AF389620

Sustainability in Buildings: New Trends in the Management of Construction and Demolition Waste—Volume I

Editors

Carlos Morón Fernández
Daniel Ferrández Vega

Basel • Beijing • Wuhan • Barcelona • Belgrade • Novi Sad • Cluj • Manchester

Editors
Carlos Morón Fernández
Universidad Politécnica de Madrid
Spain

Daniel Ferrández Vega
Universidad Politécnica de Madrid
Spain

Editorial Office
MDPI
St. Alban-Anlage 66
4052 Basel, Switzerland

This is a reprint of articles from the Topic published online in the open access journals *Sustainability* (ISSN 2071-1050), *Materials* (ISSN 1996-1944), *Sensors* (ISSN 1424-8220), *Applied Sciences* (ISSN 2076-3417), and *Processes* (ISSN 2227-9717) (available at: https://www.mdpi.com/topics/sustainability_buildings).

For citation purposes, cite each article independently as indicated on the article page online and as indicated below:

Lastname, A.A.; Lastname, B.B. Article Title. *Journal Name* **Year**, *Volume Number*, Page Range.

Volume I
ISBN 978-3-0365-8302-0 (Hbk)
ISBN 978-3-0365-8303-7 (PDF)
doi.org/10.3390/books978-3-0365-8303-7

Set
ISBN 978-3-0365-8300-6 (Hbk)
ISBN 978-3-0365-8301-3 (PDF)

Contents

About the Editors

Carlos Morón Fernández

Carlos Morón Fernández, PhD. He is a full professor at the Universidad Politécnica de Madrid and is currently the director of the research group on "Monitoring and Technological Innovation in Building". He is currently the director of the Department of Building Technology at the E.T.S. de Edificación de Madrid, has numerous scientific publications in high-impact-factor journals, and has supervised 10 doctoral theses.

Daniel Ferrández Vega

Daniel Ferrández Vega, PhD. He is an associate professor at the Universidad Politécnica de Madrid and is an internationally recognized researcher in the field of building engineering. He is currently developing his research in the field of new materials for sustainable construction and the application of new measurement techniques to buildings. He has nearly 50 publications in JCR- and SJR-indexed journals, and has supervised three doctoral theses.

Preface

The building sector is evolving towards a more efficient management of natural resources and the incorporation of circular economy criteria. In this way, the recovery and revaluation of construction and demolition waste (CDW), energy efficiency or the development of new composite materials are some of the lines that define research in building engineering in this decade. This work compiles some of the most cutting-edge research on sustainability in building, showing advances in the development of new materials, more efficient construction systems, the recycling of construction waste and other topics of interest to construction professionals.

In this sense, the Guest Editors and the research group "Monitorización e Innovación Tecnológica en Edifcación (MITE)" would like to thank all the researchers for their collaboration, and for their confidence in this MDPI Topic. We sincerely hope that the work will be useful for other professors, professionals, and researchers, and that future editions will receive the same level of participation. For all these reasons, we can only say thank you.

Carlos Morón Fernández and Daniel Ferrández Vega
Editors

Article

Improvement of Bond Strength and Durability of Recycled Aggregate Concrete Incorporating High Volume Blast Furnace Slag

Shu-Ken Lin [1] and Chung-Hao Wu [2],*

[1] Department of Civil Engineering, National Chung Hsing University, No. 145 Xingda Rd., South District, Taichung City 40224, Taiwan; sklin@nchu.edu.tw

[2] Department of Civil Engineering, Chung Yuan Christian University, No. 200, Zhongbei Rd., Zhongli District, Taoyuan City 32023, Taiwan

* Correspondence: chw@cycu.edu.tw; Tel.: +886-3-2654241

Abstract: This paper aims to experimentally investigate the effects of high volume cement replacement of blast furnace slag (BFS) on the bond, strength and durability of recycled aggregate concrete (RAC). Concrete mixtures were prepared containing 0%, 15%, 30%, 45%, 60% and 75% BFS with each of recycled aggregate and natural aggregate. Measurements of the compressive and bond strength, the resistance to chloride-ion penetration and the water permeability of concrete are reported. In addition, a microhardness test was also performed to evaluate the quality of interfacial transition zone (ITZ) in concrete. Test results of the bond strength and the compressive strength of RAC mixtures, in spite of the cement replacement amount with BFS, show that the concretes result in reduced strength when compared to natural aggregate concrete (NAC) mixtures, while the strength gains for the BFS-based concrete are higher than that of the reference mixtures without BFS at long-term ages. Incorporating BFS in concrete can inherently improve the durability properties by increasing higher resistance to chloride-ion penetration and lower water permeability. This improvement in the mechanical and durability properties of the BFS-based RAC mixture may be due to the additional pozzolanic reaction of BFS, which enhances the properties of ITZ in concrete, resulting in an improvement of the strength of concrete.

Keywords: recycled aggregated concrete; blast furnace slag; bond strength; interfacial transition zone; water permeability; chloride-ion penetration; durability

Citation: Lin, S.-K.; Wu, C.-H. Improvement of Bond Strength and Durability of Recycled Aggregate Concrete Incorporating High Volume Blast Furnace Slag. *Materials* **2021**, *14*, 3708. https://doi.org/10.3390/ma14133708

Academic Editor: F. Pacheco Torgal

Received: 17 June 2021
Accepted: 30 June 2021
Published: 2 July 2021

Publisher's Note: MDPI stays neutral with regard to jurisdictional claims in published maps and institutional affiliations.

1. Introduction

Cement and aggregate are the two main components in concrete. However, the procedures of their production emit much CO_2 and consume a large amount of energy, which is unfriendly to the natural environment. In the past decades, many researchers intended to find an effective method for producing sustainable concrete with a low amount of cement or recycled aggregate. The general measure is to crush waste concrete as coarse aggregate for producing concrete, which was regarded as recycled aggregate concrete (RAC) [1–3]. The surface of recycled coarse aggregate is generally full of pores, presenting high water absorption, which affects the interfacial transition zone (ITZ) property in concrete, leading to negative effects for the durability of concrete [4–7]. In order to improve the durability of the RAC, pozzolans such as fly ash and blast furnace slag were added to concrete; nevertheless, there are restrictions on the dosages of the pozzolans in many national standard codes [8–10].

For past decades, increasingly more research results of concrete with high volumes blast furnace slag (BFS) were reported, with most discussing natural aggregate concrete [11–15], while few dealt with RAC incorporating high replacement levels of BFS, especially on the durability aspect. Djelloul [16] investigated the improvement of workability of self-compacting concrete (SCC) mixes containing recycled aggregates with increasing BFS from

0 to 30%. They found that recycled coarse aggregate (RCA) content of 25% to 50% natural aggregate replacement and cement replacement of 15% BFS may be the best dosage to produce available SCC without any segregation or bleeding. El-Hawary [17] reported the results of experimental research on the durability and performance of various mixtures of RAC incorporating 25% BFS as cement replacement. The test results reported that incorporating mineral slag improved the mechanical properties and durability characteristics of the RACs. Addition of 25% BFS evaluated to be efficient in reducing water absorption, alkali–silica reactivity and sulfate attack in RAC. However, the experimental results also presented that introduction of 25% BFS is not effective enough to prohibit the excessive expansion of the alkali–carbonate reaction in RAC after one year. Furthermore, it has an adverse impact on resisting the drying shrinkage and water permeability of the RAC mixtures.

Hence, some studies further investigated the effect of high volume blast furnace slag. Khodair [18] examined the effect of recycled coarse aggregate (RCA) on the properties of SCC. Twenty RAC mixtures with different replacement ratio of RCA, BFS and fly ash (FA) were prepared and tested. The test results showed that the replacement ratio of recycled aggregate increased, the compressive strengths of concrete decreased at 3, 14, and 28 days. Moreover, the partial replacement of cement by pozzolans presented an adverse effect on the 28-day RAC compressive strength; however, it increased the resistance to chloride permeability.

Majhi [19] researched the effects of RCA and BFS on fresh and hardened concrete properties; 16 concrete mixtures were prepared with 0–100% replacement of natural aggregate by RCA for each 0–60% replacement of cement by BFS. The test results showed that the workability decreased with the use of RCA. The compressive, flexural and split tensile strength decreased with the increase in the replacement ratio of RCA or BFS or both. Voids and water absorption in the concrete mixtures increased with the increase of RCA content. However, the use of BFS enhanced the quality of the concrete by improving the interfacial transition zone and the bond strength between mortar and aggregate. The concrete mixture with 50% RCA and 40% BFS achieved properties similar to those of the natural aggregate concrete without BFS.

In addition, the bond strength of recycled aggregate concrete is also an interesting issue. Some studies discussed the bond strength between RAC and rebar in comparison with that of ordinary concrete [20,21]. Their test results indicated that the reduction of bond strength could be associated with the increment of recycled aggregate used in the mixture, which presented reductions of 6–8% up to 30% of bond strength. In order to improve the bond strength of RAC, some researchers added pozzolan into concrete to enhance the bond strength. Majhi [22] investigated concrete mixes containing 0%, 40% and 60% BFS with each of 50% and 100% RCA. The test results showed that the compressive strength of RAC can satisfy the 28-day compressive strength requirements of the concrete grades of M15, M20 and M25, as per IS 10262 (2009) [23]. Majhi [24] also investigated the mechanical properties and durability of RAC utilizing high volume BFS (up to 60% replacement ratio of cement) with lime activator. The results revealed that the enhancement in the mechanical properties of blast furnace slag-based RAC mixes were found to be in the range of 14.02–19.61%, 10.74–14.71%, 9.33–14.07% and 6.65–14.17% for compressive, flexural and bond, and splitting tensile strength, respectively.

Furthermore, other studies focused on discussing the microstructure at steel–concrete interface in reinforced concrete. Soylev [25] studied the influence of steel–concrete interface defects on reinforcing steel corrosion. Their test results showed that the defects related to the gaps caused by segregation, bleeding and settlement of fresh concrete under horizontal reinforcing bars (RB). These defects increased with the concrete depth below the horizontal RB and depended on the bleeding capacity of the concrete mixture. Horne [26] reported that compared with the bulk cement paste, the aggregate–cement-paste and RB–cement-paste interfaces presented more porosity and calcium hydroxide (CH) and less unreacted cement. As the hydration age increased, the porosity near the interfaces decreased, and the

CH increased with more CH close to the RB than to the aggregate. Chen [27] indicated that there was a relatively wide porous band with large voids and pores near the RB–concrete interface. They also found that size of porous band around the RB is not uniform, and that the water to cement ratio can significantly affect the distribution and size of porous band at the interface.

The abovementioned research suggests that RAC with high replacement levels of BFS may present improved durability and mechanical properties at later age. This study aims to further investigate the fresh properties, mechanical properties and durability of RAC that incorporates high replacement levels of BFS. The mechanical properties of hardened concrete were evaluated by testing the interfacial transition zone performance at the surface of aggregate and steel bars, which were measured using microhardness tests [28,29].

2. Materials and Methods

2.1. Materials

(1) Cement: Type I Portland cement obtained by Taiwan Cement Corporation Ltd., Taipei, Taiwan, with a specific gravity of 3.15. The basic properties of cement are shown in Table 1.

(2) Blast furnace slag (BFS): Grade 120 of ground granulated blast furnace slag obtained from Advanced-Tek Systems Co., Ltd., Taipei, Taiwan. The basic properties of BFS are shown in Table 1.

(3) Fine Aggregate (FA): Natural river sand with a specific gravity of 2.60 and fineness modulus of 2.40.

(4) Coarse aggregate (CA): Crushed river stone with a specific gravity of 2.61, bulk density of 1470 kg/m^3 and maximum size of 19 mm.

(5) Recycled coarse aggregate (RCA): Crushed waste concrete with a specific gravity of 2.26, bulk density of 1280 kg/m^3 and particle sizes of 5~20 mm.

(6) Superplasticizer (SP): High performance water-reducing agent obtained from HI CON Chemical Admixture Taiwan Ltd., Taipei, which meets the CNS 12283 Type G with a specific gravity of 1.2 $\pm$ 0.02 and pH 7.0 $\pm$ 1.0.

Table 1. Composition and physical properties of cement and blast furnace slag.

Components	Cement	Blast Furnace Slag
SiO_2 (%)	21.4	38.1
Al_2O_3 (%)	4.9	7.5
Fe_2O_3 (%)	3.8	0.3
CaO (%)	64.2	39.9
MgO (%)	1.1	10.6
Na_2O (%)	0.2	0.4
SO_3 (%)	2.1	0.2
Loss on ignition (%)	2.1	1.4
Specific surface area (m^2/kg)	336	538
Specific gravity	3.15	2.90

2.2. Mixture Proportion and Specimen Preparation

The mixture proportions of concrete were designed according to ACI 211.1 [30], basically, for providing the compressive strength of natural aggregate concrete (NAC) of 41 MPa at 28 days. The water to cementitious material ratio (w/cm) ranged from 0.30 to 0.50, and the BFS content ranged from 0% to 75% by weight of the total cementitious materials as cement replacement. The proportions of a dozen kinds of mixtures are shown in Table 2.

Table 2. Proportions of concrete mixtures.

Mixture	w/cm	Water	Cement	BFS [b]	FA	NCA	RCA	SP
		kg/m^3						
NS00 [a]	0.50	210	420	0	800	860	0	1.1
NS15	0.46	193	357	63	839	860	0	1.0
NS30	0.42	176	294	126	878	860	0	1.8
NS45	0.38	160	231	189	918	860	0	2.6
NS60	0.34	143	168	252	956	860	0	3.6
NS75	0.30	126	105	315	995	860	0	6.3
RS00	0.50	210	420	0	793	0	750	1.1
RS15	0.46	193	357	63	832	0	750	1.5
RS30	0.42	176	294	126	871	0	750	2.3
RS45	0.38	160	231	189	910	0	750	3.6
RS60	0.34	143	168	252	950	0	750	5.9
RS75	0.30	126	105	315	989	0	750	8.7

Note: [a] N is normal aggregate concrete, R is recycled aggregate concrete; the number following the symbol S represents the percentage of cement replacement by BFS. [b] BFS: blast furnace slag, FA: fly ash, NCA: natural coarse aggregate concrete, RCA: recycled coarse aggregate, SP: superplasticizer.

The fresh properties of concrete, including slump, air content and unit weight, were simultaneously measured for each batch. Specimens for testing were cast from each mixture: a cylinder specimen of φ100 mm × 200 mm for compressive strength test, cylinder specimen of φ150 mm × 50 mm for water permeability test and cylinder specimen of φ100 mm × 50 mm for chloride-ion permeability test. For the bond strength test, cubes of 150 mm size were prepared by embedding a reinforcing bar of #7 (nominal diameter 22.2 mm) vertically along the central axis of the specimen. A cube specimen of 50 mm × 50 mm × 50 mm in dimension with a #4 (nominal diameter 12.7 mm) bar placed horizontally in the middle was prepared for the microhardness test. Figure 1 shows the cut section of specimen for testing. After removal from the molds, all specimens were moved to a standard moist-curing room until date for testing.

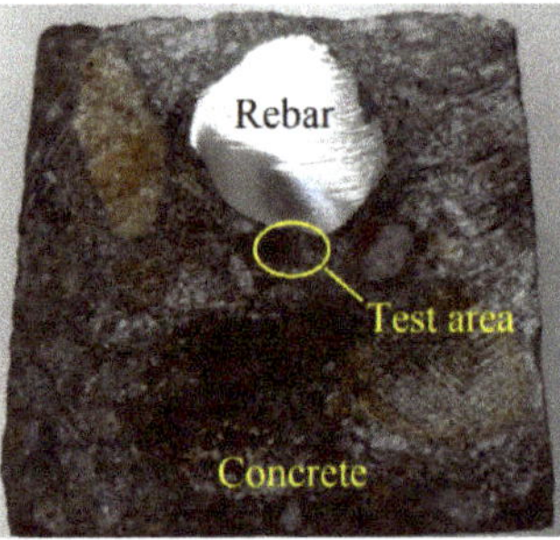

Figure 1. Vickers hardness test section of cube specimen.

2.3. Test Procedures

The compressive strength test was performed according to ASTM C 39 [31] at the ages of 7, 28, 56 and 91 days. Steel pull-out testing (ASTM C234 [32]) for bond strength determination was carried out at 7, 28, 56 and 91 days. The resistance of concrete to the penetration of chloride ion was measured using rapid chloride-ion permeability test (RCPT, ASTM C1202-12 [33]), in terms of the charge passed through the concrete in coulombs. The water permeability of concrete (IS 3085 [34]) was tested using a water permeability apparatus subjected to a water pressure of 0.29 MPa for 3 h to determine the flow through the specimen. The chloride-ion penetration and water permeability tests were carried out at 28, 56 and 91 days. Vickers hardness test was performed according to ASTM E384-17 [35] for recognizing the existence and properties of ITZ. The Vickers microhardness HV (MPa) is calculated as:

$$\text{HV} = \frac{P}{\text{As}} = 2P\frac{\sin\left(\frac{\alpha}{2}\right)}{d^2} = \frac{1.8544P}{d^2} \tag{1}$$

where As = surface area of indentation (mm^2), P = load (N), α = face of angle of indenter at 136∘ and d = mean diagonal of indentation (mm). Generally, the hardness is correlated with the strength of material tested.

3. Experimental Results and Discussion

3.1. Fresh Properties of Concrete

The measured fresh properties of concrete including slump, air content and unit weight are shown in Table 3. It shows that the concrete mixtures were mixed with super-plasticizer to produce a workable slump in the range of 210 to 250 mm. The air content of concrete ranged from 1.2% to 3.4% for natural aggregate concretes and 1.8% to 4.8% for RACs, respectively, basically increased with the increase of the cement replacement ratio of BFS. The unit weight of concrete ranged from 2215 kg/m^3 to 2314 kg/m^3 decreased as the replacement ratio of BFS increased. These results indicate that the RAC mixtures prepared for test can exhibit adequate properties of fresh concrete.

Table 3. Properties of fresh concretes.

Mixture	Slump (mm)	Unit Weight (kg/m^3)	Air Content (%)
NS00	225	2314	2.1
NS15	210	2303	1.2
NS30	220	2289	1.4
NS45	210	2285	2.3
NS60	250	2267	3.2
NS75	250	2236	3.4
RS00	215	2284	2.0
RS15	220	2273	1.8
RS30	250	2261	2.2
RS45	235	2238	3.1
RS60	230	2231	4.1
RS75	200	2215	4.8

3.2. Mechanical Properties of Concrete

3.2.1. Compressive Strength

The compressive strength of concretes measured at various ages is shown in Table 4; Figure 2 shows the development of compressive strength for each concrete mixture. It is seen that the strength development with age of the NAC and RAC mixtures present a similar trend, in which the compressive strength values of RAC series were all lower than that of the corresponding mixtures of NAC series. This indicates that despite the cement replacing amount with BFS, the mixtures of RAC result in reduced strength compared to that of the NAC.

Table 4. Compressive strengths of concretes (MPa).

Mixture	w/cm	Compressive Strength (MPa)			
		7 Days	28 Days	56 Days	91 Days
NS00	0.50	39.2 (100%)	42.1 (107%)	47.8 (122%)	56.3 (130%)
NS15	0.46	47.9 (100%)	55.4 (115%)	61.3 (128%)	64.2 (134%)
NS30	0.42	52.8 (100%)	62.9 (119%)	68.1 (129%)	71.3 (135%)
NS45	0.38	59.5 (100%)	65.6 (110%)	72.9 (123%)	78.8 (133%)
NS60	0.34	62.1 (100%)	72.3 (116%)	77.5 (125%)	82.4 (133%)
NS75	0.30	65.2 (100%)	76.2 (116%)	82.4 (126%)	86.2 (132%)
RS00	0.50	30.6 (100%)	32.6 (106%)	37.1 (121%)	38.3 (125%)
RS15	0.46	31.9 (100%)	37.1 (116%)	40.3 (126%)	44.6 (139%)
RS30	0.42	35.8 (100%)	40.3 (112%)	44.4 (124%)	49.7 (138%)
RS45	0.38	39.7 (100%)	46.7 (118%)	52.0 (130%)	54.6 (137%)
RS60	0.34	43.5 (100%)	51.1 (117%)	55.7 (128%)	58.5 (135%)
RS75	0.30	47.3 (100%)	53.6 (111%)	59.9 (127%)	64.1 (136%)

Note: The compressive strength was calculated as the average value from the test of 3 specimens for each type of mixture.

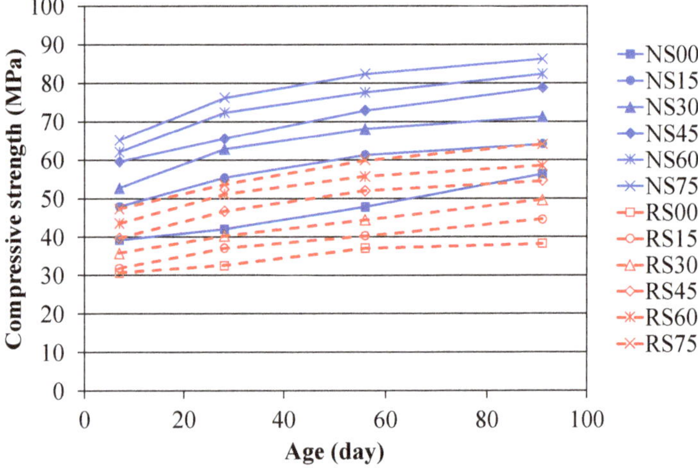

Figure 2. Compressive strength development of concrete.

It can be seen in Figure 2 that the curve slope of the BFS mixtures tends to be steeper than that of the reference mixture without BFS at later age. In other words, the strength gain for the BFS mixtures is higher than that of the reference mixture without BFS. This improvement in the long-term compressive strength may be the consequence of the pozzolanic reaction of BFS, leading to enhance the strength.

3.2.2. Properties of ITZ in Concrete

Microhardness testing has been reported to be a means for characterizing the bulk paste properties in cement pastes, especially for the properties of interfacial transition zone (ITZ). The ITZ was recognized as a significant part of the microstructural system in concrete, which plays an important role in affecting the properties of concrete, such as compressive strength, tensile strength, fracture and permeability.

Figures 3 and 4, respectively, illustrate the microhardness profiles of measured results for NACs and RACs. All curves show that in the vicinity of the rebar surface there is a gradient in microhardness, but in the outer bulk paste it tends to be constant. The width of the zone, with a varied gradient, may be referred to as ITZ. Figures 3 and 4 show that both NAC and RAC mixtures at all ages present a similar curve shape of ITZ.

The distinction from each other of any two ITZ curves displays solely in the form of depression; deeper depression represents a lower microhardness value. Both RAC and NAC mixtures incorporated with various BFS content show higher microhardness at ITZ than that of the correlated control mixture without BFS. Furthermore, for each age, concrete containing more BFS presents higher microhardness at ITZ; the concrete mixture with 75% BFS achieves the highest microhardness. These results indicate that incorporating BFS makes the ITZ in concrete denser and stronger, especially at long-term ages. This may infer to be due to the latent hydraulic activity and the additional pozzolanic reaction of BFS, which enhance the properties of ITZ and in turn the strength of BFS concrete.

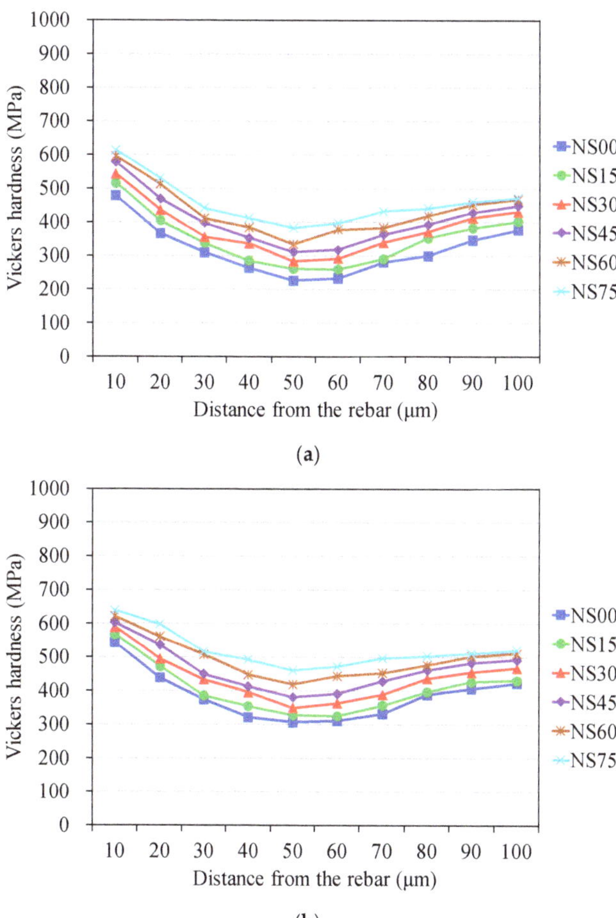

(a)

(b)

Figure 3. *Cont.*

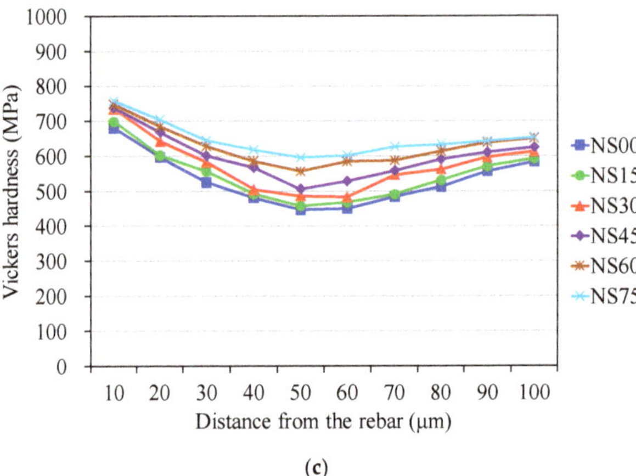

(**c**)

Figure 3. Measured Vickers hardness of normal aggregate concrete at ages of (**a**) 28 days, (**b**) 56 days and (**c**) 91 days.

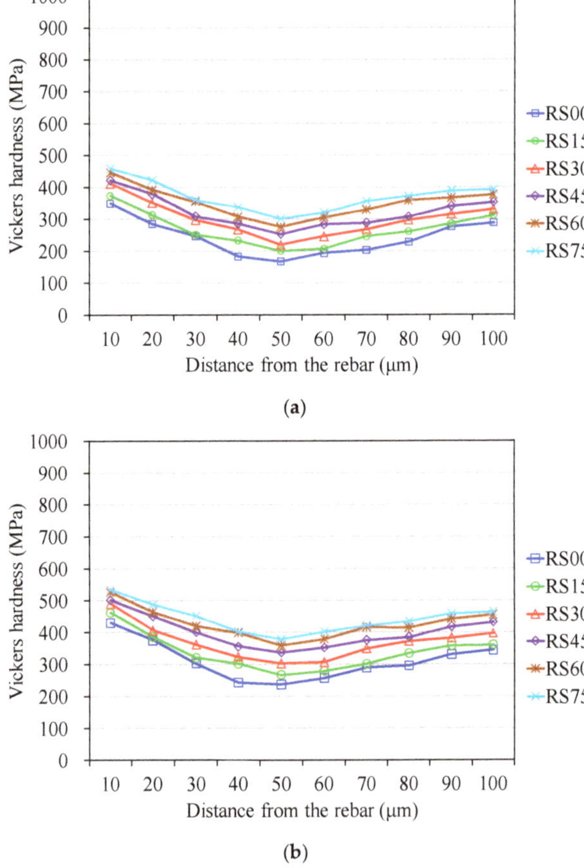

(**a**)

(**b**)

Figure 4. *Cont.*

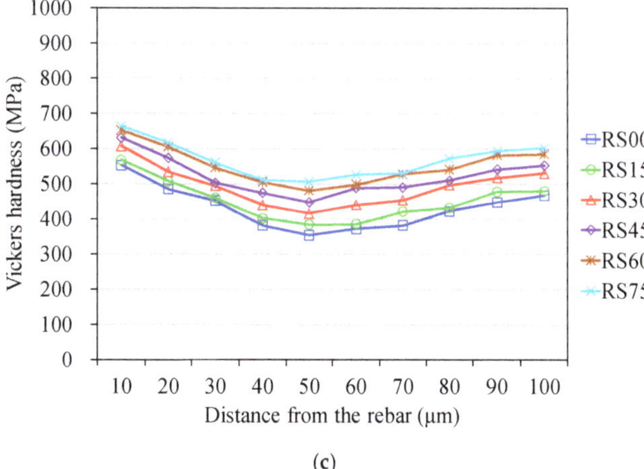

(c)

Figure 4. Measured Vickers hardness of recycled aggregate concrete at ages of (**a**) 28 days, (**b**) 56 days and (**c**) 91 days.

3.2.3. Bond Strength

The bond strength of concrete obtained from the pull-out test was calculated as:

$$\mu_{avg} = \frac{P}{\pi \cdot d_b \cdot L_e} \tag{2}$$

where μ_{avg} is the average bond stress (MPa), P is the measured maximum load (N), d_b is the diameter of rebar (mm) and L_e is the length of embedment (mm).

Table 5 summarizes the results of the pull-out test measured at the ages of 7, 28, 56 and 91 days. It is seen that the average bond stress μ_{avg} (also named bond strength) of concrete increased with the increase of compressive strength. The bond strength of BFS concretes was higher than that of the reference concrete without BFS, which increases with the increase in BFS-based content. This increase in bond strength of BFS concrete may be due to the denser and stronger ITZ at the surface of steel bars, as discussed earlier.

Table 5. Bond strength of concretes measured from pull-out tests.

Mixture	Age (day)	Concrete Strength (MPa)	Bond Strength μ_{avg} (MPa)	Age (day)	Concrete Strength (MPa)	Bond Strength μ_{avg} (MPa)
NS00		39.2	12.7 (100%)		47.8	15.1 (119%)
NS15		47.9	14.3 (100%)		61.2	17.2 (120%)
NS45		59.5	16.6 (100%)		72.9	20.5 (123%)
NS75	7	65.2	20.1(100%)	56	80.4	25.6 (127%)
RS00		30.6	9.8 (100%)		38.1	11.5 (117%)
RS15		31.9	10.1 (100%)		41.3	12.3 (122%)
RS45		39.7	11.9 (100%)		52.0	14.8 (124%)
RS75		47.3	13.2 (100%)		59.9	16.8 (127%)
NS00		42.1	13.9 (109%)		56.3	16.3 (128%)
NS15		55.4	16.0 (112%)		64.2	18.7 (131%)
NS45		65.6	18.8 (113%)		78.8	22.1 (133%)
NS75	28	76.2	23.4 (116%)	91	85.2	27.1 (135%)
RS00		32.6	10.4 (106%)		41.3	12.4 (126%)
RS15		37.1	11.3 (111%)		44.6	13.5 (134%)
RS45		46.2	13.5 (113%)		54.8	16.4 (138%)
RS75		52.6	15.0 (114%)		63.5	18.3 (139%)

To evaluate the long-term effect of BFS on the bond strength of concrete, the bond strength gains (in percentage) of each mixture series at 28, 56 and 91 days with respect to the 7-day bond strength are also calculated, as shown in Table 5. Their bond strength development with age of NAC and RAC mixtures are illustrated in Figures 5 and 6. From Table 5, it is found that the rate of bond strength gains of each BFS concrete for both NAC and RAC mixtures at later ages is higher than that of the reference concrete without BFS. The rate of strength gains at ages of 56 and 91 days increases with the increase in BFS content of the BFS-based RAC. This trend of bond strength can be also observed in Figures 5 and 6, in which the curve slope of the BFS mixtures at long-term ages tend to be steeper than that of the reference concrete without BFS. For instance, the strength gains in bond strength of the mixture RS00 at 28, 56 and 91 days with respect to 7-day strength of the same are 6%, 17% and 28%, respectively, obviously less than the RS45 mixture of 13%, 24% and 37%, and the RS75 mixture of 14%, 27% and 39%, respectively.

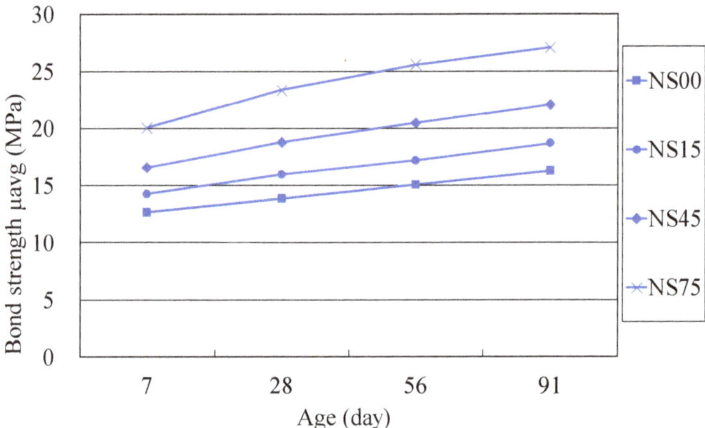

Figure 5. Bond strength of NAC versus curing age.

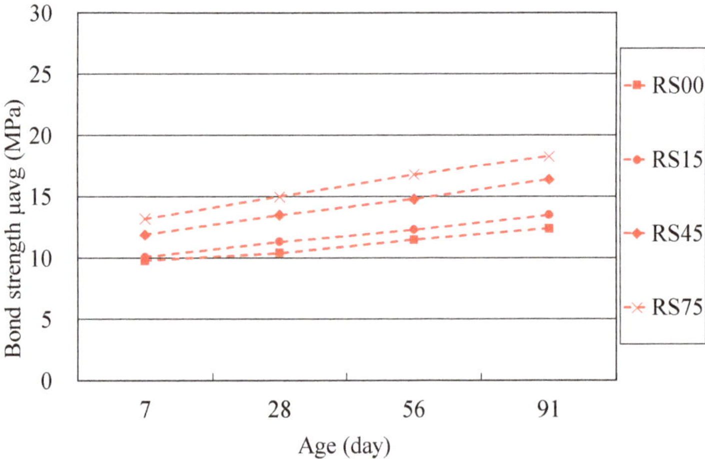

Figure 6. Bond strength of RAC versus curing age.

Based on these results, the BFS may provide the effects to enhance the long-term bond strength of RAC; the rate of strength gains at later ages beyond 56 days increase with the increase in BFS content. This may be due to the fact that the pozzolanic reaction of

BFS can improve the strength properties of ITZ at the surface of steel bars, resulting in an enhancement of the long-term bond strength of the BFS concretes.

3.3. Durability Properties

3.3.1. Resistance to Chloride-Ion Penetration

Table 6 summarizes the measured results of the RCPT test for the resistance of concrete mixture to the chloride-ion penetration; the results are also shown in Figure 7. It is found that for most concrete mixtures of the NAC and RAC series, except the mixtures NS15 and RS15, the resistance to the chloride-ion penetration for each BFS concrete is significantly higher than that of the reference concrete, without BFS at later ages. When BFS is used, the chloride-ion penetration value of concrete decreases with an increase in BFS content (Figure 7) At 91 days, the total charge passed for mixture RS75 was 581 coulombs (C) (classified as very low permeability), far less when compared with 8114 C (high permeability) for the reference mixture RS00. Even at 28 days, the total charge passed for the mixture RS 75 was 938 C (very low permeability). This indicates that incorporating high volume (higher than 45%) of BFS in RAC at later ages may greatly reduce the chloride-ion ingress, and results in obvious improvement of the durability of concrete. This is believed to be due to the contribution of the pozzolanic character of BFS, which enhanced the resistance to chloride attack at long-term ages.

Table 6. Measured results of the resistance to chloride-ion penetration of concretes.

Mixture	Total Charge Passed (Coulombs) *			Age (Days)
	28 Days	**56 Days**	**91 Days**	
NS00	9605	6567	6182	high/high/high **
NS15	6941	5589	5068	high/high/high
NS45	3562	3042	2928	moderate/moderate/moderate
NS60	2266	1478	1416	moderate/low/low
NS75	1446	1109	1076	low/low/low
RS00	10,506	8329	8014	high/high/high
RS15	6214	4813	4139	high/high/high
RS45	2973	2299	2136	moderate/moderate/moderate
RS60	1507	1385	922	low/low/very low
RS75	938	861	581	very low/very low/very low

* Average value of 3 specimens. ** Charge passed, chloride permeability, coulombs: >4000 C = high, 2000–4000 C = moderate, 1000–2000 C = low, 100–1000 C = very low and <100 C = negligible.

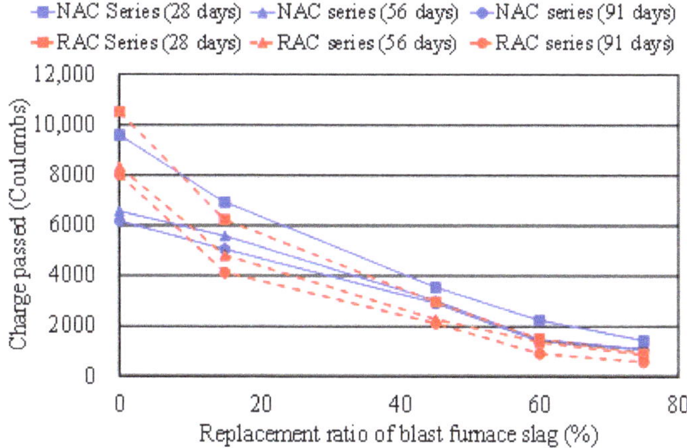

Figure 7. Total charge passed vs. BFS content of concrete.

3.3.2. Water Permeability

The water permeability (WP) of concrete was tested using a uniaxial flow apparatus performed on cylinder specimens (φ150 mm $\times$ 50 mm) subjected to 0.29 MPa pressure for 3 h. The WP was calculated with following formula:

$$\text{Water permeability} = \frac{m_2 - m_1}{m_2} \times 100\% \tag{3}$$

where m_1 = initial weight of specimen and m_2 = specimen weight after test.

Table 7 summarizes the measured WP of the concrete mixtures. It is seen that WP decreases with increasing compressive strength for both NAC and RAC mixtures. Incorporating BFS in concrete inherently reduces the WP of concrete at all ages. This is particularity found in mixtures NS 75 and RS 75 containing 75% BFS at curing age of 91 days, which has WP of 0.69% and 0.18%, respectively, comparably less with that of the reference mixture NS00 and RS00 of 0.46% and 0.79%, respectively. These results signify the fact that either NAC or RAC containing a high volume BFS at later ages may lead to lower water permeability, namely, superior durability to the concrete without BFS.

Table 7. Results of the concrete water permeability measurements.

Mixture	Compressive Strength (MPa)			Water Permeability (%)		
	28 Days	56 Days	91 Days	28 Days	56 Days	91 Days
NS00	42.1	47.8	51.3	0.52	0.47	0.46
NS15	55.4	61.3	64.2	0.46	0.41	0.40
NS45	65.6	72.9	78.8	0.36	0.28	0.21
NS60	72.3	77.5	82.4	0.25	0.22	0.16
NS75	76.2	82.4	86.2	0.15	0.13	0.09
RS00	32.6	39.1	38.3	0.84	0.81	0.79
RS15	37.1	40.3	44.6	0.71	0.68	0.57
RS45	46.7	52.0	54.6	0.58	0.49	0.38
RS60	51.1	55.7	58.5	0.43	0.32	0.21
RS75	52.6	59.9	64.1	0.30	0.23	0.18

Furthermore, the effect on the water permeability of concrete by adding BFS can also be seen from Figure 8, which illustrates the relations between water permeability and BFS content of RAC mixture at curing ages of 28 and 91 days. Note that the WP of concrete decreases with the increase of BFS content for the two curing ages, while the curve trend for the 91-day concrete presents a steeper decline than that for the 28-day concrete, indicating that BFS concrete may exhibit less WP at later ages.

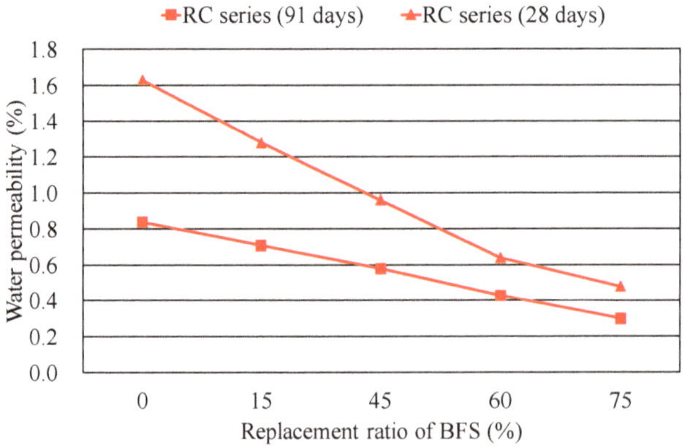

Figure 8. Influence of BFS content on the water permeability of RAC mixtures.

4. Conclusions

Based on the results and findings of the experimental work for concrete with w/cm ratio in the range of 0.30 to 0.50 and with BFS content in the range of 0% to 75% by weight of the total cementitious material as cement replacement, the following conclusions can be drawn:

1. The strength development with age of the NAC and RAC mixtures presents a similar trend, in which the compressive strength of RAC series is lower than that of the corresponding NAC series, indicating that, despite the cement replacing BFS, the mixtures of RAC result in reduced strength. In addition, the strength gain at later ages for the BFS mixtures is higher than that of the reference mixture without BFS.
2. Both of RAC and NAC mixtures incorporated with any BFS content show higher microhardness at ITZ than that of the correlated control mixture without BFS. Concrete at each age and containing more BFS presents higher microhardness at ITZ. This indicates that incorporating BFS makes the ITZ in concrete denser and stronger, especially at later ages.
3. The bond strength of BFS concretes is higher than that for the reference concrete without BFS, which increases with the increase in BFS content. The rate of bond strength gains at long-term ages with respect to the 7-day strength increases with the increase in BFS content. This may be due to pozzolanic reaction of BFS, which enhances the strength properties of ITZ at the surface of steel bars, resulting in an improvement of the long-term bond strength of BFS concretes.
4. Both RAC and NAC mixtures containing BFS present higher resistance to chloride ingress than that of the reference concrete without BFS, at all ages. The chloride-ion penetration value of the BFS-based concrete decreases with the increase in BFS content. Eventually, the mixture RS 75 comprising recycled aggregate and 75% BFS exhibits a pronounced resistance to chloride attack, revealing a chloride-ion penetration value of less than 1000 coulombs (very low permeability) at later ages.
5. Incorporating BFS in concrete can inherently reduce the water permeability of concrete at all ages. Either RAC or NAC containing high volume BFS (>45%) at later ages may lead to lower water permeability, namely, superior durability to the concrete with BFS.

Author Contributions: Conceptualization, S.-K.L. and C.-H.W.; methodology, S.-K.L. and C.-H.W.; validation, S.-K.L. and C.-H.W.; formal analysis, S.-K.L. and C.-H.W.; investigation, S.-K.L.; data curation, S.-K.L.; writing—original draft preparation, C.-H.W.; writing—review and editing, C.-H.W.; supervision, C.-H.W.; project administration, S.-K.L. and C.-H.W. Both authors have read and agreed to the published version of the manuscript.

Funding: This research received no external funding.

Institutional Review Board Statement: Not applicable.

Informed Consent Statement: Not applicable.

Data Availability Statement: Data sharing is not applicable to this article.

Conflicts of Interest: The authors declare no conflict of interest.

References

1. Yaragal, S.C.; Teja, D.C.; Shaffi, M. Performance studies on concrete with recycled coarse aggregates. *Adv. Concr. Constr.* **2016**, *4*, 263–281. [CrossRef]
2. Mao, Y.A.; Liu, J.H.; Shi, C.J. Autogenous shrinkage and drying shrinkage of recycled aggregate concrete: A review. *J. Clean. Prod.* **2021**, *295*, 126435. [CrossRef]
3. Nedeljković, M.; Visser, J.; Šavija, B.; Valcke, S.; Schlangen, E. Use of fine recycled concrete aggregates in concrete: A critical review. *J. Build. Eng.* **2021**, *38*, 102196. [CrossRef]
4. Yang, K.H.; Chung, H.S.; Ashour, A.F. Influence of type and replacement level of recycled aggregates on concrete properties. *ACI Mater. J.* **2008**, *105*, 289–296.

5.   Limbachiya, M.C.; Leelawar, T.; Dhir, R.K. Use of recycled concrete aggregate in high-strength concrete. *Mater. Struct. Vol.* **2000**, *33*, 574–580. [CrossRef]
6.   Parekh, D.N.; Modhera, C.D. Assessment of recycled aggregate concrete. *J. Eng. Res. Stud.* **2011**, *2*, 1–9.
7.   Gonza'lez-Fonteboa, B.; Martı´nez-Abella, F.; Eiras-Lo'pez, J.; Seara-Paz, S. Effect of recycled coarse aggregate on damage of recycled concrete. *Mater. Struct. Vol.* **2011**, *44*, 1759–1771. [CrossRef]
8.   *ACI 232.2R, Report on the Use of Fly Ash in Concrete*; American Concrete Institute: Farmington Hills, MI, USA, 2018.
9.   *ASTM C989, Standard Specification for Ground Granulated Blast-Furnace Slag for Use in Concrete and Mortars*; ASTM International: West Conshohocken, PA, USA, 2005.
10. *CNS 12549, Ground Granulated Blast-Furnace Slag for Use in Concrete and Mortars*; Chinese National Standards: Taipei, Taiwan, 2009.
11. Ashish, D.K.; Singh, B.; Verma, S.K. The effect of attack of chloride and sulphate on ground granulated blast furnace slag concrete. *Adv. Concr. Constr.* **2016**, *4*, 107–121. [CrossRef]
12. Kim, D.J.; Kim, C.Y.; Urgessac, G.; Choi, J.H.; Park, C.; Yeon, J.H. Durability and rheological characteristics of high-volume ground-granulated blast-furnace slag concrete containing $CaCO_3$/anhydrate-based alkali activator. *Constr. Build. Mater.* **2019**, *204*, 10–19. [CrossRef]
13. Duraman, S.B.; Richardson, I.G. Microstructure & properties of steel-reinforced concrete incorporating Portland cement and ground granulated blast furnace slag hydrated at 20 °C. *Cem. Concr. Res.* **2020**, *137*, 106193.
14. Lee, J.H.; Lee, T.Y. Durability and engineering performance evaluation of CaO content and ratio of binary blended concrete containing ground granulated blast-furnace slag. *Appl. Sci.* **2020**, *10*, 2504. [CrossRef]
15. Nicula, L.M.; Corbu, O.; Iliescu, M. Influence of blast furnace slag on the durability characteristic of road concrete such as freeze-thaw resistance. *Procedia Manuf.* **2020**, *46*, 194–201. [CrossRef]
16. Djelloul, O.K.; Menadi, B.; Wardeh, G.; Kenai, S. Performance of self-compacting concrete made with coarse and fine recycled concrete aggregates and ground granulated blast-furnace slag. *Adv. Concr. Constr.* **2018**, *6*, 103–121.
17. El-Hawary, M.; Al-Yaqout, A.; Nouh, K. Durability of recycled aggregate concrete incorporating slag. *Waste Resour. Manag.* **2019**, *172*, 107–117. [CrossRef]
18. Khodair, Y.A.; Bommareddy, B. Self-consolidating concrete using recycled concrete aggregate and high volume of fly ash, and slag. *Constr. Build. Mater.* **2017**, *153*, 307–316. [CrossRef]
19. Majhi, R.K.; Nayak, A.N.; Mukharjee, B.B. Development of sustainable concrete using recycled coarse aggregate and ground granulated blast furnace slag. *Constr. Build. Mater.* **2018**, *159*, 417–430. [CrossRef]
20. Seara-Paz, S.; González-Fonteboa, B.; Eiras-López, J.; Herrador, M.F. Bond behavior between steel reinforcement and recycled concrete. *Mater. Struct.* **2013**, *47*, 323–334.
21. Butler, L.; West, J.S.; Tighe, S.L. The effect of recycled concrete aggregate properties on the bond strength between RCA concrete and steel reinforcement. *Cem. Concr. Res.* **2011**, *41*, 1037–1049. [CrossRef]
22. Majhi, R.K.; Nayak, A.N. Bond, durability and microstructural characteristics of ground granulated blast furnace slag based recycled aggregate concrete. *Constr. Build. Mater.* **2019**, *212*, 578–595. [CrossRef]
23. *IS 10262, Concrete Mix Proportioning–Guidelines*; Indian Standard: New Delhi, India, 2009.
24. Majhi, R.K.; Nayak, A.N. Production of sustainable concrete utilising high-volume blast furnace slag and recycled aggregate with lime activator. *J. Clean. Prod.* **2020**, *255*, 120188. [CrossRef]
25. Soylev, T.A.; François, R. Quality of steel-concrete interface and corrosion of reinforcing steel. *Cement and Concrete Research* **2003**, *33*, 1407–1415. [CrossRef]
26. Horne, A.T.; Richardson, I.G.; Brydson, R.M.D. Quantitative analysis of the microstructure of interfaces in steel reinforced concrete. *Cem. Concr. Res.* **2007**, *37*, 1613–1623. [CrossRef]
27. Chen, F.J.; Li, C.Q.; Baji, H.; Ma, B.G. Quantification of steel-concrete interface in reinforced concrete using Backscattered Electron imaging technique. *Constr. Build. Mater.* **2018**, *179*, 420–429. [CrossRef]
28. Igarashi, S.B.A.; Bentur, A.; Mindess, S. Microhardness testing of cementation materials. *Adv. Cem. Based Mater.* **1996**, *4*, 48–57. [CrossRef]
29. Mindess, S.; Young, J.F.; Darwin, D. *Concrete*; Prentice Hall: Hoboken, NJ, USA, 2003; pp. 78–79.
30. *ACI 211.1, Standard Practice for Selecting Proportions for Normal, Heavyweight, and Mass Concrete*; American Concrete Institute: Farmington Hills, MI, USA, 1991.
31. *ASTM C39/C39M, Standard Test Method for Compressive Strength of Cylindrical Concrete Specimens*; ASTM International: West Conshohocken, PA, USA, 2021.
32. *ASTM C234, Standard Test Method for Comparing Concretes on the Basis of the Bond Developed with Reinforcing Steel*; ASTM International: West Conshohocken, PA, USA, 1991.
33. *ASTM C 1202, Standard Test Method for Electrical Indication of Concrete's Ability to Resist Chloride Ion Penetration*; ASTM International: West Conshohocken, PA, USA, 2012.
34. *IS 3085, Method of Test for Permeability of Cement Mortar and Concrete*; Indian Standard: New Delhi, India, 1965.
35. *ASTM E384, Standard Test Method for Microindentation Hardness of Materials*; ASTM International: West Conshohocken, PA, USA, 2017.

applied sciences

Article

Effect of C-S-H Nucleating Agent on Cement Hydration

Wenhao Zhao [1,†], Xuping Ji [2,†], Yaqing Jiang [1,*] and Tinghong Pan [1]

[1] College of Mechanics and Materials, Hohai University, Nanjing 211100, China; 191308060021@hhu.edu.cn (W.Z.); thpan@hhu.edu.cn (T.P.)

[2] State Key Laboratory of High Performance Civil Engineering Materials, Jiangsu Research Institute of Building Science, Jiangsu Sobute New Materials Co., Ltd., Nanjing 210008, China; xpji@hhu.edu.cn

* Correspondence: yqjiang@hhu.edu.cn

† Co-first authors. These authors contributed to the work equally and should be regarded as co-first authors.

Abstract: This work aims to study the effect of a nucleating agent on cement hydration. Firstly, the C-S-H crystal nucleation early strength agent (CNA) is prepared. Then, the effects of CNA on cement hydration mechanism, early strength enhancement effect, C-S-H content, 28-days hydration degree and 28-days fractal dimension of hydration products are studied by hydration kinetics calculation, resistivity test, BET specific surface area test and quantitative analysis of backscattered electron (BSE) images, respectively. The results show that CNA significantly improves the hydration degree of cement mixture, which is better than triethanolamine (TEA). CNA shortens the beginning time of the induction period by 49.3 min and the end time of the cement hydration acceleration period by 105.1 min than the blank sample. CNA increases the fractal dimension of hydration products, while TEA decreases the fractal dimension. CNA significantly improves the early strength of cement mortars; the 1-day and 3-days strength of cement mortars with CNA are more than the 3-days and 7-days strength of the blank sample. These results will provide a reference for the practical application of the C-S-H nucleating agent.

Keywords: C-S-H nucleating agent; cement hydration; early strength effect; mechanical properties

Citation: Zhao, W.; Ji, X.; Jiang, Y.; Pan, T. Effect of C-S-H Nucleating Agent on Cement Hydration. *Appl. Sci.* **2021**, *11*, 6638. https://doi.org/10.3390/app11146638

Academic Editor: Carlos Morón Fernández

Received: 21 June 2021
Accepted: 17 July 2021
Published: 20 July 2021

1. Introduction

In recent years, the use of prefabricated components has become more and more frequent in the construction industry [1–3]. Compared with traditional buildings, it has the advantages of energy saving, environmental protection and cost reduction [4,5]. However, in the process of the production of precast components, accelerating the development of early concrete strength and achieving a high turnover of mold is a key issue that needs to be solved urgently in the current precast concrete industry [6]. At present, component factories usually use high-temperature autoclave curing methods to quickly improve the early strength of prefabricated components, thereby improving the turnover efficiency of molds. However, high-temperature autoclave curing has the disadvantages of high energy consumption, heavy environmental pollution, high use cost and impact on the durability and later strength of concrete [7–9]. This is not friendly to the environment and enterprises. Therefore, seeking a method to improve the early strength of cement-based materials has become a research focus.

With the further research of various scholars on the process of cement hydration, dissolution and precipitation theory has become the mainstream theory to explain cement hydration, accelerating dissolution or accelerating precipitation has become a new way to explore early-strength admixtures [10–12]. Traditional concrete early-strength agents have certain application defects. For example, chlorine-based early-strength agents can cause corrosion of steel bars in concrete [13]. Triethanolamine (TEA) can significantly increase the early strength of cement-based materials, but cause a decrease in the flexural strength and late strength of cement [14], The mixing amount of traditional concrete early-strength is difficult to determine, and it may cause corrosion damage to concrete products due to alkali–aggregate reaction or excessive sulfate [15]. The C-S-H crystal nucleation early strength agent (CNA) as a nucleation-type early strength agent can provide nucleation sites for hydration products of the initial stage of cement hydration, accelerate the generation of hydration products and promote the early hydration of cement [16–18].

In this paper, CNA is prepared by a precipitation method, and then, the effect of CNA on cement hydration is mainly introduced and compared with traditional early strength agent TEA, based on hydration kinetic calculation, resistivity test and BSE analysis methods and combined with fractal theory, which aims to reveal the mechanism of CNA, and provides theoretical support and technical guidance for the practical application of CNA.

2. Materials and Methods

2.1. Materials

P·II portland cement (types 42.5, Nanjing Conch Cement Co., Ltd., Nanjing, China) is used as cementitious materials in this work; its chemical composition is shown in Table 1. Triethanolamine (TEA, Chengdu Kelong Chemical Reagent Field, Chengdu, China) and C-S-H crystal nucleation (CNA) are used as early strength agents of cement. Polyoxyethylene (5) nonylphenyl ether (NP5, Chengdu Kelong Chemical Reagent Field), cyclohexane and $Na_2SiO_3 \cdot 9H_2O$, and $Ca(NO)_3 \cdot 4H_2O$ are prepared to synthesis C-S-H crystal nucleation (CNA). The standard sand is used as a fine aggregate of cement mortar.

Table 1. Chemical composition of cement clinker.

Composition	SiO_2	Fe_2O_3	Al_2O_3	CaO	MgO	SO_3	f-CaO	LOI	Total
Content (%)	22.62	3.65	5.29	65.12	1.61	0.45	1.0	0.26	100

2.2. Synthesis of CNA

Firstly, NP5 and cyclohexane were mixed at a 1:10 volume ratio to form a solution A. Then, 0.2 mL $Na_2SiO_3 \cdot 9H_2O$ and 0.2 mL $Ca(NO)_3 \cdot 4H_2O$ were dropped into 0.8 mL solution A at rate of 1.6 mL/min and 0.6 mL/min, respectively. Then the pH value of the mixed solution was adjusted to about 13 by dropping NaOH solution. Finally, the mixed solution was held at 40 °C for 24 h to form the final C-S-H crystal nucleation agent (CNA). The XRD and infrared spectrum of CNA are shown in Figure 1a,b, respectively.

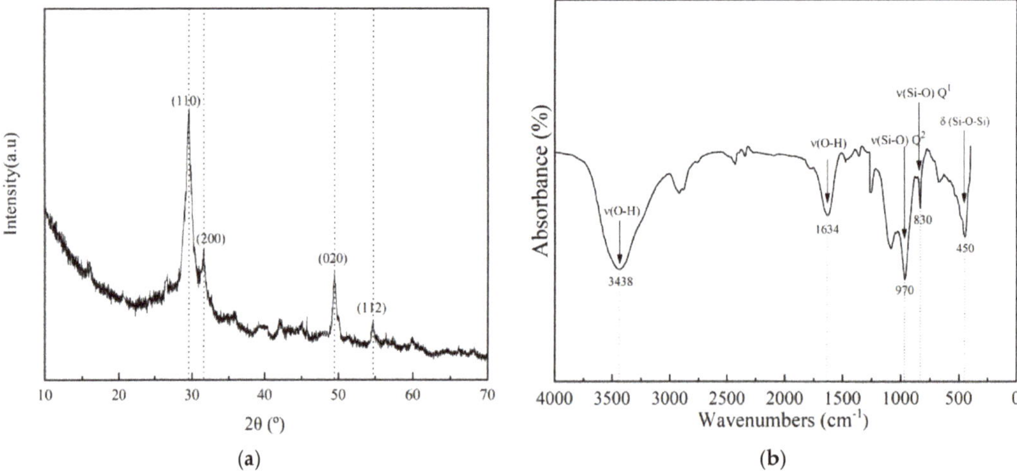

Figure 1. XRD (**a**) and infrared spectrum (**b**) of CNA.

It can be seen from Figure 1a that the characteristic peaks of C-S-H mainly appear at 2 h = 29°, 31.8°, 49.7°, and 55°, and these peaks refer to the reflection of d110, d200, d020, and d112, respectively. As can be seen from the XRD diagram, there is no other miscellaneous peak, so the reaction product obtained is C-S-H with high purity. From Figure 1b, the wide absorption peak near 970 cm^{-1} corresponds to the tensile vibration of the Si–O bond of type Q^2 Si–O tetrahedron, the absorption peak near 830 cm^{-1} corresponds to the tensile vibration of the Si–O bond of type Q^1 Si–O tetrahedron, the peak near 450 cm^{-1} corresponds to the bending vibration of Si–O–Si bond, and the absorption peaks near 1634 cm^{-1} and 3438 cm^{-1} corresponds to the O–H stretching vibration. It can be seen from Figure 1 that the synthesized C-S-H has a higher purity.

2.3. Specimen Preparation

In this work, the water–cement ratio of cement paste was set as 0.29, and the dosage of CNA and TEA was 1% and 2‰ (by mass) of the cementitious material, respectively.

BET samples preparation: The central block samples were broken into small pieces and soaked in anhydrous ethanol to stop the hydration. For about 10 days, the anhydrous ethanol was changed three times to ensure that the water in the test block was completely replaced. Then, the sample was placed in an oven for drying at 50 °C for later use.

BSE sample preparation: The hydrated samples were broken up and placed in anhydrous ethanol to stop the hydration; the soaking time was over 10 days, during which the anhydrous ethanol was changed three times, and when the soaking was completed, it was then dried. The hardened cement paste was pre-polished with 800 mesh sandpaper to form a surface to be measured, and then the resin and curing agent were stirred in a certain proportion on a magnetic stirrer for 5 min. The sample was then mixed with the inlaying resin for vacuum inlaying and then dried at 60 °C for 6 h after the inlaying was completed to cure the resin. The pre-polished surface of the inlaying sample was sanded with 125 μm, 70 μm and 30 μm sandpapers for 10 s, and the rotation speed was set to 150 revolutions/min until the hardened cement slurry was exposed. The inlaid samples were polished to further improve the flatness of the test surface. The samples were polished with polishing cloths of 9 um, 3 um and 0.25 um and a diamond polishing liquid of matching fineness, and the rotation speed was set to 100 revolutions/min.

2.4. Methods

2.4.1. Hydration Heat Test and Calculation of Hydration Kinetics

The hydration heat test was obtained by a TAM air eight-channel microcalorimeter, and the test temperature was controlled at 20 °C. The weighed water was mixed with the early strength agent to form a wet component, and then the wet component and cement were added to the 20 mL plastic bottle. This mixture was stirred with an egg beater for 60 s at a slow speed and then for 60 s at a fast speed to form a uniform cement slurry. Finally, the plastic bottle was placed into the tester, and thermal data were recorded.

The isothermal hydration heat curves are used to calculated hydration kinetic parameters based on the Krstulovic–Dabic formula. The cement hydration process can be divided into three stages, including nucleation and crystal growth stage (NG), interactions at phase boundaries diffusion stage (I) and diffusion stage (D), and the corresponding kinetic equations are [19]:

$$\text{NG stage}: \ [-\text{In}1 - \alpha]^{1/n} = K_1(t - t_0) = K_1'(t - t_0) \tag{1}$$

$$\text{I stage}: 1 - (1 - \alpha)^{1/3} = K_2 r^{-1}(t - t_0) = K_2'(t - t_0) \tag{2}$$

$$\text{D stage}: \ \left[1 - (1 - \alpha)^{1/3}\right]^2 = K_3 r^{-2}(t - t_0) = K_3'(t - t_0) \tag{3}$$

where dQ/dt represents the hydration exothermic rate of mixtures, Q is the hydration exothermic quantity and t is the hydration time; these parameters can be obtained through the hydration heat test. α is the hydration degree of cement, K_1, K_2, K_3, K_1', K_2', K_3' are the reaction rate constant, n is reaction exponent, t_0 is the end time of cement hydration induction period, r is the particle size participating in the reaction and Q_{max} is the theoretical maximum hydration heat of infinite age, which is obtained by the instrument's own software. In addition, according to Formulas (4) and (5), the hydration heat data can be converted into the hydration degree α and hydration rate $d\alpha/dt$ required by the kinetic model.

$$\alpha(t) = Q(t)/Q_{\max} \tag{4}$$

$$d\alpha/dt = dQ/dt \tag{5}$$

The hydration degree $\alpha(t)$ obtained from Equation (4) is substituted into Equation (1), draw a double logarithmic curve of $ln[-ln(1-\alpha)] - ln(t - t_0)$ (as the formula shows $ln[-ln(1-\alpha)] = nlnK_1' + nln(t - t_0)$), then the kinetic parameters n and K_1' in the NG stage are obtained by linear fitting. In the same way, the same calculation is made for Equations (2) and (3) to obtain the reaction rate constants K_2' and K_3' in the I and D stage, then, the kinetic parameters obtained are substituted into the differential expressions of Equations (1)–(3), the kinetic curves of reaction rate $F_1(\alpha)$, $F_2(\alpha)$, $F_3(\alpha)$ and reaction degree α for the NG, I and D stage are obtained, respectively.

2.4.2. Electrical Resistivity Test

A CCR-III non-contact electrodeless resistivity meter (ERM, Shanghai Lorui Instrument Equipment Co., Ltd., Shanghai, China) is used to continuously monitor the resistivity of cement slurry with different early strength agents [20]. Firstly, the petroleum jelly is used to seal the joints of the annular mold, and then the stirred cement slurry is poured into the annular mold. Gently vibrate the edge of the mold to ensure the compactness of the cement slurry. Cover the mold with plastic film and mold cover to prevent the evaporation of water. The computer automatically collects data within 48 h with an interval of 1 min. After the test, the vernier caliper is used to measure the height of the test piece and set the corresponding height on the running program to correct the resistivity data. All tests are carried out at a constant temperature of 20 °C.

2.4.3. Backscattered Electron Imaging Analysis (BSE)

The technique of backscattered electron image analysis (BSE) helps to further understand the microstructure and hydration properties of cement slurry, which can make up for the shortages of other research methods [21]. In this work, the JEOL-JXA-840A scanning electron microscope was used to obtain the SEM and BSE images. The magnification of the BSE image is 250 times, and the acceleration voltage was 15 kV. When the shooting conditions of the BSE image are the same, the gray-scale characteristic value of the phase has repeatability in different images so that BSE can be used to quantitatively count the volume content of the phase-in cement [22]. In this work, the number of pictures taken in the quantitative calculation was six. Each BSE image was analyzed to estimate the capillary porosity and hydration degree of cement. A gray-level histogram was obtained for each image. This histogram indicates the number of pixels in the image having each possible brightness value (between 0 and 255). The gray level of a phase in a BSE image is related to its backscattering coefficient η. The larger η is, the brighter phase is in the BSE image [23,24]. For hydrated cement paste, capillary pores are darkest, calcium silicate hydrate (C-S-H) and other (aluminate) hydration products are dark gray, calcium hydroxide (CH) is light gray, and unhydrated cement is the brightest. The backscattering coefficient of a single phase can be calculated by the following formula [25]:

$$\eta = \sum_i C_i \eta_i \tag{6}$$

where C_i and η_i are the mass fraction and backscattering coefficient of atoms that make up the phase, respectively. The content of each component (such as pore and unhydrated cement) in cement paste can be obtained by the threshold segmentation method [26], and the hydration degree of cement is estimated according to Equation (7).

$$\alpha = \left[1 - \frac{V_{cem}(t)}{V_{cem}(0)(1 - V_{RS})}\right] \tag{7}$$

where α is the hydration degree of the cement, $V_{cem}(t)$ is the volume (area) fraction of unhydrated cement at time t, and V_{RS} the initial volume fraction of readily soluble phases in cement. In this work, $V_{RS} = 8\%$. $V_{cem}(0)$ is calculated based on the *w/c* of the prepared paste, and the cement density is 3150 kg/m^3.

In addition, based on the BSE images, the fractal dimension of cement hydration products can be calculated. Fractal characteristics can effectively quantify and compare the complexity of hydration products, which is related to the macron performance of concrete [27]. In this work, the box-counting dimension is used to evaluate the fractal characteristics of hydration products in the BSE image. For complex structures with autocorrelation, the box-counting dimension D^b can be expressed as:

$$D^b = \log(N^k)/\log(1/k) \tag{8}$$

where N^k is the minimum number of grids that covering the surface of a complex structure with a square grid with side length k. The box-counting dimension indicates that the number of square grids N^k increases with the decrease of side length k, and the calculation algorithm of box-counting dimension on the BSE image is shown in Figure 2. Figure 2 is Calculation algorithm of box-counting dimension based on BSE image (a) and BSE image processing process. Original BSE image b)-1, BSE image after binarization b)-2, schematic diagram of meshing b)-3 and schematic diagram of fractal dimension fitting b)-4.

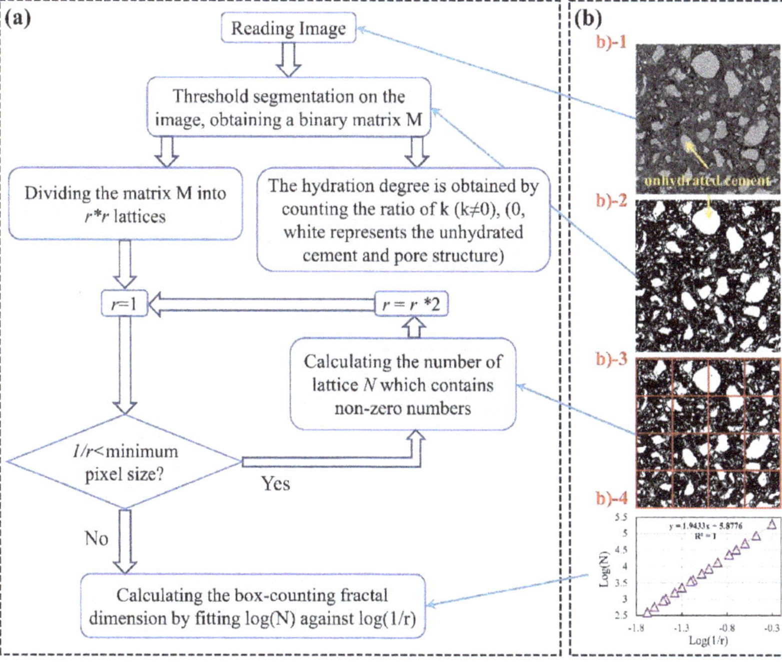

Figure 2. Calculation algorithm of box-counting dimension based on BSE image (**a**) and BSE image processing process (**b**).

2.4.4. Semi-Quantitative Analysis of C-S-H Content

It is different from quantitative C-S-H content directly due to both a lack of crystallinity and not precisely defined composition [28]. C-S-H gel obtains a higher surface area per unit volume that is at least an order of magnitude higher than any other component in hardened cement paste [29,30]. Thus, a semi-quantitative analysis method of C-S-H content is proposed, based on the BET-specific surface area testing. In the author's previous study, a quantitative relationship between the BET-specific surface area and C-S-H content of hardened cement paste is proposed [31].

$$C = 0.01129S - 0.07157 \tag{9}$$

where C is the C-S-H content of hardened cement paste, and S is the BET-specific surface area. A SA3100 nitrogen adsorption tester is used to measure the BET-specific surface area of the sample.

2.4.5. Mechanical Properties Test

In this work, the cement mortar (water–cement ratio: 0.5 and cement–sand ratio: 1:3) with different types of early strength agent was used for mechanical properties test. The dosage of CNA and TEA were 1% and 2‰ (by mass) of the cementitious material, respectively. Blocks with a size of 40 mm × 40 mm × 160 mm were selected to test the mechanical properties (flexural and compressive strength) of cement mortars with a different type of early strength agent curing at 1 day, 3 days, 7 days and 28 days, respectively. A YZH-300.10 computerized electronic universal testing machine (LUDA, Shaoxing, China) was employed. In the flexural strength test, 40 mm × 40 mm × 160 mm blocks were tested at a loading speed of 0.05 kN/s. In the compressive strength test, 40 mm × 40 mm × 40 mm blocks were utilized at a loading speed of 2.4 kN/s. For each mix, three specimens were tested and averaged.

3. Results and Discussion

3.1. Hydration Heat

The cement hydration process may be divided into five stages, including the dissolution period, the induction period, the acceleration period, the deceleration period and the slow reaction period [32]. Figure 3 shows the isothermal hydration heat curves of mixtures with and without CNA; the hydration heat emission rate and hydration heat in Figure 3 are equal to heat flow and heat, respectively. It can be found that the exothermic peak of the specimen with CNA appeared first, which is advanced by approximately 1.5 h than the blank specimen. It was also found that the cumulative heat released by the specimens with CNA is significantly higher than that without CNA. This indicates that CNA may effectively promote the early hydration process of cement. According to the instrument measurement, the theoretical maximum hydration heat (Q_{max}) of cement slurry with CNA (356.4 J·g^{-1}) is slightly higher than that without CNA (343.3 J·g^{-1}). This indicates that CNA may slightly increase the total heat release from cement hydration.

Figure 4 shows the fitting process of the hydration kinetic parameters of the CNA group and the blank group. The reaction exponent n and reaction rate constant K_1', K_2', K_3' of the mixtures with and without CNA are summarized in Table 2.

The kinetic parameters in Table 2 are substituted into the differential equations of Equations (1)–(3), the kinetic curves of reaction rate $F_1(\alpha)$, $F_2(\alpha)$, $F_3(\alpha)$ and reaction degree α for NG, I and D processes are obtained, respectively, as shown in Figure 5.

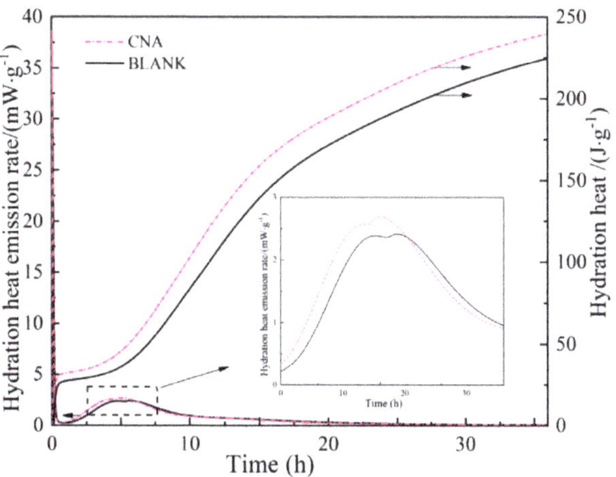

Figure 3. Cement hydration rate and heat of hydration curve.

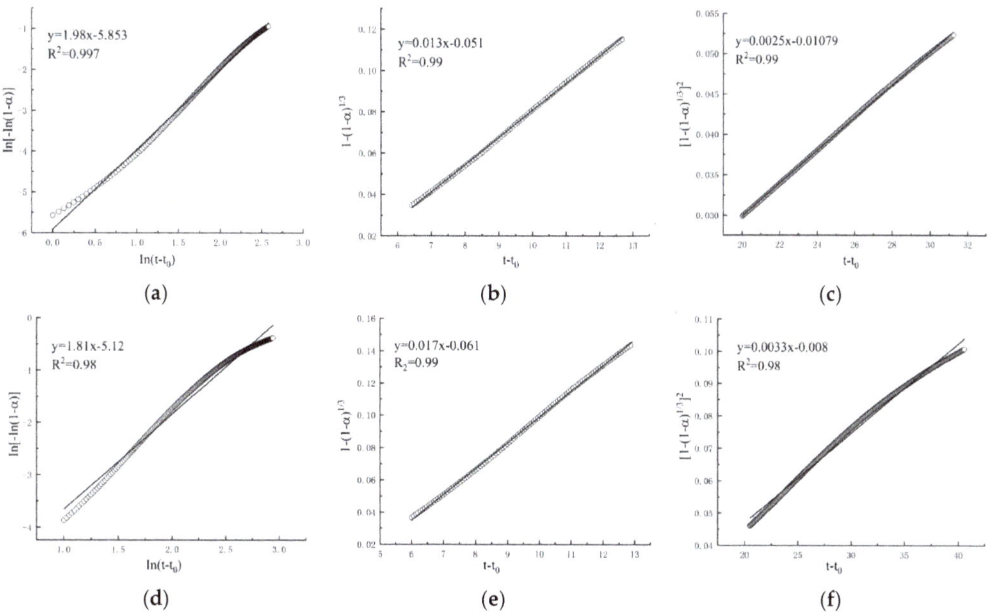

Figure 4. Fitting curves of kinetics parameters in NG stage (**a**), I stage (**b**) and D stage (**c**) of blank sample; Fitting curves of kinetics parameters in NG stage (**d**), I stage (**e**) and D stage (**f**) of cement mixture with CNA.

Table 2. Dynamic parameter solution results.

Types	Temperature of Hydration/°C	n	K_{NG}	K_I	K_D
BLANK	20.0	1.98	0.050	0.013	0.0025
CNA	20.0	1.81	0.059	0.017	0.0033

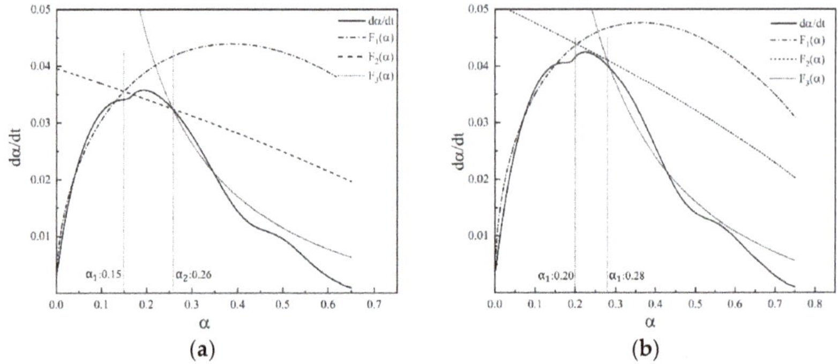

Figure 5. Hydration reaction rate curve of cement mixture with nothing (**a**) and CNA (**b**).

From Figure 5, both of the dynamic differential curves of mixtures with and without CNA can accurately fit the $d\alpha/dt$ curve obtained from the hydration heat test, which verifies that the hydration process of cement mixtures with and without CNA is dominated by the NG-I-D mechanism. In Figure 5, α_1 represents the transition point from the NG stage to the I stage, and α_2 represents the transition point from the I stage to the D stage. The mixture with CNA obtains a higher value of α_1 (0.20) than that without CNA (0.15), which indicates that CNA effectively promotes the hydration process in the NG stage. This phenomenon may be explained by that CNA sol particles with small particle size may provide a crystal nucleus for the generation of the C-S-H gel when silicate ions are saturated in the NG stage, which reduces the potential energy of C-S-H gel generation and promotes the generation of C-S-H gel. Compared with the blank group, in stage I, the CNA group's hydration degree increased by 0.02, which also promoted cement hydration.

3.2. Electrical Resistivity

Figure 6a shows the electrical resistivity curve of cement slurry with a different type of early strength agent (ρ represents electrical resistivity). All the electrical resistivity curve exhibits the same trend. Firstly, during the dissolution and precipitation processes of cement particles, more ions will be released into the system, causing a decreasing trend in the electrical resistivity curve. Then the electrical resistivity curve reaches a dynamic balance due to the competitive balance between dissolution and precipitation processes [33]. When the calcium hydroxide (CH) is supersaturated in the solution, hydration products will precipitate on the surface of the cement particles [34], which results in a rapid increase in electrical resistivity.

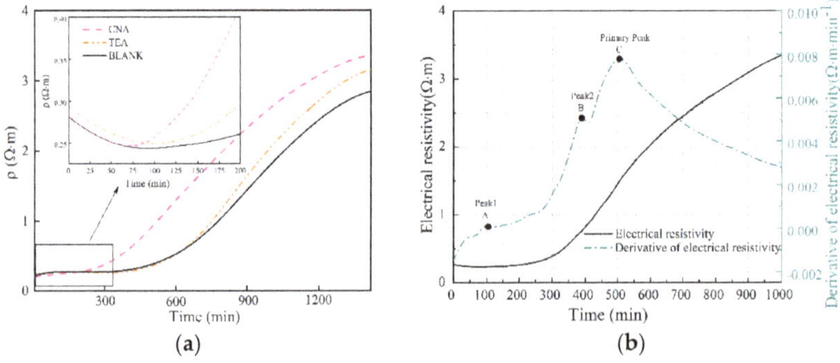

Figure 6. Resistivity development curve of cement slurry with different types of early strength agent (**a**) and the differential resistivity curve of the blank sample (**b**).

From Figure 6a, it can be found that the electrical resistivity curve with CNA increases rapidly, which is always higher than that with TEA. The electrical resistivity curve with TEA is also higher than the blank sample. This indicates that the beginning of precipitation processes of mixture with CNA is earlier than that with TEA and blank sample. Both CNA and TEA may promote the early hydration process and early strength development of cement, but CNA is better.

Peak1 appears at the beginning of the induction period, and the hydration age corresponding to Peak1 time may be understood as the beginning time of the induction period [35]. Peak2 mutation occurs during the acceleration period. At this time, the AFt content increases and the gypsum content is insufficient, and AFt starts to transform to AFm [36]. Peak2 time can be considered as the boundary point between cement setting and hardening periods. Primary Peak time corresponds to the time point where the electrical resistivity curve increases rapidly. Table 3 summarizes the characteristic points of mixtures with a different type of early strength agent.

Table 3. Characteristic time of resistivity differential curve with different early strength agents.

Types	Peak1 Time (min)	Peak2 Time (min)	Primary Peak Time (min)
BLANK	124.2	422.7	555.4
CNA	74.9	340.6	450.3
TEA	93.6	368.6	481.7

Compared with the blank sample, the Peak1 time, Peak2 time and Primary Peak time of mixtures with CNA are earlier 49 min, 82.1 min and 105 min, respectively. This indicates that CNA may effectively shorten the beginning time of the induction period, shorten the acceleration period of cement hydration and decreases the final setting time of cement. TEA may also reduce the value of Peak1 time, Peak2 time and Primary Peak time of cement mixtures, but its effect is not as well as CNA.

3.3. Amount of C-S-H

Figure 7 shows the 1-day BET adsorption diagram of cement paste with different types of early strength agents. The C-S-H content calculated based on the BET specific surface area is summarized in Table 4. The mixture with CNA obtains the maximum value of BET-specific surface area, while the mixtures with TEA possess the minimum value. The value of blank specimens is located in between these two values. Though it can be inferred that CNA may increase the compactness of the cement mixture, and TEA may decrease the compactness due to the size of the BET specific surface area being positively correlated with the compactness of the cement paste.

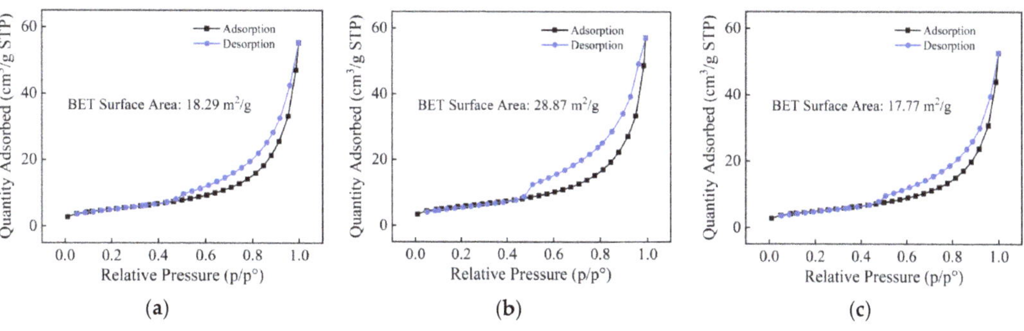

Figure 7. 1-day BET specific surface area of cement hydration products with nothing (a), CNA (b) and TEA (c).

Table 4. The C-S-H content calculation results and BET data statistics of each group.

Types	BET Surface (m^2/g)	Amount of C-S-H (%)
BLANK	18.29	13.49
CNA	28.87	25.44
TEA	17.77	12.91

From Table 4, it can be found that the mixture with CNA obtains the largest amount of C-S-H, blank sample the second, and the mixture with TEA the least. Compared with the blank sample, the mixture with TEA has lower C-S-H content. This may be explained by TEA, which can improve the early strength mainly by promoting the formation of ettringite in cement mixture rather than C-S-H gel. More ettringite is produced in cement mixture, and it even inhibits the hydration of the C_3S phase, which may be the main explanation for the lower flexural strength and late compressive strength of cement mixtures. The C-S-H content of cement mixture with CNA is higher (about twice) than the blank sample, which may be the main reason for the super early strength effect of CNA.

3.4. BSE Images Analysis

The 28-days hydration degree of cement could be obtained by the threshold segmentation of the BSE image of cement-based materials. As shown in Figure 8, the box-counting dimension of 28-days hydration products may be calculated based on the binary image, as shown in Figure 9. The fractal dimension of the 28-day hydration products and the hydration degree of the 28-day cement mixed with different early strength agents are summarized in Table 5.

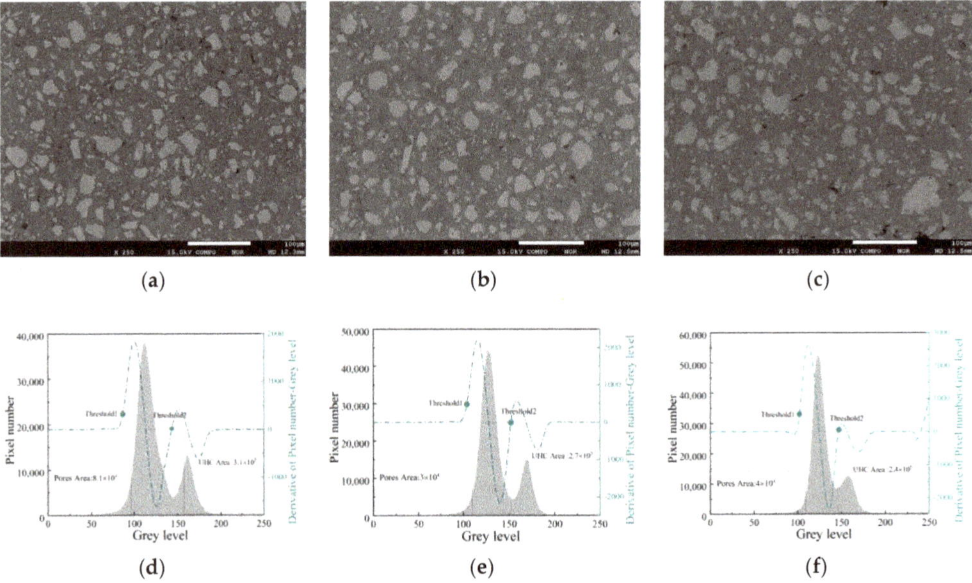

Figure 8. 28 d BSE image of cement mixture with nothing (**a**), CNA (**b**) and TEA (**c**); the gray histogram of cement mixture with nothing (**d**), CNA (**e**) and TEA (**f**).

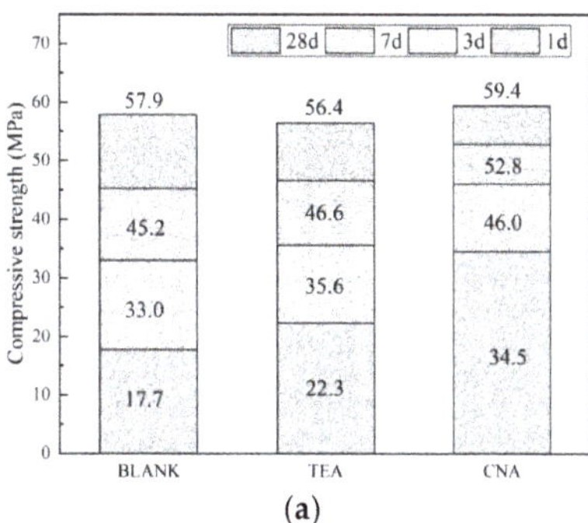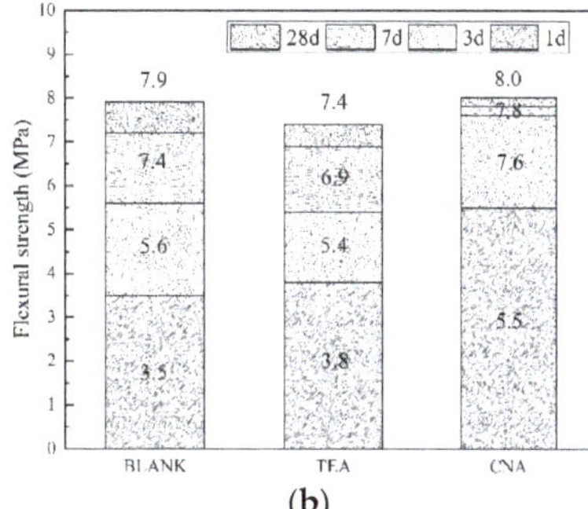

Figure 9. Compressive strength (**a**) and flexural strength (**b**) of cement mortars with different types of early strength agents.

Table 5. The hydration degree of the hardened cement slurry at 28 days.

Types	Hydration Degree	Fractal Dimension of Hydration Products
BLANK	0.632	1.62
CNA	0.641	1.67
TEA	0.615	1.56

From Table 5, it can be found that the hydration degree of the mixture with CNA (0.641) is 4.2% and 1.4%, higher than that of the mixture with TEA (0.615) and blank sample (0.632), respectively. This indicates that at the later stage of cement hydration, CNA may maintain a slightly promoting effect on the hydration of cement, but the effect is not significant. This may be explained by can, which mainly provides crystal nuclei for the formation of early cement hydration products. With the hydration process, hydration products are continuously deposited on the surface of cement particles, and the crystal nucleation effect of CNA is gradually weakened. The hydration degree of the cement mixture with TEA (0.615) is lower than the blank sample (0.632), which is a possible reason for the unsatisfactory late strength of the cement mixture with TEA [37].

It can also be found that the fractal dimension of hydration products with CNA (1.67) is 7.1% and 3.1% higher than that of mixture with TEA (1.56) and blank sample (1.62), respectively. This can be explained by the fact that more solid hydration products form and fill the pore space of cement mixture with CNA due to the higher hydration degree, which leads to a mature and complex pore network with tortuous channels [38]. Furthermore, TEA decreases the hydration degree of cement at 28-days and hinders the formation of hydration products at later ages, which is the reason that the fractal dimension of hydration products with TEA is lower than that of the blank sample.

3.5. Mechanical Properties of Cement Mortars

Figure 9 shows the compressive strength and flexural strength of cement mortars with different types of early strength agents. For all curing ages, the cement mortar with CNA obtains the largest compressive strength and flexural strength. For 1 day of age, the compressive strength of cement mortars with CNA (34.5 MPa) is 54.7% and 94.9% higher than that of mixture with TEA (22.3 MPa) and the blank sample (17.7 MPa); the flexural strength of cement mortars with CNA (5.5 MPa) is 44.7% and 57.1% higher than that of mixture with TEA (3.8 MPa) and the blank sample (3.5 MPa), respectively. For 3 days of age, the compressive strength of cement mortars with CNA (46.0 MPa) is 29.2% and 39.4% higher than that of mixture with TEA (35.6 MPa) and the blank sample (33.0 MPa); the flexural strength of cement mortars with CNA (7.6 MPa) is 40.7% and 35.7% higher than that of mixture with TEA (5.4 MPa) and the blank sample (5.6 MPa), respectively. This indicates that both CNA and TEA may significantly increase the early (1~3 days) compressive strength and flexural strength of cement mortars. In the hydration process, CNA provides nucleation sites for C-S-H and reduces the potential energy required for C-S-H nucleation, which promotes the form of C-S-H gel, fills up the remaining pores evenly and leads to a relative dense microstructure. This may explain why the cement mortars with CNA obtain the largest compressive strength and flexural strength. Meanwhile, as a traditional organic early-strength agent, TEA can chelate the aluminate phase to promote the formation of ettringite and ultimately improve the early strength of cement mortars [39].

For 28-days of age, the compressive strength of cement mortars with CNA (59.4 MPa) is 2.6% higher than the blank sample (57.9 MPa), while the compressive strength of cement mortars with TEA (56.4 MPa) is 2.7% lower than blank sample (57.9 MPa). The flexural strength has the same trend. This indicates that CNA has little effect on the mechanical properties of cement in the later stage, while TEA has certain side effects on the mechanical properties of cement in the later stage. This is consistent with the research results of Zhang et al. [40].

4. Conclusions

In this study, the influence of CNA on cement hydration mechanism, early strength enhancement effect, 28 d hydration degree and 28 d hydration product fractal dimension was mainly investigated via hydration kinetic calculation, resistivity test, BSE image analysis and mechanical properties. Triethanolamine (TEA) and the blank group are the control groups. The following results were obtained:

1. The incorporation of CNA does not change the hydration mechanism of cement, which increases the hydration degree in the nucleation and crystal growth stage (NG) by 0.5, and the interactions at phase boundaries diffusion stage(I) by 0.02, which makes the cement obtain a higher degree of hydration in the initial stage of hydration;
2. CNA shortens the beginning time of the induction period by 49.3 min and the end time of the cement hydration acceleration period by 105.1 min than the Blank sample, which can accelerate the setting and hardening of the cement;
3. CNA significantly improves the hydration degree of cement mixtures cured for 28 days, which is better than triethanolamine (TEA). CNA increases the fractal dimension of hydration products, while TEA decreases the fractal dimension;
4. CNA used in this work may provide a crystal nucleus for the generation of C-S-H gel, which reduces the potential energy of C-S-H gel generation and promotes the generation of C-S-H gel. The C-S-H content of cement mixtures with CNA is higher than that with TEA or the blank sample;
5. CNA significantly improves the early strength of cement mortars; the 1-day and 3-day strength of cement mortars with CNA are more than the 3-day and 7-day strength of the blank sample, which shows that it has an excellent early strength effect.

Author Contributions: Conceptualization, Y.J.; data curation, W.Z. and T.P.; formal analysis, W.Z., Y.J. and X.J.; funding acquisition, Y.J.; methodology, X.J. and T.P.; project administration, Y.J.; writing—original draft, W.Z.; writing—review and editing, Y.J., X.J. and W.Z. All authors have read and agreed to the published version of the manuscript.

Funding: This work was supported by the National Natural Science Foundation of China [grant numbers 51738003, 11772120] and the National Natural Science Foundation of China [grant numbers 52008191, 11772120] and the National Natural Science Foundation of China [grant numbers 51978318]; and the Primary Research and Development Plan of Jiangsu Province [grant number BE2016187].

Institutional Review Board Statement: Not applicable.

Informed Consent Statement: Not applicable.

Data Availability Statement: Data sharing is not applicable.

Conflicts of Interest: The authors declare no conflict of interest. The funders had no role in the design of the study; in the collection, analyses, or interpretation of data; in the writing of the manu-script, or in the decision to publish the results.

References

1. Scrivener, K.L.; John, V.M.; Gartner, E.M. Eco-efficient cements: Potential economically viable solutions for a low-CO_2 cement-based materials industry. *Cem. Concr. Res.* **2018**, *114*, 2–26. [CrossRef]
2. Du, Q.; Pang, Q.; Bao, T.; Guo, X.; Deng, Y. Critical factors influencing carbon emissions of prefabricated building supply chains in China. *J. Clean. Prod.* **2021**, *280*, 124398. [CrossRef]
3. Chang, Y.; Li, X.; Masanet, E.; Zhang, L.; Huang, Z.; Ries, R. Unlocking the green opportunity for prefabricated buildings and construction in China. *Resour. Conserv. Recycl.* **2018**, *139*, 259–261. [CrossRef]
4. Yu, S.; Liu, Y.; Wang, D.; Bahaj, A.S.; Wu, Y.; Liu, J. Review of thermal and environmental performance of prefabricated buildings: Implications to emission reductions in China. *Renew. Sustain. Energy Rev.* **2021**, *137*, 110472. [CrossRef]
5. Zhang, C.; Hu, M.; Laclau, B.; Garnesson, T.; Yang, X.; Tukker, A. Energy-carbon-investment payback analysis of prefabricated envelope-cladding system for building energy renovation: Cases in Spain, the Netherlands, and Sweden. *Renew. Sustain. Energy Rev.* **2021**, *145*, 111077. [CrossRef]
6. Wang, J.; Qian, C.; Qu, J.; Guo, J. Effect of lithium salt and nano nucleating agent on early hydration of cement based materials. *Constr. Build. Mater.* **2018**, *174*, 24–29. [CrossRef]
7. Zou, C.; Long, G.; Ma, C.; Xie, Y. Effect of subsequent curing on surface permeability and compressive strength of steam-cured concrete. *Constr. Build. Mater.* **2018**, *188*, 424–432. [CrossRef]
8. Castellano, C.C.; Bonavetti, V.L.; Donza, H.A.; Irassar, E.F. The effect of w/b and temperature on the hydration and strength of blastfurnace slag cements. *Constr. Build. Mater.* **2016**, *111*, 679–688. [CrossRef]
9. Zou, C.; Long, G.; Zeng, X.; Ma, K.; Xie, Y. Hydration and multiscale pore structure characterization of steam-cured cement paste investigated by X-ray CT. *Constr. Build. Mater.* **2021**, *282*, 122629. [CrossRef]
10. Sun, J.; Shi, H.; Qian, B.; Xu, Z.; Li, W.; Shen, X. Effects of synthetic C-S-H/PCE nanocomposites on early cement hydration. *Constr. Build. Mater.* **2017**, *140*, 282–292. [CrossRef]
11. Alizadeh, R.; Raki, L.; Makar, J.M.; Beaudoin, J.J.; Moudrakovski, I. Hydration of tricalcium silicate in the presence of synthetic calcium–silicate–hydrate. *J. Mater. Chem.* **2009**, *19*, 7937–7946. [CrossRef]
12. Land, G.; Stephan, D. The influence of nano-silica on the hydration of ordinary Portland cement. *J. Mater. Sci.* **2012**, *7*, 1011–1017. [CrossRef]
13. Yang, D.; Yan, C.; Zhang, J.; Liu, S.; Li, J. Chloride threshold value and initial corrosion time of steel bars in concrete exposed to saline soil environments. *Constr. Build. Mater.* **2021**, *267*, 120979. [CrossRef]
14. Lu, Z.; Kong, X.; Jansen, D.; Zhang, C.; Wang, J.; Pang, X.; Yin, J. Towards a further understanding of cement hydration in the presence of triethanolamine. *Cem. Concr. Res.* **2020**, *132*, 106041. [CrossRef]
15. Andrade Neto, J.d.S.; De la Torre, A.G.; Kirchheim, A.P. Effects of sulfates on the hydration of Portland cement—A review. *Constr. Build. Mater.* **2021**, *279*, 122428. [CrossRef]
16. John, E.; Matschei, T.; Stephan, D. Nucleation seeding with calcium silicate hydrate—A review. *Cem. Concr. Res.* **2018**, *113*, 74–85. [CrossRef]
17. Li, J.; Zhang, W.; Xu, K.; Monteiro, P.J.M. Fibrillar calcium silicate hydrate seeds from hydrated tricalcium silicate lower cement demand. *Cem. Concr. Res.* **2020**, *137*, 106195. [CrossRef]
18. Land, G.; Stephan, D. Controlling cement hydration with nanoparticles. *Cem. Concr. Compos.* **2015**, *57*, 64–67. [CrossRef]
19. Meng, T.; Hong, Y.; Wei, H.; Xu, Q. Effect of nano-SiO2 with different particle size on the hydration kinetics of cement. *Thermochim. Acta* **2019**, *675*, 127–133. [CrossRef]
20. Lu, Y.; Shi, G.; Liu, Y.; Ding, Z.; Pan, J.; Qin, D.; Dong, B.; Shao, H. Study on the effect of chloride ion on the early age hydration process of concrete by a non-contact monitoring method. *Constr. Build. Mater.* **2018**, *172*, 499–508. [CrossRef]
21. Monteiro, P.J.M.; Geng, G.; Marchon, D.; Li, J.; Alapati, P.; Kurtis, K.E.; Qomi, M.J.A. Advances in characterizing and understanding the microstructure of cementitious materials. *Cem. Concr. Res.* **2019**, *124*, 105806. [CrossRef]
22. Igarashi, S.; Kawamura, M.; Watanabe, A. Analysis of cement pastes and mortars by a combination of backscatter-based SEM image analysis and calculations based on the Powers model. *Cem. Concr. Compos.* **2004**, *26*, 977–985. [CrossRef]
23. Li, Y.; Guo, W.; Li, H. Quantitative analysis on ground blast furnace slag behavior in hardened cement pastes based on backscattered electron imaging and image analysis technology. *Constr. Build. Mater.* **2016**, *110*, 48–53. [CrossRef]
24. Shuxia, F.; Peiming, W.; Liu, X. SEM-backscattered electron imaging and image processing for evaluation of unhydrated cement volume fraction in slag blended Portland cement pastes. *J. Wuhan Univ. Technol. Mat. Sci. Ed.* **2013**, *28*, 968–972.
25. Zhao, H.; Darwin, D. Quantitative backscattered electron analysis of cement paste. *Cem. Concr. Compos.* **1992**, *22*, 695–706. [CrossRef]
26. Scrivener, K.L. Backscattered electron imaging of cementitious microstructures: Understanding and quantification. *Cem. Concr. Compos.* **2004**, *26*, 935–945. [CrossRef]
27. Gao, Y.; Wu, K.; Yuan, Q. Limited fractal behavior in cement paste upon mercury intrusion porosimetry test: Analysis and models. *Constr. Build. Mater.* **2021**, *276*, 122231. [CrossRef]
28. De Belie, N.; Kratky, J.; Van Vlierberghe, S. Influence of pozzolans and slag on the microstructure of partially carbonated cement paste by means of water vapour and nitrogen sorption experiments and BET calculations. *Cem. Concr. Res.* **2010**, *40*, 1723–1733. [CrossRef]
29. Odler, I. The BET-specific surface area of hydrated Portland cement and related materials. *Cem. Concr. Res.* **2003**, *33*, 2049–2056. [CrossRef]
30. Olson, R.A.; Jennings, H.M. Estimation of C-S-H content in a blended cement paste using water adsorption. *Cem. Concr. Res.* **2001**, *31*, 351–356. [CrossRef]
31. Chen, C.; Jiang, Y.Q.; Pan, Y.F.; Xiang, Q.D. Semi-quantitative study on hydrated calcium silicate by BET/XRD. *New Build Mater.* **2017**, *44*, 124–126.

32. Li, H.; Xue, Z.; Liang, G.; Wu, K.; Dong, B.; Wang, W. Effect of C-S-Hs-PCE and sodium sulfate on the hydration kinetics and mechanical properties of cement paste. *Constr. Build. Mater.* **2021**, *266*, 121096. [CrossRef]
33. Yousuf, F.; Xiaosheng, W. Investigation of the early-age microstructural development of hydrating cement pastes through electrical resistivity measurements. *Case Stud Constr Mater.* **2020**, *13*, e00391.
34. Scrivener, K.L.; Juilland, P.; Monteiro, P.J.M. Advances in understanding hydration of Portland cement. *Cem. Concr. Res.* **2015**, *78*, 38–56. [CrossRef]
35. Yousuf, F.; Wei, X.; Zhou, J. Monitoring the setting and hardening behaviour of cement paste by electrical resistivity measurement. *Constr. Build. Mater.* **2020**, *252*, 118941. [CrossRef]
36. Liu, L.; Yang, P.; Zhang, B.; Huan, C.; Guo, L.; Yang, Q.; Song, K.-I. Study on hydration reaction and structure evolution of cemented paste backfill in early-age based on resistivity and hydration heat. *Constr. Build. Mater.* **2021**, *272*, 121827. [CrossRef]
37. Yang, X.; Liu, J.; Li, H.; Xu, L.; Ren, Q.; Li, L. Effect of triethanolamine hydrochloride on the performance of cement paste. *Constr. Build. Mater.* **2019**, *200*, 218–225. [CrossRef]
38. Chen, X.; Yao, G.; Herrero-Bervera, E.; Cai, J.; Zhou, K.; Luo, C.; Jiang, P.; Lu, J. A new model of pore structure typing based on fractal geometry. *Mar. Petrol. Geol.* **2018**, *98*, 291–305. [CrossRef]
39. Yaphary, Y.L.; Yu, Z.; Lam, R.H.W.; Lau, D. Effect of triethanolamine on cement hydration toward initial setting time. *Constr. Build. Mater.* **2017**, *141*, 94–103. [CrossRef]
40. Yan-Rong, Z.; Xiang-Ming, K.; Zi-Chen, L.; Zhen-Bao, L.; Qing, Z.; Bi-Qin, D.; Feng, X. Influence of triethanolamine on the hydration product of portlandite in cement paste and the mechanism. *Cem. Concr. Res.* **2016**, *87*, 64–76. [CrossRef]

materials

Article

Mechanical Behavior Investigation of Reclaimed Asphalt Aggregate Concrete in a Cold Region

Wenyuan Xu, Wei Li * and Yongcheng Ji *

School of Civil Engineering, Northeast Forestry University, Harbin 150040, China; xuwenyuan@nefu.edu.cn
* Correspondence: yongchengji@126.com (Y.J.); alice091210@nefu.edu.cn (W.L.); Tel.: +86-15104571851 (Y.J.)

Abstract: Recycled construction and demolition (C&D) waste can reduce the rebuild cost, and is environmentally friendly when recycled asphalt pavement (RAP) aggregate constitutes the main part. This paper investigated the mechanical performance of RAP concrete, and the applicability of RAP in road base layers also was discussed. Several mechanical laboratory tests were selected, including the unconfined compressive-strength, splitting-strength, and compressive-resilience modulus tests. The RAP concrete had a good road performance in a cold region, which was proved by the temperature-shrinkage test, dry-shrinkage test, freeze–thaw-cycle test, and water-stability test. Based on various cement dosages from 3.5% to 5.5% in RAP concrete mix design, three RAP aggregate replacement ratios (30%, 40%, and 50%) were selected to study the variation of mechanical properties with increasing curing time, and the optimal aggregate substitute ratio was determined. A scanning electron microscope (SEM) was used to observe the inner-structure interface between the asphalt binder and cement stone. A numerical model is presented to simulate the RAP compressive strength with respect to the effect of multiple parameters. The research results can provide a technical reference for RAP use in the reconstruction and expansion of low-grade highway projects.

Keywords: recycled asphalt pavement (RAP) aggregate; mechanical performance; numerical model

Citation: Xu, W.; Li, W.; Ji, Y. Mechanical Behavior Investigation of Reclaimed Asphalt Aggregate Concrete in a Cold Region. *Materials* **2021**, *14*, 4101. https://doi.org/10.3390/ma14154101

Academic Editor: Carlos Morón Fernández

Received: 11 June 2021
Accepted: 21 July 2021
Published: 23 July 2021

Publisher's Note: MDPI stays neutral with regard to jurisdictional claims in published maps and institutional affiliations.

1. Introduction

Construction and demolition (C&D) waste management has become a worldwide concern, as up to 600 million tons of waste construction materials are produced each year. Generally, C&D waste mainly contains coarse and fine sand aggregates, aging asphalt, hardened cement hydrate, and other components [1–4]. C&D waste-recycling research began near the end of the 20th century in the United States, when the Texas Department of Transportation initially investigated the feasibility of using a waste–asphalt mixture in highway construction and maintenance application in 1994. A large number of waste materials were attempted to be used in sub-grade fillers, pavement bases, and other infrastructure construction. The United States saved 4.1 million tons of matrix asphalt and 78 million tons of natural stone in 2018 [5,6]. Similarly, the European Asphalt Pavement Association suggested all its member countries use recycled waste–asphalt materials in 2002. Over 90% of waste–asphalt mixtures have been used for pavement and pavement base materials [7–9]. However, with the development of urbanization, about 1.7 billion tons of C&D waste (2019) was generated in China, and the recycling rate is far lower than that in developed countries [10–13]. Thus, using reclaimed C&D waste in new highway construction is a promising way to solve these imminent issues.

As the largest part of ordinary concrete mixes, the excessive consumption (approximately 26 billion tons per year) of nature aggregate (NA) leads to harmful environmental pollution, and a possible solution to improve the sustainability and cost-effectiveness between C&D and NA needs to be investigated [14–17]. The reuse of recycled concrete aggregate (RCA) and recycled asphalt pavement (RAP) aggregate in highway rehabilitation has attracted extensive attention in recent years. The RAP aggregate is a mixture of aggregate and bitumen, mainly derived from old asphalt pavement. Due to the asphalt binder, the RAP aggregate has a worse environmental effect and weaker inner bond when compared with RCA.

Existing studies focus on RAP's material properties and optimal replacement ratio with NA [18–22]. Saeed and Reza [23] evaluated the performance of recycled asphalt mixtures in C&D waste materials, and the optimal binder content was determined. Test results showed that the rutting resistance was effectively improved by 30% in recycled aggregates. Akash et al. [24] used rheological and chemical methods to investigate asphalt binder and mixtures, and various recycling agents were divided into three categories. A novel parameter was developed to predict the effectiveness of various recycling agents. Wojciech [25] evaluated the fatigue life of ASP in an asphalt–concrete mixture, and several lab tests (air-void content, penetration, stiffness) were made to evaluate the mixture's mechanistic performance. A French method was presented to calculate the mixture's fatigue life. Juntao et al. [26] tested an eight-year recycled asphalt mixture bound by emulsion, and the long-term performance and interface microstructure were discussed. Test results showed that the tensile strengths and creep deformation met the performance requirement. Hassan [27] studied the behavior of reclaimed asphalt pavement (RAP) aggregate concrete, and various mixtures (natural aggregate, reclaimed coarse aggregate, reclaimed coarse and fine aggregate, and reclaimed coarse aggregate with 30% fly ash) were selected to find the optimum performance. Research results showed that the RAP aggregate partially reduced the concrete's mechanical performance (compression and tensile strength). However, the properties of ductility and microstructure were improved due to the effect of fly ash and RAP aggregate. Papakonstantinou [28] investigated the performance of recycled asphalt pavement (RAP) aggregate use in Portland cement concrete (PCC). Five weight percentages (5%, 7.5%, 10%, 12.5%, and 15%) of RAP aggregates were used in concrete mix design, and their mechanical performance was examined. The test results found that the compression strength and elastic modulus had a negative relation with increasing RAP ratio, and all mixtures met the requirements for road performance. Zaumanis et al. [29] proposed a performance-based design method to solve the asphalt mix design procedures, and the key parameters were determined to improve the asphalt mixture's performance.

Other studies investigated RAP concrete applications, including new asphalt pavement, base layers, structural members, and so on [30,31]. Giulia et al. [32] summarized the development of reclaimed asphalt pavement (RAP) material used in new asphalt construction, and the effect of RAP content was discussed. Abdulgazi [33] discussed the potential utilization of construction demolition waste (CDW) using in hot-mix asphalt pavements, and alternative CDW technical specifications and guidelines were presented that mainly considered safety, tolerability, and efficacy. Fawaz et al. [34] summarized the current knowledge about reclaimed asphalt pavement (RAP) and recycled asphalt shingles (RAS) in the United States, and the current and future challenges for reclaimed asphalt utilization were presented. Sharareh et al. [35] evaluated the effects of recycled asphalt shingles (RAS) on pavement performance. A cost analysis was conducted to assess the life-cycle cost of asphalt pavements constructed with RAS. Then, a mix with 5% post-consumer waste shingles and recycled asphalt had the lowest cost over the pavement's service life. Nasim [36] investigated axial compression behavior of concrete columns concerning four types of aggregates (NA, RCA, RAP, and RCA-RAP). Reclaimed aggregate replacement ratios from 20% to 100% also were considered. Test results indicated that the ultimate failure load had a decreasing trend with the increase of the reclaimed aggregate substitute ratio, but the structure's safety was proved.

This study's main objective was to investigate the mechanical performance of RAP concrete, and several mechanical and physical laboratory tests were selected to evaluate its road performance in a cold region. Based on the various cement dosages, from 3.5% to 5.5% in the RAP concrete mix design, three recycled aggregate mixture contents (30%, 40%, and 50%) were selected to study the variation of mechanical behavior concerning different curing times, and the optimal RAP aggregate-substitute ratio was determined. A scanning electron microscope (SEM) was used to observe the inner-structure interface between the asphalt binder and cement stone. A numerical model was created to simulate the RAP's compressive strength with respect to the effect of multiple parameters. The research results can provide a technical reference for RAP use in the reconstruction and expansion of low-grade highway projects.

2. Experimental Programs

2.1. Reclaimed Asphalt Mixture Sieve Analysis and Mix Design

Typically, asphalt pavement needs a high temperature for constant construction, and a thermal aging action occurs. In addition, the aging effect will be accelerated by light, weathering, snow/ice cover, rainwater penetration, and traffic load. In this paper, the reclaimed asphalt mixture came from the reconstruction and expansion of the National Dana Highway. The pavement structure was built in 2005, and consisted of 10 cm-thick asphalt concrete pavement. Due to the aging effect, the mechanical performance of the existing asphalt pavement cannot meet the serviceability. Therefore, the more extensive mixture was broken manually and then crushed with a small jaw crusher. Figure 1a,b show the RAP coarse and fine mixtures after being crushed by the jaw crusher, and some of them are clustered structures because of asphalt bonding. In addition, it can be observed that the asphalt and aggregate were separated due to sunlight weathering and traffic load. In order to reduce the difference of gradation composition, the RAP aggregate was sieved with an asphalt mixture centrifugal-extraction apparatus, as shown in Figure 1c,d.

Figure 1. Reclaimed asphalt mixture: (**a**) coarse aggregate; (**b**) fine aggregate; (**c**) untreated. (**d**) Extracted RAP particle-size distribution (PSD).

Table 1 shows the physical properties of the RAP and NA aggregates. The RAP aggregate exerts a higher void content and lowers apparent density, and it can be attributed to uneven surfaces and microcracks generated during the RAP crushing process. In addition, the existing asphalt and cement binder lead to the enlarging of the inner air void, eventually causing a noticeable increase in water absorption for RAP.

Table 1. Physical properties of aggregates.

Aggregate Type	Size (mm)	Apparent Density (g/cm^3)	Water Absorption (%)	Void Content (%)
RAP	5–10	2.618	7.41	44.5
	10–20	2.666	5.31	45.3
NA	5–10	2.721	0.91	42
	10–20	2.719	0.45	43

Considered with the various cement dosages from 3.5% to 5.5%, three recycled aggregate mixture contents (30%, 40%, and 50%) were selected in the RAP concrete mix design. The F30-S3.5 specimen represents 30% RAP and 70% NA, and the cement dosage is 3.5%. A total of nine combination specimen types were used, and their optimum water content and maximum dry densities are shown in Table 2.

Table 2. RAP concrete mix design.

No.	Optimum Water Content (%)	Maximum Dry Density (g/cm^3)
F30-S3.5	5.24	2.25
F40-S3.5	4.93	2.29
F50-S3.5	4.90	2.21
F30-S4.5	5.25	2.29
F40-S4.5	5.05	2.31
F50-S4.5	5.04	2.22
F30-S5.5	5.30	2.31
F40-S5.5	5.22	2.33
F50-S5.5	5.21	2.25

2.2. Axial Compressive-Strength Test

According to the Highway Engineering Inorganic Binder Stability Materials Code (JTG E51-2009), a cylinder size with 150 mm diameter by 150 mm depth was selected. Its compression test setup is shown in Figure 2. Two variable parameters were considered in the experimental test: a reclaimed aggregate replacement ratio from 30% to 50%, and cement dosage from 3.5% to 5.5%. The specimens' compressive strengths were compared with respect to various curing times, from 7 to 90 days.

Figure 2. Axial compressive-strength test.

2.3. Splitting Tensile Strength Test

According to Highway Inorganic Bond Stabilization Materials Splitting Test Method (JTG E51-2009 T0806-1994), specimens with different cement dosages and reclaimed asphalt mixtures were selected for 28- and 90-day curing times. Figure 3 shows the splitting-test setup and one typical test specimen.

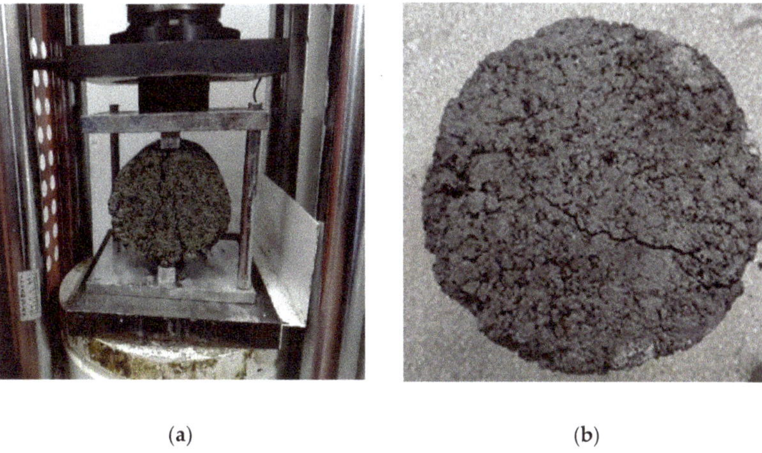

(a)   (b)

Figure 3. Splitting tensile-strength test: (**a**) test setup; (**b**) test specimen.

Similar to the splitting-strength test, specimens with two different curing times (28 and 90 days) were selected to test their compressive resilience moduli, and the relationship between multi-stage loading and their deformation was considered, as shown in Equation (1):

$$E_c = \frac{ph}{l} \tag{1}$$

where E_c is compressive resilience modulus (MPa); p is unit pressure (MPa); h is the specimen height (mm); and l is the resilient deformation (mm). Figure 4 shows the compressive-resilience modulus test setup and one typical test specimen.

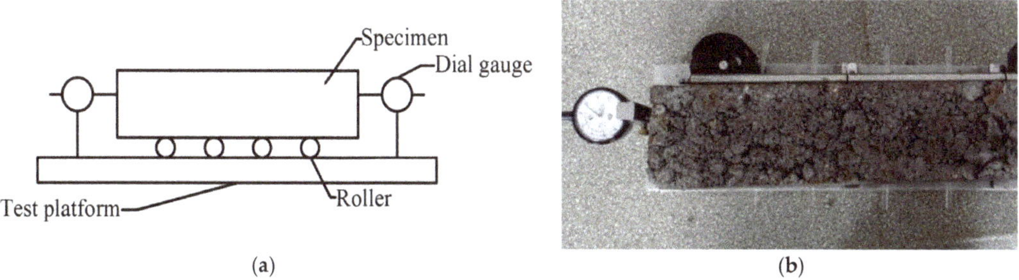

(a)   (b)

Figure 4. Compressive-resilience modulus test: (**a**) test setup; (**b**) test specimen.

2.4. Dry- and Temperature-Shrinkage Tests

According to Highway Inorganic Bond Stabilization Materials Splitting Test Method (T0854-2009), water-loss rate and dry-shrinkage strain were selected to investigate the ability of deformation resistance under the presence of pore water. The water-loss rate was calculated with the mass variation when subjected to immersion and dry conditions, as shown in Figure 5a. The dry-shrinkage strain was measured with a micrometer gauge, and the monitor period was divided into two interval times, which was recorded daily during the first week, and extended from 2 to up to 27 days.

(a)   (b)

Figure 5. Shrinkage test: (**a**) water immersion; (**b**) temperature-controlled cabinet.

Similarly, the temperature-shrinkage strain was measured with a micrometer gauge, and a lab temperature-controlled cabinet was required. Based on the on-site monitored data in Huma town (China), temperatures of 14.9 °C~18.7 °C in summer and $-30.2 \sim -20.1$ °C in winter are observed. Thus, the temperature-shrinkage measuring zone was determined to be from 20 °C to -30 °C with an interval of 10 °C, as shown in Figure 5b.

2.5. Freeze-Thaw Cycle and Water-Stability Test

Most road-base materials are macroporous, and strength loss frequently occurs during the freeze–thaw cycle. The specimens' mass and strength loss needed to be investigated through a freeze–thaw test in the seasonally frozen ground region. The 28-day and 90-day specimens were selected for the freeze–thaw test, and 10 cycles were used (frozen for 16 h at -18 °C and thawed for 8 h in 20 °C = one cycle), as shown in Figure 6a.

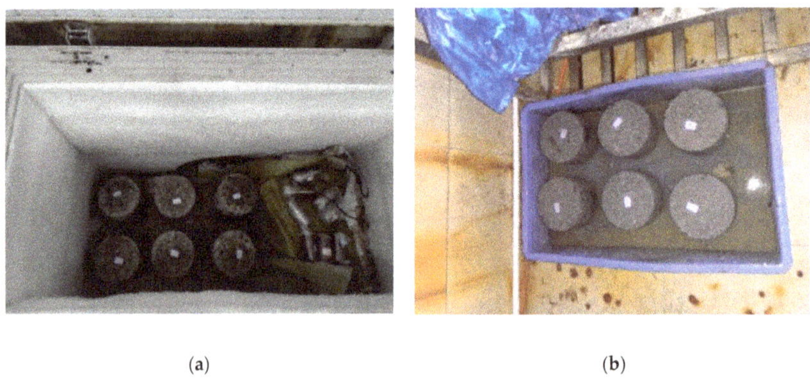

(a)   (b)

Figure 6. Freeze–thaw cycle and water-stability test: (**a**) freeze–thaw test; (**b**) water immersion.

The water-stability test examined the waterproof performance or water resistance of the road sub-base layer. For example, one water-stability cycle included initial immersion in water for 1 d and air-drying for 2 d, and then soaking in water for 1 d and air-drying for 2 d, as shown in Figure 6b. After five cycles, the compressive strength of the specimen was measured and compared with the specimen without immersion, and the calculation formula is shown in Equation (2):

$$S = \frac{R_{sc}}{R_c} \times 100 \tag{2}$$

where S is the coefficient of water stability; R_{sc} is the specimen strength after five cycles; and R_c is specimen strength without cyclic immersion.

3. Experimental Test Results

3.1. Sieve Analysis

Based on the centrifugal separation method in the Highway Engineering Asphalt and Asphalt Mixture code (JTG E20-2011), a centrifugal extraction apparatus was used to extract the waste–asphalt mixture. The 7- and 10-year RAP samples were selected, and softening point and penetration were tested to analyze the aging effects. With the increase of service life, penetration showed a decreasing trend, but the softening point showed the opposite (penetration: 65.1% for 7 years vs. 38.3% for 10 years, and softening point: 11.6% vs. 41.0%). It was observed that the coarse aggregate occupied a high portion of the waste–asphalt mixture, and a particle size below 2.36 mm accounted for about 10% of the total mass. This can be attributed to the fine aggregate bonding with asphalt to form larger particles, as shown in Figure 7a. The performance of asphalt deteriorated with time, and the sieve results of the extracted mixture are shown in Figure 7b. Compared with the untreated asphalt mixture, the fine aggregate particle-size distribution showed an apparent improvement between the upper and lower limits. It can be concluded that the untreated RAP aggregate required the extraction process before casting concrete to meet the gradation composition requirement.

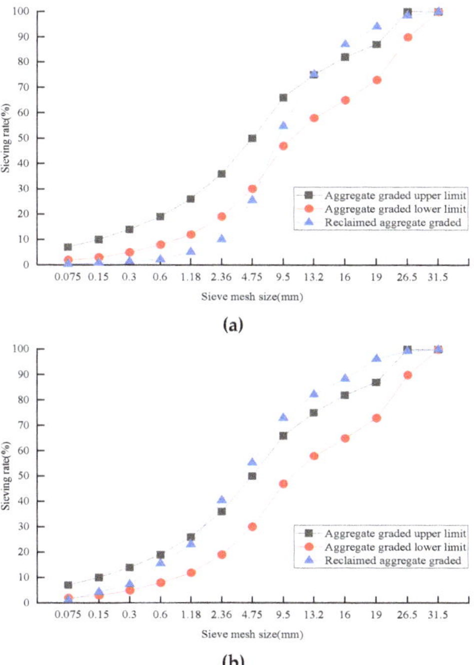

Figure 7. Reclaimed asphalt sieve analysis: (**a**) untreated RAP mixture; (**b**) extracted RAP mixture.

3.2. Axial Compressive-Strength Tests

Figures 8–10 show the RAP axial compressive-strength variation when subjected to cement dosages from 3.5% to 5.5%. It was observed that the cement dosage had a positive effect on axial strength, and the optimal cement dosage was 5.5%. Similarly, the optimal RAP substitute ratio was determined to be 40%.

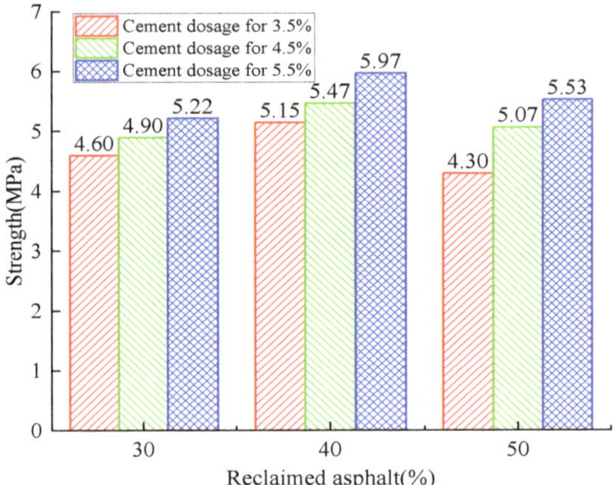

Figure 8. Analysis of the 7-day axial compressive-strength test with variation of cement dosage and RAP.

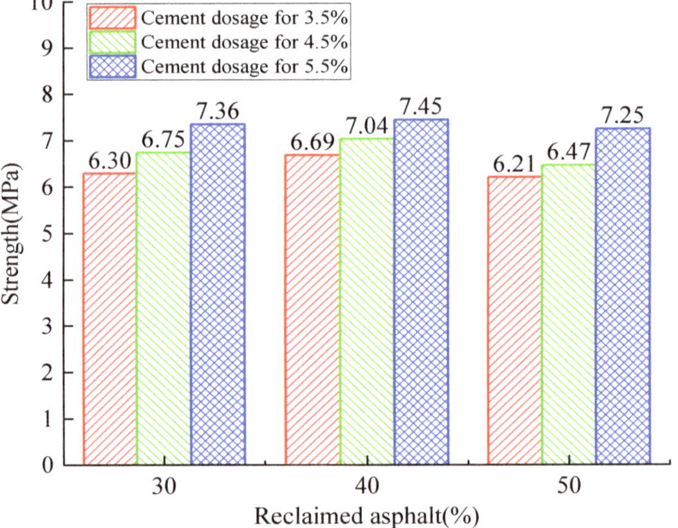

Figure 9. Analysis of the 28-day axial compressive-strength test with variation of cement dosage and RAP.

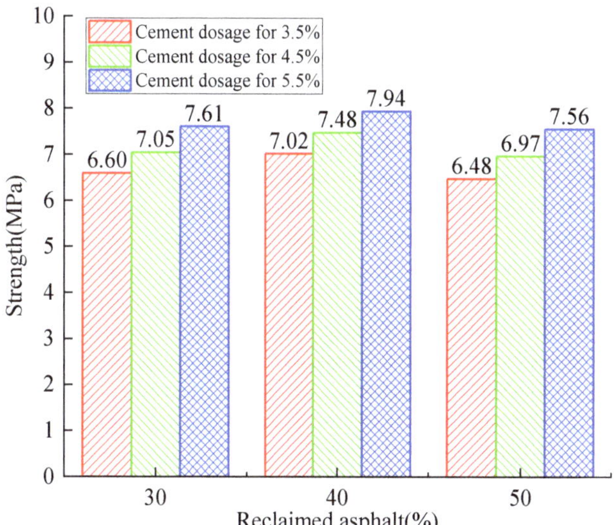

Figure 10. Analysis of the 90-day axial compressive-strength test with variation of cement dosage and RAP.

The relationship between axial compressive strength and curing time is shown in Figure 11. When the RAP replacement ratio was increased from 30% to 40%, the axial compressive strength increased by 1.22% to 14.37% as the cement dosage remained constant. On the contrary, the axial compressive strength decreased by 2.68% to 16.50% when the RAP replacement ratio reached 50%. Therefore, it can be concluded that the axial strength increases initially and decreases afterward with the increase of the RAP replacement ratio. However, the axial compressive strength exerted a continuous positive effect (4.19~17.91% for 4.5% and 5.82~12.06% for 5.5%) with the increase of cement dosage when the RAP content remained constant. All these observations show that a critical RAP substitute ratio and cement dosage exist to achieve the maximum RAP axial strength. The cost efficiency and environmental friendliness also need to be balanced.

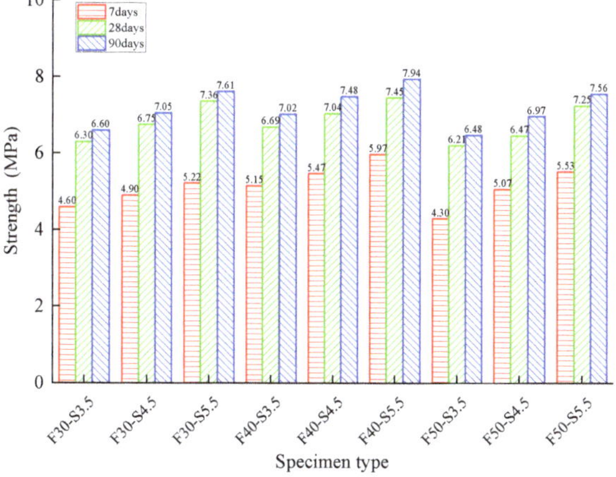

Figure 11. The relationship between axial compressive strength and curing time for various specimens.

Similarly, the compressive strength showed an increasing trend with curing time no matter the amount of cement dosage and RAP (24.79~40.99% from 7 to 28 days, and 4.28~7.73% from 28 to 90 days). The increasing trend turning flat can be explained by the hydration reaction gradually finishing.

3.3. Splitting Tensile-Strength Tests

The 28-day specimen's splitting-strength relationship between RAP and cement dosage is shown in Figure 12. It can be observed that the RAP specimen's splitting strength showed an upward trend with the increase of cement dosage. For the different RAP replacement ratios, an increasing trend was observed for all cement dosage cases, but the growth rate and maximum splitting strength varied. The maximum growth rate was 21.4% when subjected to the 5.5% cement dosage and 50% RAP replacement ratio. For the 30% RAP replacement ratio and 5.5% cement dosage, the maximum splitting strength reached 0.65 MPa. Similarly, the splitting strength showed a slowly decreasing trend with an increase of the reclaimed asphalt mixture replacement ratio. The higher cement dosage showed a better splitting resistance (16.21% for 3.5% vs. 9.1% for 5.5%). The lower RAP replacement ratio showed a higher splitting strength. It can be concluded that the RAP aggregate substitute NA had a negative influence on concrete's splitting tensile strength, and the increasing of cement dosage effectively relieved this strength degradation.

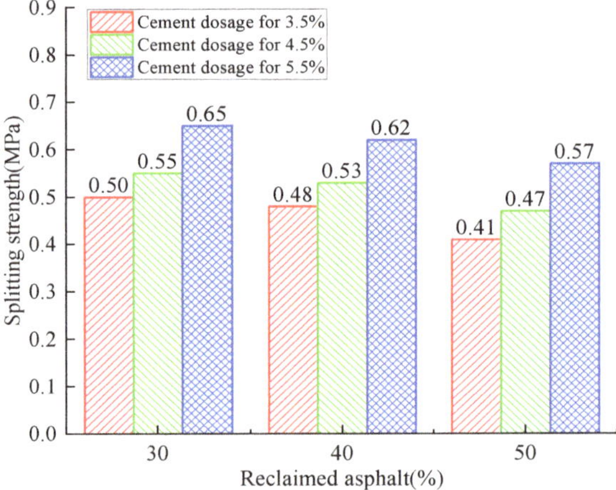

Figure 12. Analysis of the 28-day reclaimed asphalt splitting tensile-strength test with variation of cement dosage and RAP.

For 90 days of curing time, the splitting strength relationship between cement dosage and reclaimed asphalt replacement ratio is shown in Figure 13. The 90-day splitting strength had a similar increasing trend to that of 28 days, but a higher splitting strength and growth rate were observed. For the 30% reclaimed asphalt mixture, the highest growth rate of 39.1% was observed when compared with cement dosage from 3.5% to 4.5%. Therefore, the maximum splitting strength was 0.95 MPa for the 30% reclaimed asphalt mixture and 5.5% cement dosage.

Figure 14 shows the 90-day splitting strength concerning the variation of reclaimed asphalt mixture and cement dosage. The splitting strength showed a descending trend with the increase of reclaimed asphalt. The 4.5% cement dosage and 50% reclaimed asphalt exerted the highest strength-loss rate (8.8%). Compared with 28 days, specimens' splitting strength growth rate showed an increasing trend, up to 44.7% for 90 days curing time.

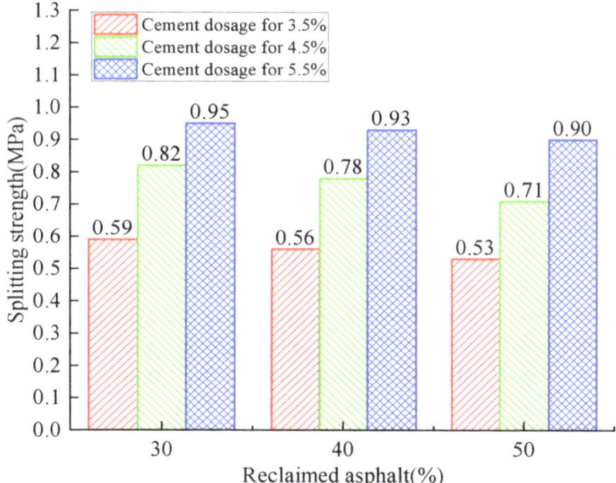

Figure 13. Analysis of the 90-day reclaimed asphalt splitting tensile-strength test with variation of cement dosage and RAP.

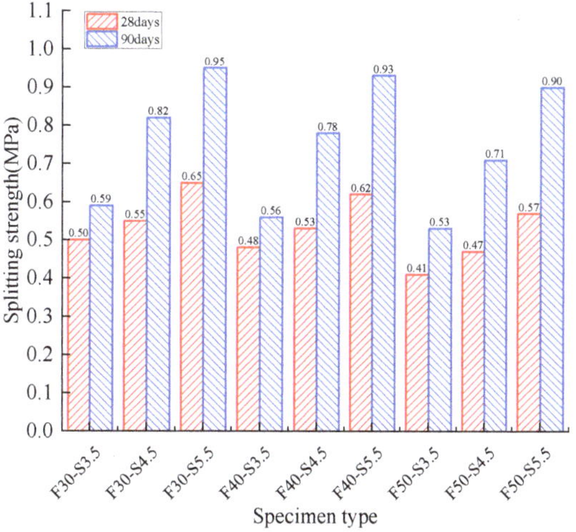

Figure 14. Reclaimed asphalt splitting tensile strength with respect to curing time.

It can be concluded that the cement dosage had a positive effect on splitting strength when the reclaimed asphalt mixture remained constant. However, reclaimed asphalt had a negative effect as the cement dosage was constant. Therefore, the splitting strength was a positively correlated function of curing time if the cement dosage and reclaimed asphalt remained constant.

The compressive resilience modulus of reclaimed asphalt with various cement dosages is shown in Figures 15–17. The compressive resilience modulus was positively correlated with cement dosage while negatively affecting the RAP replacement ratio, no matter the curing-time variation. For 28 days of curing time, the minimum compressive resilience modulus was 721.8 MPa (3.5% cement dosage and 50% reclaimed asphalt), and the maximum value was 1394.7 MPa (5.5% cement dosage and 30% reclaimed asphalt). For 3.5% cement dosage, the maximum compressive resilience modulus loss was 20.2% when the RAP content was increased from 30% to 40%. With an increased curing time of up to 90 days, the compressive resilience modulus showed an increasing trend, and the maximum value reached 1771.2 MPa. It was concluded that the increase of compressive resilience modulus mainly depended on the concrete curing time and cement dosage. However, the RAP had a significant negative effect when compared with NA.

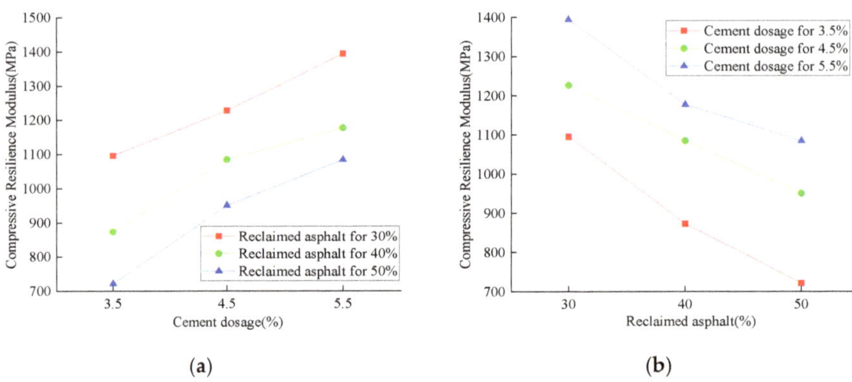

Figure 15. Results for the 28-day compressive-resilience modulus test: (**a**) cement dosage variation; (**b**) RAP replacement ratio.

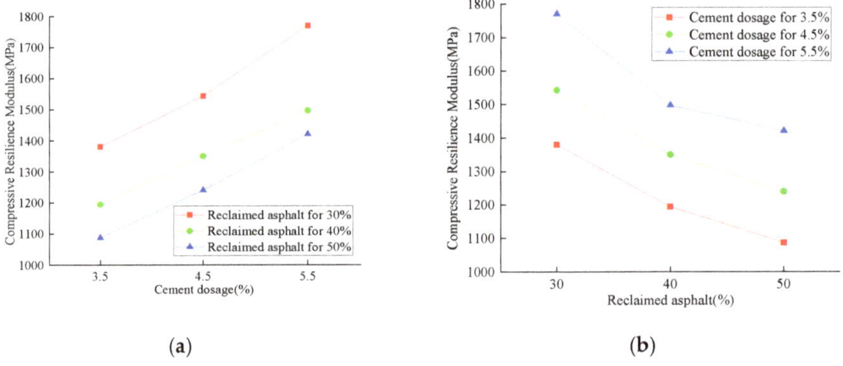

Figure 16. Analysis of the 90-day compressive-resilience modulus test: (**a**) cement dosage variation; (**b**) RAP replacement ratio.

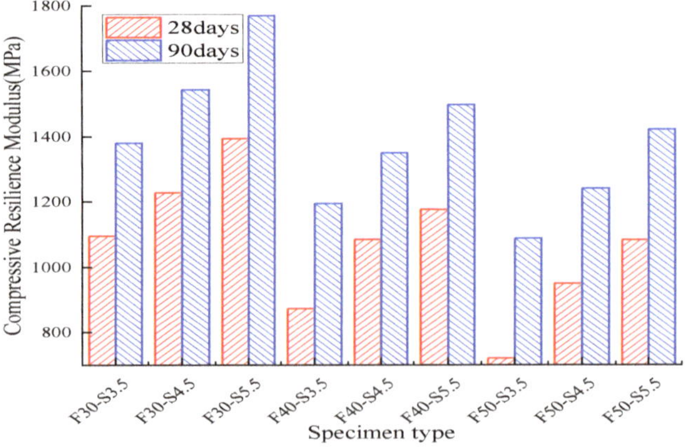

Figure 17. Compressive-resilience modulus test results with respect to curing time.

3.4. Dry- and Temperature-Shrinkage Tests

Figure 18 shows the relationship between the dry-shrinkage coefficient and water-loss rate. With the increase of water-loss rate, the drying-shrinkage coefficient showed a similar increasing trend when the cement dosage was increased from 3.5% to 5.5%. Three increase stages were observed, which included the slowly ascending stage (0–3.4% for 3.5%, 0–3.2% for 4.5%, and 0–2.6% for 5.5%), rapid-growth stage (3.4–4.5% for 3.5%, 3.2–4% for 4.5%, and 2.6–3.4% for 5.5%), and steady stage. Compared with other cases, the 5.5% cement dosage had the maximum shrinkage coefficient and minimum water-loss rate. With the increase of curing time, the drying-shrinkage strain showed a fast and then slow increasing trend. The drying-shrinkage coefficient gradually increased with the decrease of cement dosage—1.15 times for 4.5% and 1.25 times for 5.5%. The drying-shrinkage strain gradually increased with the increase of water-loss rate, and it also had a positive effect with cement dosage on constant water-loss rate. Two increase zones were observed, which included a slowly ascending zone (0–3.4% for 3.5%, 0–3.2% for 4.5%, and 0–2.6% for 5.5%) and a rapid growth zone (larger than 3.4% for 3.5%, larger than 3.2% for 4.5%, and larger than 2.6% for 5.5%). All these observations showed that the higher cement dosage facilitated an adequate cement hydration reaction, leading to higher shrinkage. Beyond that, more cement particles filled the RAP aggregate void and occupied the space that originally belonged to water.

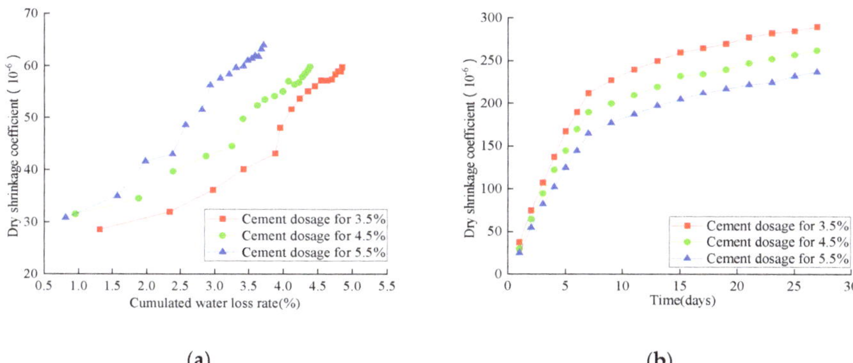

Figure 18. Analysis of the dry-shrinkage test: (**a**) dry-shrinkage coefficient related to water-loss rate; (**b**) dry-shrinkage strain related to time.

The temperature shrinkage data related to 40% reclaimed asphalt and cement dosages from 3.5% to 5.5% are summarized in Figure 19. The temperature-shrinkage coefficient showed a gradually decreasing trend with an increase of cement dosage, and all shrinkage values were located between 6.41×10^{-6} and 19.9×10^{-6}. The 3.5% cement dosage exerted the maximum temperature shrinkage coefficient, and the sharply ascend zone occurred between 0 °C and −10 °C. All the observations can be attributed to the pore water in specimens causing a freezing volume expansion between 0 °C and −10 °C, and a slight shrinkage coefficient was seen in other temperature zones.

3.5. Freeze-Thaw Cycle and Water-Stability Tests

With the increase of curing time from 28 to 90 days, both the mass and strength loss rate gradually decreased the exact cement dosage, as shown in Figure 20. For the same curing time, the increased cement dosage had a positive effect on maintaining the strength and mass of the RAP specimen. All these observations imply that the increased curing time and cement dosage effectively reduced the RAP specimen's internal porosity, which mainly depended on the degree of cement hydration reaction. It can be concluded that the higher cement dosage and longer curing time effectively improved the RAP frost resistance.

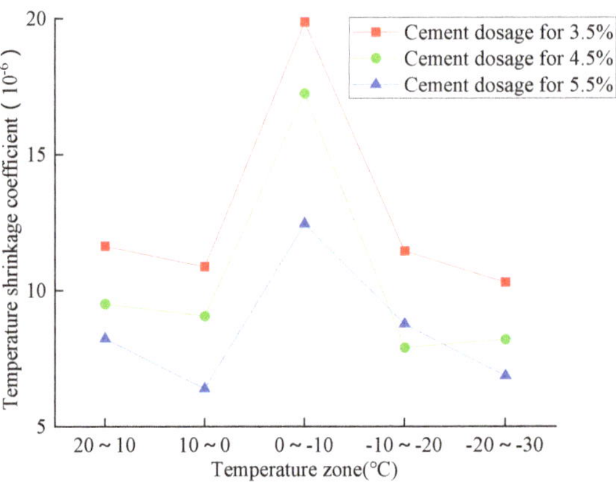

Figure 19. Analysis of the temperature shrinkage coefficient.

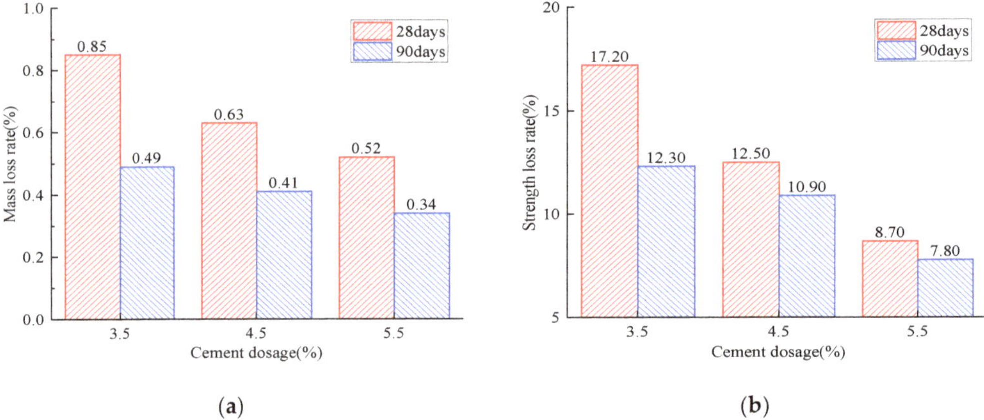

(a)   **(b)**

Figure 20. Analysis of the freeze–thaw-cycle test: (**a**) mass loss rate; (**b**) strength loss rate.

For specimens with 40% reclaimed asphalt and various cement dosages from 3.5% to 5.5%, the compressive strength related to dry–wet cycles (with and without) is summarized in Figure 21. It can be observed that the dry–wet cycles reduced the specimens' compressive strength, and the water-stability coefficient exerted a good water resistance with an increase of cement dosage.

3.6. Microcosmic Analysis

Unlike cement stone and aggregate direct adsorption, an asphalt layer exists between cement stone and aggregate in reclaimed asphalt specimens. In order to observe the interface between cement and asphalt, the scanning electron microscope (SEM) method is used to investigate the influence of existing asphalt. The asphalt sample was in a solid state in a reclaimed asphalt mixture and was not heated during recycling.

Figure 22a shows the morphology diagram of cement stone in the reclaimed asphalt mixture, and the observation area is not in contact with reclaimed asphalt. Compared with traditional cement stabilized materials, the RAP surface was uneven and angular, validated by higher void content as shown in Table 1. Figure 22b shows the interface between cement stone and asphalt in the reclaimed asphalt mixture. It can be observed that the asphalt was effectively filled with the porosity within the cement mixture, forming embedded connection strength. Due to the crushing process of RAP aggregate, the cement stone and asphalt were bonded with an irregular dividing line, and a concave and convex shape was observed. Adhesion partially retarded the hydration reaction between the cement mortar and aggregate, which caused the lower strength generation. Figure 22c shows the asphalt particle embedded in cement stone, and some small particles are also absorbed on cement stone due to crushing and mixing. Many microholes appeared on the surface of bituminous mixture particles. This can be explained by the higher water absorption than for the NA aggregate (Table 1).

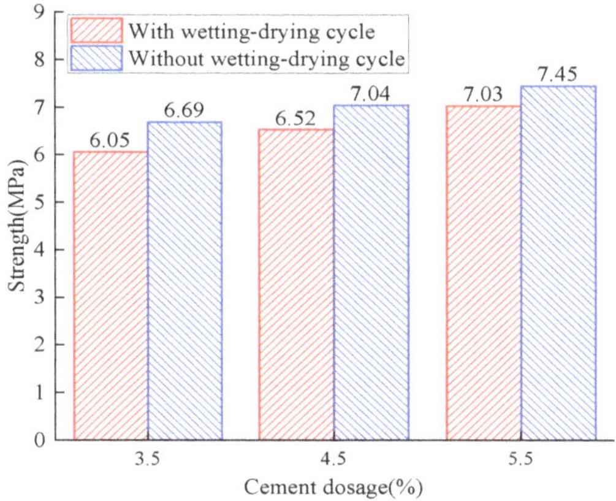

Figure 21. Analysis of the water-stability test.

(a)

(b)

(c)

Figure 22. Scanning electron microscope (SEM) method: (**a**) cement stone; (**b**) interface between cement stone and asphalt; (**c**) bituminous mixture particles bond to cement.

4. Numerical Simulation Analysis

The 7-day compressive-strength test data (Figure 8) was selected to establish a numerical analysis model, which was mainly used to predict the variation law of the reclaimed asphalt mixture's compressive strength with respect to the influence of multiple factors. Two independent variables were considered in the presented numerical analysis model: reclaimed asphalt replacement ratio (30%, 40%, and 50%) and cement dosage (3.5%, 4.5%, and 5.5%). Due to the limited test data, the Monte Carlo simulation method was used to expand the sample number to a sizeable, reasonable size (4500 samples). The simulation data distribution was assumed to be a normal distribution, and the data boundary was selected for mean value ± standard deviation, as shown in Figure 23a. Compared with the relative errors, the predicted points had good agreement with the tested values (R^2 = 0.908), which met the accuracy requirements of the model prediction analysis. A bivariate nonlinear strength-fitting equation is presented based on the nonlinear regression analysis method, as shown in Equation (3):

$$Z = -6.384 + 0.475X + 0.604Y - 0.0059X^2 - 0.0176Y^2$$
$$(20\% \leq X \leq 60\%, 3\% \leq Y \leq 6\%) \tag{3}$$

where X is the reclaimed asphalt mixture content (%); Y is cement dosage (%); and Z is the 7-day compressive strength (MPa).

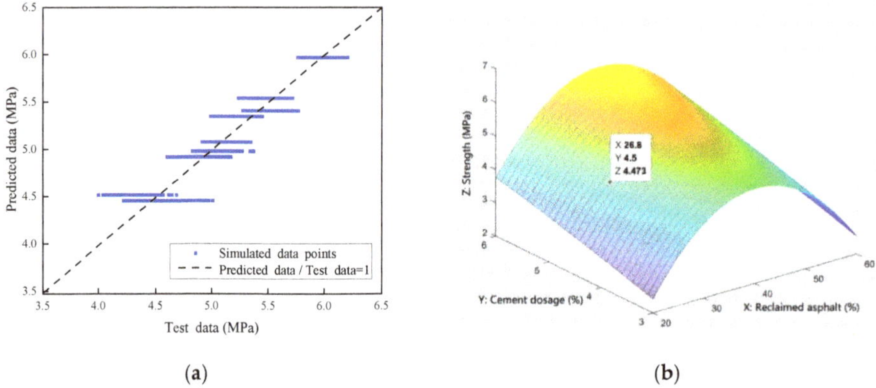

Figure 23. Numerical simulation of compressive strength: (**a**) Monte Carlo simulation; (**b**) 3D diagram of the nonlinear strength-fitting equation.

In order to exhibit the relationship among variables in the bivariate nonlinear strength equations, a three-dimensional effect diagram was constructed in MATLAB. Figure 23b shows the three-dimensional surface related to X (content of reclaimed asphalt mixture), Y (cement dosage), and Z (7-day compressive strength). For example, a point in the figure (X 26.8; Y 4.5; Z 4.473) represents a predicted 7-day compressive strength of 4.473 MPa when subjected to the reclaimed asphalt mixture content of 26.8% and a cement dosage of 4.5%. It can be concluded that the 7-day compressive strength can be predicted in the presented numerical model when subjected to various cement dosages and reclaimed asphalt mixture, and vice versa.

Similarly, considering the effect of freeze–thaw cycles, a revised numerical analysis model was established to predict the strength reduction of the reclaimed asphalt mixture. For example, two variables were considered for the 40% reclaimed asphalt mixture content: curing time (28 and 90 days) and cement dosage (3.5%, 4.5%, and 5.5%). Due to test data being limited, the Monte Carlo method was used to simulate 3000 samples, and the data-distribution type was assumed to be a normal distribution. Figure 24a shows that the simulated data points were in good agreement with the test values ($R^2 = 0.915$), which met the accuracy requirements of the model prediction analysis. The nonlinear regression analysis method was used to create a bivariate nonlinear strength-fitting equation, as shown in Equation (4):

$$Z = 28.5561 - 0.0398275X - 3.25298Y$$
$$(20 \leq X \leq 100, 0\% < Y \leq 6\%) \tag{4}$$

where X is the curing time of reclaimed asphalt mixture; Y is cement dosage (%); and Z is the compressive strength loss rate (%) of reclaimed asphalt mixture.

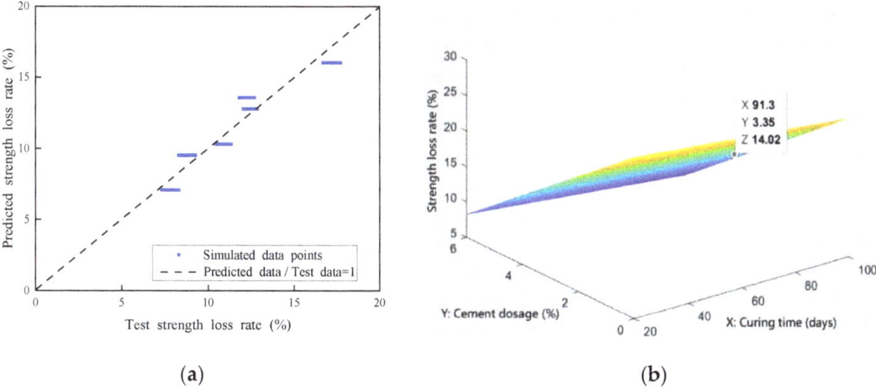

Figure 24. Numerical simulation of compressive-strength-loss rate: (**a**) Monte Carlo simulation; (**b**) 3D diagram of nonlinear strength-loss fitting equation.

Figure 24b shows the relationship among X (curing time), Y (cement dosage), and Z (compressive strength loss rate) in the 3D model. For example, a point was randomly selected in the figure (X 91.3; Y 3.35; Z 14.02), representing the predicted value of compressive strength loss rate of 14.02% for 91.3 days curing time and 3.35% cement dosage. It can be concluded that the presented numerical model could effectively predict the reclaimed asphalt mixture strength reduction when subjected to the freeze–thaw effect.

A simplified numerical model was created for practical engineering applications when considering the relationship between cement dosage and 28-day strength-loss rate. The nonlinear regression method was used to build the strength-degradation equation (Equation (5)), which included 1500 data points by Monte Carlo simulation:

$$Y = 40.558 - 8.210X + 0.439X^2 \ (3\% \leq X \leq 6\%) \tag{5}$$

where X is the cement dosage (%), and Y is the 28-day compressive-strength-loss rate (%) under the freeze–thaw effect. Compared with the relative error between the predicted value and the tested value, the 1500 simulated data points showed a good consistency with the measured value ($R^2 = 0.998$), which could meet the accuracy requirements of the model-prediction analysis, as shown in Figure 25.

In order to facilitate data selection for the construction technicians, the maximum replacement ratio of reclaimed asphalt is presented for various cement dosages, considering both with and without the freeze–thaw effect. Figure 26 shows the maximum replacement ratio of reclaimed asphalt with respect to various cement dosages. There was an apparent descending trend when considering the freeze–thaw effect, and the increasing cement dosage effectively improved the replacement ratio. It can be concluded that the maximum replacement ratio of the reclaimed asphalt mixture can be selected with different cement dosages. For specific cement dosages, the linear interpolation method can be used to obtain the maximum replacement ratio.

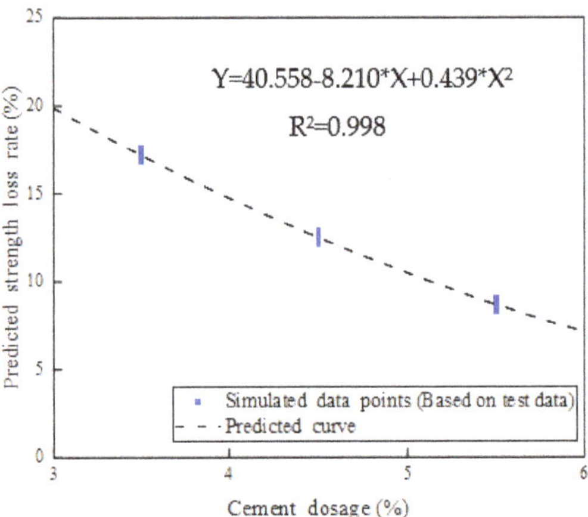

Figure 25. Nonlinear strength-loss-rate fitting equation with respect to freezing-damage effects.

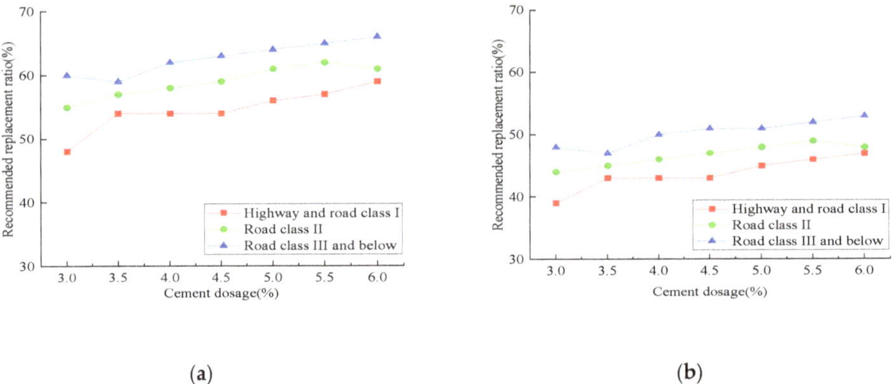

Figure 26. The maximum replacement ratio of reclaimed asphalt with respect to various cement dosage: (**a**) with the freeze–thaw effect; (**b**) without the freeze–thaw effect.

5. Conclusions

This paper investigated the mechanical performance of reclaimed asphalt mixtures. More specifically, mechanical properties such as compressive strength, splitting strength, compressive modulus of resilience, were examined, as well as the results of freeze–thaw cycling tests. The main conclusions were as follows:

According to the axial compressive and splitting tensile-strength tests, critical RAP substitute ratios (30%) and cement dosage (5.5%) exist to achieve the maximum mechanical performance. However, the cost efficiency and environmental friendliness also need to be balanced. It can be concluded that the RAP aggregate substitute NA had a negative influence on concrete compression and splitting tensile strength, and the increasing of cement dosage effectively relieved this strength degradation.

It seems that the higher cement dosage represented an adequate cement hydration reaction through dry- and temperature-shrinkage tests, which led to higher shrinkage occurring. Beyond that, a noticeable temperature-shrinkage-increase zone (between 0 °C and −10 °C) was observed due to the pore water freezing and causing a volume-expansion effect.

The 5.5% cement dosage and 90 days of curing time exerted the most robust RAP frost resistance in the freeze–thaw-cycle and water-stability tests. Therefore, it can be concluded that the increased curing time and cement dosage effectively reduced the RAP specimen's internal porosity, and mainly depended on the degree of the cement hydration reaction.

In the scanning electron microscope (SEM) analysis, an irregular dividing line between the cement stone and asphalt was observed for the RAP aggregate. The concave and convex surfaces partially retarded the hydration reaction between the cement mortar and aggregate, which caused lower strength generation and higher water absorption than NA.

A nonlinear compression-strength regression equation was presented using RAP content and cement dosage through a Monte Carlo numerical-simulation method. Considered with different highway-grade requirements, the corresponding RAP replacement ratio and cement dosage limits were determined.

Author Contributions: Conceptualization and methodology, W.X.; software and data curation, W.L. and Y.J.; validation, W.X., W.L. and Y.J.; writing—original draft preparation, W.L. and Y.J.; writing—review and editing, W.X.; funding acquisition, W.X. All authors have read and agreed to the published version of the manuscript.

Funding: This research was funded by the Transportation Department of Heilongjiang Province in China, grant number 2019HLJ0012.

Institutional Review Board Statement: Not applicable.

Informed Consent Statement: Not applicable.

Data Availability Statement: Not applicable.

Conflicts of Interest: The authors declare no conflict of interest.

References

1. Md, T.R.; Abbas, M.; Filippo, G. Recycling of Waste Materials for Asphalt Concrete and Bitumen: A Review. *Materials* **2020**, *13*, 1495.
2. Yail, J.K.; Ji, Y.C.; Troy, B. Uncertainty Modeling of Carbon Fiber-Reinforced Polymer Confined Concrete in Acid-Induced Damage. *ACI Struct. J.* **2019**, *116*, 97–108.
3. Ma, T.; Zhao, Y.L.; Huang, X.M. *Key Technology of Heat Mixing Regeneration in Asphalt Pavement Plant*; Southeast University Press: Nanjing, China, 2015.
4. Yueqin, H.; Xiaoping, J.; Jia, L.; Xianghang, L. Adhesion between Asphalt and Recycled Concrete Aggregate and Its Impact on the Properties of Asphalt Mixture. *Materials* **2018**, *11*, 2528.
5. Gilda, F.; Andrea, G.; Chiara, M.; Andrea, G. Comparing the Field and Laboratory Curing Behaviour of Cold Recycled Asphalt Mixtures for Binder Courses. *Materials* **2020**, *13*, 4697.
6. Aleksandar, R.; Ivan, I.; Michael, P.W.; Dimitrije, Z.; Marko, O.; Goran, M. The Impact of Recycled Concrete Aggregate on the Stiffness, Fatigue, and Low-Temperature Performance of Asphalt Mixtures for Road Construction. *Sustainability* **2020**, *12*, 3949.
7. Hasan, K.; Erol, T.; Sheila, B. Using Accelerated Pavement Testing to Evaluate Reclaimed Asphalt Pavement Materials for Pavement Unbound Granular Layers. *J. Mater. Civ. Eng.* **2017**, *29*, 04016205.
8. Goshtasp, C.; Augusto, C.F.; Zhanping, Y.; Siyu, C.; Yun, S.K.; Jan, W.; Ki, H.M.; Michael, P.W. Warm mix asphalt technology: An up to date review. *J. Clean. Prod.* **2020**, *268*, 122128.
9. Olivera, Đ.; Aleksandar, R.; Dimitrije, Z.; Božidar, Đ. Potential of Natural and Recycled Concrete Aggregate Mixtures for Use in Pavement Structures. *Minerals* **2020**, *10*, 744.
10. Qibo, H.; Zhendong, Q.; Jing, H.; Dong, Z.; Leilei, C.; Meng, Z.; Jinzhu, Y. Investigation on the properties of aggregate-mastic interfacial transition zones (ITZs) in asphalt mixture containing recycled concrete aggregate. *Constr. Build. Mater.* **2021**, *269*, 121257.
11. Arabani, M.; Moghadas, N.F.; Azarhoosh, A.R. Laboratory evaluation of recycled waste concrete into asphalt mixtures. *Int. J. Pavement Eng.* **2013**, *14*, 531–539. [CrossRef]
12. Merzouki, T.; Bouasker, M.; Khalifa, N.E.H. Contribution to the modeling of hydration and chemical shrinkage of slag-blended cement at early age. *Constr. Build. Mater.* **2013**, *44*, 368–380. [CrossRef]
13. Areej, A.; Nasim, S.; Ahmed, A.; Hasan, K. Influence of temperature on mechanical properties of recycled asphalt pavement aggregate and recycled coarse aggregate concrete. *Constr. Build. Mater.* **2021**, *269*, 121285.
14. Yasser, K. Self-compacting concrete using recycled asphalt pavement and recycled concrete aggregate. *J. Build. Eng.* **2017**, *12*, 282–287.

15. Le, D.; Junhui, Z.; Bowen, F.; Cheng, L. Performance Evaluation of Recycled Asphalt Mixtures Containing Construction and Demolition Waste Applicated as Pavement Base. *Adv. Civ. Eng.* **2020**, *7*, 8875402.
16. Jayakody, S.; Zimar, A.M.Z.; Ranaweera, R.A.L.M. Potential use of recycled construction and demolition waste aggregates for non-structural concrete applications. *J. Natl. Sci. Found. Sri Lanka.* **2018**, *2*, 205–216. [CrossRef]
17. Cardoso, R.; Silva, R.V.; Brito, J.D. Use of recycled aggregates from construction and demolition waste in geotechnical applications: A literature review. *Waste Manag.* **2016**, *49*, 131–145. [CrossRef] [PubMed]
18. Arulrajah, A.; Disfani, M.M.; Horpibulsuk, S. Physical properties and shear strength responses of recycled construction and demolition materials in unbound pavement base/subbase applications. *Constr. Build. Mater.* **2014**, *58*, 245–257. [CrossRef]
19. Komnitsas, K.; Zaharaki, D.; Vlachou, A. Effect of synthesis parameters on the quality of construction and demolition wastes (CDW) geopolymers. *Adv. Powder Technol.* **2015**, *26*, 368–376. [CrossRef]
20. Yang, Y.; Wang, L.; Feng, Y. Study on fractal mathematical models of pulverizing theory for ore. *Powder Technol.* **2016**, *288*, 354–359. [CrossRef]
21. Chen, R.; Fang, M.X.; Li, M. Size distribution and particle shape of iron ore powder based on the fractal theory. *Mater. Res. Innov.* **2015**, *19*, 405–409. [CrossRef]
22. Jing, H.Z.; Yun, C.S.; Shu, J.B. Experimental Study on the Loaded Property of Waste Fiber Recycled Concrete Beam-Column Joints. *Appl. Mech. Mater.* **2015**, *727*, 69–72.
23. Saeed, F.; Reza, I. Performance evaluation of recycled asphalt mixtures by construction and demolition waste materials. *Constr. Build. Mater.* **2016**, *120*, 450–456.
24. Akash, B.; Amy, E.M.; Gayle, K.; Charles, G.; Fawaz, K.; Edith, A.M. Evaluation and classification of recycling agents for asphalt binders. *Constr. Build. Mater.* **2020**, *260*, 119864.
25. Wojciech, B. Evaluation of Fatigue Life of Asphalt Concrete Mixtures with Reclaimed Asphalt Pavement. *Appl. Sci.* **2018**, *8*, 469.
26. Juntao, L.; Lin, H.; Yue, X.; Fang, X.; Pan, P. Long-term performance characteristics and interface microstructure of field cold recycled asphalt mixtures. *Constr. Build. Mater.* **2020**, *259*, 120406.
27. Hassan, K.E.; Brooks, J.J.; Erdman, M. The use of reclaimed asphalt pavement (RAP) aggregates in concrete. *Waste Mater. Constr.* **2000**, *1*, 121–128.
28. Papakonstantinou, C.G. Resonant column testing on Portland cement concrete containing recycled asphalt pavement (RAP) aggregates. *Constr. Build. Mater.* **2018**, *173*, 419–428. [CrossRef]
29. Zaumanis, M.; Poulikakos, L.D.; Partl, M.N. Performance-based design of asphalt mixtures and review of key parameters. *Mater. Des.* **2018**, *141*, 185–201. [CrossRef]
30. Mukul, R.; Martins, Z. Impact of laboratory mixing procedure on the properties of reclaimed asphalt pavement mixtures. *Constr. Build. Mater.* **2020**, *264*, 120709.
31. Chidozie, M.N.; Soon, P.Y.; Choon, W.Y.; Chiu, C.O.; Suhana, K.; Ali, M.B. Laboratory study on recycled concrete aggregate based asphalt mixtures for sustainable flexible pavement surfacing. *J. Clean. Prod.* **2020**, *262*, 121462.
32. Giulia, T.; Piergiorgio, T.; Cesare, S. The Challenges of Using Reclaimed Asphalt Pavement for New Asphalt Mixtures: A Review. *Materials* **2020**, *13*, 4052.
33. Abdulgazi, G. A review on the evaluation of the potential utilization of construction and demolition waste in hot mix asphalt pavements. *Resour. Conserv. Recycl.* **2020**, *161*, 104956.
34. Fawaz, K.; Amy, E.M.; Edith, A.M. Use of recycling agents in asphalt mixtures with high recycled materials contents in the United States: A literature review. *Constr. Build. Mater.* **2019**, *211*, 974–987.
35. Sharareh, S.; Max, A.A.; Luis, B.; Mostafa, A.E.; Samuel, C.; Louay, N.M. Mechanistic-empirical pavement performance of asphalt mixtures with recycled asphalt shingles. *Constr. Build. Mater.* **2018**, *160*, 687–697.
36. Nasim, S.; Abeer, A.A.; Hasan, K.; Mu'tasime, A.J. Investigation of axial compressive behavior of reinforced concrete columns using Recycled Coarse Aggregate and Recycled Asphalt Pavement aggregate. *Constr. Build. Mater.* **2019**, *217*, 384–393.

Article

Forecasting the Energy and Economic Benefits of Photovoltaic Technology in China's Rural Areas

Wenjie Zhang *, Yuqiang Zhao, Fengcheng Huang, Yongheng Zhong and Jianwei Zhou

School of Energy and Power Engineering, Nanjing University of Science and Technology, Nanjing 210094, China; zhao1599484022@163.com (Y.Z.); hfc_free@163.com (F.H.); 13965913060@163.com (Y.Z.); zhjwhvac@njust.edu.cn (J.Z.)
* Correspondence: zhangwenjie001@139.com

Abstract: In recent years, with the rapid development of China's economy, China's energy demand has also been growing rapidly. Promoting the use of renewable energy in China has become an urgent need. This study evaluates the potential of solar photovoltaic (PV) power generation on the roofs of residential buildings in rural areas of mainland China and calculates the area that can used for generating energy, the installed capacity, and the power generation, and conducts a comprehensive analysis of the economic benefits of investing in the construction of distributed PV systems in various provinces. The findings unveiled in this study indicate that China still has more than 6.4 billion m^2 of rural construction area available for the installation of PV modules. If this is all used for solar power generation, the annual power generation can reach up to 1.55 times the electricity consumption of urban and rural residents for the whole society. Through a comprehensive evaluation of energy efficiency and economic benefits, the Chinese mainland can be divided into three types of resource areas. The three types of resource areas have their own advantages and disadvantages. According to their own characteristics and advantages, we can reasonably formulate relevant policies to accelerate the development of PV system application in rural areas.

Keywords: solar energy; distributed PV system; energy-saving benefits; economic benefit; rural areas of China; forecasting

Citation: Zhang, W.; Zhao, Y.; Huang, F.; Zhong, Y.; Zhou, J. Forecasting the Energy and Economic Benefits of Photovoltaic Technology in China's Rural Areas. *Sustainability* **2021**, *13*, 8408. https://doi.org/10.3390/su13158408

Academic Editor: Carlos Morón Fernández

Received: 22 June 2021
Accepted: 23 July 2021
Published: 28 July 2021

Publisher's Note: MDPI stays neutral with regard to jurisdictional claims in published maps and institutional affiliations.

1. Introduction

Building a resource-conserving society is important after an in-depth study of the history of the political, economic, and social development at home and abroad, in accordance with China's social and economic development, and it is also a scientific plan for China's future social development model. Energy saving is an important part of building a resource-saving society, and as one of the three major sectors of energy consumption, building energy consumption represents about one-third of the energy consumption of the whole society, and is the biggest potential energy-saving area [1].

Energy revolution is the first step to build a resource-saving society, and the fundamental task of energy revolution is to change the energy structure dominated by fossil fuels into a low-carbon energy system dominated by renewable energy, so as to cope with the threat of climate change, reduce China's external dependence on energy, and achieve sustainable energy development. Among the many renewable energy sources, solar energy is focused on because of its unique cleanliness, low cost, high efficiency, and abundant reserves [2].

China has a vast territory, abundant solar energy resources, and huge resource potential. The maximum annual solar radiation in all regions of mainland China is 8364 MJ/m^2, the minimum is 3324 MJ/m^2, and the average is 5749 MJ/m^2; The total annual solar radiation received by the land surface of the country is about 50×10^{18} kJ, which is equivalent to 2.4×10^4 million tons of standard coal, which provides a good prerequisite for the utilization of solar energy resources in the country [3]. As a typical technology form of solar energy application, photovoltaic (PV) power generation uses the photovoltaic effect to directly convert solar radiation energy into electric energy, which is one of the most promising renewable energy technologies to realize sustainable development, and it is also a means to realize zero energy building [4]. At the same time, PV power generation, as a solution to the increasing demand for electricity, can also minimize environmental and social problems related to fossil fuels and nuclear fuels [5]. Driven by energy and environmental benefits, PV systems have developed rapidly in recent years, with an average annual growth rate approaching 50%, and have begun to gradually replace coal-fired power generation. With the advancement of technology and the increase in market demand, solar power generation has formed two more mature power generation models—centralized and distributed. Compared with the distributed PV power generation system, the centralized PV system has a relatively early development and mature technology, and its proportion in the cumulative installed capacity is higher than that of the distributed PV system [6]. In 2017, the new installed capacity of China's centralized PV power generation system reached 33.49 GW. In contrast, from 2013 to 2016, the cumulative installed capacity of the distributed PV power generation accounted for only 15% to 20% of the total PV power generation. However, in recent years, distributed PV systems have received more and more attention because of their unique advantages over remote large-scale centralized PV power plants. Its main advantages include being able to be installed on the roof near the electricity consumption area so as to obtain a lower transmission cost and power loss [7,8]. The new capacity of the distributed PV system increased significantly from 4.26 GW in 2016 to 19.44 GW in 2017, which shows the huge development potential of distributed systems [9].

There are considerable solar resources on roofs, which are also convenient for the installation of PV systems. Therefore, PV roofs have become one of the main application forms of distributed PV systems. Vardimon R et al. [10] used a complete set of GIS data covering the whole country to evaluate the available PV roof area in Israel. There are also some studies on the calculation and analysis of the roof PV power generation potential in some areas such as the United States and Austria [11,12]. In addition, Fina B [13], Senatla M [14], and Miranda R F C [15] et al. also analyzed the economic potential of rooftop PV systems in Australia, South Africa, and Brazil, respectively. However, with the rapid development of the urban economy, land value in urban areas is getting higher and higher, and there are more high-rise buildings. For high-rise buildings in urban areas, the roof area is very limited, and the area of PV systems that can be installed is also very limited. In rural areas, because the building density is smaller than in cities, and it is mostly one-story or two-story buildings, more roof area can be used to install PV systems, so there is a greater application potential for PV systems. Especially in Xinjiang, Inner Mongolia, Qinghai, Tibet, Ningxia, Gansu, Yunnan, and remote mountainous areas, where power is scarce, power transmission costs are high, but solar energy resources are abundant and land costs are low. Moreover, with the deepening of the country's new rural construction and PV precise poverty alleviation policies, the PV power generation industry has developed rapidly in the countryside, so these areas have great potential for installing PV power generation systems. On the other hand, because of the wide range of rural areas, the rural distribution network generally consists of a simple radiation chain structure, with a scattered power load, long distribution distance, and relatively unstable power supply quality. Using the performance characteristics of PV power generation, applying distributed PV power generation to rural areas according to local conditions can not only solve the impact of rural grid voltage instability, three-phase imbalance, and other problems, thus solving the power demand of rural users, but also promotes the high-quality development of the PV industry [16].

China is committed to "peak carbon dioxide emissions" and "carbon neutrality" in the future. Photovoltaic is considered to be one of the main energy situations in the future. The distributed photovoltaic system combined with buildings has many advantages, and is considered to be one of the important technical solutions to achieve "peak carbon dioxide emissions" and "carbon neutrality" in the future. At present, China needs to calculate the installed capacity and power generation that can be achieved by building integrated distributed photovoltaics. In this way, the relevant institutions can develop policies to improve data support and help. Recently, a number of provinces in China have formulated the building roof photovoltaic installation plans of cities and counties [17–21]. The calculation of the PV installed capacity in China's provinces in this study can provide assistance for the implementation of the plan.

2. Methodology

Firstly, this research analyzes the building types and converts the rural building area into the roof area. Then, the roof area is converted into the installation area of the PV modules, which can be used for power generation by considering influencing factors such as the optimal installation angle, installation spacing, roof slope, and surrounding environment. Then, according to the solar radiation data of different regions and the current PV module parameters and other data, we calculate the solar power generation potential, that is, energy efficiency. Finally, we calculate and analyze the economic benefits of the construction of distributed PV systems in rural areas under the relevant policies and measures of China. An overview of the methods used in this study is shown in Figure 1.

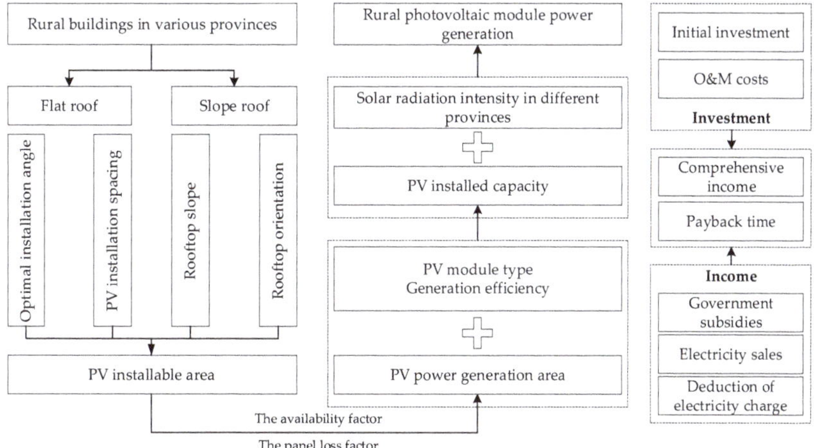

Figure 1. Overview of the methods used in the research.

3. Calculation of PV Module Power Generation Area

3.1. The Annual Solar Radiation

Because of the vast land area in China, the solar radiation intensity also varies greatly in different regions. Therefore, in the calculation process, we first divided China into several sub regions (in each partition, the intensity of solar radiation is roughly the same) according to the annual total solar radiation level, and on this basis, the installed capacity and annual power generation of PV modules in each zone were calculated. In the process of regional division, we first divided different provinces into different regions, and then divided each province into several regions with different radiation levels according to the difference of total solar radiation in each province. The annual solar radiation intensity of typical cities in each radiation area was selected as the annual total radiation calculation data of the whole area.

As shown in Figure 2, taking Jiangxi Province and Gansu Province as examples, there was no significant difference in the solar radiation intensity of each city in Jiangxi Province. The annual solar radiation amount was about 5400 MJ/m^2. Therefore, as an irradiated area, Jiangxi Province was no longer divided internally, and Ganzhou was taken as a typical city for calculation. In addition, the solar radiation resources in Gansu Province could be roughly divided into the following three areas: rich area, relatively rich area, and available area. The solar-rich areas included Jiuquan, Zhangye, and Jiayuguan in the Hexi corridor. The total annual solar radiation of the whole area is more than 6100 MJ/m^2, and the solar energy resources are stable. Jiuquan is regarded as a typical city in this area; solar energy resources are relatively rich in certain areas, including Jinchang, Wuwei, Minqin, Gulang, Tianzhu, Jingyuan, Jingtai, and Dingxi, Lanzhou, and Linxia. The total annual solar radiation in this area was between 5400~6100 MJ/m^2, and Minqin was selected as a typical city. The available areas included Tianshui, Longnan, and Gannan. The total solar radiation was between 4700~5400 MJ/m^2. Lanzhou City was used as a typical city in this area.

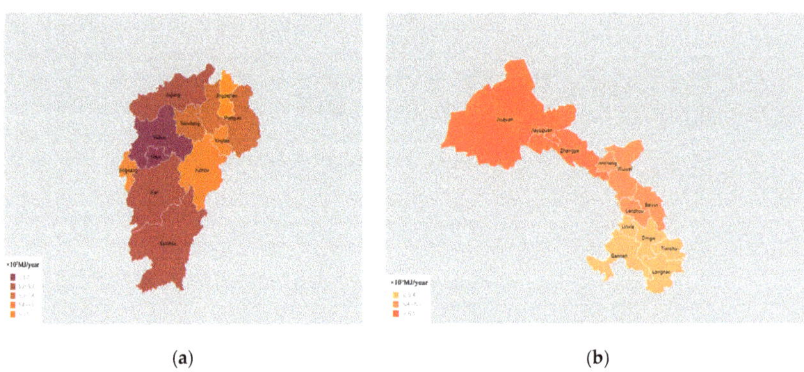

(a)   (b)

Figure 2. The solar radiation intensity distribution: (a) Jiangxi Province and (b) Gansu Province.

The annual total solar radiation data for each typical city came from the China Meteorological Data Network, and the main data come from the "Daily Data Set of Meteorological Radiation Basic Elements in China". The observation data collected in this data set include total radiation, scattered radiation, direct radiation, and net total radiation. The data used in this paper are the annual total radiation data of the radiation measurement stations in typical cities of each province, with the unit of MJ/m^2.

3.2. The Rural Building Footprint Area

Rural buildings in mainland China are generally low in height, coupled with the surrounding environment's shielding effect, resulting in a low income from installing distributed PV systems on the facade. Therefore, this calculation only considers the roof of the building as the best installation area for the distributed PV system. The calculation of the building roof area (*BFA*) needs to use rural building area (*BA*) data, which mainly comes from the sum of "township housing area at the end of the year" and "village housing area at the end of the year" in the "Statistical Yearbook of Urban and Rural Construction in China".

After obtaining the rural building area data, Formula (1) can be used to calculate the roof area of rural buildings in various regions.

$$BFA = {}^{BA}/_{f} \tag{1}$$

where f is the average number of floors. In order to obtain data on the average number of floors in each area, we conducted relevant investigations and studies, the survey results show that the existing rural buildings in various regions of mainland China are mostly one- to three-story independent buildings. The average number of floors is greatly affected by the level of local economic development; the average number of floors in areas with higher economic levels is larger. Therefore, we divided the mainland into three categories based on the current economic development.

(1) Economically developed provinces: mainly including Beijing, Shanghai, Chongqing, Tianjin, Jiangsu, Zhejiang, Shandong, and Guangdong. Most of these buildings are not less than two floors, mainly two floors, so f = 2.

(2) Medium economic areas: Heilongjiang, Jilin, Liaoning, Hebei, Shanxi, Sichuan, Hubei, Anhui, Fujian, Henan, Hunan, Hainan, Guangxi, Jiangxi, Shaanxi, and Yunnan. These provinces have a medium economy, with slightly more two-story buildings than one story buildings, so f = 1.6;

(3) Poorer economic areas: Qinghai, Ningxia, Gansu, Guizhou, Xinjiang, Tibet, and Inner Mongolia. The rural buildings are mainly on the first floor, while the second and third floors are relatively less frequent, so f = 1.2.

3.3. The Optimal Installation Angle of PV Module

Affected by cultural customs, climate, and other factors, the architectural styles of rural residential buildings in different regions of China are also different. From the perspective of the roof form, rural residential buildings can be divided into two categories: flat roof and sloping roof. For all of the provinces in China, the rural residential buildings in Henan, Anhui, Shanxi, and other northern provinces, considering the factors of water storage, the roofs are mostly flat roofs, while in Jiangsu, Zhejiang, Guangdong, and other southern provinces, due to the large amount of precipitation, in order to facilitate drainage, rural residential buildings mostly adopt a slope roof. As there are fewer sloping roofs in rural buildings in the northern provinces and less flat roofs in rural buildings in southern provinces, this work did not consider these as part of the buildings; that is, we assumed that all the residential buildings in the northern rural areas were flat roofs, while all the residential buildings in the southern rural areas were sloping roofs.

As the power output of the PV module increases with the increase in the direct sunlight intensity, when installing the PV module on a flat roof, we aimed to get the maximum amount of direct sunlight on the PV panel. There are three main installation methods for PV panels on flat roofs: horizontal installation, fixed angle installation, and tracking PV. Among them, the tracking PV can automatically adjust the height angle and azimuth angle as the sun moves, so as to increase the amount of solar radiation received, thereby increasing the amount of power generated; this method gets the most solar energy among the three installation methods. However, because of the high investment cost in the early stages and the need for certain maintenance in the later stage, the same installation capacity requires more floor space, and the cost-effectiveness is lower for large-scale installation, so this study did not consider this type of installation. The solar energy of fixed bracket installation is less than that of tracking PV, and its price is low, the structure is stable, and it is basically maintenance-free. It can also get more solar energy than horizontal installation when installed at the optimal installation angle, which is very suitable for use in rural areas. The most important thing for the fixed bracket installation to obtain the maximum solar radiation is to determine the optimal installation angle and bracket spacing in different regions. This section mainly describes the calculation of the best inclination.

The determination of the optimal inclination angle is generally obtained by comparing the annual total power generation of the PV module at different inclination angles, and the inclination corresponding to the maximum annual power generation is the optimal installation angle of the PV modules in this region. There are generally two methods for calculating the amount of power generated—experimental measurement and simulation calculation. Because of the difficulty conducting experimental measurements in different regions and the longer time period, we used the method of simulation calculation. The specific calculation steps are as follows:

(1) Calculate the hourly plane-of-array (POA) solar irradiance from the horizontal irradiance, latitude, longitude, and time in the solar resource data, and from the array type, tilt, and azimuth inputs.

(2) Calculate the effective POA irradiance to account for reflective losses from the module cover, depending on the solar incidence angle.

(3) Calculate the cell temperature based on the array type, POA irradiance, wind speed, and ambient temperature. The cell temperature model assumes a module height of 5 meters above the ground and an installed nominal operating cell temperature (INOCT) of 49 °C for the fixed roof mount option, and of 45 °C for the other array type options.

(4) Calculate the array's DC output from a DC system size at a reference POA irradiance of 1000 W/m^2, and the calculated cell temperature, assuming a reference cell temperature of 25 °C, and temperature coefficient of power of $-0.35\%/°C$ for the premium type.

(5) Calculate the system's AC output from the calculated DC output and system losses and nominal inverter efficiency input (96% by default) with a part-load inverter efficiency adjustment derived from empirical measurements of the inverter performance.

The Chinese meteorological data were taken from The ASHRAE International Weather for Energy Calculations Version 1.1 (IWEC) and the Solar and Wind Energy Resource Assessment Program (SWERA). The solar radiation data were taken from the "typical year" solar radiation values in these databases. We combined the solar resource data of the selected location with the 30-year historical temperature and wind speed data of the nearest weather station, the characteristics of the solar panel, and the direction of the solar panel relative to the sun.

In addition, the corresponding assumptions needed to be made before the calculation [22]: for a 10 kW high-quality crystalline silicon solar PV panel with a photoelectric conversion efficiency of 19%, the temperature coefficient was $-0.0035/°C$ and the efficiency loss was 10%, and we assumed that all panels were facing south with an azimuth of 180°. Based on this assumption and combined with the meteorological data, the annual power generation of the system could be calculated preliminarily. After that, the annual power generation capacity under different inclination angles was compared to obtain the optimal installation angle.

In addition, it should be pointed out that this method could only be used in the preliminary design, but it had a high reliability to determine the optimal installation angle of the PV modules.

3.4. The PV Bracket Installation Spacing

When the PV module is installed on the sloped roof, it should be installed parallel to the inclined roof. For PV modules installed at a fixed angle on a flat roof, it is necessary to determine the spacing between each row of the PV modules, because the power generation of the PV modules will be affected by the self-shadow effect, the self-shadow effect is caused by the previous row of PV modules, and is applicable to all PV modules except the first row. If the installation distance of the PV bracket is too small, the PV module cannot be fully applied because of the self-shadow effect, resulting in a waste of PV modules. However, if the distance is too large, the available roof area will be wasted, so we need to balance these two aspects and choose an optimal installation distance for the PV bracket.

In the Northern Hemisphere, the solar elevation angle is the lowest and the solar radiation is the smallest during the solstice day; at this time, if the system is required to have no self-shadow effect, theoretical maximum array spacing is required. On the day of the summer solstice, the solar altitude angle is the highest, and the solar radiation is at its maximum value; at this time, the theoretical minimum distance is calculated. In this study, in order to calculate a more conservative power generation value, the maximum array spacing was used, that is, the component spacing was greater than the maximum length of the shadow at 09:00 a.m. and 15:00 p.m. on the winter solstice. As long as no occlusion occurs during this period, there will be no occlusion within the solar module's optimal utilization range of solar radiation during the year.

The schematic diagram for calculating the array spacing is shown in Figure 3.

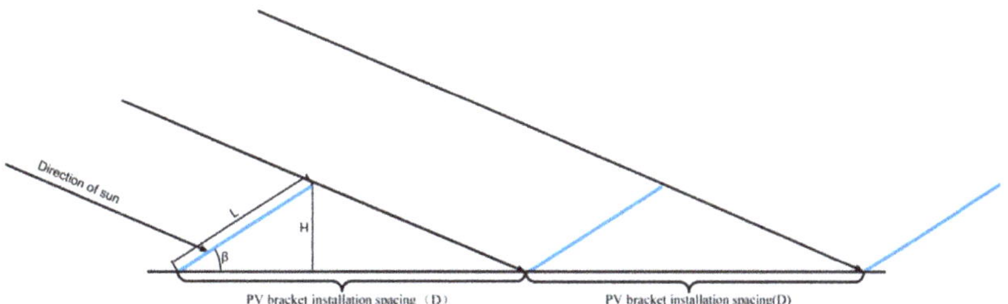

Figure 3. Schematic diagram for calculating the spacing of PV arrays.

In Figure 3, D is the distance between the two rows of brackets. The distance between the PV arrays or the vertical distance between the shield and the bottom of the array should not be less than D. The calculation formula for the array spacing D is as follows:

$$D = L\cos\beta + L\sin\beta(\cos\omega\tan\psi - \tan\delta)/(\tan\delta\tan\psi + \cos\omega) \tag{2}$$

where L is the length of the slope of the PV module. At present, most crystalline silicon PV modules on the market have a width of about one meter, so the value of L here is 1 m; β is the installation angle of the PV modules, using the optimal installation angle calculated from Section 3.3; ψ is the latitude of the location of the PV array; δ is the solar declination on the winter solstice, $\delta = -23.5°$; and ω is the hour angle, and the corresponding ω values at 09:00 a.m. and 15:00 p.m. are 45° and $-45°$, respectively. The PV array spacing calculated according to the above formula is shown in Figure 4.

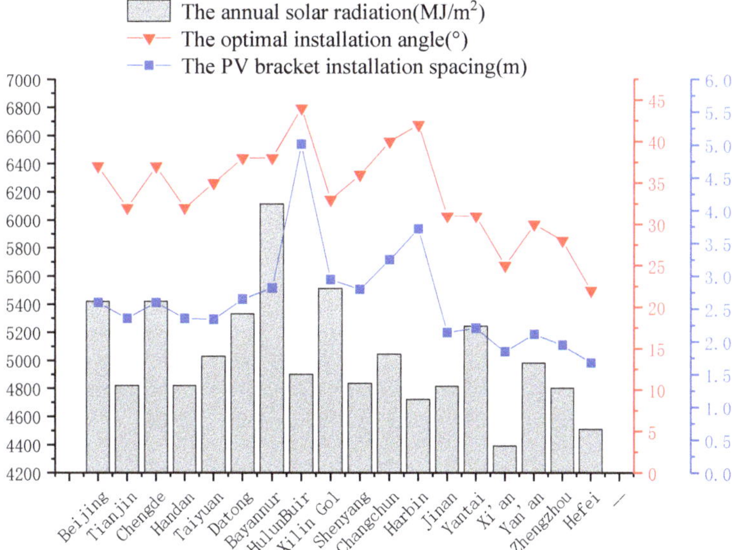

Figure 4. PV system information of representative cities in the northern provinces.

3.5. Availability Factor

In order to obtain the effective PV-available roof area (*PVA*) in China, from the rural building footprint area (*BFA*), we defined the availability factor η_a as the ratio of the effective area of solar PV panels installed on the roof to the total building footprint area.

$$\eta_a = \frac{PVA}{BFA} \tag{3}$$

The availability factor was introduced because under the influence of various factors, the entire roof area cannot be fully used to install PV systems. The main influencing factors are as follows.

(1) Shadows caused by nearby buildings, trees, or some other structures of the roof itself.

(2) The roof area is used for other purposes, such as other household facilities installed on the roof, such as air-conditioning external units.

(3) Part of the area needs to be reserved for the inspection and maintenance of the PV modules by the staff.

Cheng-Dar Yue et al. [23] evaluated the solar energy potential of Taiwan Province based on the land use situation, and used the value of coefficient η_a as 0.4–0.5 in this evaluation process. This value was based on the solar energy building experience rule of the Building Integrated PV (BIPV) potential calculation, and the applicability of the building and solar energy was considered. The suitability of the building included corrections due to construction (HVAC, elevators, terraces, etc.), historical factors, shadow effects, and available surfaces for other purposes. Scartezzini, J et al. [24] set the average value of the availability factor to 0.723 when evaluating the solar energy potential in the urban environment of Europe. J. Ordonez et al. used the sampling, analysis. and then mapping of the whole city method when analyzing the solar capacity of residential roofs in Andalusia (Spain), and set the availability factor of flat roofs in residential buildings to 0.796 and the usability coefficient of the sloped roof houses to 0.983 [25].

As the object of this study was rural buildings and the above-mentioned studies were all aimed at urban buildings, the values of the available coefficients is also different. Compared with the urban buildings, the high-rise buildings in the rural area are relatively less, so the probability of shadow shading is also small, and the proportion of equipment installed on the roof of rural residential buildings is also low, so there is a more complete space for setting up PV systems. Therefore, in the case of a reasonable arrangement of the PV system, it is only necessary to focus on the third influencing factor above, that is, to reserve part of the area for the inspection and maintenance of the PV modules by the staff. Combined with the values of the available factors in the other studies, for conservative calculations, the availability factor η_a of flat roof buildings in the northern provinces was taken as 0.8, and the availability factor of the southern slope roof buildings was taken as 0.9.

3.6. The Panel Loss Factor

The panel loss of the solar PV module refers to the part where the surface area of the module cannot be fully used to receive solar radiation because of surface contamination, wiring, nameplates, etc., so the area of the PV module's power generation area is relatively reduced.

The panel loss factor is 1 minus panel loss (expressed as a fraction), assuming that panel losses (excluding transmission and distribution losses) include contamination (2%), mismatch (2%), wiring (2%), connections (0.5%), light-induced degradation (1.5%), shielding of the nameplate (1%), availability (0.5%), and shading (0.5%).

Then, the panel loss factor is as follows:

$$\eta_p = \left[\begin{array}{l} 1 - (1-0.02) \times (1-0.02) \times (1-0.02) \times (1-0.005) \\ \times (1-0.015) \times (1-0.01) \times (1-0.005) \times (1-0.005) \end{array} \right] \times 100\% = 9.6\% \qquad (4)$$

3.7. Area of PV Modules That Can Generate Electricity

3.7.1. In the Northern Province

For flat roof buildings in rural areas in the northern provinces, the formula for the area of PV modules that can be installed to generate electricity is as follows:

$$A_{g \cdot N} = \frac{BFA}{D \times L} \times \eta_a \times (1 - \eta_p) \qquad (5)$$

where *BFA* is the floor area of the building; *D* is the calculated spacing of the PV array; and *L* is the width of the installed PV module, which is 1 m. η_a and η_p are the availability factor and the panel loss factor, respectively.

3.7.2. In the Southern Provinces

For rural residential buildings in the southern provinces of China, because of the large amount of precipitation in these provinces, in order to facilitate building drainage, the roof forms are mostly sloped roofs, and for such buildings, PV modules should be installed at an angle parallel to the roof. As the slope of the pitched roof varies from 30° to 45°, in order to obtain more conservative data, it was assumed that the slope of all pitched roofs was 30°. Moreover, only installing PV modules in the south, southeast, southwest, east, and west directions of the sloping roof could obtain greater benefits. Therefore, only one side of the slope roof was taken as the installable area of the PV module in this calculation, that is, the area conversion factor. Therefore, the formula for calculating the area of PV modules that can be installed to generate electricity on the sloped roof buildings in southern provinces is as follows:

$$A_{g \cdot S} = \frac{BFA}{\cos \theta} \times \eta_b \times \eta_a \times \left(1 - \eta_p\right) \tag{6}$$

where θ is the slope of the pitched roof, $\theta = 30°$; η_b is the area conversion factor of the buildings in southern provinces, $\eta_b = 0.55$; and η_a and η_p are the availability factor and the panel loss factor, respectively.

Table 1 shows the Solar irradiance intensity, optimum installation angle, *BA* (building area), and *BFA* of typical cities in China, and obtains the calculation results of area of PV modules that can generate electricity (A_g) of typical cities.

Table 1. Solar irradiance intensity, optimum installation angle, *BA*, and *BFA* of typical cities in China.

Province	Typical City	Annual Solar Radiation (MJ/m^2)	Optimal Installation Angle (°)	BA ($\times 10^8$ m^2)	BFA ($\times 10^8$ m^2)	A_g ($\times 10^8$ m^2)
Beijing	Beijing	5421.96	37	1.5	1	0.28
Tianjin	Tianjin	4821.84	32	0.74	0.5	0.15
Hebei	Chengde	5421.96	37	6.38	4.25	1.18
	Handan	4821.84	32	6.38	4.25	1.3
Shanxi	Taiyuan	5030.28	35	4.2	2.8	0.86
	Datong	5331.24	38	1.8	1.2	0.33
Nei Monggol	Bayannur	6113.52	38	1.41	0.94	0.24
	Hulun Buir	4900.32	44	1.05	0.7	0.1
	Xilin Gol	5510.88	33	1.05	0.7	0.17
Liaoning	Shenyang	4836.96	36	4.36	2.91	0.75
Jilin	Changchun	5047.2	40	3.15	2.1	0.47
Heilongjiang	Harbin	4721.4	42	4.66	3.11	0.6
Shandong	Jinan	4817.52	31	11.69	7.79	2.63
	Yantai	5244.48	31	5.01	3.34	1.09
Shaanxi	Xi'an	4392	25	3.89	2.6	1.01
	Yan'an	4980.96	30	2.71	1.8	0.62
Henan	Zhengzhou	4802.04	28	19.92	13.28	4.92
Anhui	Hefei	4509.72	22	14.83	9.89	4.26
Shanghai	Shanghai	4585.32		1.69	1.13	0.53
Jiangsu	Nanjing	4387.32		14.74	9.83	4.62
Zhejiang	Hangzhou	4340.52		11.35	7.56	3.55
Fujian	Fuzhou	4489.56		7.33	4.89	2.3

Table 1. *Cont.*

Province	Typical City	Annual Solar Radiation (MJ/m²)	Optimal Installation Angle (°)	BA ($\times 10^8$ m²)	BFA ($\times 10^8$ m²)	A_g ($\times 10^8$ m²)
Jiangxi	Ganzhou	4701.6		11.96	7.97	3.74
Hubei	Yichang	4030.2		11.79	7.86	3.69
Hunan	Changsha	3996.72		13.3	8.86	4.16
Guangdong	Guangzhou	4201.92		11.39	7.59	3.57
	Shantou	5167.44		1.27	0.84	0.4
Guangxi	Guilin	4229.28		11.51	7.67	3.6
Hainan	Sanya	6143.04		1.62	1.08	0.51
Chongqing	Chongqing	3189.96		7.31	4.87	2.29
Sichuan	Chengdu	3557.88		8.59	5.72	2.69
	Liangshan	5902.56		12.88	8.59	4.03
Guizhou	Guiyang	4189.32		8.08	5.39	2.53
Yunnan	Kunming	5550.48		12.18	8.12	3.81
Tibet	Lhasa	7460.64		0.86	0.58	0.27
Gansu	Jiuquan	6111		2.08	1.39	0.65
	Lanzhou	5286.96		1.56	1.04	0.49
	Minqin	6262.2		1.56	1.04	0.49
Qinghai	Gangcha	6570		0.52	0.35	0.16
	Golmud	6969.96		0.52	0.35	0.16
Ningxia	Guyuan	5599.08		0.31	0.21	0.1
	Yinchuan	5956.92		0.73	0.49	0.23
Xinjiang	Hotan	5950.8		1.26	0.84	0.39
	Northern Xinjiang	5513.76		1.26	0.84	0.39
	Urumqi	5171.4		1.26	0.84	0.39

4. The Calculation of Power Generation

4.1. Calculation of Installed Capacity

In order to calculate the power generation of the PV system, the installed capacity of the system should first be calculated first. There are generally two calculation methods for the installed capacity of a PV system. The first calculation method is based on the area of the PV modules that can be used for power generation and the photoelectric conversion efficiency of PV modules used. The calculation formula is as follows:

$$W = A_g \times 1 \text{ kW/m}^2 \times \eta \tag{7}$$

where W is the total installed capacity of the PV system, kW; A_g is the area of PV modules that can be used for power generation, m²; and η is the photoelectric conversion efficiency of the installed PV modules, %.

In addition, the installed capacity can also be calculated based on the power of the PV modules used in the PV system and the number of modules used. The calculation formula is as follows:

$$W = P \times K \div 1000 \text{ W/kW} \tag{8}$$

where p is the rated power of the PV module used, W, and K is the number of PV modules used.

As this study only has data on the area where PV modules can be installed, it is difficult was determine the power and quantity of PV modules, so formula 7 was used to calculate the installed capacity this time.

At present, the solar cells in China's PV industry are mainly divided into crystalline silicon solar cells and amorphous silicon solar cells. Furthermore, crystalline silicon PV cells are divided into monocrystalline silicon and polycrystalline silicon. The photoelectric conversion efficiency of monocrystalline silicon PV cells is higher than that of polycrystalline silicon, and the technology is more mature, but the cost is slightly higher. The conversion efficiency of the monocrystalline silicon PV cells currently on the market can reach 19%, while the conversion efficiency of polycrystalline silicon PV cells can also reach 15%. As for the amorphous silicon PV cell, because of its relatively low photoelectric conversion efficiency, only about 10%, and the global PV market is still dominated by crystalline silicon cells, and the output of amorphous silicon solar cells is only a small part, which is difficult for meeting large-scale use, so this calculation did not consider the use. Because of the differences in the economic, environmental, and social factors in different regions, the types and efficiency of PV modules installed will vary. Therefore, we calculated the change range of the installed capacity when the photoelectric conversion efficiency changed in the range of 15–19%, and the results are shown in Figure 5.

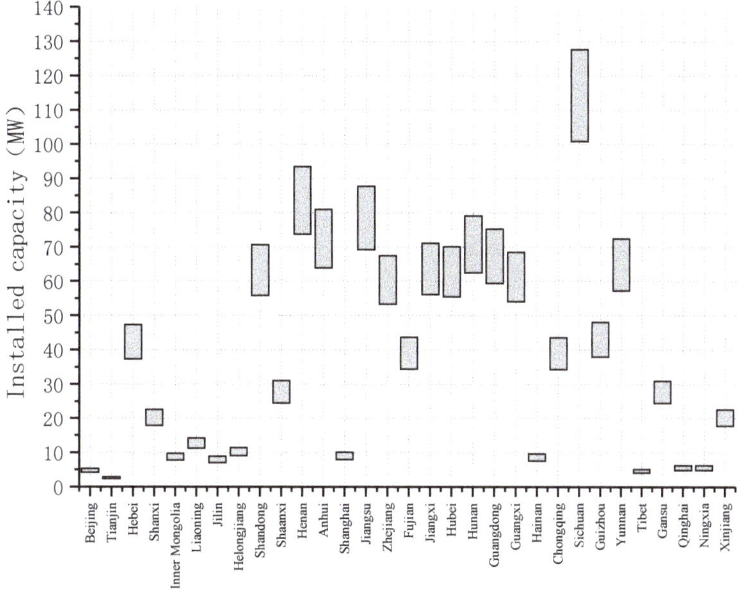

Figure 5. The installed capacity range of each province.

From the calculation results of the installed capacity, it can be seen that, with the exception of a few municipalities and more remote provinces, the installed capacity values of most areas in China can reach more than 20 GW, and some inland provinces such as Sichuan, Henan, and Anhui have a minimum of 50 GW. Across the country, several inland provinces account for a considerable proportion of the total installed capacity of the country, and are the core areas for the installation and popularization of distributed PV systems. On the whole, when different types of PV modules with different efficiencies are selected, the installed capacity of rural areas in mainland China varies within the range of 972.9~1232.34 GW. According to the "2019–2020 Annual Report on China's PV Industry" published by the China PV Industry Association, the output of PV modules in China in 2019 was 98.6 GW, an increase of 17.0% compared with 2018 [26], and it is predicted that module output will increase by 26.4% again in 2021 [27]. From this point of view, it would only take about 7 years for the output of modules to meet the needs of rural areas in China.

4.2. Calculation of Power Generation

After the installed capacity of the PV system was determined, the power generation could be calculated. At present, there are many software and tools in this industry that can be used to analyze and determine the scale, power generation, and other parameters of grid connected and independent PV systems. The main purpose of software tools related to PV system calculation can be divided into pre-feasibility analysis, capacity determination, and power generation simulation.

In this study, PVsyst was used to calculate the power generation; this software was developed by the University of Geneva, and it integrated the pre-feasibility analysis, scale calculation, and power generation simulation of a PV system. It was used to guide the design of the PV system and to simulate the power generation of the PV system. Its advantage is that it does not only conduct research, design, and data analysis on PV systems such as grid-connected systems, off-grid systems, and pumping systems, but also includes a complete PV system database and general solar energy usage tools. After entering the location and load information, we selected different PV components from the product database, and the software automatically simulated the power generation. The specific calculation process is shown in Figure 6.

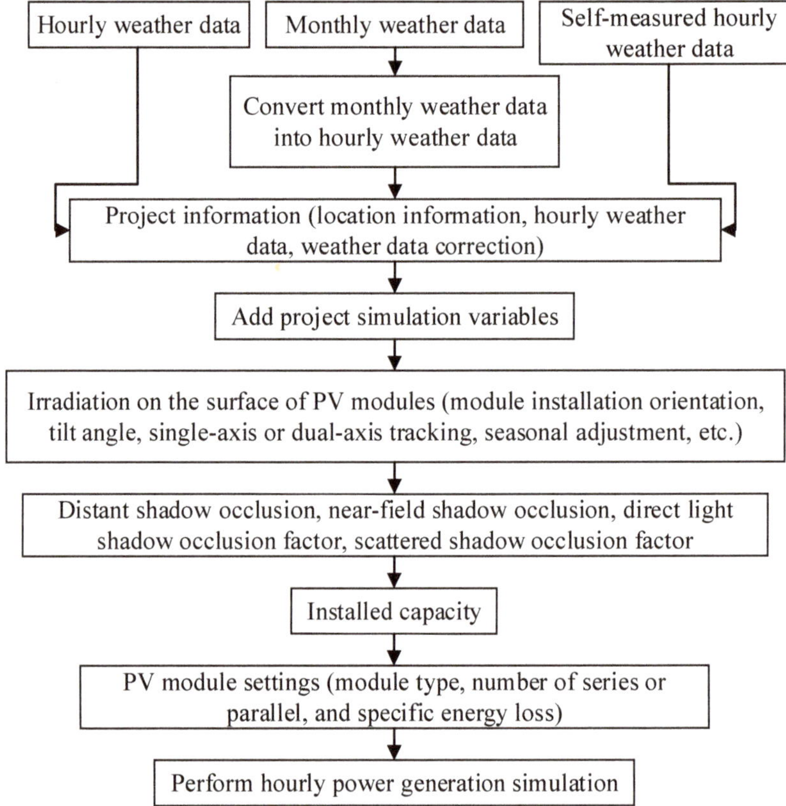

Figure 6. The specific calculation process of PVsyst software.

As the annual power generation of the PV system will gradually decrease with the aging of the PV components and related equipment, the service life of the PV system is generally 25 years. Therefore, the calculation of power generation will be calculated year by year based on the 25-year full life, and finally, an average annual system power generation will be calculated. The calculated results are shown in Figure 7.

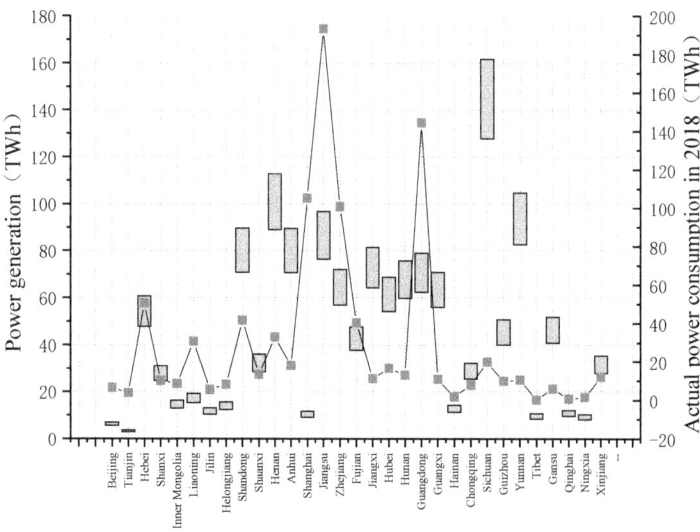

Figure 7. The specific calculation process of PVsyst software.

The simulation results show that when different types and different efficiency PV modules are selected, the total installed capacity of PV systems that can be installed on the roof of rural residential buildings in China is 972.9~1232.34 GW, and the total annual average power generation is 1158.55~1467.47 TWh. According to the data released by the National Bureau of Statistics, the electricity consumption in rural areas in 2019 was 948.293 TWh. If polysilicon PV modules with 15% photoelectric conversion efficiency are installed in all installable areas, the annual average power generation can reach 1158.55 TWh, which is 1.22 times of the total rural electricity consumption in 2019, and if 19% monocrystalline silicon PV modules are installed, the power generation will reach 1.55 times the power consumption. In other words, if all the polysilicon PV modules with a photoelectric conversion efficiency of 15% are used, each region only needs to install 82% of the installed area, and its total power generation is enough to meet the power demand of the rural areas in China. Only 65% of the installable area will be needed when all the high efficiency monocrystalline silicon PV modules with 19% efficiency are installed. It can be seen that the installation of distributed PV systems in rural areas of China has great potential. This is only the result of the roof installation for rural buildings. If we take into account the open wasteland in rural areas and the facade and roof space of urban buildings, the space resources for installing PV system will be further increased, and the power generation will also be correspondingly increased. In addition to meeting the electricity demand of urban and rural residents, the surplus power can also be supplied to other industries in China, promoting the comprehensive development of various industries.

It should be noted that although overall, the total power generation can meet the electricity demand of urban and rural residents, and there may even be some surplus, but because of the huge differences in the intensity of solar radiation in different regions of China and the different demands for electricity, the problem of mismatch between power generation and electricity consumption is prone to occur. As shown in Figure 7, the relatively developed Beijing, Tianjin, Guangdong and Jiangsu, Zhejiang, and Shanghai regions have a large annual electricity consumption, and the total amount of power generation is insufficient to meet their electricity demands. In the northwestern provinces such as Tibet, Qinghai, and Gansu, the annual maximum power generation can reach 10 times or more of the electricity consumption. In addition to the low electricity consumption in these areas, the annual solar radiation intensity in these regions is also high, the same installation areas can generate more electricity, and these are the area with the highest return rate of investment in the PV power generation system. In addition, the annual power generation of Yunnan, Jiangxi, Hunan, Guangxi, and Hainan can also reach more than five times the electricity consumption of the region with a large available power generation area but relatively small power consumption.

In addition to regional differences and economic development, the impact of seasonal differences on a PV power generation system cannot be ignored, as they are affected by geographic location, and most areas of China have higher solar radiation in summer and can generate more electricity resources. In addition to meeting their own use, a large part of electricity will be sold to the local grid. In winter, the intensity of solar radiation decreases, and the electricity generated by the PV system is insufficient to meet user requirements. Some areas may also need to purchase electricity from the grid. Therefore, the mismatch between power generation and power consumption caused by the regional and seasonal differences of solar energy requires us to be able to rationally allocate power among various regions.

5. Economic Feasibility Analysis

Since 2013, China has implemented a net measurement policy to encourage residential users install distributed PV systems; under this policy, the electricity generated by the distributed PV system can be used not only by the users themselves, but can also be sold to the power grid. A typical distributed PV system is mainly composed of five parts: PV modules, batteries, electrical loads, inverters, and power grid. The PV module converts the solar energy irradiated on its surface into electrical energy, and then the electrical energy can be stored in the battery or be directly supplied to the inverter, and the inverter converts DC power to AC power to meet the user's power needs. In addition, the connection of distributed PV systems to the grid not only means that excess electricity can be sold to the grid for profit, but the grid can also provide users with electrical support [28].

5.1. Cost Calculation of PV System

In this study, the monetary unit used in the analysis of the economy was RMB yuan.

PV modules are one of the key parts of a distributed PV system, which directly affects the operation and power generation costs of the system. In the past ten years, with the rapid development of the PV industry and the continuous expansion of the production scale, the capital cost of PV modules in China dropped from 7.4 yuan/W in 2012 to 1.75 yuan/W in 2019. In addition to PV modules, inverters and batteries are also an indispensable part of distributed PV systems. Table 2 is the current economic data and main technical characteristics of the relevant components of PV systems [28,29]. The cost of other PV system components not mentioned in Table 2 (wires, control system, and renovation costs, etc.) are not included in the cost analysis in this research. On the whole, the current cost of inverters and batteries in China is about 500 to 1000 yuan, which is still slowly decreasing.

Table 2. The economic cost and technical characteristics of each component.

Component	Technical and Economic Parameters	Value
PV system	Fixed operation and maintenance costs Service life	35.66 (yuan/year) 25 (year)
PV panel	Capital cost Replacement cost Operation and maintenance cost Service life	1750 (yuan/kW) 1750 (yuan/kW) 22.16 (yuan/kW) 25 (year)
Battery	Capital cost Replacement cost Service life	600 (yuan/kW) 1857 (yuan/kW) 5 (year)
Inverter	Capital cost Replacement cost Service life	750 (yuan/kW) 600 (yuan/kW) 15 (year)

As the design service life of the PV system of this project is 25 years, in addition to the initial investment, the replacement cost of batteries and inverters should also be added when calculating the total investment cost. For example, if the service life of the battery is 5 years, the battery needs to be replaced every five years. With economic development and the popularization of the PV industry, the cost of the battery and other related equipment will be reduced accordingly; assuming that the cost of the battery is reduced by 10% every five years, the total cost of the battery in 25 years will be 2457 yuan/kW, and assuming that the inverter price is reduced by 20% after 15 years, the total cost of the inverter will be 1350 yuan/kW. Based on comprehensive calculations, the total fixed investment cost of the PV system for 25 years of operation will be 5557 yuan/kW, and the fixed operation and maintenance cost will be 35.66 yuan/kW· year. On this basis, the initial investment cost and operation and maintenance costs of distributed PV systems in each province were calculated, and the results are shown in Figure 8.

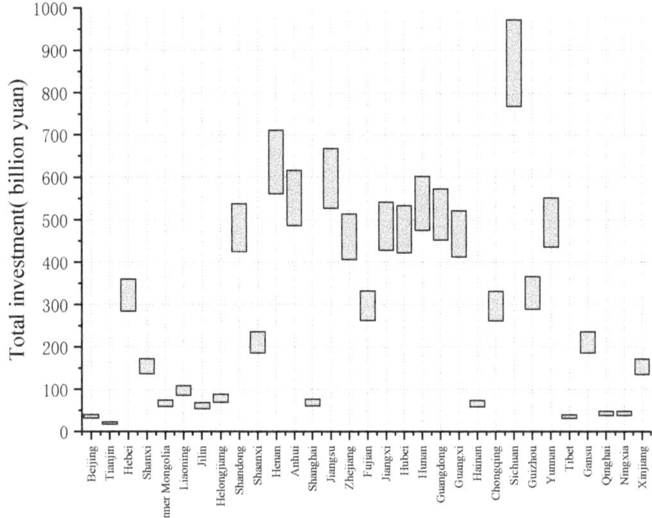

Figure 8. Total investment of the PV system.

5.2. The Benefit Analysis

If the PV module is installed on the roof of the building, because of the heat absorption of the module, the temperature in the roof and indoors can be reduced by 3–5 °C when the temperature is high in summer, achieving the effects of thermal insulation, energy saving, and emission reduction. On rainy days, PV modules can also buffer the impact of rain, effectively reducing the noise generated by rain on the roof. At the same time, it can also resist the direct rays of the ultraviolet rays on the roof and prolong the service life of the roof. In addition, users who install distributed PV systems will also receive some economic benefits.

Since the implementation of the net measurement policy in China in 2013, in addition to saving part of the electricity costs, residential users who invest in the construction of distributed PV systems can also benefit from two aspects, these are the government subsidies for distributed PV system power generation and the proceeds from the sale of surplus electricity to the State Grid, separately. In 2013, the Chinese government implemented a net measurement subsidy of 0.42 yuan/kWh for distributed PV power generation; after that, with the decline of PV power generation costs and the rapid development of the PV industry, the net metering subsidy was reduced to 0.37 yuan/kWh in 2018. In addition, the Chinese government requires grid companies to purchase renewable energy power according to the local benchmark feed-in tariff. Therefore, users of distributed PV systems can also sell excess power to the grid and make a profit from it; this is the second source of revenue [30]. At the same time, there are also the energy-saving benefits brought about by the environmental changes after the installation of a PV system to reduce the building load. However, this is greatly affected by the climate, and its income is low and unstable. Therefore, this benefit was not considered in this research. We used the following formula to represent the profits of users who invested in distributed PV systems:

$$R = \left(E_{load} + E_{grid} \right) \times P_{subsidy} + E_{grid} \times P_{grid} + E_{load} \times P_{local} \tag{9}$$

where R is the total annual income of users who invested in distributed PV systems, yuan/year; E_{load} is the part of power generation used by the users themselves, kWh/year; E_{grid} is the excess electricity sold to the national grid, kWh/year; $P_{subsidy}$ is a subsidy for the electricity implemented by the government, yuan/kWh; P_{grid} is the local coal-fired grid-connected electricity price, yuan/kWh; and P_{local} is the user's local residential electricity price, yuan/kWh. The specific price is shown in Figure 9.

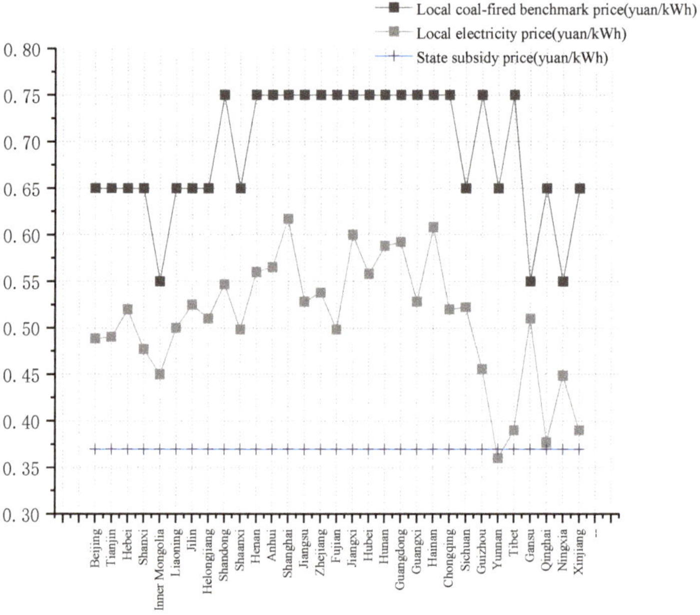

Figure 9. Electricity prices, feed-in tariffs, and subsidized prices across China in 2018.

The electrical energy generated by distributing PV systems has two main directions, namely, its own use and the sale of energy to the local power grid. Different regions will lead to different generations and demands from the PV system. The power resources used by households who install a distributed PV system consist of two parts, one of which is the electricity generated by the PV modules. When the electricity generated by the PV modules is insufficient to support the user's use, the insufficient electricity will be purchased from the local power grid. Xinyu Jia et al. [19] selected five typical cities as research objects from five regions with different solar radiation levels in China, then analyzed the benefits of China's household distributed PV system. These five cities were Chongqing, Guangzhou, Beijing, Yushu, and Lhasa. They input the PV system configuration into the HOMER software to simulate the electricity consumption of the actual residents. The simulation results show that in terms of electricity consumption, the electricity sold to the local grid in the PV systems of the five cities accounted for about 30%, which means 70% of the electricity generated by the PV systems was used by the users themselves. Therefore, in this study, 70% of the average annual power generation was taken as the self-consumption of the first year, and the remaining 30% was taken as the electricity sold to the local grid in the first year. Based on this, the profit of the users investing in the distributed PV system was calculated according to the above formula.

With the increase in operating time, the aging and wear of the relevant components of the PV system will cause the power generation to decrease year by year. Therefore, the annual power generation of PV modules were reduced by 5% per year when calculating economic benefits. The development and popularization of the PV industry will lead to a year-by-year reduction in China's subsidies for PV power generation and grid-connected benchmark electricity prices, and the reduction ratio was also set to 5%. With the development of the economy, the price of electricity for residential users will increase year by year, and the electricity consumption will increase accordingly, which will cause the proportion of electricity used by the users themselves in the power generation of PV systems to increase year by year; these three growth rates were all set to 5%. In addition, the purpose of the PV power generation subsidy implemented by the state is to stimulate the application and development of the PV industry, and the PV system has developed rapidly in recent years and has been applied in a large number of provinces, so the PV subsidy could be cancelled in the next few years. Therefore, the subsidy income will become a very uncertain income. In order to determine the impact of government subsidies on PV revenue, this study also calculated the total revenue that users who installed PV systems could get without a subsidy revenue. Table 3 shows the income and payback time of a PV system in Jiangsu Province.

Table 3. Calculation of income from PV systems in Jiangsu Province.

Years	Power Generation (TWh)	$P_{subsidy}$ (Yuan/kWh)	P_{gird} (Yuan/kWh)	P_{local} (Yuan/kWh)	Proportion of Self-Use	Revenue (No Subsidies) (Billion Yuan)	Revenue (Including Subsidies) (Billion Yuan)
1	57.31	0.37	0.75	0.5283	0.7	34.09	55.29
2	54.44	0.35	0.71	0.55	0.74	32.48	51.62
3	51.72	0.33	0.68	0.58	0.77	31.24	48.51
4	49.14	0.32	0.64	0.61	0.81	30.34	45.93
5	46.68	0.30	0.61	0.64	0.85	29.76	43.83
6	44.35	0.29	0.58	0.67	0.89	29.46	42.15
7	42.13	0.27	0.55	0.71	0.94	29.42	40.88
8	40.02	0.26	0.52	0.74	0.98	29.62	39.96
9	38.02	0.25	0.50	0.78	1	29.68	39.01
10	36.12	0.23	0.47	0.82	1	29.60	38.03
11	34.31	0.22	0.45	0.86	1	29.53	37.13
12	32.60	0.21	0.43	0.90	1	29.45	36.32
13	30.97	0.20	0.41	0.95	1	29.38	35.57
14	29.42	0.19	0.39	1.00	1	29.31	34.90
15	27.95	0.18	0.37	1.05	1	29.23	34.28
16	26.55	0.17	0.35	1.10	1	29.16	33.71
17	25.22	0.16	0.33	1.15	1	29.09	33.20
18	23.96	0.15	0.31	1.21	1	29.02	32.72
19	22.76	0.15	0.30	1.27	1	28.94	32.29
20	21.63	0.14	0.28	1.33	1	28.87	31.89
21	20.54	0.13	0.27	1.40	1	28.80	31.52
22	19.52	0.13	0.26	1.47	1	28.73	31.19
23	18.54	0.12	0.24	1.55	1	28.65	30.87
24	17.61	0.11	0.23	1.62	1	28.58	30.59
25	16.73	0.11	0.22	1.70	1	28.51	30.32

The initial investment in PV systems in Jiangsu Province was 446.6 billion Yuan. The calculation results show that when the power generation subsidy was included, the distributed PV systems installed in rural buildings in Jiangsu Province could recover the investment in the eighth year of operation, and the total income obtained over 25 years would be 125.59 billion Yuan, which is 2.8 times that of the initial investment. After removing the subsidy income, the payback period was extended to 11 years, and the total income was reduced to 98.792 billion Yuan—still more than twice the initial investment.

From a data point of view, although the payback time for installing distributed PV systems in rural areas is slightly longer, it is still a lucrative investment. Using the same method to calculate the total revenue and payback time of the other provinces in China, the calculation results are shown in Figure 10. Excluding subsidies, the distribution of the investment payback period of rural photovoltaic systems in mainland China is shown in Figure 11.

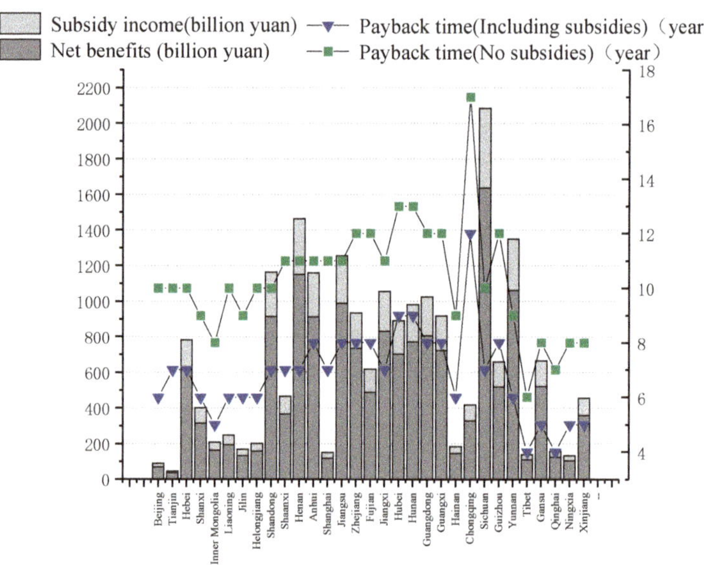

Figure 10. Total revenue and payback time of the PV system.

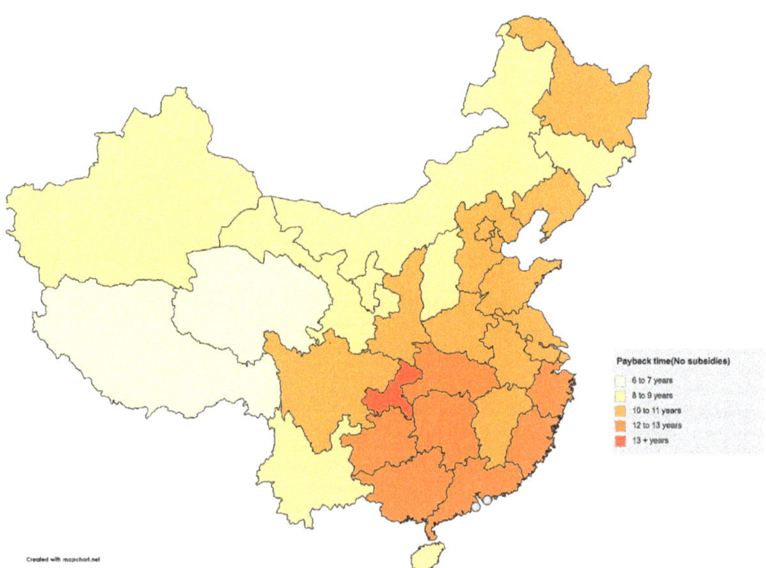

Figure 11. Payback period of the investment in various provinces (excluding subsidies).

5.3. Result Analysis

The results show that when PV subsidies were included, the payback period of investment in most areas of China was within eight years. However, Xinjiang, Tibet, Qinghai, and Gansu have a high solar radiation intensity, which is more conducive to the investment and construction of PV systems, and so the initial investment could be recovered within five years. The operating life of the PV system was set to 25 years. The first five years would be used to recover the investment, and the next twenty years would bring very considerable net profits to the users. Therefore, the shorter the payback period, the greater the benefits users can get.

Because of the great uncertainty of PV subsidies, this paper mainly analyzed the economic benefits when PV subsidies were not included. After the government subsidies for PV systems are removed, the total revenue that users can get in 25 years will shrink, the economic income of each province will be reduced by about 20%, and the payback period of investment will be extended by 3–4 years. According to the economic and energy benefits, the potential of solar energy resources in the Chinese mainland can be divided into three categories. The first is the area where the distributed PV system is recommended to be popularized, that is, the area with the highest economic benefit, which mainly includes inner Mongolia, Jilin, Hainan, Tibet, Gansu, Qinghai, and Ningxia. These provinces are rich in solar energy resources and the investment recovery period will be within 9 years; the installation of the distributed PV systems will bring considerable benefits to users in the next 25 years, which is very suitable for the investment and installation of distributed PV systems. The third kind of areas are the areas with a low economic efficiency for using solar radiation resources, mainly in Chongqing and Guizhou Province. These places are located in the basin, receive less solar radiation, and obtain less power generation under the same installed capacity compared with other provinces, so it takes a longer time to recover the initial investment, which will take 14–17 years. Compared with other provinces, the economic benefit of investment in the construction of PV system is the lowest. Besides the first and third categories of areas, the other provinces belong to the second category of resource areas, which mainly includes some central provinces that have medium solar energy resources and the investment recovery period is within 10–12 years. Although the economic benefits are not as great as those in the first category, the total income of the PV system after 25 years of operation could still reach more than twice the initial investment. In addition, these provinces have a large rural construction area and a large roof area for power generation. In terms of energy efficiency, these provinces can generate more electricity, most of the electricity in the whole country comes from these areas, and the energy efficiency is the most considerable. Therefore, these provinces will become the main force of PV power generation in China in the next few decades, and the investment and construction of distributed PV systems in these areas should be vigorously promoted.

From the perspective of overall economic benefits, even if government subsidies are not included, the total income of PV systems in most areas of China for 25 years will be more than twice the initial investment. This part of the income could not only increase the income of rural residents and reduce the gap between the rich and poor between urban and rural areas, but could also show the application potential of PV systems and stimulate the promotion and application of the PV industry in rural areas.

5.4. Social Benefit

In addition to direct economic benefits, comprehensive utilization of PV power generation can also bring other benefits, such as reducing pollutant and greenhouse gas emissions, providing jobs, ensuring energy security, and improving the safety of electricity for life and production in rural and remote areas in China.

Take Li'ao Village, Ningbo City, Zhejiang Province as an example, as shown in the Figure 12. The village invested 6 million Yuan in 2016 to install a 600 kw distributed PV power station on the roof of the village's residential houses. This practical application showed that the PV power station had an annual power generation capacity of 720,000 kWh, with an average annual income of more than 600,000 Yuan. At the same time, the installation of PV panels can also play a role in thermal insulation inside the building. It can reduce the indoor temperature by 5–7 °C in spring and summer, and can increase it by 2–3 °C in autumn and winter, saving electricity costs for heating, cooling, and air conditioning. In addition, if the power generation is calculated at 720,000 kWh per year, 288 tons of standard coal can be saved annually. The emission reduction of 576 tons of carbon dioxide, 157 tons of dust, 4.5 tons of sulfur oxides, and 1.5 tons of nitrogen oxides is equivalent to planting 16,000 trees, which would promote the sustainable development of the ecosystem.

Figure 12. Rooftop photovoltaic installation in Li'ao Village.

Therefore, the promotion of the distributed PV systems in rural areas does not only provide better investment costs and returns compared with traditional power generation, it also has extremely important social practical significance and economic value for building a new rural power supply system that is safe, reliable, energy-saving, environmentally friendly, and technologically advanced.

In addition, since the Chinese government put forward the goals of reaching a "carbon peak" and becoming "carbon neutral", adjusting the energy structure and improving the efficiency of energy use have to become the top priority of the society, and the use of renewable energy will undoubtedly be accelerated. If the PV power generation system can be widely used in rural areas of our country, this achieve a significant step forward towards the "3060 plan". In order to implement this plan, all provinces across the country have stepped up the introduction of relevant policies to stimulate PV development, and there may even be a new round of PV subsidies issued to stimulate rural residents to install PV systems.

In summary, the popularization of PV systems in rural areas can not only provide villagers with huge economic benefits, but can also play an important role in reducing carbon emissions, protecting the ecological environment, and promoting sustainable ecological development. Therefore, the popularization of PV systems can be said to be a necessary means for China to develop rural areas and to achieve carbon neutrality.

6. Conclusions

This paper takes the rural residential buildings in mainland China as the research object, carries out the solar energy resource application potential evaluation and eco-nomic benefit analysis, and draws the following conclusions.

China's rural areas are rich in space resources for the investment and installation of a distributed PV system, and the annual power generated by a PV system can bear the power load of the rural areas. The calculation results show that there are still more than 6.4 billion m^2 of building roof area in rural areas that can be used for the investment and installation of distributed PV systems, and if used rationally, the power generation will be able to reach 1.55 times the total power consumption in rural areas. The research also gives the PV installable resources in rural areas of China, which can provide some data support for China's future plan of "carbon neutrality".

Promoting the popularization of distributed PV systems in rural areas will not only bring energy benefits and alleviate environmental problems, but also bring certain economic benefits to rural farmers. Comprehensively considering economic and energy benefits, mainland China is divided into three types of resource areas, and each resource area should reasonably formulate relevant policies based on its own characteristics and advantages in order to promote the development of the PV industry in rural areas.

In this paper, the purpose of PV related prediction is to serve the formulation of relevant policies, so as to predict the total amount of energy that the distributed PVs can provide in China's rural areas in the future. It can provide assistance for many provinces in China to formulate the building roof PV installation plans of cities and counties in the near future. This study only calculated and analyzed the PV application prospects of rural building roofs in China, and lacked an analysis of the PV power generation potential of rural abandoned land in China, which would not reflect the overall PV power generation potential of China. Therefore, further work could be done to forecast the energy and economic benefits of PV technology in China.

Author Contributions: Data curation, Y.Z. (Yuqiang Zhao); Formal analysis, F.H.; Project administration, Y.Z. (Yongheng Zhong); Supervision: J.Z.; Resources: J.Z.; Writing—original draft, W.Z.; Writing—review & editing, W.Z. All authors have read and agreed to the published version of the manuscript.

Funding: This research was funded by the China National Key R&D Program (grant no. 2018YFC0704400), the National Natural Science Foundation of China (grant no. 51908287), and the National Natural Science Foundation of Jiangsu Province (grant no. BK20180484).

Institutional Review Board Statement: Not applicable.

Informed Consent Statement: Not applicable.

Data Availability Statement: Not applicable.

Acknowledgments: This study was supported by the China National Key R&D Program (grant no. 2018YFC0704400), the National Natural Science Foundation of China (grant no. 51908287), and the National Natural Science Foundation of Jiangsu Province (grant no. BK20180484).

Conflicts of Interest: The authors declare no conflict of interest.

References

1. Jiang, Y. 2020 Annual Development Research Report on building energy efficiency in China. *Build. Energy Effic.* **2020**, *48*, 166.
2. Wang, Y. The Research of Promotion & Application for PV Generation in Rural and Remote Areas of China. Master's Thesis, Chinese Academy of Agricultural Sciences, Beijing, China, 2012.
3. Shen, Y. The Spatial Distribution of Solar Energy and the Comprehensive Potential Evaluation of Regional Exploitation and Utilization in China. Master's Thesis, Lanzhou University, Lanzhou, China, 2014.
4. Li, D.H.; Yang, L.; Lam, J.C. Zero energy buildings and sustainable development implications—A review. *Energy* **2013**, *54*, 1–10. [CrossRef]
5. Tiwari, G.; Mishra, R.; Solanki, S. Photovoltaic modules and their applications: A review on thermal modelling. *Appl. Energy* **2011**, *88*, 2287–2304. [CrossRef]
6. Zhang, D.; Chai, Q.; Zhang, X.; He, J.; Yue, L.; Dong, X.; Wu, S. Economical assessment of large-scale photovoltaic power development in China. *Energy* **2012**, *40*, 370–375. [CrossRef]
7. Zhang, F.; Deng, H.; Margolis, R.; Su, J. Analysis of distributed-generation photovoltaic deployment, installation time and cost, market barriers, and policies in China. *Energy Policy* **2015**, *81*, 43–55. [CrossRef]
8. Haghdadi, N.; Copper, J.; Bruce, A.; MacGill, I. A method to estimate the location and orientation of distributed photovoltaic systems from their generation output data. *Renew. Energy* **2017**, *108*, 390–400. [CrossRef]
9. Anaya, K.L.; Pollitt, M.G. Integrating distributed generation: Regulation and trends in three leading countries. *Energy Policy* **2015**, *85*, 475–486. [CrossRef]
10. Vardimon, R. Assessment of the potential for distributed photovoltaic electricity production in Israel. *Renew. Energy* **2011**, *36*, 591–594. [CrossRef]

11. Margolis, R.; Gagnon, P.; Melius, J.; Phillips, C.; Elmore, R. Using GIS-based methods and lidar data to estimate rooftop solar technical potential in US cities. *Environ. Res. Lett.* **2017**, *12*, 074013. [CrossRef]
12. Mikovits, C.; Schauppenlehner, T.; Scherhaufer, P.; Schmidt, J.; Schmalzl, L.; Dworzak, V.; Hampl, N.; Sposato, R. A Spatially Highly Resolved Ground Mounted and Rooftop Potential Analysis for Photovoltaics in Austria. *ISPRS Int. J. Geo-Inf.* **2021**, *10*, 418. [CrossRef]
13. Fina, B.; Auer, H.; Friedl, W. Cost-optimal economic potential of shared rooftop PV in energy communities: Evidence from Austria. *Renew. Energy* **2020**, *152*, 217–228. [CrossRef]
14. Senatla, M.; Bansal, R.C.; Naidoo, R.; Chiloane, L.; Mudau, U. Estimating the economic potential of PV rooftop systems in South Africa's residential sector: A tale of eight metropolitan cities. *IET Renew. Power Gener.* **2020**, *14*, 506–514. [CrossRef]
15. Miranda, R.F.; Szklo, A.; Schaeffer, R. Technical-economic potential of PV systems on Brazilian rooftops. *Renew. Energy* **2015**, *75*, 694–713. [CrossRef]
16. Xie, Y. Key of Rural Distributed Photovoltaic Power Generation. *Appl. IC* **2020**, *37*, 48–50.
17. Notice! The National Energy Administration Released the pilot Scheme of Distributed Photovoltaic Development on the Roof of the Whole County (City, District). Available online: https://mp.ofweek.com/solar/a656714976197 (accessed on 24 June 2021).
18. 11th Province! Shanxi Province Issued the Policy of Promoting Distributed Photovoltaic in the Whole County! Available online: https://baijiahao.baidu.com/s?id=1703804914999411688&wfr=spider&for=pc (accessed on 28 June 2021).
19. Hubei Provincial Development and Reform Commission. Notice of Hubei Energy Bureau on Submitting the Pilot Scheme of Roof Distributed Photovoltaic Development for the Whole County (City, District). Available online: http://fgw.hubei.gov.cn/fbjd/xxgkml/jgzn/wgdw/nyj/xnyhkzsnyc/tzgg/202106/t20210629_3618798.shtml (accessed on 29 June 2021).
20. Notice of Fujian Provincial Development and Reform Commission on Carrying Out the Pilot Work of Centralized Promotion of Household Photovoltaic in the Whole County. Available online: https://msolar.in-en.com/html/solar-2377381.shtml (accessed on 6 June 2021).
21. Notice on Submitting the Pilot Scheme of Roof Distributed Photovoltaic Development in the Whole County (City, District) (Anhui). Available online: https://news.solarbe.com/202106/27/340661.html (accessed on 27 June 2021).
22. Jacobson, M.Z.; Jadhav, V. World estimates of PV optimal tilt angles and ratios of sunlight incident upon tilted and tracked PV panels relative to horizontal panels. *Sol. Energy* **2018**, *169*, 55–66. [CrossRef]
23. Yue, C.-D.; Huang, G.-R. An evaluation of domestic solar energy potential in Taiwan incorporating land use analysis. *Energy Policy* **2011**, *39*, 7988–8002. [CrossRef]
24. Scartezzini, J.-L.; Montavon, M.; Compagnon, R. *Computer Evaluation of the Solar Energy Potential in an Urban Environment*; EuroSun: Bologna, Italy, 2002.
25. Ordóñez, J.; Jadraque, E.; Alegre, J.; Martínez, G. Analysis of the photovoltaic solar energy capacity of residential rooftops in Andalusia (Spain). *Renew. Sustain. Energy Rev.* **2010**, *14*, 2122–2130. [CrossRef]
26. Jiang, H.; Jin, Y.; Ye, X.; Qiang, Y.; Li, J.; Han, P. Review of China's PV industry in 2019 and prospect in 2020. *Sol. Energy* **2020**, *3*, 14–23.
27. Jiang, H.; Jin, Y.; Ye, X.; Qiang, Y.; Li, J.; Zhang, D. Review of China's PV industry in the first half of 2020 and outlook in the second half of 2020. *Sol. Energy* **2020**, *9*, 5–13.
28. Zou, H.; Du, H.; Brown, M.A.; Mao, G. Large-scale PV power generation in China: A grid parity and techno-economic analysis. *Energy* **2017**, *134*, 256–268. [CrossRef]
29. Adefarati, T.; Bansal, R. Reliability and economic assessment of a microgrid power system with the integration of renewable energy resources. *Appl. Energy* **2017**, *206*, 911–933. [CrossRef]
30. Jia, X.; Du, H.; Zou, H.; He, G. Assessing the effectiveness of China's net-metering subsidies for household distributed photovoltaic systems. *J. Clean. Prod.* **2020**, *262*, 121161. [CrossRef]

 *materials*

Article

Preparation of Breathable Cellulose Based Polymeric Membranes with Enhanced Water Resistance for the Building Industry

Atif Hussain * and Pierre Blanchet

Department of Wood and Forest Sciences, Université Laval, Quebec City, QC G1V 0A6, Canada;
pierre.blanchet@sbf.ulaval.ca
* Correspondence: atif.hussain.1@ulaval.ca

Abstract: This study focuses on the development of advanced water-resistant bio-based membranes with enhanced vapour permeability for use within building envelopes. Building walls are vulnerable to moisture damage and mold growth due to water penetration, built-in moisture, and interstitial condensation. In this work, breathable composite membranes were prepared using micro-fibrillated cellulose fiber (CF) and polylactic acid (PLA). The chemical composition and physical structure of CF is responsible for its hydrophilic nature, which affects its compatibility with polymers and consequently its performance in the presence of excessive moisture conditions. To enhance the dispersibility of CF in the PLA polymer, the fibers were treated with an organic phosphoric acid ester-based surfactant. The hygroscopic properties of the PLA-CF composites were improved after surfactant treatment and the membranes were resistant to water yet permeable to vapor. Morphological examination of the surface showed better interfacial adhesion and enhanced dispersion of CF in the PLA matrix. Thermal analysis revealed that the surfactant treatment of CF enhanced the glass transition temperature and thermal stability of the composite samples. These bio-based membranes have immense potential as durable, eco-friendly, weather resistant barriers for the building industry as they can adapt to varying humidity conditions, thus allowing entrapped water vapor to pass through and escape the building, eventually prolonging the building life.

Keywords: cellulose fiber; vapor permeability; surface treatment; surfactant; bio-based materials

Citation: Hussain, A.; Blanchet, P. Preparation of Breathable Cellulose Based Polymeric Membranes with Enhanced Water Resistance for the Building Industry. *Materials* **2021**, *14*, 4310. https://doi.org/10.3390/ma14154310

Academic Editor: Carlos Morón Fernández

Received: 8 July 2021
Accepted: 29 July 2021
Published: 1 August 2021

1. Introduction

The accumulation of water in the form of excessive moisture within building envelopes can lead to the premature deterioration of building materials [1]. Buildings can be affected by three main sources of moisture: external moisture from precipitation or groundwater; internal moisture from occupant presence and their activities; and built-in moisture in the materials accumulated during manufacturing. Moisture movement in building wall cavities is facilitated by air currents, heat transfer and diffusion through materials. The penetration and accumulation of excessive moisture in building envelopes can lead to interstitial condensation, which can cause structural damages, mould growth, as well as damage to indoor materials. The moisture flow in building envelopes can be controlled by using barriers in walls, floors, ceiling, and roofs, thereby preventing interstitial condensation [2].

Weather barriers are membranes used in the exterior side of the wall system and act like a shell for buildings [3]. A premium, high-performance weather barrier has four beneficial and essential functions: air resistance, water resistance, durability during construction, and the right level of vapor permeability. Although the process of vapour permeability is least understood and heavily ignored, it can greatly affect the wall performance [4]. Vapor permeability is also discussed in terms of breathability of the material as its ability to allow moisture or water vapour to pass through it [5,6]. A good weather barrier is expected to resist bulk water (in liquid form) but should not necessarily block water vapour (in gas form) [7]. In the past, installing barriers in envelopes was not necessary as the walls had very low insulation. However, in the current scenario when the wall interior gets wet, tighter enclosures combined with high levels of thermal insulation can significantly reduce the drying potential of the wall [8]. Vapor barriers have been used with the intention to protect walls from becoming wet, however their main disadvantage is that the barrier also prevents from drying if the interior of the walls get wet [9]. Therefore, there is a need to develop a barrier with a unique structure that can specifically allow vapour to pass through but resist water entering the building wall. The most common barriers currently used in industry are non-biodegradable fossil fuel-based, which also raises the need to develop new barriers with more sustainable properties.

Bio-based materials have recently become popular in the building and construction industry due to their insulation and hygroscopic properties [10,11]. Studies have reported that using these materials in construction increases the energy efficiency of the building and provide a comfortable and healthy indoor environment [12–14]. Some bio-based materials also have the ability to absorb and release moisture with respect to changing relative humidity levels, which can reduce the load on air conditioning and have a positive impact on wellbeing of residents [15,16]. The ability of bio-based materials to capture and lock CO_2 from the atmosphere during their lifetime can be highly beneficial for the environment [17]. The Quebec building code highlights that greenhouse gas emissions from buildings account for a large share of the region's overall emissions and strives to achieve its emission reduction targets by 2030 [18]. To accomplish this goal, buildings should have a highly energy efficient design as well as lower embodied energy. Since majority of the electricity generated for use in Quebec buildings is derived from hydroelectricity, which is a renewable source, the embodied energy of buildings plays a major role in contributing towards carbon emissions during their life cycle [19]. Quebec has immense potential for developing new renewable materials from wood by-products that will result in the production extremely lower embodied energy materials.

Cellulose is the most abundant organic compound, and it can be obtained quite easily from wood pulp in the form of very thin and long fibers. Recent advancements in science have supported the development of industrial processes for the extraction of cellulose fiber from wood in large-scale volumes, having a 100% yield without the use of enzymes or chemicals [20]. The fibres have unique properties such as high specific surface area, good strength and rheological properties making them a highly versatile, biodegradable, and compostable additive for a composite as either a membrane or a coating. Recent studies have used cellulose fibres for the development of films addressing mechanical [21,22] and optical properties [23,24]. However, a few studies have reported excellent vapor barrier properties in packaging applications [25–27] and their use in the construction industry [28].

Cellulose is highly hydrophilic due to the presence of hydroxyls in its chemical structure. High moisture sensitivity in bio-based materials can lead to fungal growth and compromise the durability of the material [29]. Furthermore, the quality of the end product can be affected during the manufacturing stages if cellulosic materials encounter humid environments or unexpected water [30]. The high water absorption capacity of bio-based materials also makes them incompatible with hydrophobic polymers causing poor interfacial adhesion in the composites [31,32]. Numerous studies have investigated the effect of alkali [33], acetylation [34], ionic liquids and salts [35–37], silane [38], sol-gel [39], and surfactant [40,41] treatment of bio-based materials that improve their hydrophobicity and compatibility with polymers.

Cellulose fibers offer wide possibilities for new product development and the objective of this research is the utilization of commercially available cellulose fibers with a bio-based polymer for the development of a breathable vapor barrier for the construction industry. The work also involves the treatment of the cellulose fibers and an investigation of their compatibility with the polymer to determine the vapor permeability, morphology, as well as physical and thermal characteristics of the composite material.

2. Materials and Methods

Cellulose fibers commercially referred as cellulose filaments were received from Kruger Biomaterials Inc. (Montreal, QC, Canada), a privately held company that transforms renewable resources into sustainable essentials. The cellulose fibers used in this study were obtained in the form of suspension (2.5 wt%). The fibers were obtained by freeze-drying the suspension. PLA grade 4043D was used in this study and obtained from NatureWorks (Minnetonka, MN, USA). This particular grade of PLA is targeted for the preparation of membranes and films. Chloroform was used as the solvent for dissolving PLA and obtained from Sigma-Aldrich (Oakville, ON, Canada).

2.1. Fiber Treatment

For the treatment of cellulose fibers, an organic phosphoric acid ester-based surfactant was added to the CF suspension. The surfactant was obtained from EMCO-Inortech (Shwego-wett 6267), Montreal, QC, Canada. It is a biodegradable, VOC-free, anionic universal wetting, and dispersive additive. The concentration of surfactant used for treatment of the fibers was 20 wt% of dry fiber weight. The fiber suspension along with the surfactant was mixed using a high shear mixer at 2000 rpm for 15 min and then freeze-dried.

2.2. Film Preparation

PLA membranes were prepared using solvent casting method. Briefly, 5 g of PLA was dissolved in 100 mL of chloroform under constant stirring at 600 rpm for 2 h at room temperature and atmospheric pressure. The solution was poured into a Petri dish and left in a fume hood at room temperature for evaporating the solvent. For preparation of the CF composite membranes, varying concentrations of untreated and surfactant treated CF were the added to dissolved PLA and stirred vigorously for another 30 min using a high shear mixer at 1200 rpm. At the end of the mixing process, the PLA-CF solution was homogenous, and no bubbles were observed. The prepared PLA membranes with varying concentrations of untreated cellulose fibers (UCF) and treated cellulose fibers (TCF) used for this study are mentioned in Table 1. Micrographs showing UCF and TCF are presented in Figure 1. Photographs of all prepared films used in this experiment are shown in Figure 2. All membrane samples were vacuum dried at 40 °C for 96 h to release any remaining solvent and then sealed and stored in zipped airtight bags for characterization.

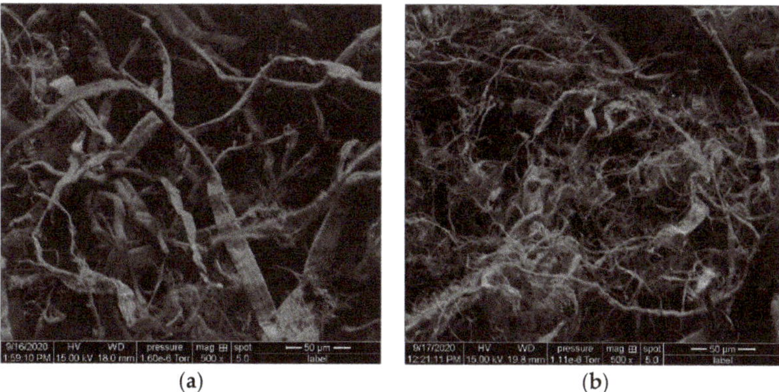

(a)   (b)

Figure 1. Micrographs of cellulose fibers (**a**) untreated, (**b**) after surfactant treatment.

Table 1. Description of film samples prepared for this study.

Sample	PLA Conc. (wt%)	CF Conc. (wt%)	Surfactant Treated
PLA	100	0	-
UCF	0	100	-
TCF	0	100	Yes
PLA-UCF1	99	1	-
PLA-UCF2	98	2	-
PLA-UCF5	95	5	-
PLA-UCF10	90	10	-
PLA-UCF15	85	15	-
PLA-UCF20	80	20	-
PLA-TCF1	99	1	Yes
PLA-TCF2	98	2	Yes
PLA-TCF5	95	5	Yes
PLA-TCF10	90	10	Yes
PLA-TCF15	85	15	Yes
PLA-TCF20	80	20	Yes

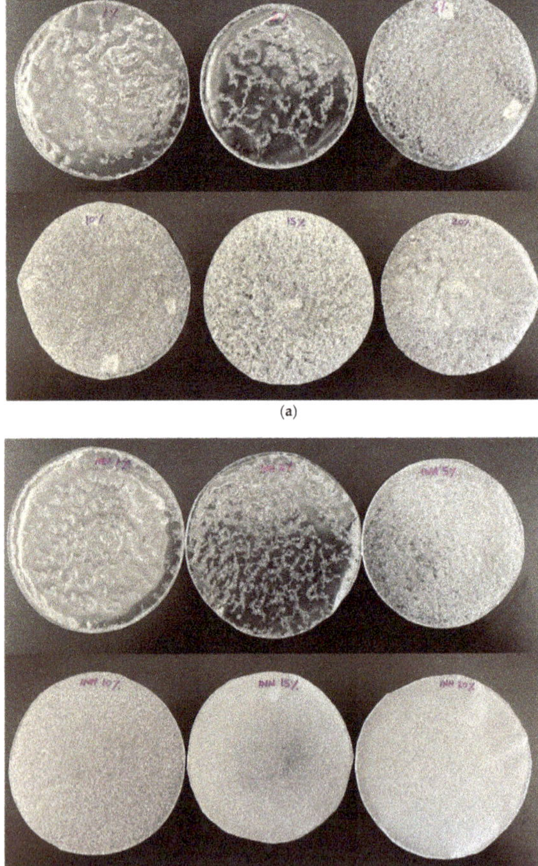

(a)

(b)

Figure 2. Photographs of (**a**) PLA-UCF films with 1–20% UCF content; (**b**) PLA-TCF films with 1–20% TCF content.

2.3. Methods

2.3.1. Vapor Permeability and Water Vapor Transmission Rate (WVTR)

The ability of a porous material to transfer moisture due to a vapour pressure gradient can be expressed by the vapour permeability of the material. The transfer of moisture during this process can take place due to three factors: diffusion (self-collision of water molecules), effusion (collision of water molecules with the pore walls), and liquid transfer (associated with capillary condensation) [42].

The vapor permeability and the WVTR of the composite membranes were determined according to ASTM standard E 96 using the wet cup method. The data used in this analysis were obtained a dynamic vapour sorption equipment DVS Advantage, SMS, London, UK. A SMS Payne Type Diffusion cell was used in this test that is specially designed to measure the permeability and moisture transmission rate of thin films in the DVS instrument. The Payne cell has two main components: cell lid that holds the test sample and a cell cup with a small reservoir. The test samples were cut to a diameter of 15.5–16.0 mm to perfectly fit in the Payne cell lid. The Payne cell cup was filled with 200 µL of water and the sample was placed between two O-rings. The bottom cell cup was screwed into the upper cell lid and placed on a metal sample pan for measurement in the DVS instrument. The samples were initially conditioned for 30 min at 95% RH and 25 °C. The test was run under 50% RH and 25 °C for 24 h and the cell was auto-weighed every minute by a DVS high mass ultra balance. The cell opening provided an active area of 113 mm^2 for moisture transport. The sample thickness was measured using a micrometer and varied between the specimens (100–500 µm). The tests were performed in triplicate and the average reading was reported.

The WVTR was determined by dividing the slope of the linear portion of the weight gain versus time curve by the tested surface area (interior cell area = 113 mm^2) of the sample using the following equation:

$$\text{WVTR} \left(\frac{\text{g}}{\text{h·m}^2} \right) = \frac{\text{slope (g/h)}}{\text{cell area (m}^2\text{)}} \tag{1}$$

WVTR describes the rate of water permeating through a test specimen into the headspace volume of a container, which differs in relative humidity (ΔRH).

Vapor permeability, P hence is represented as:

$$\text{P} \left(\frac{\text{g}}{\text{h·m·Pa}} \right) = \frac{\text{WVTR (g/h·m}^2\text{)·d (m)}}{\Delta \text{P (Pa) (m}^2\text{)}} \tag{2}$$

where d is the test specimen thickness (m)

It is simpler to compare WVTR between samples with varying thicknesses, by eliminating the thickness factor and reporting it as:

$$\text{Normalized WVTR} \left(\frac{\text{g}}{\text{h·m}} \right) = \text{WVTR} \left(\frac{\text{g}}{\text{h·m}^2} \right) \cdot \text{d (m)} \tag{3}$$

2.3.2. Water Retention

The water retention test was performed according to ASTM D-570-95 using the 24 h immersion test method. The specimen was placed in a container of water at room temperature, rested at its edge and entirely immersed. At the end of 24 h, the specimen was removed from water, wiped free of surface moisture with a dry cloth, and weighed to the nearest 0.001 g immediately. The test was performed in triplicates and the average reading was reported. The test specimens were thin films of 150 mm in diameter.

The percentage of water absorption (WA) was calculated according to the following equation:

$$\text{WA} = \frac{m_{24} - m_0}{m_0} \times 100 \tag{4}$$

where m_0 is the initial mass of the test specimen (g), m_{24} is the mass of the test specimen after partial immersion for 24 h (g).

2.3.3. Scanning Electron Microscopy

Photomicrographs of the samples were captured using a Thermo Fisher Scientific Scanning Electron Microscope (SEM), model FEI Quanta 250 (Hillsboro, OR, USA) operating at 15 kV. The samples were gold coated to obtain high magnification of morphology and texture.

2.3.4. Thermogravimetric Analysis

The thermal degradation behaviour of samples was studied using a thermogravimetric analyser TGA/DTA 851e Mettler Toledo instrument (Columbus, OH, USA). Approximately 8–10 mg of sample was placed in an uncovered 70 µL alumina crucible to determine mass loss during heating. The specimens were heated at a rate of 10 °C/min from 25 to 950 °C under nitrogen atmosphere purged at 50 mL/min. The test was performed in triplicate.

2.3.5. Differential Scanning Calorimetry

The melting point and enthalpy of the samples were determined by a differential scanning calorimeter DSC Mettler Toledo 822/e (Columbus, OH, USA). Approximately 7 mg of sample was placed in a sealed aluminium cell to determine melting and glass transition temperatures and enthalpies during heating. The samples were heated from 25 °C to 250 °C at 10 °C/min under nitrogen. The test was performed in triplicate.

2.3.6. Dynamic Mechanical Analysis

Glass transition temperature (Tg) was determined by dynamic mechanical analysis based on the maximum of the tan delta peak, which represents the ratio of storage modulus to loss modulus. The tests were conducted using a Q800 dynamic mechanical analyzer (DMA) from TA instruments, New Castle, DE, USA. The tests were conducted in tension mode at a frequency of 1 Hz and an amplitude of 10 µm. The test samples were cut from the films using a laser machine having a dimension of 20 mm (length) by 4.5 mm (wide) and approximate thickness of 0.2 mm. All specimens were initially conditioned at 25 °C in the DMA chamber, and then dynamic heating scans were performed from 25 to 150 °C at 3 °C/min. The test was performed in triplicate and the average reading was reported.

3. Results and Discussion

Moisture transfer in porous materials takes place due to the presence of a vapour pressure gradient between their top and bottom surfaces. Figure 3 represents the water vapour transmission rate (WVTR) of the prepared membranes. It was seen that, overall, the vapour transmission increased with increasing CF content in the samples. The detailed data analysis of vapour permeability of the composites is reported in Table 2.

The normalized WVTR is an important factor to consider when measuring permeability of composites with different thicknesses. The WVTR of the PLA-TCF was similar to PLA-UCF composites at lower CF concentrations. As the CF concentration increased beyond 10%, the composites with treated fibers showed significantly lower vapor transmission rates. The WVTR of PLA-TCF20 was 73% lower than that of PLA-UCF20. The increase in vapour permeability and WVTR at higher CF concentration is linked to the pore network and structure of the composite. The surfactant treatment resulted in good dispersion of fibers in the PLA matrix, thereby reducing formation of voids and limiting capillary movement of water molecules. The treatment, however, did not completely block microscopic pores in the samples that are needed for vapour diffusion.

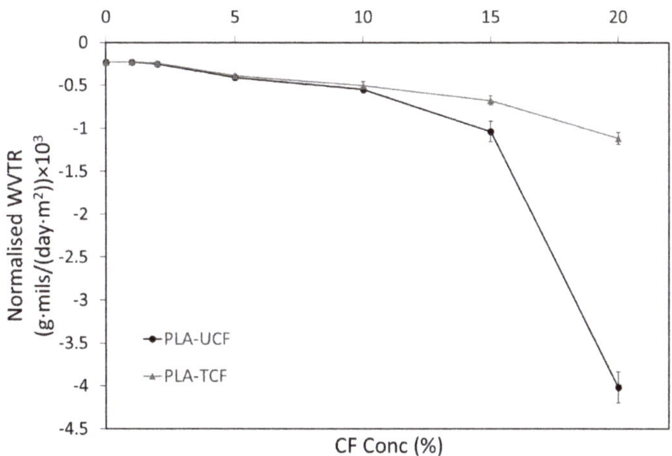

Figure 3. WVTR of the films with varying CF concentration.

Table 2. Vapor permeability results of the films.

Sample	WVTR (g/(h·m^2))	Thickness (mm)	Permeability P (g/(h·m·Pa)) $\times 10^{-7}$	Normalized WVTR (g·m/(h·m^2)) $\times 10^{-5}$	Normalized WVTR (g·mils/(day·m^2))
PLA	-2.21 ± 0.03	0.11 ± 0.01	-2.01 ± 0.12	-24.33 ± 1.47	-229.90 ± 13.96
PLA-UCF1	-1.26 ± 0.04	0.19 ± 0.01	-1.99 ± 0.03	-24.09 ± 0.34	-227.65 ± 3.26
PLA-UCF2	-1.32 ± 0.04	0.20 ± 0.01	-2.19 ± 0.02	-26.54 ± 0.25	-250.83 ± 2.38
PLA-UCF5	-1.70 ± 0.06	0.26 ± 0.02	-3.66 ± 0.13	-44.31 ± 1.59	-418.68 ± 15.05
PLA-UCF10	-1.98 ± 0.03	0.30 ± 0.01	-4.92 ± 0.14	-59.60 ± 1.74	-563.19 ± 16.49
PLA-UCF15	-3.16 ± 0.24	0.32 ± 0.02	-8.36 ± 0.54	-101.15 ± 6.71	-955.79 ± 60.04
PLA-UCF20	-10.28 ± 0.16	0.42 ± 0.02	-35.71 ± 1.58	-432.14 ± 19.12	-4083.21 ± 181.20
PLA-TCF1	-1.28 ± 0.03	0.19 ± 0.01	-2.01 ± 0.12	-24.40 ± 1.46	-230.60 ± 13.80
PLA-TCF2	-1.42 ± 0.04	0.19 ± 0.01	-2.24 ± 0.11	-27.16 ± 1.34	-256.61 ± 12.71
PLA-TCF5	-1.79 ± 0.06	0.25 ± 0.01	-3.71 ± 0.20	-44.87 ± 2.45	-423.97 ± 23.21
PLA-TCF10	-2.97 ± 0.13	0.20 ± 0.01	-4.91 ± 0.39	-59.46 ± 4.80	-561.85 ± 45.38
PLA-TCF15	-2.61 ± 0.10	0.26 ± 0.02	-5.61 ± 0.49	-67.92 ± 5.97	-641.81 ± 56.41
PLA-TCF20	-4.79 ± 0.15	0.25 ± 0.01	-9.92 ± 0.61	-119.97 ± 7.40	-1133.58 ± 69.90

The increase in the water vapour permeability at higher humidity level is related to the enhanced transport of moisture. This phenomenon is induced by the transfer of liquid in the microscopic pores of the material that are filled with water due to capillary condensation [42]. For materials that show hysteresis in their sorption isotherm, it has been reported earlier that their water vapour permeability is dependent on the moisture content [43].

The water absorption of the composites was calculated as percentage of absorption with respect to initial mass. As seen in Figure 4, the water absorption behaviour of the composites was significantly affected by the treatment of CF keeping the WA values to a minimum. PLA-UCF composites showed higher water absorption due to the absence of treatment on CF. PLA-UCF20 showed the highest absorption having a WA value of over 7% for 24 h immersion.

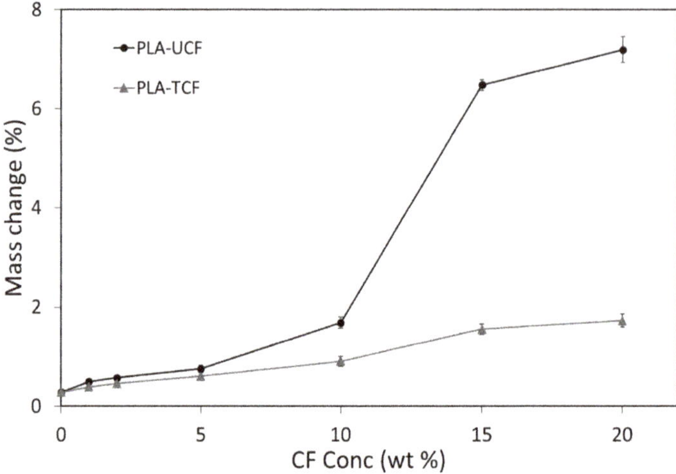

Figure 4. Water absorption behaviour of the films with varying CF concentration.

When compared to the treated fiber composite having the same CF concentration (PLA-TCF20), PLA-UCF20 showed a four-fold increase in water absorption. In addition to enhanced dispersion due to the CF treatment, it is evident that the chemical structure of the CF has also been altered, reducing available surface hydroxyl groups that are responsible for the higher water uptake in the samples.

A close examination of the composite morphology revealed that samples without CF treatment showed poor dispersion in the PLA matrix whereas the treatment was found to be effective for preparing homogenous PLA-CF samples. In Figure 5b,c, it can be seen that the untreated fibers and PLA have poor interfacial adhesion and the fibers show agglomeration at the surface. On the other hand, Figure 5d,e show that samples with surfactant treated CF are well dispersed in the PLA matrix.

The presence of the acid phosphate ester surfactant on the surface of nanocellulose facilitates their dispersion in the polymer matrix and improves the nucleation effect [40,44,45]. The surfactant acts as a stabilizing agent for obtaining a stable dispersion of cellulose fibers in the matrix [46,47]. The surfactant treatment has also been found to be effective in preventing the agglomeration of CF in solvents such as chloroform, thereby enhancing dispersibility in the polymer matrix [48].

The thermal properties of the PLA-CF films were investigated by TGA, DSC, and DMA analysis to determine the effect of surfactant treatment on the thermal stability of the polymer composites. The thermal characteristics of the composites are mentioned in Table 3.

Table 3. Thermal properties of films and fibers.

Sample	Melting Point (°C)	Melting Enthalpy (mJ)	Tg (°C)	Tmax (°C)	Residue (%)
PLA	153.6 ± 0.9	−128.6 ± 0.3	61.5 ± 0.5	368.1 ± 0.2	0.5 ± 0.01
PLA-UCF20	153.5 ± 0.6	−90.6 ± 0.2	61.3 ± 0.3	354.9 ± 0.1	2.4 ± 0.04
PLA-TCF20	153.7 ± 0.5	−120.2 ± 0.3	65.5 ± 0.5	372.5 ± 0.1	3.6 ± 0.03
UCF	-	-	-	361.2 ± 0.3	8.4 ± 0.04
TCF	-	-	-	269.3 ± 0.2	23.8 ± 0.12

(a)

(b)

(c)

Figure 5. *Cont.*

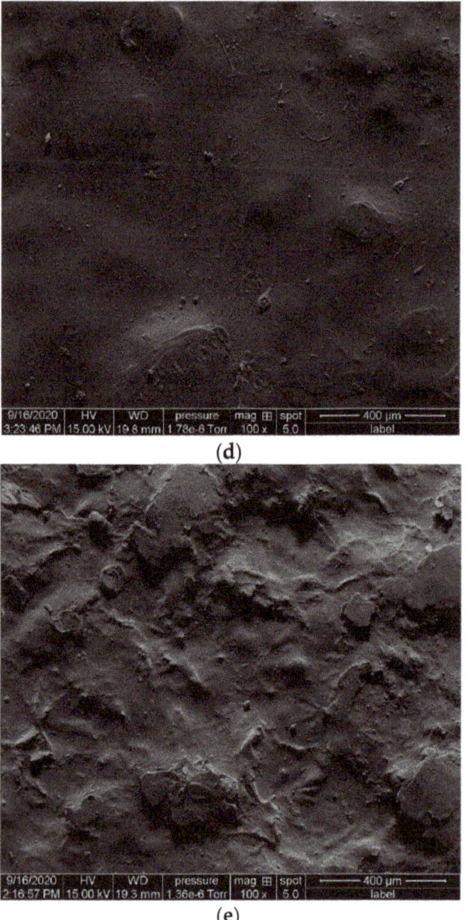

Figure 5. Micrograph of prepared films (**a**) PLA; (**b**) PLA-UCF10; (**c**) PLA-UCF20; (**d**) PLA-TCF10 and; (**e**) PLA-TCF20.

Thermogravimetric analysis of PLA, untreated and treated CF and the composite films shows a multistep degradation as seen in Figure 6. The initial step (50–100 °C) can be attributed to a loss of moisture. The treated CF shows lower thermal stability and weight loss between 250–300 °C corresponding to its respective DTG peaks in Figure 7 which are mainly associated with the effect of the surfactant on the fiber [46,48]. The PLA-TCF20 shows a smaller degradation peak corresponding to the surfactant coated fibers in the matrix.

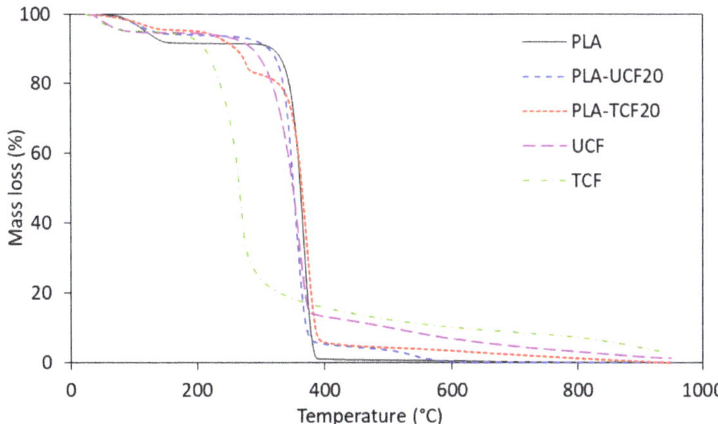

Figure 6. TGA curves of films and fibers with and without treatment.

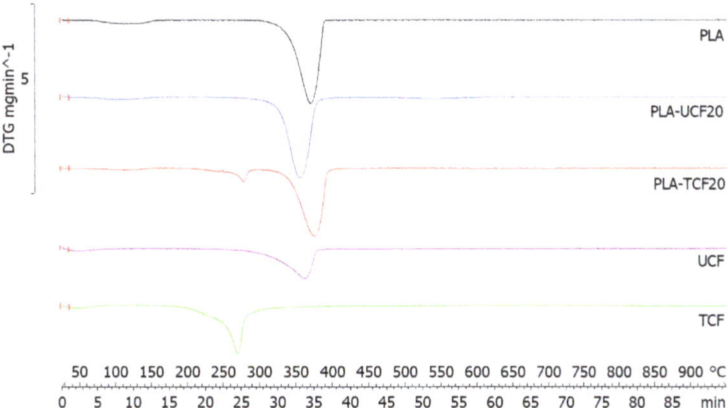

Figure 7. DTG curves of films and fibers with and without treatment.

The next main degradation step for all samples occurs between 350 and 400 °C, associated with the cellulose, hemicellulose, and thermally stable compounds in the polymer matrix. The main DTG peak corresponding to the peak degradation temperature (Tmax) for PLA was 368 °C and for the PLA-CF samples it was between 354 and 372 °C, showing that there was no degradation taking place in the composites at lower temperatures. It was observed that the addition of TCF in PLA slightly increases the thermal stability of the matrix due to better reinforcement as seen in Table 3.

Figure 8 shows the endothermic melting peaks for neat PLA, PLA-UCF20, and PLA-TCF20 composites. The three materials displayed similar curves and minor differences in their melting temperatures were observed, indicating that CF dispersion does not significantly affect the thermal properties of the polymer matrix.

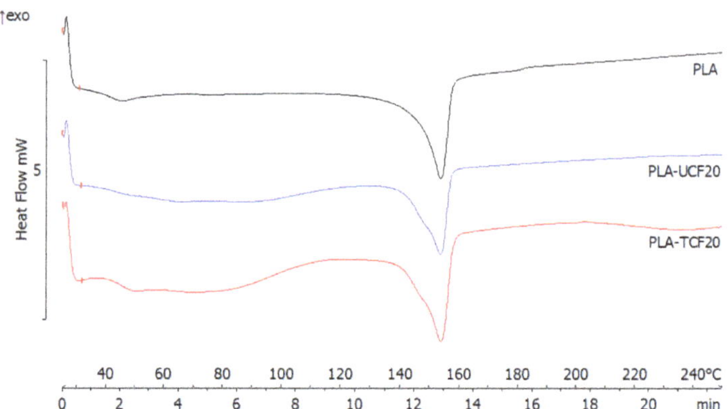

Figure 8. DSC curves of films with and without CF treatment.

From the DMA analysis, the glass transition temperatures (Tg) of the composites were evaluated using the tan δ peak temperature. It was found that both the fiber content as well as fiber treatment had an effect on the Tg, as seen in Figure 9.

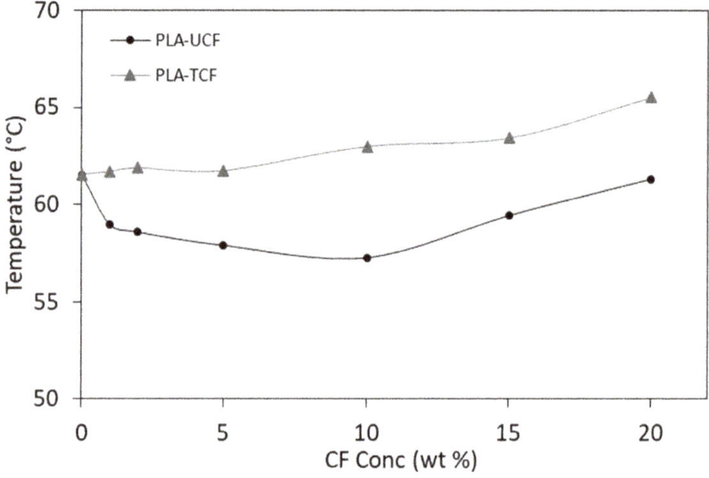

Figure 9. Glass transition temperature of films with varying CF concentration.

PLA composites prepared with surfactant treated fibers showed relatively higher Tg when compared to PLA composites containing untreated CF. The improvement in Tg (up to 10%) could be associated with the enhanced CF dispersion and interface compatibility due to surface modification. On the other hand, the PLA-UCF composites showed a lower Tg compared to neat PLA. This could be due to the presence of residual water in the system, which causes a degradation of PLA. Moreover, it was observed that the Tg increased with higher CF content in the PLA composites. The addition of acid phosphate ester-based surfactant has been reported to enhance dispersibility, mechanical, and stress-transfer properties of cellulose reinforced composites [49].

4. Conclusions

The preparation of bio-based PLA membranes incorporating micro-fibrillated cellulose fiber (CF) using a solvent casting method have been reported in this paper. The effects of CF modification, using an anionic organic phosphoric acid ester-based surfactant, on composite structure, morphology, and properties were investigated. The physical properties of the composite were significantly enhanced when prepared with treated CF. The composites containing treated CF absorbed very low amounts of water, even at 20 wt% CF loading, yet these composites were permeable to water vapour. The CF treatment resulted in a good dispersion of fibers in the PLA matrix, as seen using scanning electron microscopy, thereby reducing void formation and limiting capillary movement. However, the incorporation of treated CF did not completely block the microscopic pores in the composite samples that are needed for vapour diffusion, as seen in the vapour permeability results. The CF treatment also improved the thermal properties of the composite, increasing their glass transition temperature and peak degradation temperature. The melting point of the composites did not change before and after CF treatment. The prepared composites were breathable and showed better resistance to water. Using a relatively inexpensive treatment, smart bio-based membranes have been developed that have great potential in the building industry by increasing the hygroscopic performance as well as reducing the embodied energy of buildings. Further work is recommended considering the durability and weathering of the PLA-CF composites as well as the preparation of larger sized films and assessment of their hygroscopic performance in test cells.

Author Contributions: Conceptualization, A.H. and P.B.; methodology, A.H. and P.B.; formal analysis, A.H.; investigation, A.H.; resources, P.B.; data curation, A.H.; writing—original draft preparation, A.H.; writing—review and editing, A.H. and P.B.; visualization, A.H.; supervision, P.B.; project administration, P.B.; funding acquisition, A.H. and P.B. All authors have read and agreed to the published version of the manuscript.

Funding: The authors are grateful to Natural Sciences and Engineering Research Council of Canada for the financial support through its IRC and CRD programs (IRCPJ 461745-18 and RDCPJ 524504-18) as well as the industrial partner, Kruger Inc., of the NSERC industrial chair on eco-responsible wood construction (CIRCERB). This work was also supported by the funds received from Quebec Research Funds – Nature and Technology (FRQNT) under the Merit Scholarship Program for Foreign Students, Quebec-India Postdoctoral scholarships (PBEEE 2I, 2019-2020).

Informed Consent Statement: Not applicable.

Data Availability Statement: The data presented in this study are available on request from the corresponding author.

Conflicts of Interest: The authors declare no conflict of interest. The funders had no role in the design of the study; in the collection, analyses, or interpretation of data; in the writing of the manuscript, or in the decision to publish the results.

References

1. Viel, M.; Collet, F.; Lecieux, Y.; François, M.L.M.; Colson, V.; Lanos, C.; Hussain, A.; Lawrence, M. Resistance to mold development assessment of bio-based building materials. *Compos. Part B Eng.* **2019**, *158*, 406–418. [CrossRef]
2. Wilkinson, J.; Ueno, K.; De Rose, D.; Straube, J.F.; Fugler, D. Understanding Vapour Permeance and Condensation in Wall Assemblies. In Proceedings of the 11th Canadian Building Science & Technology Conference, National Building Envelope Council 2007, Banff, AB, Canada, 3–5 November 2007; pp. 1–14.
3. Agarwal, S.; Gupta, R.K. Plastics in Buildings and Construction. In *Applied Plastics Engineering Handbook: Processing, Materials, and Applications*, 2nd ed.; Elsevier: Amsterdam, The Netherlands, 2017; pp. 635–649. ISBN 9780323390408.
4. Lstiburek, J.W. Understanding vapor barriers. *ASHRAE J.* **2004**, *46*, 40–47.
5. Hussain, A.; Calabria-Holley, J.; Lawrence, M.; Jiang, Y. Hygrothermal and mechanical characterisation of novel hemp shiv based thermal insulation composites. *Constr. Build. Mater.* **2019**, *212*, 561–568. [CrossRef]
6. Latif, E.; Tucker, S.; Ciupala, M.A.; Wijeyesekera, D.C.; Newport, D. Hygric properties of hemp bio-insulations with differing compositions. *Constr. Build. Mater.* **2014**, *66*, 702–711. [CrossRef]
7. Stroeks, A. The moisture vapour transmission rate of block co-poly(ether- ester) based breathable films. 2. Influence of the thickness of the air layer adjacent to the film. *Polymer* **2001**, *42*, 09903–09908. [CrossRef]
8. Straube, J.; Schumacher, C. Interior insulation retrofits of load-bearing masonry walls in cold climates. *J. Green Build.* **2007**, *2*, 42–50. [CrossRef]

9. Straube, J.F. The influence of low-permeance vapor barriers on roof and wall performance. In Proceedings of the Thermal Performance of Whole Buildings VIII, Clearwater, FL, USA, 2–7 December 2001; pp. 1–12.

10. Lafond, C.; Blanchet, P. Technical performance overview of bio-based expanded polystyrene. *Buildings* **2020**, *10*, 81. [CrossRef]

11. Jiang, Y.; Lawrence, M.; Hussain, A.; Ansell, M.; Walker, P. Comparative moisture and heat sorption properties of fibre and shiv derived from hemp and flax. *Cellulose* **2019**, *26*, 823–843. [CrossRef]

12. Ansell, M.P.; Lawrence, M.; Jiang, Y.; Shea, A.; Hussain, A.; Calabria-Holley, J.; Walker, P. *Natural Plant-Based Aggregates and Bio-Composite Panels with Low Thermal Conductivity and High Hygrothermal Efficiency for Applications in Construction*; Elsevier: Amsterdam, The Netherlands, 2019; ISBN 9780081027042.

13. Segovia, F.; Blanchet, P.; Auclair, N.; Essoua Essoua, G.G. Thermo-Mechanical Properties of a Wood Fiber Insulation Board Using a Bio-Based Adhesive as a Binder. *Buildings* **2020**, *10*, 152. [CrossRef]

14. Cabral, M.R.; Blanchet, P. A state of the art of the overall energy efficiency of wood buildings—An overview and future possibilities. *Materials* **2021**, *14*, 1848. [CrossRef]

15. Maskell, D.; Thomson, A.; Walker, P.; Lemke, M. Determination of optimal plaster thickness for moisture buffering of indoor air. *Build. Environ.* **2018**, *130*, 143–150. [CrossRef]

16. Tran Le, A.D.; Maalouf, C.; Mai, T.H.; Wurtz, E.; Collet, F. Transient hygrothermal behaviour of a hemp concrete building envelope. *Energy Build.* **2010**, *42*, 1797–1806. [CrossRef]

17. Lawrence, M. Reducing the Environmental Impact of Construction by Using Renewable Materials. *J. Renew. Mater.* **2015**, *3*, 163–174. [CrossRef]

18. MELCC. *Politique-Cadre D'Électrification et de Lutte Contre Les Changements Climatiques*; Gouvernement du Québec: Québec, QC, Canada, 2020; ISBN 9782550862796.

19. Lessard, Y.; Anand, C.; Blanchet, P.; Frenette, C.; Amor, B. LEED v4: Where Are We Now? Critical Assessment through the LCA of an Office Building Using a Low Impact Energy Consumption Mix. *J. Ind. Ecol.* **2018**, *22*, 1105–1116. [CrossRef]

20. Phanthong, P.; Reubroycharoen, P.; Hao, X.; Xu, G.; Abudula, A.; Guan, G. Nanocellulose: Extraction and application. *Carbon Resour. Convers.* **2018**, *1*, 32–43. [CrossRef]

21. Abdul Khalil, H.P.S.; Bhat, A.H.; Ireana Yusra, A.F. Green composites from sustainable cellulose nanofibrils: A review. *Carbohydr. Polym.* **2012**, *87*, 963–979. [CrossRef]

22. Jonoobi, M.; Harun, J.; Mathew, A.P.; Oksman, K. Mechanical properties of cellulose nanofiber (CNF) reinforced polylactic acid (PLA) prepared by twin screw extrusion. *Compos. Sci. Technol.* **2010**, *70*, 1742–1747. [CrossRef]

23. Sharma, A.; Thakur, M.; Bhattacharya, M.; Mandal, T.; Goswami, S. Commercial application of cellulose nano-composites— A review. *Biotechnol. Rep.* **2019**, *21*, e00316. [CrossRef]

24. Erbas Kiziltas, E.; Kiziltas, A.; Bollin, S.C.; Gardner, D.J. Preparation and characterization of transparent PMMA-cellulose-based nanocomposites. *Carbohydr. Polym.* **2015**, *127*, 381–389. [CrossRef]

25. Spence, K.L.; Venditti, R.A.; Rojas, O.J.; Pawlak, J.J.; Hubbe, M.A. Water Vapor Barrier Properties of Microfibrillated Cellulose Films. *BioResources* **2011**, *6*, 4370–4388. [CrossRef]

26. Song, Z.; Xiao, H.; Zhao, Y. Hydrophobic-modified nano-cellulose fiber/PLA biodegradable composites for lowering water vapor transmission rate (WVTR) of paper. *Carbohydr. Polym.* **2014**, *111*, 442–448. [CrossRef] [PubMed]

27. Fortunati, E.; Peltzer, M.; Armentano, I.; Torre, L.; Jiménez, A.; Kenny, J.M. Effects of modified cellulose nanocrystals on the barrier and migration properties of PLA nano-biocomposites. *Carbohydr. Polym.* **2012**, *90*, 948–956. [CrossRef]

28. Hospodarova, V.; Stevulova, N. Investigation of Waste Paper Cellulosic Fibers Utilization into Cement Based Building Materials. *Buildings* **2018**, *8*, 43. [CrossRef]

29. Yi, T.; Zhao, H.; Mo, Q.; Pan, D.; Liu, Y.; Huang, L.; Xu, H. From Cellulose to Cellulose Nanofibrils—A Comprehensive Review of the Preparation and Modification of Cellulose Nanofibrils. *Materials* **2020**, *13*, 5062. [CrossRef] [PubMed]

30. Lawrence, M.; Fodde, E.; Paine, K.; Walker, P. Hygrothermal Performance of an Experimental Hemp-Lime Building. *Key Eng. Mater.* **2012**, *517*, 413–421. [CrossRef]

31. Stevulova, N.; Cigasova, J.; Schwarzova, I.; Sicakova, A.; Junak, J. Sustainable bio-aggregate-based composites containing hemp hurds and alternative binder. *Buildings* **2018**, *8*, 25. [CrossRef]

32. Cichosz, S.; Masek, A.; Rylski, A. Cellulose Modification for Improved Compatibility with the Polymer Matrix: Mechanical Characterization of the Composite Material. *Materials* **2020**, *13*, 5519. [CrossRef]

33. Kabir, M.M.; Wang, H.; Lau, K.T.; Cardona, F. Chemical treatments on plant-based natural fibre reinforced polymer composites: An overview. *Compos. Part B Eng.* **2012**, *43*, 2883–2892. [CrossRef]

34. Bledzki, A.K.; Mamun, A.A.; Lucka-Gabor, M.; Gutowski, V.S. The effects of acetylation on properties of flax fibre and its polypropylene composites. *Express Polym. Lett.* **2008**, *2*, 413–422. [CrossRef]

35. Phan-Xuan, T.; Thuresson, A.; Skepö, M.; Labrador, A.; Bordes, R.; Matic, A. Aggregation behavior of aqueous cellulose nanocrystals: The effect of inorganic salts. *Cellulose* **2016**, *23*, 3653–3663. [CrossRef]

36. Yasin, S.; Hussain, M.; Zheng, Q.; Song, Y. Large amplitude oscillatory rheology of silica and cellulose nanocrystals filled natural rubber compounds. *J. Colloid Interface Sci.* **2021**, *588*, 602–610. [CrossRef]

37. Yasin, S.; Hussain, M.; Zheng, Q.; Song, Y. Effects of ionic liquid on cellulosic nanofiller filled natural rubber bionanocomposites. *J. Colloid Interface Sci.* **2021**, *591*, 409–417. [CrossRef] [PubMed]

38. Cerny, P.; Bartos, P.; Kriz, P.; Olsan, P.; Spatenka, P. Highly Hydrophobic Organosilane-Functionalized Cellulose: A Promising Filler for Thermoplastic Composites. *Materials* **2021**, *14*, 2005. [CrossRef] [PubMed]
39. Hussain, A.; Calabria-Holley, J.; Lawrence, M.; Ansell, M.P.; Jiang, Y.; Schorr, D.; Blanchet, P. Development of novel building composites based on hemp and multi-functional silica matrix. *Compos. Part B Eng.* **2019**, *156*, 266–273. [CrossRef]
40. Fortunati, E.; Rinaldi, S.; Peltzer, M.; Bloise, N.; Visai, L.; Armentano, I.; Jiménez, A.; Latterini, L.; Kenny, J.M. Nano-biocomposite films with modified cellulose nanocrystals and synthesized silver nanoparticles. *Carbohydr. Polym.* **2014**, *101*, 1122–1133. [CrossRef] [PubMed]
41. Kaboorani, A.; Riedl, B. Surface modification of cellulose nanocrystals (CNC) by a cationic surfactant. *Ind. Crops Prod.* **2015**, *65*, 45–55. [CrossRef]
42. Collet, F.; Chamoin, J.; Pretot, S.; Lanos, C. Comparison of the hygric behaviour of three hemp concretes. *Energy Build.* **2013**, *62*, 294–303. [CrossRef]
43. Derome, D.; Derluyn, H.; Zillig, W.A.; Carmeliet, J. Model for hysteretic moisture behaviour of wood. In Proceedings of the Nordic Symposium on Building Physics 2008, Copenhagen, Denmark, 16–18 June 2008; Volume 2, pp. 959–966.
44. Shamsuri, A.A.; Md Jamil, S.N.A. A short review on the effect of surfactants on the mechanico-thermal properties of polymer nanocomposites. *Appl. Sci.* **2020**, *10*, 4867. [CrossRef]
45. Fortunati, E.; Armentano, I.; Zhou, Q.; Iannoni, A.; Saino, E.; Visai, L.; Berglund, L.A.; Kenny, J.M. Multifunctional bionanocomposite films of poly(lactic acid), cellulose nanocrystals and silver nanoparticles. *Carbohydr. Polym.* **2012**, *87*, 1596–1605. [CrossRef]
46. Bondeson, D.; Oksman, K. Dispersion and characteristics of surfactant modified cellulose whiskers nanocomposites. *Compos. Interfaces* **2007**, *14*, 617–630. [CrossRef]
47. Heux, L.; Chauve, G.; Bonini, C. Nonflocculating and chiral-nematic self-ordering of cellulose microcrystals suspensions in nonpolar solvents. *Langmuir* **2000**, *16*, 8210–8212. [CrossRef]
48. Moreno, M.; Armentano, I.; Fortunati, E.; Mattioli, S.; Torre, L.; Lligadas, G.; Ronda, J.C.; Galià, M.; Cádiz, V. Cellulose nano-biocomposites from high oleic sunflower oil-derived thermosets. *Eur. Polym. J.* **2016**, *79*, 109–120. [CrossRef]
49. Ljungberg, N.; Cavaillé, J.Y.; Heux, L. Nanocomposites of isotactic polypropylene reinforced with rod-like cellulose whiskers. *Polymer* **2006**, *47*, 6285–6292. [CrossRef]

Article

The Design and Development of Recycled Concretes in a Circular Economy Using Mixed Construction and Demolition Waste

Marcos Díaz González [1,*], Pablo Plaza Caballero [2], David Blanco Fernández [1], Manuel Miguel Jordán Vidal [3], Isabel Fuencisla Sáez del Bosque [2] and César Medina Martínez [2]

[1] Department of Construction Sciences, Metropolitan Technological University, Dieciocho, Santiago de Chile 161, Chile; dblanco@utem.cl

[2] Department of Construction, Research Institute for Sustainable Territorial Development (INTERRA), University of Extremadura, 10003 Cáceres, Spain; pcaballerop@unex.es (P.P.C.); isaezdelu@unex.es (I.F.S.d.B.); cmedinam@unex.es (C.M.M.)

[3] Department of Agrochemistry and Environment, Miguel Hernández University of Elche, 03202 Elche, Spain; manuel.jordan@umh.es

* Correspondence: mdiaz@utem.cl

Abstract: This research study analysed the effect of adding fine—fMRA (0.25% and 50%)—and coarse—cMRA (0%, 25% and 50%)—mixed recycled aggregate both individually and simultaneously in the development of sustainable recycled concretes that require a lower consumption of natural resources. For this purpose, we first conducted a physical and mechanical characterisation of the new recycled raw materials and then analysed the effect of its addition on fresh and hardened new concretes. The results highlight that the addition of fMRA and/or cMRA does not cause a loss of workability in the new concrete but does increase the amount of entrained air. Regarding compressive strength, we observed that fMRA and/or cMRA cause a maximum increase of +12.4% compared with conventional concrete. Tensile strength increases with the addition of fMRA (between 8.7% and 5.5%) and decreases with the use of either cMRA or fMRA + cMRA (between 4.6% and 7%). The addition of fMRA mitigates the adverse effect that using cMRA has on tensile strength. Regarding watertightness, all designed concretes have a structure that is impermeable to water. Lastly, the results show the feasibility of using these concretes to design elements with a characteristic strength of 25 MPa and that the optimal percentage of fMRA replacement is 25%.

Keywords: recycled concrete; recycled mixed sand; recycled mixed gravel; mechanical properties; strength; watertightness

Citation: González, M.D.; Plaza Caballero, P.; Fernández, D.B.; Jordán Vidal, M.M.; del Bosque, I.F.S.; Medina Martínez, C. The Design and Development of Recycled Concretes in a Circular Economy Using Mixed Construction and Demolition Waste. *Materials* **2021**, *14*, 4762. https://doi.org/10.3390/ma14164762

Academic Editor: Jean-Marc Tulliani

Received: 5 July 2021
Accepted: 4 August 2021
Published: 23 August 2021

Publisher's Note: MDPI stays neutral with regard to jurisdictional claims in published maps and institutional affiliations.

1. Introduction

Climate change and global warming have become important issues that directly impact the world economy. This has led to countries developing laws and regulations that control the emission of CO_2 and the appropriate management of waste. In this context, the construction industry is responsible for 12% of all greenhouse gas emissions in the European Union (EU) and for generating ~25–30% of the solid waste produced every year in the EU [1], which equates to an average annual production of 800 million tonnes. In addition, as many as 534 and 200 million tonnes of construction and demolition waste (C&DW) are generated every year in the United States and China, respectively.

Concrete is the most used material in the construction sector worldwide, with the EU and USA producing 165 and 150 million cubic metres, respectively, in 2018 [2]. This industry is characterised by requiring large amounts of natural resources, as this material is mainly composed of aggregates—60–75% of the volume of concrete—and, to a lesser extent, cement—10–15%. Its manufacturing process is responsible for ~8% of the total worldwide amount of CO_2 [3]. Three trillion tonnes of aggregate were used in the field of civil engineering in 2018, ~45% of which were used to manufacture concrete [4].

In this context, the construction industry has tried to mitigate the adverse effects of its activity in recent years by designing and developing new formulations for materials with a cement base (mortars and/or concretes) that are more sustainable [5,6]. This has been done by adding industrial by-products (agroforestry or biomass waste, ornament and ceramics [7] industry and construction and demolition waste [8,9]) as supplementary cementitious materials and/or recycled aggregate (concrete or mixed waste from C&DW). The use of recycled aggregate from C&DW is one of the most widespread strategies for simultaneously achieving the double goal of instituting the concept of circular economy and sustainability in construction.

Recycled aggregates are obtained from the appropriate treatment of C&DW in management plants, where they are classified in granulometric fractions (mainly sand, coarse and gravel) depending on their size and composition. According to this last criterion, they are classified as (i) recycled concrete aggregates (RCA) with amounts of concrete (Rc) and unbound aggregates (Ru) $\geq$90% and $\leq$10% of ceramic materials (Rb), (ii) mixed recycled aggregates (MRA) with 70% $\leq$ Rc + Ru < 90% and Rb $\leq$ 30% and (iii) recycled masonry aggregates (RMA) with Rc + Ru < 70% and Rb > 30.

RCA are recycled aggregates that have been studied in greater depth in the international literature, focusing mainly on the coarse fraction (maximum particle size $\geq$ 4 mm). In this line of work, it is worth highlighting the study conducted by Bravo et al. [10], who found that the density decreased between 4.7% and 7.7% by adding 100% of RCA, while considering that their mechanical strength decreases if their composition includes particles of a ceramic nature. Moreover, Thomas et al. [11] concluded that the density of concretes decreased by 5% with a 20% of RCA replacement due to the high porosity and the presence of adhered mortar. Etxeberría et al. [12] analysed concretes with 100% RCA, a water–cement ratio of 0.5 and 325 kg of cement/m^3, registering a decrease between 20% and 25% decrease in compressive strength compared with the reference concrete. This loss could be offset by increasing the cement content, a strategy which is of no interest economically or from the viewpoint of sustainability. Likewise, these authors observed that concretes with average compressive strength (30–45 MPa) manufactured with 25% recycled coarse aggregate had the same mechanical properties as traditional concrete. McNeil et al. [13], in their literature review, were able to summarise their findings on what concretes with RCA experience: (1) replacing natural aggregate (NA) in concrete with RCA decreases the compressive strength but yields comparable splitting tensile strength; (2) the modulus of rupture for RCA concrete was slightly lower than that of conventional concrete, likely due to the weakened interfacial transition zone (ITZ) from residual mortar; and (3) the elastic modulus is also lower than expected, caused by the more ductile aggregate.

As regards assessing the fine fraction of RCA, there have been fewer studies that have focused on studying the design of structural concretes. Evangelista et al. [14] assessed the feasibility of adding the fine fraction of RCA and observed that the compressive strength was not impacted for percentages $\leq$30% in weight. Minkwan et al. [15] found that the compressive strength of concrete with 100% of fine RCA decreased by approximately 30% and 10% compared with the reference concrete with normal strength and high strength, respectively. Bravo et al. [16] observed that (i) recycled mixes with contents of fine RCA $\leq$25% have properties that are comparable with those of the reference concrete and (ii) contents of RCA >25% cause decreases in the properties, leading to losses of up to 19% compared with traditional concrete.

Regarding the simultaneous use of coarse and fine RCA, Fernández et al. [17] concluded that the properties that are impacted the most due to replacing natural aggregate with recycled aggregate are workability (with 100% coarse aggregate replacement, it more than doubles compared with the reference concrete), the elastic modulus (decreases by 42% with 100% coarse aggregate replacement and also decreases by 7% with 30% fine aggregate replacement), the contractive deformation (obtaining deformations between 40% and 56%

after 360 days with 100% coarse aggregate replacement) and water absorption (between 8.5% and 9%, proportions that are higher than the 5% established by the Spanish Code on Structural Concrete, or Structural Concrete Instruction EHE-08).

Plaza et al. [18] report that the use of small percentages (<25%) of coarse concrete aggregate alone or with recycled concrete sand decreases the total porosity and refines the structure of the pores. The opposite effect was observed for higher percentages. Recycled concretes have less compressive strength after 28 days than the conventional material, although the decrease was less than 17% in all cases studied. The loss of strength is greater in mixes that have recycled mixed fines. Regarding tensile strength, Plaza et al. [18] revealed that it increased by 9.49% with a coarse recycled aggregate replacement rate of 100%. However, in replacement mixes (50% coarse aggregate + 50% fine aggregate) this property decreased by up to 14.13%. Furthermore, there was a 1.74% increase in strength in concretes with no coarse aggregate replacement but with 50% mixed fine aggregate compared with the reference concrete. Corinaldesi et al. [19], Malesev et al. [20] and López et al. [21] noted that the elastic modulus decreased between 15% and 44.8% compared with the reference concrete, where the recycled coarse aggregate had a greater negative influence. Lovato et al. [22] even reveals that for the same level of axial compressive strength, higher than 20 MPa, and despite the greater consumption of cement, the costs are similar to those of the reference concrete and only vary by around 20%. Agrela et al. [23] suggest the following classification for RA: Recycled concrete aggregate, with >90% concrete; Mixed recycled aggregates (MRA), with between 30% and 10% ceramic content; and recycled ceramic aggregates, with >30% ceramic content. Regarding flexural strength [18], it is similar in conventional concrete and recycled concrete with less than 75% replacement. In higher percentages, the use of recycled materials in both fractions causes up to a 15% strength loss. Recycled concretes that have both coarse aggregate recycled concrete and recycled sand are suitable for their use in structural concretes with a characteristic strength of 30 MPa.

Regarding mixed recycled aggregates (MRA), due to the volume they represent of the total amount of C&DW (~67% of the total in Spain), in the past decade they have aroused the interest of the scientific community. There is currently a scientific–technical shortcoming in the feasibility of assessing the coarse and/or fine fractions in the design of concretes. In this regard, Sáez del Bosque et al. [24] revealed that the elastic modulus of the ITZ (Interfacial Transition Zone) varies with the type of materials that are present in the recycled aggregate, with ITZs associated with organic components (such as wood, plastic and asphalt) having a lower minimum elastic modulus value depending on the content of ceramic and concrete particles. Recently, Martínez-Lage et al. [25] recorded that the addition of 100% of MRA entailed saving more than 35% of waste generation and 50% of abiotic depletion. Martínez-Lage et al. [26] revealed that the decrease in compressive strength and the deformation modulus vary from 20% to 30% and 30% to 40%, respectively, in concretes with 100% coarse recycled aggregate replacement rate. Meanwhile, Poisson's ratio, which is independent of the rate of replacement, varied from 0.14 to 0.2. Mas et al. [27] revealed that compressive strengths, 90 days later and with 30% coarse aggregate replacement, decreased by 23.8% for CEM II cements and 13.5% for CEM III/A. The strength increased, with 50% replacement, by +7.5% for CEM V/A cement (Series 1-CEM V). López et al. [28] produced non-structural concretes with 100% coarse aggregate replacement using concretes with low strength (15 MPa), using 200 kg cement/m^3 of concrete. Meanwhile, Medina et al. [29] revealed that the use of MRA in concretes with a 25% coarse aggregate replacement rate has no effect on the absorption capacity. However, at a replacement rate of 50%, the sorptivity of recycled concretes is 10–20% higher than the reference concrete. Cantero et al. [30] concluded that in concretes with coarse MRA, the higher the replacement rate, the lower the density in a fresh state and the higher the air content: in concrete with a 100% rate of replacement, the density was 7% lower and the air content 37% higher than the reference concrete. It also backs its use in concretes, with a characteristic strength of 30 MPa.

As a result of the research conducted, recycled aggregates have been added unevenly in the regulation of concrete in different countries. Table 1 shows the type of recycled aggregate, fraction and maximum percentage (minimum and maximum value) allowed by each continent, as well as the type of concrete (structural or non-structural) that can be crafted using recycled aggregates. It also shows that the common aspect is that all continents allow the use of the coarse fraction of RCA and RMA, whose maximum limit of replacement ranges from 15% to 100% for structural or non-structural concretes. Regarding the fine fraction of RCA, it reveals that, with the exception of the American continent, its use is allowed in the eco-design of new concretes. In addition, fine MRA is only allowed for non-structural uses. In this context, it is worth noting that in South American countries [31], the technical regulation does not yet allow the use of recycled aggregate (RCA/MRA), due to the lack of trust it generates and the absence of authorised managers of C&DW who obtain optimal quality recycled aggregates that are durable.

Table 1. Recycled aggregates in international regulation. Typology and maximum rate of incorporation.

Continent	Material	Fraction	% Allowed		Type of Concrete
			Minimum Value	Maximum Value	
Asia (China, Korea and Japan)	RCA	Coarse	20	100	Structural
		Fine	30	100	Structural
		Coarse	0	100	Non-structura
		Fine	0	30	Non-structural
Australia	RCA	Coarse	-	30	Structural
	MRA	Coarse	-	100	Structural
Europe (Belgium, Germany, Italy, Denmark, Holland, Portugal, Switzerland, United Kingdom, France, Spain)	RCA	Coarse	15	100	Structural
		Coarse/Fine	-	100	Structural
	MRA	Coarse	25	100	Structural
		Fine	-	20	Non-structural
America (Brazil)	RCA	Coarse/Fine	-	100	Non-structural
	MRA	Coarse/Fine	-	100	Non-structural

Note. RCA: recycled concrete aggregate; MRA: mixed recycled aggregate.

In this context, the main novelty of this research study lies in delving into the progress of the knowledge on the joint valorisation of the coarse and fine fraction of mixed recycled aggregate, as research conducted heretofore had focused mainly on the RCA fraction and, to a lesser extent, on the coarse fraction of MRA. There are no prior studies that simultaneously use fine and coarse MRA to produce structural recycled concretes. Likewise, this study contributes to the development of regulation or guides that make it possible to, depending on the quality of the fMRA (their composition and intrinsic properties), establish the applications where they could be used, such as the concrete industry, defining the minimum parameters it must meet to comply with the building code and the areas of environmental exposure where new concretes can be used, as well as other applications in the scope of the construction sector. As stated by Plaza et al. [18], the partial replacement of natural aggregate with coarse recycled aggregate alone or together with fine recycled aggregate (RCF) from concrete wastes has a generally beneficial effect on eco-efficiency (the relation between compression, tensile strength, bending and CO_2 emitted), with values that are similar or greater than those shown by HP (conventional concrete). The use of coarse recycled aggregate concrete and RMF (recycled mixed fine aggregate) has a beneficial effect on eco-efficiency to split tensile strength (strength/CO_2 emission ratio), whereas the eco-efficiency is slightly lower (<3%) than HP in terms of compressive and flexural strength. The benefits of using coarse and/or fine recycled aggregate to partially replace natural aggregate are not only a decrease in the emissions of CO_2 when manufacturing concrete but also the significant mitigation of environmental issues triggered by gathering the necessary waste.

In this line of research, the main goal of this research was to further deepen scientific and technical knowledge on the simultaneous use of the fine and coarse fractions of the mixed fraction of construction and demolition waste as components of the granular skeleton of recycled concretes. Doing so required characterising the physical (density, entrained air and consistency) and mechanical (compressive, tensile and flexural strength) characterisation, as well as the watertightness under pressure, of the new eco-concretes, which include 0%, 25% and 50% of coarse MRA and/or 0%, 25% and 50% of fine MRA, which can be used for applications in the field of civil engineering and construction.

2. Materials and Methods

2.1. Materials

The natural aggregate (NA) used was crushed siliceous greywacke with sharp edges and layered shapes (see Figure 1) which comes in two granulometric fractions: (i) natural 0/6 mm sand (fNA); and (ii) natural 6–12 mm gravel (cNA). This aggregate meets all the requirements of European standard EN 12620 for aggregates used to manufacture concretes. Table 2 shows the physical and mechanical properties of the aggregates used in the study, as well as the requirements stipulated by European standard 12620 [32] and EHE-08 for aggregates used to manufacture concretes [33].

Figure 1. View of the aggregates used to manufacture concretes. Legend: (**a**) View of natural aggregate; (**b**) View of natural aggregate after grinding; (**c**) View of recycled aggregates and (**d**) View of recycled aggregates after grinding.

Table 2. Physical and mechanical properties of the aggregates.

Property	Aggregates				EN-12620/ EHE08
	fNA	cNA	fMRA	cMRA	
Dssd (kg/m^3) [34]	2.76	2.74	2.70	2.42	-
WA$_{24}$ (wt %) [34]	1.18	0.88	5.39	6.28	<5
LC (wt %) [35]	-	16	-	32	<40
FI (wt %) [36]	-	21	-	10	<35

Note. fNA: natural sand; cNA: natural gravel; fMRA: mixed recycled sand; cMRA: mixed recycled gravel; Dssd: dry saturated surface density; WA$_{24}$: water absorption coefficient after 24 h; LC: Los Angeles coefficient; and FI: flakiness index.

The MRA used came from the ARAPLASA C&DW management plant, located in the north of the province of Cáceres (Spain), which came, like natural aggregates, in two fractions: (i) recycled mixed 0/6 mm sand (fMRA) and (ii) mixed recycled 6/12 mm gravel (cMRA). Regarding their morphology, they are characterised by preferably having rounded and slightly layered shapes (see Figure 1). Regarding their composition (see Table 3), the gravel is characterised by being comprised by ~88% debris from unbound concrete, mortar and aggregate, as well as ~11% ceramic material (tile, blocks, bathroom fittings, etc.) and other minority components (<1.3%), mainly floating particles (plastic and wood), gypsum and glass, in a percentage that makes up less than 1% of the weight. In addition, it is worth noting that fMRA was obtained from the same MRA as cMRA, with the former having a reddish colour due to the presence of ceramic fines.

Table 3. Components of the mixed recycled gravel (cMRA) (EN 933-11 classification [37]).

Class	Type	Content (% Weight)
Rc	Concrete and mortar	43.98
Ru	Natural stone	43.94
Rc + Ru		87.82
Rb	Baked clay material	10.93
Ra	Asphalt	0.87
FL	Floating particles	0.02
G	Gypsum	0.34
X + Rg	Others and glass	0.02

Figures 1 and 2 shows the granulometric distribution of the natural and recycled aggregates, revealing that they all have a continuous grain size. Likewise, it reveals that regardless of their nature, sands are within the granulometric spectrum recommended by EHE-08 for manufacturing concretes, as is the amount of particles that pass through the 0.063 mm sieve, which is less than 10% of the weight, the maximum limit stipulated by EHE-08 for crushed sands of a siliceous nature.

The Portland cement used was a CEM I 42.5 R that met all the physical, chemical and mechanical requirements established by European standard EN 197-1 [38] and was supplied by the plant of the Lafarge Holcim group located at Villaluengo de la Sagra in the province of Toledo (Spain). Lastly, we used superplasticiser additive FUCHS BRYTEN NF, a water-reducing additive made of modified polycarboxylate with a base of water and a brownish colour. It is free of chloride, has a ~20% content of solids, a density of 1.1 g/cm^3 and pH = 8.0 and was supplied by FUCHS Lubricantes S.A.U.

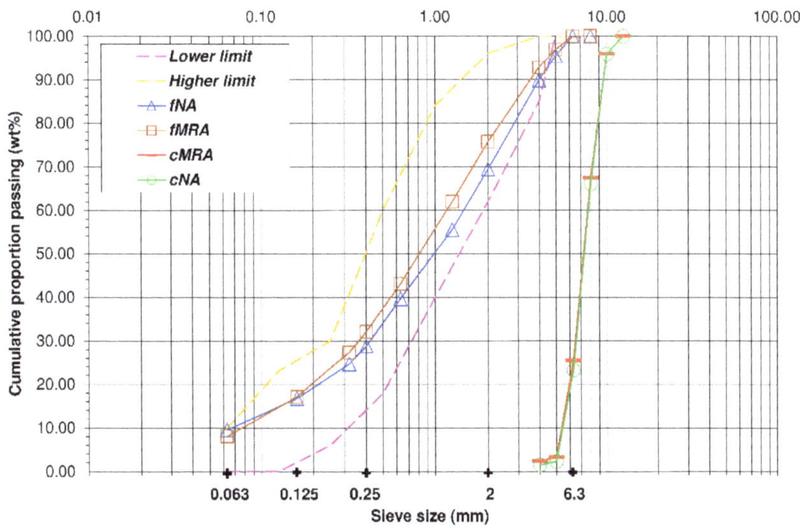

Figure 2. Granulometric distribution of aggregates (NA and MRA).

2.2. Concrete Properties Studied

Table 4 lists the properties analysed in fresh and hardened states, as well as the standard that describes the trial methodology and size of the samples used to assess the property being studied.

Table 4. The concrete properties analysed in fresh and hardened states.

Properties	Trial	Standard	Sample Size (mm)	Trial Duration (days)
Physical	Density	EN 12350-6 [39]	Cubic 150 × 150 × 150	Beginning
	Entrained air	EN 12350-7 [40]	-	Beginning
	Consistency	EN 12350-2 [41]	-	Beginning
Mechanical	Compression	EN 12390-7 [42]	Cubic 150 × 150 × 150	7, 28 and 90
	Traction	EN 12390-6 [43]	Cylindrical 100 φ × 200	28
	Bending	EN 12390-5 [44]	Prismatic 100 × 100 × 400	28
Durable	Penetration under pressure	EN 12390-8 [45]	Cylindrical 150 φ × 300	28

2.3. Mix Design

Nine mixes were manufactured: (i) Conventional concrete, with 100% NA (M1); (ii) Concrete with 25% fine MRA (M2); (iii) Concrete with 50% fine MRA (M3); (iv) Concrete with 25% coarse MRA (m4); (v) Concrete with 25% fine MRA and 25% coarse MRA (M5); (vi) Concrete with 50% fine MRA and 25% coarse MRA (M6); (vii) Concrete with 50% coarse MRA (M7); (viii) Concrete with 25% fine MRA and 50% coarse MRA (M8); (ix) Concrete with 50% fine MRA and 50% coarse MRA (M9).

In order to design and formulate the mixes we used the mix-design rules [46], taking as baseline data a design characteristic strength (f_{ck}) of 25 MPa (C25/30), a maximum aggregate size of 12.5 mm, a constant effective water–cement ratio (w/c) of 0.45 and an S2 target workability as established in EN 206-1 [47], which is equivalent to a 70 ± 20 mm slump. Likewise, dry aggregates were taken into consideration in this dosage process, as well as the water absorption of the recycled aggregates in the first 10 min of being immersed in water (~70% of the total water absorption after 24 h). With this strategy, we guaranteed that all mixes had the same amount of water available to hydrate the cement, regardless of their granular skeleton. In addition, 5.89 kg/m^3 of additive was added in order to achieve a suitable workability with this w/c ratio.

All the mixes designed met the dosing requirements (minimum cement content: 300 kg/m^3 and maximum effective w/c ratio: 0.55) stipulated by European standard EN-206-1 [47] for durability class XC2. The mixes studied (Table 5) were prepared in an 85-litre laboratory vertical shaft mixer using the following procedure: (i) the coarse aggregate (cNA and cMRA) was mixed for 30 s; (ii) the fine aggregates (fNA and fMRA) were added and the materials were mixed for 30 s; (iii) the binder (cement) was added, mixing it for 60 s; (iv) then 80% of the mixing water was added as well as the superplasticiser additive, mixing it for 45 s and (v) the remaining water was added and mixed for 240 s. Lastly, the samples were manufactured and cured following European standard EN 12390-2 [48].

Table 5. Composition of the concretes designed.

Materials (kg/m^3)	Mix								
	M1	M2	M3	M4	M5	M6	M7	M8	M9
fNA	916.8	684.0	446.4	902.4	666.0	434.4	888.0	648.0	429.6
fMRA	0.0	228.0	446.4	0.0	222.0	434.4	0.0	216.0	429.6
cNA	993.2	988.0	967.2	733.2	721.50	705.90	481.0	468.0	465.4
cMRA	0.0	0.0	0.0	244.4	240.5	235.3	481.0	468.0	465.4
Cement	380.0	380.0	380.0	380.0	380.0	380.0	380.0	380.0	380.0
Water	224.4	228.9	231.1	232.8	237.4	241.5	243.9	247.5	252.3
Additive	5.9	5.9	5.9	5.9	5.9	5.9	5.9	5.9	5.9

Figure 3 summarizes the manufacturing process and tests carried out on the studied mixtures.

Figure 3. Flow diagram corresponding to the manufacturing process and tests carried out on the studied mixtures.

3. Results and Discussion

3.1. Properties in a Fresh State

3.1.1. Consistency

Table 6 shows the results of consistency, entrained air content and density of the concretes designed. Regarding consistency, all mixes were within the target design workability (50 ≤ S2 ≤ 90 mm), which reveals that adding the mixed recycled fine and/or coarse fraction does not have a negative effect on this property. This result is in line with Plaza et al. [18] who observed that the simultaneous addition of recycled concrete aggregate and concrete or mixed sand does not lead to a decline in the workability of recycled concretes. This behaviour is in line with Agrela et al. [23] and Medina et al. [29] who respectively suggested, as strategies to mitigate the negative effect that the higher water absorption of recycled aggregates has on this property, to pre-saturate them before the mixing process or to add the water initially absorbed by these recycled aggregates to the dosage.

Table 6. Concrete properties in a fresh state.

Mix	Consistency (cm)	Entrained Air (vol %)	Density (kg/m^3)
M1	6.00	1.97	2416.37
M2	6.00	3.00	2380.97
M3	5.17	3.90	2354.57
M4	5.50	2.20	2372.79
M5	5.50	3.60	2322.27
M6	6.50	5.50	2275.41
M7	6.50	2.77	2331.01
M8	5.67	4.20	2286.24
M9	5.17	5.57	2288.47

Figure 4 shows the concrete slump test of conventional concrete (M1) and concrete with a higher content of MRA, highlighting that the individual or simultaneous use of the fractions does not have a negative effect on this property.

Figure 4. Concrete slump test: (**a**) Concrete with 100% NA (M1) and (**b**) Concrete with 50% fMRA and cMRA (M9).

3.1.2. Entrained Air

Table 6 lists the amount of entrained air in fresh concrete, revealing that the addition of fine and/or coarse MRA caused an increase in this property, with the value for M3 (50% fMRA), M7 (50% cMRA), M8 (25% fMRA + 50% cMRA) and M9 (50% fMRAf 50% cMRA) being 1.9, 1.4, 2.1 and 2.79 times higher than that registered for mix M1 (100% NA), respectively. This performance could be connected with [49,50] (i) higher water absorption by the MRA, (ii) lower density of the adhered mortar present in the recycled aggregate due to the presence of air bubbles within, (iii) the rougher texture of MRA compared with NA and (iv) the presence of microcracks inside the MRA which are not connected to the aggregate's permeable pores. These results are in line with the observations by Cantero et al. [30] and Plaza et al. [18], who registered an increase of this property when adding contents of up to 100% MRA and 100% RCA, respectively.

Figure 5 reveals the linear relationship that exists between this property and the percentage of recycled sand for different percentages of coarse aggregate replacement (0%, 25% and 50%), with all cases having an $R^2 \geq 0.99$. This trend was previously registered by Yaprak et al. [49], who added from 0% to 100% of recycled concrete sand.

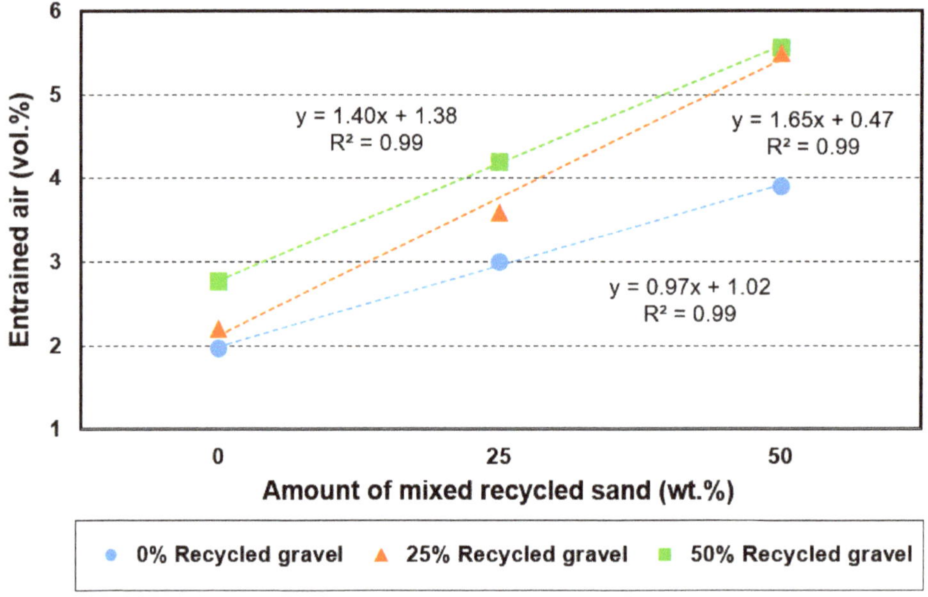

Figure 5. Connection between the percentage of mixed recycled sand and entrained air.

3.1.3. Density

Table 6 lists the density values for fresh concrete. It reveals a decrease of said property in connection with the percentage of addition of mixed recycled aggregate (fMRA and/or cMRA). These decreases reached their maximum value in mixes that had both fractions added at 50% simultaneously; as for mixes M8 and M9, these values were 5.4 and 5.3 compared with mix M1, respectively. This was also the case with the mix that added 25% cMRA + 50% fMRA (M6), with a 5.8% decrease compared with M1. This decrease is in line with Brito et al. [50], who registered decreases of up to 10% for concretes that had up to 100% of MRA. This behaviour would be connected with the lower density of recycled aggregates due to the presence of adhered mortar and ceramic material, as well as the apparent higher water/cement ratio of the new concrete mixes designed [49,51].

Regarding the values obtained, it is worth noting that they were within the 2430–2300 kg/m^3 and 2430–2220 kg/m^3 ranges of values for concretes that have different percentages of recycled concrete aggregate [52,53] and mixed recycled aggregate [51], respectively. Lastly, this observed trend is in line with research previously registered by other authors: Plaza et al. [18] registered densities of 2428.56 kg/m^3 for conventional concretes, 2367.38 kg/m^3 for concretes with a 100% coarse aggregate replacement rate and 2310.51 kg/m^3 for mixes with 100% recycled coarse aggregate replacement and 50% fMRA; César Medina et al. [52] registered a density of 2347 kg/m^3 with 25% of cMRA replacement and densities of 2335 kg/m^3 with 50% cMRA replacement. A. González et al. [51], with high-performance recycled concrete aggregates with 20% and 50% cMRA replacement, obtained densities of 2430 kg/m^3 and 2340 kg/m^3, respectively.

3.2. Properties in a Hardened State

3.2.1. Density

Table 7 lists the values of apparent density in hardened concretes analysed following 28 days of curing, once again revealing a decrease of this property with the percentage of addition of MRA (fMRA and/or cMRA), due to the lower density of this new typology of recycled aggregates compared with natural aggregates (Table 2). This decrease was between 1.4% and 5.7% compared with the reference concrete (M1). In this context, it is worth noting that said decrease is in line with the 3.3–5.0% range registered by other authors who added up to 100% of MRA [49]. Regarding the values obtained, it is worth stressing that they are within the range of values (2450–2270 kg/m^3) registered previously by other authors who added mixed recycled aggregates [10,29,30].

Table 7. Density and compressive strength evolution.

Mix	D_{28d} (kg/m^3)	a_{cs7d}	σ	a_{cs28d}	σ	a_{cs90d}	σ
M1	2412.15	33.45	0.26	40.02	0.22	46.67	0.67
M2	2377.78	31.72	0.38	41.54	1.10	55.58	0.75
M3	2330.17	33.14	0.61	40.00	0.46	50.43	0.72
M4	2368.30	30.92	0.69	38.48	0.33	48.31	0.31
M5	2321.38	31.38	1.04	37.84	0.34	47.85	0.23
M6	2274.07	31.04	0.28	38.02	0.41	47.16	0.94
M7	2319.11	30.85	0.80	37.77	0.49	50.60	0.87
M8	2283.52	30.62	0.90	36.82	0.91	52.12	0.72
M9	2278.12	29.87	0.53	39.46	0.79	52.44	0.83

Note. D_{28d}: density in a hardened state after 28 days; σ: standard deviation; a_{cs7d}: average compressive strength after seven days; a_{cs28d}: average compressive strength after 28 days; a_{cs90d}: average compressive strength after 90 days. a_{cs} average compressive strength with a cubic sample, which must be multiplied by 0.90 to turn it into a cylindrical sample of 150 $\phi \times$ 300 mm.

3.2.2. Compressive Strength

Table 7 shows the evolution of the compressive strength of the various mixes analysed after 7, 28 and 90 days. It shows that regardless of the type of concrete, (i) compressive strength increases with the curing time, an evolution which is very similar to that displayed by the reference concrete, (ii) the relative compressive strength after seven days is 75.7–83.6% of that obtained after 28 days, a similar percentage to the figure obtained (70%) in Portland cement concretes with no additions and a w/c ratio <0.45 [54] and to the 65–93% range of values registered by Bravo et al. [16] for concretes that partially added C&DW, and (iii) the average strength after 28 days is higher than the design strength of 25 MPa. Therefore, these new recycled concretes could be used as concretes for structural use.

Figure 6 shows the compressive strength variation of recycled concretes compared with conventional concrete for the various curing times. It reveals that for short times (t < 28 days), the addition of fMRA and/or cMRA caused a maximum performance loss of 10.7% compared with M1 and was suffered by the mix that had 50% fMRA + 50% cMRA (M9). However, at t = 90 days there was an increase in strength. The maximum increase was +19% over M1, observed in the mix that had 25% fMRA (M2). Likewise, concretes with 50% cMRA and 0%, 25% and 50% fMRA (M7–M9) had a better behaviour, recording increases ranging from 8.4% to 12.4% compared with M1.

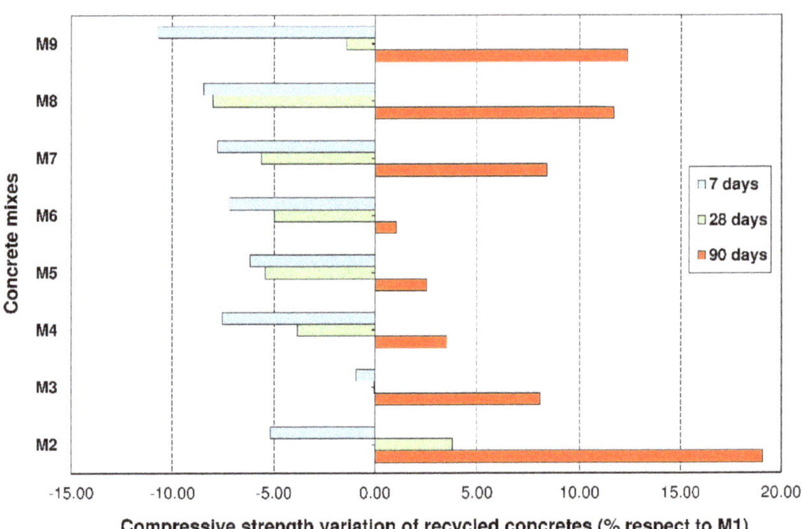

Figure 6. Compressive strength variation of recycled concretes (M2–M9) compared with conventional concrete (M1).

These results reveal that the addition of cMRA individually (M4–M7) does not have a negative effect on this attribute, registering a small decrease ($\Delta_{M4} = -3.8\%/\Delta_{M7} = -5.6\%$) and a slight increase ($\Delta_{M4} = +3.5\%/\Delta_{M7} = 8.4\%$) compared with M1 after 28 and 90 days, respectively. This performance is in line with the observations of Cantero et al. [30], Poon et al. [54], Gomes et al. [55] and Lotfy et al. [56], who revealed that for recycled coarse aggregate replacement percentages $\leq 50\%$, there are no significant differences with conventional concretes.

Regarding the individual addition of fMRA (M2 and M4), we observed that the addition of 25% in weight (M2) led to a slight increase in compressive strength, reaching +3.8% and +19.1% compared with M1 after 28 and 90 days, respectively. This behaviour is in line with the study of Bravo et al. [10], who established that type I recycled fine aggregates (Rc + Ru $\geq$ 80% and other components $\leq$20%) caused variations in compressive strength ranging from +3.8 to -12.5 after 28 days and in concretes with a replacement percentage of 25%. Likewise, this result was also obtained by other authors [57–60], who registered 3.7%, 5% and 16% increases compared with conventional concrete for 10%, 30% and 50% additions of RCA, respectively. This increase was connected with the filler effect of fine RCA and their rough texture and shape, which allowed for a better packing of the granular skeleton [61,62].

Regarding the 50% fMRA replacement (M3), the study showed that after 28 days the resistance remained constant, whereas after 90 days, there was a slight increase of +8.1% compared with M1. This behaviour is better than that established by Bravo et al. [63] for a 50% replacement with type-I sand comprised by Rc + Ru = 83.7%, Rb = 0.9% and Rg = 15.4. These authors registered an attribute loss ranging from -6.1% to -13.3% compared with conventional concrete, for a 50% percentage of replacement. This improved behaviour of M3 is linked to the better quality of fMRA (Table 2) compared with the type-I sand studied by said authors.

Likewise, Figure 5 reveals that mixes M5, M6, M8 and M9, which simultaneously incorporated cMRA (25% and 50%) and fMRA (25% and 50%) in their composition, had a worse performance after 28 days, registering a 5.5%, 5.0%, 8.0% and 1.4% decrease compared with M1, respectively. These decreases are lower than the ~7% and ~17% observed by Plaza et al. [18], who added 50% coarse RCA + 50% fine RCA and 50% coarse

RCA + 50% mixed sand, respectively. In addition, it is worth noting that similar to when MRA were added individually, an improved performance was registered for the concretes after 90 days of curing, with mix M9 (cMRA = 50% + fMRA = 50%) reaching a maximum increase of 12.4% compared with M1.

This improved behaviour of the new concretes (M2–M9) at curing times greater than 28 days could be connected with the pozzolanic activity of the fine fraction (sizes < 0.063 mm) present in fMRA, where there are mainly ceramic fines. This aspect was previously registered by Medina et al. [60,61] and Asensio et al. [62,63], who analysed the pozzolanic performance of the fine fraction of C&DW with variable compositions (26.5% $\leq$ SiO_2 $\leq$ 70.5%, 4.4% $\leq$ CaO $\leq$ 24.5%, 5.8% $\leq$ Al_2O_3 $\leq$ 18.5%), observing that they have a lime fixation capacity after 28 days of 53.3–82.1%. These values are lower than silica fume (~90%) and higher than fly ash (~45%).

In addition, this behaviour could also be linked to (i) the presence of anhydride cement particles [64–66] in the mortar adhered to the fMRA that will become hydrated, generating calcium silicate hydrate (C-S-H) which will positively contribute to the mechanical behaviour and (ii) the attributes of existing ITZs among the Rb components of MRA (cMRA and fMRA)/paste that can have a thickness equal to or lower than the one that natural aggregates normally have (e_{ITZ} = 10–50 μm), as shown previously in the studies by Medina et al. [29] and Sáez del Bosque et al. [24] in concretes that had coarse fractions of ceramic aggregate from bathroom fittings and coarse mixed recycled aggregate, respectively.

Figure 7 shows the appearance of concretes M1, M3, M6 and M9 after being subjected to the compression test after 28 days, revealing that the type of failure was similar in all of them and that their morphology can be classified as suitable according to European standard EN 12390-3 [65,67].

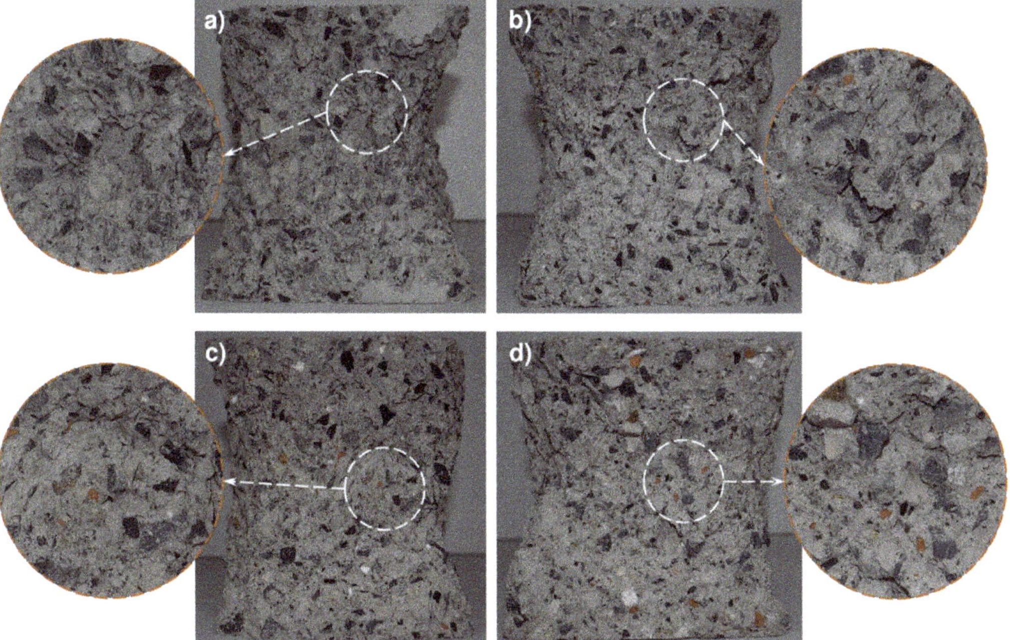

Figure 7. Type of failure of the concretes: (**a**) M1; (**b**) M3; (**c**) M6; and (**d**) M9.

Lastly, Figure 6 verifies that the aggregate/paste ITZ is the area where the cracks preferentially begin and then spread when a concrete item is subjected to an external force that surpasses its operational status limit [68]. This happens because this area is the weakest and the one where the most stress concentrates, as observed by Medina et al. [67] by simulating the stress in the coarse/paste aggregate ITZ subjected to compressive forces [69].

3.2.3. Splitting Tensile Strength

Table 8 shows the tensile strength of concretes tested after 28 days and the variations in strength compared with conventional concrete (M1) and compared with the concretes that exclusively had cMRA (M4 and M7). It reveals that the mixes that only had fMRA (M2 and M3) experienced a slight increase in tensile strength between 8.7% and 5.5% compared with M1, respectively. This behaviour is in line with the observations of Ahmed et al. [68] and Kirthika et al. [69], who revealed that the use of fine MRA in percentages lower than 50% did not lead to a significant loss of performance (<10%) in the new concretes.

Table 8. Tensile and flexural strength of concretes 28 days later.

Mix	fcmt (MPa)	Δfcmt (%) ♣	Δfcmt (%)	fcmf (MPa)	Δfcmf (%) ♣	Δfcmf (%)
M1	3.45 ± 0.09	-	-	3.82 ± 0.12	-	-
M2	3.75 ± 0.07	+8.70	-	4.16 ± 0.11	+8.90	-
M3	3.54 ± 0.13	+5.51	-	4.15 ± 0.23	+8.64	-
M4	3.21 ± 0.05	−6.96	-	4.10 ± 0.02	+7.33	-
M5	3.41 ± 0.06	−1.16	+6.23 ♠	4.28 ± 0.09	+12.04	+4.39 ♠
M6	3.31 ± 0.07	−4.06	+3.12 ♠	4.21 ± 0.06	+10.21	+2.68 ♠
M7	3.29 ± 0.06	−4.64	-	3.69 ± 0.27	−3.40	-
M8	3.40 ± 0.05	−1.45	+3.34 *	3.83 ± 0.21	+0.26	+3.79 *
M9	3.44 ± 0.06	−0.29	+4.56 *	4.06 ± 0.20	+6.28	+10.03 *

Note. fcmt: mean splitting tensile strength; fcmf: mean flexural strength; ♣ strength variation compared with M1; ♠ strength variation compared with M4; and * strength variation compared with M7.

Regarding the mixes whose composition included 25% cMRA (M4) and 50% cMRA (M7), it is worth noting that they registered a small decrease of 7.0% and 4.6% compared with M1, respectively. These losses are similar to those registered by Cantero et al. [30] in concretes with 25% and 50% of mixed recycled aggregate and lower than the 12% and 14% losses registered previously by other authors who [70–75] analysed concretes with 100% of recycled aggregates, which had 10% ≤ Rb ≤ 14%.

Regarding the mixes that include both fractions simultaneously, we observed that the addition of fMRA had a positive effect by (i) lessening the decrease in strength of the mixes (M5, M6, M8 and M9) compared with the mixes that only included cMRA (M4 and M7), with the decrease reaching 1.45% ≤ Δfcmt ≤ 4.06%, and (ii) increasing their strength compared with mixes that only had cMRA (M4 and M7), with this change ultimately being 3.12% ≤ Δfcmt ≤ 6.23%. This behaviour is better than that observed previously by Plaza et al., who registered a 7.3% and 11.0% decrease compared with conventional concrete in concretes with 25% coarse RCA + 50% fine MRA and 50% coarse RCA + 50% fine MRA, respectively. This same tendency was revealed for concretes with 50% coarse RCA + 50% fine RCA, again registering decreases ~11% compared with conventional concrete [76].

Lastly, it is worth noting that the failure mechanism was the same in all tested concretes, causing a brittle failure that led to the tested mixes splitting in half. In addition, we observed that (i) the failure surface obtained is irregular, confirming the existence of intact coarse aggregates (cNA or cMRA) and thus again revealing that the failure emerges from the ITZ of the coarse and/or fine aggregates and the cement paste, and (ii) the granular skeleton (fine and coarse aggregates) is distributed homogeneously, regardless of whether the aggregate is natural or recycled (Figure 8).

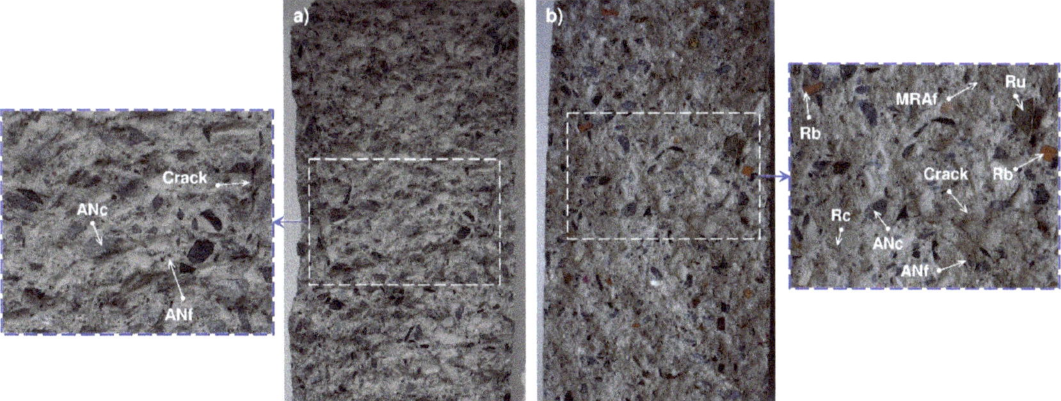

Figure 8. Appearance of the concretes when performing the tensile test 28 days after producing the mortar: (**a**) conventional concrete (M1) and (**b**) concrete with 50% cMRA + 50% fMRA (M9).

3.2.4. Flexural Resistance

Table 8 lists the flexural strength values of the tested concretes after 28 days, and the variations in strength compared both with conventional concrete (M1) and with the concretes that exclusively had cMRA (M4 and M7). It reveals that the mixes that only had fMRA (M2 and M3) experienced a slight increase in flexural strength of +8.9 and +8.6% compared with M1, respectively. This result is in line with the prior observations of Ahmed [76] and Kirthika [77], who registered 6.7% and 3.2% increases compared with conventional concrete for a fine MRA replacement percentage of 50%, respectively.

Regarding mixes with 25% cMRA (M4) and 50% cMRA (M7), it is worth noting that they displayed an uneven behaviour, with mix M4 registering a 7.3% increase compared with M1 and M7 showing a 3.4% decrease compared with M1. This behaviour was similar to that observed by Cantero et al. [30], who established that for replacement percentages lower than or equal to 50% of coarse MRA, there were no significant variations of this mechanical attribute (Δfcmf $\leq$ 10% compared with conventional concrete).

Regarding the mixes that simultaneously incorporate both fractions, we observed that the addition of fMRA had a positive effect by improving the behaviour of mixes with cMRA (M4 and M7). It is worth noting that mixes M5 and M6 registered a 4.4% and 2.7% increase compared with M4, and mixes M8 and M9 showed a 3.8% and 10.0% increase compared with M7. Likewise, all of them had a similar or improved behaviour compared with that observed in conventional concrete with 100% natural aggregate (cNA and fNA).

Lastly, all samples tested, regardless of the composition of their granular skeleton, had failures due to the formation of a crack in the middle part of the span, rising from the part being pulled (the lower part of the sample) to the part being compressed [77] (the highest part where the load is applied). Likewise, observing the cracked area, we once again saw that the failure took place along the aggregate/paste transition area, with the aggregates being detached from the matrix.

3.2.5. Water Penetration Depth under Pressure

Figure 9 shows the results obtained regarding the maximum and mean depth of the mixes analysed.

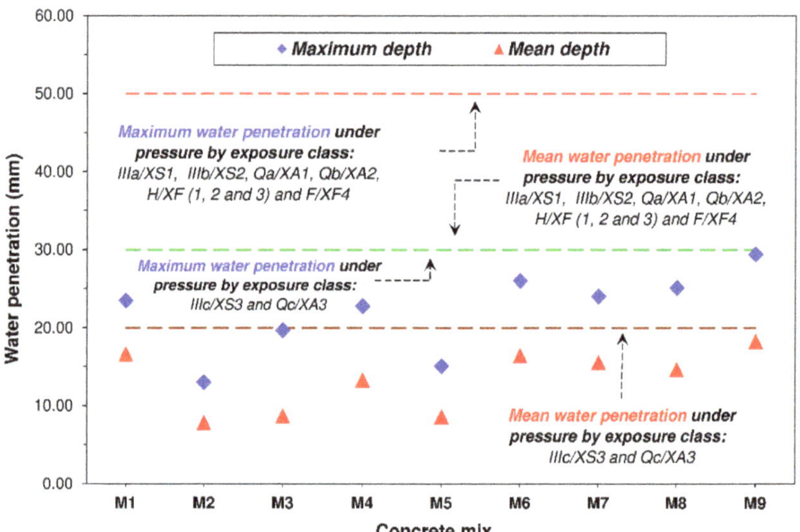

Figure 9. Water penetration of the concretes under pressure. Limits established in EHE-08 and EC-2 (Note. IIIa: marine class—subclass: aerial; IIIb: marine class—subclass: submerged; IIIc: marine class—subclass: tidal and splash zones; Qa: aggressive chemical class—subclass: weak; Qb: aggressive chemical class—subclass: average; Qc: aggressive chemical class—subclass: strong; H: with frost class—subclass: without deicing salts; and F: with frost class—subclass: without deicing salts).

Regarding the individual incorporation of fMRA (M2 and M3), we observed that these mixes experienced a larger decrease both in maximum (Pmax) and mean (Pmed) water penetration, with the decrease ending up being between 16.1% ≤ Pmax ≤ 44.6% and 47.7% ≤ Pmed ≤ 52.9% compared with conventional concrete (M1). This greater watertightness could be connected with the pozzolanic activity of ceramic fines (<0.063 mm) of the fMRA, as well as with the hydration of anhydrous cement [78] present in the fines from fMRA mortars that give them a certain hydraulic activity and thus lead to a more sealed and tortuous pore structure.

Regarding the use of cMRA, we observed that an addition of 25% (M4) and 50% (M7) caused an uneven behaviour in these two properties. In the case of M4, the table shows that Pmax and Pmed experienced a slight (2.8%) and small (20.1%) decrease compared with M1, respectively. This behaviour is in line with the prior observations of Mas et al. [77], who observed that the depth remained constant for coarse MRA replacement percentages ≤25%. Regarding mix M7, it experienced an increase of Pmax (2.4%) and a slight decrease of Pmed (6.4%). This behaviour could be connected with the fact that the microcracks present in MRA have a greater impact on Pmax than Pmed.

Regarding the simultaneous use of cMRA and fMRA as the granular skeleton of the concretes, the table shows that only the 25% cMRA + 25% fMRA mix (M5) made it possible to obtain an increase in water penetration, causing a 35.8% and 33.9% decrease in Pmax and Pmed compared with M1, respectively. This result, as happened with the other properties studied, makes it possible to say that there is no performance loss for MRA replacement percentages lower than or equal to 25%.

In this context, it is worth noting that for weight percentages of cMRA and fMRA greater than 25%, there was an increase in water penetration, which could be explained by the beneficial effect of its rougher surface texture enabling a better entry (similar ITZs) in the cement matrix, as well as the pozzolanic activity of the <0.063 mm fraction not being able to compensate for the negative effect that its intrinsic properties have on water penetration (lower density, higher water absorption and the presence of microcracks in its microstructure).

The values obtained are under the limits established in chapter VII "Durability" Section 37.3.3 "Resistance to water penetration" of the Spanish Code on Structural Concrete (EHE-08), which establishes that a concrete is watertight enough for a given type of exposure (IIIa, IIIb, IIIc, IV, Qa, E, H, F, Qb, Qc) when its maximum and mean penetration depth is lower than 50 or 30 mm and 30 or 20 mm, respectively, depending on the type of environment (see Figure 9). This highlights that recycled concretes have a porous structure that guarantees watertightness and a suitable durability throughout their useful life against this transport mechanism.

Lastly, Figure 10 shows the penetration front of the samples manufactured for concretes M1, M3, M6 and M9. As can be seen, the water penetration outlines have a similar morphology, and there are no visible differences between the mixes manufactured with NA and cMRA and/or fMRA.

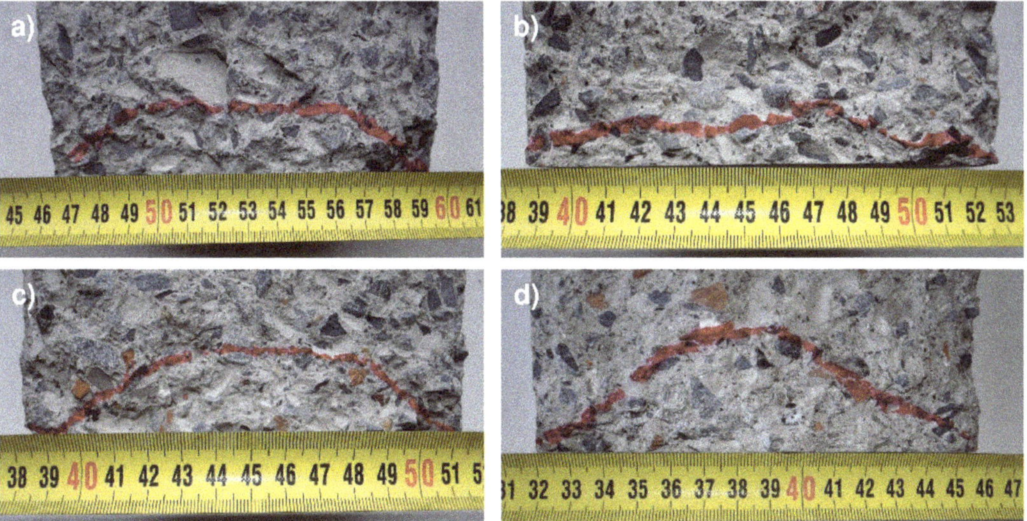

Figure 10. Penetration fronts of the concretes: (**a**) M1; (**b**) M3; (**c**) M6; and (**d**) M9.

3.2.6. Analysing the Concrete Manufacturing Costs

Table 9 lists the economic study of the cost of manufacturing the concretes analysed in order to reveal the financial aspects of this study. In this context, it is worth noting that the price of natural aggregates in Spain is lower than in other countries, as this country is characterised by having a high availability of natural resources, which enables the extraction of natural aggregates. These prices (EUR/t) will be much higher in countries with greater legal restrictions on extracting natural resources or with less availability, which would facilitate the recovery of recycled aggregates in the concrete industry from an economic point of view. This table shows that mix M9, with 50% fMRA + 50% cMRA, is the cheapest, leading to a −8.03% decrease compared with natural aggregate (M1). This result is in line with the prior observations of other authors, who registered decreases of under −50% for recycled coarse aggregate replacement percentages between 50% and 100% [78].

Table 9. Manufacturing cost of the mixes studied.

Component	Unit Price (EUR/t)	Concrete Mix								
		M1	M2	M3	M4	M5	M6	M7	M8	M9
fNA	6.79	6.23	4.64	3.03	6.13	4.52	2.95	6.03	4.40	2.92
fMRA	3.60	0.00	0.82	1.61	0.00	0.80	1.56	0.00	0.78	1.55
Can	6.54	6.50	6.46	6.33	4.80	4.72	4.62	3.15	3.06	3.04
cMRA	3.15	0.00	0.00	0.00	0.77	0.76	0.74	1.52	1.47	1.47
Cement	88.60	33.67	33.67	33.67	33.67	33.67	33.67	33.67	33.67	33.67
Water	0.50	0.11	0.11	0.12	0.12	0.12	0.12	0.12	0.12	0.13
Admixture	1.56	0.01	0.01	0.01	0.01	0.01	0.01	0.01	0.01	0.01
EUR/m^3 concrete	-	46.51	45.72	44.76	45.49	44.59	43.67	44.49	43.51	42.78

Lastly, in addition to these economic savings, one must consider the positive environmental effect, especially in terms of $kgCO_2eq/kg$, that the correct management of MRA and a decrease in the extraction of natural aggregates entail.

4. Conclusions

The conclusions drawn from this research study are:

- Coarse (cMRA) and fine (fMRA) mixed recycled aggregate meet the physical and mechanical requirements included in the relevant regulation on aggregates for manufacturing concretes.
- The workability of recycled concretes (M2–M9) is not impacted by the addition of mixed recycled aggregate, regardless of its replacement percentage, with all mixes showing an S2 consistency (50–90 mm).
- The addition of mixed recycled aggregate (cMRA and/or fMRA) causes a linear increase in the entrained air content in a fresh state of 1.9, 2.8 and 2.8 times the M1 mix in mixes M3 (50% fMRA), M6 (25% cMRA + 50% fMRA) and M9 (50% cMRA + 50% fMRA), respectively.
- The density of recycled concretes is lower than that of conventional concrete, with the loss of density increasing as the recycled aggregate replacement percentage rises and with greater intensity when using mixed aggregate. This behaviour is similar both in a fresh and hardened state.
- The compressive strength of recycled concretes is lower than that of conventional concrete, with M9 (50% cMRA + 50% fMRA) having the greatest drop (~11%) compared with M1 after 28 days of curing. After 90 days, recycled concretes have a better behaviour than conventional concretes. The addition of fMRA has a positive effect, establishing that the optimal replacement percentage is 25% in weight, regardless of the percentage of cMRA.
- The compressive strength of all concretes is higher than the design strength (f_{ck} = 25 MPa).
- The indirect tensile strength experiences a slight increase with the addition of fMRA, with the maximum increase (~9%) compared with M1 corresponding to the concrete that incorporates 25% of fMRA (M2). The addition of cMRA causes a slight loss (~7%) compared with conventional concrete (M1). This loss is softened with the simultaneous addition of fMRA, obtaining a −0.3% decrease compared with M1 for mix M9 (50% cMRA + 50% fMRA).
- The flexural strength of the new concretes is greater than conventional concrete, with the highest increases (6.3–12.0% compared with M1) belonging to mixes M2, M5 and M9. The addition of 50% of cMRA (M7) causes a −3.4% decrease compared with M1.
- All the concretes designed have a watertight structure under pressure, meeting the maximum and mean depth requirements of the relevant regulation.
- The (maximum and mean) water penetration under pressure decreases slightly for mixes M2, M3, M4 and M5, whereas an increase was registered in one (M6, M7 and M8) or both depths (M9) for all remaining concretes.

- The optimal individual or simultaneous replacement percentage of fMRA and cMRA is 25% in weight, in light of the results obtained in this research study.
- These results reveal the need for future research that addresses the behaviour of these concretes from the viewpoint of their durability properties, as well as investigating different types of mixed recycled aggregates with which to manufacture concretes.
- Lastly, this research will positively contribute to the addition of these mixed recycled aggregates to the concrete-related stipulations of structural codes.

Author Contributions: Conceptualisation, M.D.G., P.P.C., D.B.F., M.M.J.V., I.F.S.d.B. and C.M.M.; methodology, M.D.G., I.F.S.d.B. and C.M.M.; formal analysis, M.D.G., P.P.C. and I.F.S.d.B.; investigation, M.D.G., P.P.C., D.B.F., M.M.J.V., I.F.S.d.B. and C.M.M.; writing—original draft preparation, M.D.G., P.P.C. and D.B.F.; writing—review and editing, M.D.G., M.M.J.V., I.F.S.d.B. and C.M.M.; supervision, D.B.F., I.F.S.d.B. and C.M.M.; project administration, D.B.F. and C.M.M.; funding acquisition, C.M.M. All authors have read and agreed to the published version of the manuscript.

Funding: This study was funded by the Spanish Ministry for Science and Innovation under coordinated research project PID2019-107238RB-C21/AEI/10.13039/501100011033 and in conjunction with the Government of Extremadura under grant GR 18122 awarded to the MATERIA research group. The authors are grateful to the "Programa de Formación de Capital Académico Avanzado", Universidad Tecnológica Metropolitana Grant: PM-VRAC2020 UTEM, for the financial support.

Institutional Review Board Statement: Not applicable.

Informed Consent Statement: Not applicable.

Data Availability Statement: Data sharing not applicable.

Conflicts of Interest: The authors declare no conflict of interest.

References

1. Comision Europea (EU). *No 601/2012 of 21 June 2012 on the Monitoring and Reporting of Greenhouse Gas Emissions Pursuant to Directive 2003/87/EC of the European Parliament and of the Council*; European Committee: Brussels, Belgium, 2012.
2. ERMCO. Ready–Mixed Concrete Industry Statistics. 2018. [Online]. Available online: http://ermco.eu/new/wp-content/uploads/2020/08/ERMCO-Statistics-30.08.2019-R4-1.pdf (accessed on 19 March 2020).
3. Monteiro, P.J.M.; Miller, S.; Horvath, A. Towards sustainable concrete. *Nat. Mater.* **2017**, *16*, 698–699. [CrossRef]
4. European Aggregates Association. Annual Review 2019–2020. European Aggregates Association. A Sustainable Industry for a Sustainable Europe. [Online]. Available online: https://uepg.eu/mediatheque/media/UEPG-AR20192020_V13_(03082020)_spreads.pdf (accessed on 5 May 2021).
5. Tang, Y.; Feng, W.; Feng, W.; Chen, J.; Bao, D.; Li, L. Compressive properties of rubber-modified recycled aggregate concrete subjected to elevated temperatures. *Constr. Build. Mater.* **2020**, *268*, 121181. [CrossRef]
6. Rahla, K.M.; Mateus, R.; Bragança, L. Comparative sustainability assessment of binary blended concretes using Supplementary Cementitious Materials (SCMs) and Ordinary Portland Cement (OPC). *J. Clean. Prod.* **2019**, *220*, 445–459. [CrossRef]
7. Ray, S.; Haque, M.; Sakib, N.; Mita, A.F.; Rahman, M.M.; Tanmoy, B.B. Use of ceramic wastes as aggregates in concrete production: A review. *J. Build. Eng.* **2021**, *43*, 102567. [CrossRef]
8. Aghayan, I.; Khafajeh, R.; Shamsaei, M. Life cycle assessment, mechanical properties, and durability of roller compacted concrete pavement containing recycled waste materials. *Int. J. Pavement Res. Technol.* **2020**, *14*, 595–606. [CrossRef]
9. Horňáková, M.; Lehner, P. Relationship of Surface and Bulk Resistivity in the Case of Mechanically Damaged Fibre Reinforced Red Ceramic Waste Aggregate Concrete. *Materials* **2020**, *13*, 5501. [CrossRef] [PubMed]
10. Bravo, M.; de Brito, J.; Pontes, J.; Evangelista, L. Mechanical performance of concrete made with aggregates from construction and demolition waste recycling plants. *J. Clean. Prod.* **2015**, *99*, 59–74. [CrossRef]
11. Thomas, C.; Setién, J.; Polanco, J.A.; Alaejos, P.; De Juan, M.S. Durability of recycled aggregate concrete. *Constr. Build. Mater.* **2013**, *40*, 1054–1065. [CrossRef]
12. Etxeberria, M.; Vázquez, E.; Mari, A.; Barra, M. Influence of amount of recycled coarse aggregates and production process on properties of recycled aggregate concrete. *Cem. Concr. Res.* **2007**, *37*, 735–742. [CrossRef]
13. McNeil, K.; Kang, T.H.-K. Recycled Concrete Aggregates: A Review. *Int. J. Concr. Struct. Mater.* **2013**, *7*, 61–69. [CrossRef]
14. Evangelista, L.; de Brito, J. Durability performance of concrete made with fine recycled concrete aggregates. *Cem. Concr. Compos.* **2010**, *32*, 9–14. [CrossRef]
15. Ju, M.; Park, K.; Park, W.-J. Mechanical Behavior of Recycled Fine Aggregate Concrete with High Slump Property in Normal- and High-Strength. *Int. J. Concr. Struct. Mater.* **2019**, *13*, 61. [CrossRef]

16. Bravo, M.; Duarte, A.P.C.; De Brito, J.; Evangelista, L.; Pedro, D. On the Development of a Technical Specification for the Use of Fine Recycled Aggregates from Construction and Demolition Waste in Concrete Production. *Materials* **2020**, *13*, 4228. [CrossRef]
17. Fernández, S.V. Physical-mechanical characterization of the structural concrete made with recycled aggregates: Experimental comparative with substitution of fine and coarse aggregates. *An. Edif.* **2017**, *3*, 24–31.
18. Plaza, P.; del Bosque, I.S.; Frías, M.; de Rojas, M.S.; Medina, C. Use of recycled coarse and fine aggregates in structural eco-concretes. Physical and mechanical properties and CO_2 emissions. *Constr. Build. Mater.* **2021**, *285*, 122926. [CrossRef]
19. Corinaldesi, V. Mechanical and elastic behaviour of concretes made of recycled-concrete coarse aggregates. *Constr. Build. Mater.* **2010**, *24*, 1616–1620. [CrossRef]
20. Malešev, M.; Radonjanin, V.; Marinković, S. Recycled Concrete as Aggregate for Structural Concrete Production. *Sustainability* **2010**, *2*, 1204–1225. [CrossRef]
21. López-Gayarre, F.; Serna, P.; Domingo-Cabo, A.; Serrano-López, M.; López-Colina, C. Influence of recycled aggregate quality and proportioning criteria on recycled concrete properties. *Waste Manag.* **2009**, *29*, 3022–3028. [CrossRef]
22. Lovato, P.S.; Possan, E.; Molin, D.C.C.D.; Masuero, B.; Ribeiro, J.L. Modeling of mechanical properties and durability of recycled aggregate concretes. *Constr. Build. Mater.* **2012**, *26*, 437–447. [CrossRef]
23. Agrela, F.; de Juan, M.S.; Ayuso, J.; Geraldes, V.; Jiménez, J.R. Limiting properties in the characterisation of mixed recycled aggregates for use in the manufacture of concrete. *Constr. Build. Mater.* **2011**, *25*, 3950–3955. [CrossRef]
24. del Bosque, I.S.; Zhu, W.; Howind, T.; Matías, A.; de Rojas, M.S.; Medina, C. Properties of interfacial transition zones (ITZs) in concrete containing recycled mixed aggregate. *Cem. Concr. Compos.* **2017**, *81*, 25–34. [CrossRef]
25. Martínez-Lage, I.; Vázquez-Burgo, P.; Velay-Lizancos, M. Sustainability evaluation of concretes with mixed recycled aggregate based on holistic approach: Technical, economic and environmental analysis. *Waste Manag.* **2020**, *104*, 9–19. [CrossRef] [PubMed]
26. Martínez-Lage, I.; Martínez-Abella, F.; Vázquez-Herrero, C.; Ordóñez, J.L.P. Properties of plain concrete made with mixed recycled coarse aggregate. *Constr. Build. Mater.* **2012**, *37*, 171–176. [CrossRef]
27. Mas, B.; Cladera, A.; Bestard, J.; Muntaner, D.; López, C.E.; Piña, S.; Prades, J. Concrete with mixed recycled aggregates: Influence of the type of cement. *Constr. Build. Mater.* **2012**, *34*, 430–441. [CrossRef]
28. López-Uceda, A.; Ayuso, J.; Lopez, M.; Jiménez, J.R.; Agrela, F.; Sierra, M.J. Properties of Non-Structural Concrete Made with Mixed Recycled Aggregates and Low Cement Content. *Materials* **2016**, *9*, 74. [CrossRef]
29. Medina, C.; Zhu, W.; Howind, T.; De Rojas, M.I.S.; Frías, M. Influence of mixed recycled aggregate on the physical–mechanical properties of recycled concrete. *J. Clean. Prod.* **2014**, *68*, 216–225. [CrossRef]
30. Cantero, B.; del Bosque, I.S.; Matías, A.; Medina, C. Statistically significant effects of mixed recycled aggregate on the physical-mechanical properties of structural concretes. *Constr. Build. Mater.* **2018**, *185*, 93–101. [CrossRef]
31. Díaz, M.; Almendro-Candel, M.B.; Blanco, D.; Jordan, M.M. Aggregate Recycling in Construction: Analysis of the Gaps between the Chilean and Spanish Realities. *Buildings* **2019**, *9*, 154. [CrossRef]
32. European Committee for Standardization. *EN 12620. Aggregates for Concrete*; European Committee for Standardization: Brussels, Belgium, 2009.
33. Ministry of Public Works. *Code on Structural Concrete (EHE-08)*; Ministry of Public Works: Madrid, Spain, 2008; p. 704.
34. European Committee for Standardization. *EN 197. Cement. Part 1: Composition, Specifications and Conformity Criteria for Common Cements*; European Committee for Standardization: Brussels, Belgium, 2011.
35. Spanish Committee for Standardization. *UNE 83966. Concrete Durability. Test Methods. Conditioning of Concrete Test Pieces for the Purpose of Gas Permeability and Capilar Suction Tests*; Spanish Committee for Standardization: Madrid, Spain, 2008.
36. Teychenné, D.C.; Franklin, R.E. *Design of Normal Concrete Mixes*; Spanish Committee for Standardization: Madrid, Spain, 2010; p. 42.
37. European Committee for Standardization. *EN 933. Tests for Geometrical Properties of Aggregates–Part 11: Classification Test for the Constituents of Coarse Recycled Aggregate*; European Committee for Standardization: Brussels, Belgium, 2010.
38. European Committee for Standardization. *EN 12350-6. Testing Fresh Concrete–Part 6: Density*; European Committee for Standardization: Brussels, Belgium, 2020.
39. European Committee for Standardization. *EN 12350-7. Testing Fresh Concrte–Part 7: Air Content–Pressure Methods. 2020. Link: BS EN 12350-7:2019 Testing Fresh Concrete Air Content. Pressure Methods–European Standards (en-standard.eu)*; European Committee for Standardization: Brussels, Belgium, 2020.
40. European Committee for Standardization. *EN 12350-2. Testing Fresh Concrete–Part 2: Slump Test. 2020. Link: BS EN 12350-2:2019 Testing Fresh Concrete Slump Test–European Standards (en-standard.eu)*; European Committee for Standardization: Brussels, Belgium, 2020.
41. European Committee for Standardization. *EN 12390-7. Testing Hardened Concrete–Part 7: Density of Hardened Concrete*; European Committee for Standardization: Brussels, Belgium, 2020.
42. European Committee for Standardization. *EN 12390-6. Testing Hardened Concrete–Part 6: Tensile Splitting Strength of Test Specimens*; European Committee for Standardization: Brussels, Belgium, 2020.
43. European Committee for Standardization. *EN 12390-5. Testing Hardened Concrete–Part 5: Flexural Strength of Test Specimens*; European Committee for Standardization: Brussels, Belgium, 2020.
44. European Committee for Standardization. *EN 12390-8. Testing Hardened Concrete–Part 8: Depth of Penetration of Water under Pressure*; European Committee for Standardization: Brussels, Belgium, 2020.

45. European Committee for Standardization. *EN-206. Concrete–Specification, Performance, Production and Conformity*; European Committee for Standardization: Brussels, Belgium, 2018.
46. European Committee for Standardization. *EN 1239-2. Testing Hardened Concrete–Part 2: Making and Curing Specimens for Strength Tests*; European Committee for Standardization: Brussels, Belgium, 2020.
47. Kurda, R.; de Brito, J.; Silvestre, J.D. Influence of recycled aggregates and high contents of fly ash on concrete fresh properties. *Cem. Concr. Compos.* **2017**, *84*, 198–213. [CrossRef]
48. Masood, B.; Elahi, A.; Barbhuiya, S.; Ali, B. Mechanical and durability performance of recycled aggregate concrete incorporating low calcium bentonite. *Constr. Build. Mater.* **2019**, *237*, 117760. [CrossRef]
49. Yaprak, H.; Aruntas, H.Y.; Demir, I.; Simsek, O.; Durmus, G. Effects of the fine recycled concrete aggregates on the concrete properties. *Int. J. Phys. Sci.* **2011**, *6*, 2455–2461.
50. de Brito, J.; Agrela, F.; Silva, R.V. Construction and demolition waste. In *Newtrends in Eco-Efficient and Recycled Concrete*; De Brito, J., Agrela, F., Eds.; Woodhead Publishing: Amsterdam, The Netherlads, 2019; Volume 1, pp. 1–22.
51. Gonzalez-Corominas, A.; Etxeberria, M. Properties of high performance concrete made with recycled fine ceramic and coarse mixed aggregates. *Constr. Build. Mater.* **2014**, *68*, 618–626. [CrossRef]
52. Nepomuceno, M.; Isidoro, R.A.; Catarino, J.P. Mechanical performance evaluation of concrete made with recycled ceramic coarse aggregates from industrial brick waste. *Constr. Build. Mater.* **2018**, *165*, 284–294. [CrossRef]
53. Mehta, P.D.P.K.; Paulo, P.D.; Monteiro, J.M. Proportioning Concrete Mixtures. In *Concrete: Microstructure, Properties, and Materials*, 4th ed.; McGraw Hill Professional: Toronto, ON, Canada, 2014.
54. Poon, C.; Shui, Z.; Lam, L.; Fok, H.; Kou, S. Influence of moisture states of natural and recycled aggregates on the slump and compressive strength of concrete. *Cem. Concr. Res.* **2004**, *34*, 31–36. [CrossRef]
55. Gomes, M.; De Brito, J.; Bravo, M. Mechanical Performance of Structural Concrete with the Incorporation of Coarse Recycled Concrete and Ceramic Aggregates. *J. Mater. Civ. Eng.* **2014**, *26*, 04014076. [CrossRef]
56. Lotfy, A.; Al-Fayez, M. Performance evaluation of structural concrete using controlled quality coarse and fine recycled concrete aggregate. *Cem. Concr. Compos.* **2015**, *61*, 36–43. [CrossRef]
57. Evangelista, L.; de Brito, J. Mechanical behaviour of concrete made with fine recycled concrete aggregates. *Cem. Concr. Compos.* **2007**, *29*, 397–401. [CrossRef]
58. Khatib, J. Properties of concrete incorporating fine recycled aggregate. *Cem. Concr. Res.* **2005**, *35*, 763–769. [CrossRef]
59. Nedeljković, M.; Visser, J.; Šavija, B.; Valcke, S.; Schlangen, E. Use of fine recycled concrete aggregates in concrete: A critical review. *J. Build. Eng.* **2021**, *38*, 102196. [CrossRef]
60. Medina, C.; Banfill, P.; de Rojas, M.I.S.; Frías, M. Rheological and calorimetric behaviour of cements blended with containing ceramic sanitary ware and construction/demolition waste. *Constr. Build. Mater.* **2012**, *40*, 822–831. [CrossRef]
61. Medina, C.; Del Bosque, I.F.S.; Asensio, E.; Frías, M.; De Rojas, M.I.S.; De Rojas, M.I.S. New additions for eco-efficient cement design. Impact on calorimetric behaviour and comparison of test methods. *Mater. Struct.* **2016**, *49*, 4595–4607. [CrossRef]
62. Asensio, E.; Medina, C.; Frías, M.; De Rojas, M.I.S. Characterization of Ceramic-Based Construction and Demolition Waste: Use as Pozzolan in Cements. *J. Am. Ceram. Soc.* **2016**, *99*, 4121–4127. [CrossRef]
63. Asensio, E.; Medina, C.; Frías, M.; De Rojas, M.I.S. Use of clay-based construction and demolition waste as additions in the design of new low and very low heat of hydration cements. *Mater. Struct.* **2018**, *51*, 101. [CrossRef]
64. Rao, M.C. Influence of brick dust, stone dust, and recycled fine aggregate on properties of natural and recycled aggregate concrete. *Struct. Concr.* **2020**, *22*, E105–E120. [CrossRef]
65. European Committe for Standardization. *EN 12390-3:2009. Testing Hardened Concrete. Part 3: Compressive Strengh of Test Specimens*; European Committee for Standardization: Brussels, Belgium, 2009.
66. Medina, C.; Sánchez, J.; Del Bosque, I.S.; Frías, M.; De Rojas, M.S. Meso-structural modelling in recycled aggregate concrete. In *New Trends in Eco-Efficient and Recycled Concrete*; Woodhead Publishing: Amsterdam, The Netherlands, 2019; pp. 453–476. [CrossRef]
67. Medina, C.; Sánchez, J.; del Bosque, I.S.; Frías, M.; de Rojas, M.S. Modeling the interfacial transition zone between recycled aggregates and industrial waste in cementitious matrix. In *Waste and Byproducts in Cement-Based Materials*; Woodhead Publishing: Amsterdam, The Netherlands, 2021; pp. 3–27. [CrossRef]
68. Ahmed, S. Properties of Concrete Containing Recycled Fine Aggregate and Fly Ash. *J. Solid Waste Technol. Manag.* **2014**, *40*, 70–78. [CrossRef]
69. Kirthika, S.K.; Singh, S.K.; Chourasia, A. Performance of Recycled Fine-Aggregate Concrete Using Novel Mix-Proportioning Method. *J. Mater. Civ. Eng.* **2020**, *32*, 04020216. [CrossRef]
70. Kou, S.-C.; Poon, C.S.; Wan, H.-W. Properties of concrete prepared with low-grade recycled aggregates. *Constr. Build. Mater.* **2012**, *36*, 881–889. [CrossRef]
71. Yang, J.; Du, Q.; Bao, Y. Concrete with recycled concrete aggregate and crushed clay bricks. *Constr. Build. Mater.* **2011**, *25*, 1935–1945. [CrossRef]
72. Kou, S.C.; Poon, C.S.; Chan, D. Influence of Fly Ash as Cement Replacement on the Properties of Recycled Aggregate Concrete. *J. Mater. Civ. Eng.* **2007**, *19*, 709–717. [CrossRef]
73. Elhakam, A.A.; Mohamed, A.E.; Awad, E. Influence of self-healing, mixing method and adding silica fume on mechanical properties of recycled aggregates concrete. *Constr. Build. Mater.* **2012**, *35*, 421–427. [CrossRef]

74. Wang, Y.; Zhang, H.; Geng, Y.; Wang, Q.; Zhang, S. Prediction of the elastic modulus and the splitting tensile strength of concrete incorporating both fine and coarse recycled aggregate. *Constr. Build. Mater.* **2019**, *215*, 332–346. [CrossRef]
75. Kumar, R.; Gurram, S.C.B.; Minocha, A.K. Influence of recycled fine aggregate on microstructure and hardened properties of concrete. *Mag. Concr. Res.* **2017**, *69*, 1288–1295. [CrossRef]
76. Bordy, A.; Younsi, A.; Aggoun, S.; Fiorio, B. Cement substitution by a recycled cement paste fine: Role of the residual anhydrous clinker. *Constr. Build. Mater.* **2017**, *132*, 1–8. [CrossRef]
77. Mas, B.; Cladera, A.; del Olmo, T.; Pitarch, F. Influence of the amount of mixed recycled aggregates on the properties of concrete for non-structural use. *Constr. Build. Mater.* **2012**, *27*, 612–622. [CrossRef]
78. Velardo, P.; del Bosque, I.S.; Matías, A.; de Rojas, M.S.; Medina, C. Properties of concretes bearing mixed recycled aggregate with polymer-modified surfaces. *J. Build. Eng.* **2021**, *38*, 102211. [CrossRef]

 *sustainability*

Article

Integrated Cycles for Urban Biomass as a Strategy to Promote a CO_2-Neutral Society—A Feasibility Study

Nicole Meinusch [1,*], Susanne Kramer [2], Oliver Körner [2], Jürgen Wiese [3], Ingolf Seick [3], Anita Beblek [4], Regine Berges [4], Bernhard Illenberger [5], Marco Illenberger [5], Jennifer Uebbing [6], Maximilian Wolf [1], Gunter Saake [7], Dirk Benndorf [1,8], Udo Reichl [1,9] and Robert Heyer [1,*]

[1]   Bioprocess Engineering, Otto-von-Guericke-University, Universitätsplatz 2, 39104 Magdeburg, Germany; maximilian.wolf@ovgu.de (M.W.); benndorf@mpi-magdeburg.mpg.de (D.B.); ureichl@mpi-magdeburg.mpg.de (U.R.)

[2]   Leibniz Institute of Vegetable and Ornamental Crops (IGZ), Theodor-Echtermeyer-Weg 1, 14979 Grossbeeren, Germany; kramer@igzev.de (S.K.); koerner@igzev.de (O.K.)

[3]   Urban Water Management/Wastewater, Hochschule Magdeburg-Stendal, Breitscheidstrasse 2, 39114 Magdeburg, Germany; juergen.wiese@h2.de (J.W.); ingolf.seick@h2.de (I.S.)

[4]   Agrathaer GmbH, Eberswalder Street 84, 15374 Müncheberg, Germany; anita.beblek@agrathaer.de (A.B.); regine.berges@agrathaer.de (R.B.)

[5]   Jassen GmbH, Egertenweg 10, 79585 Steinen, Germany; b-illenberger@t-online.de (B.I.); marco.illenberger@gmx.de (M.I.)

[6]   Process Systems Engineering, Max Planck Institute for Dynamics of Complex Technical Systems, Sandtorstraße 1, 39106 Magdeburg, Germany; uebbing@mpi-magdeburg.mpg.de

[7]   Technical and Business Information Systems, Otto-von-Guericke-University, Universitätsplatz 2, 39194 Magdeburg, Germany; saake@iti.cs.uni-magdeburg.de

[8]   Microbiology, Anhalt University of Applied Sciences, Bernburger Strasse 55, 06354 Köthen, Germany

[9]   Bioprocess Engineering, Max Planck Institute for Dynamics of Complex Technical Systems, Sandtorstraße 1, 39106 Magdeburg, Germany

*   Correspondence: nicole.meinusch@ovgu.de (N.M.); heyer@mpi-magdeburg.mpg.de (R.H.)

Citation: Meinusch, N.; Kramer, S.; Körner, O.; Wiese, J.; Seick, I.; Beblek, A.; Berges, R.; Illenberger, B.; Illenberger, M.; Uebbing, J.; et al. Integrated Cycles for Urban Biomass as a Strategy to Promote a CO_2-Neutral Society—A Feasibility Study. *Sustainability* **2021**, *13*, 9505. https://doi.org/10.3390/su13179505

Academic Editor: Carlos Morón Fernández

Received: 22 July 2021
Accepted: 20 August 2021
Published: 24 August 2021

Publisher's Note: MDPI stays neutral with regard to jurisdictional claims in published maps and institutional affiliations.

Abstract: The integration of closed biomass cycles into residential buildings enables efficient resource utilization and avoids the transport of biowaste. In our scenario called Integrated Cycles for Urban Biomass (ICU), biowaste is degraded on-site into biogas that is converted into heat and electricity. Nitrification processes upgrade the liquid fermentation residues to refined fertilizer, which can be used subsequently in house-internal gardens to produce fresh food for residents. Our research aims to assess the ICU scenario regarding produced amounts of biogas and food, saved CO_2 emissions and costs, and social–cultural aspects. Therefore, a model-based feasibility study was performed assuming a building with 100 residents. The calculations show that the ICU concept produces 21% of the annual power (electrical and heat) consumption from the accumulated biowaste and up to 7.6 t of the fresh mass of lettuce per year in a 70 m^2 professional hydroponic production area. Furthermore, it saves 6468 kg CO_2-equivalent (CO_2-eq) per year. While the ICU concept is technically feasible, it becomes economically feasible for large-scale implementations and higher food prices. Overall, this study demonstrates that the ICU implementation can be a worthwhile contribution towards a sustainable CO_2-neutral society and decrease the demand for agricultural land.

Keywords: integrated cycles for urban biomass; biogas; carbon footprint; sustainability; renewable energy; plant cultivation; feasibility study; simulations; CO_2-neutral society

1. Introduction

One of the most demanding challenges for the future is progressive global warming caused by excessive carbon dioxide (CO_2) emissions and other greenhouse gases. To stop global warming, our society must reduce the CO_2 emission and make our entire lifestyle CO_2-neutral. While many concepts for sustainable electrical energy production already exist, CO_2-neutral agriculture and biomass circulation concepts are lacking. Additionally, since half of the world population now lives in cities, these concepts have to be also applicable to urban areas. For example, in some urban districts (e.g., the Jenfelder Au in Hamburg, Germany [1]) black water is used on-site to produce heat and electricity by anaerobic digestion (AD). Furthermore, roof-top gardens enable the production of food in the cities [2]. Fuldauer et al. [3] demonstrated that it is even feasible to connect a small-scale anaerobic digestion plant (ADP) with a hydroponic or algae cultivation system to close the biomass cycle. The benefits of closed urban biomass cycles are an efficient utilization of the resources and the avoidance of transport [4].

Although there is plenty of research about single methods for biomass recovery (e.g., biogas plants) [5–8], fertilizer production from bio waste [9], or urbane farming [10,11], there is a lack of concepts for closed biomass circles and systematic feasibility evaluations.

Therefore, this study investigates the potentials and limitations of a concept for urban biomass circulation regarding energy and food production, carbon dioxide equivalent (CO_2-eq) savings, costs, and social–cultural aspects in Germany. The concept, called Integrated Cycles for Urban Biomass (ICU), demands an in-house ADP to degrade biowaste from residential buildings to biogas and digestate. The biogas generated is converted on-site to heat and electricity through a combined heat and power plant (CHP). The remaining fermenter liquid is upgraded by a soiling®-process (EP3684909A1 [12]) and a nitrification process to refined fertilizer. Finally, the liquid fertilizer is used to produce fruits, vegetables, and ornamental plants using either in-house integrated hydroponic systems, soil-based agriculture or roof-top gardens. Finally, the residents can consume the food while the accruing plant residues are fed into the ADP to close the biomass cycle again (Figure 1). Whereas soil-based agriculture is more robust, the use of hydroponic systems for the production of vegetables enables faster growth, higher product quality and needs less space [13].

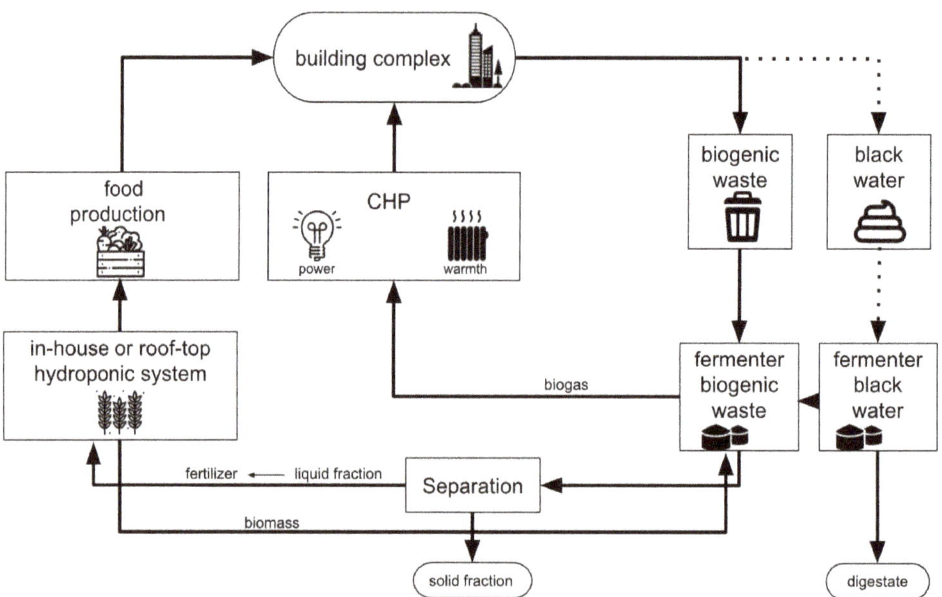

Figure 1. Process chart of ICU concept: Integrated Cycles for Urban Biomass (IUC) processes. ○: Input of biomass through the building complex into the process; outputs like solid fraction and digestate. "····": additional blackwater for biogas production.

A key challenge to close the biomass cycle between AD and agriculture is transforming the digestate into fertilizer. Digestates contain high amounts of ammonium (NH_4).

However, while NH_4 can be used as a nitrogen (N) source by plants, high NH_4 contents potentially increase N-losses by emission and can inhibit plant growth, especially in hydroponics. Therefore, NH_4 has to be oxidized via nitrite (NO_2) to nitrate (NO_3) (e.g., by the soiling®-process). In particular, for hydroponic-based crop production, fertilizer quality is of high importance [14] as, among others, its buffer capacity is very low (compared to soil). For hydroponics, under optimal conditions, synthetic or inorganic-based fertilizers are commonly applied [15–17]. With the adjustment of the correct dilution ratio and nutrient concentrations of the organic fertilizers, similar or even higher yields compared to a commercial nutrient solution are possible [18,19]. Finally, based on a Life Cycle Assessment (LCA), the reduction of CO_2-eq can be calculated [20].

2. Materials and Methods

To assess the ICU concept regarding energy and food production, the conversion of biowaste to heat and electricity using AD (Section 2.1) and agriculture (Sections 2.2 and 2.3) were simulated. CO_2-eq reductions as an indicator for the global warming potential of parts of the ICU process were evaluated by LCA (Section 2.4). In addition, the costs for the implementation of the ICU concept in real buildings were estimated (Section 2.5). Finally, social–cultural aspects of the implementation were reviewed for Germany (Section 3.5.5).

2.1. Modelling In-House Biowaste Degradation for Energy Production

Biogas production was used as a key process to model biowaste conversion to heat and electricity. The entire process was simulated using pre-implemented building blocks from the software SIMBA#Biogas (version 4.3) [21] (Figure 2) assuming a building with 100 residents.

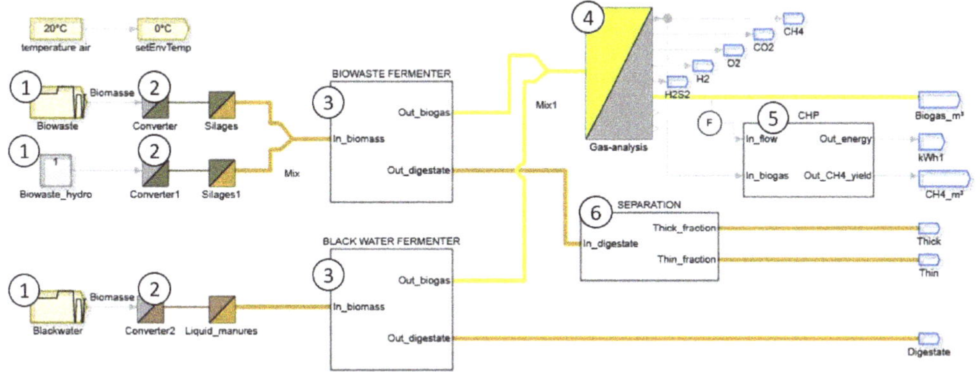

Figure 2. Simba model for in-house degradation of biowaste to heat and electricity. 1. Biomass input of biogenic waste and blackwater with 70 °C pre-treatment. 2. The Converter block determines the biomass composition. 3. The Fermenter block determines the biogas and digestate output. 4. The Gas analysis determines the biogas composition. 5. The CHP block converts biogas into energy with a calorific value of 10.5 kJ/kg 6. The Separation block determines thick (solid) and thin (liquid) fractions.

The "Biowaste block" (input) assumes homogenization at 70 °C to ensure the necessary sanitization conditions. Therefore, the biomass will be pretreated in a homogenization tank for 2–3 days before fermentation [22]. On average, residents are supposed to produce 33.6 kg of biogenic waste and 720 L of blackwater per day. Since the amount of biowaste fluctuates over the year, an increase of 22% for four months and a feeding of 30.7 kg for five days a week was applied to monitor the dynamic behavior ([23], Supplementary File 1). The "Converter block" takes into account the biomass composition based on literature values (Supplementary File 1, [23,24]). The "Biowaste Fermenter block" represents the conversion of kitchen and hydroponic biowaste into biogas and digestate. The simulations use the Anaerobic Digestion Model Number 1 da (ADM1da), which is an extension of the Anaerobic Digestion Model Number 1 (ADM1) [25]. The ADM1da comprises 32 differential and algebraic equations. They represent all relevant steps of the biomass degradation and physicochemical process parameters. Operation of the ADP at 55 °C was assumed. The "Gas analysis block" defines the biogas composition. The "CHP block" is used to determine the methane (CH_4) yield into electrical energy and heat. Here, an electrical efficiency of 38% and a thermal efficiency of 45% was used [26,27]. The electrical efficiency increases with the purity of the AD product gas [26]. The "Separation block" is used to split the digestate into a thin (liquid) and a thick (solid) fraction. The liquid effluent is further processed and nitrified by the soiling®-process into refined digestate. Soiling®-nutrient recycling fertilizer is composed of mineral N plus macro- and micronutrients. Note: only the N amount can be taken into account with Simba; for further calculations, the macro- and micronutrients were neglected.

For the "Blackwater block" (input) a second fermenter is considered as there is currently no approval for a fertilizer containing anthropogenic raw material in Germany according to the so-called Fertilizer Ordinance [28]. Therefore, blackwater fermentation is only considered for energy production but not for fertilizer production. Furthermore, the "Blackwater Fermenter block" is modeled by three different reactor block configurations to identify the most efficient one. The first scenario considers biogas production inside a continuously stirred-tank reactor (CSTR). The second scenario takes into account five CSTR blocks connected in series to simulate a plug flow bioreactor (PFR). The third scenario describes a two-stage reactor (2sR). Here, a small CSTR is used for hydrolysis and fermentation of biowaste, whereas acidogenesis and methanogenesis occur in a bigger second CSTR. The specific parameter settings for all scenarios are in the Supplementary File 2.

The ADP considered in this work was assumed to produce biogas with a calorific value of 10.5 kJ/kg Thus, neither combined cycle power plants (CCPP) nor CHP can be applied on-site because of their lower efficiency. In practice, many operators of small-scale biogas plants favor a satellite CHP over on-site power production. A satellite CHP is supplied with biogas from multiple small-scale biogas plants via a local micro gas grid [27]. The assumption is that a large number of small-scale biogas plants are nearby; an option that could also be applied here.

Suitable for the on-site power production of small-scale biogas plants are fuel cells, microturbines (MT), and engines (igniting beam engine, gas engine). Fuel cells can reach high electrical efficiencies and run quietly. However, they are comparatively large, expensive and their operation requires a high gas purity, which would make additional biogas upgrading necessary. Therefore, the use of a fuel cell was not considered in this study. MT, on the other hand, can operate with a wide range of CH_4 concentrations (30–100%) [27]. MTs reach an electrical efficiency of 25–33% with a thermal efficiency of ~49% [29], and operate silently and environmentally compatible [30]. MTs are commonly applied on a 30–550 kW scale [27,31]. Commercial 1 kW scale turbines are under development. However, at the current state, small-scale implementations are inefficient (11% electrical efficiency) at costs of around 6000 € per turbine [30,32]. Alternatively, engines (igniting beam engine, gas engine) can reach a higher electrical efficiency than MTs of 30–40% and a thermal efficiency of ~47%. Compared to MTs, engines are louder, produce noxious side products and require more maintenance. In particular igniting beam engines, which require the addition of pilot oil for combustion, produce noxious side products and soot, which inhibits the efficient use of excess heat [29]. The preferable alternative are gas engines, which operate without pilot oil, but require CH_4 concentrations above 45% [29].

In summary, gas engines were considered the most attractive option for on-site production of electrical energy from AD product gas in this study. However, if other small-scale biogas plants were available, a satellite CHP could be the more efficient alternative.

2.2. Crop Production Systems

The amount of ammonia nitrogen (NH_4-N), in the liquid effluent and the nitrification rate of the soiling process (European patent, EP3684909) was used to calculate the amounts of plant available N (NH_4-N and nitrate nitrogen (NO_3-N)) of one year. For cultivation planning (system sizing), the ratio of fresh biomass production to available N was considered. As a model, crop lettuce (*Lactuca sativa* ssp.) was used. A fresh matter N content of 0.18% was assumed (Feller et al., 2019) with a fixed dry matter fraction of 0.048%.

Four possible methods were considered for lettuce cultivation: Scenario 1 and 2 are open-air plant cultivation systems with raised beds or vertical hydroponics, respectively. The residents drive these scenarios on the roof-top with a cultivation period from April to October (vegetation period of Berlin). Both scenarios are complex as they involve the participation of community members (that are outside of the scope of the present simulations). Scenario 3 and 4 are protected cultivations with hydroponic greenhouses or plant factories, respectively. Both have to be operated year-round by trained staff and can be located on the roof-top or in the basement of buildings. Here, a pure bio-technical assessment using deterministic explanatory simulation models was applied.

A numerical simulator for controlled environments and greenhouses was used that is a further development of earlier published greenhouse simulators [33,34]. The simulator was programmed using MATLAB (MathWorks Inc., Natick, MA, USA). It was connected to a replica of commercially available climate controllers, including a setpoint generator that calculated climate setpoints for heating, ventilation, light and CO_2 concentration. The simulator was fitted to a standard Venlo-type greenhouse structure or a vertical farming hydroponics-controlled environment (scenario 3 and 4). The simulator's crop-basis is a photosynthesis-driven growth model with microclimate predictions for water and nutrient uptake according to the Penman–Monteith equation [35]. Nutrient uptake was calculated assuming that the diluted nutrients in the irrigation system are optimally taken up by the crop. As such, a perfect pH, electrical conductivity (EC), and a root environment with optimal nutrient solution composition with an optimal availability of all nutrients were assumed. In accordance with Goddek and Körner (2019) [36], all element-specific chemical, biological or physical resistances were set to zero.

For technical layout, supplementary lighting was applied with LED lamps installed either under the roof above the crop with an installed capacity of 80 W m^{-2} power and an output of 192 µmol m^{-2} s^{-1} or at an installed capacity of 110 W m^{-2} power and an output of 264 µmol m^{-2} s^{-1} in scenario 3 or 4, respectively. The light was controlled dynamically with setpoints generated using a daily light integral (DLI) of either 12 mol m^{-2} d^{-1} or 20 mol m^{-2} d^{-1} for greenhouse or vertical farming, respectively [37]. In both scenarios, CO_2 in the air was set to 700 µmol mol^{-1} and supplied according to the demand (max. at 15 g m^2 h^{-1}) during lightening when greenhouse vents were closed and at all times in the vertical farming scenario. In the greenhouse scenarios, heat exchange for cooling was calculated with passive roof ventilation while active cooling and dehumidification were used in the vertical farming-controlled environment scenarios (active cooler based on ANSI/AHRI standards 1200). Dehumidification was implemented with a commercially available dehumidification unit of the type ventilated latent heat energy converter. Further model parameters are summarized in supplementary file 3.

The simulator calculated macro- and microclimate in a time-step of 5 min, integrated hourly using controlled actuators (e.g., heating, ventilation, cooling, CO_2, light) that were re-adjusted as described by Körner and Van Straten (2008) [34]. The simulations' output included hourly biological and physical variables related to lettuce production, such as microclimate conditions, photosynthesis, yield, and resource consumption (electrical power, heating energy, water, CO_2). Input to the simulation program included, among others, physical location (latitude (LAT), longitude (LON)), humidity set point (%), set points for heating and ventilation (°C), crop planting density (plants m^{-2}) and temperature-sum related harvest time. Input climate data were hourly data sets for Berlin (Germany,

LAT 52.5N, LON 13.4) from 2009 to 2018 (Meteoblue; [38]). Calculations were performed for all scenarios for single years of each of the 10-year horizons.

Simulations were performed targeting nutrient and water uptake, yield, and energy demand for heat and lighting for either a greenhouse with a size of 70 m^2, or for a vertical farm with a four-layer system (17.5 m^2 each, in a room of 30 m^2 area and 2.50 m height). As commercially viable climate control in small greenhouses is challenging to maintain, a 500 m^2 greenhouse as minimum commercial size was modeled in addition. All simulations were done for year-round production of hydroponically grown lettuce with a fixed planting density of 36 plants m^{-2} as in commercial practice (e.g., Brechner et al., 2013 [39]).

2.3. Estimation of the CO_2 Saving Potential Using Life Cycle Assessment with openLCA

The ICU concept offers the opportunity to save CO_2 due to reduced transport of biowaste and food [40]. To quantify the amount of saved CO_2, a life cycle assessment (LCA) with the open-source software openLCA (version 1.10.3) was conducted. This software considers the total energy consumption by all components at various levels. The used database was ecoinvent35_Cut [41]. The system environment was divided into five phases: extraction of raw materials and energy sources, manufacture, use, transport and disposal (Figure 3A) [42]. The boundaries for the ecological assessment are shown in Figure 3A (grey dots). An average distance of 30 km for the transport of biowaste to the ADP in the conventional scenario was assumed. In the ICU concept, transport was neglected (Figure 3B). All flows and process data are found in supplementary file 4. The system contains specific elements providing the functional unit of 1 kg biomass for the complete life cycle. The input flow is 11 t with bio-degradable garden and park waste, food and kitchen waste from households, restaurants, caterers and retail stores, comparable waste from food processing plants, as well as forestry or agricultural residues, and manure. It does not contain sewage sludge or other biodegradable waste such as natural textiles, paper or processed wood.

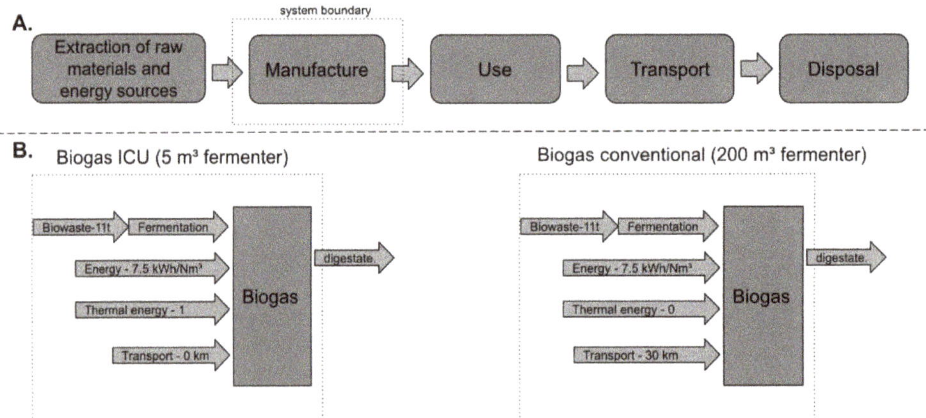

Figure 3. System boundaries of the life cycle assessment study. (**A**): System environments of the product life-cycles based on DIN 15804 in four steps and system boundary (functional unit 1 m^3 biogas). (**B**): System boundary for the ICU concept in comparison to a building with transport of biowaste to a conventional biogas plant. (Energy: Available online: https://biogas.fnr.de/daten-und-fakten/faustzahlen (accessed on 22 July 2020)).

2.4. Cost Calculation

The costs of implementing the ICU concept were estimated by a life cycle cost analysis (LCCA, Equation (1)) for the example of a building with 100 residents. Therefore, the costs for acquiring and operating the fermenters, the soiling, the hydroponic systems, the technical staff and the building's space were considered taking into account the net present value (*NPV*).

$$NPV = C + R - S + A + M + E \tag{1}$$

C = investment costs;
R = replacement costs;
S = resale value at the end of study period;
M = non-annually recurring operating, maintenance and repair cost;
E = energy costs.

Investment costs (C) were depreciated for a period of 20 years. Replacement costs were assumed with 10% of the investment costs after 20 years and maintenance costs ($A + M$) with annually 5% of the investment costs. Energy costs (E) were omitted since the system produces the required energy on its own. Additionally, the resale value (R) of the installations was assumed as "0" € as it was expected that the building's value remains at least stable. Additionally, ADP operation and plant cultivation require an experienced worker requiring at least 25 € per hour.

The production of in-house biogas generates energy in form of electricity and heat. This energy is reused inside the ICU building. If the generated energy were sold, the price for 1 kWh heat and 1 kWh electricity would be 0.024 € [43] and 0.1 € (§43 EEG 2017), respectively. A summary of results obtained is found in Table 1.

Table 1. Model output per m^2 ground area and year.

Scenario	Climate Control Energy [GJ m^{-2}]	Illumination Energy [GJ m^{-2}]	Water: Evaporation and Crop Binding [L m^{-2}]	Fresh Lettuce Yield [kg m^{-2}]
S3: Roof-top greenhouse (70 m^2 ground)	2.0	0.4	1805	91
S3 *: Roof-top greenhouse (500 m^2 ground *)	1.0	0.4	1786	90
S4: Basement vertical farming (30 m^2 ground; 70 m^2 cultivation area *)	5.2	6.7	1396	259

* 500 m^2 greenhouse is for industry the smallest size for implementation.

Against these costs, the value of the produced energy and food was considered. The benefit of the reduction in the disposal of biowaste and wastewater was neglected to avoid further complication of the calculation. Furthermore, the installation of vacuum toilets and separate black and grey water tubes is also cost-intensive. However, the cost of the installations compensates with the benefit of a reduced wastewater volume.

2.5. Overview of Important Social-Cultural Aspects Required for the Implementation

To assess social-cultural aspects for the implementation of the ICU concept in Germany, a literature survey was performed addressing the following questions:

- How great is the interest of the residents in urban agriculture and sustainable lifestyles?
- How great is the willingness of real estate owners to implement an ICU concept?
- How important is it for the government to achieve a carbon-neutral society?
- Which legal paragraphs have to be considered for implementing an ICU concept?
- Which additional social-cultural aspects might be relevant for implementing an ICU concept?

While for most of these questions results and data from literature already exist, the question on the real estate owners' willingness to implement an ICU concept has not been addressed, so far. Therefore, an online survey to collect this data was performed. To obtain a comprehensive picture of the attitude towards this new concept, 235 real estate owners, about 15 from every Federal State, were selected. All owners received a short online questionnaire containing ten questions (Supplementary File 5) to rate to which extend different aspects of the ICU concept and its implementation are important to them. In the end, only 14 answers were received.

3. Results and Discussion

3.1. Structure of Section 3

This feasibility study evaluates the amount of energy (Section 3.1) and food (Section 3.3), which could be produced by implementing an ICU concept for a building with 100 residents. Before the final simulation, the ADPs' fermenter size and configuration were evaluated and the best scenario for house-internal food production was selected. A precondition for plant growth was converting NH_4 in the digestate to NO_3 (Section 3.2).

Based on the ICU-concept's best scenario, the costs were calculated (Section 3.4) and the potential CO_2-savings (Section 3.5). Finally, social–cultural aspects were reviewed, including the laws required for implementing the concepts (Section 3.6) and potential addons for the ICU-concept (Section 3.7).

3.1.1. Utilization of Biowaste by Optimized Anaerobic Fermenters Enable to Cover 21% of the Annual Energy Demand of the Building

As the first step, the size and performance of CSTR, PFR, and 2sR ADP for processing of biowaste (Figure 4) and black water (Figure 5) were compared based on the energy content of the biogas. The volume ratio between the hydrolysis and the main fermenter of the 2sR was 1:50 as determined in the supplementary Table S6. For the PFR the sum of all five fermenters connected in series was assumed for the simulation.

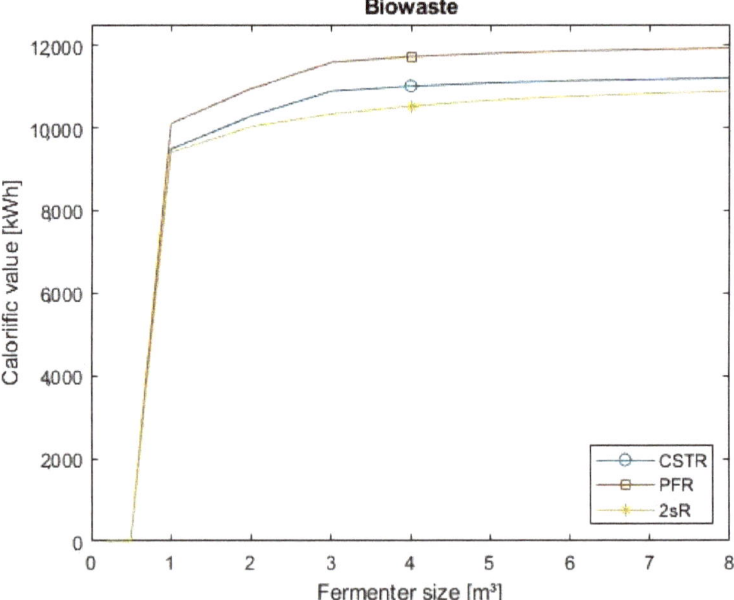

Figure 4. Evaluation of reactor scenarios with biowaste fermenters. Energy content of the biogas produced annually (without losses). Only biowaste input. For PFR reactor the sum of all 5 fermenters and for 2sR is hydrolyse + main fermenter are considered. Marker shows the chosen reactor size.

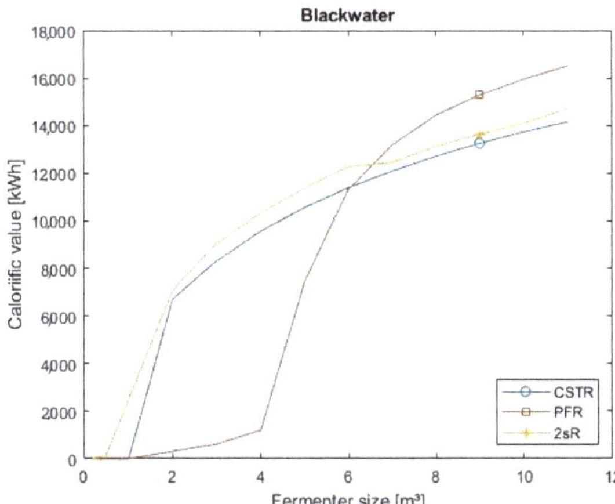

Figure 5. Evaluation of reactor scenarios with additional blackwater fermenter. Energy content of the biogas produced annually (without losses). Blackwater input. For PFR reactor the sum of all 5 fermenters and for 2sR is hydrolyse + main fermenter are considered. Marker shows the chosen reactor size.

The first scenario was the calculation of biowaste input in one fermenter with an average amount of 33.6 kg biowaste per day (Figure 4). Depending on the reactor size 1 kW energy may be produced. The second scenario calculates the additional black water fermentation in a second fermenter with an average of 720 L black water per day (Figure 5). Black water may improve the daily energy yield to 3 kW energy, fitting to the values of studies with similar substrates [44].

For the first scenario, the simulation of PFR produces about 5.5% more energy than the two-step and 22% more than the CSTR fermenter. These magnitudes between the fermenter types were also shown by Bensmann et al. [45,46]. Additionally, a PFR is more robust against contaminants like plastic material in biowaste. The shape of the power to fermenter size curve is sigmoid, reflecting that too small fermenter sizes lead to acidification, whereas too large fermenter adds no further benefit (Figure 4). As optimal biowaste fermenter sizes were chosen a 4 m^3 CSTR-fermenter, five fermenters connected in series with each 1 m^3 for the PFR-fermenter and also 4 m^3 for the main fermenter of the 2sR (Figure 4). The PFR-fermenter was selected as optimal because with 11.434 kWh calorific value annually it was able to produce the most energy. This amount of energy corresponds to 9.5% of the annual energy demand of 100 persons [47]. Since the reactors with their control units require less than 20 m^2, installation in the technical center of a building is technically feasible. Production of heat and electricity would require an additional CHP unit of about 10 m^2 size. Alternatively, the biogas can be used to cook and climatize the building. This scenario requires that the building have gas heating/heating pumps instead of an oil or electric system. Alternatively, the biogas could also be upgraded to biomethane and fed into the local gas grid [48,49].

In comparison the second scenario with an additional black water produce 25,855 kWh energy. This scenario is ecological more efficient to the first one because it can cover 21% of the yearly energy production (Figure 5). As optimal reactor size for the CSTR a 9 m^3 fermenter, for PFR five fermenters connected in series with each 1.8 m^3, and for 2sR 9 m^3 were chosen.

Due to the higher energy content, the additional fermentation of black water should be considered for the ICU-concept. For black water usage, the separation of grey and black water is needed. This required a two-pipe system and separation or vacuum toilets (e.g., Jenfelder Au, [50]). The implementation is technically demanding, but allows the reuse of greywater, which would further reduce water consumption.

3.1.2. Dynamic Behavior of the Anaerobic Digester

For the first simulation a constant supply of biowaste and black water with the chance that fluctuations occur was assumed. To assess the dynamic behavior of the system, after initial conditions (i.), a shift in the feeding rate from (ii.) constant 33.6 kg/d to (iii.) an increase of biomass of 18% for four months (iii.) to (iv.) a feeding rate of 30.7 kg/d for only five days a week was considered (Figure 6). The simulation of all three fermenter types (2sR, CSTR and PFR) shows smooth transitions for the different changes in the feeding strategy indicating an overall stable process behavior. This suggests that the ICU concept could be easily integrated into buildings. However, systems for handling fluctuations in gas production such as gas storage tanks or gas torches should be considered in case of technical problems (data not shown). Furthermore, it is known that ICU systems are prone to long periods of biomass overloading [45] and require a relatively long period for the start-up (Figure 6).

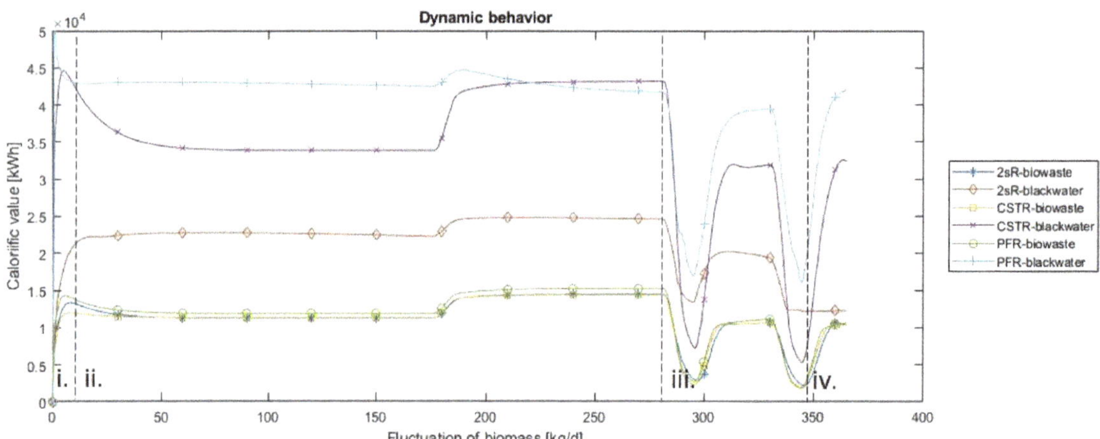

Figure 6. Dynamic behavior of the ICU model. Feed fluctuations: Energy production from biowaste (green) and additional black water (BW) (brown) for 2sR, CSTR and PFR fermenters. (**i.**): Initiation phase of the model. (**ii.**) Feeding rate of 33.6 kg/d (**iii.**) Feeding rate of 41.1kg/d (**iv.**) Feeding rate of 30.7 kg/d for five days a week.

3.1.3. The Soiling®-Process Represents an Efficient Approach to Convert Digestate to Fertilizer

Digestates contain high amounts of NH_4-N, but not NO_3-N required for plant growth in soilless cultivations. NH_4 conversion to NO_3 can be achieved by composting, by the soil microbiome and by nitrification fermenters. One efficient nitrification fermenter is the Soiling®-module from Jassen Kunststoffzentrum GmbH (EP 3684909A1 [12]). Verification of Jassen GmbH shows that before the soiling®-treatment, the NH_4 amount was 200 mg/L. After 9 days of aeration with 25 m^3/d oxygen, the NH_3 amount decreased to 15 mg/L. This means 92.5% of the NH_4-N is nitrified into NO_3-N (calculation in Supplementary File 9). This is superior to other nitrification systems, like the nitrification system of Wang et al., 2017, which has an efficiency of about 87.2%. Therefore, the soiling®-system was used for all further calculations. In addition, the project partner Jassen Kunststoffzentrum GmbH could provide exact numbers for the conversion of the digestate taken from ADP to the fertilizer. Here, a digestate yield of 2.63 L fertilizer per day containing 0.00594 kg/NO_3-N per liter. In total, this requires 535.7 kJ energy for each kg N. For comparison: the production of fertilizer using artificial nitrogen fixation process (Haber–Bosch process) requires already 10.800 kJ per kg N for the production of NH_4, which has still to be converted to nitrate. A precondition for the nitrification step is a separation of the liquid and solid components of the digestate. This separation could be achieved by a screw press that is easy to handle and has a low energy consumption (about 0.5 kJ). The remaining solid fraction could also be used to upgraded by composting, but this was not further considered here. To estimate how much N can be produced as fertilizer from a hydroponic system, the N flow according to the Simba model was considered (Figure 7).

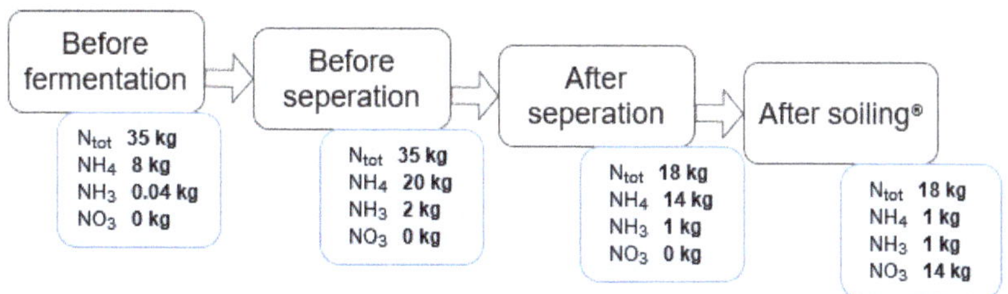

Figure 7. N flow according to the Simba model. N_{tot} = total nitrogen, NH_4 = ammonium, NH_3 = ammonia, NO_3 = nitrate. Quantities relate to one year.

3.2. Strategies and Production Potentials of Integrated Plant Growth into Building

3.2.1. Strategies to Integrate Plant Growth into Buildings

Our feasibility study considered four scenarios for house-internal gardens (plant production) and compared them regarding productivity, nutrient utilization, energy demand, required skills and social inclusion. For scenario 1 and 2, crop is produced in open roof-top gardens, while protected cultivation in greenhouses or plant factories, also called vertical farming [51], is assumed for scenario 3 and 4. Thus, the scenarios increase from low level to high-level control from scenario 1 to scenario 4 (Figure 8). Plant factories allow to cultivate crops in multiple layers with a high productivity and uniformity [52]. These systems are completely isolated from the exterior climate with the control of light, temperature, relative humidity and CO_2 concentration [51]. Especially by controlling the light quality, the yield and the nutritional value of lettuce can be increased [53]. However, with increased control from scenario 1 to scenario 4, energy consumption increases through the supply of mechanical heat in greenhouses [54] and electric light in plant factories [52,55].

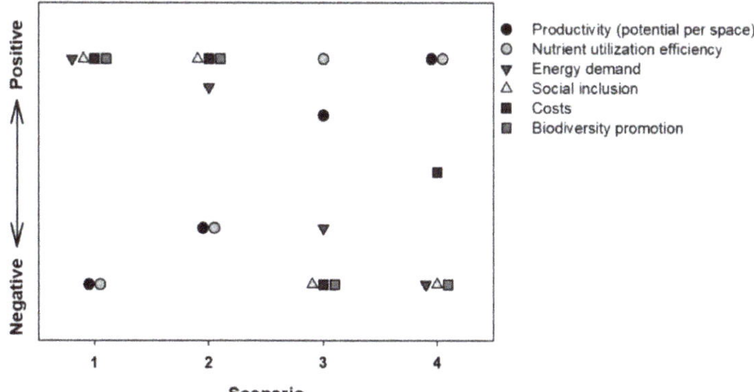

Figure 8. Relative assessment of the cultivation scenarios. Scenario 1: roof-top raised-bed, community residents; Scenario 2: roof-top vertical hydroponic system, community residents; Scenario 3: roof-top greenhouse, professional management; Scenario 4: basement vertical farm, professional management.

In scenario 1, plants are cultivated in raised-beds on the roof-top and maintained by community residents. In this scenario, the residents have a high ecological "feelgood factor". Maintenance requires only low skills, and the solid fraction of the composted digestate could be utilized, too. While food production is less efficient, a high crop diversity can be attained. With scenario 2, the production efficiency (per m^2) can be increased. Here, the residents cultivate the plants in vertical hydroponic tiers. Vertical tiers are more space-efficient, while through optimal nutrient availability, hydroponic cultivation systems allow a faster and higher crop production [56]. For example, Li et al. (2018) [57] realized in hydroponic systems about twice as much shoot fresh weight of two lettuce cultivars than for the cultivation in culture substrate. However, hydroponic systems require more experience and regular control of the composition of the nutrient solution [58].

Better suited are scenario 3 and 4 because of the controlled or semi-controlled environment as in plant factories or greenhouses. In addition, increased dynamics of the crop water and nutrient demand like in scenario 2 create unfavorable situations with the necessity to discard parts of the nutrient solution [59].

In scenario 3, the crop is cultivated in a roof-top greenhouse using hydroponics. This scenario enables very high year-round production, but with a higher demand for heating and lighting (see Table 1). The energy consumption is, in particular between autumn and spring, very high. Furthermore, a certain dynamic in water and nutrient demand, as well as in yield throughout the year, is still present. In scenario 4, this dynamic is more or less completely eliminated. Here, the plant factory is installed in the basement or rooms without natural light and the climate is fully controlled [60]. Due to the cultivation of crops in layers, even less space than for the roof-top greenhouse is needed. However, since there is no natural light, high amounts of artificial light for plant growth must be provided, and the exhaust heat needs to be removed. Thus, scenario 4 is, despite its higher productivity, a better product quality as well as a stable and constant harvest, the most energy demanding. In addition, it requires maintenance and resources all year round. As the community provides a year-round output in biowaste, continuous use of it can be best achieved with scenario 4. Thus, a decision must be made between the installation of a buffer system for seasonal provision of nutrient-rich irrigation water or a high energy demand. Next to that, non-technical issues also need to be taken into consideration, e.g., the roof-top floor is highly desired by residents and the most expensive floor in the building. In contrast, skyscrapers contain inside rooms without daylight, which must not be used for apartments or offices, at least in Germany.

3.2.2. Production Potentials and Energy Demand of Simulated Scenarios

To evaluate the potential of in-house food production, crop cultivation in protected environments with semi-closed and closed systems, i.e., greenhouses in scenario 3 and plant factories in scenario 4, a model-based simulator tuned to the respective cases was developed. Scenarios 1 and 2 were neglected since their outcome varies, among other factors, highly on the residents' skills and motivation, which complicates simulations regarding pure bio-technological scenarios significantly. As one of the most common crops in plant factories and greenhouses, lettuce was chosen as a model crop because it is relatively easy to handle during cultivation [61], and it is suitable to produce lettuce even with wastewater [62].

Based on the N content in the liquid fraction from the biowaste digestate, it was calculated that the production of 7.6 t fresh mass of lettuce per year would be theoretically possible. A challenge for implementing closed hydroponic systems in the ICU concept is the open question after what time period the fertilizer has to be replaced due to the accumulation of salts (mainly natrium and other deleterious substances. However, in comparison to open systems (the drained nutrient solution is discarded), closed re-circulating hydroponic systems reduce water and fertilizer by about 30% and 50%, respectively [63,64].

Using the ICU concept, uncertainties in the dynamic of crop water demand in scenario 3 (through seasonal fluctuations) could result in an unwanted discharge of nutrient solution. This can partly be solved by increasing the area of the cultivation system. Therefore, the area for the greenhouse in scenario 3 was with 70 m^2 partly oversized. Due to that, all available N will be taken up by the crop. This dynamic is not relevant in scenario 4 and thus a perfect sizing of the plant factory is possible in this case. Accordingly, a total annual yield of 6.4 t or 7.3 t fresh lettuce (i.e., 91 kg m^{-2} or 242 kg m^{-2}; Table 1) was predicted for the greenhouse (scenario 3) and the plant factory (scenario 4), respectively. These values correspond to the yields reported in other studies [36,65]. In these scenarios, a weekly harvest of roughly 760 or 1377 lettuce heads in a 70 m^2 greenhouse or in a plant factory can be expected. Since 1 kg of lettuce contains about 150 calories and, assuming a person requires about 2000 calories a day without burden, this would nourish about one people for one year. However, while human nutrition is not covered solely by lettuce, in a real implementation of innovative urban agriculture practitioners a broad mix of vegetables and herbs with various nutritious values should be considered in follow-up studies [66] (Table 2).

Table 2. Potential biomass production per year for different vegetable crops based on the total available N amount in the liquid effluent and respective calories.

Crop	Potential Fresh Mass [kg]	Calories [kcal/100 g]
Broccoli	3050	35 *
Bush bean	5490	25 *
Brussel sprouts	2771	43 *
Lettuce	7625	11 **
Spinach	3431	23 *

* Footspring magazine—Kalorientabelle [67]. ** wikifit—Kalorientabelle, Gemüse, Salat [68].

On the downside: The roof-top greenhouse in scenario 3 (m^2 ground) requires an annual amount of ~143 GJ (39.7 MWh) energy for heating and 25 GJ (7.1 MWh) electrical power for light. While the energy demand is highest during autumn and winter, the yield is highest in summer. Thus, the efficiency of energy use strongly drops in the cold season. Using a cultivation period of March to October requires only about 44% of the energy for heat and 10% for light. The latter, however, would require an additional storage solution for crops during that period. Furthermore, power consumption for greenhouse lighting depends on many parameters, e.g., geographical location, greenhouse cover light transmission, light source, crop, set points. Comparable results for power demand in greenhouses as in scenario 3 were reported earlier [37,69–71].

The energy demands for the plant factory case in scenario 4 are even higher. This corresponds to an annual electrical power demand of 201 GJ (56.1 MWh) per year. As each plant factory is unique and large differences exist between implementations, e.g., the numbers of layers, layer size, empty spacing (room use efficiency), light source, light set point, etc., the power consumption needs to be compared taking into account the net production area and the net installed and used light capacity. Based on our assumptions, the results obtained are in the same range as reported earlier [72].

The lettuce crop produced in scenario 4 using a full climate-controlled plant factory consumes 95% of the fertilizer of the ICU system when a (nearly) closed system is attained. About 36 m^3 water would be additionally needed if the evaporation water is not reused.

As both scenario 3 and 4 have a high energy demand, these solutions are not the best concerning the CO_2 footprint. However, to make controlled hydroponic systems more sustainable, regenerative energy could be used more intensively in this culture system, as it is already done for some urban farms [66]. Moreover, due to further advantages, like reduced water consumption and space requirements in comparison to a field cultivation, less transport and utilization of non-used spaces in the city, reliable and stable resource demand and crop yield, higher quality and food safety, there is a great potential for controlled and semi-controlled crop production in an urban scenario. Additionally, local food production is highly advantageous. In particular, it represents an alternative for highly populated mega cities lacking space or for regions lacking agricultural areas.

To stress the full potential of the ICU concept, further extrapolations were performed. The complete fertilizer produced from the biowaste (N_{tot} in liquid and solid fraction neglecting possible N-losses during further processing like composting and/or immobilization processes) allows for the cultivation of about 19.5 t lettuce. Interestingly, the 720 L of black water produced by the residents already contain 183.95 kg NH_4, which could be converted to 170.15 kg NO_3. This amount per year could theoretically allow us to produce 1.1 t/m^2 lettuce in a roof-top greenhouse, and clearly demonstrates the potential of reusing nutrients from the digestate of biowaste and black water to produce food and nourish urban populations.

3.3. Implementation of a ICU Concept for a Building with 100 Residents Saves up to 6468 kg CO$_2$-eq

Implementation of the ICU project for a building with 100 residents reduced CO_2-emission due to reduced transport of 11 tons biowaste by 693 kg CO_2. The reuse of NH_4 as fertilizer saved 2363 kg CO_2 compared to the new synthesis by the Haber–Bosch process. Usage of the produced 25,855 m^3 biogas for heating saved 3412 kg CO_2 and is comparable to the CO_2 fingerprint for heating reported in literature [73]. In total, the implementation of the ICU concept can save 6468 kg CO_2-eq. Based on a CO_2 emission price of 25€ per ton CO_2, the CO_2 saving value is currently 161.7€ (BMU, 2021 [74]) (supplementary Figure S9). Additionally, our concept has the potential to save further CO_2. For example, in the case of black water usage, less CO_2 is emitted during wastewater treatment [75]. However, further studies are required to accurately incorporate then the CO_2 emission for black water tubes and the wastewater treatment plant.

3.4. ICU Concept Becomes Economically Feasible in Large Buildings and with Growing Food Prices

Necessary for the implementation of the ICU concept is its profitability. Although the numbers for a specific implementation will differ, projection of the costs and benefits enables identifying scaling effects and targets for further optimization. To estimate the economic feasibility, yearly costs and yields of the ICU concept were estimated. Operation of an AD for biogenic waste (Figure 9, Biowaste Anaerobic Digester) costs 9535–9948 € and yields 1615–1727 € annually. In contrast, operation of an AD for blackwater utilization costs 8594–8758 € and yields 2917–3606 € annually. In consequence, the in-house use of biowaste and black water is not profitable for small buildings (Salerno et al., 2017). For large buildings, however, personal costs remain more or less the same and investment costs for larger fermenters rise only slightly. Therefore, implementation is economically feasible for large buildings or agglomeration of several buildings, favoring the ICU implementation in large cities or at the district level.

Generally, black water utilization is economically more promising than biowaste usage under the premise that the saved cost for the wastewater removal compensates for the black water system's cost. Combined usage of black water and biowaste would create the synergy that personal costs for the daily lookup and the co-generation unit can be shared.

The annual cost for a 70 m^2 greenhouse on the roof-top was 69,000 € (scenario 3), and in the basement 70,000 € (scenario 4). The annual benefit of both solutions is 11,418–14,737 €, based on a lettuce price of 0.87 €. Extrapolation showed that the system is profitable with lettuce prices of 1.80 €. This rise is possible when the agricultural space vanishes further, and the population grows. For example, in Singapore, the lettuce price is already 1.00–2.50 €. In contrast to the AD, the hydroponics' economic efficiency grows only slightly with larger systems since it scales quite well.

Whether the hydroponic should be integrated on the roof-top, in the basement, or in rooms without light depends heavily on the price. The top floor's rental price is roughly 30% more expensive. Therefore, it could bring more profit to use the top floor as a penthouse. In a prestigious skyscraper with around 100 m^2 of living space, a six-digit amount in major German cities is possible.

Description	Size/amount	Investment costs [€]	Yearly costs [€]	Comments
Biowaste Anaerobic Digester				
Fermenter (CSTR, 2sR or PFR)	5 m³	-20,000 to -25,000[1]	-2,100 to 2,625	
Co-generation unit	2 m²	-10,000	-1,050	
Technicum mini plant	15m² (2 000 (€/m²)	-30,000	-1,500	
Employee for AD	5 h/week á 25 Euro/h	---	-6,500	
Biogas (used for electricity)	11,294 - 12,080 kWh	----	1,615-1,727	refer to biogas price (14.3 cent/kWh)[2]
		sum	-9,535 to -9,948	

Description	Size/amount	Investment costs [€]	Yearly costs [€]	Comments
Blackwater Anaerobic Digester				
Fermenter (CSTR, 2sR or PFR)	7 m³	-25,000 to -30,000[1]	-2,625 to -3,150	
Co-generation unit	2 m²	-10,000	-1,050	
Technicum mini plant	20m² (2 000 (€/m²)	-30,000	-1,500	
Employee for AD	5 h/week á 25 Euro/h	---	-6,500	
Biogas (used for heating)	20,400 - 25,220 kWh	----	2,917-3,606	refer to biogas price (14.3 cent/kWh)[2]
		sum	-8,758 to -8,594	

Description	Size/amount	Investment costs [€]	Yearly costs [€]	Comments
Garden unit scenario 3				
roof-top greenhouse Greenhouse Lightning	70 m²	-140,000 -13,986	-14,700 -1,468	14 lights × 999€
Hydroponic module	70 m²	-9,200	-966	
Soiling®NRF module	1 × 300 L	-70,000	-7,350	refer to Jassen GmbH[5]
Soiling®NRF solid separator	1	-35,000	-3,675	
Building-costs	One floor (70 m² á 2 000€/m²)	-140,000	-14,700	-
Employee for garden unit	20 h/week á 25 Euro/h	---	-26,000	900 h per year[6]
Produced food (S3)	39,000 lettuce (0.87€ each)	---	34	
		sum	-34,899	

Description	Size/amount	Investment costs [€]	Yearly costs [€]	Comments
Garden unit scenario 4				
basement vertical farm Vertical farming Lightning	70 m²	-150,000 -6,540	-15,750 -686	60 lights × 109€
Hydroponic module	70 m²	-9,200	-966	
Soiling®NRF module	1 x 300 L	-70,000	-7,350	refer to Jassen GmbH[5]
Soiling®NRF solid separator	1	-35,000	-3,675	
Building-costs	One floor (70 m² á 2 000€/m²)	-140,000	-14,700	-
Employee for garden unit	20 h/week á 25 Euro/h	---	-26,000	900 h per year[6]
Produced food (S4)	70,000 lettuce (0.87€ each)	---	61	

Figure 9. Summary of cost and yields. Yields are considered from the point of the residents using Table S3. = scenario 3, S4 = scenario 4. 1. pers. communication AAT Abwasser- und Abfalltechnik GmbH, Konrad-Doppelmayr-Str. 17, 6960 Wolfurt, Austria. 2. Regulations for the expansion of renewable energies (Renewable Energies Act—EEG 2021) §43 Fermentation of biowaste (2017). 3. pers. Communication Mr. Huber, GEFOMA GmbH, Germany. 4. pers. Communication Mr. Grantham, Heliospectra AB, Sweden [76]. 5. Waste transport Magdeburg [77]. 6. Jassen Kunststoffzentrum GmbH, Germany [12]. 7. Average price of 4 Discounter 0.87 €.

3.5. High Motivation of Stakeholders for the Implementation of an ICU Concept but High Legal Barriers

3.5.1. High Motivation of Residents for a Sustainable Lifestyle

Current studies show that in Germany, a sustainable lifestyle becomes more and more important [78]. Today, 657 urban gardening projects exist in Germany [79]. However, the engagement of residents in urbane gardening might decline and require strategies to counteract. In contrast, a professional management of the urban gardening projects is more reliable, but often reduces the acceptance of residents [80]. A compromise could be combined models with hydroponic modules for the residents or a botanic garden with a cafe alongside professionally managed greenhouses.

3.5.2. Real Estate Owners would Implement the ICU Project as Long as It Is Profitable

Almost all real estate owners participating in this survey considered sustainable building as important for their business. Nevertheless, closed material cycles or exploitation of options related to the operation of AD or plant cultivation seemed less important for most of them. Some may even lose their tax benefits when they engage in another business field like urbane agriculture or operation of an AD. Therefore, it seems beneficial to outsource the operation of an ICU project to a contracting partner or a cooperative of the residents. Necessary conditions for implementing ICU concepts for real estate owners are high acceptance of the residents, lower maintenance requirements and profitability. Typically, there is no interest in pure flagship projects. Furthermore, especially for the roof-top use of buildings, ICU projects compete with other more established sustainable solutions such as the operation of photovoltaic systems (Supplementary File 8).

3.5.3. In the Framework of the Paris Agreements, Government Promote a CO_2-Society

In the Paris Agreement, 195 states including the European Union agreed to limit global warming below 2 °C [81]. Therefore, netto CO_2 emission has to be reduced to zero by 2040. In addition to the targets for the energy and building sector, new goals and measures are also being set for agriculture. In Germany, 8% (72 million tons of CO_2-eq) of the greenhouse gas emission came from agriculture in 2014. Despite the fact that of most agricultural greenhouse gas emissions are caused by natural physiological processes, the ability to reduce them is limited. The largest source with about 25 million tons of CO_2-eq is the use of N fertilizers. The use of organic nutrients considered in the ICU project, in particular the soiling®-products, are one important step towards minimization of the N use. The main goal until 2030 is to significantly reduce the emissions of mineralic fertilizers in agriculture. One option is to use financing instruments under the Common Agricultural Policy. Another option is to increase the percentage of land used for organic farming by circular economy approaches [82].

3.5.4. ICU Implementations must Fulfil High Legal Standards Favoring Large Projects or Tiny Ones for Personal Need

The implementation of ICU projects requires observing the laws of the state. In the following, the most critical regulations for implementation ICU concepts in Germany are exemplarily addressed. One important factor is to fulfil the guidelines for building security (BauGB §29–§38, in particular admissibility of projects §34, BauNVO). Additionally, urban development plans and urban planning law must be complied with. Biogas production using AD for biowaste and black water is critical since biogas is flammable and requires sufficient ventilation. Therefore, the storage of large biogas volumes should be avoided. The produced biogas should be consumed immediately, upgraded to natural gas, fed in the gas grid, or outsourced from the buildings [83]. Nevertheless, fire prevention (§§ 3 and 14 MBO) and explosion control (DGUV Regel 113-001, [84]) must be considered. For the removal of black water digestate and the solid fraction of the biowaste fermentation, the laws for sludge disposal have to be considered (AbfKlärV, [85]). Utilization of the biowaste as fertilizer requires compliance with the German laws for biowaste (Bioabfallverordnung

(BioAbfV)) and fertilizer ordinance (DüMV, [28]) as well as the EU regulations for fertilizer ordinance (EU-FPR). In particular, base materials must be allowed (DüMV, Supplementary 2, Table S7, [28]). The fertilizer has to be listed in a positive list (DüMV, Supplementary 1, Table S1, [28]) or equals any allowed fertilizer type. Furthermore, emission limits (DüMV, supplementary 2, Table S1 [28]) and minimum hygiene requirements (§ 5 DüMV, [28]) have to be fulfilled. Exception exist in the case the biowaste and the produced fertilizer are only used for personal needs. However, it is questionable if biowaste utilization in a cooperative of more than 100 residents accounts as a personal need. Due to the high legal requirements, implementation of ICU concepts seems only manageable for large projects or tiny implementations for personal needs (Supplementary File 3). Another question is liability, which is difficult to address in general and usually depends on the specific case. Therefore, no general recommendation can be given here, except to address this issue in a contract between the stakeholders (Supplementary file 6).

3.5.5. Communication and Participation Are Important for the Acceptance of Residents

Implementation of ICU projects require the participation of the residents. Residents have to separate the biowaste accurately, agree to install vacuum toilets and use the urban gardens either as gardeners or as consumers. In general, there is a high acceptance in Germany to waste separation [86], vacuum toilets [87], and urban gardening [79]. However, it is always useful to integrate all stakeholders as early as possible to successfully implement projects [88] and, in addition, to guide their participation by teaching material.

Furthermore, it is recommended to communicate potential risks [89]. For example, ICU operation has the risk of microbial contamination of the food collected. Pre-treatment of biomass at 70 °C can ensure the inactivation of harmful microorganisms. Another issue is the produced biogas, which is explosive. However, when immediately consumed, the risk is reduced to the level of a conventual gas heater.

An important cultural aspect is the utilization of black water as fertilizer. Theoretically, animal dung or manure usage and spreading it on fields are quite similar to the use of black water digestate for hydroponics. However, this is neither allowed nor accepted [90].

3.6. Strategies to Extend the ICU Concept

The ultimate goal of the ICU project is to close energy and material flow cycles in an urbane building. Additional components could increase yields and productivity and allow for a more robust operation.

For example, hydroponic modules for the balcony could be added, or food production can be elevated by aquaponic [91] or algae cultivation [92]. For an implementation, strategies to combine agriculture with photovoltaic could also be implemented [93,94]. For hot and dry areas, the ICU concept could be extended to include the water cycle [95]. For example, greywater can be reused [1] or rainwater could be used for adiabatic cooling. In order to achieve a more robust operation, a module for cleaning the biowaste, for example, via conveyor belts, can be added [96]. Additionally, storage capacities for biowaste, biogas, or fertilizers can be added. However, additional stores in the building are expensive and increase the fire load.

3.7. Overall Discussion of the ICU Project

Overall, we develop a concept for closed urbane biomass circulation and illuminated its feasibility. The final calculation for material flows and potential outcomes for our concept was straightforward. In particular, modeling frameworks such as SIMBA#Biogas for the biogas process and a greenhouse simulator [33,34] for the hydroponic enabled to compare several possibilities. However, selecting the best scenarios was challenging due to the high number of possibilities. Furthermore, besides technical and economic reasons, we recognized that social–cultural reasons such as laws and liability are essential for implementing such concepts.

In consequence of these observations, we can give three main steps to consider for decision-makers aiming to implement biomass circulation. The first step is to define criteria and their weighting (e.g., resource efficiency, economics, CO_2-emission, or social–cultural aspects). The second is to select the best technologies for their use case, depending on the climate, culture, resources, and infrastructure. Based on findings 1 and 2, the third step is to compare the different possibilities, as we have done for scenarios 1 to 4.

This will require comprehensive databases that collect the relevant information for biomass circulation (e.g., as available for LCA) [97] as well as models to calculate the best solution for each setting.

One main limitation of our concept is the high investment costs, making it only economically beneficial for large professional implementations. Another one is that our concept could be easily applied to new buildings or districts, but upgrading existing buildings is challenging.

In the future, our concept and evaluations could be improved by more advanced models, more precise parameters, scenarios, or novel technologies. However, even more important is to validate our outcome on a prototype and identify potential challenges during the implementation.

4. Conclusions

Our study proposes a concept for closed urbane biomass circulation in its entirety and comprehensively analyze its feasibility, which was not done before. Integrating biomass cycles into residential buildings, as proposed by the ICU concept, is technically feasible, reduces CO_2 emission, and is of interest to owners of urban buildings and their residents. It is profitable for implementation in large buildings or agglomeration of buildings and in case food prices further increase. However, to achieve this goal, it will require the implementation of prototypes to perfect technical details and to confirm economic and material calculations. Major challenges for the implementation come from legal aspects relate to the biowaste prescription (the German BioAbfV) and the fertilizer prescription (the German DüMV). In sum, the results of this study should bring us one step closer to a reduction in land use and to a sustainable, CO_2-neutral society.

Supplementary Materials: The following are available online at https://www.mdpi.com/article/10.3390/su13179505/s1, Figure S1: file 2,3,5,7,8, Table S1: file 1,4,6.

Author Contributions: B.I. and M.I. provided estimates for quantification of the soiling®-process and compared the conversion of NH_3 to NO_3 with oxygen feedCalculations for in-house hydroponic calculations were made by S.K. and O.K. Legal aspects for the ICU-concept were processed by A.B. and R.B. LCA-analyze was performed by M.W. Simba#Biogas simulations were created by N.M. and I.S. Data evaluation and preparation of the manuscript was conducted by N.M., R.H., S.K. and J.U., U.R., D.B., J.W. and G.S. contributed with valuable advice and by editing the manuscript. All authors have read and agreed to the published version of the manuscript.

Funding: The authors acknowledge the financial support by a funding program of the Fachagentur Nachwachsende Rohstoffe e.V. (FNR, FPNR). We additionally thank the GEFOMA GmbH for the estimation of costs of greenhouse construction. J.U. received funding from the EU-program ERDF (European Regional Development Fund) of the German Federal State Saxony Anhalt by Research Center of Dynamic Systems (CDS).

Institutional Review Board Statement: Not applicable.

Informed Consent Statement: Not applicable.

Data Availability Statement: Not applicable.

Conflicts of Interest: Jassen Kunststoff GmbH was one of the co-authors of this feasibility study. B.I and M.I. byJassen Kunststoff GmbH have a commercial interest in selling soiling®-modules. Both B.I. and M.I. confirm that they have carried out their evaluation to the best of their knowledge and judgment. All other authors declare that they have no competing interests.

References

1. Hertel, S.; Navarro, P.; Deegener, S.; Körner, I. Biogas and nutrients from blackwater, lawn cuttings and grease trap residues—Experiments for Hamburg's Jenfelder Au district. *Energy Sustain. Soc.* **2015**, *5*, 29. [CrossRef]
2. Barreca, F. Rooftop gardening. A solution for energy saving and landscape enhancement in Mediterranean urban areas. *Procedia-Soc. Behav. Sci.* **2016**, *223*, 720–725. [CrossRef]
3. Fuldauer, L.I.; Parker, B.M.; Yaman, R.; Borrion, A. Managing anaerobic digestate from food waste in the urban environment: Evaluating the feasibility from an interdisciplinary perspective. *J. Clean. Prod.* **2018**, *185*, 929–940. [CrossRef]
4. Jouhara, H.; Czajczyńska, D.; Ghazal, H.; Krzyżyńska, R.; Anguilano, L.; Reynolds, A.J.; Spencer, N. Municipal waste management systems for domestic use. *Energy* **2017**, *139*, 485–506. [CrossRef]
5. Lisowyj, M.; Wright, M.M. A review of biogas and an assessment of its economic impact and future role as a renewable energy source. *Rev. Chem. Eng.* **2020**, *36*, 401–421. [CrossRef]
6. Khalid, A.; Arshad, M.; Anjum, M.; Mahmood, T.; Dawson, L. The anaerobic digestion of solid organic waste. *Waste Manag.* **2011**, *31*, 1737–1744. [CrossRef]
7. Ardolino, F.; Parrillo, F.; Arena, U. Biowaste-to-biomethane or biowaste-to-energy? An LCA study on anaerobic digestion of organic waste. *J. Clean. Prod.* **2018**, *174*, 462–476. [CrossRef]
8. Cecchi, F.; Cavinato, C. Anaerobic digestion of bio-waste: A mini-review focusing on territorial and environmental aspects. *Waste Manag. Res.* **2015**, *33*, 429–438. [CrossRef]
9. Thomsen, M.; Seghetta, M.; Mikkelsen, M.H.; Gyldenkærne, S.; Becker, T.; Caro, D.; Frederiksen, P. Comparative life cycle assessment of biowaste to resource management systems–A Danish case study. *J. Clean. Prod.* **2017**, *142*, 4050–4058. [CrossRef]
10. Specht, K.; Siebert, R.; Hartmann, I.; Freisinger, U.B.; Sawicka, M.; Werner, A.; Thomaier, S.; Henckel, D.; Walk, H.; Dierich, A. Urban agriculture of the future: An overview of sustainability aspects of food production in and on buildings. *Agric. Hum. Values* **2014**, *31*, 33–51. [CrossRef]
11. Hui, S.C.M. Green roof urban farming for buildings in high-density urban cities. In Proceedings of the Hainan China World Green Roof Conference 2011, Hainan, China, 18–21 March 2011.
12. Kunststoff, J. GmbH. Bioreactor and Use Thereof, Method for Producing an Organic Nutrient Solution, Organic Nutrient Solution, Substrate Material and Use Thereof for Cultivating Plants (EP 3684909). U.S. Patent Application No 16/955,979, 13 December 2018.
13. Sapkota, S.; Sapkota, S.; Liu, Z. Effects of nutrient composition and lettuce cultivar on crop production in hydroponic culture. *Horticulturae* **2019**, *5*, 72. [CrossRef]
14. Thakur, N. Organic farming, food quality, and human health: A trisection of sustainability and a move from pesticides to eco-friendly biofertilizers. In *Probiotics in Agroecosystem*; Springer: Berlin/Heidelberg, Germany, 2017; pp. 491–515.
15. Stoknes, K.; Scholwin, F.; Krzesiński, W.; Wojciechowska, E.; Jasińska, A. Efficiency of a novel "Food to waste to food" system including anaerobic digestion of food waste and cultivation of vegetables on digestate in a bubble-insulated greenhouse. *Waste Manag.* **2016**, *56*, 466–476. [CrossRef] [PubMed]
16. Shinohara, M.; Aoyama, C.; Fujiwara, K.; Watanabe, A.; Ohmori, H.; Uehara, Y.; Takano, M. Microbial mineralization of organic nitrogen into nitrate to allow the use of organic fertilizer in hydroponics. *Soil Sci. Plant Nutr.* **2011**, *57*, 190–203. [CrossRef]
17. Krishnasamy, K.; Nair, J.; Bäuml, B. Hydroponic system for the treatment of anaerobic liquid. *Water Sci. Technol.* **2012**, *65*, 1164–1171. [CrossRef] [PubMed]
18. Liedl, B.E.; Cummins, M.; Young, A.; Williams, M.L.; Chatfield, J.M. Liquid Effluent from Poultry Waste Bioremediation as a Potential Nutrient Source for Hydroponic Tomato Production. *Acta Hortic* **2004**, *659*, 647–652. [CrossRef]
19. Wang, L.; Xu, X.; Zhang, M. Effects of intermittent aeration on the performance of partial nitrification treating fertilizer wastewater. *China Environ. Sci.* **2017**, *37*, 146–153.
20. Lombardi, L. Life cycle assessment comparison of technical solutions for CO_2 emissions reduction in power generation. *Energy Convers. Manag.* **2003**, *44*, 93–108. [CrossRef]
21. ifak GmbH. Simba#Biogas 4.3. 2020. Available online: https://www.ifak.eu/de/produkte/simba-biogas (accessed on 1 September 2020).
22. Luste, S.; Luostarinen, S. Anaerobic co-digestion of meat-processing by-products and sewage sludge–Effect of hygienization and organic loading rate. *Bioresour. Technol.* **2010**, *101*, 2657–2664. [CrossRef]
23. Hanc, A.; Novak, P.; Dvorak, M.; Habart, J.; Svehla, P. Composition and parameters of household bio-waste in four seasons. *Waste Manag.* **2011**, *31*, 1450–1460. [CrossRef]
24. Malakahmad, A.; Basri, N.A.; Zain, S.M. An application of anaerobic baffled reactor to produce biogas from kitchen waste. *WIT Trans. Ecol. Environ.* **2008**, *109*, 655–664.
25. Karlsson, J. Modeling and Simulation of Existing Biogas Plants with SIMBA# Biogas. Master's Thesis, Linköping University, Linköping, Sweden, 2017.
26. Liebetrau, J.; Pfeiffer, D.; Thrän, D. Collection of methods for Biogas—Methods to determine parameters for analysis purposes and parameters that describe processes in the Biogas sector. In *DBFZ Deutsches Biomasseforschungszentrum Gemeinnützige GmbH*; Fischer Druck: Leipzig, Germany, 2016.

27. Scheftelowitz, M.; Daniel-Gromke, J.; Denysenko, V.; Sauter, P.N.; Krautz, A.; Beil, M.; Beyrich, W.; Peter, W.; Schicketanz, S.; Schultze, C. Stromerzeugung aus Biomasse 03MAP250-Zwischenbericht. In *Deutsches Biomasseforschungszentrum Gemeinnützige GmbH, Leipzig*; DBFZ German Biomass Research Center gGmbH: Leipzig, Germany, 2013; p. 15.
28. Bundesministerium der Justiz und Verbraucherschutz. *Verordnung Über das Inverkehrbringen von Düngemitteln, Bodenhilfsstoffen, Kultursubstraten und Pflanzenhilfsmitteln*; DüMV: Berlin, Germany, 2012.
29. Lugmayr, R. Technisch-wirtschaftlicher Vergleich eines Gasmotors mit einer Mikrogasturbine. Master's Thesis, FH-Oberösterreich, Wels, Austria, 2010.
30. Hasemann, S. *Abschlussbericht EnBW Patenschaft*; Institut für Verbrennungstechnik: Stuttgart, Germany, 2015.
31. Lingstädt, T.; Seliger, H.; Reh, S.; Huber, A. *Technologiebericht 2.2 b Dezentrale Kraftwerke (Motoren und Turbinen) Innerhalb des Forschungsprojekts TF_Energiewende*; Wuppertal Institut für Klima, Umwelt, Enerjie: Wuppertal, Germany, 2018; p. 64.
32. Agelidou, E.; Schwärzle, A.; Hasemann, S. Mikrogasturbinen-basiertes Mikro-BHKW für den Einsatz in Einfamilienhäusern (MGT-μBHKW). Available online: https://www.researchgate.net/publication/337590819_Mikrogasturbinen-basiertes_Mikro-BHKW_fur_den_Einsatz_in_Einfamilienhausern_MGT-BHKW (accessed on 13 August 2021).
33. Körner, O.; Hansen, J.B. An on-line tool for optimising greenhouse crop production. In *IV International Symposium on Models for Plant Growth, Environmental Control and Farm Management in Protected Cultivation-957*; ISHS: Leuven, Belgium, 2012; pp. 147–154.
34. Körner, O.; Warner, D.; Tzilivakis, J.; Eveleens-Clark, B.; Heuvelink, E. (Eds.) Decision Support for Optimising Energy Consumption in European Greenhouses. *Acta Hortic.* **2008**, *801*, 803–810. [CrossRef]
35. Körner, O.; Aaslyng, J.M.; Andreassen, A.U.; Holst, N. Microclimate prediction for dynamic greenhouse climate control. *HortScience* **2007**, *42*, 272–279. [CrossRef]
36. Körner, O.; Pedersen, J.S.; Jaegerholm, J. Simulating lettuce production in a multilayer moving gutter system. *Acta Hortic.* **2018**, *1227*, 283–290. [CrossRef]
37. Körner, O.; Andreassen, A.U.; Aaslyng, J.M. Simulating dynamic control of supplementary lighting. *Acta Hortic.* **2006**, *711*, 151–156. [CrossRef]
38. Meteoblue. Meteoblue. Available online: www.meteoblue.com (accessed on 13 August 2021).
39. Brechner, M.; Both, A.J.; Staff, C.E. Hydroponic lettuce handbook. *Cornell Control. Environ. Agric.* **1996**, *834*, 504–509.
40. Finkbeiner, M.; Inaba, A.; Tan, R.; Christiansen, K.; Klüppel, H.-J. The new international standards for life cycle assessment: ISO 14040 and ISO 14044. *Int. J. Life Cycle Assess.* **2006**, *11*, 80–85. [CrossRef]
41. Wernet, G.; Bauer, C.; Steubing, B.; Reinhard, J.; Moreno-Ruiz, E.; Weidema, B. The ecoinvent database version 3 (part I): Overview and methodology. *Int. J. Life Cycle Assess.* **2016**, *21*, 1218–1230. [CrossRef]
42. McDonough, W.; Braungart, M. *Cradle to Cradle: Remaking the Way We Make Things*; North Point Press: New York, NY, USA, 2010. ISBN 1429973846.
43. Andor, M.A.; Frondel, M.; Sommer, S. Equity and the willingness to pay for green electricity in Germany. *Nat. Energy* **2018**, *3*, 876–881. [CrossRef]
44. Wriege-Bechtold, A. *Anaerobe Behandlung von Braunwasser und Klärschlamm unter Berücksichtigung von Co-Substraten*; Technische Universitaet: Berlin, Germany, 2015; ISBN 1392811880.
45. Bensmann, B.; Hanke-Rauschenbach, R.; Müller-Syring, G.; Henel, M.; Sundmacher, K. Optimal configuration and pressure levels of electrolyzer plants in context of power-to-gas applications. *Appl. Energy* **2016**, *167*, 107–124. [CrossRef]
46. Bensmann, A.; Hanke-Rauschenbach, R.; Sundmacher, K. Reactor configurations for biogas plants—A model based analysis. *Chem. Eng. Sci.* **2013**, *104*, 413–426. [CrossRef]
47. Frondel, M.; Ritter, N.; Sommer, S. Stromverbrauch privater Haushalte in Deutschland–Eine ökonometrische Analyse. *Z. Energ.* **2015**, *39*, 221–232. [CrossRef]
48. Bensmann, A.; Hanke-Rauschenbach, R.; Heyer, R.; Kohrs, F.; Benndorf, D.; Reichl, U.; Sundmacher, K. Biological methanation of hydrogen within biogas plants: A model-based feasibility study. *Appl. Energy* **2014**, *134*, 413–425. [CrossRef]
49. Ryckebosch, E.; Drouillon, M.; Vervaeren, H. Techniques for transformation of biogas to biomethane. *Biomass Bioenergy* **2011**, *35*, 1633–1645. [CrossRef]
50. Gao, M.; Zhang, L.; Florentino, A.P.; Liu, Y. Performance of anaerobic treatment of blackwater collected from different toilet flushing systems: Can we achieve both energy recovery and water conservation? *J. Hazard. Mater.* **2019**, *365*, 44–52. [CrossRef] [PubMed]
51. Carotti, L.; Graamans, L.; Puksic, F.; Butturini, M.; Meinen, E.; Heuvelink, E.; Stanghellini, C. Plant factories are heating up: Hunting for the best combination of light intensity, air temperature and root-zone temperature in lettuce production. *Front. Plant Sci.* **2020**, *11*, 2251.
52. Graamans, L.; Baeza, E.; van den Dobbelsteen, A.; Tsafaras, I.; Stanghellini, C. Plant factories versus greenhouses: Comparison of resource use efficiency. *Agric. Syst.* **2018**, *160*, 31–43. [CrossRef]
53. Cammarisano, L.; Donnison, I.S.; Robson, P.R.H. Producing enhanced yield and nutritional pigmentation in lollo rosso through manipulating the irradiance, duration, and periodicity of LEDs in the visible region of light. *Front. Plant Sci.* **2020**, *11*, 598082. [CrossRef]
54. Bailey, B.J.; Seginer, I. Optimum control of greenhouse heating. *Eng. Econ. Asp. Energy Sav. Prot. Cultiv.* **1988**, *245*, 512–518. [CrossRef]

55. Harbick, K.; Albright, L.D. Comparison of energy consumption: Greenhouses and plant factories. *Acta Hortic.* **2016**, *1134*, 285–292. [CrossRef]

56. Rorabaugh, P.; Jensen, M.; Giacomelli, G. Introduction to Controlled Environment Agriculture and Hydroponics. Available online: https://studyres.com/doc/7770673/introduction-to-controlled-environment-agriculture-and-hy (accessed on 13 August 2021).

57. Li, Q.; Li, X.; Tang, B.; Gu, M. Growth responses and root characteristics of lettuce grown in aeroponics, hydroponics, and substrate culture. *Horticulturae* **2018**, *4*, 35. [CrossRef]

58. Savvas, D.; Chatzieustratiou, E.; Pervolaraki, G.; Gizas, G.; Sigrimis, N. Modelling Na and Cl concentrations in the recycling nutrient solution of a closed-cycle pepper cultivation. *Biosyst. Eng.* **2008**, *99*, 282–291. [CrossRef]

59. Goddek, S.; Körner, O. A fully integrated simulation model of multi-loop aquaponics: A case study for system sizing in different environments. *Agric. Syst.* **2019**, *171*, 143–154. [CrossRef]

60. SharathKumar, M.; Heuvelink, E.; Marcelis, L.F.M. Vertical farming: Moving from genetic to environmental modification. *Trends Plant Sci.* **2020**, *25*, 724–727. [CrossRef] [PubMed]

61. Karimaei, M.S.; Massiha, S.; Mogaddam, M. Comparison of two nutrient solutions' effect on growth and nutrient levels of lettuce (*Lactuca sativa* L.) cultivars. In *International Symposium on Growing Media and Hydroponics 644*; ISHS: Leuven, Belgium, 2001.

62. Sikawa, D.C.; Yakupitiyage, A. The hydroponic production of lettuce (*Lactuca sativa* L.) by using hybrid catfish (*Clarias macrocephalus* × *C. gariepinus*) pond water: Potentials and constraints. *Agric. Water Manag.* **2010**, *97*, 1317–1325. [CrossRef]

63. Vogeler, I.; Thomas, S.; van der Weerden, T. Effect of irrigation management on pasture yield and nitrogen losses. *Agric. Water Manag.* **2019**, *216*, 60–69. [CrossRef]

64. van Os, E.A. Closed soilless growing systems: A sustainable solution for Dutch greenhouse horticulture. *Water Sci. Technol.* **1999**, *39*, 105–112. [CrossRef]

65. Gerbaud, A.; Andre, M. Down regulation of photosynthesis after CO_2 enrichment of lettuce; relation to photosynthetic characteristics. *Biotronics* **1999**, *28*, 33–44.

66. Armanda, D.T.; Guinée, J.B.; Tukker, A. The second green revolution: Innovative urban agriculture's contribution to food security and sustainability–A review. *Glob. Food Secur.* **2019**, *22*, 13–24. [CrossRef]

67. Footspring Magazine. Kalorientabelle. Available online: https://www.foodspring.ch/magazine/wp-content/uploads/2020/10/FS_Kalorientabelle.pdf (accessed on 13 August 2021).

68. Wikifit. Kalorientabelle Salat. Available online: https://www.wikifit.de/kalorientabelle/salat (accessed on 13 August 2021).

69. Aaslyng, J.M.; Andreassen, A.U.; Körner, O.; Lund, J.B.; Jakobsen, L.; Pedersen, J.S.; Ottosen, C.O.; Rosenqvist, E. Integrated optimization of temperature, CO_2, screen use and artificial lighting in greenhouse crops using a photosynthesis model. In *V International Symposium on Artificial Lighting in Horticulture 711*; ISHS: Leuven, Belgium, 2005; pp. 79–88.

70. Mortensen, L.M.; Strømme, E. Effects of light quality on some greenhouse crops. *Sci. Hortic.* **1987**, *33*, 27–36. [CrossRef]

71. Seginer, I.; Albright, L.D.; Ioslovich, I. Improved strategy for a constant daily light integral in greenhouses. *Biosyst. Eng.* **2006**, *93*, 69–80. [CrossRef]

72. Kozai, T. Resource use efficiency of closed plant production system with artificial light: Concept, estimation and application to plant factory. *Proc. Jpn. Acad. Ser. B* **2013**, *89*, 447–461. [CrossRef]

73. Capponi, S.; Fazio, S.; Barbanti, L. CO_2 savings affect the break-even distance of feedstock supply and digestate placement in biogas production. *Renew. Energy* **2012**, *37*, 45–52. [CrossRef]

74. Götz, T.; Adisorn, T.; Tholen, L. *Der digitale Produktpass als Politik-Konzept: Kurzstudie im Rahmen der umweltpolitischen Digitalagenda des Bundesministeriums für Umwelt, Naturschutz und Nukleare Sicherheit (BMU)*; Wuppertal Reports: Wuppertal, Germany, 2021.

75. Pilarski, G.; Kyncl, M.; Stegenta, S.; Piechota, G. Emission of biogas from sewage sludge in psychrophilic conditions. *Waste Biomass Valorization* **2020**, *11*, 3579–3592. [CrossRef]

76. Heliospectra. Industry-Leading LED Lighting for Greenhouses. Available online: https://www.heliospectra.com/ (accessed on 13 August 2021).

77. Landeshauptstadt Magdeburg. Landeshauptstadt Magdeburg. Available online: https://www.magdeburg.de/Start/B%C3%BCrger-Stadt/Leben-in-Magdeburg/Umwelt/Abfall/index.php?La=1&object=tx,37.3427.1&kat=&kuo=2&sub=0 (accessed on 13 August 2021).

78. Tölkes, C.; Butzmann, E. Motivating pro-sustainable behavior: The potential of green events—A case-study from the Munich Streetlife Festival. *Sustainability* **2018**, *10*, 3731. [CrossRef]

79. Winkler, B.; Maier, A.; Lewandowski, I. Urban gardening in Germany: Cultivating a sustainable lifestyle for the societal transition to a bioeconomy. *Sustainability* **2019**, *11*, 801. [CrossRef]

80. Specht, K.; Weith, T.; Swoboda, K.; Siebert, R. Socially acceptable urban agriculture businesses. *Agron. Sustain. Dev.* **2016**, *36*, 17. [CrossRef]

81. Paris Agreement. The Paris Agreement- Process and Meetings. Available online: www.unfccc.int/process-and-meetings/the-paris-agreement/the-paris-agreement (accessed on 13 August 2021).

82. BMU-Klimaschutzplan 2050. Klimaschutzplan 2050. Available online: https://www.bmu.de/publikation/klimaschutzplan-2050/ (accessed on 13 August 2021).

83. Schmidt-Eichstaedt, G. *Anwendungsbereich und Abgrenzungsfragen bei Entwicklungs-und Ergänzungssatzungen nach § 34 Abs. 4 Satz 1 Nr. 2 und 3 BauGB. Schaffung von Bauland*; Nomos Verlagsgesellschaft mbH & Co. KG: Berlin, Germany, 2019.

84. Frigger, A. Wie schütze ich was? Expertentipps zum explosionsschutz: Wie können welche anlagenteile optimal geschützt werden. *CITplus* **2019**, *22*, 34–35. [CrossRef]
85. Queitsch, P. Die neue klärschlamm-verordnung und die rechtsfolgen für die entsorgungspraxis. *Z. Recht Abfallwirtsch.* **2018**, *17*, 78–85.
86. Improving the quality and quantity of source-separated household food waste in areas of different socio-economic characteristics: A case study from Lübeck, Germany. In Proceedings of the 16th International Conference on Environmental Science and Technology (CEST2019), Rhodos, Greece, 14 September 2019.
87. Poortvliet, P.M.; Sanders, L.; Weijma, J.; de Vries, J.R. Acceptance of new sanitation: The role of end-users' pro-environmental personal norms and risk and benefit perceptions. *Water Res.* **2018**, *131*, 90–99. [CrossRef]
88. García-Sánchez, E.; García-Morales, V.J.; Martín-Rojas, R. Analysis of the influence of the environment, stakeholder integration capability, absorptive capacity, and technological skills on organizational performance through corporate entrepreneurship. *Int. Entrep. Manag. J.* **2018**, *14*, 345–377. [CrossRef]
89. Xia, N.; Zou, P.X.W.; Griffin, M.A.; Wang, X.; Zhong, R. Towards integrating construction risk management and stakeholder management: A systematic literature review and future research agendas. *Int. J. Proj. Manag.* **2018**, *36*, 701–715. [CrossRef]
90. Gell, K.; de Ruijter, F.J.; Kuntke, P.; de Graaff, M.; Smit, A.L. Safety and effectiveness of struvite from black water and urine as a phosphorus fertilizer. *J. Agric. Sci.* **2011**, *3*, 67. [CrossRef]
91. Chia, S.R.; Ong, H.C.; Chew, K.W.; Show, P.L.; Phang, S.-M.; Ling, T.C.; Nagarajan, D.; Lee, D.-J.; Chang, J.-S. Sustainable approaches for algae utilisation in bioenergy production. *Renew. Energy* **2018**, *129*, 838–852. [CrossRef]
92. Wongkiew, S.; Hu, Z.; Chandran, K.; Lee, J.W.; Khanal, S.K. Nitrogen transformations in aquaponic systems: A review. *Aquac. Eng.* **2017**, *76*, 9–19. [CrossRef]
93. Narvarte, L.; Fernández-Ramos, J.; Martínez-Moreno, F.; Carrasco, L.M.; Almeida, R.H.; Carrêlo, I.B. Solutions for adapting photovoltaics to large power irrigation systems for agriculture. *Sustain. Energy Technol. Assess.* **2018**, *29*, 119–130. [CrossRef]
94. Putri, R.K.; Astuti, Y.D.R.W.; Wardhany, A.K.; Hudaya, C. Building integrated photovoltaic for rooftop and facade application in Indonesia. In Proceedings of the 2018 2nd International Conference on Green Energy and Applications (ICGEA), Singapore, 24–26 March 2018.
95. Liuzzo, L.; Notaro, V.; Freni, G. A reliability analysis of a rainfall harvesting system in southern Italy. *Water* **2016**, *8*, 18. [CrossRef]
96. Verma, S. Anaerobic digestion of biodegradable organics in municipal solid wastes. *Columbia Univ.* **2002**, *7*, 98–104.
97. Frischknecht, R.; Rebitzer, G. The ecoinvent database system: A comprehensive web-based LCA database. *J. Clean. Prod.* **2005**, *13*, 1337–1343. [CrossRef]

Article

Characterization and Hydration Mechanism of Ammonia Soda Residue and Portland Cement Composite Cementitious Material

Dong Xu [1,2,3], Pingfeng Fu [3,4], Wen Ni [2,3,4,*], Qunhui Wang [1,2] and Keqing Li [3,4]

[1] School of Energy and Environmental Engineering, University of Science and Technology Beijing, Beijing 100083, China; sdlycsxd@163.com (D.X.); wangqh59@163.com (Q.W.)

[2] Beijing Key Laboratory on Resource-Oriented Treatment of Industrial Pollutants, University of Science and Technology Beijing, Beijing 100083, China

[3] Key Laboratory of the Ministry of Education of China for High-Efficient Mining and Safety of Metal Mines, University of Science and Technology Beijing, Beijing 100083, China; pffu@ces.ustb.edu.cn (P.F.); Lkqing2003@163.com (K.L.)

[4] School of Civil and Resource Engineering, University of Science and Technology Beijing, Beijing 100083, China

* Correspondence: niwen@ces.ustb.edu.cn

Citation: Xu, D.; Fu, P.; Ni, W.; Wang, Q.; Li, K. Characterization and Hydration Mechanism of Ammonia Soda Residue and Portland Cement Composite Cementitious Material. *Materials* **2021**, *14*, 4794. https://doi.org/10.3390/ma14174794

Academic Editor: Carlos Morón Fernández

Received: 21 July 2021
Accepted: 20 August 2021
Published: 24 August 2021

Publisher's Note: MDPI stays neutral with regard to jurisdictional claims in published maps and institutional affiliations.

Abstract: The use of ammonia soda residue (ASR) to prepare building materials is an effective way to dispose of ASR on a large scale, but this process suffers from a lack of data and theoretical basis. In this paper, a composite cementitious material was prepared using ASR and cement, and the hydration mechanism of cementitious materials with 5%, 10%, and 20% ASR was studied. The XRD and SEM results showed that the main hydration products of ASR-cement composite cementitious materials were an amorphous C-S-H gel, hexagonal plate-like $Ca(OH)_2$ (CH), and regular hexagonal plate-like Friedel's salt (FS). The addition of ASR increased the heat of hydration of the cementitious material, which increased upon increasing the ASR content. The addition of ASR also reduced the cumulative pore volume of the hardened paste, which displayed the optimal pore structure when the ASR content was 5%. In addition, ASR shortened the setting time compared with the cement group, and the final setting times of the pastes with 5%, 10%, and 20% ASR were 30 min, 45 min, and 70 min shorter, respectively. When the ASR content did not exceed 10%, the 3-day compressive strength of the mortar was significantly improved, but the 28-day compressive strength was worse. Finally, the hydration mechanism and potential applications of the cementitious material are discussed. The results of this paper promote the use of ASR in building materials to reduce CO_2 emissions in the cement industry.

Keywords: ammonia soda residue; hydration mechanism; cementitious material; resource utilization; CO_2 emissions

1. Introduction

Ammonia soda residue (ASR) is a solid waste produced during the preparation of soda ash by the ammonia-soda method. Its main components are $CaCO_3$, $CaSO_4$, NaCl, and $CaCl_2$, and its pH value is generally in the range of 10–12 [1]. In China, the annual growth rate of ASR exceeds 5 million tons, and the accumulated amount exceeds 50 million tons due to the lack of effective treatment methods (the current treatment efficiency is only 3%–4%) [2]. The accumulation of such a large amount of ASR poses a potential environmental pollution hazard due to the storage area required, and the impact on groundwater and air increases each day [3]. Because the production of soda ash by the ammonia method requires large amounts of chloride salts as raw materials, most soda plants are located near the sea to conveniently obtain chloride salts; therefore, ASR is piled on beaches, which threatens the ocean environment and also poses hidden dangers in the form of dam breaches [4]. Thus, the treatment and utilization of ASR have become a popular research topic.

Many studies have investigated the treatment and utilization of ASR, such as extracting calcium components [5], and using it to prepare desulfurizers [6] and soil regulators (pH regulators, heavy metal solidification) [7,8], but these methods produce secondary pollution or face small-scale issues. Some researchers have studied the use of ASR to prepare building materials. Zhao [9] prepared a geopolymer from ASR and fly ash that displayed good volume stability and found that ASR improved the comprehensive performance of the geopolymer, and when the mass ratio of ASR to fly ash was 1:4, the 60-day and 180-day compressive strength reached 13.5 MPa and 18 MPa, respectively. Wang [10] studied the application of ASR for the preparation of cement and prepared cement that satisfied the requirements of 52.5-grade cement; however, the impermeability and freeze–thaw resistance of the cement were poor. Guo [11] utilized soda solid waste ASR and calcium carbide slag to synergistically activate ground granulated blast-furnace slag (GGBS) and fly ash (FA). The 28-day compressive strength of the binder system reached 17.5–43.2 MPa. Xu [12] used ASR, GGBS, steel slag, and flue gas desulfurization gypsum as cementitious materials and waste iron ore tailings as an aggregate to prepare clinker-free concretes. The compressive strength of the optimized concrete reached 36.29 MPa and 66.31 MPa after 3 and 360 days, which were, respectively, 32% and 27% higher than the control specimen.

From the perspective of the treatment scale, the use of ASR to prepare building materials is the only way to effectively reuse ASR. At present, cement is the main artificial building material throughout the world, but its production consumes natural resources and produces CO_2 emissions. In fact, the cement industry has always been one of the main sources of CO_2 emissions [13]. According to statistics, each ton of produced cement emits 900 kg of CO_2, and the total amount of CO_2 emissions from cement production accounts for more than 5% of the world's total annual CO_2 emissions [14]. Methods to reduce the production of cement is a current and future research hotspot, so it is particularly important to select appropriate materials to completely or partially replace cement to prepare building materials, such as using supplementary cementitious materials (SCMs) (e.g., GGBS, SS, and FA) [15]. In this regard, the use of solid waste to replace cement to prepare building materials can reduce the amount of cement, thereby reducing CO_2 emissions associated with cement production processes. Therefore, the use of ASR to replace part of cement to prepare building materials can reuse ASR on a large scale and also reduce the amount of cement, which is highly feasible in terms of technology, costs, and policy. In particular, two policies are taken into account: (1) environmental protection tax (25 yuan per ton of solid waste) in China since 2018; (2) the Chinese government has made a solemn commitment to strive for the peak of carbon dioxide emissions by 2030 and to achieve carbon neutrality by 2060.

Despite these goals, there are some deficiencies in the current research on the preparation of building materials from ASR, such as the high costs of cement burning, or the complex preparation process; therefore, a simple inexpensive treatment process is needed to solve the current problem of ASR stacking. The purpose of this paper was to provide theoretical data support for the preparation of cementitious materials from ASR via a simple and inexpensive method to reuse ASR and reduce CO_2 emissions of the cement industry. To this end, an ASR-cement composite cementitious material was prepared. The hydration mechanism and hydration characteristics of the composite cementitious material were analyzed by XRD, SEM, TG-DSC, and MIP. The physical properties of the mortar prepared by the ASR-cement composite cementitious material were studied. Finally, the potential applications of the ASR-cement cementitious material were discussed.

2. Raw Materials and Methods

2.1. Raw Materials

The cement used in this work was P.O 42.5 cement produced by the China United Cement Group Co., Ltd. (Beijing, China), with a Blaine fineness of 351 m^2/kg. The ASR was obtained from Tangshan Sanyou Chemical Co., Ltd. (Tangshan, China), which is one of the largest soda ash manufacturers in China. The ASR had a moisture content of 25%–50%, dried and ground to a powder with a Blaine fineness of 370 m^2/kg. The chemical compositions of ASR and cement are shown in Table 1. The main chemical constituents of ASR are CaO, MgO, and SiO_2, with a chloride content of 11.5%.

Table 1. Chemical composition of raw materials (mass fraction/%).

Composition	Cement	ASR
CaO	62.38	52.88
SiO_2	22.11	10.19
MgO	2.28	8.35
SO_3	2.62	8.87
Na_2O	1.73	1.84
Al_2O_3	4.43	3.25
Fe_2O_3	3.13	1.23
K_2O	0.26	0.38
TiO_2	0.35	0.19
MnO	0.3	0.06
Cl	0.012	11.95

Figure 1 shows the morphology and XRD pattern of ASR, which is gray-white in appearance. The mineral phases are mainly $CaCO_3$, NaCl, $CaSO_4 \cdot 2H_2O$ and $CaSO_4 \cdot 0.5H_2O$. The aggregate used in this study was an ISO reference sand with a particle size smaller than 2 mm according to ISO R679 standard. It was purchased from Xiamen ISO Standard Sand Co., Ltd. (Xiamen, China).

2.2. Mix Proportions and Specimen Preparation

After the ASR was crushed and dried, it was ground to a specific surface area of 370 m^2/kg using a grinder mill. The ASR and cement were weighed according to the cementitious material ratio in Table 2 and then mixed with water. The water-binder (*w/b*) ratio was 0.5. The mixture was put into a planetary mixer (NJ-160) for cement paste and mixed at a low speed for 120 s, stopped for 15 s, and then mixed at a high speed for 120 s. The paste sample was poured into a 30 mm × 30 mm × 50 mm iron mold for trial molding and then transferred to a constant-temperature and humidity curing box for storage. The curing temperature was 20 ± 2 °C, and the relative humidity was ≥90%. At 3 d and 28 d, the paste samples were taken out and then crushed and put into anhydrous ethanol to terminate the hydration reaction and characterize the hydration mechanism. Cementitious material (450 g) was weighed according to the ASR-and-cement ratios in Table 2. Then, 1350 g standard sand and 225 g water were added, and a planetary mixer (JJ-5) was used to mix the mortars evenly. The mixing time was 30 s at low-speed, high-speed stirring for 30 s, pausing for 90 s, and high-speed stirring for 60 s. After the mortar was prepared, it was immediately poured into an iron mold with a size of 40 mm × 40 mm × 160 mm. After curing for 1 d at 20 ± 2 °C and a relative humidity of ≥90%, the mortar was demolded and then placed in 20 ± 1 °C water to cure for 3 d, 7 d, and 28 d for compressive strength testing. The setting time of the cementitious material was tested in accordance with the Chinese standard GB/T 1346-2011, and the compressive strength tests were carried out in accordance with GB/T 17671-2020.

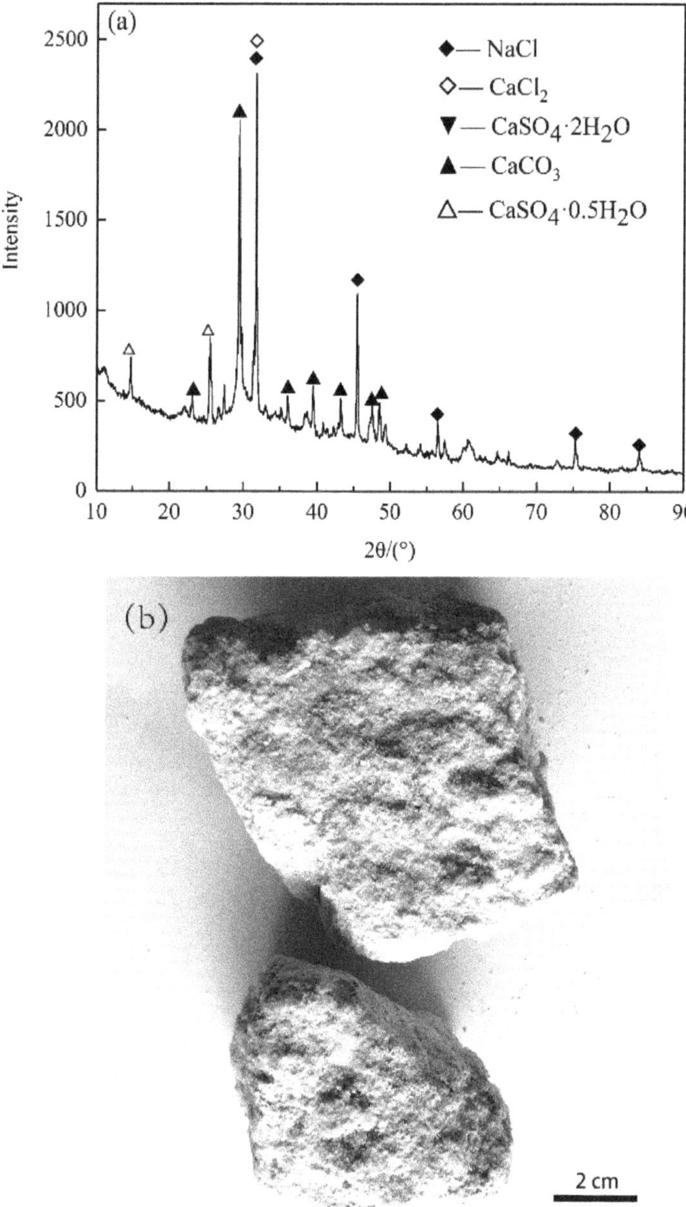

Figure 1. (**a**) XRD pattern and (**b**) photo of ASR.

Table 2. Mix proportions of the pastes with different amounts of ASR (g).

Mix Code	Cement	ASR	Water
OPC	100	0	50
ASR-5	95	5	50
ASR-10	90	10	50
ASR-20	80	20	50
ASR-30	70	30	50

2.3. Testing Methods

The hydration mechanism of the composite cementitious materials was characterized by XRD, SEM, TG-DSC, and MIP. The mineral phases of materials were characterized by an Ultima IV X-ray diffractometer equipped with a copper target working at 40 kV and 40 mA. Scanning was carried out for 2θ values from $10°$ to $90°$ at a scanning rate of $10°$ min^{-1}. The microstructure of pastes was analyzed by a SUPRA55 field emission scanning electron microscope. The sample preparation method for SEM is as follows: the paste sample was crushed into thin slices with a diameter of about 1–2 mm and a thickness of 0.3–0.5 mm. Then, the sample was placed in a drying oven and dried to a constant weight at a constant temperature of 65 °C. Finally, conductive glue was used to fix the sample to the sample stage, and the surface of the sample was evenly sputtered with gold to render it conductive. TG-DSC analysis was carried out by an STA409C/CD differential scanning calorimeter, and the testing temperature was increased from 20 °C to 1000 °C at a rate of 10 °C/min under an inert nitrogen (N_2) atmosphere. The pore structures of hardened pastes were measured by a Quantachrome Autoscan-33 mercury intrusion porosimeter. The heat of hydration of composite cementitious materials was measured by a TAM-Air isothermal calorimeter at a constant temperature of 25 °C with a 0.5 w/b ratio according to the ASTM C1702 standard test method.

3. Results

3.1. XRD Analysis

The XRD patterns of the paste samples with 0, 5%, 10%, 20%, and 30% of ASR contents at 3 days and 28 days are shown in Figure 2. The results show that the hydration products of OPC are mainly C-S-H gel and $Ca(OH)_2$ (CH). The C-S-H gel has no characteristic diffraction peaks in the XRD pattern because it is amorphous. A new hydration product, Friedel's salt (FS), appeared in the ASR-cement composite cementitious material control OPC. According to the classification of substances with typical characteristic XRD peaks, there are mainly three types of substances: (1) Hydration products—mainly dicalcium silicate and tricalcium silicate, which are hydrated to form CH, and the new hydration product FS; (2) The active mineral components that were not hydrated—C_3S and C_2S; (3) Materials that do not possess pozzolanic activity and cannot be hydrated—$CaCO_3$ and NaCl.

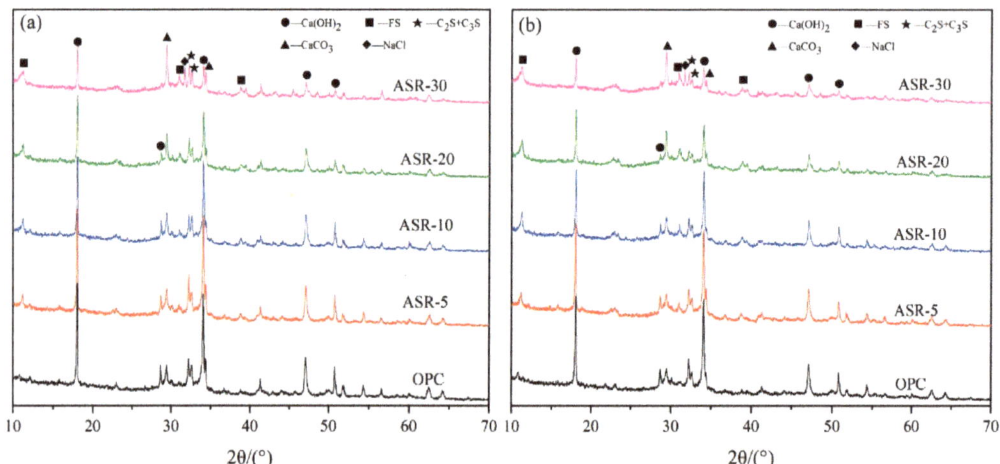

Figure 2. XRD patterns of ASR-cement composite cementitious materials: (**a**) 3 days; (**b**) 28 days.

From Figure 2a, after curing for 3 days, the amount of CH changed significantly upon increasing the amount of ASR, especially when the ASR content exceeded 10% (ASR-10). The intensity of the CH characteristic diffraction peaks gradually decreased, indicating that the CH content in the hardened paste gradually decreased. The intensity of the $CaCO_3$ diffraction peak gradually became stronger upon increasing the ASR content due to overlap of the $CaCO_3$ diffraction peak in the ASR and the $CaCO_3$ diffraction peak generated from carbonization during curing. Similarly, the characteristic diffraction peak of NaCl also gradually increased upon increasing the ASR content, indicating that a large amount of NaCl remained without participating in the hydration reaction. In Figure 2b, after curing for 28 days, the characteristic diffraction peaks of various minerals showed roughly the same trend as those in the sample cured for 3 days, indicating that the hydration products of ASR-cement composite cementitious materials do not change significantly during the later stage of hydration.

3.2. SEM Analysis

The XRD results showed that the early and late hydration products of the ASR-cement composite were basically the same, and the types of hydration products did not change with the ASR content; therefore, the samples with an ASR content of 10% (ASR-10) and a hydration age of 3 days were selected for hydration product morphology observations. Figure 3 shows SEM images of the ASR-10 hardened paste of the ASR-cement composite. In Figure 3a there are many regular hexagonal plate-shaped FS particles with a diameter of about 1–2 μm, while there are also hexagonal plate-shaped substances in Figure 3b; however, they have a coarser appearance with a diameter of 5–10 μm, which is a typical CH crystal morphology. Around the CH crystals, there is an amorphous, honeycomb-like substance, which was the typical morphology of a C-S-H gel in the early stage. The formation of CH and FS has an important effect on the density of the paste, because they will fill the pores of the C-S-H gel with unreacted particles and form a dense structure, which plays an important role in the strength development of the paste.

Figure 3. SEM image of hardened paste of ASR-10 cementitious material, the morphology of Friedel's salt (**a**); and Ca(OH)$_2$ (**b**).

3.3. TG-DSC Analysis

The TG-DSC curves of the hardened pastes of ASR-cement composite cementitious material after aging for 3 days are shown in Figure 4. From the DSC curve in Figure 4a, the exothermic peak near 120 °C represents the dehydration of the C-S-H gel. The endothermic peak near 200 °C represents the decomposition of monocarbonate (Mc), and the exothermic peak near 450 °C is caused by the decomposition of CH, while the exothermic peak above 600 °C is caused by the decomposition of CaCO$_3$. It can be seen from Figure 4b,c that the endothermic decomposition peak of FS appeared in the DSC curves of ASR-10 and ASR-20 from 230–410 °C, which is consistent with the XRD and SEM results. The CaCO$_3$ in OPC samples were generated by carbonization of the hardened paste during curing, while the CaCO$_3$ in ASR-10 and ASR-20 samples were not only carbonized but also introduced from ASR.

It can be seen from the TG curves in Figure 4 that the weight loss rates of OPC, ASR-10, and ASR-20 samples in the temperature range of 25–150 °C were 2.48%, 4.67%, and 4.52%, respectively; however, the results do not mean that the addition of ASR increased the amount of C-S-H gel in the pastes, because the CaSO$_4 \cdot$2H$_2$O in ASR will also be dehydrated in this temperature range. The total weight loss rates of the three groups of samples in the temperature range of 25–1000 °C were 15.11%, 18.79%, and 18.68%, respectively. These do not indicate the number of hydration products in the samples, because the amounts of C-S-H gel, CH, FS, and CaCO$_3$ are different, and the weight loss of each substance during decomposition is also different.

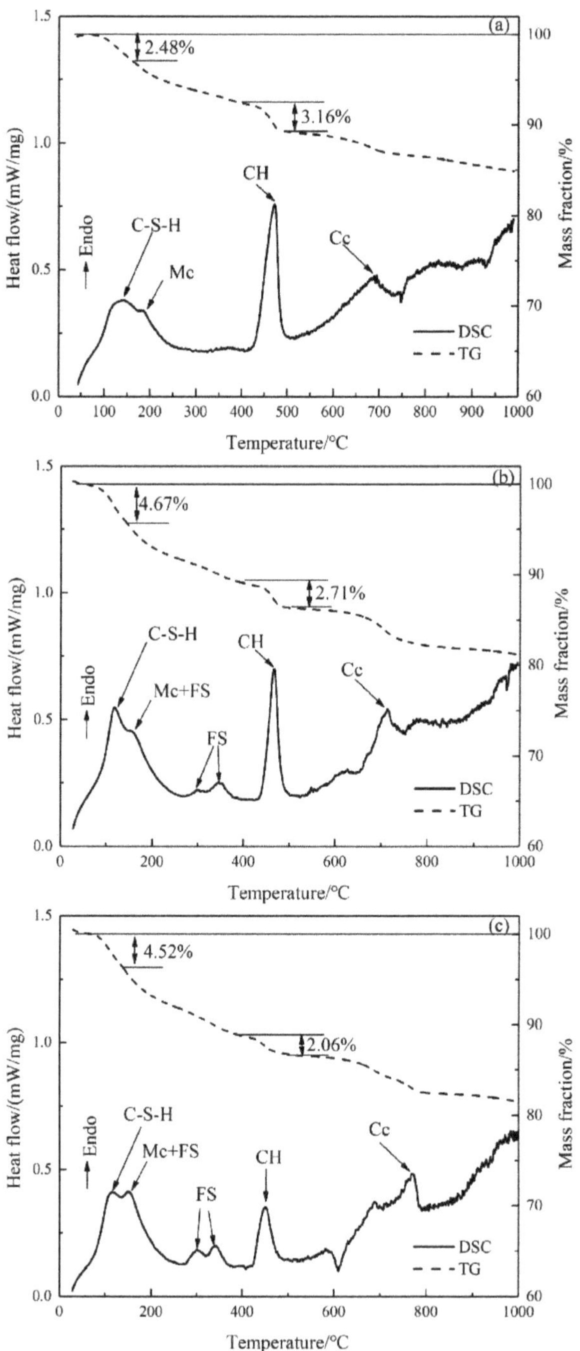

Figure 4. DSC and TG curves of (**a**) OPC paste, (**b**) ASR-10 paste, and (**c**) ASR-20 paste (Mc: mono-carbonate; FS: Friedel's salt; CH: Ca(OH)$_2$; Cc: CaCO$_3$).

In cement-based materials, the CH quantitative analysis method is often used to characterize the degree of hydration of cementitious materials. The weight loss in the temperature range of 400–500 °C is caused by the dehydration of CH, so the weight loss within this range can be used to calculate the CH content of the paste [16]. The FS content in the paste can be determined using the weight loss value in the interval of 230–410 °C, because FS generally loses six interlayer water molecules in this interval [17]. The amount of CH ($P_{Ca(OH)_2}$, in wt.%) formed in the paste can be calculated from the DSC curves using Equation (1) [18], and the Friedel's salt content formed in the paste can be calculated from the DSC curves using Equation (2) [19]:

$$P_{Ca(OH)_2} = WL_{Ca(OH)_2} \frac{MW_{Ca(OH)_2}}{MW_{H_2O}} \tag{1}$$

where $WL_{Ca(OH)_2}$ is the mass loss from the hydroxylation of $Ca(OH)_2$ (wt.%), and MW_{H_2O} and $MW_{Ca(OH)_2}$ are the molar masses of H_2O and $Ca(OH)_2$, respectively.

$$M_{FS} = \frac{M_{FS}}{6M_{H_2O}} m_{H_2O} \tag{2}$$

where m_{H_2O} is the mass loss from the DSC curves of the main-layer water in Friedel's salt (wt.%), and M_{FS} and m_{H_2O} are the molar masses of Friedel's salt and water, with values of 561.3 g/mol and 18 g/mol, respectively.

Figure 5 shows the CH and FS contents of the hardened paste of ASR-cement composite cementitious material calculated by Equations (1) and (2). Assuming that ASR acts as an inert material and does not participate in the hydration reaction, in theory, the CH content in ASR-10 and ASR-20 should be 90% and 80% of the OPC, respectively; however, the amounts of CH in ASR-10 and ASR-20 were 85.7% and 65.15% of the OPC, respectively, which are lower than the theoretical values. This indicates that the addition of ASR consumed CH in the paste, but it does not mean that the hydration degree of cement was reduced. Similarly, the theoretical FS content of the ASR-20 group should be twice that of the ASR-10 group, but the FS contents of ASR-10 and ASR-20 pastes in Figure 5 were 3.83% and 6.17%, respectively. The latter value is much lower than the theoretical value, which is due to the amount of aluminate (C_3A) in cement, because the formation of FS is related to the aluminate-to-chloride ratio.

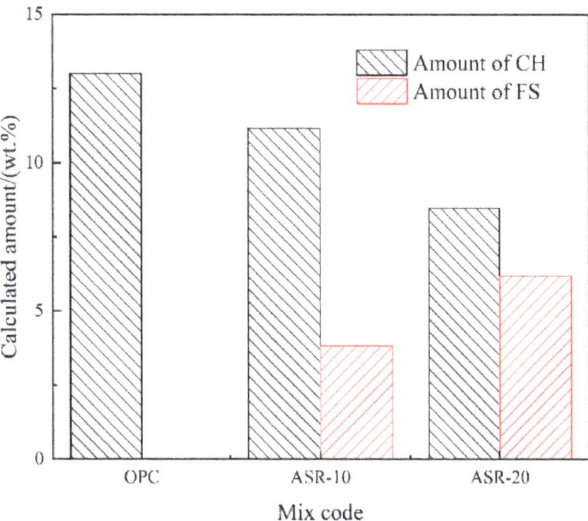

Figure 5. Calculated amounts of $Ca(OH)_2$ and Friedel's salt in the ASR-cement hardened pastes.

3.4. Hydration Heat

The hydration heat of cementitious materials can reflect the hydration rate and has an important influence on the setting time and early strength of the slurry. Figure 6 shows the hydration heat release curves of ASR-cement composite cementitious materials over 72 h. Generally, the hydration heat release of cementitious materials can be divided into five stages [20]: (I) the rapid exothermic reaction, (II) the dormant period, (III) the acceleration, (IV) the deceleration, and (V) the steady state.

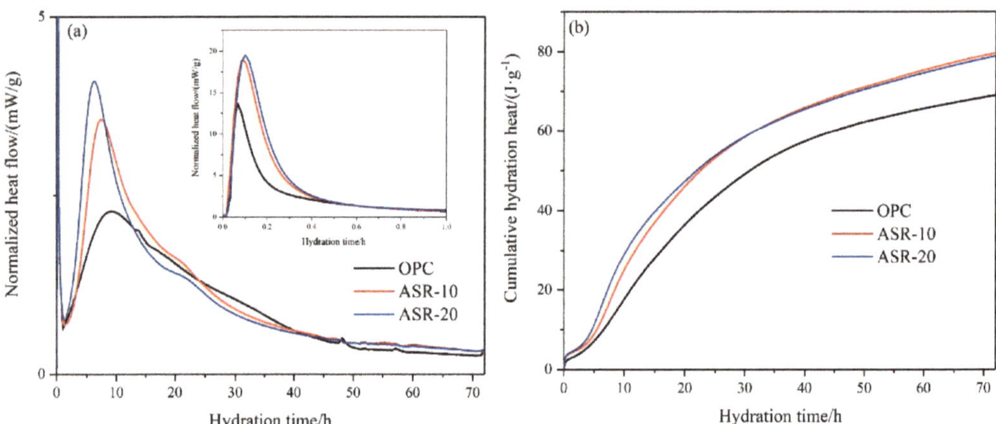

Figure 6. The hydration heat of varying pastes within 72 h: (**a**) normalized heat flow; (**b**) cumulative hydration heat.

It can be seen from Figure 6a that the cementitious material reacted rapidly with water and released heat, indicating the rapid exothermic reaction stage (I). The released heat was mainly caused by the dissolution of raw material minerals in the liquid phase. After the addition of ASR, the exothermic heat of the rapid exothermic reaction stage increased significantly, indicating that the addition of ASR accelerated the dissociation of raw material particles. The dissociation rate increased at higher amounts of ASR. With the continuous hydration reaction, the raw materials released more heat after the dormant period (II) and entered the acceleration period (III). The addition of ASR caused the fast exothermic peak to become narrow and the maximum exothermic point to advance. The maximum exothermic peaks of OPC, ASR-10, and ASR-20 appeared at 9.18 h, 7.88 h, and 6.16 h, respectively, which indicates that ASR significantly promoted the hydration of cementitious materials and accelerated the hydration reaction.

In Figure 6b, the cumulative hydration heat release of the three groups of cementitious materials over 24 h followed the order ASR-20 > ASR-10 > OPC. This indicates that the addition of ASR increased the hydration heat release of cementitious materials in the early stages of hydration, and a higher ASR content increased the hydration heat release; however, after 72 h of hydration, the order of the cumulative hydration heat release became ASR-10 > ASR-20 > OPC, whose values were 78.89 J·g⁻¹, 79.11 J·g⁻¹, and 69.23 J·g⁻¹, respectively. The main reason for this phenomenon was that the ASR does not have pozzolanic activity, so the main body of the hydration reaction is cement; thus, the later hydration reaction will lose its material support when the amount of cement is reduced.

3.5. Pore Structure

The porosity and pore distribution of the hardened pastes of cementitious materials can be used to study the hydration reaction of cement-based materials. In this respect, the cumulative pore volume reflects the overall compactness of a paste, which is positively related to the compressive strength of the paste, and the different pore distributions will affect the durability of pastes by affecting their shrinkage and impermeability [21]. The quantity, shape, and size of hydration products have important influences on the porosity and pore distribution of hardened pastes.

From the cumulative pore volume of the hardened pastes with different ASR contents at 3 days in Figure 7a, the order of the cumulative pore volume from small to large is ASR-5, OPC, ASR-10, and ASR-20. This means that when 5% ASR was added, the cumulative pore volume of the hardened paste decreased. When the ASR content exceeded 10%, the cumulative pore volume of the hardened paste was greater than that of the blank control group (OPC), and the greater the ASR content, the greater the pore volume; therefore, from the point of view of cumulative pore volume, the ASR content in cementitious materials should not exceed 10%.

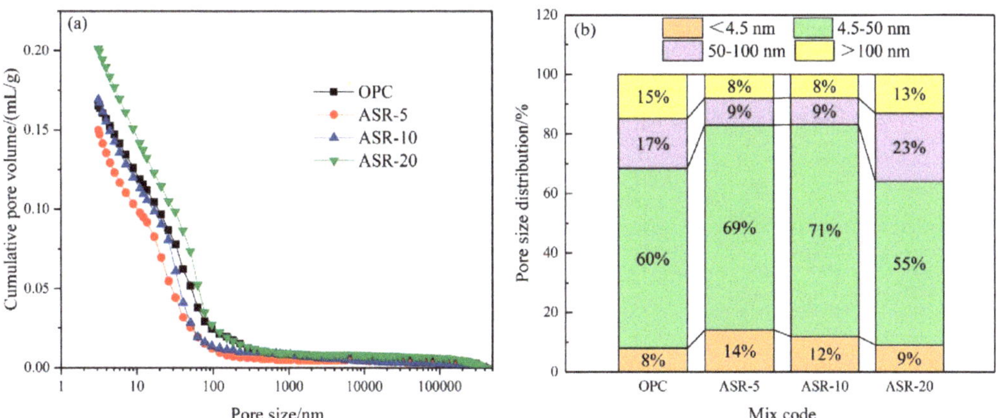

Figure 7. Pore structures of the ASR-cement hardened pastes after 3 days. Cumulative pore volume (**a**); pore size distributions (**b**).

Figure 7b shows the pore distribution of the hardened pastes at 3 days. Many studies have shown that the pore distribution of pastes can be divided into four size ranges [22,23]: gel micropores (<4.5 nm), mesopores (4.5–50 nm), middle capillary pores (50–100 nm), and large capillary pores (>100 nm). A pore size larger than 50 nm can also be called harmful pores [24]. It can be seen from Figure 7b that when the alkali slag content is ≤ 10%, the volume of gel micropores and mesopores in the paste increased significantly, while the volume of middle capillary pores and large capillary pores decreased. In this respect, an appropriate amount of ASR can be added to reduce the proportion of harmful pores (>50 nm), which can improve the penetration resistance of the paste and optimize its durability.

There are two main effects of ASR on the pore structure of the pastes. One is the formation of a new hydration product, FS, with a 1–2 μm diameter. The second is that ASR is formed by the pressure filtration of very fine particles (<1 μm particle size), which play a physical filling role, thereby reducing the pore volume of the paste; however, the addition of ASR will reduce the amount of cement, thereby reducing the C-S-H gel and CH contents in the hydration products, especially when the ASR content exceeds 10%. The TG analysis also supports this. The pore structure of the paste is the result of the accumulation of C-S-H gels, FS, and CH, which explains why ASR-5 displayed the smallest cumulative pore volume and the best pore distribution.

3.6. Setting Time and Compressive Strength

Figure 8 shows the setting time of ASR-cement composite cementitious materials and the compressive strength of the resulting mortars. According to Figure 8a, the addition of ASR significantly shortened the setting time of the cementitious materials—the higher the

ASR content, the shorter the initial setting time and final setting time. This is mainly because the addition of ASR accelerated the hydration rate of the cement, which is supported by the normalized heat flow of the cementitious materials (Figure 6a). Compared with the OPC control group, the final setting time of cementitious materials with 5%, 10%, and 20% ASR decreased by 30 min, 45 min, and 70 min, respectively. This was mainly because the addition of ASR accelerated the hydration rate of cement, which is also supported by the hydration heat results (Figure 6b).

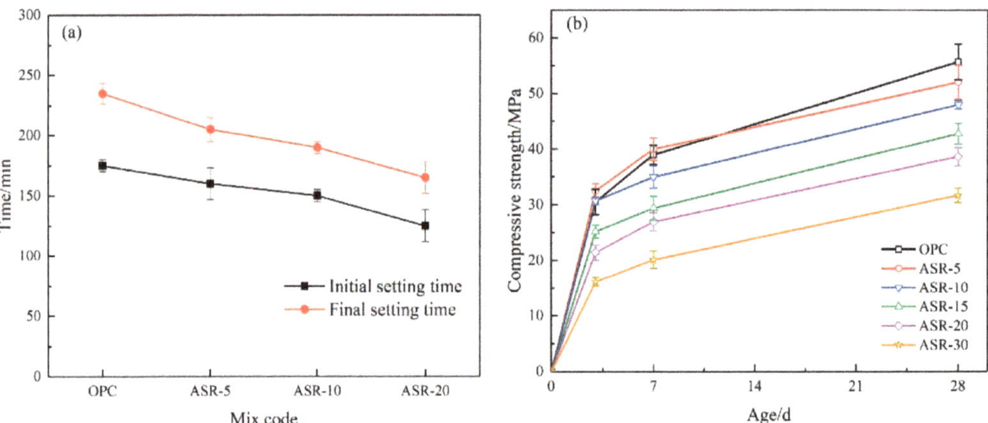

Figure 8. The setting time (**a**) of ASR-cement composite cementitious material and (**b**) mortar compressive strength.

The compressive strengths of the mortars test at 3 d, 7 d, and 28 d are shown in Figure 8b. At the 3 days age, the mortar sample with 5% ASR content had the highest compressive strength of 32.5 MPa, which is 6.6% higher than the OPC control group. The compressive strength of the mortar sample with an ASR content of 10% was basically the same as that of the OPC group. When it exceeded 10%, it was lower than the blank group OPC. The results mean that the addition of ASR improved the 3-day compressive strength of the mortar when its content did not exceed 10%. After curing for 7 days, only the ASR-5 group had a higher compressive strength than OPC, which was 40.02 MPa, an increase of 7.09%. The 28-day compressive strength of all samples containing ASR was lower than that of the OPC control group, indicating that the addition of ASR did not improve the long-term compressive strength of the mortar. A higher ASR content had a more negative effect on the long-term compressive strength.

4. Further Discussion

4.1. Hydration Mechanism

The mineral phase composition of ASR indicates that it has no pozzolanic activity; therefore, the hydration mechanism of ASR-cement composite cementing material mainly involves the activation of cement by ASR. According to the XRD and SEM analysis, the hydration products of ASR-cement composite cementitious materials were mainly C-S-H gel, CH, and FS. Among them, the C-S-H gel and CH formed mainly due to the reaction of silicate in cement and water (as shown in Equation (3)) [25], while the formation of FS was mainly caused by the reaction of aluminate in cement and chloride in ASR (as shown in Equation (4)) [26] or the reaction of sulfoaluminate and chloride salt in the liquid phase (as shown in Equation (5)) [27].

$$3CaO{\cdot}SiO_2 + nH_2O \rightarrow xCaO{\cdot}SiO_2{\cdot}yH_2O + (3-x)Ca(OH)_2 \tag{3}$$

$$3CaO{\cdot}Al_2O_3 + 2Ca(OH)_2 + 2NaCl + 10H_2O \rightarrow 3CaO{\cdot}Al_2O_3{\cdot}CaCl_2{\cdot}10H_2O + 2NaOH \tag{4}$$

$$AFm + 2NaCl \rightarrow 3CaO \cdot Al_2O_3 \cdot CaCl_2 \cdot 10H_2O + 2NaSO_4 + 2H_2O \tag{5}$$

Equation (5) indicates that a small amount of gypsum in cement reacts with aluminate to form monosulfate (AFm, $3CaO \cdot Al_2O_3 \cdot CaSO_4$). In the presence of Cl^- in a liquid environment, the Cl^- will replace SO_4^{2-} in AFm to generate FS [28].

Figure 5 shows that the addition of ASR reduces the theoretical content of CH in the paste, indicating that part of the CH is consumed during hydration. This possibly occurred via its consumption during FS formation (Equation (4)). Another chloride-containing hydration product, calcium oxychloride ($3CaO \cdot CaCl_2 \cdot 15H_2O$), is also likely to exist in the paste (Equation (6)) [29]. The formation of calcium oxychloride also consumes CH, but its formation conditions generally require a higher $CaCl_2$ concentration. There is also a competitive relationship between the formation of calcium oxychloride and the formation of FS, so the amount of calcium oxychloride is lower, so it was not observed in the characterization results, but many studies have proved its existence [30,31].

$$12H_2O + Ca^{2+} + 2Cl^- + 3Ca(OH)_2 \rightarrow 3CaO \cdot CaCl_2 \cdot 15H_2O \tag{6}$$

According to the hydration heat and TG-DSC results, ASR has an obvious effect on the early activation of cement, which is mainly dominated by chloride salts and $CaCO_3$. The activation effect of chloride salts on cement is mainly reflected in two aspects: firstly, it can accelerate the dissociation rate of silicates (C_3S and C_2S), thus accelerating the hydration reaction rate. Secondly, the formation of FS consumes the aluminate ions in the liquid phase and, therefore, accelerates the dissociation rate of aluminate in the liquid phase. The activation effect of $CaCO_3$ on cement is also mainly reflected in two aspects. One is that $CaCO_3$ can accelerate the hydration rate of silicate, and the other is that $CaCO_3$ particles can serve as nucleation sites in the liquid phase. The hydration products can quickly adhere to its surface after forming, so that the hydration reaction can continue in the liquid phase. In this respect, a faster hydration rate means a higher hydration heat release and shorter setting time, which is consistent with the results in Figures 4 and 8a.

4.2. Applications

The characteristics of ASR-cement composite cementitious materials prevent their use in many fields. First, they cannot be used to prepare reinforced concrete because the chloride salts in ASR can corrode steel [32]. Secondly, they cannot be used to prepare mass concrete because the hydration heat is higher than that of OPC, which increases the concrete temperature and the cracking risk. Finally, ASR contains carbonates, which decrease the sulfate resistance of concrete [33].

The results of this study shared some similarities and differences with those of other studies [1,9–12]. The common point is that ASR shortened the setting time of cementitious materials, whether it is used with cement or with GGBS or FA materials. There are two main differences. One is that when ASR is used together with cement, it improves the pore structure of the paste because the volume of pores >50 nm was reduced, but this result is not clear in other studies. Secondly, when ASR was used with cement, its 28-day compressive strength was poor. Li [34] obtained the same result when using ASR to burn magnesium oxychloride cement, but, when used with GGBS-based materials, its 3- and 28-day compressive strength was significantly improved.

Overall, the applications of ASR-cement composite cementitious materials have several potential development directions. One is the preparation of special-purpose concrete, such as for prefabricated components, because they can greatly improve the production efficiency. They can also be used to produce road concrete, slope stones, and artificial reef concrete, because soda plants are located near the sea [35]. Second, they can be used in general building materials, such as building mortar and bricks. Third, they can be used as a binder for mine filling materials. Finally, the costs of using ASR are almost negligible, which gives ASR-cement composite cementing materials good economic benefits—the higher the amount of ASR used, the better the economic benefits.

In addition, the application scope of ASR-cement composite cementitious materials can be broadened if it is used together with SCMs with high aluminate content (e.g., GGBS and FA). Some researchers [35,36] have studied the properties of ASR-GGBS materials, and the results showed that ASR activates GGBS; however, it also has significant disadvantages, such as low compressive strength in the early stage (3 days) and a high proportion of harmful pore sizes in the hardened paste. In Figures 7 and 8, when ASR and cement are used together, volume of the harmful pore is reduced when using 10% ASR, and the 3-day compressive strength was also significantly increased. It seems that a more optimal paste pore structure can be obtained if the ASR-cement cementitious material is used with GGBS. In this respect, using ASR-cement composite cementitious materials and SCMs with a high aluminate content can be used to prepare cementitious materials with better performance, but this requires further research.

5. Conclusions

In this paper, a composite cementing material was prepared using ASR and cement, and its hydration mechanism and hydration characteristics were studied using XRD, SEM, TG-DSC, and MIP. The main conclusions are as follows:

The main hydration products of ASR-cement were C-S-H gel, CH, and FS. The shape of FS was hexagonal plate-like CH but smaller in size (1–2 μm). The amount of FS produced depends on the ASR content in the cementitious material, and more FS was produced at a higher ASR dosage. Meanwhile, the addition of ASR also reduced the theoretical content of CH in the paste, mainly because CH was consumed during the hydration reaction.

The addition of ASR improved the pore structure of ASR-cement paste by decreasing the number of harmful pores (>50 nm). The paste displayed the optimal pore structure at an ASR content of 5%. Meanwhile, ASR increased the heat of hydration of cementitious materials, mainly because chlorides and carbonates accelerated the hydration of silicates and aluminates. Correspondingly, the setting time of the cementitious material also decreased upon increasing the ASR content.

The early compressive strength of the mortar was significantly improved when the ASR content in the cementitious material exceeded 10%, but the long-term compressive strength was poor regardless of the ASR content; therefore, there are disadvantages in the mechanical properties of ASR-cement cementitious materials, but they have advantages in their working performance.

ASR-cement composite cementing materials can be applied in plain concrete without steel bars and are especially suitable for preparing marine concrete, road concrete, etc. When used with GGBS or fly ash, they will have better performance and can also broaden their applications. Overall, the ASR-cement composite cementitious materials have significant advantages in terms of costs and reducing CO_2 emissions of the cement industry.

Author Contributions: D.X.: Conceptualization, Methodology, Validation, Investigation, Resources, Writing—original draft, Writing—review and editing; P.F.: Resources, Project administration, Funding acquisition; W.N.: Methodology, Validation, Writing—original draft, Supervision; Q.W.: Conceptualization, Writing—original draft; K.L.: Project administration, Funding acquisition. All authors have read and agreed to the published version of the manuscript.

Funding: This study was financially supported by National Key R & D Program of China (NO.2018Y-FC1900603) and Fundamental Research Funds for the Central Universities (NO. FRF-MP-20-01).

Institutional Review Board Statement: Not applicable.

Informed Consent Statement: Not applicable.

Data Availability Statement: The data presented in this study are available on request from the corresponding author. The data are not publicly available due to the privacy restrictions.

Conflicts of Interest: There are no conflict to declare.

References

1. Lin, Y.; Xu, D.; Zhao, X. Effect of soda residue addition and its chemical composition on physical properties and hydration products of soda residue-activated slag cementitious materials. *Materials* **2020**, *13*, 1789. [CrossRef]
2. Yang, Y.B.; Yong-Qiang, P.U.; Yan, W.J.; Guo, W.Y.; Wang, H.C. Microstructure and chloride ion dissolution characteristics of soda residue. *J. South China Univ. Technol. Nat. Sci. Ed.* **2017**, *45*, 82–89.
3. Wang, X.-B.; Yan, X.; Li, X.-Y. Environmental risk for application of ammonia-soda white mud in soils in China. *J. Integr. Agric.* **2020**, *19*, 601–611. [CrossRef]
4. Lin, S.; Luo, H.J. Preparation of soil nutrient amendment using white mud produced in ammonia-soda process and its environmental assessment. *T. Nonferr. Metal. Soc.* **2009**, *19*, 1383–1388.
5. Li, Y.; Song, X.; Chen, G.; Sun, Z.; Xu, Y.; Yu, J. Preparation of calcium carbonate and hydrogen chloride from distiller waste based on reactive extraction–crystallization process. *Chem. Eng. J.* **2015**, *278*, 55–61. [CrossRef]
6. Kasikowski, T.; Buczkowski, R.; Igliński, B.; Peszyńska-Białczyk, K.; Lemanowska, E. Utilization of distiller waste from am-monia-soda processing. *J. Clean. Prod.* **2004**, *12*, 759–769. [CrossRef]
7. He, J.; Shi, X.-K.; Li, Z.-X.; Zhang, L.; Feng, X.-Y.; Zhou, L.-R. Strength properties of dredged soil at high water content treated with soda residue, carbide slag, and ground granulated blast furnace slag. *Constr. Build. Mater.* **2020**, *242*, 118126. [CrossRef]
8. Liu, J.; Zha, F.; Long, X.; Kang, B.; Zhang, J. Strength and microstructure characteristics of cement-soda residue solidified/stabilized zinc contaminated soil subjected to freezing–thawing cycles. *Cold Reg. Sci. Technol.* **2020**, *172*, 102992. [CrossRef]
9. Zhao, X.; Liu, C.; Wang, L.; Zuo, L.; Zhu, Q.; Ma, W. Physical and mechanical properties and micro characteristics of fly ash-based geopolymers incorporating soda residue. *Cem. Concr. Compos.* **2019**, *98*, 125–136. [CrossRef]
10. Wang, Q.; Li, J.; Yao, G.; Zhu, X.; Hu, S.; Qiu, J.; Chen, P.; Lyu, X. Characterization of the mechanical properties and microcosmic mechanism of Portland cement prepared with soda residue. *Constr. Build. Mater.* **2020**, *241*, 117994. [CrossRef]
11. Guo, W.; Zhang, Z.; Bai, Y.; Zhao, G.; Sang, Z.; Zhao, Q. Development and characterization of a new multi-strength level binder system using soda residue-carbide slag as composite activator. *Constr. Build. Mater.* **2021**, *291*, 123367. [CrossRef]
12. Xu, D.; Ni, W.; Wang, Q.; Xu, C.; Li, K. Ammonia-soda residue and metallurgical slags from iron and steel industries as cementitious materials for clinker-free concretes. *J. Clean. Prod.* **2021**, *307*, 127262. [CrossRef]
13. Hasanbeigi, A.; Menke, R.; Price, R. The CO_2 abatement cost curve for the thailand cement industry. *J. Clean. Prod.* **2010**, *18*, 1509–1518. [CrossRef]
14. Benhelal, E.; Zahedi, G.; Shamsaei, E.; Bahadori, A. Global strategies and potentials to curb CO2 emissions in cement industry. *J. Clean. Prod.* **2013**, *51*, 142–161. [CrossRef]
15. Lothenbach, B.; Scrivener, K.; Hooton, R.D. Supplementary cementitious materials. *Cem. Concr. Res.* **2011**, *41*, 1244–1256. [CrossRef]
16. Vedalakshmi, R.; Raj, A.S.; Srinivasan, S.; Babu, K.G. Quantification of hydrated cement products of blended cements in low and medium strength concrete using TG and DTA technique. *Thermochim. Acta* **2003**, *407*, 49–60. [CrossRef]
17. Birnin-Yauri, U.; Glasser, F. Friedel's salt, Ca2Al(OH)6(Cl,OH)·2H2O: Its solid solutions and their role in chloride binding. *Cem. Concr. Res.* **1998**, *28*, 1713–1723. [CrossRef]
18. Qiao, C.; Ni, W.; Wang, Q.; Weiss, J. Chloride diffusion and wicking in concrete exposed to NaCl and MgCl2 solutions. *J. Mater. Civ. Eng.* **2018**, *30*, 04018015. [CrossRef]
19. Shi, Z.; Geiker, M.R.; Lothenbach, B.; Weerdt, K.D.; Garzón, S.F.; Enemark-Rasmussen, K.; Skibsted, J. Friedel's salt profiles from thermogravimetric analysis and thermodynamic modelling of Portland cement-based mortars exposed to sodium chloride solution. *Cem. Concr. Comp.* **2017**, *78*, 73–83. [CrossRef]
20. Wang, Q.; Shi, M.; Jun, Y. Influence of classified steel slag with particle sizes smaller than 20 μm on the properties of cement and concrete. *Constr. Build. Mater.* **2016**, *123*, 601–610.
21. Das, B.; Kondraivendhan, B. Implication of pore size distribution parameters on compressive strength, permeability and hydraulic diffusivity of concrete. *Constr. Build. Mater.* **2012**, *28*, 382–386. [CrossRef]
22. Mehta, P.K.; Monteiro, P. *Concrete: Microstructure, Properties, and Materials*; McGraw Hill Education: New York, NY, USA, 2013.
23. Zeng, Q.; Li, K.; Teddy, F.; Dangla, P. Pore structure characterization of cement pastes blended with high-volume fly-ash. *Cem. Concr. Res.* **2012**, *42*, 194–204. [CrossRef]
24. Zhao, Y.; Qiu, J.; Xing, J.; Sun, X. Chemical activation of binary slag cement with low carbon footprint. *J. Clean. Prod.* **2020**, *267*, 121455. [CrossRef]
25. Bullard, J.W.; Jennings, H.M.; Livingston, R.; Nonat, A.; Scherer, G.; Schweitzer, J.S.; Scrivener, K.; Thomas, J. Mechanisms of cement hydration. *Cem. Concr. Res.* **2011**, *41*, 1208–1223. [CrossRef]
26. Suryavanshi, A.K.; Scantlebury, J.D.; Lyon, S.B. Mechanism of friedel's salt formation in cements rich in tri-calcium alu-minate. *Cem. Concr. Res.* **1996**, *26*, 717–727. [CrossRef]
27. Baroghel-Bouny, V.; Wang, X.; Thiery, M.; Saillio, M.; Barberon, F. Prediction of chloride binding isotherms of cementitious materials by analytical model or numerical inverse analysis. *Cem. Concr. Res.* **2012**, *42*, 1207–1224. [CrossRef]
28. Matschei, T.; Lothenbach, B.; Glasser, F. The AFm phase in Portland cement. *Cem. Concr. Res.* **2007**, *37*, 118–130. [CrossRef]
29. Jones, C.; Ramanathan, S.; Suraneni, P.; Hale, W.M. Calcium oxychloride: A critical review of the literature surrounding the formation, deterioration, testing procedures, and recommended mitigation techniques. *Cem. Concr. Compos.* **2020**, *113*, 103663. [CrossRef]

30. Qiao, C.; Suraneni, P.; Ying, T.N.W.; Choudhary, A.; Weiss, J. Chloride binding of cement pastes with fly ash exposed to $CaCl_2$ solutions at 5 and 23 °C. *Cem. Concr. Compos.* **2019**, *97*, 43–53. [CrossRef]
31. Qiao, C.; Suraneni, P.; Weiss, J. Flexural strength reduction of cement pastes exposed to $CaCl_2$ solutions. *Cem. Concr. Compos.* **2018**, *86*, 297–305. [CrossRef]
32. Glass, G.; Buenfeld, N. The influence of chloride binding on the chloride induced corrosion risk in reinforced concrete. *Corros. Sci.* **2000**, *42*, 329–344. [CrossRef]
33. Tosun, K.; Felekoğlu, B.; Baradan, B.; Altun, İ.A. Effects of limestone replacement ratio on the sulfate resistance of Portland limestone cement mortars exposed to extraordinary high sulfate concentrations. *Constr. Build. Mater.* **2009**, *23*, 2534–2544. [CrossRef]
34. Li, C.; Liang, Y.; Jiang, L.; Zhang, C.; Wang, Q. Characteristics of ammonia-soda residue and its reuse in magnesium oxychloride cement pastes. *Constr. Build. Mater.* **2021**, *300*, 123981. [CrossRef]
35. Xu, D.; Ni, W.; Wang, Q.; Xu, C.; Jiang, Y. Preparation of clinker-free concrete by using soda residue composite cementitious material. *J. Harbin Inst. Technol.* **2020**, *52*, 151–160.
36. Lin, Y.; Xu, D.; Zhao, X. Properties and hydration mechanism of soda residue-activated ground granulated blast furnace slag cementitious materials. *Materials* **2021**, *14*, 2883. [CrossRef]

Article

Investigation on Control Burned of Bagasse Ash on the Properties of Bagasse Ash-Blended Mortars

Redeat Seyoum [1], Belay Brehane Tesfamariam [1,*], Dinsefa Mensur Andoshe [1], Ali Algahtani [2,3], Gulam Mohammed Sayeed Ahmed [4,5] and Vineet Tirth [2,3]

[1] Department of Materials Science and Engineering, Adama Science and Technology University, Adama 1888, Ethiopia; redeatseyoumseyoum@gmail.com (R.S.); dinsefa.mensur@astu.edu.et (D.M.A.)

[2] Department of Mechanical Engineering, College of Engineering, King Khalid University, Abha 61413, Asir, Saudi Arabia; alialgahtani@kku.edu.sa (A.A.); vtirth@kku.edu.sa (V.T.)

[3] Research Center for Advanced Materials Science (RCAMS), King Khalid University, Abha 61413, Asir, Saudi Arabia

[4] Department of Mechanical Design and Manufacturing Engineering, Adama Science and Technology University, Adama 1888, Ethiopia; drgmsa786@gmail.com

[5] Center of Excellence (COE) for Advanced Manufacturing Engineering, Department of Mechanical Design and Manufacturing Engineering, Adama Science and Technology University, Adama 1888, Ethiopia

* Correspondence: bellove22@gmail.com

Abstract: In recent years, partial replacement of cement with bagasse ash has been given attention for construction application due to its pozzolanic characteristics. Sugarcane bagasse ash and fine bagasse particles are abundant byproducts of the sugar industries and are disposed of in landfills. Our study presents the effect of burning bagasse at different temperatures (300 °C and 600 °C) on the compressive strength and physical properties of bagasse ash-blended mortars. Experimental results have revealed that bagasse produced more amorphous silica with very low carbon contents when it was burned at 600 °C/2 h. The compressive strength of mortar was improved when 5% bagasse ash replaced ordinary portland cement (OPC) at early curing ages. The addition of 10% bagasse ash cement also increased the compressive strength of mortars at 14 and 28 days of curing. However, none of the bagasse ash-blended portland pozzolana cement (PPC) mortars have shown improvement on compressive strength with the addition of bagasse ash. Characterization of bagasse ash was done using XRD, DTA-TGA, SEM, and atomic absorption spectrometry. Moreover, durability of mortars was checked by measuring water absorption and apparent porosity for bagasse ash-blended mortars.

Keywords: sugarcane bagasse ash; pozzolanic; amorphous silica; OPC and PPC; mortar

Citation: Seyoum, R.; Tesfamariam, B.B.; Andoshe, D.M.; Algahtani, A.; Ahmed, G.M.S.; Tirth, V. Investigation on Control Burned of Bagasse Ash on the Properties of Bagasse Ash-Blended Mortars. *Materials* **2021**, *14*, 4991. https://doi.org/10.3390/ma14174991

Academic Editor: Carlos Morón Fernández

Received: 2 August 2021
Accepted: 29 August 2021
Published: 1 September 2021

Publisher's Note: MDPI stays neutral with regard to jurisdictional claims in published maps and institutional affiliations.

1. Introduction

Partial replacement of cement with waste materials has been given great attention in recent years due to high CO_2 emissions from cement industries, the high cost of cement, and the need to improve cement properties [1,2]. Typically, ashes of rice husk [3,4], bagasse [5–8], coffee husk [9], cob corn [10], fly ash [11], and silica fume [12] were studied as their pozzolanic nature has highly reactive amorphous siliceous and aluminous materials. Five (5)% rice husk ash with an average particle size of 95 μm blended OPC have enhanced durability of concrete as well as compressive strength from 36.8 MPa to 38.7 MPa at 28 days of curing [13]. Ash-blended cement also has the potential to reduce the energy consumption of cement manufacturing. Silica fume has also improved high-performance concrete due to its ultra-fine particles, which leads to reducing the porosity of concrete and the formation of calcium silicate hydrate (C–S–H) gel [14–16]. Sabir, B.B also investigated high-curing-temperature (50 °C) results in higher strengths of silica-fume-embedded concrete when compared to a lower curing temperature (20 °C) at early ages [14]. Compressive strength, as well as flexural strength of OPC, was enhanced with the addition of silica fume up to 12% [16].

OPC with partial bagasse ash improved the compressive strength compared to ordinary concretes at 28 days curing age [17]. Certain bagasse ash-blended concretes can also enhance the durability of concrete [18] and decrease the heat of hydration [8]. However, industrial bagasse ash has a high carbon content and unburned organic matter [19] which negatively affects concrete properties and also lowers the workability of concrete. Trifunovic, P.D. et al. explained the negative effect of carbon on the compressive strength of bottom ash-blended mortar samples [20]. Thus, we expected an optimum amount of control-burned bagasse ash replacement to have no contrary influence on the properties of Ordinary Portland Cement (OPC).

Therefore, we focused on investigating the BA amount of a control-burned bagasse ash at a high temperature (600 °C/2 h) on the properties of bagasse ash-blended mortar. Mineralogical composition and properties of bagasse ash differ due to diversity of sugarcane plants, firing temperature and time, cooling rate, quality of bagasse and collection techniques, and ash particle sizes [21]. Specific-gravity and specific-area of BA also differ with firing temperature. Thus, we compared compressive strength and physical properties of bagasse ash-blended OPC/PPC mortar samples with conventional mortar samples. We selected bagasse ash as OPC replacement instead of other waste material ashes because bagasse ash has higher pozzolanic reactivity and is available in large amounts, free of cost in many countries. It was predicted that partial replacement of OPC with well-burnt bagasse ash will increase the compressive strength and durability of blended mortars.

2. Experimental Procedure

2.1. Preparation and Characterization of Bagasse Ash

Bagasse ash (BA) was collected from Wonji-Shoa sugar factory (8°27′14.4″ N, 39°13′48″ E) of Oromia, Ethiopia after burned at a temperature around 300 °C (a lower temperature) and the ash leaves as a landfill as shown in Figure 1a. The dark (char) color of ash was an indication of higher carbon content due to incomplete combustion, which was also explained in previous studies [20,22]. Large amounts of bagasse particles similarly leave as landfills in the factory as shown in Figure 1b. We also collected bagasse particles and then pretreated them in distilled water to remove all undesirable materials. After that, we dried bagasse at 100 °C for 24 h and then burned it at a controlled temperature from 300 °C to 750 °C for 1–2 h (Table 1). Bagasse ash's color was changed to white as burning temperatures increased due to the elimination of carbon. We analyzed bagasse ash burned at 300 °C (low temperature (LT–BA)) and 600 °C (higher temperature (HT–BA)) using Differential Thermal Analysis-Thermogravimetry Analysis (Shimadzu, DTG-60H). Scanning Electron Microscopy (SEM, Shimadzu, COXIEM-30) and X-ray Diffractometer (Shimadzu, XRD 7000, λ CuK α = 1.5418 Å) were used to confirm the presence of phase and morphology of bagasse ashes. Atomic Absorption Spectrometry (AAS-model, spectra AA-20 plus) was used to analyze bagasse ash's major and minor oxide compositions. Moreover, we measured the compressive strength bagasse ash-blended mortar samples. Lastly, water absorption and apparent porosity tests for HT–BA blended mortars were performed to check the durability of the specimens according to ASTM C20-00 (2015) [23].

(**a**) Bagasse ash

(**b**) Bagasse

Figure 1. Byproducts as landfills at Wonji Sugar factory, Ethiopia, (**a**) Burned sugarcane bagasse ash at a temperature $\geq$300 °C for 4–5 min and inset image of bagasse ashes; (**b**) Unburned sugarcane bagasse and inset image of the dry bagasse particles.

Table 1. Bagasse ashes (BA) were prepared by burning 300–750 °C for 1–2 h.

Burning Time	Burning Temperature			
	300 °C	450 °C	600 °C	750 °C
1 h	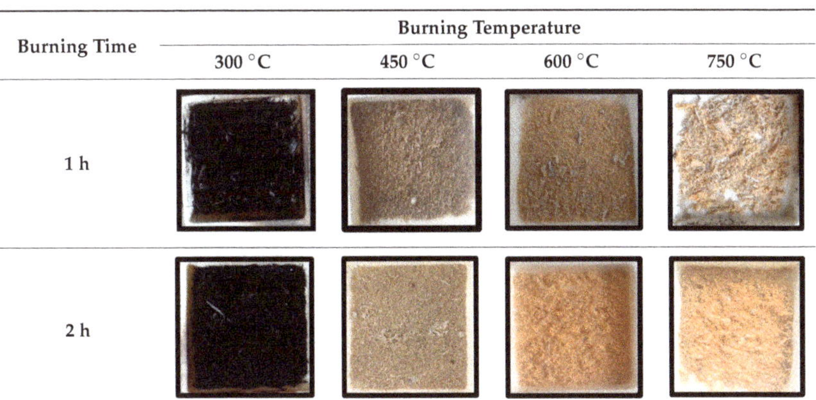			
2 h				

We calcinated bagasse at a controlled temperature from 300 °C to 750 °C for 1–2 h after the bagasse was dried at 100 °C for 24 h. We obtained BA having black to white color/appearance as shown in Table 1. As the burning temperature and duration increase, BA color gradually changed from black to white ash.

2.2. Preparation and Analysis of BA-Blended Mortar Samples

Mortar samples with sizes of $70 \times 70 \times 70$ mm were prepared using bagasse ashes, portland cement, sand, and water as shown in Table 2. Water to cement plus BA (0%, 5%, or 10%) ratio was 0.53. However, as bagasse ash content increased (above 10%), the workability of mixing proportion BA blended mortars decreased a lot. Thus, extra water (20 mL) was added, while BA 15%, 20%, and 25% was added to cement for mortar preparation. The samples were then de-molded after 24 h and measured the compressive strength at the curing age of 3, 7, 14, and 28 days. We used bagasse ash after burning at a high temperature (HT–BA), 600 °C/2 h, and bagasse ash after burning at a low temperature, 300 °C/2 h (LT–BA) as revealed in Table 2 below.

Table 2. Mixing proportion of BA-blended PC mortars, Cement : Sand = 1 : 3.

Amount BA (%)	PC-Portland Cements (gm)	LT-BA (gm)	HT-B (gm)	Sand (gm)	Water (mL)
BA 0	150.0	0	0	450	80
LT-BA 5	142.5	7.5	0	450	80
LT-BA 10	135.0	15.0	0	450	80
LT-BA 15	127.5	22.5	0	450	100
LT-BA 20	120.0	30.0	0	450	100
LT-BA 25	112.5	37.5	0	450	100
HT-BA 5	142.5	0	7.5	450	80
HT-BA 10	135.0	0	15.0	450	80
HT-BA 15	127.5	0	22.5	450	100
HT-BA 20	120.0	0	30.0	450	100
HT-BA 25	112.5	0	37.5	450	100

3. Results and Discussion

3.1. Mineralogical Compositions of Bagasse Ash (BA)

Table 3 displays mineralogical compositions of BA collected from the sugar industry. We used an Atomic Absorption Spectrometry measurement to reveal the major and minor oxide composition of the bagasse ash, as well as the loss on ignition (LOI) percent to check its pozzolan oxide contents (SiO_2 + Al_2O_3 + Fe_2O_3), alkaline oxides, and loss on ignition. The amount of pozzolan oxides (SiO_2 + Al_2O_3 + Fe_2O_3) in BA is higher (80.02%) than the minimum content of these oxides (70%) required by ASTM C-618 [24]. The sum of pozzolan oxides of OPC is 31.6%, but bagasse ash has 80.02%, as shown in Table 3. Thus, the result testified the pozzolanic nature of BA as per ASTM C-618 specifications and taken as pozzolan materials for cement replacement. Moreover, loss on ignition (LOI) of BA is 4.75% which is less than 10%. Thus, the pozzolanic activity would upsurge with partial BA replacement cement as stated in a previous study [25]. Chemical analysis data has also indicated BA has a three-times higher silica content (65.06%) than OPC (22.82%). However, bagasse ash has a high alkali content (Na_2O + K_2O = 8.66%), implying a high potential for the alkali–silica reaction, which might have an adverse effect.

Table 3. Chemical compositions of sugarcane bagasse ash (SCBA) and Dangote OPC cement.

Amount (%)	SiO_2	Al_2O_3	Fe_2O_3	CaO	MgO	Na_2O	K_2O	MnO	P_2O_5	TiO_2	H_2O	LOI
Bagasse Ash	65.06	10.88	4.08	1.14	1.30	2.06	6.60	0.10	0.79	0.24	0.66	4.75
OPC [26]	22.82	5.41	3.37	66.32	1.46	-	-	-	-	-	-	-

3.2. Thermal Analysis of Bagasse Ash

We used DTA–TGA to measures the thermal stability of bagasse at different temperatures, as illustrated in Figure 2. From the DTA measurement, a small endothermic peak was observed at nearly 70 °C, which is attributed to water evaporation (4.35%). Endothermic peaks between 200 to 300 °C and 350 to 500 °C represent the decomposition of organic matters/compositions of bagasse [27]. During the combustion process, a strong exothermic peak was detected at 356 °C and 446 °C. The first strong peak at 356 °C indicated the oxidation of the volatile product [28]. The second strong absorption peak was at approximately 550 °C, representing carbon removal from bagasse as stated in previous studies; i.e., at nearly 600 °C for prolonged heating, carbon can be removed from BA [19,22]. TGA analysis has shown the mass loss between 200 to 300 °C and 350 to 500 °C is more than 80%, which is caused by the decomposition of organic matter and the combustion of unburned carbon. The second mass-loss is mainly associated with the thermal decomposition of hemicellulose with much less cellulose and lignin, while the third region is especially associated with the cellulose with much less hemicellulose and lignin.

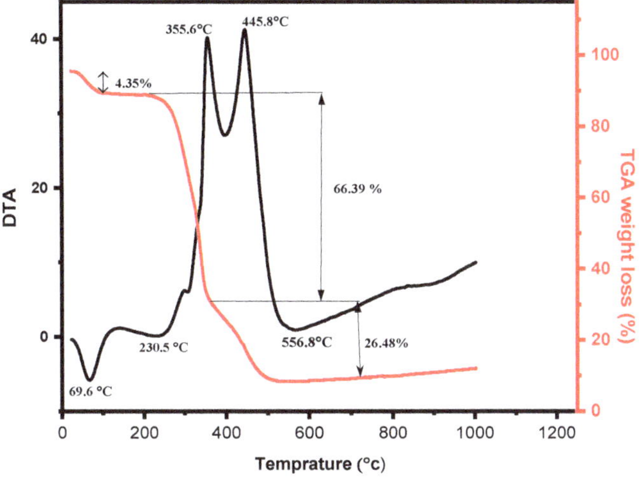

Figure 2. TGA and DTA curves for sugarcane bagasse heating up to 1000 °C.

3.3. Compressive Strength of Bagasse Ash (BA)-Blended OPC Mortar Sample

The compressive strength of HT–BA blended and LT–BA blended OPC mortar samples are shown in Figure 3a,b, respectively. A higher compressive strength of mortar was obtained at 5% HT–BA replacement OPC, compared to conventional mortar for all curing ages. The 10% HT–BA replacement also gave high strength to OPC mortars at curing ages of 14 and 28 days (Table 4). Voids among the cement and sand in mortars were probably filled with fine BA, which caused an increase in the BA-blended mortar's strength at an early age. Moreover, the formation of additional Calcium Silicate Hydrated (C–S–H) gel as a result of the reaction of the active silica of BA and the $Ca(OH)_2$ from cement hydration is necessary to improve the strength of mortars. However, the compressive strength of OPC mortars decreased with a high amount of BA replacement at an early age. This is most likely related to the reduction of the $3CaO \cdot SiO_2$ (C_3S) alite phase amount in a mortar with BA replacement, as the alite phase is responsible for giving early-age strength. Moreover, the $3CaO \cdot Al_2O_3$ (C_3A) aluminite phase is also expected to decrease with increasing BA content. Thus, the heat of the hydration rate most probably decreased with BA, as explained in a previous study [8]. In our experimental works, the workability of BA-blended mortar was decreased with increasing BA content, which is related to high water absorption and a larger specific surface area of BA. Thus, we added an extra 20 mL of water as 15%, 20%, and 25% BA-blended mortars were prepared. This might cause higher overall porosity of the mortars. Higher compressive strength was obtained at HT–BA blended OPC mortars compared to LT–BA blended OPC mortars at early ages.

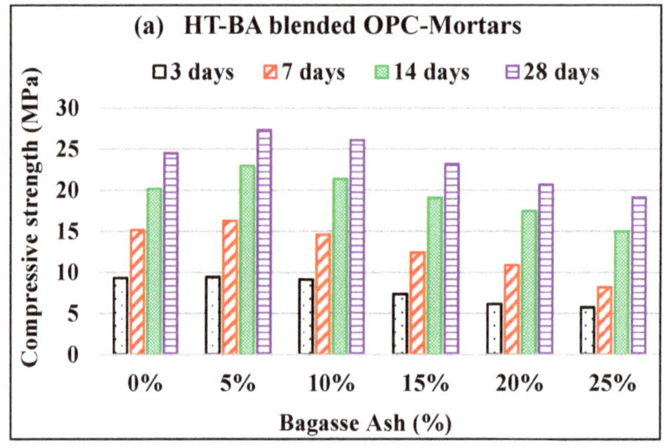

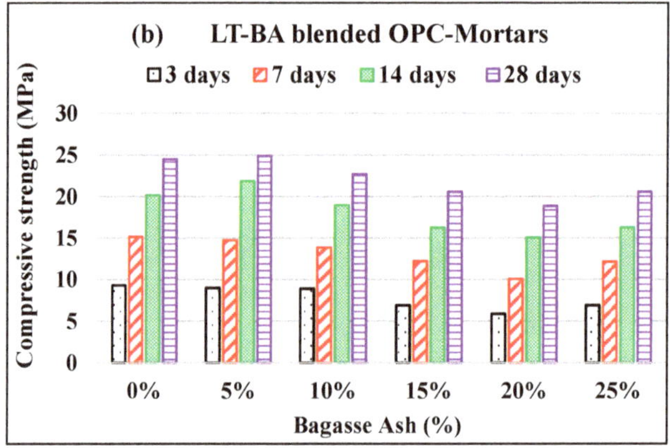

Figure 3. Compressive strength of bagasse ash-blended OPC mortars: (**a**) BA burned at 600 °C/2 h (HT-BA); (**b**) BA burned at 300 °C/2 h (LT-BA).

Table 4. Compressive strength of bagasse ash (HT–BA and LT–BA) blended OPC mortar samples. BA amount from 0 to 25% of cement replacement.

No.	Bagasse Ash (BA) (%)	Compressive Strength (MPa) of BA Blended OPC–Mortars							
		3 Days		7 Days		14 Days		28 Days	
		LT-BA	HT-BA	LT-BA	HT-BA	LT-BA	HT-BA	LT-BA	HT-BA
1	BA 0	9.26		15.20		20.2		24.5	
2	BA 5	8.97	9.40	14.8	16.29	21.9	23.0	24.9	27.3
3	BA 10	8.88	9.10	13.9	14.61	19.0	21.4	22.7	26.1
4	BA 15	6.89	7.32	12.2	13.50	16.3	19.1	20.6	23.2
5	BA 20	5.85	6.12	10.1	10.90	15.1	17.5	18.9	20.7
6	BA 25	5.30	5.71	7.71	8.21	14.3	15.0	17.30	19.1

3.4. Compressive Strength of Bagasse Ash-Blended PPC Mortar Samples

Table 5 illustrates the compressive strength of BA-blended PPC mortars decreased with the increase of BA at early ages up to 28 days. None of the LT–BA and HT–BA addition has shown an improvement in the compressive strength of PPC mortars. This is most likely related to the fact that the alite phase amount decreased with an increasing BA in BA-blended PPC mortars. Nevertheless, the compressive strength variation between PPC mortars and 5% HT–BA-blended PPC mortars was less than 1.5 MPa for all curing ages of 3, 7, 14, and 28 days. The compressive strength of HT–BA- (600 °C/2 h) blended PPC mortars is greater than LT–BA- (300 °C/2 h) blended PPC mortars for all BA replacements and all early curing days, as shown in Table 5. This is probably due to the high carbon amount found at LT–BA. Further studies are required on the effect of HT–BA blended PPC mortars on the properties of mortars at later curing ages.

Table 5. Compressive strength of bagasse ash (BA)-blended PPC mortars at early curing ages, 3–28 days; BA amount from 0 to 25% of PPC replacement.

No.	Bagasse Ash (%)	Compressive Strength (MPa) of BA Blended PPC–Mortar Samples							
		3 Days		7 Days		14 Days		28 Days	
		LT-BA	HT-BA	LT-BA	HT-BA	LT-BA	HT-BA	LT-BA	HT-BA
1	BA 0	2.71		3.53		7.05		16	
2	BA 5	1.73	1.83	2.61	3.08	5.10	6.08	12.5	14.7
3	BA 10	1.81	2.34	2.78	3.02	2.42	4.31	6.69	9.36
4	BA 15	1.77	2.04	2.46	2.77	2.12	3.28	4.67	6.91
5	BA 20	1.73	1.79	1.83	1.98	1.48	2.67	4.34	5.95
6	BA 25	1.38	1.42	1.48	1.77	1.44	2.30	2.42	3.97

3.5. Water Absorption and Apparent Porosity of HT–BA-Blended OPC Mortar Samples

Water absorption of mortars increased from 15.08% to 19.82% with the addition of HT–BA from 0% to 25% at 3 days curing, as shown in Figure 4a, due to a higher specific surface area of bagasse ash (4716 cm^2/gm), compared to the surface area of cement (3394 cm^2/g) [29]. BA-blended OPC mortars have higher water absorption at a curing age of 3 days compared to the 7 days of curing for OPC mortars. This may be attributed to an increase in pores spaces in the blocks as the BA percentage increases. Apparent porosity of HT–BA-blended mortars also increased from 22.6% to 26.8% with the increment of BA from 0 to 25% at 3 days curing, as shown in Figure 4b. Apparent porosity is expressed as a percentage of the volume of the internal open pores in the specimen to its exterior volume.

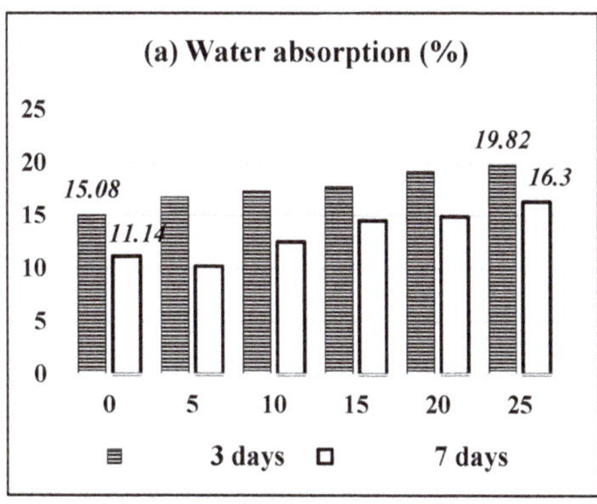

Figure 4. *Cont.*

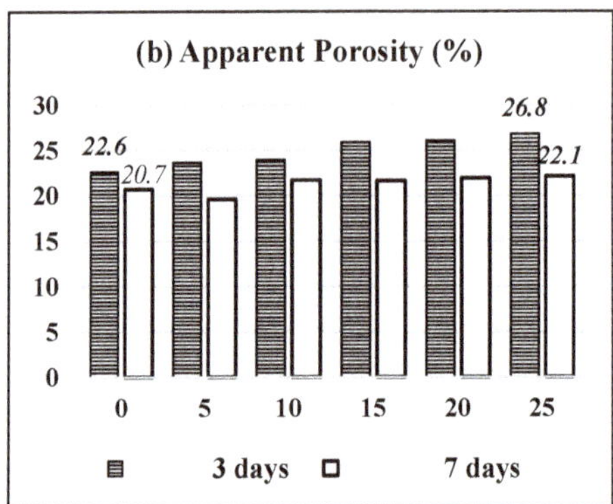

Figure 4. HT–BA blended OPC-mortar samples; (**a**) Water absorption after 3 and 7 days of curing, (**b**) Apparent porosity after 3 and 7 days of curing.

3.6. Microstructure Characterization of Bagasse Ash and Bagasse Ash-Blended Mortar

We prepared HT–BA (bagasse burned at 600 °C) and LT–BA (bagasse burned at 300 °C) for XRD analysis. XRD results for HT–BA illustrate a wider hump between 15° and 30° at 2θ which is an indication of the occurrence of the amorphous silica and also Quartz (Q) and Cristobalite (C) phases (Figure 5). A similar observation was also explained in previous studies [17,30]. This amorphous character contributed to the pozzolanic activity of the material to be added to portland cement. Katare, V.D. et al., stated that the optimal temperature for producing pozzolanic bagasse ash is 600 °C [22]. At this firing temperature of bagasse, it has mainly amorphous silica, which is more reactive and has a high pozzolanic activity index [17,22]. Modification of silica assists in the development of C–S–H, as stated in the previous study [31].

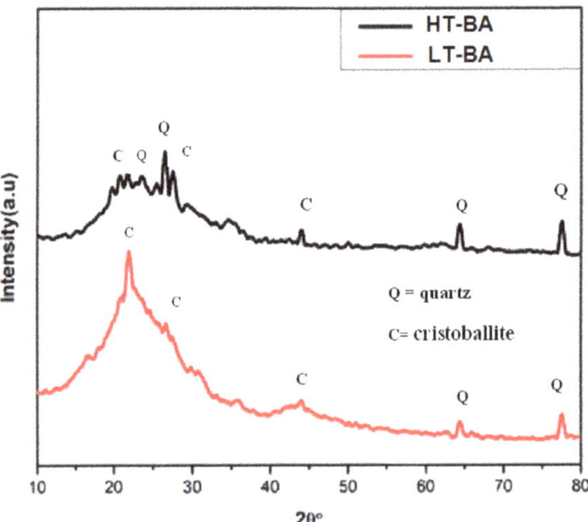

Figure 5. XRD pattern of bagasse burned at 600 °C (top image) and bagasse burned at 300 °C (bottom); Q = Quartz, C = Cristoballite.

From XRD measurement, we also confirmed the formation of Calcium Silicate Hydrate (C–S–H) in both mortar samples with BA-blended (Figure 6a) and without BA-blended (Figure 6b) mortar. However, BA-blended mortar revealed extra C–S–H peaks at 2θ diffracted angles and more intense C–S–H peaks compared to mortar without BA. This is due to a high proportion of reactive amorphous silica of BA reacted with Ca(OH)$_2$ from cement hydration, which is crucial to form C–S–H that assists to increase the strength of samples. In Figure 6a, at 2θ between 25° and 30°, it is shown that CH peak intensity decreased for BA-blended mortar. Similarly, a previous study also used XRD to detect and analyze the C–S–H [32,33]. C–S–H is a nanoscale material, which is mainly responsible for the compressive strength of cement.

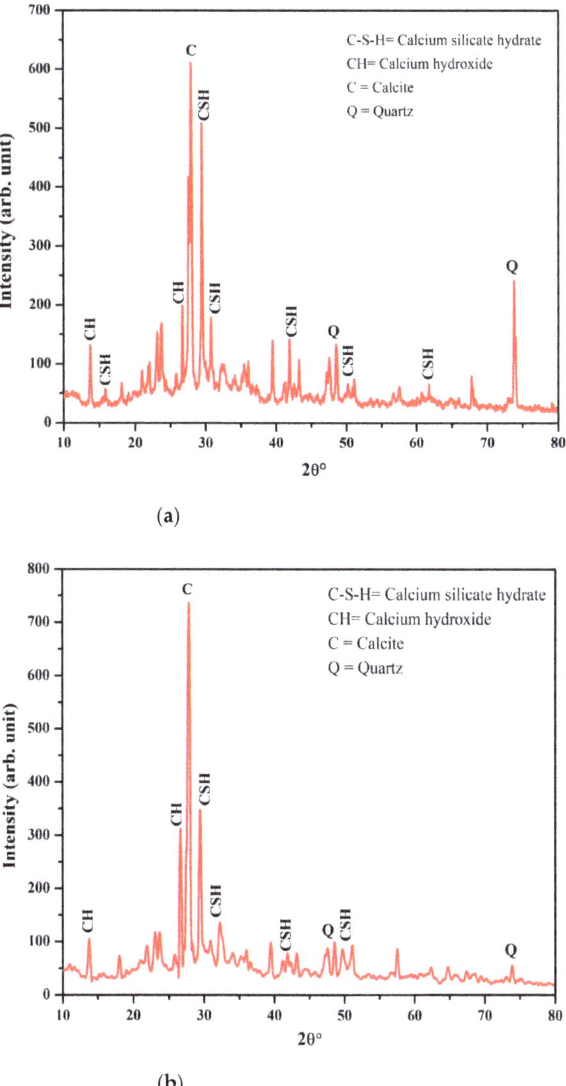

Figure 6. XRD results; (**a**) BA-blended mortar, (**b**) Mortar without BA-blend.

SEM images of HT–BA and LT–BA are shown that particles with different sizes and morphology. HT–BA (600 °C/2 h) has porous microstructures (Figure 7a) compared to LT–BA (Figure 7b). It is probably due to the removal of carbon and other components from bagasse ash at a higher temperature (600 °C). We confirmed the effect using DTA–TGA analysis (i.e., at a temperature above 500 °C), and BA mass decreased due to the escape of carbon and other components from BA, as shown in Figure 2. A similar explanation was also provided in previous papers, and they exhibited a decrement of carbon content in sugarcane bagasse ash with an increment of calcination temperature [34].

(**a**)

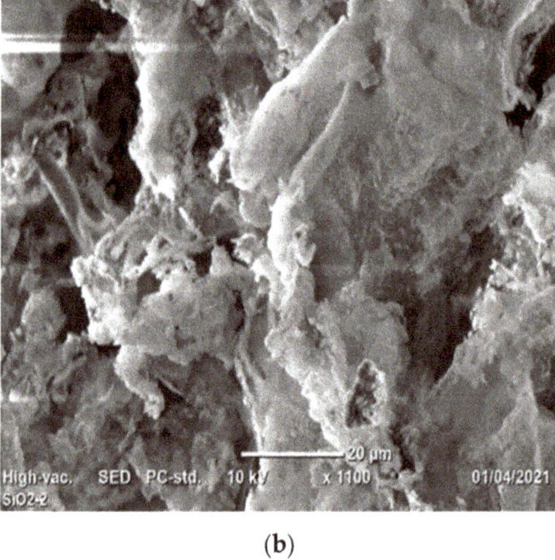

(**b**)

Figure 7. SEM images: (**a**) Controlled-burned (600 °C/2 h) bagasse ash (HT–BA); (**b**) Uncontrolled-burned (300 °C/5 min) bagasse ash (LT–BA).

4. Conclusions

Based on our study, the following conclusions can be drawn:

Compressive strength (σ_c) of OPC mortar samples increased with an up-to-10% bagasse ash (HT–BA) addition for early curing ages. The strength of mortars increased from 9.26 MPa without BA to 9.40 MPa with HT–BA 5% after 3 days of curing. At 28 days of curing, OPC mortars have σ_c = 24.5 MPa without BA and σ_c = 27.3 MPa with BA 5%. Enhancement of the compressive strength of OPC mortars with 5–10% HT–BA are most likely related to extra C–S–H formation in cement paste as a result of reactive amorphous silica of BA reacting with $Ca(OH)_2$. It might also relate to the fact that BA probably filled the voids in the mortar. However, the compressive strength of OPC mortars decreased with the addition of BA above 10% at early an curing age, which is likely related to a reduction of the 3CaO.SiO2 (C3S) phase in cement paste. None of the bagasse ash-blended PPC mortars have shown an enhancement in the compressive strength.

Water absorption in bagasse ash-blended OPC mortars increased from 15.08% (without BA) to 19.82% (with the addition of BA 25%) after a 3-day curing age, which might lead to an increase in the pore spaces in dry mortars as the BA percentage increases. When the samples were cured for 7 days, water absorption of BA-blended OPC mortars decreased compared to the samples cured for 3 days. Apparent porosity of BA blended OPC mortars increased from 22% (BA = 0%) to 26.8% (with BA = 25%) after 3 days of curing.

The pozzolanic activity of bagasse ashes was improved for controlled burning temperature and duration (600 °C/2 h), which facilitated the formation of more C–S–H in mortars. Thus, we improved compressive strength, water absorption, and apparent porosity of bagasse ash-blended OPC mortars by controlling BA amounts, firing temperatures, and duration of bagasse.

Author Contributions: Conceptualization, R.S. and B.B.T.; methodology, B.B.T.; software, D.M.A. and G.M.S.A.; validation, B.B.T. and D.M.A.; formal analysis, R.S.; investigation, R.S. and B.B.T.; resources, D.M.A.; writing—original draft preparation, B.B.T. and R.S.; writing—review and editing, G.M.S.A.; visualization, G.M.S.A.; supervision, B.B.T.; project administration, V.T. and A.A.; funding acquisition, A.A. All authors have read and agreed to the published version of the manuscript.

Funding: The authors gratefully acknowledge the funding and support provided by the Deanship of Scientific Research, King Khalid University (KKU), Abha-Asir, Saudi Arabia, with grant number G.R.P/93/42, under a general research program to complete the research work.

Institutional Review Board Statement: Not applicable.

Informed Consent Statement: Not applicable.

Data Availability Statement: The data presented in this study are available on request from the corresponding author.

Acknowledgments: The authors extend their appreciation to the Department of Materials Science and Engineering, Adama Science and Technology University, Ethiopia, for Technical expertise and financial assistance in doing experiments. The authors also extend their appreciation to the Deanship of Scientific Research at King Khalid University for funding this work through the general research program under grant number (G.R.P-/93/42).

Conflicts of Interest: The authors declare no conflict of interest.

References

1. Worrell, E.; Price, L.; Martin, N.; Hendriks, C.; Meida, L.O. Carbon dioxide emissions from the global cement industry. *Annu. Rev. Energy Environ.* **2001**, *26*, 303–329. [CrossRef]
2. Fairbairn, E.M.; Americano, B.B.; Cordeiro, G.C.; Paula, T.P.; Toledo Filho, R.D.; Silvoso, M.M. Cement replacement by sugar cane bagasse ash: CO_2 emissions reduction and potential for carbon credits. *J. Environ. Manag.* **2010**, *91*, 1864–1871. [CrossRef] [PubMed]
3. Papadakis, V.; Antiohos, S.; Tsimas, S. Supplementary cementing materials in concrete: Part I: Efficiency and design. *Cem. Concr. Res.* **2002**, *32*, 1525–1532. [CrossRef]
4. Zain, M.F.M.; Islam, M.N.; Mahmud, F.; Jamil, M. Production of rice husk ash for use in concrete as a supplementary cementitious material. *Constr. Build. Mater.* **2011**, *25*, 798–805. [CrossRef]
5. Zhang, P.; Liaob, W.; Kumar, A.; Zhang, Q.; Ma, H. Characterization of sugarcane bagasse ash as a potential supplementary cementitious material: Comparison with coal combustion fly ash. *J. Clean. Prod.* **2020**, *277*, 123834. [CrossRef]
6. Deepika, S.; Anand, G.; Bahurudeen, A.; Santhanam, M. Construction products with sugarcane bagasse ash binder. *J. Mater. Civil Eng.* **2017**, *29*, 04017189. [CrossRef]
7. Cordeiro, G.C.; Toledo Filho, R.D.; Tavares, L.M.; Fairbairn, E.M.R. Pozzolanic activity and filler effect of sugar cane bagasse ash in Portland cement and lime mortars. *Cem. Concr. Compos.* **2008**, *30*, 410–418. [CrossRef]
8. Chusilp, N.; Jaturapitakkul, C.; Kiattikomol, K. Utilization of bagasse ash as a pozzolanic material in concrete. *Constr. Build. Mater.* **2009**, *23*, 3352–3358. [CrossRef]
9. Lin, L.-K.; Kuo, T.-M.; Hsu, Y.-S. The application and evaluation research of coffee residue ash into mortar. *J. Mater. Cycles Waste Manag.* **2016**, *18*, 541–551. [CrossRef]
10. Adesanya, D.A.; Raheem, A.A. Development of corn cob ash blended cement. *Constr. Build. Mater.* **2009**, *23*, 347–352. [CrossRef]

11. Tan, H.; Nie, K.; He, X.; Deng, X.; Zhang, X.; Su, Y.; Yang, J. Compressive strength and hydration of high-volume wet-grinded coal fly ash cementitious materials. *Constr. Build. Mater.* **2019**, *206*, 248–260. [CrossRef]
12. Song, H.-W.; Pack, S.-W.; Nam, S.-H.; Jang, J.-C.; Saraswathy, V. Estimation of the permeability of silica fume cement concrete. *Constr. Build. Mater. Tech.* **2010**, *23*, 315–321. [CrossRef]
13. Givi, A.N.; Rashid, S.A.; Aziz, F.N.A.; Mohd-Salleh, M.A. Assessment of the effects of rice husk ash particle size on strength, water permeability and workability of binary blended concrete. *Constr. Build. Mater.* **2010**, *24*, 2145–2150. [CrossRef]
14. Sabir, B.B. High-strength condensed silica fume concrete. *Mag. Concr. Res.* **1995**, *47*, 219–226. [CrossRef]
15. Elsayed, A.A. Influence of silica fume, fly ash, super pozz and high slag cement on water permeability and strength of concrete. *Jordan J. Civ. Eng.* **2011**, *159*, 1–13.
16. Kumar, R.; Dhaka, J. Review paper on partial replacement of cement with silica fume and its effects on concrete properties. *Int. J. Eng. Res.* **2016**, *4*, 2347–4718.
17. Bahurudeen, A.; Kanraj, D.; Dev, G.V.; Santhanam, M. Performance evaluation of sugarcane bagasse ash blended cement in Concrete. *Cem. Concr. Compos.* **2015**, *59*, 77–88. [CrossRef]
18. Lima, S.A.; Sales, A.; Almeida, F.D.C.R.; Moretti, J.P.; Portella, K.F. Concretes made with sugarcane bagasse ash: Evaluation of the durability for carbonation and abrasion tests. *Ambient Constr.* **2011**, *11*, 201–212. [CrossRef]
19. Embong, R.; Shafiq, N.; Kusbiantoro, A.; Nuruddin, M.F. Effectiveness of low-concentration acid and solar drying as pre-treatment features for producing pozzolanic sugarcane bagasse ash. *J. Clean Prod.* **2016**, *112*, 953–962. [CrossRef]
20. Trifunovic, P.D.; Marinkovic, S.R.; Tokalic, R.; Matijasevic, T.D. The effect of the content of unburned carbon in bottom ash on its applicability for road construction. *Thermochim. Acta* **2010**, *498*, 1–6. [CrossRef]
21. Xu, Q.; Ji, T.; Gao, S.-J.; Yang, Z.; Wu, N. Characteristics and Applications of Sugar Cane Bagasse Ash Waste in Cementitious Materials. *Materials* **2019**, *12*, 39. [CrossRef]
22. Katare, V.D.; Madurwar, M.V. Experimental characterization of sugarcane biomass ash—A review. *Constr. Build. Mater.* **2017**, *152*, 1–15. [CrossRef]
23. ASTM C20-00(2015). *Standard Test Methods for Apparent Porosity, Water Absorption, Apparent Specific Gravity, and Bulk Density of Burned Refractory Brick and Shapes by Boiling Water*; ASTM International: West Conshohocken, PA, USA, 2015.
24. *ASTM C618: Standard Specification for Coal Fly Ash and Raw or Calcined Natural Pozzolan for Use in Concrete*; ASTM International: West Conshohocken, PA, USA, 2019.
25. Chusilp, N.; Chai, J.; Kiattikomol, K. Effects of LOI of ground bagasse ash on the compressive strength and sulfate resistance of mortars. *Constr. Build. Mater.* **2009**, *23*, 3523–3531. [CrossRef]
26. Gebreyouhannes, E.; Geremew, M. Bagasse Ash as a Partial Substitute of Cement in Concrete Rigid Pavement. Master's Thesis, AAU, Addis Ababa, Ethiopia, 2017.
27. Maliger, V.R.; Doherty, W.; Frost, R.L.; Mousavioun, P. Thermal Decomposition of Bagasse: Effect of Different Sugar Cane Cultivars. *Ind. Eng. Chem. Res.* **2011**, *50*, 791–798. [CrossRef]
28. Garcia-Perez, M.; Chaala, A.; Yanga, J.; Roy, C. Co-pyrolysis of sugarcane bagasse with petroleum residue. Part I: Thermogravimetric analysis. *Fuel* **2001**, *80*, 1245–1258. [CrossRef]
29. Bahurudeen, A.; Santhanam, M. Influence of different processing methods on the pozzolanic performance of sugarcane bagasse ash. *Cem. Concr. Compos.* **2015**, *56*, 32–45. [CrossRef]
30. Batool, F.; Masood, A.; Ali, M. Characterization of Sugarcane Bagasse Ash as Pozzolan and Influence on Concrete Properties. *Arab. J. Sci. Eng.* **2020**, *45*, 3891–3900. [CrossRef]
31. Baltakys, K.; Jauberthie, R.; Siauciunas, R.; Kaminskas, R. Influence of modification of SiO_2 on the formation of calcium silicate hydrate. *Mater. Sci.-Pol.* **2007**, *25*, 663–669.
32. Nasir, M.; Al-Kutti, W. Performance of Date Palm Ash as a Cementitious Materials by Evaluating Strength, Durability, and Characterization. *Buildings* **2019**, *9*, 6. [CrossRef]
33. Bergold, S.T.; Goetz-Neunhoeffer, F.; Neubauer, J. Quantitative analysis of C–S–H in hydrating alite pastes by in-situ XRD. *Cem. Concr. Res.* **2013**, *53*, 119–126. [CrossRef]
34. Ribeiro, D.V.; Morell, M.R. Effect of Calcination Temperature on the Pozzolanic Activity of Brazilian Sugar Cane Bagasse Ash (SCBA). *Mater. Res.* **2014**, *17*, 974–981. [CrossRef]

 *sustainability*

Article

Identification of Methods of Reducing Construction Waste in Construction Enterprises Based on Surveys

Marta Białko [1],* and Bożena Hoła [2]

[1] Department of Architectural Engineering, College of Engineering, University of Sharjah, Sharjah P.O. Box 27272, United Arab Emirates

[2] Department of Civil Engineering, Wrocław University of Science and Technology, ul. Wybrzeże Wyspiańskiego 27, 50-370 Wrocław, Poland; bozena.hola@pwr.edu.pl

* Correspondence: mbialko@sharjah.ac.ae

Abstract: The article presents the analysis of the dependence between methods of reducing construction waste and the size of the construction enterprise. The analysis was carried out for the following construction products: steel, concrete, wood, and small-sized (ceramic, concrete) and finishing (ceramic and stone tiles) products. Based on the literature review, the 13 most frequently used methods of reducing construction waste were identified. Surveys were then conducted among 140 construction enterprises. The research was conducted in Sharjah in the United Arab Emirates. In order to test whether there is a relationship between the used waste-reduction method for a given construction product and the size of the enterprise, the Pearson chi-square test of independence was used. The null hypothesis and the alternative hypothesis were formulated, and the critical level of significance $\alpha = 0.05$ was adopted. The results were statistically significant for 7 methods of reducing construction waste. The identified methods include appropriate storage, the training of employees in the field of waste management, the use of monitoring systems, the appropriate transport and unloading of products, the appropriate involvement of subcontractors, the use of prefabricated elements, and the reuse of products on the construction site. Based on the conducted research, it was found that these methods are more often used with an increase in the size of the enterprise. The presented analysis emphasizes the urgent need to improve, integrate, and adjust the promotion of both the reduction of construction waste and the benefits of this reduction in construction enterprises, especially those of the smallest size.

Keywords: construction enterprise; construction waste; methods of reducing construction waste; survey research; chi-square test

Citation: Białko, M.; Hoła, B. Identification of Methods of Reducing Construction Waste in Construction Enterprises Based on Surveys. *Sustainability* **2021**, *13*, 9888. https://doi.org/10.3390/su13179888

Academic Editors: Carlos Morón Fernández and Daniel Ferrández Vega

Received: 24 July 2021
Accepted: 29 August 2021
Published: 2 September 2021

1. Introduction

The natural environment is constantly being exploited. In order to protect natural resources from destructive human activity, the concept of sustainable development was developed. It was first presented on 26 May 1969, by the United Nations (UN) in the report "Problems of the human environment". This report is considered to be a turning point in the perception of the devastating impact of humans on the environment [1,2]. Sustainable development aims to prevent the deepening destruction of the environment while at the same time satisfying the needs of mankind and enabling unlimited progress. The concept of sustainable development should be applied in all areas of human life [3]. The current assumptions of sustainable development were presented during the UN summit in New York on 25–27 September 2015, in a document entitled "Transforming our world: Agenda for Sustainable Development—2030". Sustainable Development Goals have been incorporated into the legislation of UN member states. In the regulation of the European Parliament and the Council of the European Union from 2011, a sustainable construction was introduced as a new basic requirement [4]. Since then, reusing and recycling construction products has not only become a choice but also a necessity. Countries that changed their law according to the new requirements reduced the production of construction waste, which was proven by the analysis of statistics from 2016 performed by Wing-Yan Tam and Lu [5]. Appropriate waste management is included in the 2008/98/EC Directive, which is based on the latest UN assumptions from the 2030 Agenda. The hierarchy of proceeding with waste is written in such a way that the first and most desirable method (reduction) minimizes the amount of generated waste by reducing the use of construction products that cause this waste. The second method (re-consumption) involves the reuse of products that were originally produced as disposable but can still have auxiliary functions. The third principle (recycling) emphasizes the need to recycle the waste that can be processed and used in the production of new construction products [6]. For proper waste management to be as effective as possible, the activities of industrial, research, civic, and public authorities should be combined [7,8]. It has been also proven that the appropriate management of construction waste not only brings environmental, but also economic benefits [9,10].

In the subject literature, construction waste is defined in many ways. One of them is the definition of waste as materials produced "in the process of the production, construction, renovation, or demolition of structures" [11]. A more detailed definition was created by A. Denmark [12]. It reads: "Construction and demolition waste is a complex waste stream that consists of a wide variety of materials such as rubble, earth, concrete, steel, wood and a mixture of materials resulting from various construction activities, including soil removal, demolition, road construction and the modernization of buildings." A European Union (EU) report from 1999 defines construction waste to be a wide range of materials resulting from the complete or partial demolition of buildings or roads, the construction of buildings/roads, the removal of soil, construction works, and building/road restoration works [13]. According to the works of B. Kourmpanis (2008), construction waste differs in individual countries due to the economic and cultural situation of these countries, the characteristics of waste classified as construction waste, and the type of recorded data [14]. Due to divergences in the definition of construction waste in different countries, the EU has developed a European Waste Catalog for its Member States [15]. In the United Arab Emirates (UAE), federal law generally defines waste as all toxic and non-toxic waste, including nuclear waste, which must be disposed of and recycled in accordance with the law. These wastes include solid waste, such as municipal, industrial, agricultural, medical, and construction waste [16]. To sum up, construction waste can be defined as the difference "between the materials ordered and those used to construct a building" [17].

Construction waste can be classified according to the type of material that was used in the production of a construction product [11,13,14,18]. One example includes the division of construction waste in the European Union into concrete, bricks, ceramic tiles, ceramics and gypsum-based materials, wood, glass, plastics, asphalt, tar and tarred products, metals (including metal alloys), soil and earth, insulating materials, mixed construction waste, and hazardous construction waste [15]. Construction waste can also be distinguished according to the properties of the materials, e.g., recyclable and non-recyclable waste, potentially biodegradable waste, waste that is potentially suitable for disposal at a landfill, and waste that is potentially suitable for incineration [11]. Another way to classify construction waste involves the consideration of its origin, e.g., from the construction of a new building, reworks or demolition [11,19,20], or related to the function of a building [21,22] or the source where this waste was generated [23–28].

The knowledge concerning the sources of generating waste facilitates the identification of methods for reducing this waste. The sources of construction waste in the life cycle of a building were first identified by Gavilan and Bernold (1994) during their research carried out in the Netherlands. They distinguished the following sources where construction waste is generated in the production process: (1) design, (2) procurement, (3) handling of materials, (4) operation, (5) residual, including scrap and nonconsumables, and (6) other sources [17]. In 2000, Lingard, on the basis of the research conducted among employees of general contractors, classified four sources of construction waste and added a behavioral theme to the existing knowledge: (1) production and delivery, (2) transport and storage, (3) construction, and (4) culture related [23].

The influence of the behavioral factor on the production of construction waste has been more widely studied and also confirmed in other scientific studies [24,25,29,30]. In 2004, a survey conducted in Singapore among general contractors identified four main sources of construction waste: (1) design, (2) production and delivery, (3) material management, and (4) construction [26]. The results of these both studies, which were carried out in 1994 and 1996, and following ones confirmed that the maximum number of sources of the construction waste occur at the design stage [23,27–29]. Based on a subject literature review [21–28,31–33], 13 methods of reducing the amount of construction waste used in construction companies were found. These methods are discussed in detail in Table 1 in the next chapter.

Table 1. The methods of reducing construction waste used in construction enterprises.

Lp.	Method of Reducing Construction Waste	Benefits
1	Appropriate storage	Protection against mechanical damage and weather conditions;
2	Ordering products of an appropriate size	Minimizing the need for cutting to size elements; eliminating waste;
3	Training employees in the field of waste management	Reduction of losses caused by the inadequate processing of products;
4	Use of systems for monitoring the flow of products on the construction site	Reducing the risk of making mistakes in the management of construction products;
5	Appropriate transport and unloading	Damage prevention;
6	Appropriate involvement of subcontractors	Reduction of the amount of waste on the construction site;
7	Security of the construction site	The prevention of theft, vandalism, and double-ordering;
8	Use of prefabricated elements	Minimizing the amount of waste related to the production of elements on the construction site;
9	Waste segregation on the construction site	Preventing contamination of products by providing containers for each type of waste. Non-contaminated waste can be recycled or reused;
10	Designation of a place for waste segregation	Recovering products for reuse in the designated area, e.g., removing nails from wooden elements or crushing concrete elements;
11	Reuse of products on the construction site	E.g., formwork timber used several times; use of concrete or ceramic and stone waste as rubble for temporary roads and pavements;
12	Delivery of products according to the schedule	Reduction of storage time and the risk of damage;
13	Development of a waste-disposal plan	Easier management of construction waste.

The aim of the conducted research and analyses is to find out whether the application of construction waste-reduction methods with regards to selected construction materials depends on the size of the enterprise. In terms of the number of people employed, the enterprises were classified into five groups: (1) from 1 to 9 employees, (2) from 10 to 49 employees, (3) from 50 to 99 employees, (4) from 100 to 249 employees, and (5) 250 employees and more. The analyses were based on the results of a survey conducted among engineers employed in construction companies. The research was conducted in Sharjah, United Arab Emirates (UAE). The analyzes were performed with the use of the SPSS 26 computer program.

2. Materials and Methods

The methods of reducing construction waste that are used in construction companies, which were identified based on the literature review, are presented in Table 1. The benefits of using each of them are also listed. Research was carried out for the following construction products: steel, concrete, wood, and small-sized (ceramic, concrete) and finishing (ceramic and stone tiles) elements.

2.1. Size and Structure of the Studied Population

The population of the studied enterprises consists of five subpopulations. Each of them includes enterprises with a certain number of employees. The groups have been derived from the population as per the characteristic of construction enterprises. In Sharjah, construction companies employ a limited number of people to keep low insurance and municipality fees, and they outsource work to subcontractors. The survey was conducted using the technique of personal interviews and telephone interviews due to the possibility of obtaining the most accurate data and immediate clarification of ambiguities in the obtained answers. The research was carried out in 140 enterprises of general contractors. The structure of the population is presented in Table 2.

Table 2. Structure of the studied population of enterprises.

Number of Employees Hired in the Assessed Enterprises	Number of Enterprises n_i	Percentage Share	
1–9 employees	42	30%	
10–49 employees	41	29%	
50–99 employees	15	11%	
100–249 employees	20	14%	
250 employees and more	22	16%	
Total	140	100%	

Table 2 presents the number and percentage share of the enterprise sizes in the studied population. In the surveyed representative group, 42 enterprises (30%) employ from 1 to 9 employees, 41 enterprises (29%) employ from 10 to 49 employees, 15 enterprises (11%) employ from 50 to 99 employees, 20 enterprises (14%) employ between 100 and 249 employees, and 22 enterprises (16%) employ 250 employees or more. The largest sub-population includes enterprises that employ the least workers.

Research was also carried out with regards to the companies' experience in the market of construction works. The results of the surveys are presented in Table 3.

Table 3. Experience in the construction market among the surveyed companies.

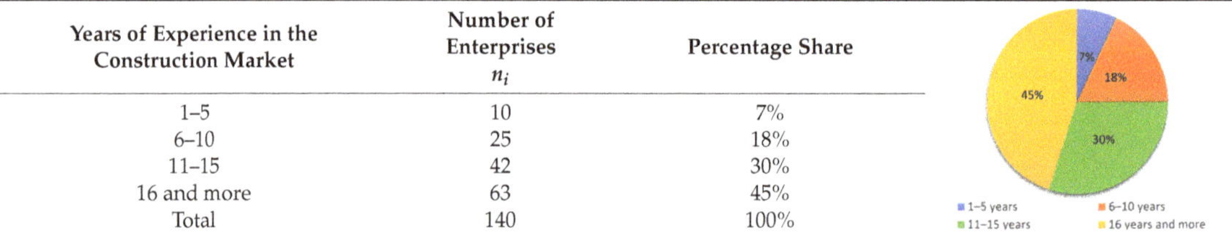

Years of Experience in the Construction Market	Number of Enterprises n_i	Percentage Share	
1–5	10	7%	
6–10	25	18%	
11–15	42	30%	
16 and more	63	45%	
Total	140	100%	

Among the surveyed companies, 63 companies (45%) had 16 years or more of experience in the construction market, 42 companies (30%) had 11 to 15 years of experience, 25 companies (18%) had 6 to 10 years of experience, and 10 enterprises (7%) had between 1 and 5 years of experience. To sum up, the most numerous group were the oldest enterprises, with 16 years of experience or more.

2.2. Methodology of Identifying Methods of Reducing Construction Waste with Regards to the Size of the Enterprise

In the subject literature review, the correlation analysis of two variables is especially popular [34–36]. In the presented paper, the subject of the study is to determine the relationship between the method of reducing construction waste in relation to a given construction product and the size of the construction enterprise. For this purpose:

- The answers of the respondents concerning the applied methods of waste reduction were qualified into five groups with regards to particular construction products. Each of these groups represented a certain size of an enterprise. In each group, the number of positive answers (YES) and the number of negative answers (NO) were determined;

- In order to test whether there is a relationship between the waste-reduction method used in the case of a given construction product and the size of the enterprise, the Pearson chi-square (χ^2) test of independence was used [37]. This test is used to check the relationship between the two nominal variables X and Y. In the conducted research, the nominal variable X is the size of the enterprise, while the nominal variable Y is the answer Yes/No in relation to the tested reduction method.
- The Pearson chi-square test is based on comparing the values obtained in the study (the so-called observed or empirical frequencies) with theoretical values calculated based on the assumption that there is no relationship between variables X and Y. The chi-square test statistic has the form of formula (1):

$$\chi^2 = \sum_{i=1}^{r}\sum_{j=1}^{c} \frac{\left(O_{ij} - E_{ij}\right)^2}{E_{ij}},\tag{1}$$

where:

χ^2—chi-square statistic,
O_{ij}—observed counts obtained from surveys,
E_{ij}—theoretical counts,
r—number of levels of variable X (X = 5) (number of enterprise groups), and
c—number of levels of variable Y (Y = 2) (number of possible answers).

Chi-square statistics were calculated using the SPSS-26 computer program. The chi-square statistic has a distribution of χ^2 with $(r - 1)(c - 1)$ degrees of freedom. In the analyzed case, the number of degrees of freedom is 4. The p value determined for the chi-square test statistic is compared with the significance level α. The critical significance level of $\alpha = 0.05$ was adopted in the analyzes.

The null hypothesis H_0 and the alternative hypothesis H_1 were formulated:

- H_0: Variables X and Y are independent if $p > \alpha$
- H_1: Variables X and Y are not independent if $p \leq \alpha$

where:

p—the probability (the value of p is compared to the theoretical value of α)
α—the significance level.

If $p > \alpha \Rightarrow$ it can be assumed that there are no reasons to reject hypothesis H_0. This means that there is no significant relationship between the size of the enterprise and the use of the analyzed method of reducing construction waste. The result is statistically insignificant.

If $p \leq \alpha \Rightarrow$ it can be assumed that there are reasons for rejecting hypothesis H_0. Based on the tested sample, it can be assumed that there is a relationship between the size of the enterprise and the use of the analyzed method of reducing construction waste. The result is statistically significant.

3. Results

Calculations of the chi-square test were performed for all the tested methods of reducing construction waste in relation to all the analyzed construction products. Table 4 only presents those results that are statistically significant, namely the statistics of frequency and percentage rates of YES and NO responses as well as the calculated values of the χ^2 (4) test and probability p. Other calculation results, which are not included in the table below, show that there is no significant relationship between the size of the enterprise and the use of the analyzed method of reducing construction waste.

Table 4. The statistics of the chi-square test and probability p of the occurrence of methods of reducing construction waste that were used by construction companies, with their size also provided.

Methods of Reducing Construction Waste	Construction Products	Answer	Enterprise Size (Number of Employees)												$Chi2(4)$; Probability p
			1–9		10–49		50–99		100–249		250 and More		Total		
			n	%	n	%	n	%	n	%	n	%	n	%	
Appropriate storage	small-sized products	Yes	20	47.6%	24	58.5%	8	53.3%	16	80.0%	18	81.8%	86	61.4%	$Chi2(4) = 10.711$; $p = 0.03$
		No	22	52.4%	17	41.5%	7	46.7%	4	20.0%	4	18.2%	54	38.6%	
	ceramic and stone tiles	Yes	23	54.8%	25	61.0%	8	53.3%	18	90.0%	18	81.8%	92	65.7%	$Chi2(4) = 11.433$; $p = 0.022$
		No	19	45.2%	16	39.0%	7	46.7%	2	10.0%	4	18.2%	48	34.3%	
	wood	Yes	25	59.5%	27	65.9%	10	66.7%	18	90.0%	20	90.9%	100	71.4%	$Chi2(4) = 11.179$; $p = 0.025$
		No	17	40.5%	14	34.1%	5	33.3%	2	10.0%	2	9.1%	40	28.6%	
Training employees in the field of waste management	small-sized products	Yes	24	57.1%	26	63.4%	9	60.0%	19	95.0%	13	59.1%	91	65.0%	$Chi2(4) = 9.60$; $p = 0.048$
		No	18	42.9%	15	36.6%	6	40.0%	1	5.0%	9	40.9%	49	35.0%	
Use of monitoring systems	steel	Yes	24	57.1%	19	46.3%	11	73.3%	17	85.0%	18	81.8%	89	63.6%	$Chi2(4) = 13.751$; $p = 0.008$
		No	18	42.9%	22	53.7%	4	26.7%	3	15.0%	4	18.2%	51	36.4%	
	concrete	Yes	24	57.1%	21	51.2%	11	73.3%	17	85.0%	18	81.8%	91	65.0%	$Chi2(4) = 11.272$; $p = 0.024$
		No	18	42.9%	20	48.8%	4	26.7%	3	15.0%	4	18.2%	49	35.0%	
	small-sized products	Yes	19	45.2%	18	43.9%	8	53.3%	17	85.0%	16	72.7%	78	55.7%	$Chi2(4) = 13.754$; $p = 0.008$
		No	23	54.8%	23	56.1%	7	46.7%	3	15.0%	6	27.3%	62	44.3%	
	ceramic and stone tiles	Yes	18	42.9%	18	43.9%	9	60.0%	17	85.0%	17	77.3%	79	56.4%	$Chi2(4) = 16.369$; $p = 0.003$
		No	24	57.1%	23	56.1%	6	40.0%	3	15.0%	5	22.7%	61	43.6%	
	wood	Yes	22	52.4%	16	39.0%	10	66.7%	17	85.0%	16	72.7%	81	57.9%	$Chi2(4) = 14.996$; $p = 0.005$
		No	20	47.6%	25	61.0%	5	33.3%	3	15.0%	6	27.3%	59	42.1%	
Appropriate transport and unloading of products	small-sized products	Yes	34	81.0%	31	75.6%	9	60.0%	19	95.0%	21	95.5%	114	81.4%	$Chi2(4) = 10.777$; $p = 0.029$
		No	8	19.0%	10	24.4%	6	40.0%	1	5.0%	1	4.5%	26	18.6%	
Appropriate involvement of subcontractors	steel	Yes	20	47.6%	20	48.8%	11	73.3%	16	80.0%	16	72.7%	83	59.3%	$Chi2(4) = 10.671$; $p = 0.031$
		No	22	52.4%	21	51.2%	4	26.7%	4	20.0%	6	27.3%	57	40.7%	
	small-sized products	Yes	11	26.2%	18	43.9%	9	60.0%	12	60.0%	17	77.3%	67	47.9%	$Chi2(4) = 17.855$; $p = 0.001$
		No	31	73.8%	23	56.1%	6	40.0%	8	40.0%	5	22.7%	73	52.1%	
	wood	Yes	17	40.5%	17	41.5%	10	66.7%	12	60.0%	17	77.3%	73	52.1%	$Chi2(4) = 11.495$; $p = 0.022$
		No	25	59.5%	24	58.5%	5	33.3%	8	40.0%	5	22.7%	67	47.9%	
The use of prefabricated elements	steel	Tak	7	16.7%	22	53.7%	7	46.7%	9	45.0%	9	40.9%	54	38.6%	$Chi2(4) = 13.259$; $p = 0.01$
		No	35	83.3%	19	46.3%	8	53.3%	11	55.0%	13	59.1%	86	61.4%	
	small-sized products	Yes	1	2.4%	6	14.6%	3	20.0%	7	35.0%	5	22.7%	22	15.7%	$Chi2(4) = 12.315$; $p = 0.015$
		No	41	97.6%	35	85.4%	12	80.0%	13	65.0%	17	77.3%	118	84.3%	
Reuse of products on the construction site	wood	Yes	34	81.0%	20	48.8%	10	66.7%	17	85.0%	14	63.6%	95	67.9%	$Chi2(4) = 13.027$; $p = 0.011$
		No	8	19.0%	21	51.2%	5	33.3%	3	15.0%	8	36.4%	45	32.1%	
	concrete	Yes	34	81.0%	25	61.0%	9	60.0%	18	90.0%	13	59.1%	99	70.7%	$Chi2(4) = 9.862$; $p = 0.043$
		No	8	19.0%	16	39.0%	6	40.0%	2	10.0%	9	40.9%	41	29.3%	
	small-sized products	Yes	33	78.6%	21	51.2%	9	60.0%	19	95.0%	15	68.2%	97	69.3%	$Chi2(4) = 14.825$; $p = 0.005$
		No	9	21.4%	20	48.8%	6	40.0%	1	5.0%	7	31.8%	43	30.7%	
	ceramic and stone tiles	Yes	30	71.4%	19	46.3%	10	66.7%	17	85.0%	12	54.5%	88	62.9%	$Chi2(4) = 11.056$; $p = 0.026$
		No	12	28.6%	22	53.7%	5	33.3%	3	15.0%	10	45.5%	52	37.1%	

4. Discussion

The analysis of the results of the calculations included in Table 4 helped to indicate methods of reducing waste, the application of which depends on the size of the construction company, to be indicated with a probability greater than 0.95. A significant statistical dependence was found for seven methods of reducing construction waste, namely appropriate storage, employee training in the field of waste management, use of monitoring systems, appropriate transport and unloading of products, appropriate involvement of subcontractors, use of prefabricated elements, and the reuse of products on the construction site. In all these cases, the statistic $Chi2(4) > 9.487$, and $p < 0.05$. No statistically significant correlation was found for the other six methods of reducing construction waste, namely ordering products to size and in the appropriate quantity, the security of the construction site, waste segregation on the construction site, the designation of a place for waste segregation on the construction site; timely delivery, and having a waste disposal plan. In all these cases, the value of the statistic $Chi2(4) > 9.487$, and $p > 0.05$. This means that the application of a given reduction method does not depend on the size of the enterprise. A detailed summary of the test results is provided in Table 5.

Table 5. Summary of research results.

#	Method of Reducing Construction Waste	The Results of The Calculations Confirm:	
		Statistically Significant Correlation between the Size of the Enterprise and:	Non-Statistically Significant Correlation between the Size of the Enterprise and:
1	Appropriate storage	• small-sized products • ceramic and stone tiles • wood	• steel • concrete
2	Training employees in the field of waste management	• small-sized products	• steel • concrete • ceramic and stone tiles • wood
3	Use of monitoring systems	• steel • concrete • small-sized products • ceramic and stone tiles • wood	
4	Appropriate transport and unloading of products	• small-size products	• steel • concrete • ceramic and stone tiles • wood
5	Appropriate involvement of subcontractors	• steel • small-sized products • wood	• concrete • ceramic and stone tiles
6	The use of prefabricated elements	• steel • small-sized products	• concrete • ceramic and stone tiles • wood
7	Reuse of products on the construction site	• concrete • small-sized products • ceramic and stone tiles • wood	• steel

Table 6 lists the construction products for which a statistically significant correlation was found between the method of reducing construction waste and the size of the enterprise. Moreover, for each construction product, the strength of this relationship (PW) was determined. The frequency of the affirmative answer (YES) indicated by the respondents was adopted as a measure of this strength. The following designations were adopted:

$PW = 1$ when the frequency is $\leq 60\%$,
$PW = 2$ when the frequency is between 61% and 75%, and
$PW = 3$ when the frequency is between 76% and 100%

Table 6. Construction products for which a statistically significant correlation was found between the method of reducing construction waste and the size of the construction enterprise.

#	Construction Product	Method of Reducing Construction Waste	Enterprise Size (Number of Employees)				
			1–9	10–49	50–99	100–249	250 and More
1	steel	Appropriate storage	-	-	-	-	-
	concrete		-	-	-	-	-
	small-sized products		1	1	1	3	3
	ceramic and stone tiles		1	2	1	3	3
	wood		1	2	2	3	3
2	steel	Training employees in the field of waste management	-	-	-	-	-
	concrete		-	-	-	-	-
	small-sized products		1	2	1	3	3
	ceramic and stone tiles		-	-	-	-	-
	wood		-	-	-	-	-
3	steel	Use of monitoring systems	1	1	2	3	3
	concrete		1	1	2	3	3
	small-sized products		1	1	1	3	2
	ceramic and stone tiles		1	1	1	3	3
	wood		1	1	2	3	2

Table 6. *Cont.*

#	Construction Product	Method of Reducing Construction Waste	Enterprise Size (Number of Employees)				
			1–9	10–49	50–99	100–249	250 and More
4	steel	Appropriate transport and unloading of products	-	-	-	-	-
	concrete		-	-	-	-	-
	small-sized products		3	2	1	3	3
	ceramic and stone tiles		-	-	-	-	-
	wood		-	-	-	-	-
5	steel	Appropriate involvement of subcontractors	1	1	2	3	2
	concrete		-	-	-	-	-
	small-sized products		1	1	1	1	3
	ceramic and stone tiles		-	-	-	-	-
	wood		1	1	2	1	3
6	steel	Use of prefabricated elements	1	1	1	1	1
	concrete		-	-	-	-	-
	small-sized products		1	1	1	1	1
	ceramic and stone tiles		-	-	-	-	-
	wood		-	-	-	-	-
7	steel	Reuse of products on the construction site	-	-	-	-	-
	concrete		3	2	1	3	1
	small-sized products		3	1	1	3	2
	ceramic and stone tiles		2	1	2	3	1
	wood		3	1	2	3	2

Based on the results of the research, the following conclusions were drawn:

1. Out of the 13 analyzed methods, a statistically significant correlation between the size of the enterprise and the method of reducing construction waste was found in the case of seven methods. These include appropriate storage, the training of employees in the field of waste management, the use of monitoring systems, the appropriate transport and unloading of products, the appropriate involvement of subcontractors, the use of prefabricated elements, and the reuse of products on the construction site (justification in Table 4). As the size of the enterprise grows, these methods are used more frequently;

2. Each group of analyzed methods of waste reduction includes construction products for which no statistically significant correlation was found between their use and the size of the enterprise. No such dependence was found with regards to the method of:

    - Appropriate storage in the case of steel and concrete;
    - Training of employees in the field of waste management in the case of steel, concrete, and wood;
    - Appropriate transport and unloading of products in the case of steel, concrete, and wood;
    - Appropriate involvement of subcontractors in the case of concrete and ceramic and stone tiles;
    - Use of prefabricated elements in the case of concrete, wood, and ceramic and stone tiles; and
    - Reuse of products on site in the case of steel.

3. The use of seven separate methods of reducing construction waste in enterprises of certain sizes is as follows:

    3.1. In enterprises employing 250 or more employees, the following methods are used:

    3.1.1. Most often (PW = 3):

    - Appropriate storage in relation to small-sized products, wood, and ceramic and stone tiles;
    - Training of employees in the field of waste management in the case of small-sized products;

- Use of monitoring systems in the case of steel, concrete, ceramic and stone products;
- Appropriate transport and unloading of products in relation to small-sized products; and
- Appropriate involvement of subcontractors in the case of small-sized products and wood.

3.1.2   Often (PW-2):

- Use of monitoring systems in the case of small-sized products and wood;
- Appropriate involvement of subcontractors in the case of steel products; and
- Reuse of products on the construction site in the case of small-sized products and wood.

3.1.3.   Rare (PW = 1):

- Use of prefabricated elements in relation to steel and small-sized products;
- Reuse of products on the construction site in the case of concrete products and ceramic and stone tiles.

3.2.   In enterprises employing from 100 to 249 employees, the following methods are used:

3.2.1.   Most often (PW = 3):

- Appropriate storage with regards to small-sized products, wood, and ceramic and stone tiles;
- Training employees in the field of waste management with regards to small-sized products;
- Use of monitoring systems in relation to all the groups of analyzed construction products;
- Appropriate transport and unloading of products with regards to small-sized products;
- Appropriate involvement of subcontractors with regards to steel products; and
- Reuse of products on the construction site with regards to concrete, small-sized products, wood, and ceramic and stone tiles;

3.2.2.   Often (PW-2):

- No such cases were observed.

3.2.3.   Rare (PW = 1):

- Appropriate involvement of subcontractors with regards to small-sized products and wood;
- Use of prefabricated elements with regards to steel and small-sized products.

3.3.   In enterprises employing from 50 to 99 employees, the following methods are used:

3.3.1.   Most often (PW = 3):

- No such cases were observed.

3.3.2.   Often (PW-2):

- Appropriate storage with regards to wooden products;
- Use of monitoring systems with regards to steel, concrete and wooden products;
- Appropriate involvement of subcontractors with regards to steel and wooden products; and

- Reuse of products on the construction site with regards to wood, and ceramic and stone tiles.

3.3.3. Rare (PW = 1):

- Appropriate storage with regards to small-sized products and ceramic and stone tiles;
- Training employees in the field of waste management with regards to small-sized products;
- Use of monitoring systems with regards to small-sized products and ceramic and stone tiles;
- Appropriate transport and unloading of products with regards to small-sized products;
- Appropriate involvement of subcontractors with regards to small-sized products;
- Application of prefabricated elements with regards to steel and small-sized products; and
- Reuse of products on the construction site with regards to concrete and small-sized products.

3.4. In enterprises employing from 10 to 49 employees, the following methods are used:

3.4.1. Most often (PW = 3):

- No such cases were observed.

3.4.2. Often (PW-2):

- Appropriate storage with regards to wood and ceramic and stone tiles;
- Training employees in the field of waste management with regards to small-sized products;
- Appropriate transport and unloading of products with regards to small-sized products; and
- Reuse of products on the construction site with regards to concrete products.

3.4.3. Rare (PW = 1):

- Appropriate storage with regards to small-sized products;
- Use of monitoring systems with regards to all the groups of analyzed construction products;
- Appropriate involvement of subcontractors with regards to steel, small-sized, and wooden products;
- Use of prefabricated elements with regards to steel and small-sized products; and
- Reuse of products on the construction site with regards to small-sized products, wood, and ceramic and stone tiles.

3.5. In enterprises employing from 1 to 9 employees, the following methods are used:

3.5.1. Most often (PW = 3):

- Appropriate transport and unloading of products with regards to small-sized products;
- Reuse of products on the construction site in relation to concrete, small-sized, and wooden products.

3.5.2. Often (PW-2):

- Reuse of products on the construction site with regards to ceramic and stone tiles.

3.5.3. Rare (PW = 1):

- Appropriate storage with regards to small-sized products, wood, and ceramic and stone tiles;
- Training employees in the field of waste management with regards to small-sized products;
- Use of monitoring systems with regards to all the groups of analyzed construction products;
- Appropriate involvement of subcontractors with regards to steel, small-sized, and wooden products; and
- Use of prefabricated elements with regards to steel and small-sized products.

5. Conclusions

The subject of the study was to determine the relationship between 13 methods of reducing construction waste and the size of the construction enterprise in relation to selected construction products. The selected construction products included steel, concrete, small-sized products, wood, and ceramic and stone tiles. Enterprises were divided into groups according to the number of employees, namely from 1 to 9 employees, from 10 to 49 employees, from 50 to 99 employees, from 100 to 249 employees, and for 250 employees and more. Employee surveys were conducted in enterprises belonging to the designated groups. The values of the chi-square test for the significance level of 0.05 and the degree of freedom 4 confirmed a statistically significant correlation between the size of the enterprise and seven methods of reducing construction waste, which included appropriate storage, the training of employees in waste management, the use of monitoring systems, the appropriate transport and unloading of products, the appropriate involvement of subcontractors, the use of prefabricated elements, and the reuse of products on the construction site. The dependence between the use of waste-reduction methods and the size of the enterprise did not always apply to all tested construction products, e.g., no statistically significant correlation was found in relation to steel and concrete in the case of the appropriate storage method. For the remaining six methods of reducing construction waste, no statistical correlation was found between the application of these methods and the size of the enterprise, but this does not mean that these methods were not used. The use of these methods or their non-application may be influenced by other factors that are not included in these studies.

In further research, it is recommended to focus on behavioral motives that can have a large impact on the use of methods that reduce construction waste in construction enterprises.

Studies presented in this paper make a significant contribution to the existing research concerning the reduction of construction waste. In conclusion, based on the conducted research, it was found that the bigger the enterprise, the more methods of reducing construction waste were applied. It can be assumed that larger construction enterprises have more human resources and financial support to plan, organize, and implement more methods of reducing construction waste than smaller enterprises. Therefore, it is crucial that governmental bodies will support reduction of construction waste by providing necessary trainings and financial support and will effectively require it. Thus, the presented analysis emphasizes the urgent need to improve, integrate, and adjust the promotion of the reduction of construction waste and the benefits of this reduction in construction enterprises, especially those of the smallest size.

Author Contributions: Conceptualization, M.B.; methodology, M.B. and B.H.; software, B.H.; validation, M.B. and B.H.; formal analysis, M.B. and B.H.; investigation, M.B.; resources, M.B.; data curation, M.B.; writing—original draft preparation, M.B.; writing—review and editing, B.H.; supervision, B.H. All authors have read and agreed to the published version of the manuscript.

Funding: This research received no external funding.

Institutional Review Board Statement: Not applicable.

Informed Consent Statement: Informed consent was obtained from all subjects involved in the study.

Data Availability Statement: All data included in this study are available upon request by contact with the corresponding author.

Conflicts of Interest: The authors declare no conflict of interest.

References

1. United Nations. *Economic and Social Council, Problems of the Human Environment, 47 Session, E/4667*; United Nations: San Francisco, CA, USA, 1969.
2. Sertyesilisik, B.; Remiszewski, B.; Al-Khaddar, R. Sustainable waste management in the UK construction industry. *Int. J. Constr. Proj. Manag.* **2012**, *4*, 173–188.
3. United Nations. *Report of the World Commission on Environment and Development: Our Common Future, Annex to A/42/427*; United Nations: San Francisco, CA, USA, 1987.
4. United Nations. *General Assembly, Transforming Our World: The 2030 Agenda for Sustainable Development, 70 Session, A/RES/70/1*; United Nations: San Francisco, CA, USA, 2015.
5. Wing-Yan Tam, V.; Lu, W. Construction Waste Management Profiles, Practices, and Performance: A Cross-Jurisdictional Analysis in Four Countries. *Sustainability* **2016**, *8*, 190. [CrossRef]
6. European Commission. *Directive 2008/98 / EC of the European Parliament and of the Council of November 19, 2008 in the Reports and Repealing Certain Wastes*; European Commission: Brussels, Belgium, 2008.
7. Ghaffar, S.H.; Burman, M.; Braimah, N. Pathways to circular construction: An integrated management of construction and demolition waste for resource recovery. *J. Clean. Prod.* **2020**, *244*, 118710. [CrossRef]
8. Wu, H.; Zuo, J.; Zillante, G.; Wang, J.; Yuan, H. Status quo and future directions of construction and demolition waste research: A critical review. *J. Clean. Prod.* **2019**, *240*, 118163. [CrossRef]
9. Islam, R.; Nazifa, T.H.; Yuniarto, A.; Uddin, A.S.M.S.; Salmiati, S.; Shahid, S. An empirical study of construction and demolition waste generation and implication of recycling. *Waste Man.* **2019**, *95*, 10–21. [CrossRef]
10. Wang, B.; Yan, L.; Fu, Q.; Kasal, B. A Comprehensive Review on Recycled Aggregate and Recycled Aggregate Concrete. *Res. Cons. Rec.* **2021**, *171*, 105565. [CrossRef]
11. Yeheyis, M.; Hewage, K.; Shahria, A.M.; Eskicioglu, C.; Sadiq, R. An overview of construction and demolition waste management in Canada: A lifecycle analysis approach to sustainability. *Clean Tech. Environ. Policy* **2013**, *15*, 81–91. [CrossRef]
12. Dania, A.; Kehinde, J.; Bala, K. A study of construction material waste management practices by construction firms in Nijeria. In Proceedings of the 3rd Scottish Conference for Postgraduate Research of the Built and Natural Environment, Glasgow, Scotland, UK, 20–22 November 2007; pp. 121–129.
13. Symonds Group Ltd.; ARGUS; COWI; PRC Bouwcentrum. *Construction and Demolition Waste Management Practices, and Their Economic Impacts*; Report to DGXI, European Commission February; Symonds Group Ltd.: Somerset, UK, 1999.
14. Kourmpanis, B.; Papadopoulos, A.; Moustakas, K.; Stylianou, M.; Haralambous, K.J.; Loizidou, M. Preliminary study for the management of construction and demolition waste. *Waste Manag. Res.* **2008**, *26*, 267–275. [CrossRef]
15. European Commission. *COUNCIL DIRECTIVE 2000/43/EC of 29 June 2000 Implementing the Principle of Equal Treatment between Persons Irrespective of Racial or Ethnic Origin*; European Commission: Brussels, Belgium, 2000.
16. The UAE Federal Environmental Agency. *Federal Law No. (24) of 1999 for the Protection and Development of the Environment*; The Federal Environmental Agency: Abu Dhabi, UAE, 1999.
17. BREEAM International New Construction 2016. Reference: SD233, Issue 2.0. 2017. Available online: https://www.breeam.com/BREEAMInt2016SchemeDocument/ (accessed on 20 May 2021).
18. Gavilan, R.M.; Bernold, L.E. Source Evaluation of Solid Waste in Building Construction. *J. Constr. Eng. Manag.* **1994**, *120*, 536–555. [CrossRef]
19. Tongo, S.O.; Oluwatayo, A.A.; Adeboye, A.B. Correlation of material waste types with life cycle stages in building projects across the states in southwest Nigeria. In Proceedings of the IOP Conference Series: Materials Science and Engineering, Ota, Nigeria, 10–14 August 2020; 1107, p. 012164. [CrossRef]
20. Mahamid, I. Impact of rework on material waste in building construction projects. *Int. J. Constr. Manag.* **2020**. [CrossRef]
21. Saez, P.V.; Del Río Merino, M.; Gonzalez, A.S.A.; Porras-Amores, C. Best practice measures assessment for construction and demolition waste management in building constructions. *Resour. Conserv. Recycl.* **2013**, *75*, 52–62. [CrossRef]
22. Saez, P.V.; Del Río Merino, M.; Porras-Amores, C. Estimation of construction and demolition waste volume generation in new residential buildings in Spain. *Waste Manag. Res.* **2012**, *30*, 137–146. [CrossRef]
23. Lingard, H.; Graham, P.; Smithers, G. Employee perceptions of the solid waste management system operating in a large Australian contracting organization: Implications for company policy implementation. *Constr. Manag. Econ.* **2000**, *18*, 383–393. [CrossRef]
24. Teo, M.M.M.; Loosemore, M.; Marosszeky, M.; Gardner, K.K. Operatives attitudes towards waste on construction project. In Proceedings of the 16th Annual ARCOM Conference, Glasgow, Scotland, UK, 6–8 September 2000; Akintoye, A., Ed.; Association of Researchers in Construction Management: Leeds, UK, 2000; Volume 2, pp. 509–517.
25. Srour, I.; Chong, W.K.; Zhang, F. Sustainable Recycling Approach: An Understanding of Designers' and Contractors' Recycling Responsibilities Throughout the Life Cycle of Buildings in Two US Cities. *Sustain. Dev.* **2012**, *20*, 350–360. [CrossRef]
26. Ekanayake, L.L.; Ofori, G. Building waste assessment score: Design-based tool. *Build. Environ.* **2004**, *39*, 851–861. [CrossRef]

27. Bossink, B.A.G.; Brouwers, H.J.H. Construction waste: Quantification and source evaluation. *J. Constr. Eng. Manag.* **1996**, *122*, 55–60. [CrossRef]
28. Al-Hajj, A.; Hamani, K. Material Waste in the UAE Construction Industry: Main Causes and Minimization Practices. *Arch. Eng. Des. Manag.* **2011**, *7*, 221–235. [CrossRef]
29. Olanrewaju, S.D.; Ogunmakinde, O.E. Waste minimisation strategies at the design phase: Architects' response. *Waste Manag.* **2020**, *118*, 323–330. [CrossRef] [PubMed]
30. Liu, J.; Yi, Y.; Wang, X. Exploring factors influencing construction waste reduction: A structural equation modelling approach. *J. Clean. Prod.* **2020**, *276*, 123185. [CrossRef]
31. Chen, J.; Hua, C.; Liu, C. Considerations for better construction and demolition waste management: Identifying the decision behaviours of contractors and government departments through a game theory decision-making model. *J. Clean. Prod.* **2019**, *212*, 190–199. [CrossRef]
32. Jin, R.; Yuan, H.; Chen, Q. Science mapping approach to assisting the review of construction and demolition waste management research published between 2009 and 2018. *Resour. Conserv. Recycl.* **2019**, *140*, 175–188. [CrossRef]
33. Bekr, G.A. Study of the Causes and Magnitude of Wastage of Materials on Construction Sites in Jordan. *J. Constr. Eng.* **2014**. [CrossRef]
34. Gu, Q.; Wang, L.; Li, Y.; Deng, X.; Lin, C. Multi-scale response sensitivity analysis based on direct differentiation method for concrete structures. *Compos. Part B* **2019**, *157*, 295–304. [CrossRef]
35. Zhou, S.; Jia, Y.; Wang, C. Global Sensitivity Analysis for the Polymeric Microcapsules in Self-Healing Cementitious Composites. *Polymers* **2020**, *12*, 2990. [CrossRef] [PubMed]
36. Zimmermann, T.; Lehký, D.; Strauss, A. Correlation among selected fracture-mechanical parameters of concrete obtained from experiments and inverse analyses. *Struct. Concr.* **2016**, *17*, 1094–1103. [CrossRef]
37. Pearson, K. On the criterion that a given system of deviations from the probable in the case of a correlated system of variables is such that it can be reasonably supposed to have arisen from random sampling. *Lond. Edinb. Dublin Philos. Mag. J. Sci.* **1900**, *50*, 157–172. [CrossRef]

Article

The Use of Modal Analysis in Addition Percentage Differentiation, and Mechanical Properties of Ordinary Concretes with the Addition of Fly Ash from Sewage Sludge

Gabriela Rutkowska [1,*], Mariusz Żółtowski [2] and Michał Liss [3]

[1]   Institute of Civil Engineering, Warsaw University of Life Sciences (SGGW), 02-776 Warsaw, Poland
[2]   Water Centrum, Warsaw University of Life Sciences (SGGW), 02-776 Warsaw, Poland; mariusz_zoltowski@sggw.edu.pl
[3]   Faculty of Mechanical Engineering, UTP-University of Science and Technology in Bydgoszcz, 85-796 Bydgoszcz, Poland; michal.liss@utp.edu.pl
*   Correspondence: gabriela_rutkowska@sggw.edu.pl

Abstract: Production cost reduction and constraints on natural resources cause the use of waste materials as substitutes of traditional raw materials to become increasingly important. The dynamic development of sewerage systems and sewage treatment plants leads to increases in the produced sewage sludge. According to the Waste Law, municipal sewage sludge can be used if it is properly stabilized. This process results in significant quantities of fly ash that must be utilized. This paper presents investigation results of partial cement replacement influence by the fly ash from sewage sludge on concrete parameters. The results confirm the possibility of fly ash waste applications as a cement substitute in concrete manufacturing. In the later parts of the publication, a pilot study was conducted using the modal analysis methodology and aimed at checking the hypothesis of whether vibration methods can be used in the assessment of the amount of the admixture used in concrete and the effect it has on concrete properties. This is the first time that vibration tests have been used to determine the diversity of the concrete mix composition and to distinguish the percentage of ash added. There are no studies using modal analysis to distinguish the composition of a concrete mix in the scientific literature. The article shows that the vibration test results show the differentiation of concrete composition and can be further improved as a method for determining the composition of mixtures and for distinguishing their mechanical properties. These are only pilot studies, which, in order to develop the target cognitive inference, should be performed in the future on a significantly enlarged number of the studied samples.

Keywords: concrete; sewage sludge; fly ash; cement; vibrations; modal analysis

Citation: Rutkowska, G.; Żółtowski, M.; Liss, M. The Use of Modal Analysis in Addition Percentage Differentiation, and Mechanical Properties of Ordinary Concretes with the Addition of Fly Ash from Sewage Sludge. *Materials* **2021**, *14*, 5039. https://doi.org/10.3390/ma14175039

Academic Editors: Carlos Morón Fernández and Daniel Ferrández Vega

Received: 28 July 2021
Accepted: 29 August 2021
Published: 3 September 2021

1. Introduction

Concrete, sometimes referred to as the stone of the present day, is by far the most widely used composite material among man-made materials, second only to water in the entire complex of used materials and without which modern construction would not function. It is an ecological composite that is often made of local raw materials—aggregate (as a filler), cement (as a binder), water, admixtures, and possibly mineral additives. It is a safe product that guarantees the stability and load-bearing capacity of a structure.

One of the most important issues in the development of the construction sector is the continued efforts to make concrete an ecological, even more environmentally friendly material for the environment. The search is undoubtedly indispensable in the case of developing the composition of a concrete mix, the two components of which—aggregate and cement—contribute to the anthropopressive effects at their production and acquisition stages. The big problem is still the fact that during the production of 1 ton of cement, 0.5 to 1 ton of greenhouse gases are produced, which, according to various data, constitutes 6–8%

of the total anthropogenic emission. Research shows that the production and processing of this binder is responsible for 85% of the CO_2 production in the entire concrete industry. To reduce the environmental impact of the produced concrete, its two main components, aggregate and cement, can be replaced by more sustainable alternatives. Natural aggregate can be replaced with recycled aggregate, while Portland cement can be partially replaced with binding additives such as ground granulated blast furnace slag or silica fly ash. Adding a binding additive to the concrete mix has the potential to reduce CO_2 emissions and the cost of concrete production. As reported by Jones et al., replacing cement with 20, 35, and 55% fly ash can lead to a carbon dioxide reduction of 19.4%, 45.8%, and 53.7%, respectively [1–4].

The carbon dioxide emission limits introduced by the European Union (target: 55% emission reduction by 2030) encourage research into next-generation materials containing less clinker. Currently, siliceous fly ash from hard coal combustion is widely used in cement technology, mainly in concrete technology. The wide application siliceous fly ash is mainly determined by its high fineness, similar to cement, their chemical and phase composition, and their pozzolanic activity. The use of silica fly ash to produce concrete is only possible when the requirements specified in PN-EN 450-1:2012 are met [5–7].

The development of sewage networks and thus the municipal sewage treatment plants observed in recent years causes the formation of an increasing amount of sewage sludge. Municipal wastewater supplied to wastewater treatment plants is primarily a mixture of domestic and industrial wastewater, which is also supplied with rainwater and water infiltrating from the ground. The qualitative and quantitative characteristics of such wastewater depend on the type and condition of the sewage system, the amount of water used, the industrialization of the city, or the standard of living of the inhabitants. Both the quantity and composition of the wastewater flowing into wastewater treatment plants generally change in the daily, weekly, monthly, and yearly cycles [8]. Considering the ban on the storage of sludge since 1 January 2016, their management has become not only a technical and economic problem, but also an ecological problem [9]. Until now, the generated sewage sludge was landfilled or, after initial stabilization (aerobic, anaerobic, with lime), was released into the environment, e.g., as fertilizers or for earthworks. However, the sanitary danger, high weight, and hydration of generated sewage sludge, have always been a technical problem. Considering that the presence of toxic substances and heavy metals in generated sewage sludge limits the possibility of its use in agriculture, thermal methods are the most appropriate methods of sewage sludge disposal [10–14]. As a result of this process, the volume of waste (sludge) is reduced, the heat or electricity is obtained, and the content of sulfur and nitrogen compounds in the exhaust gas is reduced. The secondary waste material generated in sewage sludge thermal treatment installations is waste (fly ash—Fly ash sewage sludge—SSA) with the code 19 01 14, which also requires appropriate management. According to the idea of circular economy, a near-zero emission economy, ash from sewage sludge should be treated as a potential product [15].

The indication of appropriate waste management methods after the thermal treatment of sewage sludge is of particular importance in the building materials industry. There are five main trends in the literature on the use of such fly ash. They mainly concern the use of fly ash from sewage sludge as an active additive to inorganic cement binders—concrete and mortars, a component of light sintered aggregate, and a component of excess raw materials to produce construction ceramics as well as a component of a raw material mixture to produce cement or a substitute for cement/sand in earth road structures [16–21]. It should be noted, however, that their application is associated with limitations resulting from the provisions in force in Poland (Journal of Laws 2016, item 108), which in terms of their regulation, implement the Directive of the European Parliament and of the Council [EU/2010/75], which enforce rules regarding ash with a different (from fly ashes from hard coal and brown coal combustion) physico-chemical composition and, most importantly, with the fact that such ash also contains pollutants—heavy metals [22,23].

The research on fly ash from the thermal treatment of sewage sludge presented in the literature proves that the introduction of a certain amount of SSA into the concrete mix does not adversely affect the strength parameters and frost resistance of mortars and concretes. Fly ash is treated as a partial binder replacement or as a mineral additive [24–27]. In studies conducted by Monzo et al. [28], it was shown that mortars containing up to 30% SSA and that were cured in water at a temperature of 40 °C do not show a reduction in compressive strength compared to with the addition of ash. The changes the in strength of mortars are related to the pozzolanic properties of the SSA, maturation conditions, and the content of C3A. Ing et al. [29] found that cement mortars in which cement was replaced with ash in an amount of up to 10% were characterized by a similar or higher compressive strength than conventional mortars. In the studies of Cyr et al. [30], it was estimated that the fly ash introduced into the mortars in the amount of 25% and 50% caused a decrease in the compressive and bending strength parameters compared to the reference mortars. Lin et al. [31] analyzed the influence of fly ash on the compressive strength of building materials. They found that the additive with a finer graining is characterized by a higher index of pozzolanic activity, which resulted in obtaining higher compressive strength from the tested composites. In the studies of Baeza-Brotons et al. [32], ash was added to the cement composite in the amounts of 5, 10, 15 and 20%. It was found that ash influences the absorption of water contained in the composite. The pozzolanic properties of fly ash from the tarmac treatment of sewage sludge and their chemical composition are similar to traditional mineral additives. Fly ash from sewage sludge, considering the fulfillment of both environmental and technical criteria, as stated by other authors [25,26,33–36] in the amount of 5–20%, can be successfully used as an active additive for cement production. In concrete, they show an analogy to traditional mineral additives.

Test methods using modal analysis have been used for many years in technical mechanics. In construction, this methodology is less widespread. In the works of many authors [37,38], the possibilities of using this methodology in the evaluation of research on mechanical properties and its composition in various structures and elements of technical mechanics have been described. Later authors such as [39,40] organized research to adapt this research methodology to civil engineering. Many years of laboratory tests and tests conducted on real building structures have resulted in the creation of a methodology dedicated to the use of materials and building elements, including concrete elements, in testing.

The research shown in first part of the article shows the possibility of reusing fly ash from the thermal conversion of sewage sludge as an additive to concrete and to analyze the impact of the physico-chemical properties of this ash on the compressive strength of concretes produced with their participation, in accordance with the assumptions of the circular economy. The obtained test results were compared to the blank sample, which did not contain any fly ash. The main goal in this article was to conduct experimental laboratory research with the use of the modal analysis vibration methodology to determine the possibility of a non-invasive concrete mix composition by distinguishing the percentage content of plastic ash in the mix.

2. Materials and Methods

For laboratory tests, a concrete mix of C20/25 class with a K2 consistency was designed according to PN-EN 206 + A1: 2016-12 using the computational and experimental method according to Bukowski [41,42]. The mixture was prepared using CEM I 42.5R Portland cement (Cement Ożarów S.A., Ożarów, Poland), 0–2 mm fine aggregate (Warsaw, Poland) and 2–16 mm coarse aggregate (KIM Sp.z o.o. Warsaw, Poland), water, and an additive—fly ash from thermal treatment of sewage sludge at the "Płaszów" sewage treatment plant in Krakow (Poland). In each composition of the concrete mix, the same composition of the fine aggregate selected by the sieve analysis method and the same composition of the course (gravel) selected by the iteration method (Table 1) were adopted.

Table 1. Percentage of the contents of the aggregates selected by iterations.

Fraction	Fraction Mixing Percentage Ratio (for Sand and Gravel)			Grain Composition of (%)	
	I Stage	II Stage	III Stage	Sand	Gravel
0.0–0.125				1.26	0.38
0.0125–0.25				12.33	3.70
0.25–0.50			30	36.35	10.91
0.50–1.0				33.12	9.94
1.0–2.0				16.94	5.07
2.0–4.0		35			24.50
4.0–8.0	49	65	70		22.29
8.0–16.0	51				23.21

To determine the effect of fly ash on the selected mechanical properties of ordinary concretes, two types of samples were prepared:

CO—no additive;

FA—with the addition of fly ash from thermal treatment of sewage sludge in the amounts of 5% to 20%.

The physical and chemical properties as well as the phase composition of CEM I 42.5R Portland cement in accordance with the requirements of PN-EN 197-1:2012 [43] are presented in Tables 2 and 3.

Table 2. Physical properties and phase composition of cement CEM I 42.5R.

Beginning of Binding Time (min)	Compressive Strength after 2 Days (MPa)	Compressive Strength after 28 Days (MPa)	Blaine's Specific Surface Area (cm^2/g)
218	22	49.8	3330
The share of mineral phases CEM I (5 mass)			
C_3S—61.8	C_2S—12.3	C_3A—7.6	C_4AF—4.1

Table 3. Chemical properties of cement CEM I 42.5R.

SiO_2 (%)	Sulfate Content SO_3 (%)	Roasting Loss (%)	Chloride Content Cl (%)	Alkali Content Na_2O_{eq} (%)
20.20	3.20	3.20	0.06	0.73
CaO	Al_2O_3	Fe_2O_3	MgO	CaO_w
65.21	4.34	2.37	1.51	1.76

The composition of the prepared concrete townspeople per 1 m^3 is presented in Table 4.

To assess the effect of fly ash from the thermal transformation of sewage sludge on the properties of concrete mix and concrete, the following were tested: apparent density (PN-EN 12350-6: 2011), air content (determined by the pressure method (PN-EN 12350-7: 2011)), and consistency (determined by the method fall cone (PN-EN 12350-2: 2011)). Compressive strength was tested in a hydraulic testing machine H011 Matest (Italy) in accordance with the guidelines contained in PN-EN 12390-3: 2019-07 after 28, 56, and 356 days of maturation [44–48]. For each series of tests, 15 samples with the dimensions $10 \times 10 \times 10$ cm were prepared.

Table 4. Concrete mix proportions by weight.

Specification	Mass of Concrete Ingredients (kg/m^3)			
	Aggregate	Water	Cement	Fly Ash
Concrete CO	1812.15	193.55	380.12	-
FA 2.5% Concrete with quantity 5% of fly ash	1812.15	193.55	370.62	9.50
FA 5% Concrete with quantity 5% of fly ash	1812.15	193.55	361.11	19.01
FA 7.5% Concrete with quantity 5% of fly ash	1812.15	193.55	351.61	28.51
FA 10% Concrete with quantity 10% of fly ash	1812.15	193.55	342.11	38.01
FA12. 5% Concrete with quantity 5% of fly ash	1812.15	193.55	332.60	47.52
FA 15% Concrete with quantity 15% of fly ash	1812.15	193.55	323.10	57.02
FA 17.5% Concrete with quantity 5% of fly ash	1812.15	193.55	313.60	66.52
FA 20 % Concrete with quantity 20% of fly ash	1812.15	193.55	308.10	72.02

The municipal wastewater that is delivered to the treatment plant is a mixture of industrial and domestic wastewater, which is additionally supplied with rainwater and infiltrating water from the ground. The amount and composition of the wastewater flowing into the wastewater treatment plant is subject to changes during daily, weekly, monthly, and yearly cycles. Additionally, their qualitative and quantitative characteristics primarily depend on the condition and type of a given sewage system, the standard of living of the inhabitants, the amount of water used, or the industrialization of the city. An important rule is that there is no typical composition and quality of municipal wastewater; thus, there is no typical composition of sewage sludge—fly ash generated during thermal transformation—Figure 1. To determine the physico-chemical properties, tests were conducted on fly ash from the thermal transformation sewage sludge.

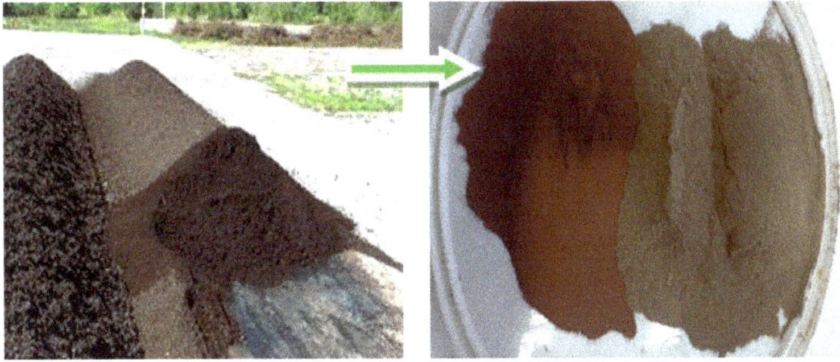

Figure 1. Converted sludge to fly ash from the thermal treatment of sewage sludge—SSA.

Pozzolanic activity determination was conducted in accordance with PN-EN 450-1: 2012 and ASTM C379-65T. The chemical composition of the material was determined by the X-ray energy dispersion fluorescence (XRF) method on an Epsilon 3 spectrometer (Panalytical, Eindhoven, The Netherlands). The study was conducted in the measuring range of Na-Am using an apparatus equipped with an X-ray Rh 9W lamp, working at 50 kV and 1 mA, a 4096-channel spectrum analyzer (Panalytical, Eindhoven, The Netherlands), six measurement filters (Panalytical, Eindhoven, The Netherlands) (Cu-500, Cu-300, Ti, Al-50, Al-200, Ag) and a high-resolution semiconductor SDD detector (Panalytical, Eindhoven, The Netherlands) (Be window, 50 μm thick) cooled with a Peltier cell. Particle size distribution analysis was performed based on the laser diffraction phenomenon using the Mastersizer 3000 analyzer (Malvern Panalytical, Malvern, UK). The measurement was conducted in a dispersing liquid (demineralized water) in the presence of an ultrasonic probe in order to break up the larger aggregates from the tested samples. Grains with equivalent diameters ranging from 0.1 μm to 1000 μm were analyzed. Morphology and chemical composition in the micro-area determination was conducted using the SEM Quanta 250 FEG scanning microscope (Panalytical, Eindhoven, Netherlands) by FEI, which was equipped with the chemical composition analysis system based on the energy dispersion of X-rays—EDS (Energy Dispersive X-Ray Spectroscopy), by EDAX. Mineral composition was determined using X-ray phase analysis (XRD). Measurements were made using the powder method using a Panalytical X'pertPRO MPD X-ray diffractometer with a PW 3020 goniometer (Panalytical, Eindhoven, Netherlands. A Cu copper lamp (CuKa = 1.54178 Å) was used as the X-ray emission source. X'Pert Highscore software (Panalytical, Eindhoven, Netherlands) was used for processing the diffraction data. The identification of the mineral phases was based on the PDF-2 release 2010 database formalized by JCPDS-ICDD [7,27,49].

After conducting the strength tests for the purpose of vibration tests using the modal analysis methodology, standard samples were prepared. In the first case, the cube was made from an ordinary concrete cube, while in the second case, the material composition was modified by adding 20% fly ash from sewage sludge ash. A characteristic feature of this change is the color of both samples, which is shown in Figures 2 and 3. In both cases, the tested cube had the same external dimensions (10 cm edge) and weight (approx. 2.47 kg). This was the first time that a pilot study was conducted using the modal analysis methodology and aimed to confirm the hypothesis of whether vibration methods can be used in the assessment of the amount of the admixture used in concrete and the effect that it has on the properties of concrete.

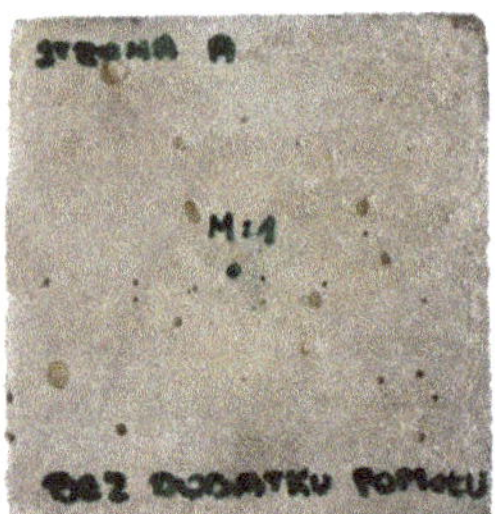

Figure 2. Ordinary concrete cube.

Figure 3. Ordinary concrete cube with the addition of 20% fly ash from sewage sludge.

2.1. Types of Modal Analysis

The modal analysis of mechanical systems is a method of examining dynamic properties consisting of tracking changes in the parameters of the modal structure or wall element model under test resulting from wear, damage, or failure and is based on current observations of the object. In this method, a modal model is created in the form of the natural frequencies set, the vibration modes, and the damping coefficients for an object without damage as a pattern. Then, during operation, the modal model is identified, and its correlation with the model for an undamaged object is examined. When such a correlation occurs, it can be stated that the object is fit. In the absence of correlation, the object is in a state of unfitness, caused, for example, by aging or damage [38,39,50–52].

Based on the knowledge of the modal model, it is also possible to predict the response of an object to any disturbance, both in the time and frequency domains. It is implemented as a theoretical, experimental, or operational modal analysis.

Theoretical modal analysis is defined as the intrinsic problem of a matrix, which is dependent on the matrix of masses, stiffness, and damping. Theoretical modal analysis requires solving the problem for the adopted structural model of the structure under study [37,40,53]. The determined sets of eigenfrequencies, damping coefficients for the eigenfrequencies, and the modes of free vibrations allow for the simulation of structure behavior with any input, the selection of the controls, structure modification, among other features. It is used in the design process when it is not possible to conduct research on a similar object.

It is especially useful in the analysis of the resistance of structures to various disturbances and enables the assessment of skyscrapers, bridges, railway wagons, vibrations, and tectonic movements. An example of the above may be a model of a skyscraper structure—a system of heavy plates connected to each other by relatively flexible girders. So far, in practical application, modal analysis has been used to diagnose lattice structures (masts, antennas, cranes), diagnose a turbine set, and diagnose the quality of bridge structures. In most of these applications, it is assumed that because of failure, the stiffness of the structure changes locally, which causes changes in the parameters of the modal model. By tracking changes in the mode of natural vibrations, it is possible to determine the area in which significant destruction occurs [54].

The sets of the eigenfrequencies, the damping coefficients for the eigenfrequencies, and the modes of free vibrations determined in the eigenvalues problem allow for the simulation of the behavior of the structure under any excitation, control selection, structure modification, among other factors.

The analysis of the eigenfrequencies and eigenvectors is obtained from the equation of motion (after omitting the terms containing the damping matrix and the vector of external loads). Then, the equation of the natural vibration motion has the following form:

$$B\ddot{q} + Kq = 0 \tag{1}$$

For a system with one degree of freedom, the solution has the form of a function:

$$q(t) = \vec{q}\,sin(\omega t + \varphi) \tag{2}$$

where q is the vector of the natural vibration amplitudes.

Substituting the above equation and the second derivative into the equation of motion, we achieve

$$\left(-\omega^2 B + K\right)\vec{q}\,sin(\omega t + \varphi) = 0 \tag{3}$$

The above equation must be fulfilled for any moment t—then, the system of algebraic equations is obtained:

$$\left(K - \omega^2 B\right)\vec{q} = 0$$

$$\left(k_{11} - \omega^2 m_{11}\right)q_1 + \left(k_{12} - \omega^2 m_{12}\right)q_2 + \ldots + \left(k_{1n} - \omega^2 m_{1n}\right)q_n = 0$$
$$\left(k_{21} - \omega^2 m_{21}\right)q_1 + \left(k_{22} - \omega^2 m_{22}\right)q_2 + \ldots + \left(k_{2n} - \omega^2 m_{2n}\right)q_n = 0$$
$$\left(k_{n1} - \omega^2 m_{n1}\right)q_1 + \left(k_{n2} - \omega^2 m_{n2}\right)q_2 + \ldots + \left(k_{nn} - \omega^2 m_{nn}\right)q_n = 0$$

$$(4)$$

The result is a linear system of homogeneous algebraic equations that has a non-zero solution only when

$$\det\left(K - \omega^2 B\right) = 0 \tag{5}$$

After transformations, we obtain an nth degree polynomial with respect to ω^2. Multiple roots can occur among the roots, and the vector made up of frequencies set arranged in the increasing values order is called the frequency vector, and the first frequency is called the fundamental frequency [39].

$$\omega = [\omega_1, \omega_2 \ldots, \omega_n] \tag{6}$$

Theoretical modal analysis is mainly used in the design process when it is not possible to conduct research on the object.

Experimental modal analysis is one of the techniques that is used for identifying structure modal parameters. Experimental modal analysis is a technique that is often used in practice for examining the properties of objects, both at the stage of construction and in operation. The identification experiment in experimental modal analysis consists of forcing the vibrations of the object with the simultaneous measurement of the exciting force and the response, most often in the form of a vibration acceleration spectrum. The modal model is obtained from the stabilization diagram and the animation of the vibration form presented in the software.

The identification experiment in experimental modal analysis (Figure 4) consists of forcing the vibrations of the object with the simultaneous measurement of the exciting force and the response of the system, most often in the form of vibration acceleration amplitude.

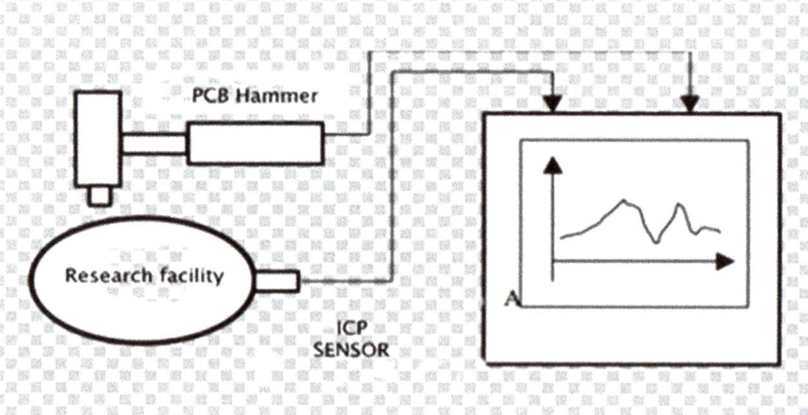

Figure 4. The essence of research in experimental and operational modal analysis.

The measurement results based on the adopted geometric model are processed in the software system (LMS, MATLAB, VIOMA), obtaining the vibration spectrum of the driving force at the input of the system, the spectrum of the vibration acceleration amplitude at the output of the system, and the stabilization diagram—from which the parameters of the modal model can be estimated [39,40]. Stabilization diagram is simply a plot of different model orders or damping ratios versus the frequencies identified at each of model orders. Generally, the typical stabilization diagram is a graph where the frequency is plotted as the X-axis and the model order as the Y -axis

The next Figure, Figure 5, shows example stabilization diagrams with the distinguished frequencies of the free vibrations.

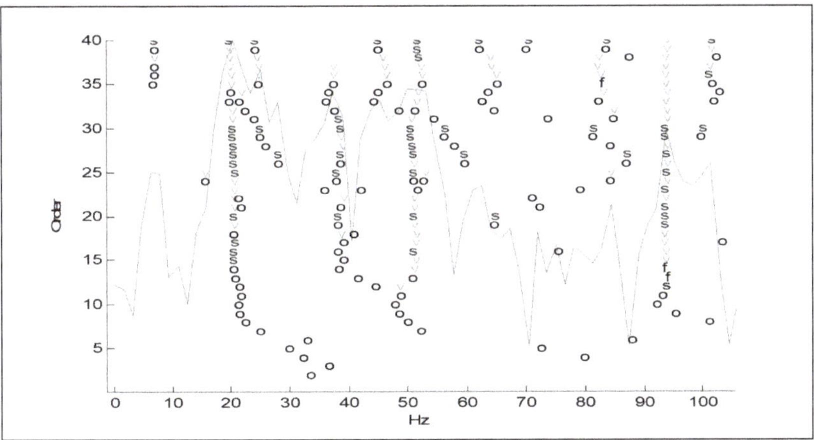

Figure 5. Stabilization diagrams. Notations: order—order of the pole, o—unstable pole, f—pole has a constant frequency. The v-pole has a constant frequency and modal vector; s—stable pole.

By subjecting the recorded vibration signal and the response of the system to the FFT (Fast Fourier Transform) analysis (fast Fourier analysis), it is possible to obtain the measure of the dynamic state FRF after the transformations. The function (FRF- Frequency Response Function) is a basic measure that describes the global dynamic properties of the structure under study. The FRF describes an input–output relationship between two points on a structure (material) as a function of frequency.

The experimental parameters of the modal model (natural frequency, damping, and vibration modes) as well as many other important vibration estimators can additionally be obtained without major difficulty in the procedures accompanying the determination of the FRF.

Operational modal analysis is used to identify objects of large spatial dimensions and large masses. It is based on the measurement of the response to operational forces, resulting from external forces or kinematic excitations and the process of destruction of the building elements [40].

Operational modal analysis:

- Enables the analysis of large-sized objects for which laboratory tests are difficult;
- Models objects more correctly because the inputs correspond to real loads;
- Enables the identification of nonlinear models.

For determining the vibration form, the operational analysis method is based on a multi-channel response measurement at the modal points of a real object. The system enables a graphic representation of the object's behavior under operating conditions. The input data to the system are the time courses of mechanical vibrations occurring at the modal points of the object that are related to one of them (with the highest amplitude). The essence of applying operational modal analysis is shown in Figure 6.

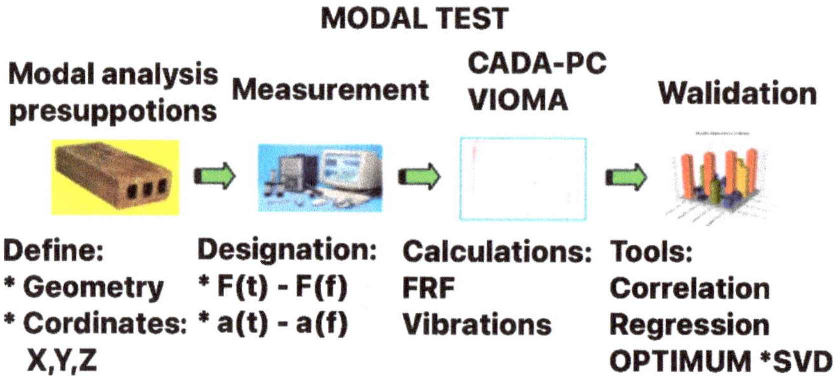

Figure 6. The essence and scope of research in operational modal analysis.

Operational modal analysis is a technique based on the measurement of the system's response to unknown operational forces resulting from the action of technological process forces or kinematic excitations and the process of structural element destruction. During the tests, the measurement points and reference points should be determined and should be constant during the measurements. In this method, based on the measured signals at the facility exit obtained during the measurements at the selected reference and measurement points for unknown system inputs, the estimation of the modal parameters is performed. As a result of the estimation, the poles of the system are identified, and then, on their basis, free vibration modes are estimated. In this method, the selection of the number and the location of the reference points becomes important.

2.2. Measurement Systems

The LMS SCADAS Recorder system (SIEMENS LMS TEST EXPRESS, Leuven, Belgium) is one of the most advanced measurement systems used in research on object condition degradation. A LMS SCADAS Recorder is a device that combines the features of an analyzer and a classic recorder. Figure 5 shows the front panel of the LMS SCADAS system. Depending on the specifics of the tests, the user configures the device with the necessary set of measurement cards with the appropriate number of measurement channels. Contrary to classic LMS recorders, a SCADAS Recorder is fully automatic and does not need to be computer or remote controlled to conduct the process of recording the measurement signals, and the data are saved on a Compact Flash card. This device is fully compatible with professional engineering software, and it is shown in picture below (Figure 7).

LMS Test.Lab is a system for measurement, data acquisition, analysis, and reporting. The system includes dedicated procedures for structural and acoustic testing, environmental testing, vibration control, and quality testing.

LMS Test.Lab is used to initially provide data collected on real objects and to integrate them into the simulation process. It can be used to provide simulation software with data on dependent models that are either too difficult to implement or that would take too long to create. These data can be, for example, time waveforms, spectral FRF transition functions, cross power waveforms, etc. After testing the prototype created based on the simulation, we can test it, and LMS Test.Lab will provide us with data for its modification and improvement. The only way to obtain the best results is a combination of simulations and tests on real objects that were based on them [37,40].

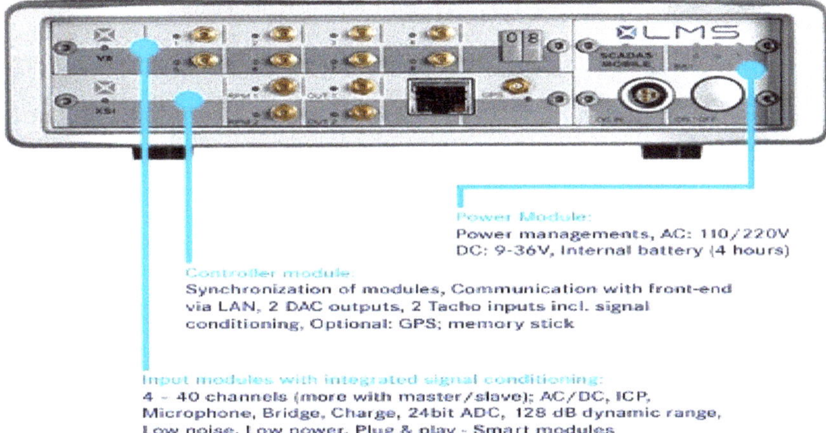

Figure 7. LMS SCADAS Recorder.

Test.Lab is a multi-functional software built modularly so that the program can be adapted to specific measurements. The software configured during the modal analysis allows you to create geometry, calibrate sensors, and determine appropriate measurement ranges; control inductors; record the inputs, responses, and estimates; select modal parameters; and visualize modal forms.

The LMS Test.Lab measurement system is complemented by the LMS Virtual.Lab system, which allows you to combine the most important aspects of simulation and tests on existing facilities. It offers a unique, hybrid approach to simulation—input data from real objects are combined with data from simulated objects.

Presented issues indicate that the use of technical diagnostics is important at all stages of the existence of facilities and is particularly useful at the stage of their operation, providing information and data for the study of the constructed models of operational strategies.

The condition of actual building structures or masonry elements should be tested in a simple and effective method, using the minimum number of measurements. In laboratory tests of building elements, experimental modal analysis is used for this purpose, and in tests of entire structures, operational modal analysis is used. In both modal analysis cases, the most modern measuring equipment, purchased for research projects conducted by means of LMS—under the name LMS TEST.XPRESS—was used to measure the time courses of the excitation and the system response to the given excitation and to determine the main measures of quality (FRF and Coherence functions). This software allows you to easily conduct, according to the developed degradation state analysis algorithm, modal analysis of masonry elements and any other building structures.

The LMS Test. Xpress 4A software (SIEMENS LMS TEST EXPRESS, Leuven, Belgium) allows for the registration of the input force signal and the response in the form of time waveforms and can generate a cross-correlation function, which, in turn, can generate a stabilization diagram with natural vibration frequencies for each structure element.

Correct measurement depends on obtaining the level of the exciting force that was previously determined and on the appropriate level of the response signal. Repeatability tests of waveforms concerning measures of general changes in the state of material destruction, the construction of stabilization diagrams, and the determination of the values of the areas under the FRF and coherence curves are just some of the advantages of the LMS program.

3. Results and Discussion

3.1. Properties of the Fly Ash

Fly ash from coal combustion is currently an important and valuable raw material used in the production of cement or concrete. The applicable standards define the requirements for the physical properties and chemical composition of ash that can be used as an additive to cement or concrete [7]. An important ash parameter that determines its use in concrete technology is its phase and chemical composition, pozzolanic activity index, and fineness. According to the requirements of the standard, the total content of silicon dioxide, iron oxide, and aluminum should be a min. of 65% by weight, with a reactive SiO_2 content of at least 25% by weight. Additionally, the content of reactive CaO should not exceed 10% MgO 4%, and the total alkali content, calculated as Na2O content (equivalent), should not exceed 5% by weight. The content of soluble phosphate (P2O5) should not exceed 100 mg/kg. The activity index after 28 days of maturation should reach the value of $\geq$75%, and after 90 days, the value of $\geq$85% should be reached. The pozzolanic activity of fly ash from the thermal treatment of sewage sludge after 28 days of maturation was 66.4%, while after 90 days, it was 77.3%. Additionally, it should be emphasized that the PN-EN 450-1: 2012 standard refers to ash from hard coal combustion—silica fly ash [7].

The results of the analysis of the oxide composition of fly ash from the thermal transformation of sewage sludge are presented in Figure 8.

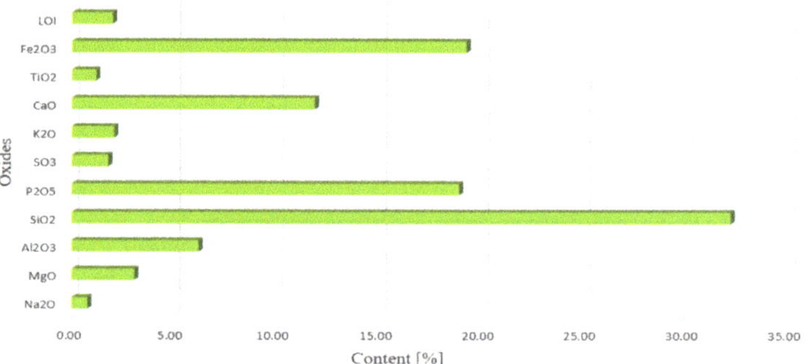

Figure 8. Oxide composition of fly ash from thermal treatment of sewage sludge.

It was observed that the sum of the content of aluminum oxide (Al_2O_3), silicon dioxide (SiO_2), and iron oxide (Fe_2O_3) in the fly ash from sewage sludge was at the level of 57.71%. The largest percentage of the ash was silicon dioxide (32.21%), iron oxide (19.25%), phosphorus (18.91%), and calcium (18.64%). The loss on ignition, which expresses the content of unburned carbon in the tested sample of fly ash in the fluidized bed furnace at a temperature above 850 °C, was only 0.5%. The phosphate content was also much higher than that of conventional ash. This is due to the removal of phosphorus from the wastewater and its accumulation in the sludge.

Figure 9 shows the volume distribution of the individual particle fractions of the additive used. At the largest percentage share above 25%, there were grains with a diameter of 20–250 μm, while at the smallest percentage share, 1.45%, there were grains with a diameter of 0.01–2.0 μm. For comparison, in the research conducted by Kosior-Kazeruk, the content of particles (0.25 mm) in ash from the thermal treatment of sewage sludge was 7% [55]. According to Joshi and Loftia, the diameter of the ash grains ranges from 1 to 150 μm and is similar to the grains of the binder—cement [56].

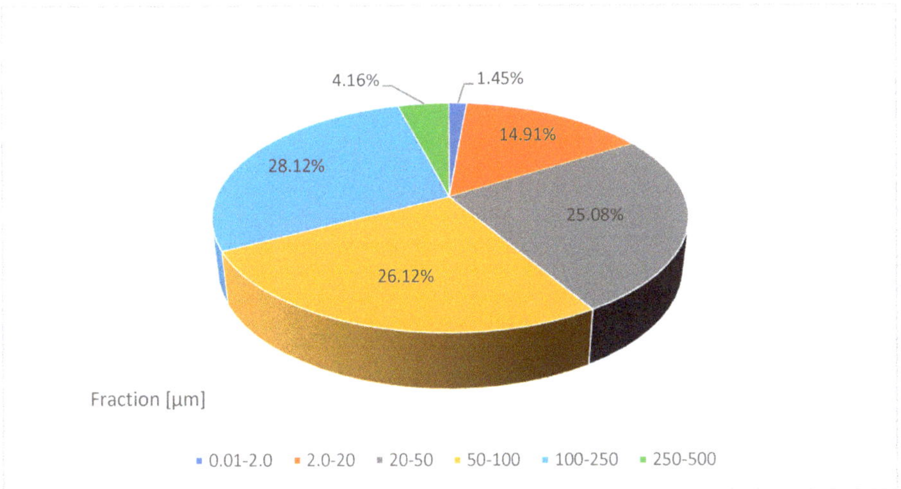

Figure 9. Volumetric distribution of the particle fractions of fly ash from thermal treatment of sewage sludge.

Figure 10 shows a SEM photo of fly ash from the thermal treatment of sewage sludge. Chemical analysis in the SEM-EDS micro-area showed the presence of grains with the chemical composition of calcium, aluminum, iron, and phosphorus as well as potassium, magnesium, and sodium. However, considering the mineral composition of ash from sewage sludge, hematite, quartz, and anhydrite were dominant (Figure 11). The fineness of the ash from sewage sludge determined according to PN-EN 451-2: 2017-06 was 50.7%, while the specific density tested according to PN-EN 1097-07: 2008 was 2534 kg/m^3 [57,58].

(A) **(B)**

Figure 10. SEM photo of fly ash from thermal treatment of sewage sludge in enlargement. (**A**)— 4000 times, (**B**)—30,000 times.

Fly ash samples are predominantly irregular grains with variable size and a strongly developed surface, showing a loose and rough structure of the material that also has high porosity. It leads to higher water absorption, which is connected to the increased water demand of concrete containing SSA. This finding is consistent to that of Alonso and Wesche [59], who wrote that ash grains with a diameter of over 125 µm are highly porous and that the Blaine number (SSA) of these materials falls into the range 250–550 m^2/kg. As Malhotra and Ramezanianpour demonstrated, the grain size and SSA of fly ash is not associated with the formation source. Spherical and cuboidal forms are very rare. Grains of fly ash produced in coal combustion are usually spherical, but they can be irregular or angular [60].

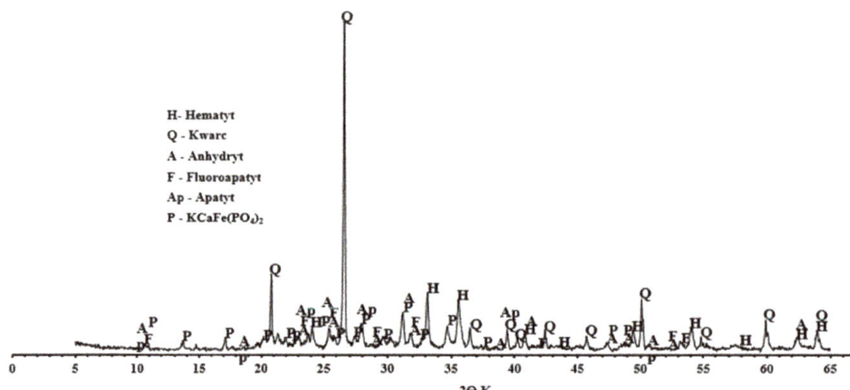

Figure 11. Mineral composition of fly ash from the thermal treatment of sewage sludge.

3.2. Properties of the Concrete Mix

The basic properties of each of the prepared concrete samples were checked: consistency, density, and air content. Based on the results, it was found that the fly ash from the thermal treatment of sewage sludge affects its individual parameters. The lowest air content was observed for a sample of the reference concrete mixture CO, with the air content being equal to 1.8%, while the highest air content occurred in the mixture in which 20% of the cement was exchanged with ash, FA20%, which had an air content equal to 2.9%. The density of the concrete mix ranged from 2322 to 2394 kg/m^3. Concrete mix without additive obtained a plastic consistency, similar to concrete mixes containing different amounts of additive.

3.3. Compressive Strength

The results of the compressive strength test in the four maturation periods (28, 56, 90, and 365 days) of concrete samples with different fly ash content from the thermal treatment of sewage sludge are shown in Figure 12.

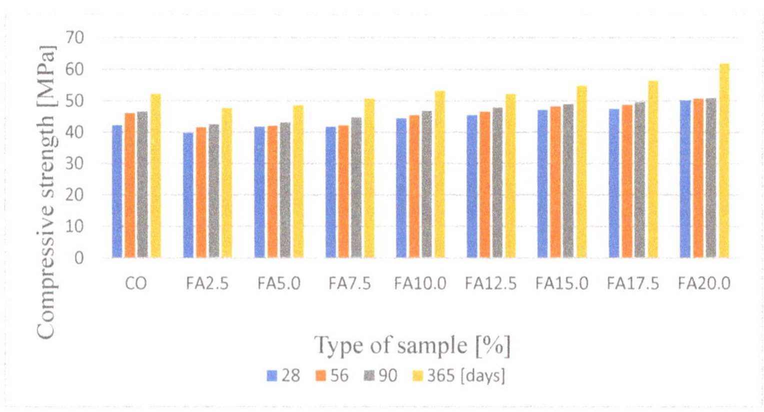

Figure 12. Average compressive strength in different periods of concrete maturation.

Fly ash from the thermal treatment of sludge does not meet the requirements of PN-EN 450-1: 2012, but replacing cement with this additive in the amount of up to 20% had a positive effect on the compressive strength. The highest compressive strength in the first maturation period (28 days), equal to 50.1 MPa, was achieved by the FA20% concrete samples, while the lowest strength, equal to 39.7 MPa, was achieved by the samples in which the cement was replaced with 2.5% ash. Compared to the reference concrete, the decrease in concrete FA5% was 5.6%, while the increase in concrete strength FA 20% was 18.7%. Taking into account the subsequent maturing periods, i.e., after 56, 90, and 365 days, a further increase in the compressive strength of all of the samples was observed. After 28 courses, the compressive strength FA20% was the highest in all of the studied maturation periods. These values were 50.6 MPa, 50.8 MPa, and 61.9 MPa. The increase in endurance between days 28 and 365 of puberty was 23.6%. The lowest compressive strength was observed for the FA2.5% concrete samples. After 56 days, the strength of these samples was 41.6 MPa and 42.5 MPa after 56 days and 47.7 MPa after 365 days. The increase in strength between the first and last tested maturation period for these samples was 12.6%. In addition, it was observed that the reference concrete showed a lower compressive strength than the FA20% samples by a value equal to 18.8% after one year of maturation. Our own experimental research conducted on the influence of fly ash from the thermal transformation of sewage sludge on the compressive strength in different maturation periods confirms the positive influence of this type of ash [25–27].

According to Williams [61], a low loss of ignition and the P2O5 content affect the strength of concrete. In addition, it is assumed that the presence of phosphate ions may cause a slow increase in the strength of concretes containing ash due to the delay in the cement hydration process [62]. The diversified chemical composition of different types of fly ash (Figure 8) allows for the production of ash concretes that meet the standard strength requirements. The pozzolanic and hydraulic properties as well as the chemical composition (silica, iron, aluminum, calcium, magnesium, phosphorus, and oxygen) of sewage sludge ash used in concrete as a replacement for Portland cement are analogous to traditional mineral additives [63]. Chang et al. [64] showed that the addition of ash from the thermal treatment of sewage sludge increases the water absorption capacity and reduces the workability and compressive strength. The most favorable conditions were obtained with the incorporation of 10% ash into the mixture. In our own research, similar results were obtained. Thus, the compressive strength of concrete with the 10% addition of sewage sludge ash was higher than for the samples with 5% and 15% ash content. It should be added here that the compressive strength of this sample was only lower than that of concrete with a 15% share of silica ash [26]. The optimal ash content from sediment, according to information provided by other authors, in cement materials ranges from 5% to 20% [34,35]. Taking environmental and technical criteria into account, Chen et al. [65] analyzed the possibilities of using ash as a substitute for cement and/or a sand substitute in building materials. It was confirmed that mortar containing 10% ash has a compressive strength that is similar to that of a conventional mortar. Replacing cement with ash in the amount of 25% delays the process of setting the slurry and slows the increase in the compressive strength of mortars and concretes compared to reference composites without the addition of ash. However, by extending the maturation time of the composites, it is possible to obtain the strength required for construction concretes [55,66]. An important and binding rule is that there is no typical composition and quality of municipal wastewater; thus, there is no typical composition of the fly ash that is formed during the thermal transformation of sewage sludge, which hinders the common use of ash as active additives for concrete.

3.4. Modal Analysis Results of Ordinary Concretes with the Addition of Fly Ash from Sewage Sludge

The object of the current research is a construction material in the shape of a cubic cube. In the case of the objects that were tested, the composition of the material used to make the cube changed. During the modal test experiment, four cubes were tested (two that were made of ordinary concrete and two that were concrete cubes with the 20% addition of fly ash from the sewage sludge ash addition) (Figure 13).

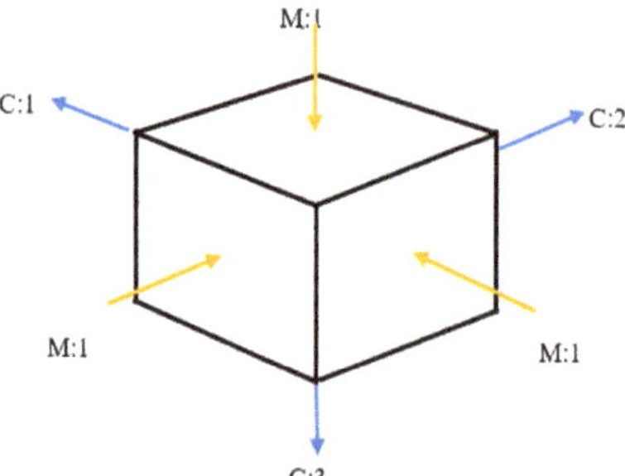

Figure 13. Distribution of excitations and signal response points on the research object.(C and M: the names of axels in software).

During the experiment, two time courses were recorded: one from the modal hammer and the other from the vibration sensor attached to the test object. In modal studies, the frequency range from 0.7 Hz to 4100 Hz was adopted. Due to the symmetry of the cube, it was decided that the modal study would be conducted in three directions, and it was because of this that three spectral transition functions were obtained for each cube. Figures 14 and 15 show the FRF plots obtained from the modal study performed on both cubes.

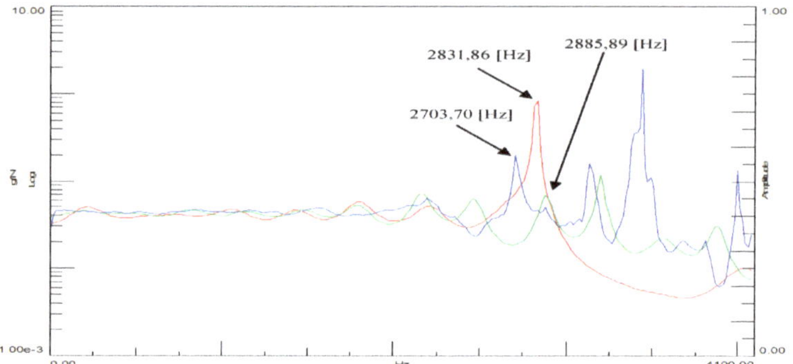

Figure 14. Spectral functions of the FRF transition for ordinary concrete cube. Legend: ——— spectral transition function FRF (Side A), ——— spectral transition function FRF (Side B), ——— spectral transition function FRF (Page C).

The initial analysis of the obtained spectra focuses on the case of the cube without the addition of ash around the three natural frequencies marked in Figure 15. The situation is completely different in the case of the spectral transition functions determined for the cube with the 20% ash addition. In addition to much more convergent spectra, three characteristic vibration frequencies were also obtained. A particularly interesting situation is that in the predetermined natural frequencies, there is a very large discrepancy between the cube with and without the additive. This would initially indicate a significant differentiation of these two building materials in terms of their dynamic properties. Additionally, the analysis of individual phase shifts indicates the occurrence of the structural vibrations of at least one of the three predetermined natural frequencies for the cube without the addition of ash.

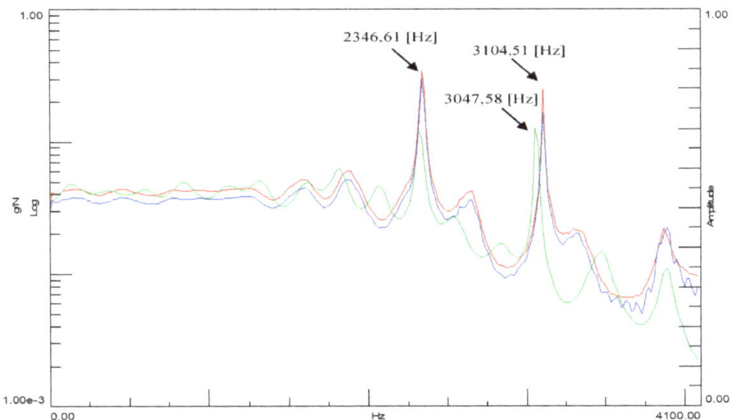

Figure 15. Spectral functions of the FRF transition for the ordinary concrete cube with the 20% addition of fly ash from the sewage sludge ash addition. Legend: ——— spectral transition function FRF (Side A), ——— spectral transition function FRF (Side B), ——— spectral transition function FRF (page C).

Figure 16 shows the spectral transition function of the FRF in the amplitude form for a cube without the addition of ash with a marked natural frequency of 2831.86 [Hz]. It should be clearly emphasized that the determined natural frequency occurs simultaneously with the phase shift, which may be interpreted as a typical characteristic of structural vibrations (Figure 17). In Figures 16, 19 and 21 on the Spectral function of the FRF transition is shown the relevant numbers of sensors connection to the measuring apparatus (C2), the connection point of the modal hammer to inputs (M1) and the measurement directions (X and -X axes), and the time in which the measurements were performed (SUM in seconds)

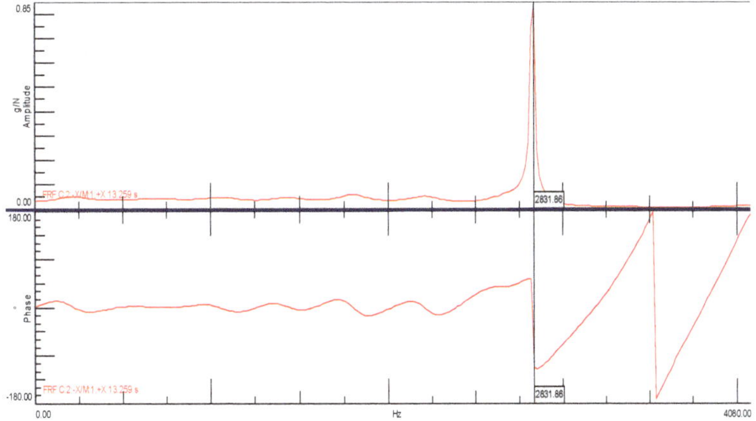

Figure 16. Spectral function of the FRF transition for the ordinary concrete cube with the first natural frequency marked.

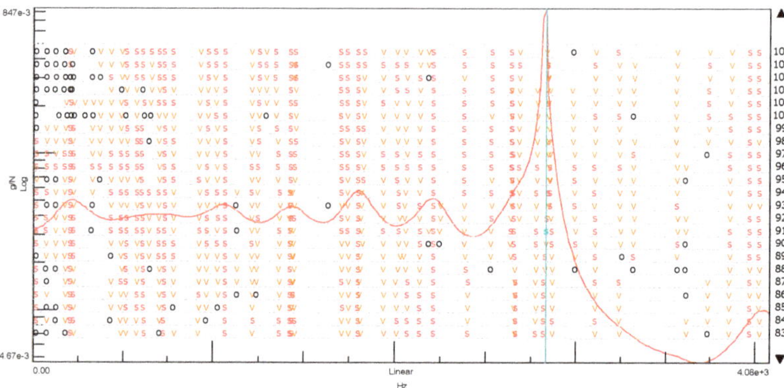

Figure 17. Stabilization diagram made of the spectral transition function FRF (Side B) for ordinary concrete cube (s—stable frequency field, v—modal vector, o—empty field).

We can easily find this frequency of vibrations in the stabilization diagram (PolyMAX estimation method—pLSCFD) created from the spectral transition function determined on the B side of the cube without the addition of ash (Figure 18).

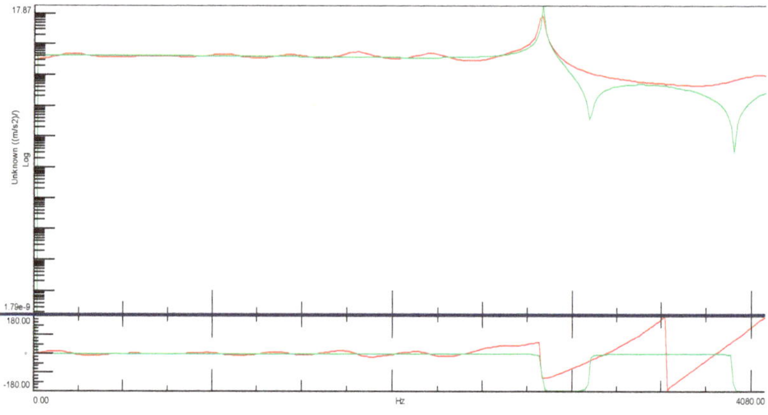

Figure 18. Verification of the measured spectral transition function FRF with its synthesized form.

To verify the correctness of the selected natural frequency of the cube without the addition of ash, the method of comparing the synthesized FRF spectra with the corresponding original FRF spectra was used, thus obtaining a correlation at the level of 84.77%. This indicates a very high level of correctness for the selected model. The effect of the method outside of the selected band was compensated for with the upper and lower residuals.

In another modal study, a stabilization diagram was created considering all three spectral transition functions of the FRF corresponding to each of the three sides of the cube without the addition of ash. For this purpose, a function in the form of the SUM indicator was used, which strengthens their exposure on the chart in the places where peaks occur. This situation is shown in Figure 19 by comparing the averaged spectral function of the FRF transition with phase shifts. The graph clearly shows that the phase shifts, which are a kind of determinant in the case of selecting the appropriate natural frequencies, are not as clear as in the previous case. On this basis, a stabilization diagram was created, which is shown in Figure 20.

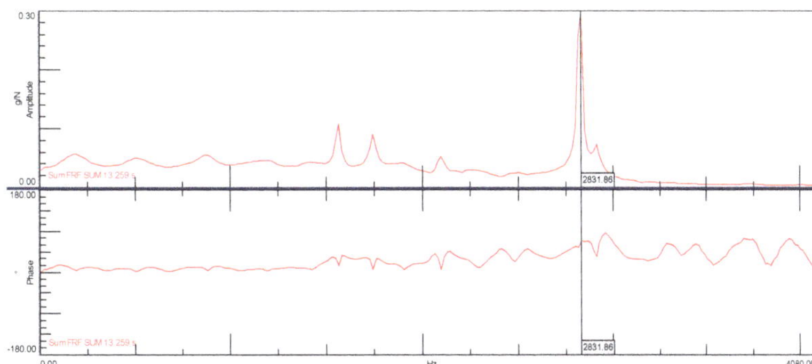

Figure 19. Spectral function of the FRF transition created with the use of the SUM indicator for the ordinary concrete cube.

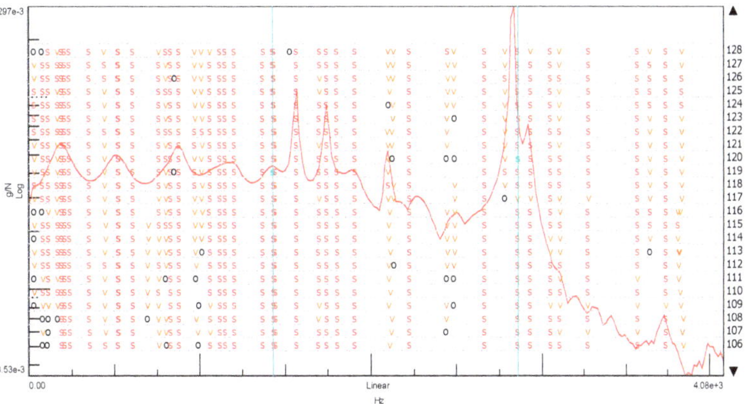

Figure 20. Stabilization diagram made of the spectral transition function FRF (using the SUM index) for the ordinary concrete cube (s—stable frequency field, v—modal vector, o—empty field).

Additionally, in this case, the most significant natural vibration frequency is the one with the value of 2852.83 (Hz). This characteristic eigenfrequency occurs, however, in the vicinity of several other frequencies, and after multiple selection, the frequency of 1427.99 (Hz) was specified only for further analysis. Ultimately, however, due to the very high modal damping factor (exceeding 35%), it was necessary to eliminate this form from further analysis. Thus, the effect of considering all of the spectral FRF transition functions of the cube without the addition of ash in the modal analysis resulted in obtaining one characteristic natural vibration frequency. Its value, compared to the modal analysis based on one spectral transition function FRF, is shifted by approximately 21 (Hz). Therefore, it can be assumed that the first and the same time characteristic frequency of the cube's natural vibrations without the addition of ash is at the value of 2831.86 (Hz).

Another modal analysis concerned a cube with a 20% ash addition. Considering the very high convergence of the determined spectral functions of the FRF transition for each side of the cube with the 20% ash addition, it was decided to analyze all of the FRF spectra using the SUM index. Its form is shown in Figure 21 together with the phase shifts. The diagram presented in Figure 22 shows the frequencies of the natural vibrations, which were assumed to be characteristic for the cube without the addition of ash. Thanks to this, it is clearly visible that there is a very clear difference in the characteristic frequencies of the natural vibrations of both of these objects. This difference is at the level of about 495.17 (Hz) in the direction of the beginning of the graph's coordinate system and is 274.2 (Hz) in the direction of higher frequencies. This is a significant discrepancy beyond the tolerance associated with the measurement error.

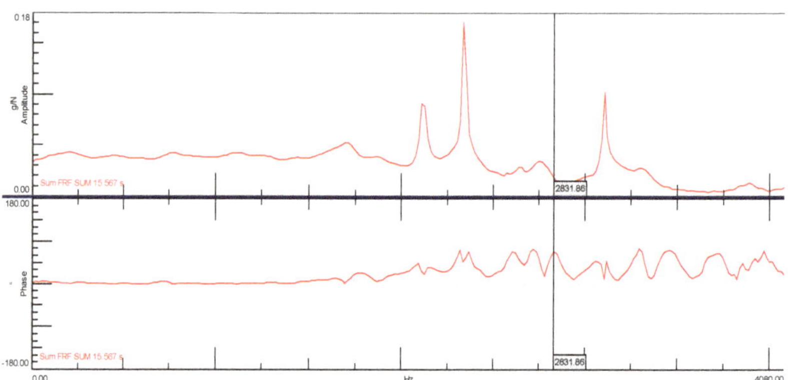

Figure 21. Spectral function of the FRF transition created with the use of the SUM indicator for the ordinary concrete cube with the addition of fly ash from sewage sludge.

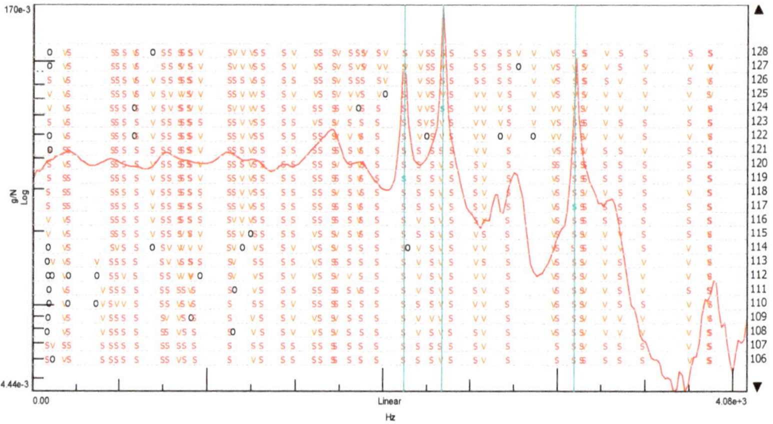

Figure 22. Stabilization diagram made of the spectral transition function FRF (using the SUM index) for the ordinary concrete cube with the 20% addition of fly ash from sewage sludge. (s—stable frequency field, v—modal vector, o—empty field).

Based on the stabilization diagram created in this way, three natural frequencies were initially determined, which describe the dynamic properties of the cube with the addition of ash. The pre-selected vibration frequencies are shown in Table 5.

Table 5. Pre-determined natural frequencies of the ordinary concrete cube with the 20% addition of fly ash from sewage sludge ash.

Ordinary Concrete Cube with 20% Addition of Fly Ash from Sewage Sludge		Mode 1	Mode 2	Mode 3
		2119.214	2336.438	3094.247
Mode 1	2119.214	100	27.697	61.491
Mode 2	2336.438	27.697	100	1.322
Mode 3	3094.247	61.491	1.322	100

For selected free vibration frequencies, a validation was conducted using the AutoMAC method. The result of this validation is presented in a tabular form in Table 6 and in a graphic form in Figure 23. The validation shows that the first selected form of vibration is inappropriate for describing the dynamic properties of the cube with the addition of ash.

Table 6. Determined natural frequencies of the ordinary concrete cube with the 20% addition of fly ash from sewage sludge ash.

Ordinary Concrete Cube with 20% Addition of Fly Ash from Sewage Sludge		Mod 2336.438	Mod 3094.247
Mode 2	2336.438	100	1.464
Mode 3	3094.247	1.464	100

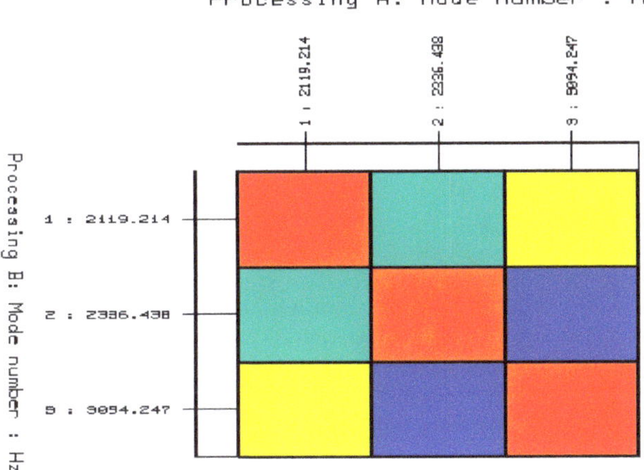

Figure 23. Validation of the selected stable poles and the corresponding modes of the free vibrations using the AutoMAC function (convergence percentage between the modes of on-vibrations determined from the stabilization diagrams: blue = 4% yellow = 28% red = 100% turquoise = 15%).

As a result, the correct and final number of natural frequencies describing the dynamic properties of the cube with ash added was obtained, as shown in Table 6. Additionally, Figure 24 shows the result of the validation with the AutoMAC method.

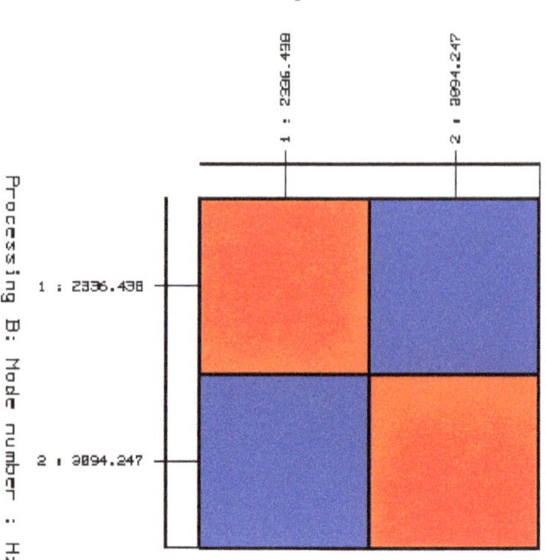

Figure 24. Validation of selected stable poles and the corresponding modes of the free vibrations using the AutoMAC function. (convergence percentage between the modes of on-vibrations determined from the stabilization diagrams: blue = 4%, red =100%).

4. Conclusions

The use of waste—fly ash from the thermal treatment of sewage sludge with the code 19 01 14 in building materials creates great economic benefits. The experimental studies conducted in this research showed that it is possible to use fly ash from sewage sludge to produce concrete as a partial replacement for cement. The obtained results and their analysis allow for the following conclusions:

1. There is no typical quality and typical composition of municipal wastewater supplied to the treatment plant; thus, there is no typical physico-chemical composition formed during the thermal treatment of sewage sludge—fly ash;
2. Waste-generated fly ash from the thermal transformation of sewage sludge used to produce concrete has a positive effect on its compressive strength;
3. Fly ash from the thermal transformation of sewage sludge does not meet the requirements of PN-EN 450-1: 2012. It has a different physico-chemical composition compared to the fly ash from coal combustion used in concrete technology. The highest percentage of the collected fly ash samples were oxides of silicon, calcium, phosphorus, and aluminum;
4. Concrete containing fly ash from the incineration of sewage sludge in its composition was characterized by a compressive strength comparable to that of the reference concrete without the additive. With an ash content up to 20%, they can be used as a cement substitute. The average compressive strength of concrete containing 20% fly ash from the thermal transformation of sewage sludge was after 28, 56, and 365 days of maturation was 50.1 MPa, 50.8 MPa, and 61.9 MPa, respectively;
5. The results of the conducted modal tests of cubes without and with the addition of ash showed that it is possible to distinguish both products using the method of experimental modal analysis. For the cube without the addition of ash, one natural frequency was determined at the level of 2831.86 (Hz), while for the cube with the ash addition, two natural frequencies were established at level I—2336.44 (Hz) and II—3094.25 (Hz). The obtained results of the analysis clearly indicate that the dynamic state in the case of the cube without the addition of ash and the cube with the 20% ash addition is different. This is part of an additional research study that will be continued in the future. The results of these studies are assumed to result in the development

of a model for the non-invasive diagnosis of the mechanical properties of concrete containing fly ash from the incineration of sewage sludge.

Author Contributions: Conceptualization, G.R.; methodology, G.R.; validation, G.R.; formal analysis, G.R.; investigation, M.L.; resources, G.R. and M.Ż.; writing—original draft preparation, G.R. and, M.L.; writing—review and editing, M.L and M.Ż.; visualization, G.R. All authors have read and agreed to the published version of the manuscript.

Funding: This research received no external funding.

Institutional Review Board Statement: Not applicable.

Informed Consent Statement: Not applicable.

Data Availability Statement: Data are not publicly available. The data may be made available upon request from the corresponding author.

Conflicts of Interest: The authors declare no conflict of interest.

References

1. International Statistics Yearbook. Available online: https://stat.gov.pl/en/topics/statisticalyearbooks/statistical-yearbooks/concise-statistical-yearbook-of-poland-2020,1,21.html (accessed on 12 February 2021).
2. IEA—Shaping a Secure and Sustainable Energy Future for All. Available online: www.iea.org (accessed on 12 February 2021).
3. Kuzielová, E.; Slaný, M.; Žemlička, M.; Masilko, J.; Palou, T.M. Composition of Silica Fume—Portland Cement Systems Formed under Hydrothermal Curing Evaluated by FTIR, XRD, and TGA. *Materials* **2021**, *14*, 2786. [CrossRef]
4. Jones, R.; McCarthy, M.; Newlands, M. Fly ash route to low embodied CO_2 and implications for concrete construction. In Proceedings of the 2011 World of Coal Ash Conference, Denver, CO, USA, 9–12 May 2011; pp. 1–14.
5. Gupta, S.M. Support vector machines-based modelling of concrete strength. *Int. J. Intel. Technol.* **2007**, *3*, 12–18.
6. Kim, J.I.; Kim, D.K. Application of neural networks for estimation of concrete strength. *KSCE J. Civ. Eng.* **2002**, *6*, 4. [CrossRef]
7. *PN-EN 450-1:2012 Fly Ash for Concrete. Part 1: Definition, Specifications and Conformity Criteria*; Polish Standardization Committee: Warsaw, Poland, 2012.
8. *Strategy for Dealing with Municipal Sewage Sludge for 2019–2022*; Ministry of the Environment: Warsaw, Poland, 2018.
9. Ordinance of the Minister of Economy of 16 July 2015 on Admitting Waste for Storage in Landfills (Dz.U. 2015 poz. 1277). Warsaw, Poland. Available online: https://www.gov.pl (accessed on 10 July 2021).
10. Sadecka, Z.; Myszograj, S.; Suchowska-Kisielewicz, M. Legal aspects of natural use of sewage sludge. *Zesz. Nauk. Inż. Śr. Univ. Ziel. Góra* **2011**, *144*, 5–17.
11. Bień, J.; Wystalska, K. Sludge management—The need to change strategic decisions. *Environ. Eng. Prot.* **2014**, *17*, 357–361.
12. Bień, J.; Neczaj, E.; Neczaj, M.; Worwąg, A.; Grosser, D.; Nowak, M.; Milczarek, M.; Janik, S. Directions of sediment management in Poland after 2013. *Environ. Eng. Prot.* **2011**, *14*, 378.
13. Borowski, G.; Gajewska, M.; Haustein, E. Possibilities of utilization of ashes from thermal transformation of sewage sludge in fluidized bed boilers. *Eng. Environ. Prot.* **2014**, *17*, 393–402.
14. Pająk, T. Thermal treatment of sewage sludge in the face of the challenges of 2016. *Inż. Ochr. Śr.* **2014**, *17*, 363–376.
15. Circular Economy, i.e., a Circular Economy. Available online: https://www.euractiv.pl/section/energia-i-srodowisko/news/circular-economy-czyli-ekonomia-zrownowazonego-rozwoju (accessed on 22 July 2021).
16. Park, Y.J.; Park, S.; Moon, O.; Heo, J. Crystalline phase control of glass ceramics obtained from sewage sludge fly ash. *Ceram. Int.* **2003**, *29*, 223–227. [CrossRef]
17. Merino, I.; Arévalo, L.F.; Romero, F. Preparation, and characterization of ceramic products by thermal treatment of sewage sludge ashes mixed with different additives. *Waste Manag.* **2007**, *27*, 1829–1844. [CrossRef]
18. Lin, D.F.; Luo, H.L.; Cheng, J.F.; Zhuang, M.L. Strengthening tiles manufactured with sewage sludge ash replacement by adding micro carbon powder. *Mater. Struct.* **2016**, *49*, 3559–3567. [CrossRef]
19. Lin, D.F.; Luo, H.L.; Lin, K.L.; Liu, Z.K. Effects of waste glass and waste foundry sand additions on reclaimed tiles containing sewage sludge ash. *Environ. Technol.* **2017**, *38*, 1679–1688. [CrossRef] [PubMed]
20. Piasta, W.; Lukawska, M. The effect of sewage sludge ash on properties of cement composites. *Proc. Eng.* **2016**, *161*, 1018–1024. [CrossRef]
21. Vouk, D.; Nakic, D.; Stirmer, N.; Cheeseman, C.R. Use of sewage sludge ash in cementitious materials. *Rev. Adv. Mater. Sci.* **2017**, *49*, 158–170.
22. Ordinance of the Minister of Development of 21 January 2016, on the Requirements for the Thermal Waste Treatment Process and the Methods of Handling Waste Resulting from this Process. (Dz.U. 2016 poz. 108). Available online: http://isap.sejm.gov.pl (accessed on 22 July 2021).
23. UE/75/2010 Directive of the European Parliament and the Council of 24 November 2010 on Industrial Emissions (Integrated Pollution Prevention and Control). Available online: http://isap.sejm.gov.pl (accessed on 10 July 2021).

24. Rutkowska, G.; Fronczyk, J.; Filipczuk, S. Influence of flight ash properties from thermal transformation of sewage sludge on ordinary concrete parameters. *Acta Sci. Pol. Archit.* **2020**, *19*, 43–54.
25. Rutkowska, G.; Wichowski, P.; Franus, M.; Mendryk, M.; Fronczyk, J. Modification of Ordinary Concrete Using Fly Ash from combustion of Municipal Sewage Sludge. *Materials* **2020**, *13*, 487. [CrossRef]
26. Rutkowska, G.; Wichowski, P.; Fronczyk, J.; Franus, M.; Chalecki, M. Use of fly ashes from municipal sewage sludge combustion in production of ash concretes. *Constr. Build. Mater.* **2018**, *188*, 874–883. [CrossRef]
27. Rutkowska, G.; Ogrodnik, P.; Fronczyk, J.; Bilgin, A. Temperature influence on ordinary concrete modified with fly ash from thermally conversed municipal sewage sludge strength parameters. *Materials* **2020**, *13*, 5259. [CrossRef]
28. Monzó, J.; Paya, J.; Borrachero, M.V.; Peris-Mora, E. Mechanical behavior of mortars containing sewage sludge ash (SSA) and Portland cements with different tricalcium aluminate content. *Cem. Concr. Res.* **1999**, *29*, 87–94. [CrossRef]
29. Ing, D.S.; Chin, S.C.; Guan, T.K.; Suil, A. The use of sewage sludge ash (SSA) as partial replacement of cement in concrete. *ARPN J. Eng. Appl. Sci.* **2016**, *11*, 3771–3775.
30. Cyr, M.; Coutand, M.; Clastres, P. Technological and environmental behavior of sewage sludge ash (SSA) in cement-based materials. *Cem. Concr. Res.* **2007**, *37*, 1278–1289. [CrossRef]
31. Lin, K.L.; Chang, W.C.; Lin, D.F.; Luo, H.L.; Tsai, M.C. Effects of nano-SiO_2 and different ash particle sizes on sludge ash-cement mortar. *J. Environ. Manag.* **2008**, *88*, 708–714. [CrossRef]
32. Baeza-Broton, F.; Garces, P.; Paya, J.; Saval, J.M. Portland cement systems with addition of sewage sludge ash. Application in concretes for the manufacture of blocks. *J. Clean. Prod.* **2014**, *82*, 112–124.
33. Lin, K.-L.; Lin, C.-Y. Hydration characteristics of waste sludge ash utilized as raw cement material. *Cem. Concr. Res.* **2005**, *35*, 1999–2007. [CrossRef]
34. Yen, C.L.; Tseng, D.H.; Lin, T.T. Characterization of eco-cement paste produced from waste sludges. *Chemosphere* **2011**, *84*, 220–226. [CrossRef]
35. Ferreira, C.; Ribeiro, A.; Ottosen, L. Possible applications for municipal solid waste fly ash. *J. Hazard. Mater.* **2003**, *96*, 201–216. [CrossRef]
36. Chen, Z.; Li, J.S.; Poon, C.S. Combined use of sewage sludge ash and recycled glass cullet for the production of concrete blocks. *J. Clean. Prod.* **2018**, *171*, 1447–1459. [CrossRef]
37. Uhl, T. Computer-aided identification of mechanical structure models. *WNT Sci. Tech. Publ.* **1997**, *3*, 45–57.
38. Roberts, G.W.; Meng, X.; Dodson, A.H. Integrating a Global Positioning System and Accelerometers to Monitor the Deflection of Bridges. *J. Surv. Eng.* **2004**, *130*, 65–72. [CrossRef]
39. Li, B.; Cai, H.; Mao, X.; Huang, J.; Luo, B. Estimation of CNC Machine-tool Dynamic Parameters Based on Random Cutting Excitation Through Operational Modal Analysis. *Int. J. Mach. Tools Manuf.* **2013**, *71*, 26–40. [CrossRef]
40. Żółtowski, M.; Liss, M. The use of modal analysis in the evaluation of welded steel structures. *Stud. Proc. Pol. Assoc. Knowl. Manag.* **2016**, *79*, 233–248.
41. PN-EN 206+A1:2016-12 Concrete. Part 1. Requirements, Properties, Manufacturing and Conformity. Available online: https://sklep.pkn.pl (accessed on 10 April 2021).
42. Jamroży, Z. *Concrete and Its Technologirs*, 3rd ed.; PWN: Warsaw, Poland, 2015. (In Polish)
43. PN-EN 197-1: 2012 Cement—Part 1: Composition, Requirements, and Compliance Criteria for Common Cements. Available online: http://www.puntofocal.gov.ar/notific_otros_miembros/mwi40_t.pdf (accessed on 10 April 2021).
44. PN-EN 12350-6:2011 Testing of Fresh Concrete. Part 6: Density. Available online: https://sklep.pkn.pl (accessed on 10 March 2021).
45. PN-EN 12350-7:2019 Testing of Fresh Concrete. Part 7: Air Content—Pressure Method. Available online: https://standards.iteh.ai/catalog/standards/cen/444b4a93-2e0f-41e7-96e9-7d25505d78bd/en-12350-7-2019 (accessed on 10 March 2021).
46. PN-EN 12350-2:2011 Testing of Fresh Concrete. Part 2. Slump Test. Available online: https://sklep.pkn.pl (accessed on 10 April 2021).
47. PN-EN 12390-3: 2019-07. Testing of Hardened Concrete. Part 3: Compressive Strength of Test Specimens. Available online: https://standards.iteh.ai/catalog/standards/cen/7eb738ef-44af-436c-ab8e-e6561571302c/en-12390-3-2019 (accessed on 10 March 2021).
48. PN-B-06265:2004 Plain Concrete. Available online: https://sklep.pkn.pl (accessed on 10 April 2021).
49. ASTM C379-65T Specification for Fly Ash for Use as a Pozzolanic Material with Lime. Available online: https://www.astm.org/Standards/C379.htm (accessed on 12 February 2021).
50. Jain, H.; Rawat, A.; Sachan, A.K. A Review on Advancement in Sensor Technology in Structural Health Monitoring System. *J. Struct. Eng. Manag.* **2015**, *2*, 1–7.
51. Żółtowski, B.; Żółtowski, M. *Vibration Signals in Mechanical Engineering and Construction*; ITE—PIB: Radom, Poland, 2015.
52. He, C.; Xing, J.C.; Zhang, X. A New Method for Modal Parameter Identification Based on Natural Excitation Technique and ARMA Model in Ambient Excitation. *Adv. Mater. Res.* **2015**, *1065*, 1016–1019. [CrossRef]
53. Żółtowski, M. Investigations of harbour brick structures by using operational modal analysis. *Pol. Marit. Res.* **2014**, *21*, 42–53. [CrossRef]
54. Le, T.P.; Paultre, P. Modal identification based on the time-frequency domain decomposition of unknown-input dynamic tests. *Int. J. Mech. Sci.* **2013**, *71*, 41–50. [CrossRef]

55. Kosior-Kazberuk, M. Application of SSA as partial replacement of aggregate in concrete. *Pol. J. Environ. Stud.* **2011**, *20*, 365–370.
56. Joshi, R.C.; Lohtia, R.P. Fly ash in concrete: Production properties and uses. In *Advances and Concrete Technology*; Malhot, V.M., Ed.; Gordon and Breach Science Publishers: Ottawa, ON, Canada, 1997; Volume 2, p. 269.
57. PN-EN 451-2:2017-06 Fly Ash Test Method—Determination of Fineness by Wet Sieving. Available online: https://standards.iteh.ai/catalog/standards/cen/39f1ef7f-e230-41c3-8335-b3a377bab125/en-451-2-2017 (accessed on 10 March 2021).
58. PN-EN 1097-07:2008 Determination of the Filler Density. Available online: https://sklep.pkn.pl (accessed on 10 April 2021).
59. Alonso, J.L.; Wesche, K. Characterization of fly ash. Fly ash in concrete properties and performance. In *Report of Technical Committee 67-FAB—Use Fly Ash in Building*; Wesche, K., Ed.; RILEM. E&FN SPON: London, UK, 1991.
60. Malhotra, V.M.; Ramezanianpour, A.A. *Fly Ash in Concrete*, 2nd ed.; CANMT—Canada Centre for Mineral and Energy Technology: Ottawa, ON, Canada, 1994.
61. Williams, P.T. *Waste Treatment and Disposal*, 2nd ed.; John Wiley & Sons: London, UK, 2005.
62. Małolepszy, J.; Tkaczewska, E. Influence of fly ashes from the co-combustion of coal and biomass on the hydration process and properties of cement. *Concr. Days* **2006**, *23*, 591–601. (In Polish)
63. Yusur, R.O.; Noor, Z.Z.; Din, M.D.F.M.D.; Abba, A.H. Use of sewage sludge ash (SSA) in the production of cement and concrete-a review. *Int. J. Glob. Environ. Issues* **2012**, *12*, 214–228.
64. Chang, F.; Lin, J.; Tsail, C.; Wang, K. Study on cement mortar and concrete made with sewage sludge ash. *Water Sci. Technol.* **2010**, *62*, 1689–1693. [CrossRef]
65. Chen, M.; Blanc, D.; Gautier, M.; Mehu, J.; Gourdon, R. Environmental and technical assessments of the potential utilization of sewage sludge ashes (SSAs) as secondary raw materials in construction. *Waste Manag.* **2013**, *33*, 1268–1275. [CrossRef]
66. Kosior-Kazberuk, M. New mineral additives foe concreto. *Civil Environ. Eng.* **2011**, *2*, 47–55.

Article

An Evaluation of the Physical and Chemical Stability of Dry Bottom Ash as a Concrete Light Weight Aggregate

Jinman Kim [1], Haseog Kim [2] and Sangchul Shin [2,*]

[1] Department of Architectural Engineering, Kongju National University, 275 Cheonan-daero, Cheonan City 330-717, Chungcheongnam-do, Korea; jmkim@kongju.ac.kr
[2] Environment-Friendly Concrete Research Institute, Kongju National University, 275 Cheonan-daero, Cheonan City 330-717, Chungcheongnam-do, Korea; bravo3po@kongju.ac.kr
* Correspondence: hiykhj@kongju.ac.kr; Tel.: +82-41-521-9338

Abstract: Compared to the bottom ash obtained by a water-cooling system (wBA), dry process bottom ash (dBA) makes hardly any unburnt carbon because of its stay time at the bottom of the boiler and contains less chloride because there is no contact with seawater. Accordingly, to identify the chemical stability of dBA as a lightweight aggregate for construction purposes, the chemical properties of dBA were evaluated through the following process of the reviewing engineering properties of a lightweight aggregate (LWA). Typically, river gravel and crushed gravel have been used as coarse aggregates due to their physical and chemical stability. The coal ash and LWA, however, have a variety of chemical compositions, and they have specific chemical properties including SO_3, unburnt coal and heavy metal content. As the minimum requirement to use the coal ash and lightweight aggregate with various chemical properties for concrete aggregate, the loss on ignition, the SO_3 content and the amount of chloride should be examined, and it is also necessary to examine heavy metal leaching even though it is not included in the standard specifications in Korea. Based on the results, it is believed that there are no significant physical and chemical problems using dBA as a lightweight aggregate for concrete.

Keywords: bottom ash; coal ash; dry process; light weight aggregate

Citation: Kim, J.; Kim, H.; Shin, S. An Evaluation of the Physical and Chemical Stability of Dry Bottom Ash as a Concrete Light Weight Aggregate. *Materials* **2021**, *14*, 5291. https://doi.org/10.3390/ma14185291

Academic Editors: Carlos Morón Fernández and Daniel Ferrández Vega

Received: 23 August 2021
Accepted: 6 September 2021
Published: 14 September 2021

Publisher's Note: MDPI stays neutral with regard to jurisdictional claims in published maps and institutional affiliations.

1. Introduction

Coal ash is produced by drying and pulverizing bituminous coal in a pulverizer at a power plant and burning the pulverized coal in a boiler. About 15 to 45% of the pulverized coal is mainly collected from a dust collection system or the bottom of the boiler. Bottom ash refers to the ash that has fallen into the clinker hopper at the bottom of the boiler. After it falls to the bottom of the boiler and becomes accumulated in the hopper, it is ground by a grinder. The amount of the bottom ash produced accounts for around 10 to 25% of the total coal ash [1,2].

Approximately 40% of the bottom ash produced from Korea is recycled, while the remaining 60% is landfilled. Most of the bottom ash is recycled into landfill aggregates, which have low added value. Bottom ash has a high chloride concentration, a high unburned carbon content, and a high moisture content due to the use of seawater in the cooling process, and the chloride contained in the bottom ash causes corrosion of the embedded reinforcing bars. Moreover, the non-constant condition causes technical difficulties when using bottom ash and acts as an impediment to adding value as it cannot be used in a dry state [3,4].

The unburned carbon in bottom ash is partially used as a low value-added material such as a filling material or a subsidiary material for cement, whereas its uses in high value-added products such as ready-mixed concrete (unburned carbon content of less than 5%) are only marginal. As for domestic technologies related to bottom ash, there have mostly been studies on its uses as a raw material for cement, binder, or construction material, the pretreatment of bottom ash, and its catalytic potential and other applications. Due to the characteristics of bottom ash, there are limits to its use as a cement, binder, an aggregate or a lightweight aggregate (LWA) material. Accordingly, it is mainly used as an aggregate (low value-added use) for filling and covering sports fields and drainage areas. In order to promote the recycling of bottom ash, research has been carried out to determine its dissolution potential based on the content of harmful substances and dissolution characteristics so that it could be used as a fill and cover material, but a negative perception of bottom ash is prevalent due to the uncertainty of the potential environmental impact. For the purpose of resolving the issue of unstable quality, domestic power plants have switched from the wet process to the dry process when it comes to bottom ash treatment so as to lower the content of moisture, salt and unburned coal, and are striving to boost the recycling rate and expand the scope of recycling. Dry bottom ash (dBA) is an industrial byproduct emitted from thermal power plants, and a detailed description is provided in Section 2.1.

As of 2018, the amount of bottom ash is 1.35 million tons in Korea, of which dBA is estimated to be 30%, which is about 400,000 tons. The outer section of dry bottom ash is characterized by an open-cell porous structure, so it is lightweight with a high absorption rate, high hardness, and excellent physical properties, making it suitable as a permeable aggregate [5,6].

In this study, the physical and chemical properties of dry bottom ash produced in Korea using the dry method were analyzed, and the possibility of using dry bottom ash as LWA for lightweight concrete was experimentally examined.

2. Preceding Research on the dBA

2.1. The Discharging Process of dBA

Dry bottom ash (hereinafter referred to as dBA) is an industrial by-product discharged from thermal power plants in Korea, as shown in Figure 1. dBA discharged from a dry process is moved without coming into contact with water by the primary clinker conveyor. After passing through the primary conveyor, it is pulverized to a certain size and then finally cooled on the secondary conveyor and discharged after the secondary crushing process. The passage time of the primary and secondary conveyors is 40 to 60 min each, and most of the unburned coal is burned during residence on the conveyor. The particle size of the discharged dBA varies greatly in range, from more than 100 mm to less than 1 mm. It is quite irregular in shape, as shown in Figure 2 [7].

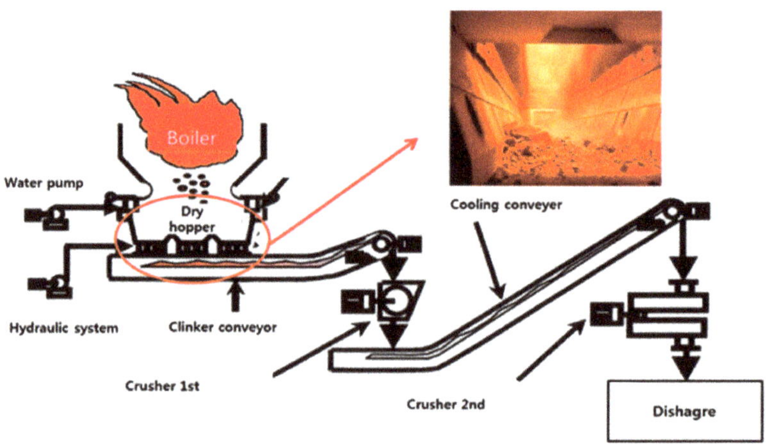

Figure 1. Bottom ash discharging system by the air-cooling process.

Figure 2. Shape of dBA [7].

2.2. Pore Properties of dBA

Figure 3 is a photograph showing the surface and internal pore properties of dBA observed with a scanning electron microscope (SEM). It shows that the dBA surface is composed of pores on a thin film and that the outside and the inside are connected due to partial destruction. On the inside, there are pores of various sizes that exist independently or continuously. Figure 4 is a graph showing the porosity according to the particle size. While the total porosity varies depending on the size of dBA, it has been found to be 50 to 60%, and there was a positive correlation between the size of the aggregate and the number of closed pores.

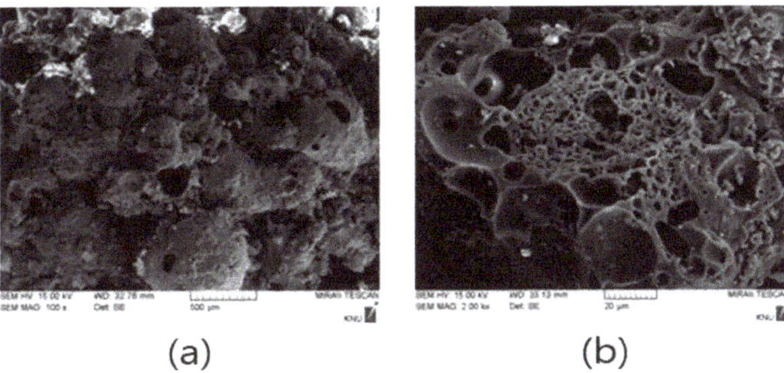

(a)   (b)

Figure 3. Pore properties of dBA; (**a**) Surface porosity of dBA; (**b**) Internal porosity of dBA.

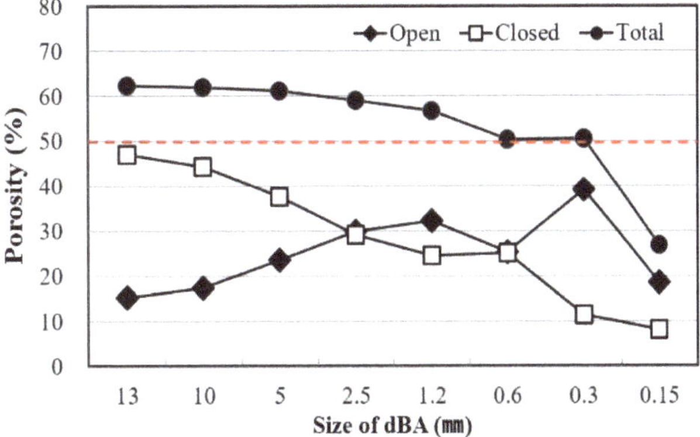

Figure 4. Pore distribution of dBA.

2.3. Physical Properties of dBA

Table 1 shows the density, absorption rate, unit volume mass, and the percentage of absolute volume as an examination of the physical properties by dBA size. The density and absorption rate of dBA were measured using the standard test method for the density and absorption of a coarse aggregate (KS F 2503) and the standard test method for the density and absorption of a fine aggregate (KS F 2504). The unit volume mass and percentage of the absolute volume of dBA were measured by using the standard test method for the bulk density and the percentage of absolute volume in an aggregate (KS F 2505) [8–10].

Table 1. Physical properties of dBA according to size.

Aggregate Size (mm)	Density (g/cm^3)		Absorption (%)	Bulk Density (kg/m^3)	Percentage of Absolute Volume (%)
	OD *	SSD **			
13	1.09	1.14	4.21	437.7	40.1
10	1.05	1.17	11.38	437.2	41.6
5	1.08	1.22	12.20	432.0	39.8
2.5	1.34	1.52	13.29	497.1	37.1
1.2	1.51	1.70	12.79	585.1	38.8

* OD: Oven Dry condition of dBA. ** SSD: Saturated Surface-dry condition of dBA.

An analysis of the correlation between density and absorption rate shows that a decrease in particle size is accompanied by an increase in density and absorption rate, as shown in Figure 5. This is opposite to the inverse relationship between density and absorption rate in general. Figure 6 shows the correlation between the density and absorption rate of general aggregates. The difference between dBA and general aggregates is deemed to be due to the large number of pores present in the matrix in the former. In the case of density, the larger the particle, the lower the number of pores and the higher the absorption rate, and this is judged to be because there are a relatively larger number of continuous pores when there are smaller dBA particles, resulting in an increased amount of moisture penetrating the dBA. Moreover, while the unit mass increases, the percentage of the absolute volume tends to decrease. It is thought that this is because the number of open pores increases as the particle size decreases, as shown in Figure 4. Furthermore, it has been indicated that the rate of sphericity or flatness of the aggregate does not improve even when the aggregates become smaller [11,12].

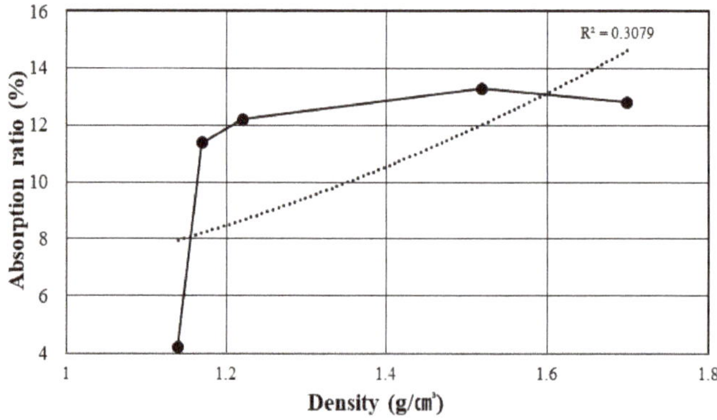

Figure 5. Relationship between the density and the absorption rate of dBA.

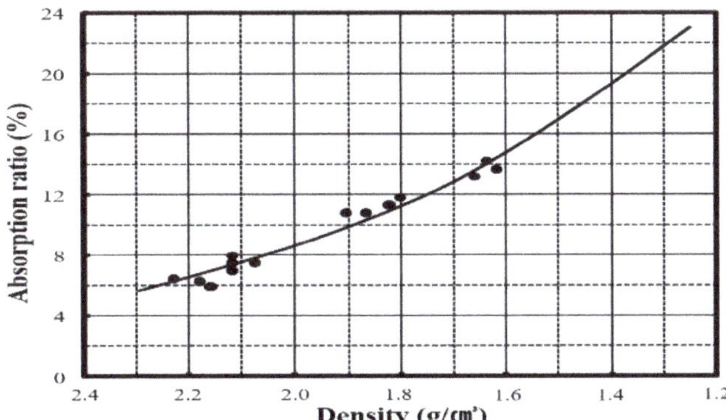

Figure 6. Relationship between the density and the absorption rate of a general aggregate (natural river aggregate or crushed aggregate). Source: Kim, H.S. A Study on the Quality Improvement of Recycled Fine Aggregate using Neutralization and Low Speed Wet Abraser. Notification on 2011-02; Kongju National University: Cheonan, South Korea, 2011.

2.4. Feasibility of dBA as a Construction Material in Relation to its Physical Properties

In a preceding study, the following conclusions were obtained as a result of examining the physical properties of dBA, a by-product in the electric power industry, as a LWA (Light Weight Aggregate) for construction [13].

As for the shape of dBA, it is structurally weak due to the sharp and angular edges and the flat and elongated shape, but unlike the surface, where there are open cells, the inside features closed cells and is relatively high in hardness.

In addition, although the particle density decreases along with a decrease in the particle size, the apparent density tends to increase slightly. This is because large cells are destroyed as the particles become smaller. Since closed cells that are larger than 100 μm are destroyed and turn into open cells as a result of the reduced particle size, there is an increase in the open porosity, a decrease in the closed porosity and total porosity, and an increase in the absorption rate.

Therefore, based on the findings of the previous study, it appears that dBA can be used as a LWA and has excellent properties in terms of weight and thermal insulation if the relatively weak surface is removed and it is processed into a near-spherical shape [14].

Table 2 shows the advantages and drawbacks of various aggregates including dBA used in this study. The wBA discharged from the wet process is difficult to recycle due to problems such as high unburned carbon, high chloride, and moisture contents. Natural aggregates are relatively high-quality aggregates that have been used for a long time, but they are related to environmental problems and resource depletion. In the case of the artificial lightweight aggregate, it is a product with a constant quality with a spherical shape and a uniform particle size because it is produced in a factory; however, it has the disadvantage of high manufacturing costs and greenhouse gas emissions due to the calcination process. In addition, in the case of Korea, the cost is high because most of it depends on imports. On the other hand, dBA has the potential to be applied as an alternative material for lightweight aggregates, owing to the low content of unburned carbon, chloride and SO_3. Nevertheless, the structural weakness caused by the irregular shape and high absorption is a problem to be overcome.

Table 2. Comparison of various aggregates.

Type of Aggregates	Advantages	Drawbacks
dBA	• Low unburned carbon content • Low chloride content • Low SO_3 content • Open pore → can be applied to artificial ground or insulation materials	• Irregular shape • Structural weakness • High absorption • Large particles → need to crush
wBA	-	• High absorption • Low unburned carbon content (~25%) • High chloride content → steel corrosion
Artificial light weight aggregate	• Spherical shape • Uniform particle size	• High manufacturing cost • Greenhouse gas emissions
Natural aggregate	• Long-term usage experience • Relatively high quality	• Environmental problems • Resource depletion

3. Experimental Plan and Method

3.1. Experimental Plan

Table 3 shows the experimental plan of this research. This study was conducted with the aim of determining the possibility of using dBA, which has been confirmed to have appropriate physical properties as LWA, for concrete manufacturing by evaluating its chemical properties.

Table 3. Experimental plan.

ID	Full Name of Light Weight Aggregate	Test Items
dBA	Bottom ash discharged from dry process	- Oxide composition by XRF *
wBA	Bottom ash discharged from wet process	- Mineralogical analysis by XRD ** - Chloride content
FA	Fly ash	- Unburnt carbon
LWA-1	Artificial light aggregate from Korea	- Potential of hydrogen - Heavy metal leaching test
LWA-2	Artificial light aggregate from USA	- Minor inorganic compounds by ICP ***
LWA-3	Artificial light aggregate from Japan	- Heating loss by DT-TGA ****
LWA-4	Artificial light aggregate from China	

* XRF: X-ray fluorescence, ** XRD: X-ray Diffraction. *** ICP: Inductively Coupled Plasma, **** DT-TGA: Thermogravimetry-Differential thermal analysis.

The chemical properties of dBA were analyzed based on X-ray fluorescence (XRF) to check the oxide composition and SO_3 content, X-ray diffractometry (XRD) to examine the mineralogical properties, loss on ignition to measure the amount of unburned coal as well as chloride content and the pH, and the heavy metal leaching test.

The dBA evaluated in this study was bottom ash discharged from a dry process at the B Thermal Power Plant operated by J Power in Korea. For a comparison, fly ash (FA), wet bottom ash (wBA), and four domestic and foreign artificial LWAs were examined as well. The experimental plan is shown in Table 2.

3.2. Shapes of Various Artificial Aggregates

Figure 7 shows digital camera and SEM images of the samples used in this study.

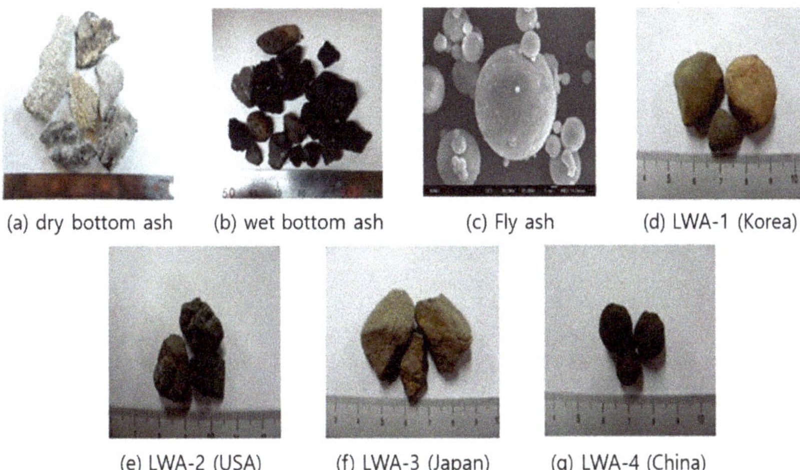

(a) dry bottom ash   (b) wet bottom ash   (c) Fly ash   (d) LWA-1 (Korea)

(e) LWA-2 (USA)   (f) LWA-3 (Japan)   (g) LWA-4 (China)

Figure 7. Shape of the test samples.

As described above, dBA has sharp edges and is flat, elongated, and irregular in shape. In the case of wBA, it has the appearance of a crushed aggregate (crushed stone), and the corners are relatively round compared to dBA. On the other hand, FA has a near-perfect spherical shape.

LWA-1 is an artificial LWA produced in Korea that has a generally round shape, while LWA-2 is a product from the United States and has a relatively angular shape compared to LWA-1. In the case of LWA-3, it is a product made in Japan and is an intermediate between LWA-1 and 2 in terms of shape, while LWA-4 is a product made in China and its shape is closest to a sphere.

3.3. Test Method

Aggregates used as coarse aggregates for concrete are generally made from river gravel and crushed gravel with stable physical and chemical properties. Coal ash and LWA, on the other hand, have various chemical compositions and have specific chemical properties such as SO_3, unburned carbon, and heavy metal content. As such, in order to use coal ash and a LWA with various chemical properties as aggregates for concrete production, the loss on ignition, sulfur trioxide content and chloride content among other factors are reviewed as a minimum requirement in Korea. Plus, in order to use dBA as industrial waste generated from power plants, it is necessary to determine its environmental impact. Therefore, this paper presents a review of the leaching of heavy metals from dBA, and the experimental items and evaluation methods were as follows.

3.3.1. Oxide Composition by XRF

This is a method of analyzing the characteristic energy emitted from the samples using a Rigaku ZSX Primus X-ray Fluorescence Spectrometer (XRF-WDX) after grinding the samples to 80 μm or less in size in order to examine their chemical composition. The composition of the constituent elements was quantitatively evaluated.

3.3.2. Mineralogical Analysis by XRD

In order to determine the mineral phase of coal ash and the artificial lightweight aggregate and the bonding state of minerals, the samples pulverized to 8 μm or less in size were examined by X-ray diffraction using Rigaku DMAX2000. This method involves analyzing the diffraction image obtained when the X-ray is incident on the crystal surface at a specific angle and scattered by the atomic layer in the crystal surface which satisfies Bragg's law.

3.3.3. Chloride Content

The chloride analysis was conducted in accordance with KS F 2713—Testing method for analysis of chloride in concrete and concrete raw materials [15].

3.3.4. Unburned carbon

To test the amount of unburned carbon in the test specimen, an experiment was conducted based on KS L 5405—fly ash [16]. According to the instructions, 1 g of the sample was measured to 0.1 mg in a crucible, ignited for 15 min in an electric furnace at 975 ± 25 °C, and cooled in a desiccator before the mass was measured. Ignition was repeated for 15 min until a constant mass was reached. Moreover, the PerkinElmer Pyris 1 TGA equipment was used to examine the rate of change in weight according to temperature.

3.3.5. Potential of Hydrogen

The pH test was conducted according to KS M 0011—the method for the determination of the pH of aqueous solutions, and a pH meter with a precision of ± 0.01 was used [17]. In this test, 200 g of the sample was immersed in 200 g of normal water and distilled water and then the pH of the aqueous solution was measured for 48 h. Moreover, in order to examine the initial pH and leaching characteristics, the samples were stirred for 30 s before measurement.

3.3.6. Heavy Metal Leaching Test

The heavy metal leaching test was conducted by applying two test methods in parallel: the Waste Process Test Method (Ministry of Environment of Korea Notice No. 2011-160 [18]) and the Official Soil Pollution Test Method (Ministry of Environment of Korea Notice No. 2013-113 [19]). Both tests use the ion detection method and the methods of preparing the test solution are identical. To be more specific, the sample and the solvent were mixed in a 1:10 ratio to create at least 500 mL of the mixture, after which the mixture was shaken for about 6 h at a rate of around 200 times per minute in a 4–5 cm shaker. Then it was filtered for the filtrate to be used as the test solution.

3.3.7. Minor Inorganic Compounds by ICP

In order to measure the concentrations of inorganic elements in coal ash and artificial LWA, an inductively coupled plasma optical emission spectrometer, Optima 5000 DV from GE, was used. A total of 0.25 g of the sample was pretreated with 0.5 mL of hydrofluoric acid (HF) and 5 mL of nitric acid (HNO_3) each and diluted by $200\times$.

3.4. The Physical Properties of Various Artificial Aggregates

Table 4 shows a prior study that measured the physical properties of fly ashes and LWAs. The absolute dry density was 2.1 g/cm^3 of FA which was the highest among the aggregates in the analysis. LWA 2, 3, and 4 showed similar values of 1.50~1.53 g/cm^3, and LWA-1 was 1.70 g/cm^3 which was relatively slightly higher. dBA showed an absolute density of 1.72 g/cm^3 and among the overall aggregates in analysis it showed moderate values.

As for the absorption, LWA-2 was 2.36%, which was relatively very low compared to the other aggregates with values of around 10%.

From the measurements of the unit volume mass and performance rate, dBA had a relatively low performance while LWA-1 showed the highest performance. This is due to the high density of LWA-1 and the fact that it is spherical with a smooth surface. Therefore, it has a high-performance rate while dBA has a relatively low sphericity, and the pores are exposed on the surface.

Table 4. Experimental plan and the physical properties of various artificial aggregates.

ID	Density (g/cm^3)		Absorption (%)	Bulk Density (kg/m^3)	Percentage of Absolute Volume (%)
	OD	SSD			
dBA	1.72	1.76	12.11	732.8	45.52
wBA	1.67	1.73	12.34		
FA	2.10	-	-	-	-
LWA-1	1.70	1.88	10.16	1029.1	60.54
LWA-2	1.47	1.51	2.36	766.2	52.12
LWA-3	1.38	1.50	8.89	738.4	53.50
LWA-4	1.39	1.53	10.15	767.0	55.18

Source: Kim, J.M. Development for Environment Friendly Construction Material having Low Density Using the Dry Processed Bottom Ash. Notification on 2011-06; Ministry of Environment: Seoul, South Korea, 2016.

4. Test Results and Discussion

4.1. Oxide Composition

The chemical composition of coal ash varies according to the type, quality, and combustion characteristics of coal, and it is as shown in Figure 8. SiO_2, which is the primary component, accounts for 50%, while Al_2O_3 accounts for 20%, CaO accounts for 2~3%, and MgO accounts for 1~2%. From the perspective of utilizing coal ash, alkali cations, such as Ca, Mg, K, and Na, in addition to Fe and S, are important evaluation factors, and coal ash is classified according to the content ratio of these elements [20]. As for the chemical composition of typical bituminous coal, a type of coal ash generated from domestic power plants, SiO_2 and Al_2O_3 account for at least 85%, CaO accounts for 2~3%, MgO accounts for around 1%, and SO_3 accounts for around 1~2% [21]. Soluble SiO_2 in coal ash combines with $Ca(OH)_2$ generated during cement hydration at room temperature to form insoluble calcium silicate ($CaO \cdot SiO_2 \cdot nH_2O$) that is stable in form, thereby causing a pozzolanic reaction that increases the long-term compressive strength of concrete. Moreover, when fly ash is used as a subsidiary material in cement, it behaves as a fusing agent (Al_2O_3, Fe_2O_3) that lowers the melting point during the clinker reaction, which is why it is a crucial raw material in the manufacturing process of cement [22–25].

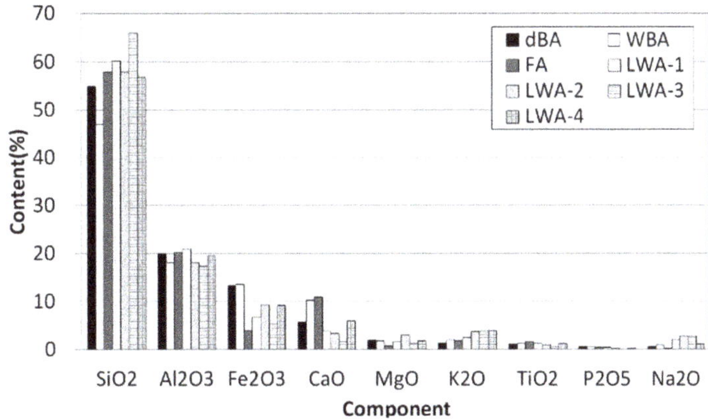

Figure 8. Major oxide composition of the test materials.

In this study, an XRF analysis was carried out to determine the chemical composition of each sample, and the results are shown in Table 5. The composition analysis of dBA showed that the largest amount of compound found was SiO_2, followed by Al_2O, Fe_2O_3,

CaO, and MgO at 54.9, 19.9, 13.3, 5.7, and 1.9 wt.%, respectively, with a total content of 96%.

Table 5. Trace elements of coal ash and LWA (unit: %).

	MnO	SrO	BaO	Cl	Cr_2O_3	ZrO_2	NiO
dBA	0.209	0.181	0.129		0.066	0.066	0.018
wBA	0.113	0.112	0.134	0.013	0.034	0.050	0.014
FA	0.190	0.025	0.081		0.038	0.024	0.009
LWA-1	0.059	0.019			0.020	0.023	0.006
LWA-2	0.143	0.029	0.062	0.009	0.033	0.037	0.013
LWA-3	0.145	0.099	0.140	0.038	0.042	0.035	0.031
LWA-4		0.061		0.098		0.076	

In the case of wBA, FA, and LWA-1 to 4, the composition ratio was found to be similar to that of dBA with SiO_2, Al_2O, Fe_2O_3, CaO, and MgO accounting for more than 90% of the content, despite slight differences. These compositions were found to be similar to that of clay, which is about 50 to 60% SiO_2, 15 to 18% Al_2O, 5 to 7% Fe_2O_3, and 1 to 7% CaO.

4.2. Mineralogical Analysis

4.2.1. Quartz

The mineral crystal form of coal ash provides important information on the reactivity and hardening necessary for the effective utilization of coal ash. The quartz and alumina components of coal are transferred to mullite ($3Al_2O_3 \cdot 2SiO_2$) or cristobalite (SiO_2) at high temperatures to transition into a crystalline state. When there is a crystalline phase such as Fe_2O_3, only 20 to 30% of the coal ash is in a crystalline state and the rest is present in a glassy state. Except for quartz, which does not undergo chemical deformation at high temperatures and instead mostly retains its original shape, most of the minerals in coal are generally broken down during combustion [26,27]. In other words, clay minerals lose water and are dissociated into various types of glassy or amorphous components in the cooling process. Among coal ash, quartz particles that are the size of silt and clay contained in coal remain as they are, irrespective of the melt that forms together with other inorganic matter during combustion, and the quartz content remains in all coal ash without being significantly affected by the type or composition of the coal ash. Mullite, which is directly crystallized from the melt when coal ash particles start to cool or are formed as a result of the crystallization of the glassy material at high temperatures in the furnace, is the main crystalline state of the low-calcium coal ash and is independent of the cement reaction [28].

Figure 9 shows the X-ray diffraction analysis of dBA, wBA, FA, and LWA-1 to 4. It was found that quartz (SiO_2), not cristobalite (SiO_2), was the main mineral of both the coal ash and the four kinds of artificial LWA. This is because quartz (SiO_2) can be generated at low temperatures of 200 °C or below and at high temperatures of 2000 °C or above depending on the pressure, but cristobalite (SiO_2) is only generated at low pressures of 1 atm and in a temperature range of 1500 to 1700 °C. In the case of boilers used at domestic thermal power plants, the pressure is relatively high, and the temperature is maintained in the range of 1500 ± 200 °C, which is probably why the quartz (SiO_2) state is maintained. Even in the case of the artificial LWA, it is fired and foamed using a rotary kiln, and it is believed that the quartz state is maintained because the firing temperature is around 1100~1300 °C during this process [29,30].

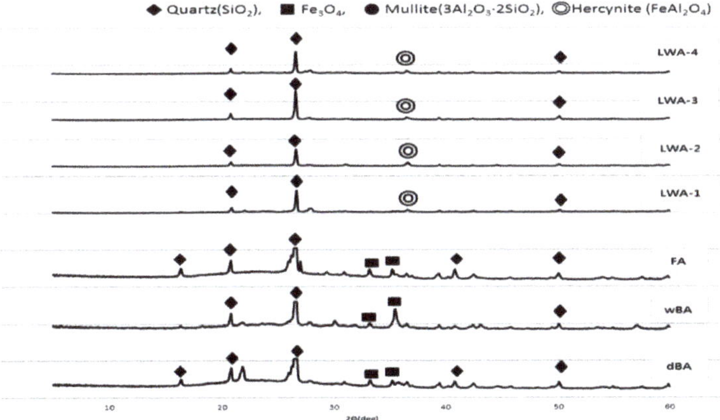

Figure 9. Mineral composition of the test materials.

4.2.2. Sulfate

In the case of bottom ash, there is a difference in the content of sulfur trioxide (SO_3) depending on the coal type and combustion efficiency. Generally, oxides of sulfur are considered to be stable because they are dispersed in the glass aggregate that has been rapidly cooled, as in the case of bottom ash, but if sulfur oxides are present in large amounts, they react with C_3A ($3CaO·Al_2O_3$) among the minerals constituting cement and generate ettringite, the expandable hydration product. This in turn causes sulfate erosion, which can be a problem in concrete structures [31].

According to KS F 2534:2009—Lightweight aggregate for structural concrete, there are restrictions on the sulfur trioxide content in the case of artificial and natural LWAs, but in the case of bottom ash LWA, it must be no more than 0.8%. This is because some of the coal ash discharged from bituminous coal-fired power plants contains sulfur trioxide. As shown in Figure 10, the sulfur oxide content was found to be 0.06 wt.% in the case of dBA and 2.9 wt.% in FA, while wBA recorded the highest sulfur oxide content at 4.29 wt.%. As such, it can be seen that dBA conforms to KS F 2534, whereas wBA does not. In addition, the fired artificial LWA was found to have a relatively stable sulfur oxide content of 0.1 to 1.0 wt.%.

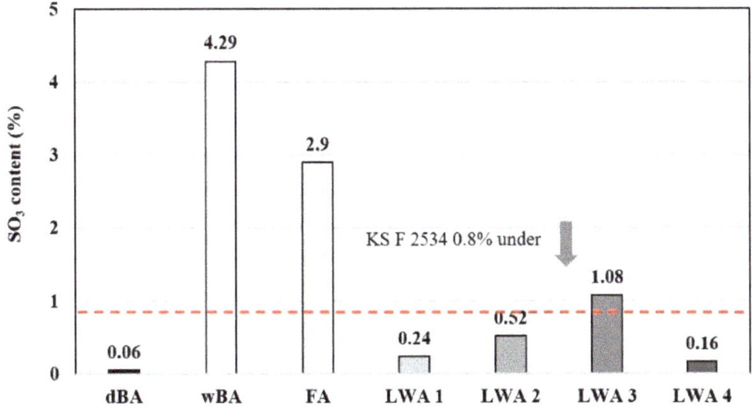

Figure 10. SO_3 contents of the test materials.

4.2.3. Chloride

The sulfur trioxide (SO_3) content was analyzed using the XRF result values of each sample. In Korea, dBAs are generally combusted in a boiler and settled in a tank that is filled with seawater at the bottom, after which they are crushed to a certain size and then transferred to an ash treatment plant using a water pipe to be buried. This is why dBAs have a substantially high salt content.

Due to the high salt content, dBAs can cause corrosion of rebars and can seriously reduce the durability of the reinforced concrete structure. For this reason, there is a limit to the amount of chloride that can be contained in the raw materials used in structural concrete.

KS F 2526—Aggregates for concrete and KS F 4570—Bottom ash aggregate for the precast concrete product, for example, require that the salt content be no more than 0.04% and 0.025%, respectively [32,33]. In addition, KS F 2534—Lightweight aggregate for structural concrete, limits the salt content of fine aggregates made from bottom ash to no more than 0.025% to ensure quality.

As a result of measuring the chloride content of the various test materials, it was found that dBA generated by the dry process does not contain any chloride, as shown in Figure 11. However, in the case of wBA generated by the wet process, the chloride content was measured to be 0.038%, exceeding the limit specified in the KS. This is believed to be because wBA is cooled and transported using seawater, as described above.

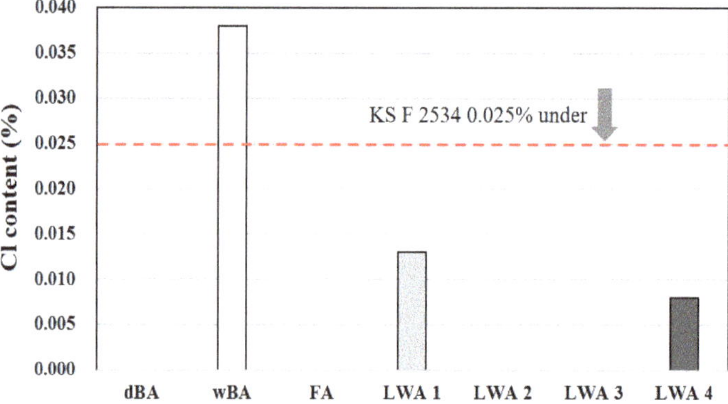

Figure 11. Cl contents of the test materials.

Moreover, Cl ions (approx. 0.01%) were detected even in the artificial LWA, which suggests the need to pay attention to the chloride content even in the case of fired artificial LWA.

4.2.4. Unburned Carbon

Bottom ash produced by the wet process contains a large amount of unburned carbon because it is cooled immediately when it falls to the bottom of the boiler. If unburned carbon is evaluated based on loss on ignition, the loss on ignition of the bottom ash from the wet process in Korea varies depending on the coal type and combustion efficiency, but it is generally known to be about 1.5 to 15%. If the loss on ignition is high, the unburned carbon adsorbs the entraining air introduced to improve the durability of the concrete, and this adversely affects the durability of the concrete. Therefore, the unburned carbon content is limited to 5% for concrete materials used in powder form. In the case of aggregate-like materials, there is no limit when it comes to the unburned carbon content, but in case of using bottom ash aggregates with high loss on ignition, more admixtures are required to ensure the same workability and air volume, making concrete production less economical. Unburned carbon powder with large particles may cause concrete delamination and thus may serve as a factor deteriorating the quality of the concrete surface, causing problems in ensuring quality [34,35].

A comparison of the unburned carbon content of dBA and FA is shown in Figure 12. The samples used were FA and dBA discharged from Facilities 7 and 8 of the B Thermal Power Plant operated by J Power in Boryeong, Chungcheongnam-do Province. It was found that the unburned carbon content of FA exceeded the requirement (5%) specified in KS L 5405—Fly ash. On the other hand, dBA was found to have an unburned carbon content of no more than 2% during the 6-month period and was found to exhibit stable qualities satisfying the requirement (5% or less) specified in KS F 2534:2007—Lightweight aggregate for structural concrete.

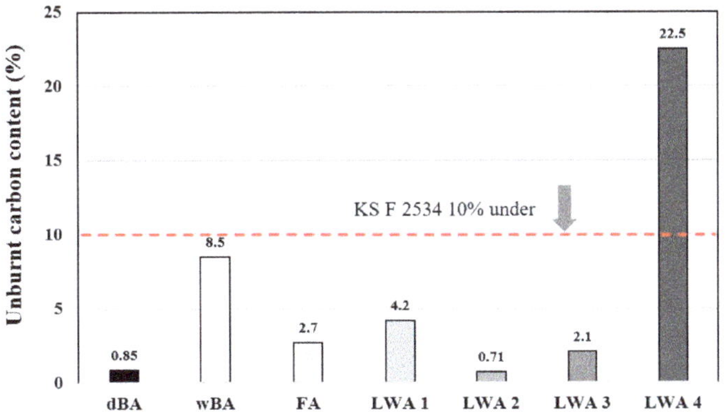

Figure 12. Unburnt carbon of dBA.

Figure 13 shows the results of a gravimetric analysis of dBA coal ash and LWAs using DT-TGA. dBA was found to have a low unburned carbon content with a weight reduction rate of less than 1%, as was the case during the 6-month tracking period. Likewise, in the case of artificial LWAs, although LWA-4 recorded a relatively high value, all of the samples underwent a small degree of weight reduction of less than 1%. This is due to the fact that artificial LWA is put through a sintering process for the foaming of the aggregates at around 1100 to 1300 °C, and most of the unburned carbon is burned in this temperature range [36,37].

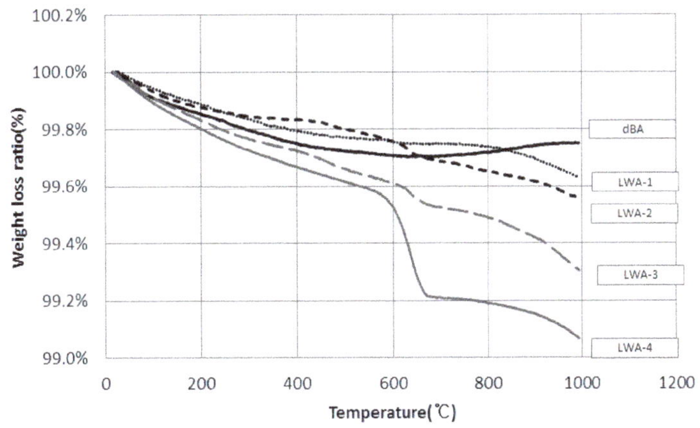

Figure 13. DT-TGA of coal ash and LWA.

4.2.5. Hydrogen Exponent (pH)

Aggregates used in the construction industry are mostly stored outdoors, as they are used in large amounts and are large in volume. Aggregates stored outside in this way lead to leachates due to rainwater, etc., which makes it necessary to consider the impact on the surrounding environment. Unlike natural aggregates, crushed stone aggregates are usually chemically stable and do not cause environmental problems. It has been pointed out that recycled aggregates and slag aggregates may result in strong alkaline leachates and may cause environmental problems [38,39].

Therefore, the possibility of environmental problems caused by leachate was examined by measuring the pH of domestic and foreign artificial LWAs and bottom ash. For the experimental method, KS F 2013—Method of measuring the pH of soil was referenced, and the prescribed test method was modified to suit the purpose of this study [40]. Two types of water were used in the measurement: tap water and distilled water. The artificial LWA and dBA samples were immersed in the two types of water in a weight ratio of 1:1, and the changes in pH over 48 h were measured.

Figures 14 and 15 show the results of the measurement of the pH using tap water and distilled water, respectively. Figure 14 shows that the initial pH of the dBA aggregate sample was similar to that of the artificial LWA, but the pH increased after 3 h of immersion before it stabilized at 8.61 after 48 h. Generally, ash represents basicity. WBA has a high unburnt carbon content and dBA has a low unburnt carbon content. Therefore, dBA is relatively highly basic and shows a somewhat higher pH value. As such, dBA has a low level of alkalinity, but this is not expected to adversely affect the surrounding environment. However, when dBA was pulverized into powder, the pH was found to be the highest at 9.4 at the beginning of immersion, and although it decreased somewhat to 9.15 after 48 h, it was still the highest among all the samples. This is believed to be due to the fact that the specific surface area was increased by pulverization. Generally, when the particle size is smaller and the specific surface area is high, ion elution becomes more active, so the pH value can vary according to the ionization rate according to the specific surface area of the same material. Therefore, care should be taken not to store dBA outdoors when it is in powder form.

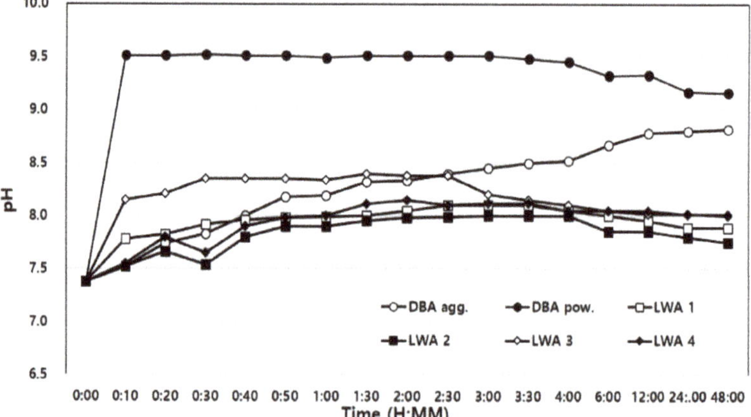

Figure 14. Time-dependent pH variance in tap water.

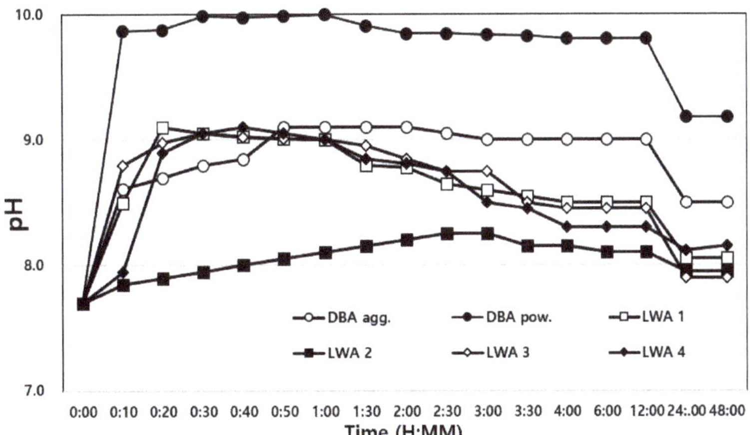

Figure 15. Time-dependent pH variance in distilled water.

In the case of the four types of artificial LWA under comparison, their pH levels were found to be stable at the beginning of the measurement and after 48 h. The pH was 7.4 to 7.5 initially and increased to about 7.8 to 8.6 after 48 h.

The results of the experiment in which the samples were immersed in distilled water for pH measurement are shown in Figure 15. Since the impurities that could interfere with the leaching of ions were removed, the initial pH measurements were all higher than those of the samples immersed in service water. However, the results were similar to their counterparts after 48 h due to ion stabilization.

Moreover, the reason that the pH increased after immersion for all the test specimens was deemed to be due to the leaching of alkali ions, Na^+, Mg^{++}, K^+, Ca^{++}, etc.

4.2.6. Heavy Metal Leaching Test

In order to use bottom ash as a structural aggregate, it is necessary to measure the environmental pollutants that can possibly be leached prior to understand their impact on the environment before applying them in the field. In each country, there are laws regulating the amount of pollutant that can be leached during recycling and specifying whether bottom ash can be recycled. Bottom ash is composed of SiO_2, Al_2O_3, CaO, Fe_2O_3, etc., but it also contains trace amounts of heavy metals [41,42]. Column leaching tests have shown that some of the heavy metals tested according to the drinking water quality standards of Korea, the USA, the EPA, and the WHO were below the baseline or the detection limit. On the other hand, the Pb and Zn levels exceeded the baseline, but the possibility of causing water pollution is low, as they have recorded no more than 1 PVE in all dissolution experiments. Clause 2.3 of KS F 2534:2009—Lightweight aggregate for structural concrete prescribes the content requirements for Pb, Cd, $Cr6^+$, Cu, Hg, As, etc. Accordingly, in this study, a heavy metal leaching test was conducted on the bottom ash discharged from the dry process. In the case of artificial LWAs that are distributed in the market, the heavy metal leaching test was not performed.

The heavy metal leaching test was carried out using two methods specified in the Official Waste Test Standards (Ministry of Environment of Korea Notice No. 2011-160, Type–I) and the Official Soil Pollution Test Standards (Ministry of Environment of Korea Notice No. 2013-113, Type– II) and test results are shown in Tables 6 and 7. When the official waste test method was applied, dBA was found to be stable, as Pb, Cd, $Cr6^+$, Cu, Hg and As were not detected. Also, when the official soil pollution test method was applied, there was partial leaching of five heavy metals Pb, Cd, Cu, Hg, and As, but not $Cr6^+$, but leaching occurred in very small amounts compared to the level that would cause concerns of soil pollution.

Table 6. Heavy metal content analysis type—I.

ID	Test Items (mg/L)					
	Pb	**Cd**	**Cr6**$^-$	**Cu**	**Hg**	**As**
Standard *	3	0.3	1.5	3	0.005	1.5
dBA	N.D	N.D	N.D	N.D	N.D	N.D

* Official Wastes Test Method; Korea ministry of environment Regulation.

Table 7. Heavy metal content analysis type—II.

ID	Test Items (mg/L)					
	Pb	**Cd**	**Cr6**$^-$	**Cu**	**Hg**	**As**
Standard	200	4	5	150	4	25
dBA—2.5 mm size	8.04	0.27	N.D	27.7	0.01	1.61
dBA—5 mm under	6.86	N.D	N.D	6.23	N.D	N.D

4.2.7. Minor Inorganic Compounds by ICP

Heavy metals and pollutants may be present on the surface of the sample in a chemically or physically adsorbed state or sometimes in a form in which the pollutant itself is mixed in the sample. Ultimately, in order to protect the soil environment and prevent pollution, there needs to be information on the chemical reactions that can occur between the sample and heavy metals and the characteristics of adsorption to the surface of the sample [43].

Table 8 shows the results of an ICP analysis of coal ash and an artificial LWA. Among the six heavy metals, Pb, Cr, and Cu were found at 0.1, 1.0, and 515 mg/L, respectively, in dBA, whereas Cd, Hg, and As were not leached [44]. Moreover, alkali metals and alkaline earth metals, Na, K, Mg, and Ca, were found at concentrations of 1.2, 34,770, 10,775, and 22.4 mg/L, respectively, and the leaching of these alkali ions was considered to be the cause of the increased pH levels examined above.

Table 8. ICP analysis.

	Alkaline Metals and Earth Metals (mg/L)				Heavy Metal (mg/L)					
	K	**Mg**	**Ca**	**Na**	**Pb**	**Cd**	**Cr**	**Cu**	**Hg**	**As**
dBA	1.2	34,770	10,775	22.4	0.1	ND *	1.0	515	ND	ND
wBA	0.2	5552	2223	6.3	1.3	ND	0.1	56	ND	ND
FA	0.1	9235	9552	4.9	0.0	ND	0.3	178	ND	ND
LWA-1	0.1	458	134	0.9	0.1	ND	0.6	260	ND	ND
LWA-2	4.5	673	2942	1.2	0.1	ND	0.9	174	ND	ND
LWA-3	5.1	4277	2419	1.3	0.1	ND	0.4	76	ND	ND
LWA-4	5.1	12,052	5880	28.1	0.2	ND	0.9	294	ND	ND
ST1 **	-	-	-	-	200	4	5	150	4	25
ST2 ***	-	-	-	-	400	10	15	500	10	50
ST3 ****	-	-	-	-	700	60	40	2000	20	200

* ND: Not Detected, ** ST1: Soil pollution concern Standard 1 area, *** ST2: Soil pollution concern Standard 2 area, **** ST3: Soil pollution concern Standard 3 area.

Even in the heavy metal content analysis, Cd, Hg, and As were not detected, while the Pb and Cr content was similar to that of an artificial LWA. In the case of the Cu content, it was found to be relatively high, but the amount leached in the leaching test was below the baseline. Thus, it was deemed that there will not be any major problems in using dBA as aggregate for concrete manufacture.

5. Conclusions

In this study, the chemical properties of bottom ash produced by the dry process, which is a by-product of the electric power industry, were examined following an analysis of its physical properties to check the possibility of its use as a lightweight aggregate, and the following conclusions were reached as a result:

(1) An oxide analysis of dBA showed a composition similar to that of wBA, FA, and four types of LWA, and a mineral analysis showed that the primary component was quartz (SiO_2), and in the case of dBA, wBA, and FA, the presence of Fe_3O_4 was clearer.

(2) As for the SO_3 content, it was found in significantly lower amounts in dBA compared to wBA, FA, and LWA-1 to 4, and the chloride content was also lower in dBA compared to the existing artificial LWAs. Thus, it is believed that there will not be any concerns of expansion caused by SO_3 or corrosion of rebar by chloride.

(3) As a result of measuring the unburned carbon content, it was found that dBA would be a stable aggregate for concrete production, as it has a relatively very low unburned carbon content compared to wBA, FA, and artificial LWAs.

(4) In a leaching test carried out in accordance with the Official Waste Test Standards, the six major heavy metals were not detected. In contrast, Pb, Cd, Du, Hg, and As were detected when the test was conducted according to the Official Soil Pollution Test Standards, but they were all below the baseline.

(5) An ICP analysis showed that some of the alkali metals and alkaline earth metals were found in larger amounts in dBA than in artificial LWAs, wBA, and FA, thereby resulting in higher pH levels, and Pb, Cr, and Cu were detected in a heavy metal leaching test. However, the heavy metals detected were found at levels below the baseline, based on which it was judged that there will not be any problems in real-life application.

(6) Based on the above results, it is believed that there are no significant physical and chemical problems in using dBA as a lightweight aggregate.

(7) This paper physically and chemically examined the possibility of using dBA as aggregate for concrete and it is intended that its characteristics by using it in mortar and concrete in the future will be examined.

Author Contributions: S.S., H.K. and J.K. conceived and designed the experiments; H.K. performed the experiments; S.S. and J.K. analyzed the data; H.K. contributed materials/analysis tools; and S.S. and J.K. wrote the paper. All authors have read and agreed to the published version of the manuscript.

Funding: This research was funded by the National Research Foundation of Korea(NRF), grant number 2020R1I1A1A01074492.

Institutional Review Board Statement: Not applicable.

Informed Consent Statement: Not applicable.

Data Availability Statement: The data presented in this study are available on request from the corresponding author.

Acknowledgments: This work was supported by the National Research Foundation of Korea(NRF) grant funded by the Ministry of Education (No.2020R1I1A1A01074492) and the research grant of the Kongju National University in 2020.

Conflicts of Interest: The authors declare no conflict of interest.

References

1. Korea Energy Economics. *Yearbook of Energy Statistic*; Notification on 2011; Korea Energy Economics Institute: Seoul, Korea, 2011; pp. 172–179.
2. Korea Energy Economics. *Yearbook of Energy Statistic*; Notification on 2011; Korea Energy Economics Institute: Seoul, Korea, 2013; pp. 345–349.
3. Choi, S.-J.; Kim, M.-H. A Study on the Durabilities of High Volume Coal Ash Concrete by the Kinds of Coal Ash. *J. Korean Inst. Build. Constr.* **2009**, *9*, 73–78. [CrossRef]
4. Abis, M.; Bruno, M.; Simon, F.-G.; Grönholm, R.; Hoppe, M.; Kuchta, K.; Fiore, S. A Novel Dry Treatment for Municipal Solid Waste Incineration Bottom Ash for the Reduction of Salts and Potential Toxic Elements. *Materials* **2021**, *14*, 3133. [CrossRef]
5. Kim, J.M.; Choi, H.B.; Lee, M.J.; Yu, J.S.; Sun, J.S. Properties of Fresh Concrete with Dry Bottom Ash Processed by Various Method. *J. Korean Inst. Build. Const.* **2014**, *l14*, 92–93.
6. Kim, M.H.; Yeon, N.K. Properties of Mortar Using Lightweight Fine Aggregate Made by Bottom Ash Discharged Air Cooling Process according to Grading. *J. Korean Inst. Build. Const.* **2016**, *l14*, 92–93.
7. Kim, J.M. *Development for Environment Friendly Construction Material Having Low Density Using the Dry Processed Bottom Ash*; Notification on 06-2011; Ministry of Environment: Seoul, Korea, 2016.
8. *Standard Test Method for Density and Absorption of Coarse Aggregate*; KS F 2503; Korea Standard Committee: Seoul, Korea, 2019.
9. *Standard Test Method for Density and Absorption of Fine Aggregate*; KS F 2504; Korea Standard Committee: Seoul, Korea, 2020.
10. *Standard Test Method for Bulk Density and Percentage of Absolute Volume in Aggregate*; KS F 2505; Korea Standard Committee: Seoul, Korea, 2017.
11. U.S. Environmental Protection Agency (EPA). *Wastes from the Combustion of Coal by Electric Utility Power Plants*; Notification on February 1988; U.S. Environmental Protection Agency: Washington, DC, USA, 1988.
12. Cai, Y. Laboratory Investigation of Pulverized Coal Combustion Bottom Ash as a Fine Aggregate in Roller Compacted Concretes. Master's Thesis, Department of Civil Engineering, Southern Illinois University, Carbondale, IL, USA, 1996.
13. Kim, J.M.; Choi, S.M.; Sung, J.H.; Bok, Y.J.; Sun, J.S. Engineering Properties of Lightweight Aggregate Concrete Using Dry Bottom Ash as Coarse Aggregate. *J. Korea Instit. Build. Constr.* **2013**, *13*, 166–167.
14. Kim, H.-S.; Kim, J.-M.; Kim, B. Quality improvement of recycled fine aggregate using steel ball with the help of acid treatment. *J. Mater. Cycles Waste Manag.* **2017**, *20*, 754–765. [CrossRef]
15. *Testing Method for Analysis of Chloride in Concrete and Concrete Raw Materials*; KS F 2713; Korea Standard Committee: Seoul, Korea, 2002.
16. *Fly Ash*; KS L 5405; Korea Standard Committee: Seoul, Korea, 2009.
17. *Surface Active Agents—Determination of pH of Aqueous Solutions—Potentiometric Method*; KS M ISO 4316; Korea Standard Committee: Seoul, Korea, 2009.
18. Korea Ministry of Environment Regulation. *Official Wastes Test Method*; Notification on 2011-160; Ministry of Environment: Seoul, Korea, 2011.
19. Korea Ministry of Environment Regulation. *Official Wastes Test Method*; Notification on 2013-113; Ministry of Environment: Seoul, Korea, 2013.
20. Ministry of Trade. *Development of Adsorbent Using Highly Unburned Carbon*; Notification on 2000; Ministry of Industry & Energy: Seoul, Korea, 2000.
21. Jung, M.Y. Beneficiation of Domestic Anthracite Fly Ash by Reverse Flotation. *J. Korean Inst. Miner. Energy Resour. Eng.* **2000**, *37*, 72–79.
22. Ministry of Trade. *Coal Ash Ceramic Body Manufactured by Low Temperature Processing*; Notification on 2002; Ministry of Industry & Energy: Seoul, Korea, 2002.
23. Jeon, D.H.; Yum, W.S.; Song, H.M.; Yoon, S.Y.; Bae, Y.H.; Oh, J.E. Use of Coal Bottom Ash and CaO-CaCl$_2$-Activated GGBFS Binder in the Manufacturing of Artificial Fine Aggregates through Cold-Bonded Pelletization. *Materials* **2020**, *13*, 5598. [CrossRef] [PubMed]
24. Risdanareni, P.; Villagran, Y.; Schollbach, K.; Wang, J.; De Belie, N. Properties of Alkali Activated Lightweight Aggregate Generated from Sidoarjo Volcanic Mud (Lusi), Fly Ash, and Municipal Solid Waste Incineration Bottom Ash. *Materials* **2020**, *13*, 2528. [CrossRef]
25. Kurda, R.; Vasco Silva, R.; de Brito, J. Incorporation of Alkali-Activated Municipal Solid Waste Incinerator Bottom Ash in Mortar and Concrete: A Critical Review. *Materials* **2020**, *13*, 3428. [CrossRef] [PubMed]
26. Yang, I.H.; Park, J.H. A Study on the Thermal Properties of High-Strength Concrete Containing CBA Fine Aggregates. *Materials* **2020**, *13*, 1493. [CrossRef] [PubMed]
27. Sun, Q.; Wei, X.; Li, T.; Zhang, L. Strengthening Behavior of Cemented Paste Backfill Using Alkali-Activated Slag Binders and Bottom Ash Based on the Response Surface Method. *Materials* **2020**, *13*, 855. [CrossRef] [PubMed]
28. Menéndez, E.; Argiz, C.; Sanjuán, M.Á. Chloride Induced Reinforcement Corrosion in Mortars Containing Coal Bottom Ash and Coal Fly Ash. *Materials* **2019**, *12*, 1933. [CrossRef] [PubMed]
29. Van den Heede, P.; Ringoot, N.; Beirnaert, A.; Van Brecht, A.; Van den Brande, E.; De Schutter, G.; De Belie, N. Sustainable High Quality Recycling of Aggregates from Waste-to-Energy, Treated in a Wet Bottom Ash Processing Installation, for Use in Concrete Products. *Materials* **2016**, *9*, 9. [CrossRef] [PubMed]

30. Holm, T.A. Performance of structural lightweight concrete in a marine environment. *J. Am. Concr. Inst.* **1980**, *65*, 589–608.
31. Tuan, B.L.A.; Tesfamariam, M.G.; Chen, Y.-Y.; Hwang, C.-L.; Lin, K.-L.; Young, M.-P. Production of Lightweight Aggregate from Sewage Sludge and Reservoir Sediment for High-Flowing Concrete. *J. Constr. Eng. Manag.* **2014**, *140*, 04014005. [CrossRef]
32. *Lightweight Aggregate for Structural Concrete*; KS F 2534; Korea Standard Committee: Seoul, Korea, 2009.
33. *Concrete Aggregate*; KS F 2526; Korea Standard Committee: Seoul, Korea, 2007.
34. *Bottom Ash Aggregate for the Precast Concrete Product*; KS F 4570; Korea Standard Committee: Seoul, Korea, 2007.
35. Berry, E.E.; Hemming, R.T. Beneficiation of Fly Ash. An Overview of a Resource. In Proceedings of the 2nd International Conference on the Use of Fly Ash, Silica Fume, Slag and Natural Pozzolans in Concrete, Madrid, Spain, 21–25 April 1986; pp. 241–274.
36. Cheeseman, C.; Makinde, A.; Bethanis, S. Properties of lightweight aggregate produced by rapid sintering of incinerator bottom ash. *Resour. Conserv. Recycl.* **2005**, *43*, 147–162. [CrossRef]
37. Bhatty, J.I.; Reid, K.J. Lightweight aggregates from incinerated sludge ash. *J. Waste Manag. Res.* **1986**, *7*, 363–376. [CrossRef]
38. Yip, W.; Tay, J. Aggregate Made from Incinerated Sludge Residue. *J. Mater. Civ. Eng.* **1990**, *2*, 84–93. [CrossRef]
39. Park, J.-B.; Lee, B.-C.; Jang, M.-H.; Na, H.-H. Environmental Characteristics of Leachates from Steel Slag. *J. Korean Geosynth. Soc.* **2012**, *11*, 31–38. [CrossRef]
40. *Test Method for pH of Soils*; KS F 2103; Korea Standard Committee: Seoul, Korea, 2013.
41. Wang, W.; Qin, Y.; Song, D.; Wang, K. Column leaching of coal and its combustion residues, Shizuishan, China. *Int. J. Coal Geol.* **2008**, *75*, 81–87. [CrossRef]
42. Hong, Y.-K.; Kim, J.-W.; Kim, H.-S.; Lee, S.-P.; Yang, J.-E.; Kim, S.-C. Bottom Ash Modification via Sintering Process for Its Use as a Potential Heavy Metal Adsorbent: Sorption Kinetics and Mechanism. *Materials* **2021**, *14*, 3060. [CrossRef] [PubMed]
43. Ministry of Trade. *Quality Standard for Drinking Water*; Notification on 2012; Ministry of Environment: Seoul, Korea, 2012.
44. Brigatti, M.F.; Corradini, F.; Franchini, G.C.; Mazzoni, S.; Medici, L.; Poppi, L. Interaction between montmorillonite and polluants from industrial waste-water exchange of Zn^{2+} and Pb^{2+} from aqueous solution. *Appl. Clay Sci.* **1995**, *9*, 383–395. [CrossRef]

Review

Research Trends of Human–Computer Interaction Studies in Construction Hazard Recognition: A Bibliometric Review

Jiaming Wang [1,2], Rui Cheng [2], Mei Liu [3] and Pin-Chao Liao [2,*]

1   School of Economics and Management, Tongji University, Shanghai 200092, China; 1551288@tongji.edu.cn
2   Department of Construction Management, Tsinghua University, Beijing 100084, China; chengr19@mails.tsinghua.edu.cn
3   School of Urban Economics and Management, Beijing University of Civil Engineering and Architecture, Beijing 100084, China; liumei0311@126.com
*   Correspondence: pinchao@tsinghua.edu.cn

Abstract: Human–computer interaction, an interdisciplinary discipline, has become a frontier research topic in recent years. In the fourth industrial revolution, human–computer interaction has been increasingly applied to construction safety management, which has significantly promoted the progress of hazard recognition in the construction industry. However, limited scholars have yet systematically reviewed the development of human–computer interaction in construction hazard recognition. In this study, we analyzed 274 related papers published in ACM Digital Library, Web of Science, Google Scholar, and Scopus between 2000 and 2021 using bibliometric methods, systematically identified the research progress, key topics, and future research directions in this field, and proposed a research framework for human–computer interaction in construction hazard recognition (CHR-HCI). The results showed that, in the past 20 years, the application of human–computer interaction not only made significant contributions to the development of hazard recognition, but also generated a series of new research subjects, such as multimodal physiological data analysis in hazard recognition experiments, development of intuitive devices and sensors, and the human–computer interaction safety management platform based on big data. Future research modules include computer vision, computer simulation, virtual reality, and ergonomics. In this study, we drew a theoretical map reflecting the existing research results and the relationship between them, and provided suggestions for the future development of human–computer interaction in the field of hazard recognition from a practical perspective.

Keywords: human-computer interaction; construction; hazard recognition; bibliometric review

Citation: Wang, J.; Cheng, R.; Liu, M.; Liao, P.-C. Research Trends of Human–Computer Interaction Studies in Construction Hazard Recognition: A Bibliometric Review. *Sensors* **2021**, *21*, 6172. https://doi.org/10.3390/s21186172

Academic Editors: Carlos Morón Fernández and Daniel Ferrández Vega

Received: 17 July 2021
Accepted: 4 September 2021
Published: 15 September 2021

1. Introduction

According to the Encyclopedia Britannica, human–computer interaction is usually defined as "the science concerned with designing effective interaction between users and computers and the construction of interfaces that support this interaction" [1], i.e., the process of exchanging information between a human and a computer in a certain manner, using some kind of conversational language to accomplish a defined task [2]. As an interdisciplinary area of research involving many fields such as computer science, psychology, sociology, graphic design, and industrial design, human–computer interaction has evolved from the early stage of manual work to the stage of the web user interface and multimodal intelligent interaction, with user customization, embedded computing, augmented reality, social computing, knowledge-driven human–computer interaction, emotion interaction, and brain–computer interfaces as the focus of research [3], which has dramatically improved human life.

In the field of civil engineering, "hazard" is usually defined as "the source of energy that, if released and results in exposure, could cause injury or death" [4]. Due to the specificity of the construction industry, the overall hazard recognition rate for construction projects is relatively low compared with other industries, at 66.5%. At the individual level, in general, the recognition rate of hazards among construction workers with more than 10 years of experience is less than 80% [5]. Therefore, effective recognition of potential hazards is of great significance to reduce the accident rate in the construction industry and ensure the safety of construction workers. However, the traditional hazard recognition technology is single-modal and relies too much on humans' subjective feelings. Consequently, the hazard recognition technology has progressed slowly and failed to meet the needs of construction industry development to date, which is one of the main reasons why the global number of casualties in the construction industry has still not clearly decreased [5]. For example, in the United States, the number of construction fatalities was 1102 in 2019, revealing an increase of 6.17% compared to 2018 [6]. Today, the effective recognition of potential hazards has become a major strategic issue related to the sustainable development of the construction industry, in addition to the safety of workers.

As a consequence, due to the pace of the fourth industrial revolution, human–computer interaction technologies are increasingly being applied to the construction industry, effectively driving advances in hazard recognition technologies. For instance, the modeling, measurement, and enhancement of the effectiveness of various types of interfaces between computer applications and construction workers, and maximization of the accuracy of the mapping of data from one modality to another, are academic frontiers that are currently being addressed by scholars [7]. Therefore, it can be inferred that the application of human–computer interaction technology in construction hazard recognition has a solid research foundation and a broad development prospect.

In this paper, we define CHR-HCI as the research related to both human–computer interaction and construction hazard recognition. Although some scholars have deeply explored this field, limited research has concentrated on providing a comprehensive overview for these studies. Thus, to systematically summarize the research related to this topic and identify the directions for future research, this study aimed to: (1) collect peer-reviewed papers and conference articles associated with CHR-HCI from 2000 to 2021 to ensure the literature's completeness and representativeness; (2) extract the characteristics of the research in this field including the numbers, types, and country sources of articles by analyzing the basic information of the papers; (3) undertake keyword co-occurrence analysis and time-zone analysis to identify the research content and evolution trend; (4) summarize the research modules through cluster analysis and propose valuable potential research topics for scholars; (5) establish a research framework and a theoretical map of CHR-HCI to show the existing research progress and shortcomings in this field, and provide practical guidance and assistance for construction management.

The following sections of this paper are organized in accordance with the order of the bibliometric research. Section 2 discusses the research methodology and process, and Section 3 presents the basic information analysis. Next, the keyword co-occurrence network in Section 4 reveals the research content and the association between keywords over the past 20 years. Section 5 clusters the terms to present the research modules composed of keywords with a high association, and then the timeline in Section 6 analyzes the evolution of research themes and trends. Section 7 illustrates a theoretical framework for research in this field and provides an outlook on research topics to be further explored. Finally, the theoretical contributions, practical contributions, and limitations of this paper are summarized.

2. Research Process

2.1. Paper Retrieval

As shown in Figure 1 [8], the paper retrieval mainly included the following steps.

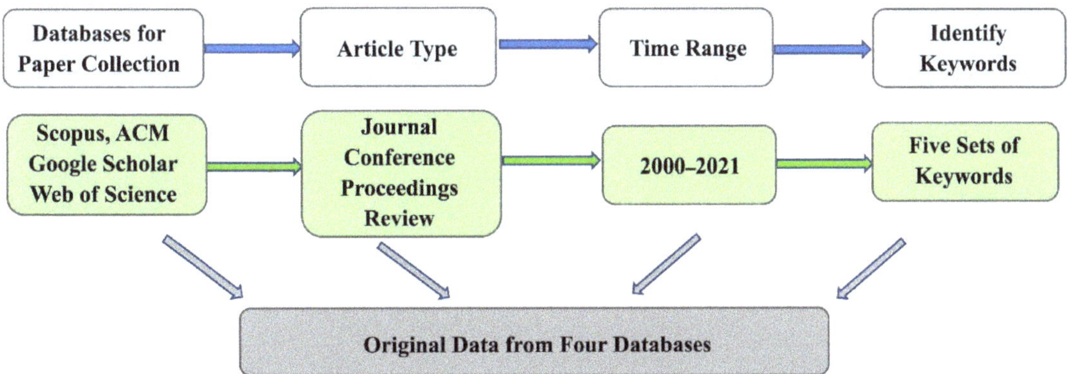

Figure 1. Paper retrieval process.

Firstly, we identified data sources. After careful consideration and repeated comparison, we selected Scopus, ACM Digital Library, Web of Science, and Google Scholar databases for the literature search. As important navigation tools, the four databases, which globally cover the most extensive abstracts, references, and indexes of academic literature, can help researchers understand the most advanced developments regarding CHR-HCI, thus explicitly identifying the current research topics and future trends.

Second, we selected the types of literature. The main literature resources of this study are journal papers about human–computer interaction technology and hazard recognition. In addition, because the research of hazard recognition and HCI depends on practical applications, conference papers should also be a necessary part of the literature resources, because academic conferences are an important channel for scholars to exchange research progress and solve scientific problems encountered in this field. Because these journal and conference papers have undergone strict peer review and selection before publication, the quality of these documents in the Scopus database is sufficiently high to represent the main body of the CHR-HCI research field.

Finally, we searched the literature based on the restrictions. The constraints of the paper retrieval mainly related to keywords and the time range. Regarding the time range, HCI, established in the 1950s [9], was first applied to hazard recognition considerably later. Thus, we selected 2000, when HCI was initially applied to the field of hazard recognition, as the starting year of the retrieval. In addition, before 2000, the quantity of relevant literature was minimal. In summary, we searched the papers published from 2000 to 2021. Regarding the keywords, we identified "construction", "hazard", "recognition", "human-computer" and "interaction," and searched the dictionary for synonyms and near-synonyms of each keyword. To ensure the completeness and comprehensiveness of the literature search, the following methods were used: Boolean operators were adopted to connect synonyms and near-synonyms, and the results were searched for in the databases. We added the synonyms and near-synonyms that had not been found previously, based on the keywords, abstracts, and papers with high relevance in the search results. After several iterations, if the number of documents we searched did not change significantly, we could determine the final search strategies. Taking Scopus as an example, the retrieval methods based on keywords and Boolean operators are shown in Table 1. When we used the other three databases to retrieve papers, we adjusted the Boolean operators according to the rules of each database.

Table 1. An example of paper retrieval.

Search Attributes	Values Used in the Search
Database	Scopus
Keywords	Construction hazard recognition; Human-computer interaction
Boolean operators	(TITLE-ABS-KEY(construct*) OR TITLE-ABS-KEY(build*) OR TITLE-ABS-KEY(erect)) AND (TITLE-ABS-KEY(hazard*) OR TITLE-ABS-KEY(hazard*) OR TITLE-ABS-KEY(peril) OR TITLE-ABS-KEY(risk*) OR TITLE-ABS-KEY(threat*)) AND (TITLE-ABS-KEY(recogni*) OR TITLE-ABS-KEY(identif*) OR TITLE-ABS-KEY(supervis*) OR TITLE-ABS-KEY(detect*) OR TITLE-ABS-KEY (inspect*) OR TITLE-ABS-KEY(realiz*) OR TITLE-ABS-KEY(cogni*) OR TITLE-ABS-KEY(notice*) OR TITLE-ABS- KEY(perceiv*) OR TITLE-ABS-KEY(verif*)) AND (TITLE-ABS-KEY(computer) OR TITLE-ABS-KEY(machine) OR TITLE-ABS-KEY(robot) OR TITLE-ABS-KEY(sensor)) AND (TITLE-ABS-KEY(collaborat*) OR TITLE-ABS-KEY(cooperat*) OR TITLE-ABS-KEY(interact*) OR TITLE-ABS-KEY(combin*)) AND PUBYEAR > 1999

Note: The "*" means searching the words with the same letters before "*". For instance, "identif*" means searching the words including "identify", "identification", "identified" and so on, i.e., any word starting with the letters "identif".

2.2. Bibliometric Analysis Method

Human–computer interaction (HCI) is an emerging interdisciplinary subject, encompassing numerous fields, such as computer science, industrial design, psychology, behavioral science, organizational behavior, and physiology. After HCI was introduced into construction hazard recognition, many new research topics requiring multidisciplinary knowledge were spawned. As a result, due to the large number of academic papers, it is not feasible to conduct a manual analysis of such a complex interdisciplinary field. The large number of documents creates a tremendous workload, leading to the waste of substantial effort by researchers to identify the research focus and correctly classify the literature. Furthermore, literature classification based on researchers' subjective interpretation is prone to human error, leading to a significant deviation between the conclusions drawn in the literature review and the facts.

Bibliometrics usually refers to the science that applies mathematical and statistical tools to quantitatively analyze literature resources and obtain relatively accurate and objective statistical results. Compared with the traditional literature classification method, bibliometric analysis can not only reduce the workload of researchers and shorten the research cycle, but also help researchers draw more scientific conclusions. As a consequence, the introduction of the bibliometric method is indispensable for the analysis of the literature in this field.

Bibliometric software is a fundamental tool for conducting bibliometric analysis. To explore the research progress of CHR-HCI and improve the efficiency of analysis, this study adopted a combination of CiteSpace and Vosviewer to undertake the bibliometric research. CiteSpace and Vosviewer, as typical bibliometric analysis software with an automatic calculation function, can be used by the researcher to visualize the theoretical rules, knowledge structure, and advanced content of the research field, and clearly reveal the relationship between two research topics to finally generate a visual chart called a "scientific knowledge map".

Before using the software for bibliometric analysis, we needed to identify the final research sample. Because the aim of the paper search is to comprehensively retrieve relevant papers and avoid the exclusion of relevant articles, all of the literature obtained by the search may not be relevant to this study. Furthermore, the literature search will inevitably encounter the problem of duplicate research. For example, some papers are first accepted in conferences and then published in journals, thus generating duplicate publications. To address the above issues, the study sample was identified through the following three-level screening process.

First, the research team exported four datasets from four databases and then compared them by reading the titles carefully to identify an initial literature list that did not contain any apparent duplicates. Second, based on reading the publication titles, the literature abstracts were carefully checked, and thus irrelevant or duplicate papers were eliminated.

Third, we browsed the general content of the literature and further eliminated the non-compliant literature on the basis of the first two steps. Finally, 274 eligible papers were selected as the study sample. Figure 1 shows the framework of the study methodology.

After identifying the sample, bibliometric analysis was conducted through CiteSpace and Vosviewer. Because the abstract and keywords are a powerful representation of the main idea of a paper, in this research we systematically identified the current research status and future development trends in this field using basic information analysis, cluster analysis, keyword co-occurrence analysis, and keyword timeline analysis. In addition, it should be noted that the corresponding indexes of cluster analysis should meet certain credibility criteria to obtain reliable research modules. Among the indicators describing credibility, the mean silhouette value P was used to measure the homogeneity of clusters, and modularity Q was used to represent the strength of connections between nodes. The results are usually considered to be reliable when the values of the two indicators are not lower than 0.7 and 0.3, respectively [10]. Figure 2 shows the process of bibliometric analysis [8].

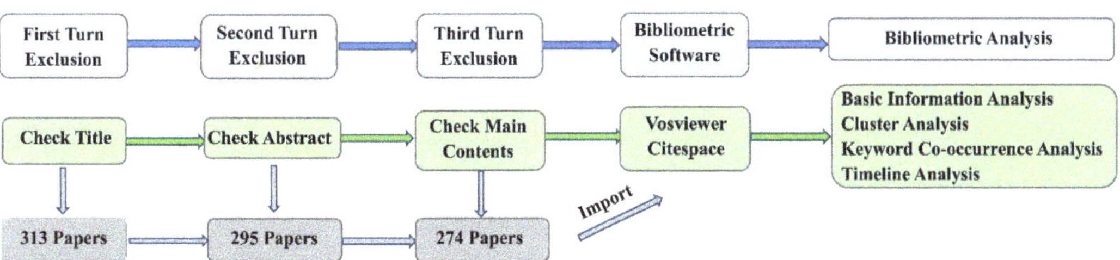

Figure 2. Bibliometric analysis method.

3. Basic Information Analysis

After identifying the sample, we first analyzed the basic information of the 274 publications. Similar to the descriptive statistics in some experimental studies, the main purpose of this section is to help readers understand the basic information including the number of annual publications, the composition of literature types, and publication countries or regions in this field. Analysis of the number of annual publications aims to explore the trends of the literature, concentrating on CHR-HCI and its future development potential; the statistics relating to the literature types are used to reflect the research progress through the ratio of articles, conference papers, and reviews; and the analysis of publication countries/regions can reveal the global distribution of research and provide advice on location selection for scholars interested in this field.

3.1. Number of Annual Publications

The trend of annual publications in this field from 2000 to 2020 is shown in Figure 3. Before 2009, the number of relevant papers published in most years was small. Since 2011, particularly since 2015, publications have shown a significant upward trend, surging from nine papers in 2015 to 48 papers in 2020. Because the current year has not yet ended, the papers published in 2021 are not plotted in Figure 3 to avoid any visual misrepresentation for the readers. This figure shows that, despite the impact of the COVID-19 pandemic, the number of publications in this field is considerable.

Moreover, using the least-squares method, a regression model taking the number of publications as the dependent variable and the year as the independent variable was applied to reveal the trend of publications; the resulting slope is positive, as shown by the dashed line in Figure 3. In addition, we calculated the Price Index as the percentage of the number of publications in the last five years (defined as 2016–2020 because 2021 has not yet ended) divided by the total number of publications (defined as 2000–2020). The value of the Price Index was 0.594, showing that the literature in this field, rather than aging, has excellent research prospects. To conclude, the number of annual publications indicates that the research related to CHR-HCI has attracted a significant amount of attention and has been a burgeoning research area in recent years.

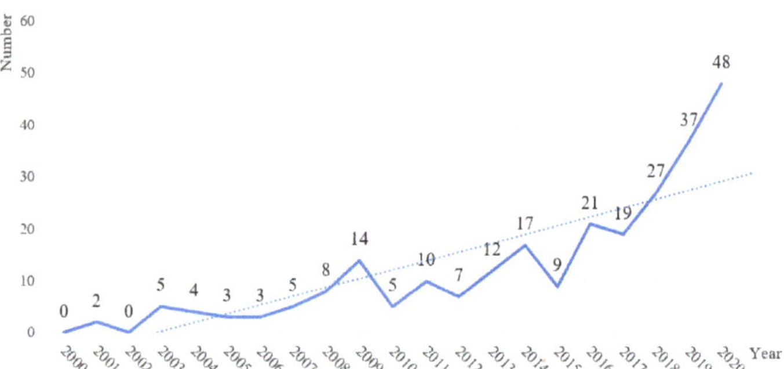

Figure 3. Distribution of related papers from 2000 to 2020.

3.2. Composition of Literature Types

The types of the 274 documents are shown in Figure 4. The proportion of different literature types can reflect different research dimensions in the field. First, journal articles accounted for more than half of the articles, at 56%. Because journal articles usually represent basic research related to experiments, this reflects a certain amount of basic research in the field, which is still in a vigorous development stage. Second, conference papers also account for a large proportion, at 42%. The percentage of conference papers usually shows the excellence of the academic communication mechanism in this field; thus, the ratio of more than 40% proves that the area has a high research value and scholars have a strong will to communicate and improve together. Hence, more communication platforms have been established through academic conferences. Third, review articles are relatively scarce, accounting for only 2%, which indicates that the literature review has not kept pace with the rapidly emerging fundamental research, and further reveals that few scholars have systematically summarized the research in this field.

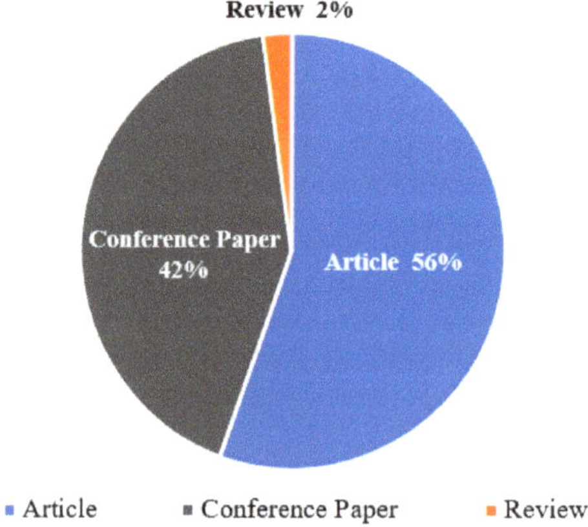

Figure 4. Distribution of literature types.

3.3. Publication Country/Region

As shown in Figure 5, 95 papers were published in this field in China and 72 in the United States, which is in line with the construction industry's comprehensive strength and development level in both countries. As a country with an early start in the technological revolution, the USA has many advantages in developing emerging technologies, including a rich technology reserve, high-quality human resources, and good innovation performance. As the world's largest market in the construction industry, China, which has numerous projects under construction, is able to provide researchers with rich practical opportunities. Thus, China has also made significant progress in the field of CHR-HCI. In addition, the UK, Germany, Korea, and Italy have also published more than 10 articles, showing that they are also essential contributors to the development of human–computer interaction applications in the construction industry.

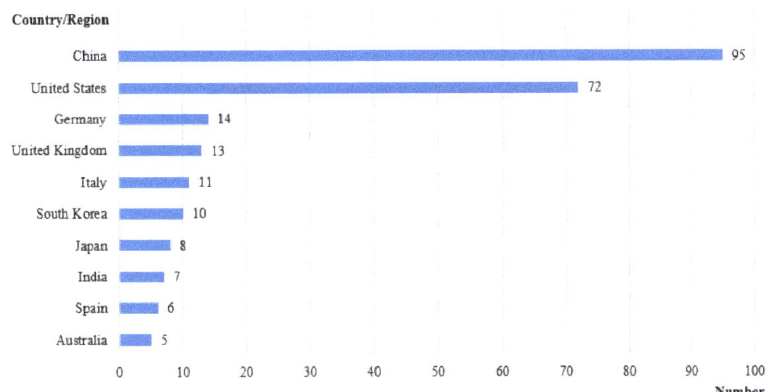

Figure 5. Publication country/region distribution.

In summary, we believe that future research in this field is expected to see greater cooperation among countries. At the national level, taking China and the United States as an example, China's emerging infrastructure construction provides good opportunities for practice in the field of construction engineering safety. However, the rich technical reserves and high-quality human resources in the United States are also needed in China; thus, cooperation between the two countries can compensate for the disadvantages of each and allow the two to work together for a better future in hazard identification. At the personal level, scholars devoted to the field of hazard identification may consider going to UK, Germany, USA, China, or other suitable countries/regions to start research or actively cooperate with scholars in those countries/regions.

4. Keyword Co-Occurrence Network

The keyword co-occurrence network diagram shows, in detail, the centrality, importance, and connections of the research terms in a field. The keyword co-occurrence network consists of three basic elements: nodes, node markers, and links between nodes. The color of the node represents when the corresponding term was first explored. The cooler the node's color, the earlier the node appeared. The radius of each node shows the frequency of the term, and the distance of its position from the network's center represents its centrality. The width of the line between two nodes represents their connection strength, and the line's color reveals the first time at which the two terms were first co-occurred [11]. Similarly, the cooler the link's color, the earlier the two terms were connected.

First, after keyword co-occurrence analysis was completed using Vosviewer, we obtained the scientific knowledge graph shown in Figure 6 [8], which identifies the research progress of CHR-HCI. Second, the mean silhouette (P) value obtained after cluster analysis of the keyword co-occurrence network is 0.7533 and the modularity (Q) is 0.796, both of which meet the corresponding credibility criteria. Consequently, we can directly adopt the keyword co-occurrence network shown in Figure 6 to make clustering analysis. Finally, the research terms in Figure 6 can be divided into three categories according to the levels they belong to. The first level is the research goal, i.e., keywords related to construction safety and hazard recognition; the second level is the general method to achieve the goal, i.e., keywords related to human–computer interaction; and the third level is the specific technology adopted by researchers, i.e., keywords related to machine learning and specific algorithms.

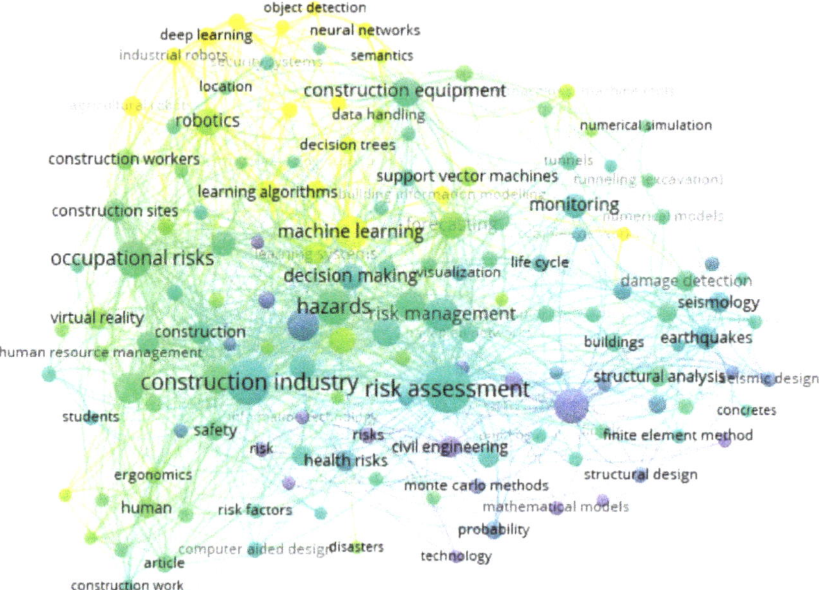

Figure 6. Keyword cooccurrence networks.

4.1. Terms Related to Construction Safety and Hazard Recognition

In the past 20 years, an important shift has occurred in the research field of CHR-HCI, i.e., the guiding ideology has changed from post-accident analysis to accident prevention and hazard prediction [12]. The term "forecasting" in Figure 6 reveals this important shift. In contrast to post-accident analysis, accident prevention and hazard prediction emphasize that construction-related staff should accurately identify potential risks and take appropriate measures to avoid accidents before an accident occurs. The emergence of key words such as "risk perception" and "risk analysis" is closely related to the change in this guiding ideology, in that effective risk perception and analysis are essential for staff to accurately recognize hazards in complex construction environments [13].

Furthermore, in the field of risk prediction and hazard recognition, research on earthquakes, a major natural hazard, have attracted the continuous attention of scientists. Scholars have launched useful explorations from the perspectives of seismic design, building planning, and new materials [14]. Terms such as "earthquakes", "seismic design", "seismology", "architectural design", and "reinforced concrete" reflect the research progress of this subject, whose main research directions include earthquake prediction, real-time intelligent monitoring of earthquakes, design of new seismic structures, and development of new seismic resistant materials [15].

The keyword co-occurrence network also reflects the changes in the organizational management in this field. A revolution in organizational management and safety technologies is required to ensure the purpose of accident prevention and hazard prediction becomes reality [16]. Because construction safety is closely connected to organizational management, researchers are continuously hoping to improve safety performance through innovations in management. Terms such as "decision making", "monitoring", "safety training", and "risk management" are vivid reflections of the changes in this field. In the past 20 years, the concepts of risk management, risk decision making, engineering structural health, and safety training in engineering construction have been generally accepted and have become important research topics of current scholars [17].

4.2. Terms Related to Human-Computer Interaction

In the practice of construction safety, it is not enough to merely rely on correct concepts and high-quality management to make breakthroughs in improving the hazard recognition rate—it is imperative to make changes in safety technology, which has been significantly improved by the introduction of human–computer interaction technology.

Human–computer interaction, usually occurring in the implementation of a specific automated task, refers to a mode in which humans and computer-related equipment share an operating space [18]. It has triggered a significant technological change in hazard recognition. Current HCI research in hazard recognition can be divided into three main categories: key technologies, typical products, and product performance.

Key technologies refer to basic or breakthrough technologies needed in the formation of HCI-related products applied to hazard recognition. Basic technologies generally include sensor technology, positioning and map construction technology, robot operating systems, 3D modeling technology and virtual simulation technology [19], whereas breakthrough technologies refer to computer vision, computer simulation, neural network, and high-performance material manufacturing technology [20]. As shown in Figure 6, terms such as "virtual reality", "three-dimensional computer graphics", "computer simulation", and "computer vision" reflect the researchers' focus on technology.

To date, researchers have developed typical products for HCI with certain hazard recognition functions, mainly including construction robots for specific scenarios and automated construction systems for integrated scenarios. Robots used for a single scene are capable of repeatedly completing specified tasks, such as excavation robots, handling robots, and painting robots that can complete hazard recognition in a specific scene [21]. Automated construction systems used for integrated scenarios usually have the ability to integrate multiple single-task robots, such as ABCS systems and SMART systems with more complete hazard recognition functions [22].

Product performance of HCI applied to hazard recognition refers to the product attributes, product cost, operation efficiency, operation quality, operation safety, etc. The performance can be evaluated either by horizontal comparison of typical HCI products and traditional operation methods in terms of product cost, operation efficiency, operation quality, and operation safety, or by vertical comparison of different HCI products in terms of human resources, building material consumption, machine quality, machine power, machine load, movement speed, operation accuracy, etc. [23]. In the future, HCI products applied to hazard recognition are expected to pay more attention to the directions of integrating design and construction, improving the mobility of humanoid robots, and enhancing the load capacity and the positioning accuracy of intelligent machinery [24].

4.3. Terms Related to Machine Learning and Specific Algorithms

Machine learning is the foundation of artificial intelligence, and deep learning is the further development and extension of machine learning. Compared with traditional neural networks, deep learning sets multiple implicit layers and thus has higher training efficiency and a better learning effect in the fields of computer vision, audio recognition, and natural language processing [25]. Under the CHR-HCI framework, deep learning is mainly used for data processing in the development experiments of HCI devices, writing of internal programs for electronic HCI devices, and risk prediction for hazard recognition systems [26]. To further explore deep learning, researchers are focusing on algorithms such as convolutional neural networks, stacked autoencoder network models, deep belief networks, etc. [27].

As shown in Figure 6, among the algorithms related to machine learning, the support vector machine (SVM) is widely applied to the research and development of devices related to hazard recognition. SVM, proposed by Soviet scientists in 1964, is a generalized linear classifier for binary classification of data using the kernel method in a supervised learning manner [28]. Since the 1990s, SVM has been rapidly popularized with the breakthrough of HCI-related technology, and thus a series of improved and extended algorithms have been derived. Because SVM can solve pattern recognition problems, such as portrait recognition, action recognition, emotion recognition, text classification, handwritten character recognition, etc., it has also been applied to hazard recognition. Recently, the improved algorithms of SVM applied to hazard recognition have encompassed improved algorithms for skewed data, probabilistic SVM, multiclassification SVM, least-squares SVM, structured SVM, and multi-kernel SVM. The extended algorithms include support vector regression, support vector clustering, and semi-supervised SVM [29]. As an algorithm still in the optimization phase, there are three main future research directions for SVM: the improvement of kernel functions, classification of big data, and combination of models [30].

As a multi-disciplinary subject, machine learning also involves multiple disciplines such as probability theory, statistics, approximation theory, convex analysis, and algorithmic complexity theory. Keywords such as "numerical/mathematical models" and "Monte Carlo methods" show the importance of numerical simulations represented by Monte Carlo simulations. Depending on the computer technology, the modern Monte Carlo simulation has two main advantages, i.e., simplicity and speed, and has become an important technical basis in construction project management [31]. Monte Carlo simulation is now an important tool for computer simulations in hazard recognition experiments.

5. Cluster Analysis

Because keyword co-occurrence networks are too detailed to simplify research subjects and identify research modules, we used cluster analysis to summarize the main trends of CHR-HCI.

Cluster analysis is an analysis mode in which text data are processed with optimized computational methods in statistics to obtain potential research themes. In this study, a combination of Vosviewer and CiteSpace was selected for cluster analysis, i.e., CiteSpace was used to optimize Vosviewer's cluster analysis results. In CiteSpace, there are three usual methods for determining module names: log-likelihood ratio, mutual information, and highest word frequency [32]. Due to the representativeness of the module names, we chose the highest word frequency method to identify the modules.

After analysis and optimization, we obtained four modules with no obvious containment relationship, as shown in Figure 7 [8]: computer vision, ergonomics, computer simulation, and virtual reality. Because the keywords are not clearly shown in this figure, we listed the top three most frequently occurring keywords in each cluster (except the cluster label) in Table 2.

5.1. Cluster 1: Computer Vision

Among the 251 articles retrieved, 177 articles were related to this keyword, indicating that computer vision played a pivotal role in hazard recognition experiments. Advances in computer vision technology are based on the continuous optimization of deep learning algorithms, such as convolutional neural networks, stacked autoencoder network models, and deep belief networks [33,34]. The vital research topics include content-based image extraction, pose evaluation, multimodal data recognition, autosomal motion, image tracking, scene reconstruction, image recovery, and system integration [35].

In the field of hazard recognition, computer vision is divided into two research themes.

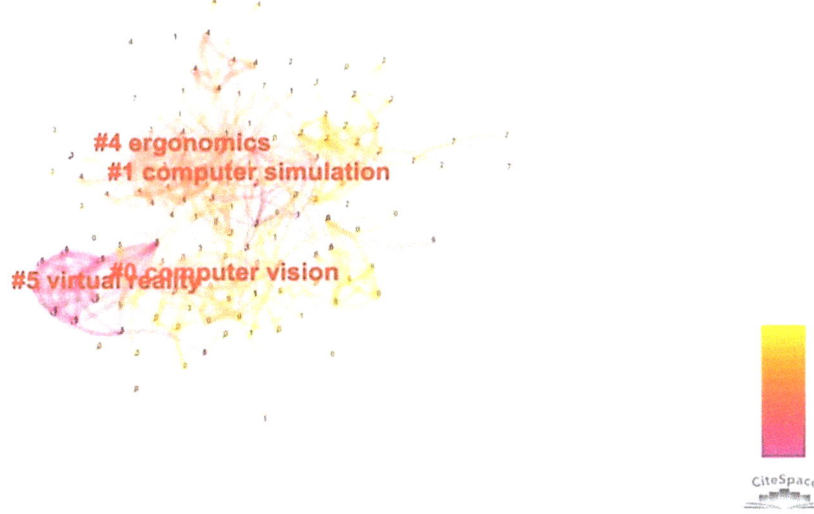

Figure 7. Cluster analysis.

Table 2. The top three most frequently occurring keywords in each cluster except the cluster label.

Keywords	Occurrence Times	Cluster Label
Accident Prevention	24	Computer Vision
Support Vector Machine	20	Computer Vision
Three-Dimensional Computer Graphics	15	Computer Vision
Risk Perception	20	Ergonomics
Risk Management	13	Ergonomics
Human Resource Management	9	Ergonomics
Machine Learning	31	Computer Simulation
Neural Networks	28	Computer Simulation
Monte Carlo Methods	22	Computer Simulation
Safety Training	14	Virtual Reality
Augmented Reality	13	Virtual Reality
Forecasting	10	Virtual Reality

One is the development of models. The algorithms related to computer vision are based on the cognitive psychological models represented by the template matching model and the recognition-by-component theory model [36]. However, the algorithms mentioned above can only distinguish the low-visual-complexity hazard from a fixed perspective, which is challenging to adapt to the dynamic construction scenes with time-variant characteristics. In recent years, feature matching models have also been proposed to take into account the dynamic nature of the temporal dimension [37], but due to the lack of sufficient data to train the models and the inaccuracy of inter-modal mapping, the effectiveness of hazard recognition remains unreliable.

The other is the analysis of cognitive associations. Driven by recognition, human attention can be differentially distributed into different regions and shifted over time. Computers lack this ability to exploit and learn human attentional cues, which is one of the important factors restricting the development of computer vision techniques [38]. Moreover, the existing computer vision's hazard judgment logic is based only on real-time construction scenarios and cannot combine existing cues to make predictions and inferences about future safety status. Therefore, future researchers should not ignore the time-variant nature of visual behavior, but are expected to focus on the logical association between existing cues and determine potential hazards.

5.2. Cluster 2: Ergonomics

Since 2015, ergonomics has become closely associated with CHR-HCI, and has developed in the directions of diversification, humanization, and intelligence. Currently, researchers are trying to use physiological and psychometric means to study the rational coordination relationship between the structural-functional, psychological, and mechanical aspects of the human body and computers, which aims at improving the performance of hazard recognition [39]. Sixty-five of the 251 articles retrieved were connected with this keyword, confirming that the interaction between construction hazard recognition and ergonomics is adequate and a large number of researchers have deeply explored the technical means. At present, research in this field focuses on task evaluation and quantification, brain-computer interfaces, and experimental paradigms in engineering psychology [40].

Akanmu et al. described a cyber-physical postural training environment where workers could perform work with reduced ergonomic risks [41]. Inyang et al. illustrated a proposed methodology to assess and quantify the ergonomic hazard effects of wall framing tasks on the back, legs, neck, shoulder, hands, and wrists of residential construction workers involved in manufacturing wood framing activities in a construction factory [42]. The focus was placed on risks associated with awkward work postures, force and static loading, contact stress, hand-arm vibration, repetitive tasks, and environmental factors. In addition, the contributions of organizational factors and the total daily duration of exposure to each risk factor were presented. The proposed methodology has been being incorporated into a computer program, "ErgoCheck", which has shown success in quantifying work-related ergonomic hazards.

5.3. Cluster 3: Computer Simulation

Computer simulation, also known as computer emulation, is a computer program used to simulate an abstract model of a specific system [43]. Ninety-seven of the 251 articles retrieved were related to this keyword. Currently, computer simulation research related to hazard recognition is oriented towards discrete simulation, analogous simulation, simulation-based on probe elements, and simulation of stochastic processes or deterministic models, with the primary purpose of simulating hazards in construction scenarios through simulation software and external parameters. This research involves the development of source code and the optimization of existing programs [44]. Numerous scholars have continuously optimized discrete event simulation languages, such as GPSS, SIMSCRIPT, GASD, CSL, and SIMULA, in addition to continuous system simulation languages represented by DARE, ACSL, CSS, and CSSL, which have laid a solid foundation for human–computer interaction technology and promoted the development of the field of hazard recognition [45].

5.4. Cluster 4: Virtual Reality

Virtual reality (VR) technology is a human–computer interaction system with the function of creating a virtual world, which can immerse the users in the virtual environment [3]. Among the 251 articles retrieved, 52 were related to this keyword, indicating that virtual reality has a broad development prospect after being introduced into the field of hazard recognition. From the perspective of technology development, scholars are working on optimizing four key technologies: dynamic environment modeling technology, real-time 3D graphics generation technology, stereo display and sensor technology, and system integration technology [46]. From the perspective of technology application, virtual reality technology is mainly developed for safety training of construction workers and building risk assessment systems [47]. From the standpoint of technical disadvantages, the main problems faced by virtual reality include the high cost of production and unstable user visual experience [45].

For example, Huang et al. noted that a virtual reality system composed of a brain–computer interface (BCI) and electroencephalogram (EEG) neural network was constructed to collect users' EEG signals and evaluate the accident susceptibility of construction workers. Based on the EEG and physiology data, a statistical model is used in the safety assessment framework to establish the risk standard. In addition, augmented reality has also been used to provide safety training for workers by superimposing virtual scenes onto the real world, thus enhancing the visual experience and reducing the production expense [48,49]. Therefore, using augmented reality to circumvent some of the drawbacks of virtual reality technology will also be a meaningful research topic in the field of CHR-HCI in the future.

6. Keyword Timeline Analysis

For this section, we undertook keyword timeline analysis to obtain a dynamic graph reflecting the evolution of popular research topics. The timeline analysis diagram consists of three elements: nodes, node markers, and lines between nodes. Each element in this graph represents the same meaning as that in the keyword co-occurrence network graph, except that this graph shows the year in which each keyword appears via a timeline [50].

The results of the timeline analysis are shown in Figure 8. Because the small number of publications prior to 2010 resulted in a blank between 2000 and 2010 in the final analyzed image, Figure 8 mainly shows the research dynamics between 2010 and 2021 [8]. The portion of literature published from 2000 to 2010, as a percentage of the total literature, is 17.88%. After reading and analyzing the papers, we found that the keywords that appeared relatively frequently in the research during this period were "computer simulation", "project management", "civil engineering", etc. At that time, the computer simulation technology used for hazard recognition was not mature, and the discussion of hazard recognition in project management was not detailed, indicating that the research on CHR-HCI was in an initial stage during this period. Although the research published between 2000 and 2010 was relatively superficial, it provided a solid foundation for subsequent developments. Additionally, this study categorized the keywords in the timeline analysis based on the cluster analysis results, aiming at presenting the most valuable conclusions to readers.

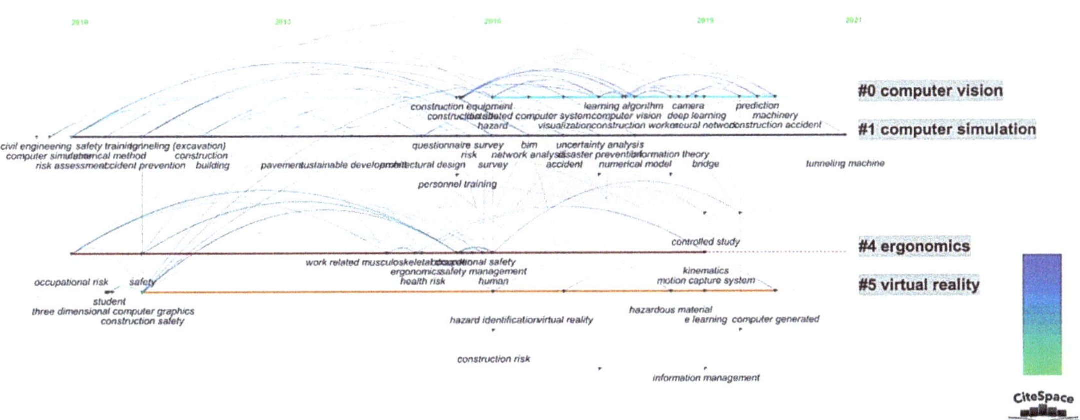

Figure 8. Timeline analysis.

From 2010 to 2011, terms such as occupational risk, construction safety, accident prevention, and risk assessment emerged, confirming our judgment in Section 4.1 that hazard recognition and construction safety gained researchers' attention, and the opinion of accident prevention was gradually established in academic circles. Moreover, at this time, computer simulation, safety training, and 3D computer graphics were also initially applied to the study of hazard recognition to undertake accident prevention and risk prediction.

In 2013, terms such as sustainable development became a popular research topic, which symbolized the further maturation of guiding ideas in construction hazard recognition. Researchers combined the goal of achieving accident prevention with the new concept of green, healthy, safe, and sustainable development, which has been commonly advocated globally, to drive progress in hazard recognition. Moreover, terms such as "pavement" also emerged in 2013, which was mainly because the HCI research in intelligent driving could consistently provide meaningful references for hazard recognition in the construction industry: improving the design of the interaction interface including buttons, lights, and displays; setting up enclosures to divide the operating space between humans and computers; developing convenient mode-switching buttons to quickly start, change, and stop tasks; placing light curtains, laser sensors, and pressure-sensing mats to reduce the risk of irrelevant personnel entering the workspace unintentionally [51].

The year of 2016 was a landmark during which keywords containing safety management, health monitoring, construction equipment, virtual reality, construction informatics, and ergonomics emerged and established strong links with the preceding keywords of occupational risks, construction safety, and accident prevention. This progress reveals that, subsequent to the accumulation of the previous 15 years, an increasing number of scholars engaged in this research and published numerous research results in this year. Research at this stage focused on studying methods and models, mainly including safety training based on individual characteristics, questionnaire surveys, work-related musculoskeletal system monitoring, and wearable devices and sensors [52].

During 2018–2020, research on numerical models, deep learning algorithms, uncertainty analysis and neural networks became popular. These keywords representing specific algorithms and analysis methods mainly address computer vision and computer simulation, which is an inevitable continuation of the research growth that occurred around 2016. Many results emerged in this phase, such as motion capture systems, e-learning, information integration management systems, and improved support vector machines, representing a further deepening of algorithmic research in the field of CHR-HCI [53].

7. Framework Development, Future Directions and Discussion

7.1. Framework Development

To reflect the overview of the research field from 2000 to 2021, we mapped the research framework and knowledge structure of CHR-HCI based on bibliometric analysis, as shown in Figures 9 and 10 [8,36].

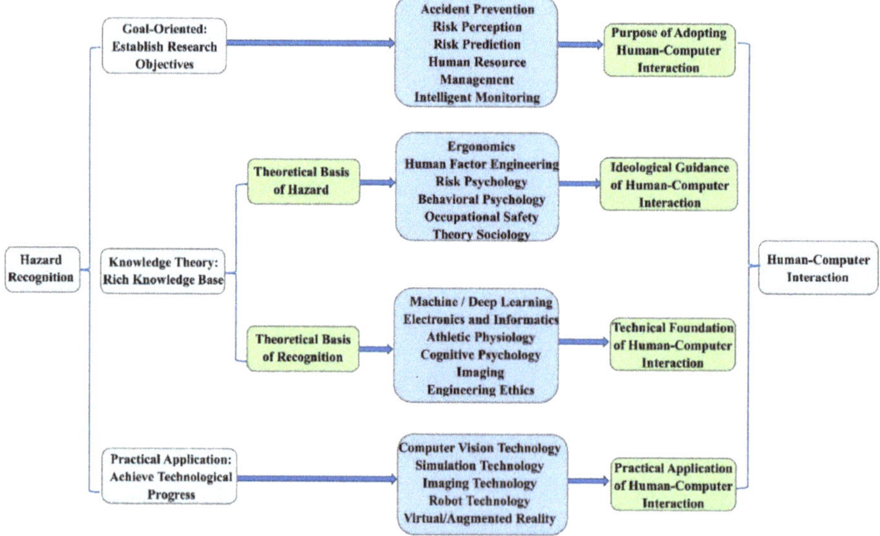

Figure 9. Research framework: from hazard recognition to HCI.

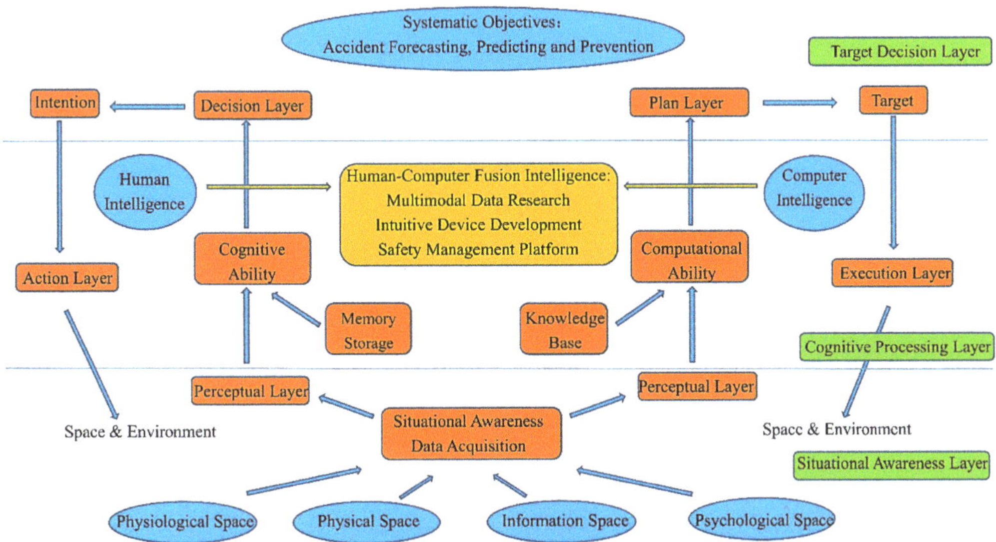

Figure 10. Hazard recognition based on human–computer fusion intelligence.

Because human–computer interaction is an emerging interdisciplinary field encompassing various fields, and hazard recognition also involves complex theoretical knowledge and practical techniques, a new framework was proposed in this paper to divide the CHR-HCI field into three levels:

(I) In terms of goal orientation, hazard recognition aims to study the process of identifying, perceiving, and recognizing hazards, and their influencing factors, in the construction environment for the purpose of risk perception, accident avoidance, potential hazard prevention, accident prediction, and intelligent monitoring. The change from post-accident analysis to pre-accident prediction and prevention is the core advancement in engineering safety guiding ideology within the last 20 years with the development of human–computer interaction technology. Therefore, this is also the aim of introducing HCI technology.

(II) From the aspect of theoretical foundation, hazard recognition mainly contains two parts: hazard-related/risk-related theory and recognition-related/identification-related theory. Engineering hazard-related theories include risk psychology, ergonomics, human factors engineering, behavioral psychology, sociology, and other disciplines [54], which are theoretical guides for the application of HCI technology. For example, the concepts of people-oriented and sustainable development in ergonomics provide theoretical directions for HCI technology. Recognition-related theories, as the technical basis of conducting HCI-related studies, mainly cover cognitive psychology, machine learning, deep learning, athletic physiology, imaging, electronics, informatics, etc. [55]. Furthermore, in addition to the development of science and technology, engineering ethics has become essential, and its supervisory role in scientific experiments has attracted significant attention from academic circles [56]. Thus, engineering ethics should also be regarded as a crucial guiding theory for hazard recognition.

(III) Regarding practical applications, the practical application of hazard recognition should focus on computer simulation technology, computer vision technology, virtual reality technology, augmented reality technology, and robotics [57]. We believe the emphasis of future research on hazard recognition should be placed on three issues. The primary goal of researchers is to explore appropriate methods to process multimodal data in hazard recognition experiments, and then to develop intuitive devices for hazard recognition using the appropriate data processing methods, as noted above. The ultimate goal is to establish an integrated safety management platform based on the development of suitable multimodal data processing methods and intuitive devices. Therefore, these three research aspects have not only been applied to a certain extent, but also represent explicit directions for the future.

In summary, as shown in Figure 9, we can construct the relationship between hazard recognition and human–computer interaction. Starting from the three dimensions of hazard recognition, namely, goal orientation, theoretical basis, and practical application, we can divide the CHR-HCI research field into three aspects, among which the theoretical basis can be further classified into the two aspects of hazard-related theory and recognition-related theory. These aspects eventually support human–computer interaction, illustrating the close and complex relationship between hazard recognition and human–computer interaction.

As a static schematic diagram, although Figure 9 provides a clear description of the research framework in this area, to better illustrate the implications of Figure 9 from a dynamic perspective, we drew Figure 10 to show the framework of hazard recognition from the perspective of human–computer fusion intelligence. The system consists of three layers: the situational awareness layer, cognitive processing layer, and target decision layer [36].

(I) The situational awareness layer is mainly used to collect information. In general, there are three elements in human–computer interaction systems: humans, computers, and environments. The humans provide the physiological and psychological space, the computers offer the information space, and the environments afford the physical space [36]. Through situational awareness, human–computer fusion intelligence can collect information from the above four kinds of spaces and import it into the perception layer of the humans and computers. The design of the perception layer requires the use of technologies such as wearable devices, combined with the disciplines of athletic physiology, imaging, electronics, and informatics, as shown in Figure 9.

(II) The cognitive processing layer, between the situational awareness layer and target decision layer, has the function of analyzing the data obtained from the perception layer by fusing the intelligence of the human and computer. Hazard recognition with human–computer fusion intelligence is expected to enable the processing of multimodal data, the development of intuitive devices, and the construction of integrated safety management platforms, which echoes the practical applications in Figure 9. Humans have the cognitive ability to analyze data combined with memory storage, whereas computers have the computational capability to process data in combination with a knowledge base. As a result, human–computer fusion intelligence has both the superb computing power of computers and the intelligent cognitive ability of humans, thus forming a dynamic closed loop of "perception"—"analysis"—"decision"—"execution"—"feedback", which continuously optimizes the process of hazard recognition. In this process, cognitive and computational analysis corresponds to machine learning and deep learning in Figure 9.

(III) The target decision layer is able to make judgements subject to data analysis. Through cognition, humans can make decisions and form intentions, leading to corresponding actions. Similarly, through computation, computers can complete planning and determine goals, leading to the execution of feedback [36]. The systematic goals in Figure 10, also highlighting accident prediction and prevention, are consistent with the goal orientation in Figure 9, and the process of making decisions is closely related to the risk psychology, ergonomics, human factor engineering, behavioral psychology, and sociology in Figure 9.

In addition, to further explain the framework in Figures 9 and 10, we list the traditional research areas of hazard recognition and the new research subjects arising from the introduction of HCI in Figure 11 [8]. Traditional hazard recognition mainly depends on manual monitoring, traditional human source management, and post-accident analysis, whereas current hazard recognition concentrates on risk prediction, accident prevention, deep learning, intuitive device based on brain-wave and eye movement, multimodal data process, etc.; thus, the subjects initiated by human–computer interaction relate to virtual reality, augmented reality, computer vision, computer simulation, etc. After identifying the framework of CHR-HCI, we are able to define the direction of future research.

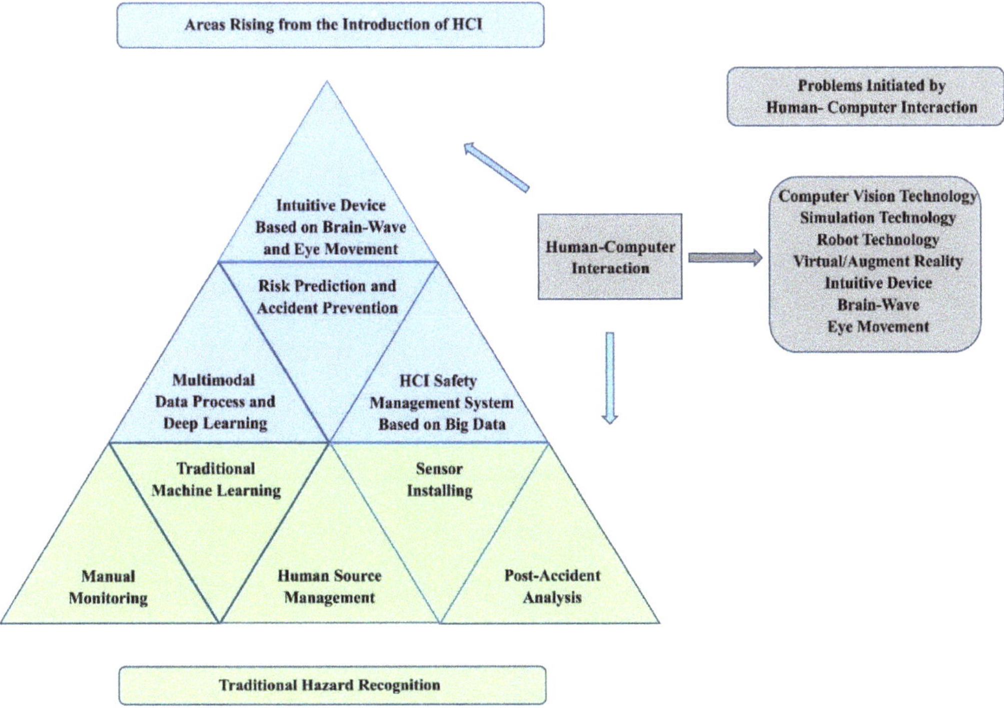

Figure 11. Interaction of hazard recognition and HCI.

7.2. Future Directions

As mentioned above, we summarized three main directions with recursive relationships: improving algorithms to process multimodal data, developing intuitive devices for hazard recognition, and building integrated safety management platforms.

7.2.1. Improve Algorithms to Process Multimodal Data in Hazard Recognition Experiments

To effectively recognize hazards, a large number of studies have monitored workers' physiological data in construction scenarios and explored the relationship between physiological data and hazard recognition performance. The application of multimodal methods has energized the research on hazard recognition, overcome the restrictions caused by unimodal data, and generated a series of new research topics following the combination with human–computer interaction technology.

(I) Multimodal representation learning. The data obtained from hazard recognition experiments generally have more than three modalities; however, current specific representation learning is limited to the case of two modalities. Consequently, it is recommended that researchers expand the number of modalities to which specific representation learning can apply. Furthermore, because dynamic real-time monitoring is an important future direction for hazard recognition experiments, and the mainstream methods of representation learning are often confined to static conditions, identifying approaches for the use of multimodal physiological data for dynamic deep learning is also a frontier issue [58].

(II) Translation among modalities. Because the physiological signals obtained from the hazard recognition experiments are of various types, including brain waves, visual pathways, heart rate, blood pressure, respiratory rate, speech, audio, and video, translation among modalities is needed in the studies. However, there is usually no single correct answer or optimal solution in this process, and the subjective consciousness of researchers can also have a large impact on the results; thus, the final results of hazard recognition experiments often fail to confirm the representation of the same entity between different modalities. It remains a challenge for a series of evaluation systems proposed by academia, such as BLEU, ROUGE, Meteor, and CIDEr, to address the subjectivity problem of the results [59].

(III) Multimodal semantic alignment. In the study of hazard recognition, multimodal semantic alignment is an important step in the analysis the data, and the problems in this field are mainly as follows. First, semantic alignment faces the problem of temporal effects because most data in hazard recognition experiments have temporal relationships. Next, it is difficult to design indicators to measure the similarities between different modalities, and manual design is time consuming and laborious. Furthermore, because experimental data may be missing or redundant, elements in different models may have one-to-many or many-to-one relationships, or may not be able to be matched if significant data are missing. When the elements are incorrectly matched, the performance of the model will be seriously degraded [60]. Finally, the performance of multimodal semantic alignment can be significantly influenced by noise.

(IV) Multimodal data fusion. Similar to multimodal semantic alignment, noise also affects the performance of multimodal data fusion to a large extent, and each modality may suffer from noise interference at different moments. When the data are affected by noise, the results will not be able to accurately represent the original features [61]. In addition, the effectiveness of multimodal fusion will also be significantly impacted if the modalities are not well aligned with each other, as in the case in which data from brain waves and eye tracking are not synchronized. To summarize, researchers in this field should pay attention to addressing the interference of time-variant effects and noise on multimodal data fusion.

7.2.2. Develop Intuitive Devices for Hazard Recognition

As a tool for effective hazard recognition, intuitive devices can not only monitor multimodal physiological data, but also make judgments and give feedback according to the obtained data. To date, intuitive devices have been developed based on two kinds of data: brain activity states (such as EEG) and eye-tracking (such as scan path). The studies of this subject are closely combined with psychology knowledge [62].

The main issue in intuitive device development based on eye-tracking technology is that there is no consensus on the processing of eye-tracking data in the construction industry. Some studies measured workers' cognitive load or safety attention through cross-sectional visual signal concentration (such as gaze duration and times) [40], whereas others used pupil diameter [63]. Additionally, a small number of studies used time-varying trajectory quantitative analysis as evidence of the mental state [64].

(I) In principle, however, different kinds of information processing can produce the same results. For instance, a portion of studies interpreted the average fixation duration as the degree of concentration and found that the duration of fixation reflects the time of individual understanding and processing, which means that the shorter the time, the less attention, leading to recognition errors [65]. However, partial studies interpreted the same indicator as the experience of subjects; that is, the shorter the fixation time, the less time spent in the scenario, resulting in better recognition performance [66]. Moreover, regarding visual trajectories, some scholars claimed that successful hazard recognizers usually spread their attention within the working environment [67]. On the contrary, certain researchers ignored unnecessary distractions in the environment, which is considered to be the key to successful hazard recognition [68]. Obviously, there is no consensus on the interpretation of eye movement data in the construction industry. Therefore, a single indicator cannot satisfy the demonstration of the process of hazard recognition, and research is required to investigate different indicators to describe the physiological process of hazard recognition.

(II) The development of intuitive devices based on electrophysiological signals faces three main problems.

(i) First, numerous researchers collected EEG signals and calculated the voltage value, frequency, vibration, and other parameters to describe the mental state of the participants, such as fatigue, emotion, motivation, and load. Researchers then demonstrated the correlation among these indicators of the participants and hazard recognition performance [69,70]. However, these studies ignored the complex characteristics of EEG signals: each person's subtle actions in the process of performing tasks may be related to different EEG signals, so the conclusions are likely to be misunderstood due to the action interference required by the experiment [71]. For example, although researchers adopted a longitudinal experiment design to control the differences between groups, allowing the participants to perform the hazard visual search task in a VR environment, they did not specify the participants' executive actions (squatting, raising hands, raising head, etc.), and such physical actions would significantly interfere with the interpretation of EEG [72].

(ii) Second, the current research on hazard recognition is only based on the distributed brain activity to reveal behavior [73]. However, according to connectionism theory, human behavior should be on the basis of more complex brain region cooperation, rather than a single brain region in a linear manner [74]. Based on the data of EEG analysis, Wang D. et al. noted that it is possible to recognize and quantify a participant's brain cognition changes by analyzing short time intervals of the raw EEG signal and the combination of different rhythms [75]. Other scholars suggested that the spectral dynamics of EEG in the posterior brain is strongly related to the decrease in vigilance [76]. In this manner, the construction industry's research on hazard recognition appears to oversimplify the cognitive mechanism of the brain.

(iii) Third, the existing experiments need to reflect on the impact of the amount of data on the results. With regard to the amount of data, both the number of trials and samples were generally too small [77]. Regarding trial times, the number was mainly about 30, which only met the minimum standards of basic statistics and could not easily demonstrate the validity of the results [31].

(III) In light of the shortcomings of the above two series of studies, cross-validation by connecting the two studies through psychological methods is an important means to advance the research on hazard recognition. In previous studies, participants were either equipped with eye-tracking devices to record oculomotor features or quantitative psychological monitoring technologies, such as electroencephalography (EEG) and near-infrared spectrum instrument (NIRS), were used to measure the mental workload evoked by hazardous environments [78]. Because it is difficult to fully demonstrate the cognitive process of potential hazards using unimodal physiological data, other physiological signals are needed to accurately describe the hazard recognition process. Although these attempts were innovative and meaningful for occupational safety research, they failed to identify the psychological mechanisms of cooperation between different types of cognitive functions, such as brain activity and pupillary responses. Accordingly, future researchers may consider making a contribution in this area. Michail et al. proposed that the study requires cross-validation of multiple physiological signals to describe the psychological process [79]; for instance, revealing the cognitive rules of eye activity demands the comparison with specific brain activities, reflecting the adequacy of hypothesis to be tested [80]. Moreover, psychological experimental paradigms applied to hazard recognition need to be further enriched, and the current paradigms mainly include the oddball paradigm, spatial cueing task paradigm, and visual search task paradigm [81].

7.2.3. Establish Integrated Safety Management Platforms

In the research related to organization management, the safety management platform based on big data is a highly promising direction. Traditional construction safety management relies excessively on regular safety inspections and an individual's ability to recognize potential hazards. However, because of the complex construction scenarios in the construction industry and the difficulty of unification and standardization, traditional safety management tends to have a low rate of hazard recognition. In consequence, building a safety management platform for the construction industry based on big data technology is a highly significant research topic.

At present, automated construction systems designed in Japan and other countries, such as the ABCS and SMART systems, have certain safety management functions; nonetheless, some shortcomings remain. First, the cost is high. This expense makes it difficult to popularize the approach, and a large number of studies have shown the concern about the lack of funds and the high cost of construction machinery manufacturing. Second, it is hard to ensure the stability and safety of the system. Furthermore, the system, whose monitoring technology still needs to be improved, is not specifically developed for hazard recognition and the hazard recognition rate cannot meet users' requirements. Finally, the lack of interoperability among various information systems and the strengthening of mutual perception among different devices are also problems that need to be solved [82].

Based on the above deficiencies, we indicate that the system should be improved from both internal and external aspects.

(I) Strengthen the stability of the safety management system itself. In terms of industrial robots, the International Organization for Standardization has issued a series of safety management standards for industrial and collaborative robots, and proposed four types of human–robot interaction safety management: safety-related monitoring stop, hand guidance, speed and separation monitoring, and power and force limitation [83], which differ in key control variables, direct human–robot contact, and simultaneous human-robot motion. However, their common essence is to protect human safety during human–computer interaction. Compared with the manufacturing industry, construction scenarios are complex and dynamic, and the construction process is non-standardized, indicating that HCI safety management methods applicable to the manufacturing industry need to be adapted to the construction industry. For this purpose, researchers can propose HCI safety management methods and principles to meet the needs of the construction industry by referring to those methods applicable to the manufacturing industry.

(II) Improve the ability of safety management systems to recognize potential hazards in the surrounding environment. The improvement of this capability depends largely on the crucial technologies involved in CHR-HCI. Most HCI safety management systems based on different technologies are composed of an environmental monitoring system including sensing elements configured at the mechanical stage and an HCI safety warning system configured at the remote-control stage. At present, research in this field focuses on three areas: wearable technologies, virtual technologies, and image sensing technologies.

(i) HCI safety management based on wearable technology can enhance the realism and accuracy of interaction and control. Wearable devices are usually used for visual-assisted control, haptic-assisted control, and optimizing the human–computer interface of construction machinery in the case of insufficient information [84]. Moreover, the integration of wearable devices and remote operation control platforms can increase the realism of manipulation during human–computer interaction [85].

(ii) HCI safety management based on virtual technology mainly includes virtual reality and augmented reality technologies. Compared with sensors and cameras, virtual reality can visualize the blind area of human vision, improve the depth and safety of HCI, and guide operators to complete their tasks [86]. Augmented reality technology can fuse virtual 3D models and the camera surveillance field of view to show operators the invisible construction environment beyond the camera surveillance. Operators can easily identify the depth information and positioning information of the construction site with a fixed view of the camera [87]. In addition, augmented reality can provide a more immersive experience than virtual reality.

(iii) HCI security management based on image sensing technology can identify entities and relationships between them in the environment to enhance the security of HCI [88]. Deep neural network algorithms are widely used in construction resource detection, action recognition, surveillance, and other environmental perception studies [89]. Existing research mostly centers on entity recognition and lacks the comprehension of the relationships between human and machine entities, which is one of the challenges in the field of computer vision [90].

8. Conclusions

In this paper, 274 related papers published between 2000 and 2021 were collected from four databases, and a bibliometric method was adopted to analyze the research progress, development, and future trends in this field.

In general, in this study we undertook basic information analysis, keyword co-occurrence analysis, keyword timeline analysis, and cluster analysis. First, we identified the basic information of relevant studies. Second, we constructed a static network with keywords to describe the interrelationships among concepts. Third, term clustering analysis revealed four modules representing the research areas. Fourth, we conducted a timeline analysis and elucidated the changes in research subjects. Finally, we proposed a research framework and three future directions.

The theoretical contributions of this paper are as follows. First, we established a framework showing the current research area by carefully studying the relevant documents, and provided a systematic summary and overview of the current research status. In addition, we explored the relationship between hazard recognition and human–computer interaction, enriched the theory related to CHR-HCI, identified the current research results and their interrelationships, and then identified more helpful directions for future researchers.

In terms of practical contributions, many conclusions drawn from the bibliometric approach were demonstrated. The introduction of human–computer interaction technology has injected new vitality into the field of construction engineering safety and introduced a series of new research topics. As a result, possible essential research topics for future practical research include "how to make multimodal methods better serve experimental data processing of hazard recognition", "how to develop intuitive sensors and devices", and "how to build a safety management platform for human-computer interaction based on big data".

The limitations of this paper include two main points: one is the limitation of the data. In spite of the broad coverage of the four databases, follow-up research will be undertaken to select documents from more academic databases. The other is the limitation of the method. Future researchers are recommended to use quantitative methods, empirical data, and mathematical models to explore the impact of human–computer interaction on the research progress in the field of construction safety.

Author Contributions: Conceptualization, J.W., M.L., R.C. and P.-C.L.; methodology, J.W., R.C. and M.L.; software, J.W., R.C. and M.L.; validation, J.W. and R.C.; formal analysis, J.W., R.C. and M.L.; investigation, J.W., M.L. and P.-C.L.; resources, J.W.; data curation, J.W. and R.C.; writing—original draft preparation, J.W.; writing—review and editing, J.W. and R.C.; visualization, J.W. and R.C.; supervision, M.L. and P.-C.L.; project administration, J.W., R.C. and M.L.; funding acquisition, P.-C.L. All authors have read and agreed to the published version of the manuscript.

Funding: This work was supported by the National Natural Science Foundation of China (Grant Number: 51878382) and a grant from the Institute for Guo Qiang, Tsinghua University (Grant Number: 2020GQC0007).

Institutional Review Board Statement: Not applicable.

Informed Consent Statement: Not applicable.

Data Availability Statement: The data presented in this study are available on request from the corresponding author.

Acknowledgments: We are grateful for Qin Ding and Chen Xie for their help in English editing.

Conflicts of Interest: All the authors declare no conflict of interest.

References

1. Encyclopedia Britannica. Available online: https://www.britannica.com/topic/human-machine-interaction (accessed on 14 July 2021).
2. Kim, D.; Goyal, A.; Newell, A.; Lee, S.; Deng, J.; Kamat, V.R. Semantic relation detection between construction entities to support safe human-Robot collaboration in construction. In Proceedings of the Computing in Civil Engineering 2019: Data, Sensing, and Analytics, Atlanta, GA, USA, 17–19 June 2019; pp. 265–272.
3. Zhou, Z.; Goh, Y.M.; Li, Q. Overview and analysis of safety management studies in the construction industry. *Saf. Sci.* **2015**, *72*, 337–350. [CrossRef]
4. Albert, A.; Hallowell, M.R.; Skaggs, M.; Kleiner, B. Empirical measurement and improvement of hazard recognition skill. *Saf. Sci.* **2017**, *93*, 1–8. [CrossRef]
5. Ladewski, B.J.; Al-Bayati, A.J. Quality and safety management practices: The theory of quality management approach. *J. Saf. Res.* **2019**, *69*, 193–200. [CrossRef]
6. Aniekwu, N. Accidents and Safety violations in the Nigerian construction industry. *J. Sci. Technol.* **2007**, *27*, 81–89. [CrossRef]
7. Schulte, A.; Donath, D.; Lange, D.S.; Gutzwiller, R.S. A Heterarchical Urgency-Based Design Pattern for Human Automation Interaction. In Proceedings of the 15th International Conference on Engineering Psychology and Cognitive Ergonomics, Las Vegas, NV, USA, 15–20 July 2018.
8. Ganbat, T.; Chong, H.P.; Liao, P.-C.; Wu, Y.D.; Zhao, X.B. A Bibliometric Review on Risk Management and Building Information Modeling for International Construction. *Adv. Civ. Eng.* **2018**, *2018*, 8351679. [CrossRef]
9. Marvel, J.A.; Falco, J.; Marstio, I. Characterizing task-based human-robot collaboration safety in manufacturing. *IEEE Trans. Syst. Man Cybern. Syst.* **2014**, *45*, 260–275. [CrossRef]
10. Blashfield, R.K.; Kaufman, L.; Rousseeuw, P.J. Finding groups in data—An introduction to cluster-analysis. *J. Classif.* **1991**, *8*, 277–279.
11. Chen, C.; Chen, Y.; Hou, J.; Liang, Y. CiteSpace II: Detecting and Visualizing Emerging Trends and Transient Patterns in Scientific Literature. *J. China Soc. Sci. Tech. Inf.* **2009**, *28*, 401–421. [CrossRef]
12. Borys, D. The role of safe work method statements in the Australian construction industry. *Saf. Sci.* **2012**, *50*, 210–220. [CrossRef]
13. Zhang, M.; Cao, T.; Zhao, X. Applying sensor-based technology to improve construction safety management. *Sensors* **2017**, *17*, 1841. [CrossRef]
14. Farahani, S.; Tahershamsi, A.; Behnam, B. Earthquake and post-earthquake vulnerability assessment of urban gas pipelines network. *Nat. Hazards* **2020**, *101*, 327–347. [CrossRef]
15. Joyner, M.D.; Sasani, M. Building performance for earthquake resilience. *Eng. Struct.* **2020**, *210*, 110371. [CrossRef]
16. Fu, G.; Xie, X.C.; Jia, Q.S.; Tong, W.Q.; Ge, Y. Accidents analysis and prevention of coal and gas outburst: Understanding human errors in accidents. *Process. Saf. Environ. Prot.* **2020**, *134*, 1–23. [CrossRef]
17. Yeo, C.J.; Yu, J.H.; Kang, Y. Quantifying the Effectiveness of IoT Technologies for Accident Prevention. *J. Manag. Eng.* **2020**, *36*, 04020054. [CrossRef]
18. Vähä, P.; Känsälä, K.; Heikkilä, R. Use of 3-D product models in construction process automation. *Autom. Constr.* **1997**, *6*, 69–76. [CrossRef]
19. Saidi, K.S.; Bock, T.; Georgoulas, C. Robotics in construction. In *Springer Handbook of Robotics*; Springer: Cham, Switzerland, 2016; pp. 1493–1520.
20. Guo, H.; Yu, Y.; Skitmore, M. Visualization technology-based construction safety management: A review. *Autom. Constr.* **2017**, *73*, 135–144. [CrossRef]
21. Ikeda, Y. The automated building construction system for high-rise steel structure building. In Proceedings of the Council on Tall Buildings and Urban Habitat (CTBUH) 2004, Seoul, Korea, 10–13 October 2004.
22. Yamazaki, Y.; Maeda, J. The SMART system: An integrated application of automation and information technology in production process. *Comput. Ind.* **1998**, *35*, 87–99. [CrossRef]
23. Wakisaka, T.; Furuya, N.; Inoue, Y. Automated construction system for high-rise reinforced concrete buildings. *Autom. Constr.* **2000**, *9*, 229–250. [CrossRef]
24. Bock, T. Construction robotics. *Auton. Robot.* **2007**, *22*, 201–209. [CrossRef]
25. Schmidhuber, J. Deep learning in neural networks: An overview. *Neural Netw.* **2015**, *61*, 85–117. [CrossRef]
26. Lee, S.; Moon, J.I. Introduction of human-robot cooperation technology at construction sites. In Proceedings of the 31st International Symposium on Automation and Robotics in Construction and Mining (ISARC 2014), Sydney, Australia, 9–11 July 2014; Volume 31, pp. 1–6.
27. Sobotka, A.; Pacewicz, K. Mechanisation and automation technologies development in work at construction sites. *IOP Conf. Ser. Mater. Sci. Eng.* **2017**, *251*, 012046. [CrossRef]
28. Nurmikko, A.V.; Donoghue, J.P.; Hochberg, L.R.; Patterson, W.R.; Song, Y.K.; Bull, C.W.; Shang, D.A.; Zhang, H.X.; Zhu, L.; Sun, J.D. Adversarial cross-modal retrieval based on dictionary learning. *Neurocomputing* **2019**, *355*, 93–104.
29. Ma, C.; Luo, G.N.; Wang, K.Q. Concatenated and Connected Random Forests with Multiscale Patch Driven Active Contour Model for Automated Brain Tumor Segmentation of MR Images. *IEEE Trans. Med. Imaging* **2018**, *37*, 1943–1954. [CrossRef]

30. Lantzanakis, G.; Mitraka, Z.; Chrysoulakis, N. X-SVM: An Extension of C-SVM Algorithm for Classification of High-Resolution Satellite Imagery. *IEEE Trans. Geosci. Remote. Sens.* **2021**, *59*, 3805–3815. [CrossRef]
31. Zhang, H.; Wang, T.; Dai, G. Semi-supervised cross-modal common representation learning with vector-valued manifold regularization. *Pattern Recognit. Lett.* **2020**, *130*, 335–344. [CrossRef]
32. Waltman, L.; Eck, N.J.; Noyons, E.C.M. A unified approach to mapping and clustering of bibliometric networks. *J. Inf.* **2010**, *4*, 629–635. [CrossRef]
33. Guo, B.H.W.; Zou, Y.; Fang, Y.; Goh, Y.M.; Zou, P.X.W. Computer vision technologies for safety science and management in construction: A critical review and future research directions. *Saf. Sci.* **2021**, *135*, 105130. [CrossRef]
34. Luo, H.; Wang, M.; Wong, P.K.Y.; Tang, J.; Cheng, J.C.P. Construction machine pose prediction considering historical motions and activity attributes using gated recurrent unit (GRU). *Autom. Constr.* **2021**, *121*, 103444. [CrossRef]
35. Hossain, M.M.; Prybutok, V.R. Towards developing a business performance management model using causal latent semantic analysis. *Int. J. Bus. Perform. Manag.* **2016**, *17*, 161–183. [CrossRef]
36. Wang, H.T.; Song, L.H.; Xiang, T.T.; Liu, L.J. New Development Direction of Artificial Intelligence—Human Cyber Physical Ternary Fusion Intelligence. *Chin. Comput. Sci.* **2020**, *47*, 1–5+22.
37. Yan, X. Development of ergonomic posture recognition technique based on 2D ordinary camera for construction hazard prevention through view-invariant features in 2D skeleton motion. *Adv. Eng. Inform.* **2017**, *34*, 152–163. [CrossRef]
38. Kowalski-Trakofler, K.M.; Barrett, E.A. The concept of degraded images applied to hazard recognition training in mining for reduction of lost-time injuries. *J. Saf. Res.* **2003**, *34*, 515–525. [CrossRef] [PubMed]
39. Zhang, H.; Yan, X.; Li, H. Ergonomic posture recognition using 3D view-invariant features from single ordinary camera. *Autom. Constr.* **2018**, *94*, 1–10. [CrossRef]
40. Hasanzadeh, S.; Esmaeili, B.; Michael, D. Impact of Construction Workers' Hazard Identification Skills on Their Visual Attention. *J. Constr. Eng. Manag.* **2017**, *143*, 04017070. [CrossRef]
41. Akanmu, A.A.; Olayiwola, J.; Ogunseiju, O.; McFeeters, D. Cyber-physical postural training system for construction workers. *Autom. Constr.* **2020**, *117*, 103272. [CrossRef]
42. Inyang, N.; Al-Hussein, M. Ergonomic hazard quantification and rating of residential construction tasks C3. In Proceedings of the Annual Conference Canadian Society for Civil Engineering, Ottawa, ON, Canada, 14–17 June 2011.
43. Kim, K.; Chen, J.; Cho, Y.K. Evaluation of Machine Learning Algorithms for Worker's Motion Recognition Using Motion Sensors. In Proceedings of the Computing in Civil Engineering 2019: Data, Sensing, and Analytics, Atlanta, GA, USA, 17–19 June 2019.
44. Asadzadeh, A.; Arashpour, M.; Li, H.; Ngo, T.; Bab-Hadiashar, A.; Rashidi, A. Sensor- based safety management. *Autom. Constr.* **2020**, *113*, 103128. [CrossRef]
45. Hu, Z.; Zhang, J.; Zhang, X. Construction collision detection for site entities based on 4-D space time model. *J. Tsinghua Univ. Sci. Technol.* **2010**, *50*, 820–825.
46. Zhang, S.; Teizer, J.; Lee, J.K. Building information modeling (BIM) and safety: Automatic safety checking of construction models and schedules. *Autom. Constr.* **2013**, *29*, 183–195. [CrossRef]
47. Benjaoran, V.; Bhokha, S. An integrated safety management with construction management using 4D CAD model. *Saf. Sci.* **2010**, *48*, 395–403. [CrossRef]
48. Huang, D.; Wang, X.; Li, J.; Tang, W. Virtual Reality for Training and Fitness Assessments for Construction Safety. In Proceedings of the 2020 International Conference on Cyberworlds (CW), Caen, France, 29 September–1 October 2020.
49. Skibniewski, M.J.; Jang, W.S. Simulation of accuracy performance for wireless sensor-based construction asset tracking. *Comput.-Aided Civ. Infrastruct. Eng.* **2009**, *24*, 335–345. [CrossRef]
50. Newman, M.E.J. Modularity and community structure in networks. *Proc. Natl. Acad. Sci. USA* **2006**, *103*, 8577–8582. [CrossRef] [PubMed]
51. Moon, H.; Dawood, N.; Kang, L. Development of workspace conflict visualization system using 4D object of work schedule. *Adv. Eng. Inform.* **2014**, *28*, 50–65. [CrossRef]
52. Zhang, J.P.; Hu, Z.Z. BIM-and 4D-based integrated solution of analysis and management for conflicts and structural safety problems during construction: 1. Principles and methodologies. *Autom. Constr.* **2011**, *20*, 155–166. [CrossRef]
53. Pradhananga, N.; Teizer, J. Automatic spatio-temporal analysis of construction site equipment operations using GPS data. *Autom. Constr.* **2013**, *29*, 107–122. [CrossRef]
54. Montaser, A.; Moselhi, O. RFID indoor location identification for construction projects. *Autom. Constr.* **2014**, *39*, 167–179. [CrossRef]
55. Yang, J.J.; Ye, G.; Xiang, Q.T.; Kim, M.; Liu, Q.J.; Yue, H.Z. Insights into the mechanism of construction workers' unsafe behaviors from an individual perspective. *Saf. Sci.* **2021**, *133*, 105004. [CrossRef]
56. Li, H.; Chan, G.; Wong, J.K.W. Real-time locating systems applications in construction. *Autom. Constr.* **2016**, *63*, 37–47. [CrossRef]
57. Maalek, R.; Sadeghpour, F. Accuracy assessment of Ultra-Wide Band technology in tracking static resources in indoor construction scenarios. *Autom. Constr.* **2013**, *30*, 170–183. [CrossRef]
58. Wu, F.; Lu, X.; Song, J.; Yan, S.; Zhang, Z.M.; Rui, Y.; Zhuang, Y. Learning of Multimodal Representations with Random Walks on the Click Graph. *IEEE Trans. Image Process.* **2016**, *25*, 630–642. [CrossRef] [PubMed]

59. Yao, T.; Pan, Y.W.; Li, Y.H.; Mei, T. Incorporating Copying Mechanism in Image Captioning for Learning Novel Objects. In Proceedings of the 30th IEEE Conference on Computer Vision and Pattern Recognition, Honolulu, HI, USA, 22–25 July 2017; pp. 5263–5271.
60. Xia, Y.J.; Zhang, L.M.; Liu, Z.G.; Nie, L.Q.; Li, X.L. Weakly Supervised Multimodal Kernel for Categorizing Aerial Photographs. *IEEE Trans. Image Process.* **2017**, *26*, 3748–3758. [CrossRef] [PubMed]
61. Li, H.B.; Sun, J.; Xu, Z.B.; Chen, L.M. Multimodal 2D + 3D Facial Expression Recognition with Deep Fusion Convolutional Neural Network. *IEEE Trans. Multimed.* **2017**, *19*, 2816–2831. [CrossRef]
62. Meng, X.R.; Xu, X.Z.; Zhao, G.M. True three-dimensional coal mine personnel positioning system based on 3D visualization and Zigbee technology. *J. China Coal Soc.* **2014**, *39*, 603–608.
63. Jeelani, I.; Albert, A.; Han, K.; Azevedo, R. Are Visual Search Patterns Predictive of Hazard Recognition Performance? Empirical Investigation Using Eye-Tracking Technology. *J. Constr. Eng. Manag.* **2018**, *145*, 111–113. [CrossRef]
64. Sharafi, Z.; Soh, Z.; Guéhéneuc, Y. A systematic literature review on the usage of eye-tracking in software engineering. *Inf. Softw. Technol.* **2015**, *67*, 79–107. [CrossRef]
65. Hasanzadeh, S.; Esmaeili, B.; Dodd, M.D. Examining the Relationship between Construction Workers' Visual Attention and Situation Awareness under Fall and Tripping Hazard Conditions: Using Mobile Eye Tracking. *J. Constr. Eng. Manag.* **2018**, *144*, 04018060. [CrossRef]
66. Dolezalova, J.; Popelka, S. Evaluation of the User Strategy on 2D and 3D City Maps Based on Novel Scan-Path Comparison Method and Graph Visualization. *Int. Arch. Photogramm. Remote Sens. Spat. Inf.* **2016**, *XLI-B2*, 637–640. [CrossRef]
67. Xu, Q.; Chong, H.-Y.; Liao, P.-C. Exploring eye-tracking searching strategies for construction hazard recognition in a laboratory scene. *Saf. Sci.* **2019**, *120*, 824–832. [CrossRef]
68. Xu, Q.; Chong, H.-Y.; Liao, P.-C. Collaborative information integration for construction safety monitoring. *Autom. Constr.* **2019**, *102*, 120–134. [CrossRef]
69. Siddharth, B.; Alex, A.; Matthew, R. Construction Hazard Recognition: Themes in Scientific Research. In Proceedings of the Construction Research Congress 2020: Safety, Workforce, and Education, Tempe, AZ, USA, 8–10 March 2020.
70. Bhandari, S.; Hallowell, M.R.; Correll, J. Making construction safety training interesting: A field-based quasi-experiment to test the relationship between emotional arousal and situational interest among adult learners. *Saf. Sci.* **2019**, *117*, 58–70. [CrossRef]
71. Lingard, H. Occupational health and safety in the construction industry. *Constr. Manag. Econ.* **2013**, *31*, 505–514. [CrossRef]
72. Albert, A.; Hallowell, M.R.; Kleiner, B.M. Enhancing construction hazard recognition and communication with energy-based cognitive mnemonics and safety meeting maturity model: Multiple baseline study. *J. Constr. Eng. Manag.* **2013**, *140*, 04013042. [CrossRef]
73. Haslam, R.A.; Hide, S.A.; Gibb, A.G.F.; Gyi, D.E.; Pavitt, T.; Atkinson, S.; Duff, A.R. Contributing factors in construction accidents. *Appl. Ergon.* **2005**, *36*, 401–415. [CrossRef]
74. Tixier, A.J.; Hallowell, M.R.; Albert, A.; Boven, L.; Kleiner, B.M. Psychological antecedents of risk-taking behavior in construction. *J. Constr. Eng. Manag.* **2014**, *140*, 04014052. [CrossRef]
75. Wang, D.; Chen, J.; Zhao, D.; Dai, F.; Zheng, C.; Wu, X. Monitoring workers' attention and vigilance in construction activities through a wireless and wearable electroencephalography system. *Autom. Constr.* **2017**, *82*, 122–137. [CrossRef]
76. Domdouzis, K.; Kumar, B.; Anumba, C. Radio-Frequency Identification (RFID) applications: A brief introduction. *Adv. Eng. Inform.* **2007**, *21*, 350–355. [CrossRef]
77. Jiang, S.; Skibniewski, M.J.; Yuan, Y. Ultra-wide band applications in industry: A critical review. *J. Civ. Eng. Manag.* **2011**, *17*, 437–444. [CrossRef]
78. Hasanzadeh, S.; Esmaeili, B.; Dodd, M. Measuring construction workers' real-time situation awareness using mobile eye-tracking. In Proceedings of the Construction Research Congress, San Juan, Puerto Rico, 31 May–2 June 2016; pp. 2894–2904.
79. Aryal, A.; Ghahramani, A.; Becerik-Gerber, B. Monitoring fatigue in construction workers using physiological measurements. *Autom. Constr.* **2017**, *82*, 154–165. [CrossRef]
80. Brito, R.; Pradhananga, N.; Jianu, R.; Orabi, W. Eye-Tracking Technology for Construction Safety: A Feasibility Study. In Proceedings of the International Symposium on Automation and Robotics in Construction, Auburn, AL, USA, 18–21 July 2016; pp. 282–290.
81. Chen, J.; Song, X.; Lin, Z. Revealing the "invisible gorilla" in construction: Estimating construction safety through mental workload assessment. *Autom. Constr.* **2016**, *63*, 173–183. [CrossRef]
82. Wang, D.; Li, H.; Chen, J. Detecting and measuring construction workers' vigilance through hybrid kinematic-EEG signals. *Autom. Constr.* **2019**, *100*, 11–23. [CrossRef]
83. International Organization for Standardization. *Robots and Robotic Devices-Safety Requirements for Industrial Robots—Part 2: Robots Systems and Integration*; International Organization for Standardization: Geneva, Switzerland, 2011.
84. Jamil, S.M.; Albert, A.; Alsharef, A.; Pandit, B.; Patil, Y. Hazard Recognition Patterns Demonstrated by Construction Workers. *Int. J. Environ. Res. Public Health* **2020**, *17*, 7788.
85. Derlukiewicz, D.; Ptak, M.; Koziołek, S. Proactive failure prevention by human-machine interface in remote-controlled demolition robots. *New Adv. Inf. Syst. Technol.* **2016**, *1*, 711–720.
86. Tanzini, M.; Tripicchio, P.; Ruffaldi, E. A novel human-machine interface for working machines operation. In Proceedings of the IEEE RO-MAN, Gyeongju, Korea, 26–29 August 2013; Volume 1, pp. 744–750.

87. Patrinostro, S.; Tanzini, M.; Satler, M. A haptic-assisted guidance system for working machines based on virtual force fields. In Proceedings of the XXV International Conference on Information, Communication and Automation Technologies, Sarajevo, Bosnia and Herzegovina, 29–31 October 2015; Volume 1, pp. 1–6.
88. Jacinto-Villegas, J.M.; Satler, M.; Filippeschi, A. A novel wearable haptic controller for teleoperating robotic platforms. *IEEE Robot. Autom. Lett.* **2017**, *2*, 2072–2079. [CrossRef]
89. Moon, J.; Son, Y.; Park, S. Development of immersive augmented reality interface for construction robotic system. In Proceedings of the International Conference on Control, Automation and Systems, Seoul, Korea, 17–20 October 2007; Volume 1, pp. 1192–1197.
90. Zhu, Z.; Ren, X.; Chen, Z. Integrated detection and tracking of workforce and equipment from construction jobsite videos. *Autom. Constr.* **2017**, *81*, 161–171. [CrossRef]

Article

Study of Pavement Micro- and Macro-Texture Evolution Due to Traffic Polishing Using 3D Areal Parameters

Yiwen Zou [1,*], Guangwei Yang [2,*], Wanqing Huang [3], Yang Lu [1], Yanjun Qiu [1] and Kelvin C. P. Wang [2]

[1] School of Civil Engineering, Southwest Jiaotong University, Chengdu 610031, China; luyang@home.swjtu.edu.cn (Y.L.); publicqiu@vip.163.com (Y.Q.)

[2] School of Civil and Environmental Engineering, Oklahoma State University, Stillwater, OK 74078, USA; wangchenping@swjtu.cn

[3] Sichuan Communication Surveying & Design Institute Co., Ltd., Chengdu 610041, China; hangwq1978@163.com

* Correspondence: zouyiwen@my.swjtu.edu.cn (Y.Z.); guangwy@okstate.edu (G.Y.)

Abstract: Pavement micro- and macro-texture have significant effects on roadway friction and driving safety. The influence of traffic polish on pavement texture has been investigated in many laboratory studies. This paper conducts field evaluation of pavement micro- and macro-texture under actual traffic polishing using three-dimensional (3D) areal parameters. A portable high-resolution 3D laser scanner measured pavement texture from a field site in 2018, 2019, and 2020. Then, the 3D texture data was decomposed to micro- and macro-texture using Fourier transform and Butterworth filter methods. Twenty 3D areal parameters from five categories, including height, spatial, hybrid, function, and feature parameters, were calculated to characterize pavement micro- and macro-texture. The results demonstrate that the 3D areal parameters provide an alternative to comprehensively characterize the evolution of pavement texture under traffic polish from different aspects.

Keywords: traffic polishing; pavement micro-texture; pavement macro-texture; high-resolution 3D texture image; 3D areal texture parameters

Citation: Zou, Y.; Yang, G.; Huang, W.; Lu, Y.; Qiu, Y.; Wang, K.C.P. Study of Pavement Micro- and Macro-Texture Evolution Due to Traffic Polishing Using 3D Areal Parameters. *Materials* **2021**, *14*, 5769. https://doi.org/10.3390/ma14195769

Academic Editor: Carlos Morón Fernández

Received: 12 September 2021
Accepted: 30 September 2021
Published: 2 October 2021

Publisher's Note: MDPI stays neutral with regard to jurisdictional claims in published maps and institutional affiliations.

1. Introduction

1.1. Background

Pavement skid resistance has positive effects on reducing traffic accidents in dry or wet conditions [1]. For example, the skidding risk would increase rapidly when the pavement skid number (SN) from locked-wheel skid testers is below 50 and decrease significantly when the value of SN is over 65 [2]. Additionally, traffic accidents increased 60% when the value of SN decreased from 48 to 33 [3]. Therefore, it is critical to design pavements surface with high skid resistance and wear resistance to ensure driving safety during design life [4].

Skid resistance of pavement surface varies under traffic polishing throughout the design life: It typically increases to a peak at the initial stage and decreases continuously at the following stages of pavement life [5,6]. However, the level of skid resistance is dependent on the wear process of pavement surface texture [7], as the skid resistance of pavement comes directly from the contact between vehicle tires and pavement surface micro- and macro- texture [8]. Pavement surface texture is recognized as the dominating factor influencing pavement skid resistance [9].

Therefore, many studies have been performed to investigate how traffic wear affects pavement texture over time in order to evaluate pavement skid resistance under traffic polish. Pavement texture can be categorized into macro-texture (0.5 mm < wavelength < 50 mm) and micro-texture (wavelength < 0.5 mm) based on the wavelength of its components [10]. The pavement macro-texture provides drainage channels when it rains and comprises the hysteretic component of friction, while the pavement micro-texture provides actual contact with the tire and comprises the adhesion part of friction [11].

Typically, measurement of pavement macro-texture adopts the sand patch method, the outflow meter, or the circular texture meter (CTM) using a two-dimensional (2D) texture profile [12]. Indicators like mean texture depth (MTD), mean profile depth (MPD), or root mean square depth (RMSD) is customarily applied to characterize pavement macro-texture [13]. Besides, the pavement micro-texture is evaluated by indirect friction measurement devices testing at low speed, such as the British portable tester (BPT), the dynamic friction tester (DFT), and the locked-wheel skid trailer [12].

Recently, pavement surface micro-texture was measured with high-resolution cameras in the laboratory to obtain more texture details [14]. Moreover, the advanced high-resolution laser device can conveniently collect three-dimensional (3D) surface texture data from the field and achieve enough accuracy to characterize pavement micro- and macro-texture [15]. The acquisition of high-resolution surface texture information can significantly assist the investigation of the micro- and macro-texture contributions to skid resistance [16].

Further, 3D areal surface texture parameters have been utilized extensively in modern manufacturing industries to control and evaluate the surface finishing of products [17]. The 3D areal texture parameters contain aspects of surface height, spatial, hybrid, function, and feature information, whereas the traditional parameters only contain height information [17]. The areal texture parameters can characterize surface texture functionality and understand texture characteristics in different perspectives that the traditional texture parameter fails to achieve [18]. Therefore, some recent studies attempted to evaluate pavement texture using 3D areal parameters and correlate them with skid resistance [19–21]. The study described in this paper used 3D areal parameters to evaluate pavement texture changes under traffic loading.

Many studies have been performed to investigate how traffic polish affects pavement texture over time so that pavement can be constructed with desired texture features to maintain good skid resistance [22]. Several devices were developed to study the wear-resisting feature of pavements in the laboratory, such as the Wehner/Schulze device (W/S) [23], the Aachen polishing machine (APM) [24], and other accelerated polishing machines [25–28]. These devices evaluate the evolution of pavement texture under simulated traffic polishing in controlled laboratory conditions rather than actual traffic polishing from various vehicles.

Further, Kane et al. proposed a polishing model to predict the surface variation with polishing cycles based on laboratory testing using the W/S machine and adopted the roughness parameters (R_q) to validate the model [29]. Druta et al. conducted accelerated polish testing on stone matrix asphalt (SMA) species and found that MPD had completely different changing trends with BPN during the polish process [30]. Wang et al. tried to quantify the effect of aggregate size with W/S device on polishing resistance, and the texture variation characterized by power spectral density (PSD) showed that the coarser aggregate had a significantly rougher texture [5]. Wu and Abadie simulated the wearing process with an accelerated polishing machine and measured MPD with a CTM, and the results indicated that the MPD values tended to remain constant under different polishing cycles [31]. Plati and Pomoni investigated the long-term field data of skid resistance and macro-texture and found that the MPD and grip number (GN) presented a contrary trend under traffic polish [32].

1.2. Research Need

Several limitations in previous studies have been identified and summarized as follows:

(1) Many previous studies evaluated pavement wear performance in the laboratory using polishing machines. In laboratory studies, it is challenging to repeat the actual pavement polish process in the field involving traffic polishing from various vehicle types under different environmental conditions such as temperature, precipitation, or freeze-thaw cycles.

(2) In previous studies, 2D texture profiles were typically collected at the macro-texture level to evaluate pavement texture variation under traffic polish. With the advance-

ment of pavement data collection, high-resolution 3D texture data should be applied to understand better the evolution of asphalt pavement micro- and macro-texture under traffic polish.

(3)   Traditional pavement texture parameters only consider texture height distribution while miss other texture characteristics (such as spatial, hybrid, and so on). Hence, different categories of 3D areal parameters should be explored to characterize pavement micro- and macro-texture under traffic polishing from different aspects.

Therefore, it is necessary to conduct field studies to understand pavement micro- and macro-texture evolution under actual traffic polish using different 3D areal texture parameters.

1.3. Objective

In this study, a field asphalt pavement site was monitored from 2018 to 2020 to study the influence of traffic polish on the evolution of pavement micro- and macro-texture. Pavement 3D texture images were collected using an LS-40 3D laser scanner (HyMIT Measurement Instrument Technology, Austin, TX, USA) for three years. Then, pavement micro- and macro-texture were separated from the obtained 3D texture data using two-dimensional discrete Fourier Transform method and Butterworth filters method 21. Next, twenty 3D areal parameters from five categories (height, spatial, hybrid, function, and feature) were calculated for pavement 3D micro- and macro-texture. The obtained micro- and macro-texture parameters in three years were analyzed to describe the evolution of surface texture under actual traffic polish from different perspectives.

2. Field Data Collection

This study selected the field site on a suburb road paved with dense-graded asphalt mixture (HMA-13) in Yongning avenue, located in an industrial district of Chengdu, China. There were about 3 million passage cars on this site in 2019, and the traffic volume had an 11.9% growth in 2020. The site was constructed in 2018 and monitored until 2020 for pavement texture variations under actual traffic polishing.

Figure 1 shows example pavement images from this site in 2018, 2019, and 2020, individually. Most of the aggregates were coated with bitumen in 2018 when the pavement was just constructed (Figure 1a). After traffic polish from 2018 to 2019, the bitumen layer was gradually removed, and the coarse aggregate was exposed to field environmental effect and traffic polishing and compacting, as shown in Figure 1b. Further, the texture of coarse aggregates in 2020 looked smoother and more aging than that in 2019 due to traffic polish (see Figure 1c).

Figure 2a shows the LS-40 Portable Surface Analyzer (LS-40, HyMIT Measurement Instrument Technology, Austin, TX, USA) that was used to record 3D texture images on this site to quantify pavement texture evolution due to traffic polish. The LS-40 scans a 102.4×102.4 mm pavement surface area with height resolution (z) at 0.01 mm and lateral resolution (x, y) at 0.05 mm. From 2018 to 2020, a total number of 42 3D texture images were obtained from the wheel path during each data collection from previously marked locations on this site.

The 3D texture data collected by LS-40 was denoised (see Figure 2b) by a Gaussian smoothing filter with a kernel size of 5×5. Then the Fourier transform converted the texture height data into the frequency domain, and the Butterworth filter separated texture components in the frequency domain into micro-texture and macro-texture at a boundary of 2 Hz. Subsequently, the inverse Fourier transform converted the frequency domain micro-texture and macro-texture data back to texture height data, respectively, as shown in Figure 2c,d. The detailed procedure of texture data processing was published in a previous research [21].

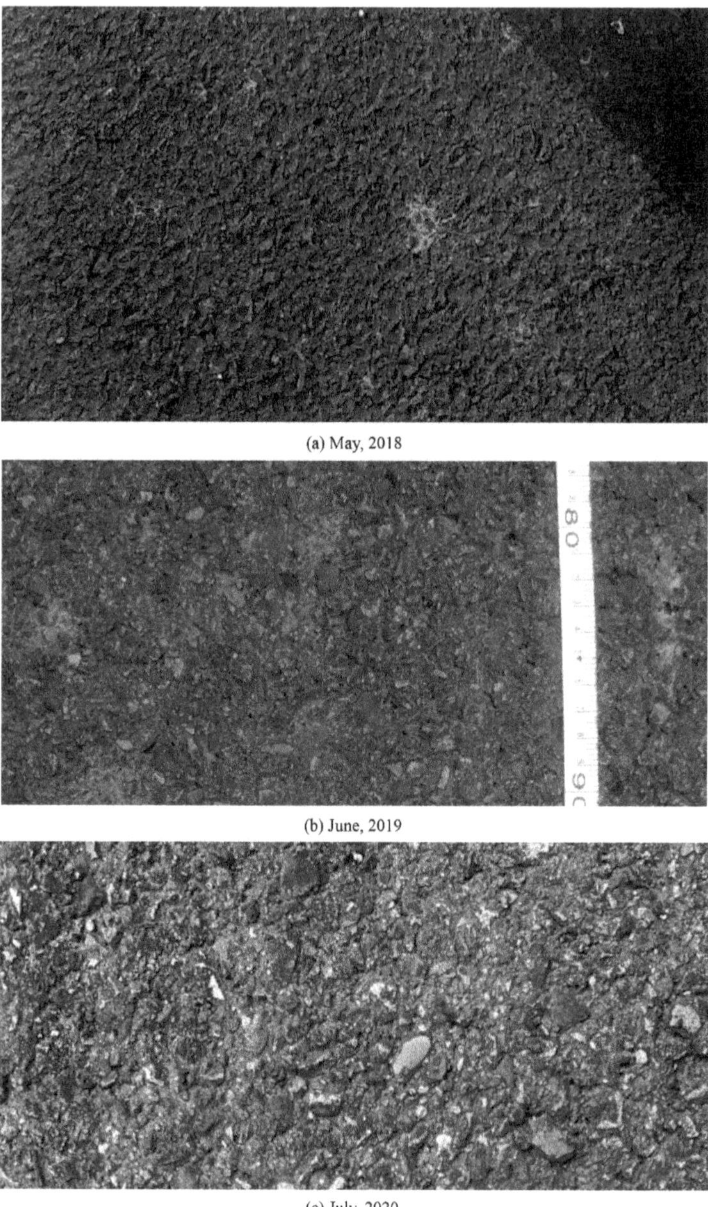

(a) May, 2018

(b) June, 2019

(c) July, 2020

Figure 1. Pavement surface pictures from the field site: (**a**) May 2018, (**b**) June 2019, and (**c**) July 2020.

Figure 3 shows an example of how 3D pavement micro- and macro-texture changed from 2018 to 2020. It is noteworthy that the height of macro-texture was decreasing over time and tiny stripes formed on the micro-texture along driving direction in 2020. To characterize the evolution of micro- and macro-texture, quantitative analysis is conducted in the following section via 3D areal parameters.

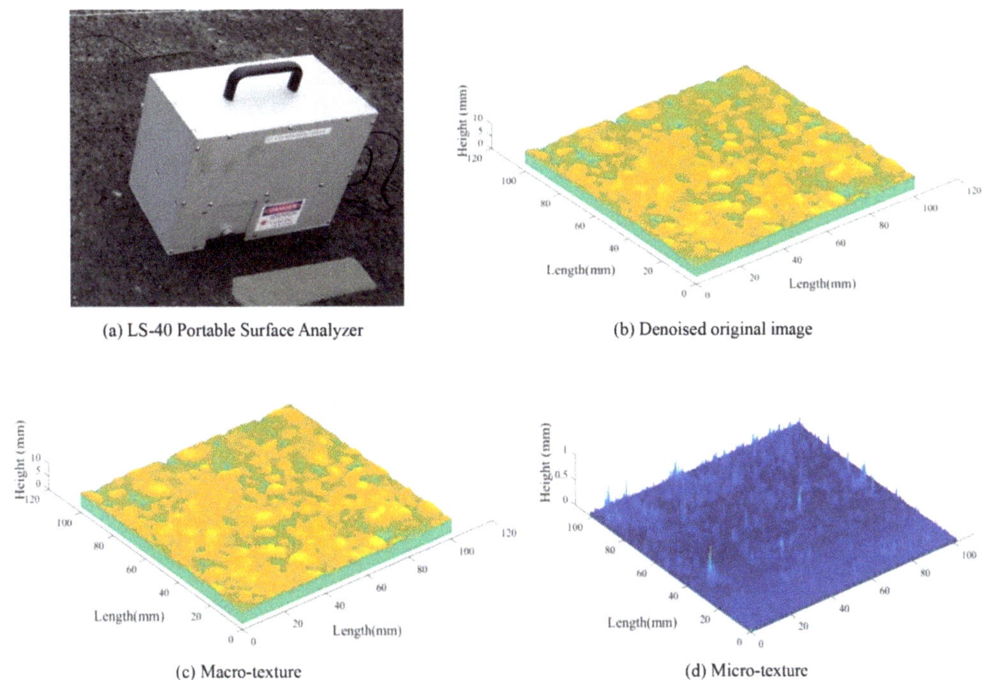

(a) LS-40 Portable Surface Analyzer

(b) Denoised original image

(c) Macro-texture

(d) Micro-texture

Figure 2. LS-40 and examples 3D texture data: (**a**) LS-40 Portable Surface Analyzer, (**b**) Denoised original image, (**c**) Macro-texture, and (**d**) Micro-texture.

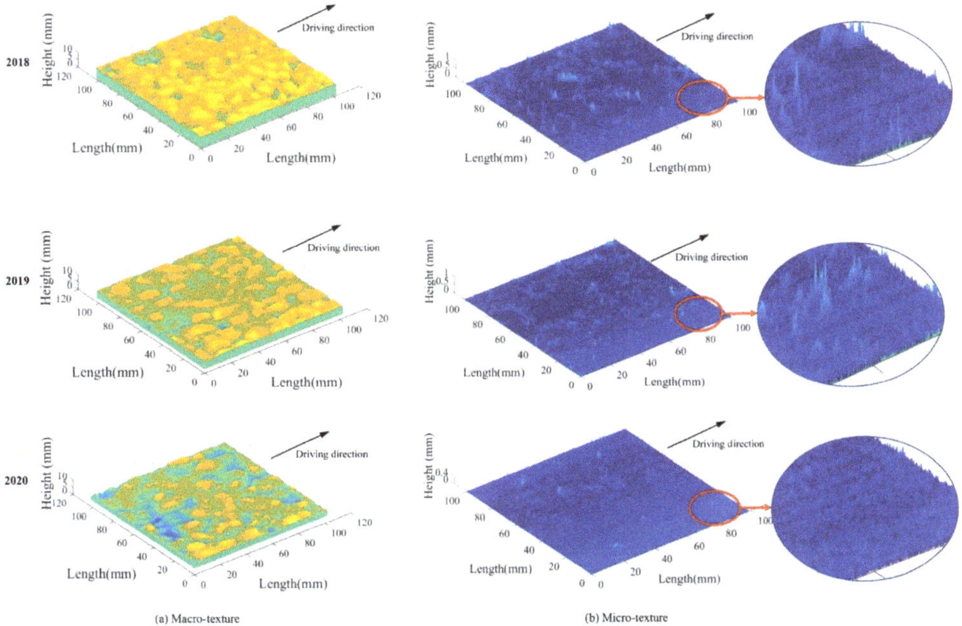

Figure 3. Evolution of pavement micro- and macro-texture: (**a**) Macro-texture, and (**b**) Micro-texture.

3. Three Dimensional Areal Parameters

In this section, twenty 3D areal texture parameters from five categories (height, spatial, hybrid, functional, and feature) were calculated for both 3D micro- and macro-texture to investigate the evolution of pavement surface texture under actual traffic polishing and environmental impacts. The category, name, and unit of these 3D areal parameters are summarized in Table 1. The detailed definition of these parameters is introduced as follows.

Table 1. Summary of 3D areal parameters.

Category			Parameters	Unit
	Height Parameters		arithmetic mean height (S_a)	mm
			root mean square height (S_q)	mm
			skewness (S_{sk})	-
			kurtosis (S_{ku})	-
	Spatial Parameters		autocorrelation length (S_{al})	mm
			texture aspect ratio (S_{tr})	-
			texture direction (S_{td})	rad
	Hybrid Parameters		root mean square gradient (S_{dq})	-
			developed interfacial area ratio (S_{dr})	%
Function Related Parameters	Material Ratio Parameters		peak extreme height (S_{xp})	mm
			surface section difference (S_{dc})	mm
	Stratified Parameters		reduced peak height (S_{pk})	mm
			core height (S_k)	mm
			reduced dale height (S_{vk})	mm
	Volume Parameters		peak material volume (V_{mp})	mm^3
			core material volume (V_{mc})	mm^3
			core void volume (V_{vc})	mm^3
			dales void volume (V_{vv})	mm^3
	Feature Parameters		peak density (S_{pd})	mm^{-2}
			peak curvature (S_{pc})	mm^{-1}

3.1. Height Parameters

The height parameters consider the surface height information, but neglect the horizontal input. In this section, four height parameters, including arithmetic mean height (S_a), root mean square height (S_q), skewness (S_{sk}), and kurtosis (S_{ku}), were calculated per Equations (1)–(4) [10]. The S_a and S_q measure the overall height of a surface and correlate intensely with each other, and the S_{sk} and S_{ku} describe the shape of the surface probability density [10]. The S_{sk} indicates the symmetry of the height probability density curve, and the S_{ku} characterizes the kurtosis of the probability density curve. Pavement surface with positive S_{sk} would have spike structure, and surface with negative S_{sk} would have valley structure. Moreover, a higher S_{ku} implies more significant height variation of surface peaks or valleys. Significantly, the S_{sk} is 0.0 and the S_{ku} is 3.0 when the surface probability density function is Gaussian distribution [33].

$$S_a = \sqrt{\frac{1}{A} \iint |z(x,y)| \, dxdy} \tag{1}$$

$$S_q = \sqrt{\frac{1}{A} \iint z(x,y)^2 \, dxdy} \tag{2}$$

$$S_{sk} = \frac{1}{A \, S_q^3} \iint z(x,y)^3 \, dxdy \tag{3}$$

$$S_{ku} = \frac{1}{A S_q^4} \iint z(x,y)^4 \, dxdy \tag{4}$$

where A is the area of a 3D image; z is the height value of pixels in a 3D image; x and y are the horizontal coordinates of pixels in a 3D image.

3.2. Spatial Parameters

The calculation of spatial parameters involves the autocorrelation function (ACF) of a 3D texture surface. The ACF calculates the similar degree of a surface z(x, y) and the duplicate surface z(x-τ_x, y-τ_y) with a horizontal shift (τ_x, τ_y) [17]. Equation (2) shows the function to calculate ACF, and Figure 4a shows the ACF of an obtained LS-40 data as an example. For instance, the ACF is 1.0 when the LS-40 data has a horizontal shift (0, 0), the ACF equals 0.2 when the LS-40 data has a horizontal change along the red circle highlighted in Figure 4a.

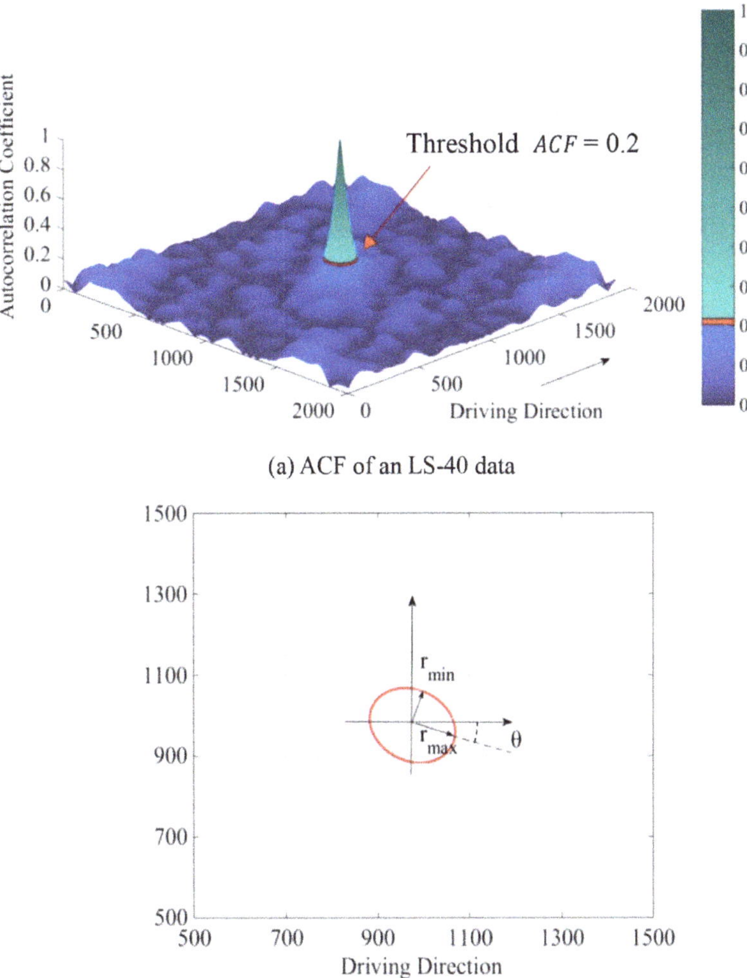

(a) ACF of an LS-40 data

(b) r_{min}, r_{max}, and θ when $ACF = 0.2$

Figure 4. Calculation of spatial parameters: (**a**) ACF of an LS-40 data, and (**b**) r_{min}, r_{max}, and θ when ACF = 0.2.

The spatial parameters, including autocorrelation length (S_{al}), texture aspect ratio (S_{tr}), and texture direction (S_{td}), were obtained for each LS-40 data per Equations (5)–(8) using 0.2 as the threshold of ACF. Figure 4b illustrates an example of how to calculate r_{min}, r_{max}, and θ from the red circle when ACF = 0.2 [10]. The S_{al} is defined as the horizontal distance r_{min} that has the fastest decay to ACF = 0.2 [34]. Additionally, the S_{tr} is calculated as the ratio of the fastest decay distance r_{min} to the slowest decay distance r_{max}, which is the most critical indicator to characterize isotropy of surface texture in the horizontal direction. The surface is isotropic when S_{tr} equals 1, and the surface is anisotropic when S_{tr} equals 0. Further, the S_{td} gives the angle of r_{max} for a surface texture, as shown in Figure 4b.

$$ACF(\tau_x, \tau_y) = \frac{\iint z(x,y)z(x - \tau_x, y - \tau_y)dxdy}{\iint z^2(x,y)dxdy} \tag{5}$$

$$S_{al} = \min_{\tau_x, \tau_y \in R} \sqrt{\tau_x^2 + \tau_y^2} = r_{min} \tag{6}$$

where $R = \{(\tau_x, \tau_y) : ACF(\tau_x, \tau_y) \leq 0.2\}$.

$$S_{tr} = \frac{r_{min}}{r_{max}} \tag{7}$$

$$S_{td} = \theta \tag{8}$$

where τ_x and τ_y are the shifting along and perpendicular to the driving direction, respectively; r_{max} and r_{min} are the slowest and fastest decay distance to ACF = 0.2, respectively; θ is the angle between r_{max} and driving direction.

3.3. Hybrid Parameters

The hybrid parameters describe the height and spacing information of a surface texture. They measure the angular slope of the 3D profile and are handy for assessing friction, adhesion, vibration, etc. The root mean square gradient (S_{dq}) and developed interfacial area ratio (S_{dr}) were calculated per Equations (9) and (10) as the hybrid parameters based on the surface local gradient. The S_{dq} and S_{dr} can be utilized to assess surface cosmetic flatness and correlate to adhesion property [35]. A flat surface would have both S_{dq}, and S_{dr} value equals 0. Besides, a 45° inclined surface would have a S_{dq} value of 1 and a S_{dq} value of 41.4%.

$$S_{dq} = \sqrt{\frac{1}{A} \iint \left(\frac{\partial z}{\partial x}\right)^2 + \left(\frac{\partial z}{\partial y}\right)^2 dxdy} \tag{9}$$

$$S_{dr} = \frac{1}{A}\left\{\iint \left[\sqrt{1 + \left(\frac{\partial z}{\partial x}\right)^2 + \left(\frac{\partial z}{\partial y}\right)^2} - 1\right]dxdy\right\} \times 100\% \tag{10}$$

3.4. Function Parameters

The function parameters are strongly associated with surface functions, such as wearing, bearing, and hydroplaning. Three sub-categories of functional parameters, material ratio, stratified, and volume, were calculated for 3D micro- and macro-texture data based on the cumulative height distribution curve or material ratio curve. The dashed line in Figure 5 shows examples of the cumulative height curve of the 3D texture. The ordinate is the surface height, and the abscissa is the cumulative probability above a certain height. The function parameters characterize peak, core, and valley features of pavement micro- and macro-texture. Details of how to calculate stratified and volume parameters per the cumulative height curve are introduced as follows.

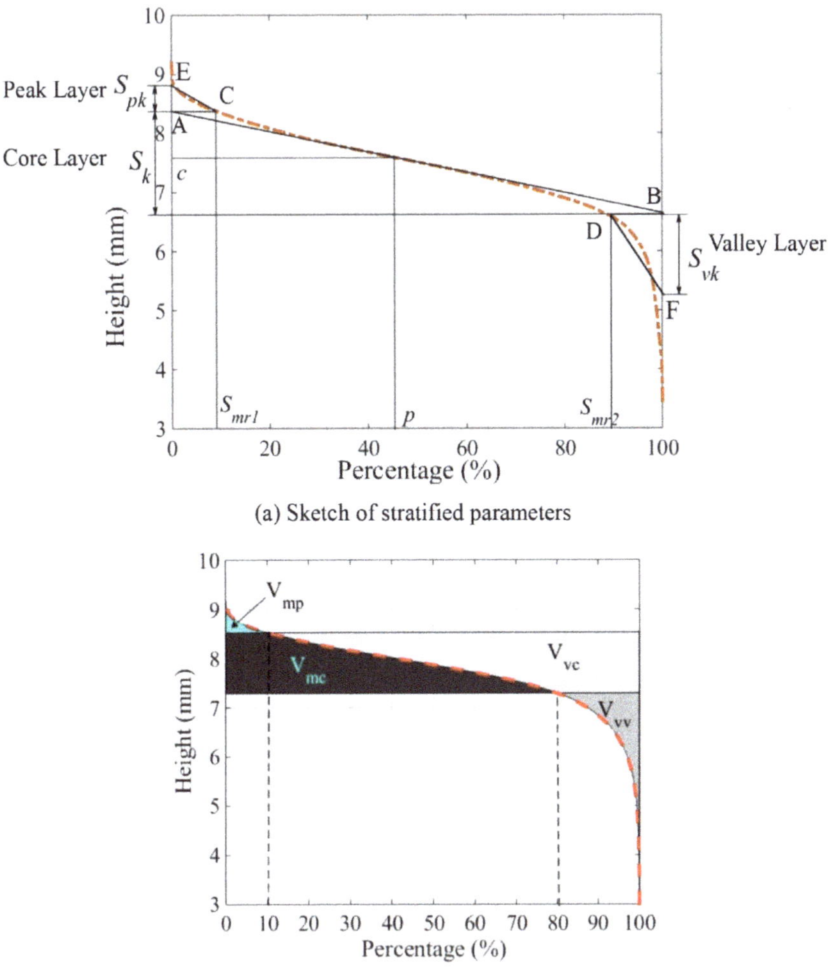

(a) Sketch of stratified parameters

(b) Sketch of volume parameters

Figure 5. Calculation of function parameters: (**a**) Sketch of stratified parameters, and (**b**) Sketch of volume parameters.

3.4.1. Material Ratio Parameters

The areal material ratio parameters employ the peak extreme height (S_{xp}) and the surface section difference (S_{dc}) to characterize the upper half part and the general height of a surface, respectively [18]. The calculation of S_{xp} considers the surface part among the mean plane (50%) and the summit (2.5%). The parameter S_{dc} defines the general height difference of the surface without taking the highest peaks (below 2%) and the lowest valleys (above 98%) into account, as shown in Equations (11) and (12).

$$S_{xp} = S_{mc}(2.5\%) - S_{mc}(50\%) \tag{11}$$

$$S_{dc} = S_{mc}(2\%) - S_{mc}(98\%) \tag{12}$$

where $S_{mc}(p)$ is the height value c corresponding to a material ratio p in Figure 5a.

3.4.2. Stratified Parameters

According to the cumulative height distribution curve, the surface topography is stratified into three parts: peak layer, core layer, and valley layer. The three parts of a surface texture are represented by reduced peak height (S_{pk}), core height (S_k), and reduced dale height (S_{vk}), as shown in Figure 5a [10]. The calculation of stratified parameters in this study requires the following steps, as illustrated in Figure 5a:

- First, a straight line is plotted tangent to the middle part of the dashed cumulative height distribution curve. The tangent line intersects the vertical axes of percentage 0% and 100% at two points, A and B. The corresponding height of A and B are peak and valley thresholds.
- Then, points C and D are projected on the cumulative height distribution curve for heights A and B to define percentages of Smr1 and Smr2. The height difference of A and B is defined as S_k.
- The area enclosed below the cumulative height distribution curve and above AC is represented by the triangle ACE that has an equivalent area. The height difference of E and A is defined as S_{pk}.
- The area enclosed above the cumulative height distribution curve and below BD is represented by the triangle BDF that has an equivalent area. The height difference of F and B is defined as S_{vk}.

Generally, the S_{pk} measures the equivalent height of the surface summit, which is the primary and the most worn surface height. The S_k evaluates the long-term contact height of a surface. The S_{vk} measures the equivalent height of deep grooves, which would hold debris from the upper surface [36].

3.4.3. Volume Parameters

The peak material volume (V_{mp}), core material volume (V_{mc}), core void volume (V_{vc}), and dales void volume (V_{vv}) were calculated as per Equations (13)–(16) [10] as volume parameters and illustrated in Figure 5b. The material ratios, 10% and 80%, are specified as thresholds of the accumulated height to define peak and void of a surface texture [10]. The V_{mp} represents the material volume that is most likely to be removed by traffic polish. Moreover, the V_{mc} measures the material volume polished by traffic but not as much as the V_{mp} is. The V_{vc} is the surface void volume opposite to the V_{mc}. The V_{vv} indicates the void volume with a cumulative height distribution of the lowest 20%.

$$V_{mp} = V_m(10\%) \tag{13}$$

$$V_{mc} = V_m(80\%) - V_m(10\%) \tag{14}$$

$$V_{vc} = V_v(10\%) - V_v(80\%) \tag{15}$$

$$V_{vv} = V_v(80\%) \tag{16}$$

where $V_m(mr)$ is material volume above the height corresponding to a material ratio mr to the highest peak; $V_v(mr)$ is void volume below the height corresponding to a material ratio mr to the lowest valley.

As shown in Figure 5, stratified parameters and volume parameters divide the surface texture into peak, core, and valley with a different method based on the cumulative height distribution curve. To define surface peak and valley, volume parameters use 10% and 80% material ratios, whereas stratified parameters utilize the tangent line of the cumulative height distribution curve to determine mr1 and mr2. Further, volume parameters calculate these layers' material or void volume, and the stratified parameters estimate equivalent height for surface peak or valley layer.

3.5. Feature Parameters

The feature parameters can be used to characterize specified features of surface texture. The peak density, S_{pd}, is calculated by dividing the number of peaks by the unit area, and the peak curvature, S_{pc}, is the arithmetic mean curvature of significant peaks. A peak is selected as the highest pixel within a 16 by 16 nearest neighbors. These two feature parameters can be applied in surface contact models [37].

4. Evolution of Micro- and Macro-Texture

The evolution of pavement micro- and macro-texture was evaluated by comparing 3D areal texture parameters from the three years' data collection on the field site. The Figure 6, Figure 7, Figure 9, and Figures 11–14 summarize the variations of height, spatial, hybrid, function, and feature parameters for pavement micro- and macro-texture under actual traffic polishing. In each figure, the lines with markers display the actual 3D parameters from each data collection, while the bar chart in the upper-right corner shows the average number and standard deviation of each 3D parameter.

4.1. Evolution of Height Parameters

The variations of height parameters for pavement micro- and macro-texture under actual traffic polish are shown in Figure 6. For macro-texture from 2018 to 2020, (1) both S_a and S_q had no significant distinction in mean value and standard deviation, indicating that traffic polishing was not decreasing the macro-texture's height. This result corresponds to a previous study that the MPD values tended to remain constant under different polishing cycles [31]; (2) the negative S_{sk} indicates that pavement macro-texture had valley structure; (3) the declined average S_{ku} means that the height variation of surface peaks or valleys was decreasing.

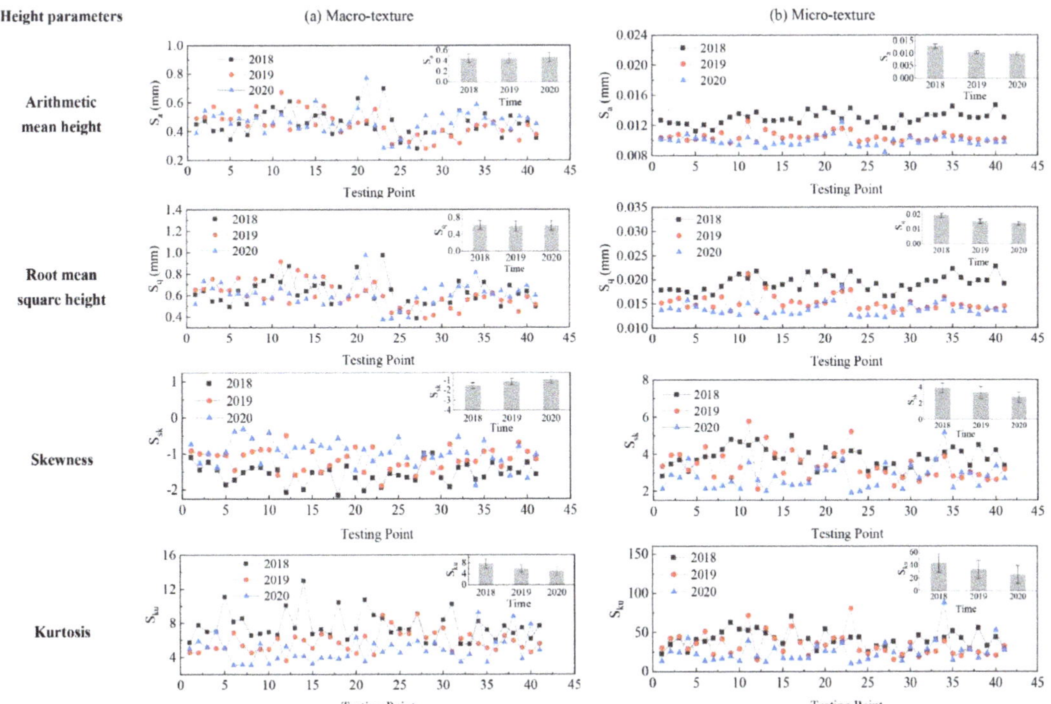

Figure 6. Pavement texture variations via height parameters: (**a**) Macro-texture, and (**b**) Micro-texture.

For micro-texture from 2018 to 2020, (1) the S_a and S_q had an approximate reduction of 20% from 2018 to 2019, and 5% from 2019 to 2020; (2) the S_{sk} were positive and decreased year after year, suggesting the spike structure of micro-texture was decreasing; (3) the S_{ku} was greater than that of macro-texture and gradually reduced, which means the considerable height variation of micro-texture was decreasing as well. The evolution of these height parameters means the spike structure of pavement micro-texture was gradually polished under traffic, as illustrated in Figure 3b.

4.2. Evolution of Spatial Parameters

The variation of spatial parameters for pavement micro- and macro-texture is displayed in Figure 7. For macro-texture from 2018 to 2020, (1) the S_{al} had a 19.5% growth from 2018 to 2019 and stabilized from 2019 to 2020; (2) the S_{tr} was around 0.76 during polish, which means the isotropy of macro-texture was unchanged; (3) the S_{td} was fluctuating around zero. Examples of ACF = 0.2 for macro-texture from 2018 to 2020 are shown in Figure 8a: the shape was stable, meaning the spatial characteristics of macro-texture were not changed from 2018 to 2020 under traffic polish.

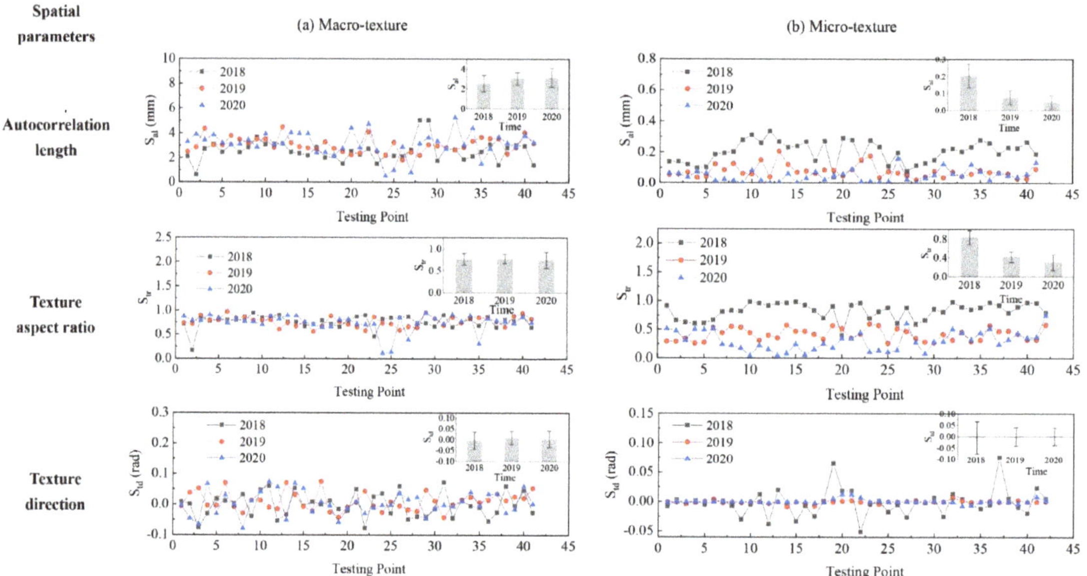

Figure 7. Pavement texture variations via spatial parameters: (**a**) Macro-texture, and (**b**) Micro-texture.

For micro-texture from 2018 to 2020, (1) the S_{al} had a descent of 62.3% from 2018 to 2019, and 36.7% from 2019 to 2020; (2) the S_{tr} decreased year after year, indicating the texture changed from isotropic to anisotropic under traffic polishing; (3) the S_{td} was fluctuating around zero, and its deviation decreased year after year. As shown in Figure 8b, the shape of ACF = 0.2 was round in 2018 and became long and thin in 2020, which corresponded to $S_{tr} = r_{min}/r_{max}$ decreased from 1.0 to 0.

The spatial evolution of micro-texture can be seen intuitively from Figure 3b. The micro-texture asperities were isotropically distributed in 2018, corresponding to $S_{tr} = 1$. The micro-texture asperities were anisotropic distributed along the driving direction: stripes appeared in 2020, and the S_{tr} equals 0. Thus, the spatial parameters successfully characterize how pavement micro-texture evolved from isotropic to anisotropic along driving direction under traffic polish.

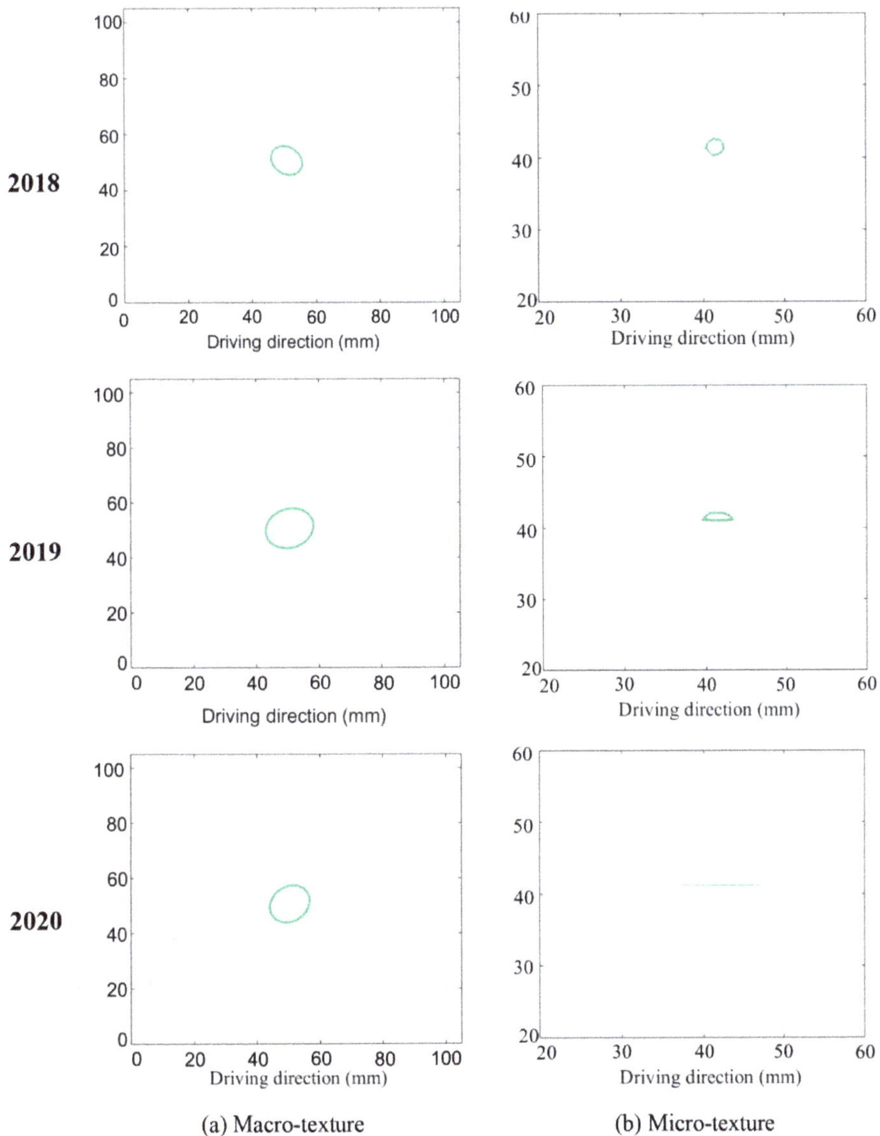

(a) Macro-texture   (b) Micro-texture

Figure 8. ACF = 0.2 for pavement macro- and micro-texture: (**a**) Macro-texture, and (**b**) Micro-texture.

4.3. Evolution of Hybrid Parameters

The variation of hybrid parameters for pavement micro- and macro-texture is displayed in Figure 9. Similar decreasing treads were observed for S_{dq} and S_{dr} from 2018 to 2020. For S_{dq}, a reduction of 46.1% and 32.8% were observed for macro- and micro-texture from 2018 to 2019, and another 16.0% and 11.5% of reduction were observed for macro- and micro-texture from 2019 to 2020. As S_{dq} is getting closer to 0, it means the texture surface is getting close to flat under traffic polish with angular slope decreased. For S_{dr}, reductions of 43.4% and 32.8% were observed for macro- and micro-texture from 2018 to 2019, and another 16.0% and 113.4% of reduction for macro- and micro-texture from 2019 to 2020. The evolution of hybrid parameters suggests that the steepness and the developed interfacial area of pavement micro- and macro-texture were decreased year after year under traffic polish.

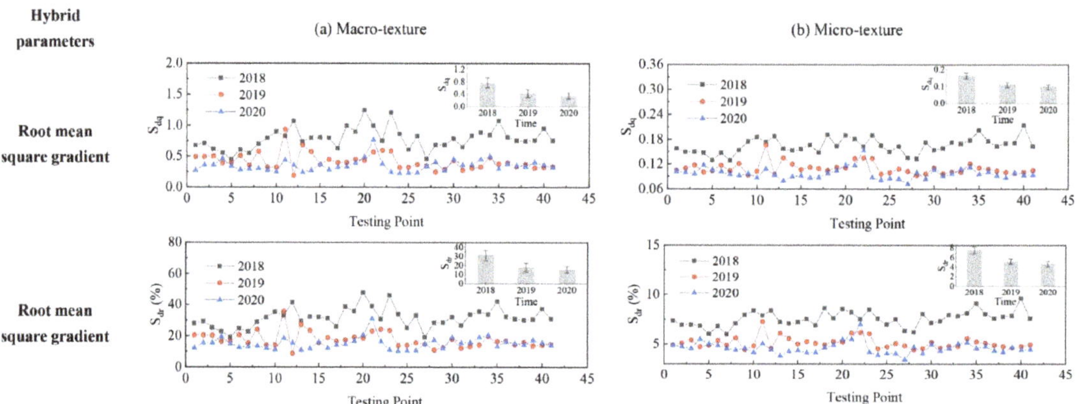

Figure 9. Pavement texture variations via hybrid parameters: (**a**) Macro-texture, and (**b**) Micro-texture.

4.4. Evolution of Function Parameters

Under traffic polishing, the peak and valley of pavement texture change over time. The cumulative height distribution curve of pavement texture provides an ideal tool to visualize how the texture profile changes due to polishing. Figure 10 shows examples of cumulative height distribution curves for macro- and micro-texture over the years. For example, for macro-texture, the material ratio corresponding to height 8 mm were 41.0% in 2018, 30.1% in 2019, and 16.5% in 2020; for micro-texture, the material ratio corresponding to height 0.05 mm were 4.9% in 2018, 3.4% in 2019, and 2.4% in 2020. This implies that the material of pavement texture was worn due to traffic polish.

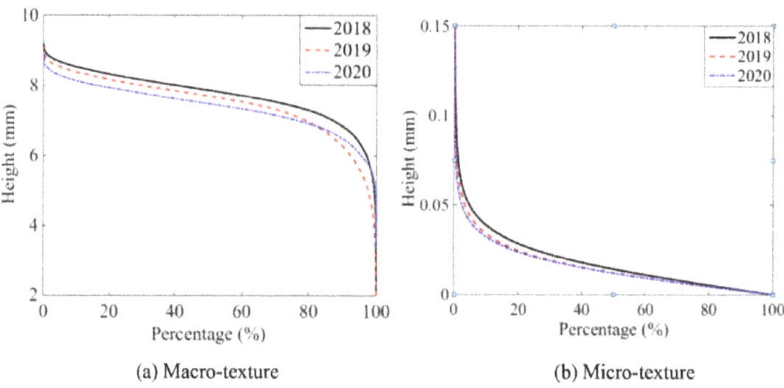

Figure 10. Cumulative height distribution curve of pavement macro- and micro-texture: (**a**) Macro-texture, and (**b**) Micro-texture.

Notably, the cumulative height distribution curve of macro-texture in 2019 was lower than that of 2018. It means that the texture material was worn, and the texture valley was increased from 2018 to 2019, which should correspond to the bitumen removal process. The cumulative height distribution curve of macro-texture in 2020 was lower at the peak layer and core layer but higher at the valley layer than that of 2019. This phenomenon illustrates that the upper part of macro-texture was removed by traffic polishing and field environmental erosion, and the valley void collected dust, debris, or chipping under traffic polishing. Besides, micro-texture's cumulative height distribution curve was getting lower year after year, suggesting micro-texture was consistently polished by traffic.

4.4.1. Evolution of Material Ratio Parameters

The variation of material ratio parameters for pavement micro- and macro-texture is displayed in Figure 11. For macro-texture from 2018 to 2020, (1) the S_{xp} slightly increased from 2018 to 2019 and remained stable after the second polish year; (2) the S_{dc} kept almost unchanged. For micro-texture from 2018 to 2020, (1) the S_{xp} had a 20% decrement from 2018 to 2019 and another 7.5% decrement from 2019 to 2020; (2) the S_{dc} decreased by20% and 6.9%, respectively, after the first and second years of polishing. The material ratio parameters of micro-texture changed more than that of macro-texture by traffic polish, suggesting traffic polish mainly affects materials of micro-texture.

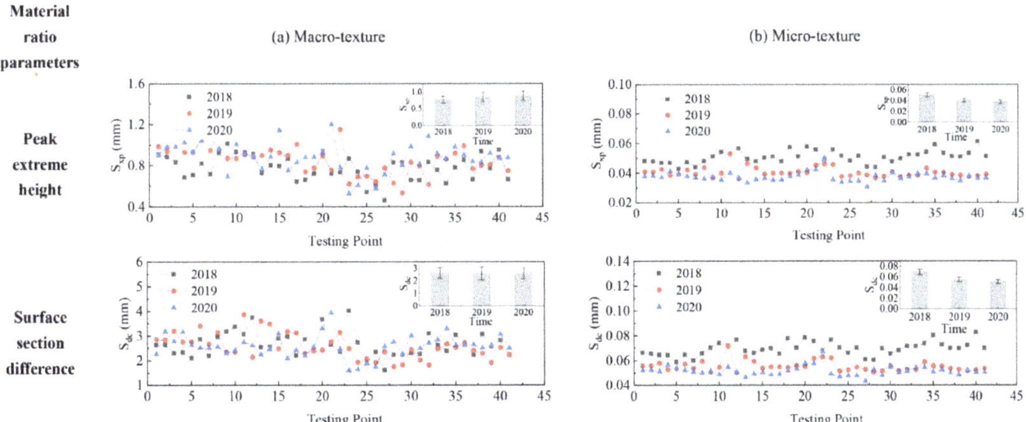

Figure 11. Pavement texture variations via material ratio parameters: (**a**) Macro-texture, and (**b**) Micro-texture.

4.4.2. Evolution of Stratified Parameters

The variation of stratified parameters for pavement micro- and macro-texture is displayed in Figure 12. For macro-texture from 2018 to 2020, (1) the S_{pk} increased by 17.6% from 2018 to 2019 and remained unchanged roughly after the second year of polishing, indicating that the peak layer remained unchanged after the bitumen layer was removed; (2) the S_{vk} of macro-texture decreased by 10% and 5.7% for each polishing year, which implies that the valley structure of macro-texture was gradually filled by dust, debris, or residue under traffic polishing; (3) the S_k showed minor variance after two years of traffic polish, which means the core layer of macro-texture was stable under traffic polish. Therefore, the variation of stratified parameters for macro-texture reveals that the traffic polish mainly affects the peak and valley layers but not the core layer of pavement macro-texture.

For micro-texture from 2018 to 2020, (1) the S_{pk} and S_k had significant decrement after the first year's polish and minor change after the second year's polish; (2) the mean value and standard deviation of micro-texture S_{vk} was almost zero, because the dale stratification did not exist in the cumulative height distribution curve of micro-texture, as shown in Figure 10b. It means traffic polish affects peak, core, and valley layers of pavement micro-texture.

4.4.3. Evolution of Volume Parameters

Figure 13 shows the variation of volume parameters in three years for pavement micro- and macro-texture. For macro-texture from 2018 to 2020, (1) the V_{mp} increased 9.9% and 5.9% after each polishing year, suggesting more material from the peak layer was exposed under traffic polish; (2) the V_{mc} and V_{vc} remained unchanged, indicating the material and void volume of core layer was unaffected by traffic polish; (3) the V_{vv} slightly decreased 8.5% and 3.9% sequential under traffic, meaning the void volume of valley layer was gradually reduced by collecting dust, debris, or residue under traffic polishing.

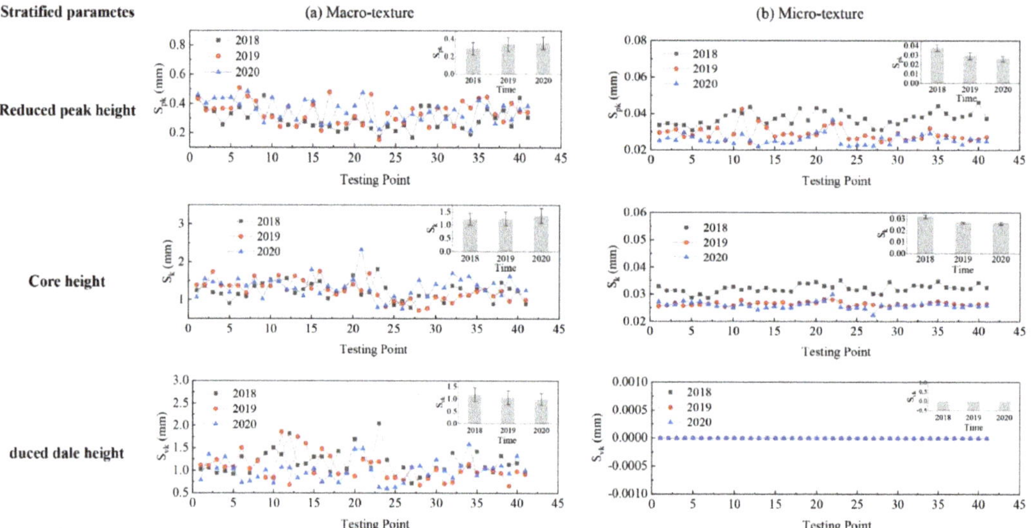

Figure 12. Pavement texture variations via stratified parameters: (**a**) Macro-texture, and (**b**) Micro-texture.

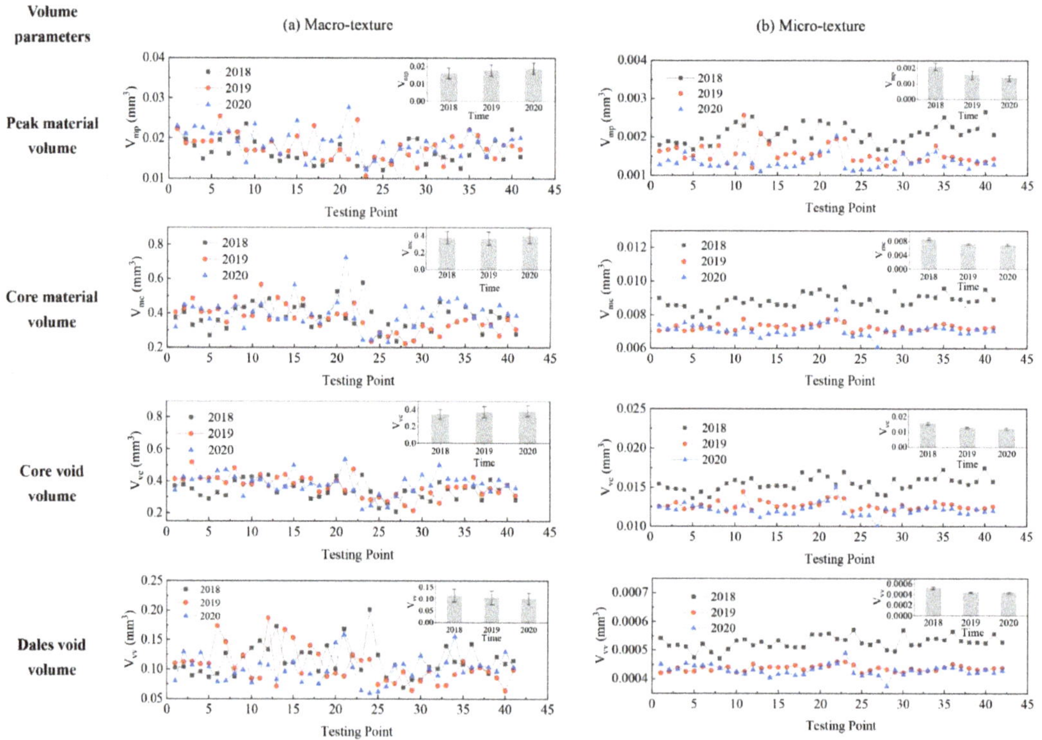

Figure 13. Pavement texture variations via volume parameters: (**a**) Macro-texture, and (**b**) Micro-texture.

For micro-texture from 2018 to 2020, (1) the V_{mp} decreased by 42.9% and 14.0%; (2) the V_{mc} decreased by 16.7% from 2018 to 2019, and changed minor (2.9%) after the second year's traffic polish; (3) the V_{vc} had a large descend of 17.8% from 2018 to 2019, and minor change (4.7%) from 2019 to 2020; (4) the V_{vv} also had consecutive drops of 16.0% and 1.9%. This result implies that traffic polish affects the material and void volume of pavement micro-texture.

Therefore, the volume parameters suggest that traffic polish influences pavement macro-texture in the following aspects: (1) exposed more material from the peak layer into contact; (2) filled up the valley layer with dust, debris, or residue; (3) had a minor impact on the core layer. Additionally, traffic polish consistently reduced the height or volume of pavement micro-texture peak, core, and valley layers.

4.5. Evolution of Feature Parameters

The evolution of feature parameters for macro- and micro-texture is shown in Figure 14. Generally, the S_{pd} of macro-texture had a tiny descend of around 5%, which means the number of contact peaks was reduced by abrasion. The average number of S_{pc} dropped 26.3% after the first year's polish, corresponding to the removal of the bitumen layer and fine aggregate. Then the S_{pc} had only a 7% drop from 2019 to 2020, because the coarse aggregate in pavement structure was gradually exposed and was harder to get worn than bitumen layer under traffic polish.

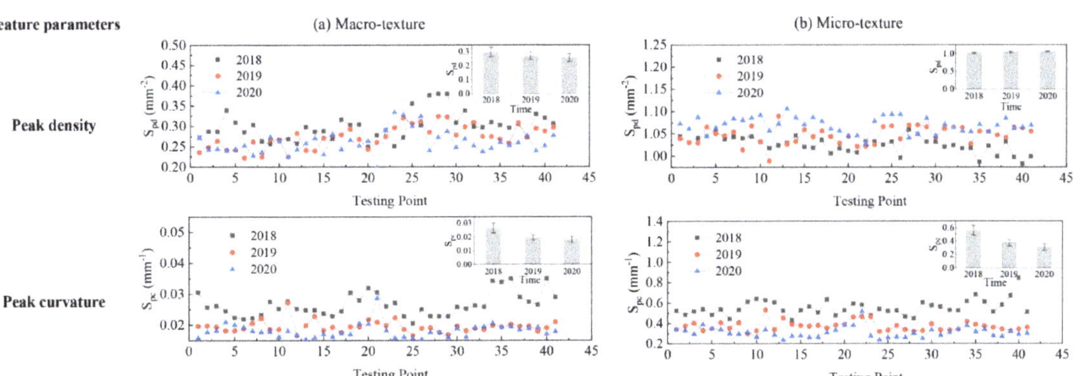

Figure 14. Pavement texture variations via feature parameters: (**a**) Macro-texture, and (**b**) Micro-texture.

Unlike the macro-texture, the S_{pd} of micro-texture was slightly increased year after year, as displayed in Figure 14b. The enlarged micro-texture in Figure 3b also shows more peaks existed on micro-texture over time due to traffic polish. The coarse aggregate exposure from 2018 to 2019 and the new micro-texture generated in the wearing process of coarse aggregates from 2019 to 2020 may contribute to the increased S_{pd}. However, the S_{pc} was lessening by 33% and 17% after each year's polish, which means the pavement micro-texture was gradually rounded by polish.

5. Conclusions

This paper applies 3D areal parameters to investigate asphalt pavement micro- and macro-texture evolution under actual traffic polish and environmental conditions. The portable 3D laser scanner LS-40 collected high-resolution 3D pavement texture data from predefined locations on a field site in 2018, 2019, and 2020, respectively. The obtained LS-40 data was decomposed into pavement micro- and macro-texture data sets to calculate 3D areal texture parameters. A total number of twenty parameters under five categories (height, spatial, hybrid, functional, and feature) were calculated to study the evolution of pavement micro- and macro-texture under actual traffic polish. The conclusions are summarized as follows:

1. the traffic polish and environmental conditions change the pavement micro-texture as follows: (1) the spike structure was gradually shrunk; (2) the spatial characterization evolved from isotropic to anisotropic; (3) the steepness and the developed interfacial area were decreased; (4) the height or volume of the peak, core, and valley layers reduced consistently; and (5) the peak density increased but peak curvature decrease.

2. the traffic polish and environmental conditions change the pavement macro-texture as follows: (1) the had valley structure and the height variation of surface peaks decreased; (2) the spatial characteristics were not changed under traffic polish; (3) its steepness and the developed interfacial area were decreased; (4) the material of the peak layer removed, and the valley layer filled up with dust, debris, or residue, and (5) the peak density and peak curvature were all decreased.

The results demonstrate the advantage of 3D areal parameters to describe the evolution characterization of pavement micro- and macro-texture under traffic polish. However, this paper only recorded texture data from one asphalt mixture in three years. Thus, it is expected that more asphalt pavement texture categories could be collected for a longer time frame in a future study to understand how traffic polish affects pavement micro- and macro-texture for different pavements. Furthermore, the relationship of texture wear and skid resistance should be studied in the future as well.

Author Contributions: Conceptualization, Y.Z.; methodology, Y.Z. and G.Y.; software, Y.Z.; validation, Y.Z. and G.Y.; formal analysis, Y.Z. and G.Y.; investigation, Y.Z.; resources, W.H., Y.L., Y.Q., K.C.P.W.; data curation, Y.Z. and G.Y.; writing—original draft preparation, Y.Z.; writing—review and editing, Y.Z. and G.Y.; visualization, Y.Z.; supervision, W.H., Y.L., Y.Q., K.C.P.W.; project administration, Y.Z.; funding acquisition, W.H., Y.L. and Y.Q. All authors have read and agreed to the published version of the manuscript.

Funding: This research was funded by [National Natural Science Foundation of China] grant number [51778541 and 51878574], [Sichuan Province Science and Technology Project] grant number [2020YFS0362], and [Sichuan Province Youth Science and Technology Innovation Team] grant number [2021JDTD0023]. The APC was funded by [National Natural Science Foundation of China] grant number [51778541].

Institutional Review Board Statement: Not applicable.

Informed Consent Statement: Not applicable.

Data Availability Statement: Not applicable.

Conflicts of Interest: The authors declare no conflict of interest.

References

1. Fwa, T. Skid resistance determination for pavement management and wet-weather road safety. *Int. J. Transp. Sci. Technol.* **2017**, *6*, 217–227. [CrossRef]
2. Giles, C.G.; Sabey, B.E. Skidding as a factor in accidents on the roads of Great Britain. In Proceedings of the First International Skid Prevention Conference, Charlottesville, VA, USA, 1959; pp. 27–40.
3. Xiao, J.; Kulakowski, B.T.; Ei-Gindy, M. Prediction of Risk of Wet-Pavement Accidents: Fuzzy Logic Model. *Transp. Res. Rec.* **2000**, *1717*, 28–36. [CrossRef]
4. Chu, L.; Fwa, T.F. Incorporating pavement skid resistance and hydroplaning risk considerations in asphalt mix design. *J. Transp. Eng.* **2016**, *142*, 04016039. [CrossRef]
5. Wang, D.; Liu, P.; Xu, H.; Kollmann, J.; Oeser, M. Evaluation of the polishing resistance characteristics of fine and coarse aggregate for asphalt pavement using Wehner/Schulze test. *Constr. Build. Mater.* **2018**, *163*, 742–750. [CrossRef]
6. Chu, L.; Zhou, B.; Fwa, T.F. Directional characteristics of traffic polishing effect on pavement skid resistance. *Int. J. Pavement Eng.* **2021**, 1–17. [CrossRef]
7. Do, M.-T.; Tang, Z.; Kane, M.; de Larrard, F. Evolution of road-surface skid-resistance and texture due to polishing. *Wear* **2009**, *266*, 574–577. [CrossRef]
8. Kanafi, M.M.; Kuosmanen, A.; Pellinen, T.K.; Tuononen, A.J. Macro- and micro-texture evolution of road pavements and correlation with friction. *Int. J. Pavement Eng.* **2015**, *16*, 168–179. [CrossRef]

9. Ueckermann, A.; Wang, D.; Oeser, M.; Steinauer, B. Calculation of skid resistance from texture measurements. *J. Traffic Transp. Eng. (Engl. Ed.)* **2015**, *2*, 3–16. [CrossRef]

10. ISO. *Geometrical Product Specifications (GPS)—Surface Texture: Areal—Part 2: Terms, Definitions and Surface Texture Parameters*; ISO 25178-2:2012(en); ISO: London, UK, 2012.

11. Gheni, A.; Abdelkarim, O.I.; AbdulAzeez, M.M.; ElGawady, M. Texture and design of green chip seal using recycled crumb rubber aggregate. *J. Clean. Prod.* **2017**, *166*, 1084–1101. [CrossRef]

12. Flintsch, G.W.; De León, E.; McGhee, K.K.; Ai-Qadi, I.L. Pavement Surface Macrotexture Measurement and Applications. *Transp. Res. Rec.* **2003**, *1860*, 168–177. [CrossRef]

13. Abe, H.; Tamai, A.; Henry, J.J.; Wambold, J. Measurement of Pavement Macrotexture with Circular Texture Meter. *Transp. Res. Rec.* **2001**, *1764*, 201–209. [CrossRef]

14. Chen, J.; Huang, X.; Zheng, B.; Zhao, R.; Liu, X.; Cao, Q.; Zhu, S. Real-time identification system of asphalt pavement texture based on the close-range photogrammetry. *Constr. Build. Mater.* **2019**, *226*, 910–919. [CrossRef]

15. Meegoda, J.N.; Gao, S. Evaluation of pavement skid resistance using high speed texture measurement. *J. Traffic Transp. Eng. (Engl. Ed.)* **2015**, *2*, 382–390. [CrossRef]

16. Zuniga-Garcia, N.; Prozzi, J.A. Contribution of Micro- and Macro-Texture for Predicting Friction on Pavement Surfaces (No. CHPP Report-UTA# 3-2016). Available online: https://www.chpp.egr.msu.edu/wp-content/uploads/2018/03/CHPP-Report-UTA3-2017_Texture.pdf (accessed on 12 September 2021).

17. Leach, R. *Characterisation of Areal Surface Texture*; Springer Science & Business Media: Berlin/Heidelberg, Germany, 2013. [CrossRef]

18. Li, L.; Wang, K.C.; Li, Q. Geometric texture indicators for safety on AC pavements with 1 mm 3D laser texture data. *Int. J. Pavement Res. Technol.* **2016**, *9*, 49–62. [CrossRef]

19. Li, Q.; Yang, G.; Wang, K.C.P.; Zhan, Y.; Wang, C. Novel Macro- and Microtexture Indicators for Pavement Friction by Using High-Resolution Three-Dimensional Surface Data. *Transp. Res. Rec.* **2017**, *2641*, 164–176. [CrossRef]

20. Yang, G.; Yu, W.; Li, Q.J.; Wang, K.; Peng, Y.; Zhang, A. Random Forest–Based Pavement Surface Friction Prediction Using High-Resolution 3D Image Data. *J. Test. Eval.* **2019**, *49*, 1141–1152. [CrossRef]

21. Zou, Y.; Yang, G.; Cao, M. Neural network-based prediction of sideway force coefficient for asphalt pavement using high-resolution 3D texture data. *Int. J. Pavement Eng.* **2021**. [CrossRef]

22. Dunford, A.M.; Parry, A.R.; Shipway, P.H.; Viner, H.E. Three-dimensional characterisation of surface texture for road stones undergoing simulated traffic wear. *Wear* **2012**, *292*, 188–196. [CrossRef]

23. *BS EN 12697-49: Bituminous Mixtures–Test Methods for Hot Mix Asphalt. Part 49: Determination of Friction after Polishing*; BSI: London, UK, 2014.

24. Wang, D.; Chen, X.; Xie, X.; Stanjek, H.; Oeser, M.; Steinauer, B. A study of the laboratory polishing behavior of granite as road surfacing aggregate. *Constr. Build. Mater.* **2015**, *89*, 25–35. [CrossRef]

25. Lei, C. *Study on Pavement Surface Function Accelerated Loading System*; Graduate Faculty of South China University of Technology: Guangzhou, China, 2010.

26. Liang, R.Y. *Long Term Validation of an Accelerated Polishing Test Procedure for HMA Pavements*; Technical Report; Ohio Department of Transportation and the U.S. Department of Transportation, Federal Highway Administration: Columbus, OH, USA, 2013.

27. Ren, W.; Han, S.; He, Z.; Li, J.; Wu, S. Development and testing of a multivariable accelerated abrasion machine to characterize the polishing wear of pavement by tires. *Surf. Topogr. Metrol. Prop.* **2019**, *7*, 035006. [CrossRef]

28. Vollor, T.W.; Hanson, D.I. Development of laboratory procedure for measuring friction of HMA mixtures–PHASE I. In *Final Report of NCAT*; North Carolina Agricultural and Technical State University: Greensboro, NC, USA, 2006.

29. Kane, M.; Do, M.T.; Piau, J.M. On the Study of Polishing of Road Surface under Traffic Load. *J. Transp. Eng.* **2010**, *136*, 45–51. [CrossRef]

30. Druta, C.; Wang, L.; Lane, D.S. Evaluation of the MMLS3 for Accelerated Wearing of Asphalt Pavement Mixtures Containing Carbonate Aggregates. Available online: https://www.virginiadot.org/vtrc/main/online_reports/pdf/14-r17.pdf (accessed on 12 September 2021).

31. Wu, Z.; Abadie, C. Laboratory and field evaluation of asphalt pavement surface friction resistance. *Front. Struct. Civ. Eng.* **2018**, *12*, 372–381. [CrossRef]

32. Plati, C.; Pomoni, M. Impact of Traffic Volume on Pavement Macrotexture and Skid Resistance Long-Term Performance. *Transp. Res. Rec.* **2019**, *2673*, 314–322. [CrossRef]

33. Blunt, L.; Jiang, X. *Advanced Techniques for Assessment Surface Topography: Development of a Basis for 3D Surface Texture Standards "Surfstand"*; Elsevier: Amsterdam, The Netherlands, 2003.

34. Barányi, I.; Czifra, Á.; Kalácska, G. Height-independent topographic parameters of worn surfaces. *Int. J. Sustain. Constr. Des.* **2011**, *2*, 35–40. [CrossRef]

35. Franco, L.A.; Sinatora, A. 3D surface parameters (ISO 25178-2): Actual meaning of Spk and its relationship to Vmp. *Precis. Eng.* **2015**, *40*, 106–111. [CrossRef]

36. Salguero, J.; Del Sol, I.; Vazquez-Martinez, J.; Schertzer, M.; Iglesias, P. Effect of laser parameters on the tribological behavior of Ti6Al4V titanium microtextures under lubricated conditions. *Wear* **2019**, *426–427*, 1272–1279. [CrossRef]

37. Bush, A.; Gibson, R.; Thomas, T. The elastic contact of a rough surface. *Wear* **1975**, *35*, 87–111. [CrossRef]

Review

Waste Mineral Wool and Its Opportunities—A Review

Zhen Shyong Yap [1], Nur Hafizah A. Khalid [2,*], Zaiton Haron [1], Azman Mohamed [1], Mahmood Md Tahir [3], Saloma Hasyim [4] and Anis Saggaff [4]

[1] School of Civil Engineering, Faculty of Engineering, Universiti Teknologi Malaysia, Skudai 81310, Johor, Malaysia; zsyap2@graduate.utm.my (Z.S.Y.); zaitonharon@utm.my (Z.H.); azmanmohamed.kl@utm.my (A.M.)

[2] Centre for Advanced Composite Materials (CACM), School of Civil Engineering, Faculty of Engineering, Universiti Teknologi Malaysia, Skudai 81310, Johor, Malaysia

[3] UTM Construction Research Centre, Universiti Teknologi Malaysia, Skudai 81310, Johor, Malaysia; mahmoodtahir@utm.my

[4] Civil Engineering Department, Faculty of Engineering, Sriwijaya University, Kota Palembang 30128, Sumatera Selatan, Indonesia; salomaunsri@gmail.com (S.H.); anissagaf@yahoo.com (A.S.)

* Correspondence: nur_hafizah@utm.my

Abstract: Massive waste rock wool was generated globally and it caused substantial environmental issues such as landfill and leaching. However, reviews on the recyclability of waste rock wool are scarce. Therefore, this study presents an in-depth review of the characterization and potential usability of waste rock wool. Waste rock wool can be characterized based on its physical properties, chemical composition, and types of contaminants. The review showed that waste rock wool from the manufacturing process is more workable to be recycled for further application than the post-consumer due to its high purity. It also revealed that the pre-treatment method—comminution is vital for achieving mixture homogeneity and enhancing the properties of recycled products. The potential application of waste rock wool is reviewed with key results emphasized to demonstrate the practicality and commercial viability of each option. With a high content of chemically inert compounds such as silicon dioxide (SiO_2), calcium oxide (CaO), and aluminum oxide (Al_2O_3) that improve fire resistance properties, waste rock wool is mainly repurposed as fillers in composite material for construction and building materials. Furthermore, waste rock wool is potentially utilized as an oil, water pollutant, and gas absorbent. To sum up, waste rock wool could be feasibly recycled as a composite material enhancer and utilized as an absorbent for a greener environment.

Keywords: composite behavior; material composition; materials recycling; reuse; sustainable resource; waste characterization

Citation: Yap, Z.S.; Khalid, N.H.A.; Haron, Z.; Mohamed, A.; Tahir, M.M.; Hasyim, S.; Saggaff, A. Waste Mineral Wool and Its Opportunities—A Review. *Materials* **2021**, *14*, 5777. https://doi.org/10.3390/ma14195777

Academic Editor: Carlos Morón Fernández

Received: 5 September 2021
Accepted: 30 September 2021
Published: 2 October 2021

Publisher's Note: MDPI stays neutral with regard to jurisdictional claims in published maps and institutional affiliations.

1. Introduction

The volume of global solid waste generation is estimated at 1.3 billion tons per year [1] and it is expected to climb to about 70%—2.2 billion tons by 2025 [2,3]. One of the major global solid wastes is mineral wool waste which is generated at 2.54 million tons per year [4]. This volume is measured based on rock wool productivity over a 30-year lifespan [5] and assuming 5% wastage during installation [6]. The volume of rock wool was estimated to have grown linearly from 2.25 million in 2010 to 2.54 million in 2020 [5]. It is expected to grow to 2.82 million in 2030, as shown in Figure 1 based on the researchers' calculation. It is notable that an increasingly massive amount of fibrous waste will become a pinnacle issue to the environment [7,8] and economy [9,10] if left unresolved for a long period.

Rock wool or stone wool is the main mineral wool with fine and intertwined fibers produced by spinning molten rocks at high speed that is akin to make cotton candy. The mass of fibers is shaped into sheets with two prominent characteristics—low density and high porosity, making rock wool an excellent construction choice for sound absorption, thermal insulation [11], and fire resistance [12,13]. In the light of these prominent characteristics, rapid development and industrialization of rock wool insulation material started in the early 20th century [14]. To date, rock wool has become commonplace as a building material [15] and pipe insulation material [16] where it makes up more than 50% of the world's insulation material in the market [17,18]. Alongside this development is the massive amount of degraded rock wool. It is generated and discarded as waste [19].

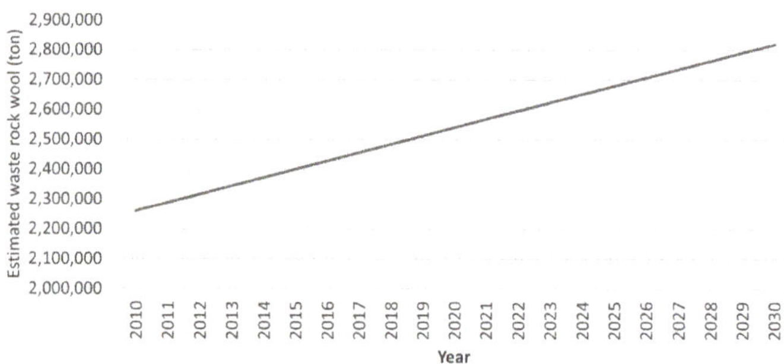

Figure 1. Annual waste mineral wool estimation in European Union Countries. (Modified from [5]).

The landfill and dumping areas are practically the most common solutions to waste rock wool. However, it comes with a series of common environmental issues [20–22], and most are connected to greenhouse-gas emissions [23–25]. The second major problem concerns the larger space and logistics arrangement required to store the loose and bulky waste rock wool. This issue further aggravates the difficulties related to landfill shortages [26], natural resources shortages [27,28], and an increase in waste disposal costs [29,30]. These remain a crucial issue for the wool fiber industry throughout the world.

Several studies have been published to overcome the conundrums on the reclamation of waste rock wool into the rock wool production process [31,32]. It encompasses introducing waste rock wool as recycled fillers into different materials, including pozzolanic material [33], geopolymer concrete [34], alkali-activated material [35], polymer composites [36], and wood-fiber composites [37] as well as reutilizing the waste as a pollutant sorbent to remove split oil [38] and hexavalent chromium Cr(VI) [39] from water and filter hydrogen sulfide (H_2S) gas. However, none has performed a thorough review of comparing the available recycling techniques of waste rock wool to understand its feasibility for the above functions.

As such, this paper reviews the available recycling methods and the inherent properties of rock wool before concluding with some recommendations for future research development. Hence, the objectives of this review are: (i) review the characterizations such as physical characterization, chemical composition, and possibly contaminants of waste rock wool, (ii) review both the advantages and disadvantages on the performance of the waste rock wool reinforced composites as well as other recycling approaches, and (iii) propose critical perspectives or knowledge gaps for future research. With much confidence, this review would benefit both academia and industry to formulate more scientific and practical strategies to handle waste rock wool in the future.

2. Research Approach

This review takes a comprehensive approach to analyzing the research outputs in the field of waste rock wool that have been published in Scopus, ScienceDirect, and Google Scholar, which are the widely used search engines for academic outputs. The keywords used in the literature search were "rock*wool" or "mineral wool" or "man-made mineral fiber" or "man-made vitreous fiber", as well as other related terms such as "waste*" or "recycle" or "used". These keywords were applied with Boolean operators, quotation marks, wildcards, and query sets.

The complete workflow is shown in Figure 2, which includes the literature searching and screening procedures, and the scope of the qualitative discussion. As illustrated in Figure 2, the literature search was divided into three sub-steps to weed out articles that were either out of scope or did not focus on the recycling of waste rock wool. The literature was manually screened by first reading the abstract and findings.

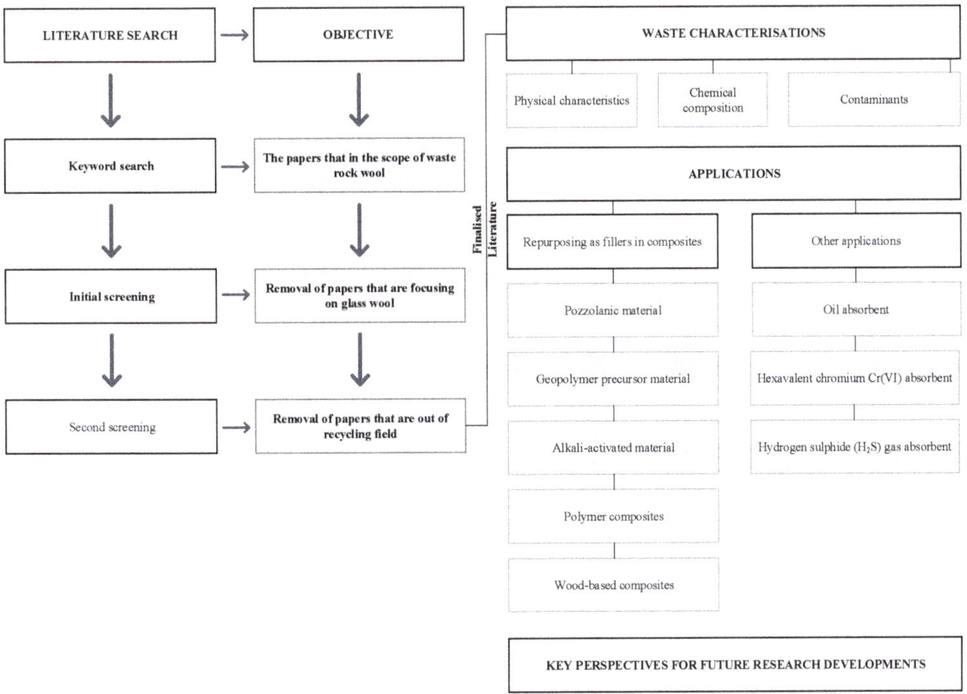

Figure 2. Literature search and scopes of the review.

Following literature searching and screening, a qualitative discussion was conducted. After reviewing the finalized literature, the review structure was categorized into three main groups for in-depth qualitative discussion. The three important scopes are:

- Identify the characterization of waste rock wool (physical properties, chemical composition, and contaminants).
- Review the performance of composites after incorporated waste rock wool.
- Investigate other applications and their effectiveness.

3. Waste Characterizations

Waste rock wool is commonly generated from two primary sources: manufacturing and post-consumer waste [40,41]. The characteristics and volume of waste from both sources are significantly different. Manufacturing waste is cleaner with a known composition that is commonly reintroduced into its manufacturing process. This method is an effective way to lower the raw material cost by replacing coke usage. Moreover, a recent study compared the melting behavior of virgin raw materials and waste rock wool and concluded that re-melting the waste rock wool is 23% more energy-efficient on differential scanning calorimetry (DSC) [31].

For post-consumer waste, a study pointed out that the inappropriate sorting of used waste rock wool gives rise to unknown quality and contaminations [42]. Thus, post-consumer rock wool is hard to reintroduce as raw material to its manufacturing production.

This section aims to identify and quantify predominantly the physical characteristics and chemical composition present in post-consumer wastes [43] before an appropriate treatment method can be assigned.

3.1. Physical Characteristics

Waste rock wool has a fibrous-open porous structure and may differ in densities and strengths [44] due to degradation when exposed to moisture in the air or acid rain [19]. It is generally regarded as unrecyclable [5,34] until modern technologies have made its blending with other materials possible.

Comminution is currently the preferred method to produce a homogeneous mixture of particles [45]. It involves crushing, grinding, cutting, or other similar reduction methods to break a solid material into smaller sizes to mix with other composites. Depending on the actual process adopted, the processed rock wool's bulk density, fiber length, and width would change [46]. Table 1 summarises the different approaches that have been adopted to produce homogenous fiber particles.

Table 1. The different approaches to comminution for waste rock wool.

Machine Used	Average Fibre Length (μm)	Average Fibre Width (μm)	Average Particle Length Versus Width	Percentage of Particles <20 μm (%)	Citation
Raw rock wool	424	7.6	55.7	1.2	
Granulator 200 series	387	7.6	50.6	1.2	
100 UPZ fine-impact mill	365	8.1	45.0	1.5	
ZRI homogenizer	191	7.7	24.7	2.0	[46]
Hydraulic press 30 ton	76	10.3	7.4	3.0	
Vibratory disc mill RS200	37	11.7	3.2	19.0	
	33.6	–	–	–	[4]
Laboratory scale Agitator	4760	500	9.5	–	[47]

3.2. Chemical Composition

The chemical composition of rock wool is highly dependent on its manufacturing process. Each rock wool factory uses different raw materials based on their preference and intended use. Nevertheless, the raw material is typically from the rock crystalline group and may include anorthosite, dolomite, diabase, olivine sand, or even slag [31]. Table 2 shows the compositions of chemicals of ordinary waste rock wool from different research studies.

Table 2. Composition of chemicals of waste rock wool from different studies.

Composition (%)									Sources
SiO_2	CaO	Al_2O_3	Fe_2O_3	MgO	K_2O	Na_2O	SO_3	TiO_2	
38.70	20.90	18.60	5.30	7.00	2.00 *		NA	NA	[48]
16.90	46.90	5.40	16.20	2.60	NA	NA	NA	NA	[47]
40.6	3.52	2.14	6.91	11.1	6.34	6.71	2.41	0.23	[49]
40.40	17.40	1.80	9.20	12.60	0.40	NA	NA	0.80	[4]
42.00	14.70	16.60	11.30	12.20	0.50	1.60	0.03	0.90	[34]
44.06	16.36	15.94	11.93	5.68	0.57	3.71	NA	1.58	[50]
27.04	9.57	6.31	8.03	5.85	5.45	3.43	5.05	NA	[51]
40.76	21.09	11.48	11.30	NA	NA	NA	NA	NA	[52]
60.10	22.60	0.76	1.36	1.39	0.58	5.29	1.36	0.20	[33]
42.60	18.30	18.10	7.10	8.40	0.60	2.10	NA	0.80	[31]
40.00	14.50	16.00	9.00	14.50	2.00	NA	NA	NA	[46]
44.1	16.6	14.3	5.5	14.7	0.3	1.2	0.0	0.2	[37]
51.54	13.94	1.07	0.36	2.92	0.56	8.20	0.3	0.04	[42]
40–52	10–12	8–13	5.5–6.5	8–15	0.8–2.0	0.8–3.3	0–0.2	1.5–2.7	[44]

* the summation of K_2O and Na_2O.

A general categorization of rock wool properties may seem almost impossible. However, a closer look at Table 2 would reveal that most waste rock wool consists of three major components—silicon dioxide (SiO_2), calcium oxide (CaO), and aluminum oxide (Al_2O_3). SiO_2 is the major component that accounts for around 38.7–60.1% of the total weight in most cases except in the two studies [47,51]. For CaO and Al_2O_3, the majority weight ratio may vary from 10 to 46.9% and 0.8 to 18.6%, respectively.

The difference in chemical composition is also apparent between industrial and construction usage based on the content, contamination level, and purity level. For instance, the rock wool used as piping insulation is often mingled with water, oil, or an unknown chemical. The wire mesh inside the rock wool panel used for pipes tends to rust over time due to water or moisture in the air. These contaminants or dirt in the porous rock wool pose the biggest hurdle to recycling efforts, making it the most apparent reason behind million tons of waste rock wool piling up every year in landfills and adversely affecting the environment. All in all, to diminish the contaminants of the waste, the initial stage of source separation is essential to getting the waste recyclable [53].

A point worth noting is that no research has performed a physical measurement on either post-consumer waste or manufacturing waste. This problem may reveal a vast difference in waste quality and its effect on the choice of recycling methods and the resultant product.

3.3. Contaminants

Waste rock wool is contaminated in two phases. The first phase is during the manufacturing process, where hazardous materials such as phenol-formaldehyde binders are coated onto the fiber surface [54]. The second phase is during the application process, where the waste rock wool becomes contaminated with other pollutants such as mycological substances, impurity dust, or chemical substances. With the contaminants, waste rock wool's composition becomes complex and more difficult to be recycled [55].

The resin application as a binder during the production process is unavoidable as it is required to bind the rock wool fiber together after the molten rock fiber has been cooled down by air [56]. The amount of resin is small, typically in 1:10 wt% of the rock wool, but it can affect the properties of waste rock wool and increase the difficulty of recycling waste rock wool [57]. Another common contaminant is mycological substances, for example, fungi, present in a humid environment [58]. Fortunately, recent research shows that resin or other organic matter can be burnt off by heating the waste to a specific temperature [31]. This finding also infers that if the resin is the only contaminant present, the chance to recycle the waste rock wool as raw material is high. However, in post-consumer waste, the resin should be neutralized using hydrogen peroxide, some catalysts, and a mixture of diammonium hydro phosphate and carbamide [59].

Apart from organic contaminants, other impurities such as dust or chloride are commonly diffused in the air and easily found on the surface or even in the porous structure of waste rock wool [60]. Chloride stored in the rock wool pores forms hydrochloric acid, which is a corrosive and volatile liquid. The transformation of chloride particles to acid usually occurs when the rock wool is hot and humid. This problem can be resolved by washing ground waste rock wool with tap water at a water-to-solid ratio of 10 to significantly reduce chlorine content [51]. In the study, the chloride content was reduced from 4.901 to 0.612 wt% after stirring the waste in water for a few minutes.

Overall, the characterization of waste rock wool in terms of the physical, chemical composition, and contaminants reveals that waste rock wool from the manufacturing process has a higher potential to be recycled than the mineral waste rock wool from other industries. This is attributed to the known composition and purity content that is easier for recycling processes. Among the recycling methods, "comminution" is preferred as its recycled product has better mechanical strength and chemical resistance for application. Meanwhile, the aim of reducing contaminants can be achieved by using "washing" with water, mainly to reduce the chlorine content that is undesirable for incorporating into composites. The feasibility and cost-effectiveness of pre-treatment and recycling methods are one of the major concerns in reprocess waste. However, a thorough and systematic investigation of these two aspects is yet to be conducted. Therefore, more studies on the practicability and costing should be done to develop efficient and cost-effective waste material pre-treatment and recycling methods.

4. Repurposing as Fillers in Composites

Repurposing waste materials in composites are the common trend of utilizing waste and other materials to increase the property of the composites [61]. This section reviews the five potential solutions for waste rock wool to mix with other materials to form a better performance composite.

4.1. Pozzolanic Material

Pozzolanic material is widely used in concrete making to reduce cement content; improve the workability of fresh concrete; increase the concrete strength; enhance the durability of hardened concrete [62]; and lower the overall production cost and impact on the environment. In studies of pozzolanic material, the incorporation of waste material to drive down overall production costs [63,64] is an exciting topic with studies already done extensively on silica fume, fly ash, ground granulated blast slag [65], rice husk ash, textile fiber [66], and palm oil fuel ash [67]. Many continued to perform with a proven track record in industrial applications [68–70]. The study on rock wool is considered a more recent venture to ultimately achieve the same objective [71,72].

The significant enablers for rock wool to become a pozzolanic material are the high SiO_2 (up to 70%) and CaO (up to 20%) content. In the amorphous phase, silica is necessary to initiate and maintain pozzolanic activity. Since the pozzolan reaction only occurs on the silica particle surface, the specific surface area is the most significant factor in the effectiveness of pozzolanic activity [71–73]. A study has been conducted on fine rock wool particles to facilitate the early hydration of cement, however, the result showed a prolonged dormant period that is undesirable [51]. Table 3 below summarises other methods used to mill the thin and soft rock wool fiber with their respective composition and its incorporation into concrete-making.

Table 3. Physical and chemical characteristics, and mixing method of incorporating waste rock wool into concrete.

Fiber Length (μm)	Density (kg/m³)	Major Chemical Composition	Mixing Methods	Mixing Proportion	Sources
75	206	SiO_2 (38.7 wt%) CaO (20.9 wt%)	Replace cement	10%, 20%, 30% & 40% (mass)	[74]
24.4	-	SiO_2 (40.6 wt%)	Additive	1%, 2% and 3%	[49]
-	-	-	Replace aggregate	30% (mass)	[75]
1–50	400	SiO_2 (30.0 wt%) CaO (10.35 wt%)	Additive	5%, 15% and 20%	[51]
45	742	SiO_2 (53.51 wt%) CaO (22.5 wt%)	Replace cement	25% (mass)	[76]
500–1000 μm	-	SiO_2 (70.0 wt%) CaO (20%)	Replace sand	30% (mass)	[33]
500–1000 μm	-	-	Replace aggregates	30%, 40%, 50% (mass)	[77]

One of the studies replaced 20% of cement with waste rock wool and successfully obtained a high increment in fresh concrete flow spread [71]. The ground waste rock wool had also prolonged the induction period of cement hydration at 5 wt% (1 h 17 min), 15 wt% (4 h 34 min), and 20 wt% (6 h 27 min). Beyond the 20% replacement of cement with waste rock wool, the flow spread decreased due to an unfavorable demand for water to moisten the particles' surface [76].

In hardened concrete, waste rock wool's fibrous nature had shown to slightly increase the porosity and decrease the density (-0.09 g/cm³) of the final product [33]. These could be promising to improve the sound [78] and thermal insulating properties [79]. However, in terms of mechanical properties, the observed outcomes vary for different researchers. Some studies found that the inclusion of waste rock wool had caused a little decrease in the mechanical properties of concrete [15,33,75]. However, there are some studies that reported a growth in overall strength [49,72]. The latter also reported enhancement in other properties such as compressive strength, splitting tensile strength, absorption, resistivity, chloride-ion penetration resistance, and abrasion resistance. At 50% replacement of aggregates with waste rock wool, the compression strength of concrete reduced while the flexural strength was enhanced up to 12% [77].

The contradictory findings have been addressed [51], whereby the paper attributes the differences to the presence of contaminants commonly found in waste rock wool, namely halite (NaCl) and sylvite (KCl) [33]. Without proper cleaning of waste rock wool, the NaCl and KCl content might have increased the early hydration of cement and promoted the generation of a large amount of Friedel's salt that would consume the portlandite ($Ca(OH)_2$) in cement and inhibit C_3H gel formation needed for the development of compressive strength. Consistent with this inference, the usage of waste rock wool with negligible NaCl and KCl content would mean a positive increase in the mechanical properties of concrete [49,51,72]. It should also be highlighted that these three studies documented the washing of waste rock wool with tap water to leach off and reduce the overall chloride content (from 4.901 wt% to 0.612 wt%) before mixing with cement.

4.2. Geopolymer Precursor Material

The geopolymer precursor material may be an appropriate replacement to the traditional Portland cement [80,81]. It contains geological source material with silicon and aluminum content and a caustic activator to form a binder [82,83]. In 2017, a study revealed that waste rock wool is an ideal replacement for slag and fly ash from iron and metal to produce geopolymer concrete due to its high silicon and aluminum content [34]. This finding was confirmed using the selective solubility tests at pH 11.6 [84,85], which revealed that the selective solubility of milled waste rock wool (SiO_2—73.0%, Al_2O_3—87.3%) was almost twice higher than fly ash's (SiO_2—32.9.0%, Al_2O_3—35.8%) [34].

After incorporating waste rock wool, the mechanical properties of geopolymer concrete showed that, with 33% rock wool and 47% fly ash inclusion in geopolymer concrete, the compressive strength after 28 days was 12 MPa [34]. In one of the studies [4], both concrete compressive and flexural strengths had increased 30 and 20.1 MPa, respectively, after the inclusion of waste rock wool. Recently, a study compared the new and waste rock wool and revealed that both geopolymer precursor materials achieve 15–20 MPa compressive strength after 1 day of curing, and resulted in 44 and 36 MPa for new and waste rock wool, respectively, after 28 days [37]. The compressive strength difference might be due to the impurities or microorganisms in the waste rock wool, as previously stated. Overall, the adoption of waste rock wool as a geopolymer shows the potential to replace cement with more sustainable material.

4.3. Alkali-Activated Material

The alkali activation method enables industrial by-products to be recycled [86]. Waste rock wool contains a high amount of SiO_2, Al_2O_3, and CaO which thus are suitable as the precursor for producing alkali-activated materials [87,88].

Attempts to establish alkali-activated material as a more environmental alternative to Portland cement [89] are not new. In fact, some had shown promising performance in the enhancement of early and late strength development [90,91], high-temperature resistance [92,93], durability under chemical attack [94–96] as well as adhesion to metallic and non-metallic surfaces [97]. Alkali-activated material is identical to geopolymer but different in the chemical reaction. It has lower calcium content than geopolymer, higher mechanical strength but is not as durable.

Milled waste rock wool is a promising precursor to produce alkali-activated cementitious binder because of its favorable chemical composition and high surface area [35,98]. The highest compressive and flexural strength is currently recorded at 30.0 and 20.1 MPa, respectively [4]. The activation process in waste rock wool with cement starts when Mg-O, Ca-O, Si-O-Si, Al-O-Si, and Al-O-Al bonds are destroyed, and a layer of Si-Al is formed on the grain's surface to form hydration products [88]. The rock wool-based alkali-activated binder comprises an amorphous sodium and aluminum substituted calcium silicate hydrate (C-(N-)A-S-H) gel, the main constituent in concrete. This gel increases when $Na_2O.2SiO_2$ is used as the alkali activator to dissolve more Si [99]. Increased soluble Si content generally increases the rock wool-based alkali-activated binder's density to allow more aluminosilicate chain-crosslinking in the C-(N-)A-S-H-type gel and the creation of an extra Al-rich N-A-S-H gel product during the N-A-S-H gel reaction [100]. The major challenge lies in controlling efflorescence formation during concrete mix design [101].

4.4. Polymer Composites

A polymer composite is a multi-phase material with its polymer matrix integrated with reinforcing fillers to enhance the final product's mechanical properties [102,103]. The incorporation of low-cost recycled fillers such as fiber [104,105], carbon black, and silica in rubber [106,107] into the polymer is currently a hot research topic of reducing conventional fillers' usage. For waste rock wool, a study depicted agglomerated milled waste rock wool (with diameter 1–3 μm and length 8–200 μm) as a particular filler into a different type of polyethylene polymer: high-density polyethylene, low-density polyethylene, recycled high-density polyethylene, and recycled low-density polyethylene [108].

Polyethene is a high-flammable material. By introducing waste rock wool as filler to the polyethylene, the polyethylene ignition time could increase 6–8 s, and the mass loss of the composite after heated could be halved [108]. This performance indicates that the developed composites are less liable to combust and is, in fact, a commercially attractive property to produce high-temperature resistant products.

The introduction of 40% weight of waste rock wool into the polyethylene has shown an increase in bending stress by 5 to 6 MPa, impact energy by 5 to 7 kJ/m^2, and Brinell hardness by 12 to 18 MPa alongside a decrease in tensile stress by 10 to 15 MPa for all types of polyethylene. Composites with virgin rock wool fillers performed better than those containing recycled rock wool. This result could be explained by the reinforcing influence of the activated non-hardened resin.

However, another study recorded a significant drop of 93% in impact strength for the composite with 40 to 60 % waste rock wool fillers [36] similar to the study on recycled wood-polyurethane composite [109]. The composite's Brinell hardness was not increased by the addition of waste rock wool fillers even up to 60%. The authors concluded that incorporating micro-fillers into polymer had increased the porosity of the composite beyond favorable conditions. A point to note is that the study did not mention the fineness of the fillers. It also included another variation of the study, which was adding 3% coupling agent and 3% of processing additive. The effect of these variations on the overall result remained unknown [109]. As such, it was stating that waste rock wool is not an effective filler may not be an objective conclusion in this case.

Undoubtedly, more research on the incorporation of waste rock wool into polyethylene is required to validate the properties of the resultant polymer composites. A more standard approach is also required and with due consideration given to the approach's economics and carbon footprint to understand its applicability better.

4.5. Wood-Based Composites

The potential of waste rock wool in wood-based composites is worth highlighting through the study, incorporating the waste into particleboards to reduce ignitability [110]. The study showed 20 to 30% addition of rock wool had shortened the smoldering path by 19 to 30%. Although this reduction in ignitability is desirable, it was accompanied by a deterioration in the modulus of rupture (up to 65%) and internal bonding (up to 71%) due to the low cohesion of the rock wool domains within the board.

A study noted a remarkable increase in durability performance after adding rock wool filler into a fiber composite [111]. This result is attributed to the high tensile strength, modulus, chemical resistance, and dimensional stability of the rock wool fiber mineral [112]. Profound enhancement of thermal insulation was also noted [37] and moisture resistance properties [111]. However, there was still a decrease in mechanical properties.

For plaster composite, the reinforcement of recycled rock wool at a maximum percentage of 10% (in weight) and a waste-plaster ratio of 0.6 to 0.8 [15] had enhanced the flexural strength by 26.58% [15]. This finding was slightly higher than the results obtained in previous studies pertaining to plaster or gypsum reinforced with short sisal fibers [113]. Moreover, adding 4% of waste rock wool enhanced 14.64% of the surface hardness—a feature essential to reduce the defects induced during the transportation and installation processes. However, the plaster composite's compressive strength had decreased due to the increment of the pore in the material that had also increased its density by 6.75%.

Current studies have generally depicted a reasonable enhancement in the ignitability and durability of wood-based material incorporated with rock wool fillers, but deterioration of mechanical properties cannot be neglected. As such, future studies in this aspect are inevitable.

5. Other Applications

5.1. Oil Absorbent

Porous materials can be used to absorb oil contaminants from water [38]. Common sorbent materials include birch bark, glass wool, cork, polyurethane foam [114], aerogels, and polystyrene [115]. The major problem with these absorbents is the limited capability to separate high-viscosity oils. Waste rock wool, in this case, has recently been reported to have overcome this limitation, providing a lower cost and environment-friendly alternative [38].

Fibers' efficiency in reclaiming water contaminated with crude oil is dependent on their surface qualities [116]. Waste rock wool is associated with oil and water due to its functional group (-OH groups on the surface). To turn waste rock wool into an effective oil sorbent, a study adopted the dip-coating method (PDMS/SiO$_2$ nanoparticle mixture) to turn the waste rock wool into superhydrophobic material [38]. With a thin layer coating of PDMS and SiO$_2$ nanoparticles on rock wool fiber, as illustrated in Figure 3, the superhydrophobic properties over a water contact angle from 0 to 152.9° have been enhanced, which is excellent in repelling water.

In terms of absorption effectiveness, a kilogram of waste rock wool can absorb 8 to 13 kg of spilled oil, depending on the oil density. The modified rock wool's volume absorption ability could reach up to 90% [38]. Furthermore, waste rock wool may absorb spilled oil in a circular fashion and reload after combustion or squeezing without deforming its shape, which is advantageous for transportation and post-treatment activities. Due to its inorganic composition, it is also mechanically stronger than other absorbents.

The most notable property of rock wool is its exceptional thermal stability. It allows application in an extremely hot environment, indicating that high viscosity oil absorption continuity during combustion is possible.

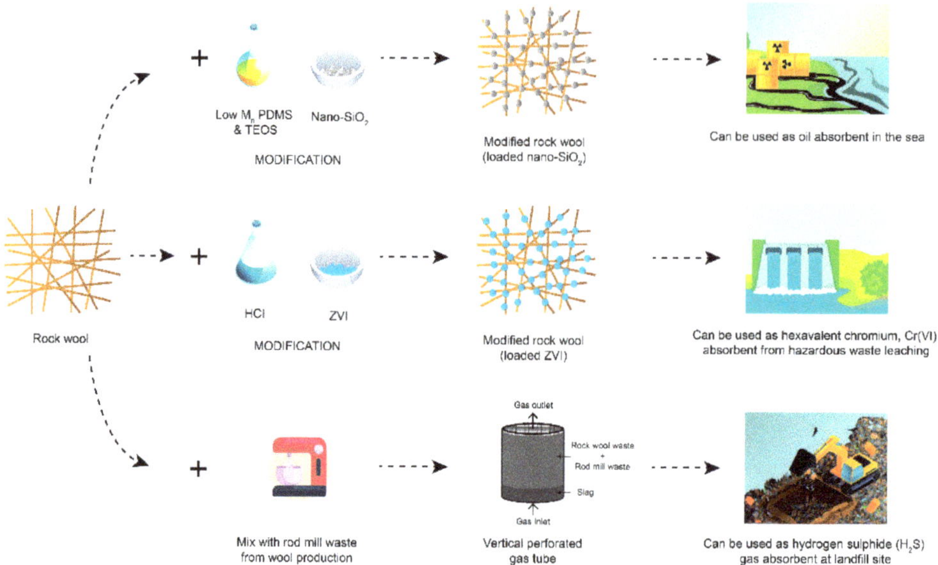

Figure 3. The applications of waste rock wool and its modification methods.

5.2. Hexavalent Chromium Cr(VI) Absorbent

Another study has succeeded in modifying waste rock wool by coating zero-valent iron (ZVI) nanoparticles onto its fiber surface to remove hexavalent chromium, Cr(VI), from water through adsorption and reduction [39]. Cr(VI) is a form metallic element that enters drinking water sources via dye discharges and paint pigments [117], wood preservatives, chrome plating wastes, and leaching from hazardous waste sites [118]. In this case, the waste rock wool was modified using hydrochloric acid to load more Zero-valent Iron (ZVI) nanoparticles onto the surface (Figure 3). Due to the steric hindrance effect, acid-modified waste rock wool could effectively limit the aggregation and oxidation of ZVI nanoparticles and successfully removed 197.69 mg/g of Cr(VI) in 30 min.

By comparing with other trendy absorbents such as bamboo biochar with a maximum absorption capacity of 61.88 mg/g [119], ethylenediaminetetraacetic acid—chitosan modified Fe$_3$O$_4$@SiO$_2$ (47.27 mg/g) [120], and diethylenediamine functionalized magnetic carbon-based adsorbents (22.24 mg/g) [121], the acid-modified waste rock wool has the most outstanding absorption performance. In conclusion, modified waste rock wool has great potential as an absorbent to remove Cr(VI); however, the removing efficiency on higher concentration pollutants could be investigated in future studies.

5.3. Hydrogen Sulfide (H2S) Gas Absorbent

For gas absorbance, at the early stage, a few studies discovered rock wool as a gas absorbent to remove wastes such as polycyclic aromatic hydrocarbons [122] and sulphur gas [123]. Recently, a few studies successfully established waste rock wool as a promising absorbent for toxic and corrosive biogas—hydrogen sulfide (H$_2$S) gas [124–126]. Biogas usually contains carbon dioxide, methane, and a small amount of hydrogen sulfide, which people usually utilize to generate energy. However, toxic H$_2$S gas will cause damage or corrosion to the combustion engines. Hence, contaminants in biogas, particularly hydrogen sulfide, must be eliminated [127].

The waste rock wool was added into rod mill waste at a ratio of 19:1, designed as a vertical perforated gas tube (Figure 3). H$_2$S gas was administered with a downward flow, and the rock wool filter succeeded in removing 87.9% [128] and 98% of the sulfide gas over 80 days. Another study emphasized that hydrogen sulfide gas's actual gas flow at the landfill is different from the study [129]. Because of this, the actual scale of H_2S gas elimination should be studied further in actual condition as well as its efficiency as compared to typical approaches such as electrochemical precipitation [130], ion exchange [131], membrane separation [132], and typical adsorption [133,134].

6. Conclusions

This paper has broadly reviewed the characteristics of waste rock wool, its recycling methods, and finally, its potential usages based on various studies. Several remarks are concluded as follow:

- Waste rock wool constitutes a high content of SiO_2 (38.7–60.1%), CaO (10 to 46.9%), and Al_2O_3 (0.8 to 18.6%) generally. This contributes to the enhancement of cementitious properties such as strength, chemical resistivity, and durability. Therefore, waste rock wool is regarded as a strong potential alternative to fillers in primarily concrete production.
- Incorporating waste rock wool into wood-based composites resulted in an increase in ignitability and durability, but deterioration in mechanical properties.
- The performance of composite materials after being incorporated with milled waste rock wool is disputable. This is attributed to the presence of various contaminants such as resin, dust, chemical substances, and mycological substances, depending on their source and handling method. It has posed a significant hurdle to formulate a standardized method to treat or recycle waste rock wool.
- Various pre-treating and reprocessing strategies such as heating, washing, and comminution are adopted to maximize the usability of waste rock wool. Heating can burn off the resin while washing could significantly reduce chlorine content. Comminution can produce a homogeneous mixture of particles that allow waste rock wool to be incorporated into different types of composites to improve desired properties.
- Waste rock wool is also promisingly utilized as an absorbent for spilled oil, hexavalent chromium contaminants, and hydrogen sulfide (H_2S) gas. The high heat stability and inorganic composition of rock wool are its notable features since it is mechanically stronger and may be used in high-temperature environments. However, the pre-treatments such as $PDMS/SiO_2$ nanoparticle dip-coating and Zero-valent Iron (ZVI) nanoparticles coating on the fiber surface are needed, which are neither energy nor cost-effective to the recycling process.

Results of the pertaining studies are more consistent in this regard and are worth exploring beyond current development. Nevertheless, comparison with other existing methods is encouraged to ensure its commercial viability and applicability in the field beyond laboratory setup.

7. Key Perspectives for Future Research Development

Generally, research on fibrous wastes is not as profound as other waste materials. To be more efficient and commercially viable in the recycling, repurposing, and reusing of waste rock wool, the following are suggested for future development:

Purity classification of waste rock wool from different sources needs to be performed prior to determining the suitable approach to handle the waste. This classification should primarily address the type of rock wool, source, prior usage, types of contaminants, physical condition, and chemical composition.

The current comminution process is neither energy nor cost-effective to pre-treat waste rock wool. This process is seen as the critical factor in driving down the overall cost and sparking more interest in research and development.

Author Contributions: Conceptualization, Z.S.Y., N.H.A.K., Z.H. and A.M.; methodology, Z.S.Y.; formal analysis, Z.S.Y., N.H.A.K.; writing—original draft preparation, Z.S.Y.; writing—review and editing, N.H.A.K., Z.H., A.M., M.M.T., S.H. and A.S.; funding acquisition, N.H.A.K., A.M., M.M.T., S.H. and A.S. All authors have read and agreed to the published version of the manuscript.

Funding: This work was supported by the Ministry of Higher Education Malaysia [R.J130000.7351.
4B525, Q.J130000.3051.01M94, Q.J130000.3651.03M24, Q.J130000.2451.04G54].

Institutional Review Board Statement: Not applicable.

Informed Consent Statement: Not applicable.

Data Availability Statement: Not applicable.

Conflicts of Interest: The authors declare no conflict of interest.

References

1. Al-Fakih, A.; Mohammed, B.S.; Liew, M.S.; Nikbakht, E. Incorporation of waste materials in the manufacture of masonry bricks: An update review. *J. Build. Eng.* **2018**, *21*, 37–54. [CrossRef]
2. Kaza, S.; Yao, L.; Bhada-Tata, P.; Van Woerden, F. *What a Waste 2.0: A Global Snapshot of Solid Waste Management to 2050*; The World Bank: Washington, DC, USA, 2018.
3. Vijayan, D.; Parthiban, D. Effect of Solid waste based stabilizing material for strengthening of Expansive soil- A review. *Environ. Technol. Innov.* **2020**, *20*, 101108. [CrossRef]
4. Yliniemi, J.; Kinnunen, P.; Karinkanta, P.; Illikainen, M. Utilization of Mineral Wools as Alkali-Activated Material Precursor. *Materials* **2016**, *9*, 312. [CrossRef]
5. Väntsi, O.; Kärki, T.; Ka, T.; Väntsi, O.; Kärki, T.; Ka, T. Mineral wool waste in Europe: A review of mineral wool waste quantity, quality, and current recycling methods. *J. Mater. Cycles Waste Manag.* **2014**, *16*, 62–72. [CrossRef]
6. Stephan, A.; Crawford, R.; de Myttenaere, K. A comprehensive assessment of the life cycle energy demand of passive houses. *Appl. Energy* **2013**, *112*, 23–34. [CrossRef]
7. Marinković, S.; Radonjanin, V.; Malešev, M.; Ignjatović, I. Comparative environmental assessment of natural and recycled aggregate concrete. *Waste Manag.* **2010**, *30*, 2255–2264. [CrossRef] [PubMed]
8. Duan, H.; Wang, J.; Huang, Q. Encouraging the environmentally sound management of C&D waste in China: An integrative review and research agenda. *Renew. Sustain. Energy Rev.* **2015**, *43*, 611–620.
9. Sharif, A.; Arshian, R.; Najmi, A.; Tseng, M.-L.; Lim, M.K. Dynamic and causality interrelationships from municipal solid waste recycling to economic growth, carbon emissions and energy efficiency using a novel bootstrapping autoregressive distributed lag. *Resour. Conserv. Recycl.* **2021**, *166*, 105372.
10. Ghisellini, P.; Ripa, M.; Ulgiati, S. Exploring environmental and economic costs and benefits of a circular economy approach to the construction and demolition sector. A literature review. *J. Clean. Prod.* **2018**, *178*, 618–643. [CrossRef]
11. Ntimugura, F.; Vinai, R.; Harper, A.B.; Walker, P. Environmental performance of miscanthus-lime lightweight concrete using life cycle assessment: Application in external wall assemblies. *Sustain. Mater. Technol.* **2021**, *28*, e00253.
12. Desai, D.A.; Sadiku, R.; Dunne, R.K. A review of porous automotive sound absorbers, their environmental impact and the factors that influence sound absorption. *Int. J. Veh. Noise Vib.* **2017**, *13*, 137. [CrossRef]
13. Shi, X.; Wu, J.; Wang, X.; Zhou, X.; Xie, X.; Xue, Z. Novel sound insulation materials based on epoxy/hollow silica nanotubes composites. *Compos. Part B Eng.* **2017**, *131*, 125–133. [CrossRef]
14. Kiss, B.; Manchón, C.G.; Neij, L. The role of policy instruments in supporting the development of mineral wool insulation in Germany, Sweden and the United Kingdom. *J. Clean. Prod.* **2013**, *48*, 187–199. [CrossRef]
15. Piñeiro, S.R.; Merino, M.D.R.; García, C.P. New Plaster Composite with Mineral Wool Fibres from CDW Recycling. *Adv. Mater. Sci. Eng.* **2015**, *2015*, 854192.
16. Ricciardi, P.; Belloni, E.; Cotana, F. Innovative panels with recycled materials: Thermal and acoustic performance and Life Cycle Assessment. *Appl. Energy* **2014**, *134*, 150–162. [CrossRef]
17. Islam, S.; Bhat, G. Environmentally-friendly thermal and acoustic insulation materials from recycled textiles. *J. Environ. Manag.* **2019**, *251*, 109536. [CrossRef] [PubMed]
18. Asdrubali, F. Survey on the Acoustical Properties of New Sustainable Materials for Noise Control. *Euronoise* **2006**, *30*, 1–10.
19. Li, P.; Yuan, Z.N.; Chen, X.M.; Xu, K.H.; Wu, Q.F. Corrosion Mechanism of Rock Wool in Sulfuric Acid. *Appl. Mech. Mater.* **2013**, *448–453*, 1260–1264. [CrossRef]
20. Wei, M.-S.; Huang, K.-H. Recycling and reuse of industrial wastes in Taiwan. *Waste Manag.* **2000**, *21*, 93–97. [CrossRef]
21. Chen, C.; Huang, R.; Wu, J.; Yang, C. Waste E-glass particles used in cementitious mixtures. *Cem. Concr. Res.* **2006**, *36*, 449–456. [CrossRef]
22. Sohn, J.L.; Kalbar, P.; Banta, G.T.; Birkved, M. Life-cycle based dynamic assessment of mineral wool insulation in a Danish residential building application. *J. Clean. Prod.* **2017**, *142*, 3243–3253. [CrossRef]
23. Ntimugura, F.; Vinai, R.; Harper, A.; Walker, P. Mechanical, thermal, hygroscopic and acoustic properties of bio-aggregates—Lime and alkali—Activated insulating composite materials: A review of current status and prospects for miscanthus as an innovative resource in the South West of England. *Sustain. Mater. Technol.* **2020**, *26*, e00211. [CrossRef]
24. Ding, Z.; Yi, G.; Tam, V.W.; Huang, T. A system dynamics-based environmental performance simulation of construction waste reduction management in China. *Waste Manag.* **2016**, *51*, 130–141. [CrossRef] [PubMed]

25. Wang, J.; Wu, H.; Duan, H.; Zillante, G.; Zuo, J.; Yuan, H. Combining life cycle assessment and Building Information Modelling to account for carbon emission of building demolition waste: A case study. *J. Clean. Prod.* **2017**, *172*, 3154–3166. [CrossRef]
26. Tao, H.; He, P.; Zhang, Y.; Sun, W. Performance evaluation of circulating fluidized bed incineration of municipal solid waste by multivariate outlier detection in China. *Front. Environ. Sci. Eng.* **2017**, *11*, 2704. [CrossRef]
27. Enyoghasi, C.; Badurdeen, F. Resources, Conservation & Recycling Industry 4.0 for sustainable manufacturing: Opportunities at the product, process, and system levels. *Resour. Conserv. Recycl.* **2021**, *166*, 105362.
28. Hossain, U.; Poon, C.S.; Lo, I.M.C.; Cheng, J.C. Comparative environmental evaluation of aggregate production from recycled waste materials and virgin sources by LCA. *Resour. Conserv. Recycl.* **2016**, *109*, 67–77. [CrossRef]
29. Aslam, M.S.; Huang, B.; Cui, L. Review of construction and demolition waste management in China and USA. *J. Environ. Manag.* **2020**, *264*, 110445. [CrossRef]
30. Kim, H.; Jang, Y.-C.; Hwang, Y.; Ko, Y.; Yun, H. End-of-life batteries management and material flow analysis in South Korea. *Front. Environ. Sci. Eng.* **2018**, *12*, 3. [CrossRef]
31. Schultz-Falk, V.; Agersted, K.; Jensen, P.A.; Solvang, M. Melting behaviour of raw materials and recycled stone wool waste. *J. Non-Cryst. Solids* **2018**, *485*, 34–41. [CrossRef]
32. Gualtieri, A.; Foresti, E.; Lesci, I.; Roveri, N.; Gualtieri, M.L.; Dondi, M.; Zapparoli, M. The thermal transformation of Man Made Vitreous Fibers (MMVF) and safe recycling as secondary raw materials (SRM). *J. Hazard. Mater.* **2009**, *162*, 1494–1506. [CrossRef] [PubMed]
33. Ramírez, C.P.; Sanchez, E.A.; Merino, M.D.R.; Arrebola, C.V.; Barriguete, A.V. Feasibility of the use of mineral wool fibres recovered from CDW for the reinforcement of conglomerates by study of their porosity. *Constr. Build. Mater.* **2018**, *191*, 460–468. [CrossRef]
34. Kinnunen, P.; Yliniemi, J.; Talling, B.; Illikainen, M. Rockwool waste in fly ash geopolymer composites. *J. Mater. Cycles Waste Manag.* **2017**, *19*, 1220–1227. [CrossRef]
35. Yliniemi, J.; Walkley, B.; Provis, J.; Kinnunen, P.; Illikainen, M. Nanostructural evolution of alkali-activated mineral wools. *Cem. Concr. Compos.* **2019**, *106*, 103472. [CrossRef]
36. Sormunen, P.; Kärki, T. Compression Molded Thermoplastic Composites Entirely Made of Recycled Materials. *Sustainability* **2019**, *11*, 631. [CrossRef]
37. Yliniemi, J.; Luukkonen, T.; Kaiser, A.; Illikainen, M. Mineral wool waste-based geopolymers. *IOP Conf. Ser. Earth Environ. Sci.* **2019**, *297*, 012006. [CrossRef]
38. Hao, W.; Xu, J.; Li, R.; Zhao, X.; Qiu, L.; Yang, W. Developing superhydrophobic rock wool for high-viscosity oil/water separation. *Chem. Eng. J.* **2019**, *368*, 837–846. [CrossRef]
39. Zhou, L.; Li, R.; Zhang, G.; Wang, D.; Cai, D.; Wu, Z. Zero-valent iron nanoparticles supported by functionalized waste rock wool for efficient removal of hexavalent chromium. *Chem. Eng. J.* **2018**, *339*, 85–96. [CrossRef]
40. Latif, E.; Lawrence, R.M.H.; Shea, A.D.; Walker, P. Energy & Buildings An experimental investigation into the comparative hygrothermal performance of wall panels incorporating wood fibre, mineral wool and hemp-lime. *Energy Build.* **2018**, *165*, 76–91.
41. Gnip, I.; Vaitkus, S.; Vėjelis, S. An estimated prediction of the deformability of mineral wool (MW) slabs under long-term compressive stress. *Constr. Build. Mater.* **2013**, *38*, 675–680. [CrossRef]
42. Ji, R.; Zheng, Y.; Zou, Z.; Chen, Z.; Wei, S.; Jin, X.; Zhang, M. Utilization of mineral wool waste and waste glass for synthesis of foam glass at low temperature. *Constr. Build. Mater.* **2019**, *215*, 623–632. [CrossRef]
43. Kohl, C.A.; Gomes, L.P. Physical and chemical characterization and recycling potential of desktop computer waste, without screen. *J. Clean. Prod.* **2018**, *184*, 1041–1051. [CrossRef]
44. Yörükoğlu, A.; Akkurt, F.; Çulha, S. Investigation of boron usability in rock wool production. *Constr. Build. Mater.* **2020**, *243*, 118222. [CrossRef]
45. Diani, M.; Pievatolo, A.; Colledani, M.; Lanzarone, E. A comminution model with homogeneity and multiplication assumptions for the Waste Electrical and Electronic Equipment recycling industry. *J. Clean. Prod.* **2018**, *211*, 665–678. [CrossRef]
46. Yliniemi, J.; Laitinen, O.; Kinnunen, P.; Illikainen, M. Pulverization of fibrous mineral wool waste. *J. Mater. Cycles Waste Manag.* **2018**, *20*, 1248–1256. [CrossRef]
47. Alves, J.O.; Espinosa, D.C.R.; Tenório, J.A.S. Recovery of Steelmaking Slag and Granite Waste in the Production of Rock Wool. *Mater. Res.* **2015**, *18*, 204–211. [CrossRef]
48. Lin, W.-T.; Han, T.Y.; Wu, Y.C.; Cheng, A.; Huang, R.; Weng, T.L. Engineered Properties of Sustainable Cement-Based Composites Containing Recycled Rock Wool and Supplementary Cementitious Materials. *Appl. Mech. Mater.* **2013**, *284–287*, 113–117. [CrossRef]
49. Stonys, R.; Kuznetsov, D.; Krasnikovs, A.; Škamat, J.; Baltakys, K.; Antonovič, V.; Černašėjus, O. Reuse of ultrafine mineral wool production waste in the manufacture of refractory concrete. *J. Environ. Manag.* **2016**, *176*, 149–156. [CrossRef]
50. Menon, S.; Naranje, V.G. Experimental Investigation of Recycling of Rock-Wool Insulation as Insulator in Concrete Blocks. *Int. J. Eng. Appl. Sci.* **2017**, *4*, 71–74.
51. Kubiliute, R.; Kaminskas, R.; Kazlauskaite, A. Mineral wool production waste as an additive for Portland cement. *Cem. Concr. Compos.* **2018**, *88*, 130–138. [CrossRef]
52. Namchan, P.; Denpetkul, T.; Pheinsusom, P.; Khaodhiar, S. Rock Wool and Printed Circuit Boards (PCBs) as Partial Fine Aggregate Replacements in Cement Mortar. *Key Eng. Mater.* **2018**, *775*, 567–575. [CrossRef]
53. Liu, J.; Yu, S.; Shang, Y. Toward separation at source: Evolution of Municipal Solid Waste management in China. *Front. Environ. Sci. Eng.* **2020**, *14*, 36. [CrossRef]

54. Hjelmgaard, T.; Thorsen, P.A.; Bøtner, J.A.; Kaurin, J.; Schmücker, C.M.; Nærum, L. Towards greener stone shot and stone wool materials: Binder systems based on gelatine modified with tannin or transglutaminase. *Green Chem.* **2018**, *20*, 4102–4111. [CrossRef]

55. Binici, H.; Aksogan, O. Eco-friendly insulation material production with waste olive seeds, ground PVC and wood chips. *J. Build. Eng.* **2016**, *5*, 260–266. [CrossRef]

56. Pedroso, M.; de Brito, J.; Silvestre, J. Characterization of eco-efficient acoustic insulation materials (traditional and innovative). *Constr. Build. Mater.* **2017**, *140*, 221–228. [CrossRef]

57. Okhrimenko, D.; Thomsen, A.; Ceccato, M.; Johansson, D.; Lybye, D.; Bechgaard, K.; Tougaard, S.; Stipp, S. Impact of curing time on ageing and degradation of phenol-urea-formaldehyde binder. *Polym. Degrad. Stab.* **2018**, *152*, 86–94. [CrossRef]

58. Dujardin, N.; Feuillet, V.; Garon, D.; Ibos, L.; Marchetti, M.; Peiffer, L.; Pottier, D.; Séguin, V.; Theile, D. Impacts of environmental exposure on thermal and mycological characteristics of insulation wools. *Environ. Impact Assess. Rev.* **2018**, *68*, 66–80. [CrossRef]

59. Kizinievič, O.; Balkevičius, V.; Pranckevičienė, J.; Kizinievič, V. Investigation of the usage of centrifuging waste of mineral wool melt (CMWW), contaminated with phenol and formaldehyde, in manufacturing of ceramic products. *Waste Manag.* **2014**, *34*, 1488–1494. [CrossRef]

60. Müller, A.; Leydolph, B.; Stanelle, K. Recycling Mineral Wool Waste—Technologies for the Conversion of the Fiber Structure, Part 1. *Interceram* **2009**, *58*, 378–381.

61. Utekar, S.; More, S.V.K.N.; Rao, A. Comprehensive study of recycling of thermosetting polymer composites—Driving force, challenges and methods. *Compos. Part B Eng.* **2021**, *207*, 108596. [CrossRef]

62. Sigvardsen, N.M.; Kirkelund, G.M.; Jensen, P.E.; Geiker, M.R.; Ottosen, L.M. Impact of production parameters on physiochemical characteristics of wood ash for possible utilisation in cement-based materials. *Resour. Conserv. Recycl.* **2019**, *145*, 230–240. [CrossRef]

63. Loginova, E.; Schollbach, K.; Proskurnin, M.; Brouwers, H. Municipal solid waste incineration bottom ash fines: Transformation into a minor additional constituent for cements. *Resour. Conserv. Recycl.* **2020**, *166*, 105354. [CrossRef]

64. Zhang, C.-Y.; Yu, B.; Chen, J.-M.; Wei, Y.-M. Green transition pathways for cement industry in China. *Resour. Conserv. Recycl.* **2020**, *166*, 105355. [CrossRef]

65. Mastali, M.; Dalvand, A.; Sattarifard, A.; Abdollahnejad, Z.; Nematollahi, B.; Sanjayan, J.; Illikainen, M. A comparison of the effects of pozzolanic binders on the hardened-state properties of high-strength cementitious composites reinforced with waste tire fibers. *Compos. Part B Eng.* **2018**, *162*, 134–153. [CrossRef]

66. Barrera, M.D.M.B.; Pombo, O.; Navacerrada, M. Textile fibre waste bindered with natural hydraulic lime. *Compos. Part B Eng.* **2016**, *94*, 26–33. [CrossRef]

67. Ayobami, A.B. Performance of wood bottom ash in cement-based applications and comparison with other selected ashes: Overview. *Resour. Conserv. Recycl.* **2020**, *166*, 105351. [CrossRef]

68. Deepankar, K.A. Concrete made with waste marble powder and supplementary cementitious material for sustainable development. *J. Clean. Prod.* **2019**, *211*, 716–729.

69. Juenger, M.C.; Siddique, R. Recent advances in understanding the role of supplementary cementitious materials in concrete. *Cem. Concr. Res.* **2015**, *78*, 71–80. [CrossRef]

70. Aprianti, E.S. A huge number of artificial waste material can be supplementary cementitious material (SCM) for concrete production—A review part II. *J. Clean. Prod.* **2017**, *142*, 4178–4194. [CrossRef]

71. Lin, W.-T.; Cheng, A.; Huang, R.; Wu, Y.-C.; Han, T.-Y. Rock wool wastes as a supplementary cementitious material replacement in cement-based composites. *Comput. Concr.* **2013**, *11*, 93–104. [CrossRef]

72. Cheng, A.; Lin, W.-T.; Huang, R. Application of rock wool waste in cement-based composites. *Mater. Des.* **2011**, *32*, 636–642. [CrossRef]

73. Acordi, J.; Luza, A.; Fabris, D.; Raupp-Pereira, F.; De Noni, A., Jr.; Montedo, O. New waste-based supplementary cementitious materials: Mortars and concrete formulations. *Constr. Build. Mater.* **2019**, *240*, 117877. [CrossRef]

74. Lin, W.-T.; Cheng, A.; Huang, R.; Zou, S.-Y. Improved microstructure of cement-based composites through the addition of rock wool particles. *Mater. Charact.* **2013**, *84*, 1–9. [CrossRef]

75. Kosior-Kazberuk, M.; Krassowska, J.; Ramirez, C.P. Post cracking behaviour of fibre reinforced concrete with mineral wool fibers residues. *MATEC Web Conf.* **2018**, *174*, 02016. [CrossRef]

76. Silva, K.D.C.; Silva, G.C.; Natalli, J.F.; Mendes, J.C.; Silva, G.J.B.; Peixoto, R.A.F. Rock Wool Waste as Supplementary Cementitious Material for Portland Cement-Based Composites. *ACI Mater. J.* **2018**, *115*, 653–662. [CrossRef]

77. Ramírez, C.P.; Merino, M.D.R.; Arrebola, C.V.; Barriguete, A.V.; Kosior-Kazberuk, M. Analysis of the mechanical behaviour of the cement mortars with additives of mineral wool fibres from recycling of CDW. *Constr. Build. Mater.* **2019**, *210*, 56–62. [CrossRef]

78. Mohammad, N.Z.; Shyong, Y.Z.; Haron, Z.; Ismail, M.; Mohamed, A.; Khalid, N.H.A. The Feasibility of Rock Wool Waste Utilisation in a Double-Layer Concrete Brick for Acoustic: A Conceptual Review. *J. Comput. Theor. Nanosci.* **2020**, *17*, 635–644. [CrossRef]

79. Ding, Y.; Zhang, X.; Wu, B.; Liu, B.; Zhang, S. Highly porous ceramics production using slags from smelting of spent automotive catalysts. *Resour. Conserv. Recycl.* **2020**, *166*, 105373. [CrossRef]

80. Lazorenko, G.; Kasprzhitskii, A.; Yavna, V.; Mischinenko, V.; Kukharskii, A.; Kruglikov, A.; Kolodina, A.; Yalovega, G. Effect of pre-treatment of flax tows on mechanical properties and microstructure of natural fiber reinforced geopolymer composites. *Environ. Technol. Innov.* **2020**, *20*, 101105. [CrossRef]

81. Zhang, P.; Zheng, Y.; Wang, K.; Zhang, J. A review on properties of fresh and hardened geopolymer mortar. *Compos. Part B: Eng.* **2018**, *152*, 79–95. [CrossRef]

82. Silva, G.; Kim, S.; Aguilar, R.; Nakamatsu, J. Natural fibers as reinforcement additives for geopolymers—A review of potential eco-friendly applications to the construction industry. *Sustain. Mater. Technol.* **2019**, *23*, e00132. [CrossRef]

83. Kozub, B.; Bazan, P.; Gailitis, R.; Korniejenko, K.; Mierzwiński, D. Foamed Geopolymer Composites with the Addition of Glass Wool Waste. *Materials* **2021**, *14*, 4978. [CrossRef]

84. Buruberri, L.; Tobaldi, D.M.; Caetano, A.; Seabra, M.P.; Labrincha, J. Evaluation of reactive Si and Al amounts in various geopolymer precursors by a simple method. *J. Build. Eng.* **2018**, *22*, 48–55. [CrossRef]

85. Hajimohammadi, A.; van Deventer, J.S. Dissolution behaviour of source materials for synthesis of geopolymer binders: A kinetic approach. *Int. J. Miner. Process.* **2016**, *153*, 80–86. [CrossRef]

86. Mehta, A.; Siddique, R. An overview of geopolymers derived from industrial by-products. *Constr. Build. Mater.* **2016**, *127*, 183–198. [CrossRef]

87. Khan, M.I.; Siddique, R. Utilization of silica fume in concrete: Review of durability properties. *Resour. Conserv. Recycl.* **2011**, *57*, 30–35. [CrossRef]

88. Moghadam, M.J.; Ajalloeian, R.; Hajiannia, A. Preparation and application of alkali-activated materials based on waste glass and coal gangue: A review. *Constr. Build. Mater.* **2019**, *221*, 84–98. [CrossRef]

89. Bernal, S.A.; Provis, J.L.; Walkley, B.; Nicolas, R.S.; Gehman, J.D.; Brice, D.G.; Kilcullen, A.R.; Duxson, P.; van Deventer, J.S.J. Gel nanostructure in alkali-activated binders based on slag and fly ash, and effects of accelerated carbonation. *Cem. Concr. Res.* **2013**, *53*, 127–144. [CrossRef]

90. Singh, G.B.; Subramaniam, K.V. Influence of processing temperature on the reaction product and strength gain in alkali-activated fly ash. *Cem. Concr. Compos.* **2018**, *95*, 10–18. [CrossRef]

91. Chen, T.-A.; Chen, J.-H.; Huang, J.-S. Effects of activator and aging process on the compressive strengths of alkali-activated glass inorganic binders. *Cem. Concr. Compos.* **2017**, *76*, 1–12. [CrossRef]

92. Sarker, P.; Mcbeath, S. Fire endurance of steel reinforced fly ash geopolymer concrete elements. *Constr. Build. Mater.* **2015**, *90*, 91–98. [CrossRef]

93. Salahuddin, M.M.; Norkhairunnisa, M.; Mustapha, F. A review on thermophysical evaluation of alkali-activated geopolymers. *Ceram. Int.* **2015**, *41*, 4273–4281. [CrossRef]

94. Albitar, M.; Ali, M.M.; Visintin, P.; Drechsler, M. Durability evaluation of geopolymer and conventional concretes. *Constr. Build. Mater.* **2017**, *136*, 374–385. [CrossRef]

95. Aiken, A.T.; Sha, W.; Kwasny, J.; Soutsos, M.N. Resistance of geopolymer and Portland cement based systems to silage effluent attack. *Cem. Concr. Res.* **2016**, *92*, 56–65. [CrossRef]

96. Van Deventer, J.S.; Nicolas, R.S.; Ismail, I.; A Bernal, S.; Brice, D.G.; Provis, J.L. Microstructure and durability of alkali-activated materials as key parameters for standardization. *J. Sustain. Cem. Mater.* **2014**, *4*, 116–128. [CrossRef]

97. Borges, P.H.R.; Banthia, N.; Alcamand, H.A.; Vasconcelos, W.; Nunes, E.H. Performance of blended metakaolin/blastfurnace slag alkali-activated mortars. *Cem. Concr. Compos.* **2016**, *71*, 42–52. [CrossRef]

98. Nguyen, H.; Kaas, A.; Kinnunen, P.; Carvelli, V.; Monticelli, C.; Yliniemi, J.; Illikainen, M. Fiber reinforced alkali-activated stone wool composites fabricated by hot-pressing technique. *Mater. Des.* **2019**, *186*, 108315. [CrossRef]

99. Walkley, B.; Nicolas, R.S.; Sani, M.-A.; Rees, G.; Hanna, J.V.; van Deventer, J.S.; Provis, J.L. Phase evolution of C-(N)-A-S-H/N-A-S-H gel blends investigated via alkali-activation of synthetic calcium aluminosilicate precursors. *Cem. Concr. Res.* **2016**, *89*, 120–135. [CrossRef]

100. Bernal, A.S.; Nicolas, R.S.; Myers, R.; DE Gutierrez, R.M.; Puertas, F.; van Deventer, J.S.; Provis, J. MgO content of slag controls phase evolution and structural changes induced by accelerated carbonation in alkali-activated binders. *Cem. Concr. Res.* **2014**, *57*, 33–43. [CrossRef]

101. Luukkonen, T.; Abdollahnejad, Z.; Yliniemi, J.; Kinnunen, P.; Illikainen, M. One-part alkali-activated materials: A review. *Cem. Concr. Res.* **2018**, *103*, 21–34. [CrossRef]

102. Adeniyi, A.; Onifade, D.; Ighalo, J.O.; Adeoye, S. A review of coir fiber reinforced polymer composites. *Compos. Part B Eng.* **2019**, *176*, 107305. [CrossRef]

103. Gharde, S.; Kandasubramanian, B. Mechanothermal and chemical recycling methodologies for the Fibre Reinforced Plastic (FRP). *Environ. Technol. Innov.* **2019**, *14*, 100311. [CrossRef]

104. Dobránsky, J.; Běhálek, L.; Baron, P.; Kočiško, M.; Dulebová, L.; Doboš, Z. The influence of the use of technological waste on the mechanical behavior of fibrous polymer composite. *Compos. Part B Eng.* **2019**, *166*, 162–168. [CrossRef]

105. Khan, A.; Jagdale, P.; Castellino, M.; Rovere, M.; Jehangir, Q.; Mandracci, P.; Rosso, C.; Tagliaferro, A. Innovative functionalized carbon fibers from waste: How to enhance polymer composites properties. *Compos. Part B Eng.* **2018**, *139*, 31–39. [CrossRef]

106. Senthivel, K.; Manikandan, K.; Prabu, B. Studies on the Mechanical Properties of Carbon Black/Halloysite Nanotube Hybrid Fillers in Nitrile Rubber Nanocomposites. *Mater. Today: Proc.* **2015**, *2*, 3627–3637. [CrossRef]

107. Ivanoska-Dacikj, A.; Bogoeva-Gaceva, G.; Valić, S.; Wießner, S.; Heinrich, G. Benefits of hybrid nano-filler networking between organically modified Montmorillonite and carbon nanotubes in natural rubber: Experiments and theoretical interpretations. *Appl. Clay Sci.* **2017**, *136*, 192–198. [CrossRef]

108. Bredikhin, P.; Kadykova, Y. Waste Stone Wool as an Effective Filler for Polyethylene. *Int. Polym. Sci. Technol.* **2017**, *44*, 41–44. [CrossRef]

109. Sommerhuber, P.F.; Welling, J.; Krause, A. Substitution potentials of recycled HDPE and wood particles from post-consumer packaging waste in Wood–Plastic Composites. *Waste Manag.* **2015**, *46*, 76–85. [CrossRef]

110. Mamiński, M.; Król, M.E.; Jaskółowski, W.; Borysiuk, P. Wood-mineral wool hybrid particleboards. *Eur. J. Wood Wood Prod.* **2011**, *69*, 337–339. [CrossRef]

111. Väntsi, O.; Kärki, T. Utilization of recycled mineral wool as filler in wood–polypropylene composites. *Constr. Build. Mater.* **2014**, *55*, 220–226. [CrossRef]

112. Brum Dutra da Rocha, E.; de Sousa, A.M.F.; Furtado, C.R.G. Properties Investigation of novel nitrile rubber composites with rockwool fibers. *Polym. Test.* **2020**, *82*, 106291. [CrossRef]

113. Deng, Y.-H.; Furuno, T. Properties of gypsum particleboard reinforced with polypropylene fibers. *J. Wood Sci.* **2001**, *47*, 445–450. [CrossRef]

114. Gołub, A.; Piekutin, J. Use of porous materials to remove oil contaminants from water. *Sci. Total Environ.* **2018**, *627*, 723–732. [CrossRef] [PubMed]

115. Wang, J.; Lou, H.; Cui, Z.-H.; Hou, Y.; Li, Y.; Zhang, Y.; Jiang, K.; Shi, W.; Qu, L. Fabrication of porous polyacrylamide/polystyrene fibrous membranes for efficient oil-water separation. *Sep. Purif. Technol.* **2019**, *222*, 278–283. [CrossRef]

116. Al Zubaidi, I.A.; Al Tamimi, A.K.; Ahmed, H. Remediation of water from crude oil spill using a fibrous sorbent. *Environ. Technol. Innov.* **2016**, *6*, 105–114. [CrossRef]

117. Khamparia, S.; Jaspal, D.K. Adsorption in combination with ozonation for the treatment of textile waste water: A critical review. *Front. Environ. Sci. Eng.* **2017**, *11*, 8. [CrossRef]

118. Han, Y.; Xie, H.; Liu, W.; Li, H.; Wang, M.; Chen, X.; Liao, X.; Yan, N. Assessment of pollution of potentially harmful elements in soils surrounding a municipal solid waste incinerator, China. *Front. Environ. Sci. Eng.* **2016**, *10*, 7. [CrossRef]

119. Liu, W.; Yu, Y. Removal of recalcitrant trivalent chromium complexes from industrial wastewater under strict discharge standards. *Environ. Technol. Innov.* **2021**, *23*, 101644. [CrossRef]

120. Wang, J.; Mao, M.; Atif, S.; Chen, Y. Adsorption behavior and mechanism of aqueous Cr(III) and Cr(III)-EDTA chelates on DTPA-chitosan modified Fe3O4@SiO2. *React. Funct. Polym.* **2020**, *156*, 104720. [CrossRef]

121. Wang, J.; Chen, Y.; Sun, T.; Saleem, A.; Wang, C. Enhanced removal of Cr(III)-EDTA chelates from high-salinity water by ternary complex formation on DETA functionalized magnetic carbon-based adsorbents. *Ecotoxicol. Environ. Saf.* **2020**, *209*, 111858. [CrossRef]

122. Gerde, P.; Scholander, P. Adsorption of polycyclic aromatic hydrocarbons on to asbestos and man-made mineral fibres in the gas phase. In *Non-occupational Exposure to Mineral Fibers*; Bignon, J., Peto, J., Saracci, R., Eds.; IARC Sci. Publ: Lyon, France, 1989.

123. Chan, A.A.; Grennberg, K. Attempted biofiltration of reduced sulphur compounds from a pulp and paper mill in Northern Sweden. *Environ. Prog.* **2006**, *25*, 152–160. [CrossRef]

124. De Crisci, A.G.; Moniri, A.; Xu, Y. Hydrogen from hydrogen sulfide: Towards a more sustainable hydrogen economy. *Int. J. Hydrogen Energy* **2018**, *44*, 1299–1327. [CrossRef]

125. Rubright, S.L.M.; Pearce, L.L.; Peterson, J. Nitric Oxide Environmental toxicology of hydrogen sulfide. *Nitric Oxide* **2017**, *71*, 1–13. [CrossRef]

126. Ali, M.; Zhang, J.; Raga, R.; Lavagnolo, M.C.; Pivato, A.; Wang, X.; Zhang, Y.; Cossu, R.; Yue, D. Effectiveness of aerobic pretreatment of municipal solid waste for accelerating biogas generation during simulated landfilling. *Front. Environ. Sci. Eng.* **2018**, *12*, 5. [CrossRef]

127. Kougias, P.G.; Angelidaki, I. Biogas and its opportunities—A review Keywords. *Front. Environ. Sci. Eng.* **2018**, *12*, 1–22. [CrossRef]

128. Galera, M.M.; Cho, E.; Tuuguu, E.; Park, S.J.; Lee, C.; Chung, W.J. Effects of pollutant concentration ratio on the simultaneous removal of NH3, H2S and toluene gases using rock wool-compost biofilter. *J. Hazard. Mater.* **2008**, *152*, 624–631. [CrossRef] [PubMed]

129. Bergersen, O.; Haarstad, K. Treating landfill gas hydrogen sulphide with mineral wool waste (MWW) and rod mill waste (RMW). *Waste Manag.* **2014**, *34*, 141–147. [CrossRef] [PubMed]

130. Sun, T.R.; Pamukcu, S.; Ottosen, L.M.; Wang, F. Electrochemically enhanced reduction of hexavalent chromium in contaminated clay: Kinetics, energy consumption, and application of pulse current. *Chemica* **2015**, *262*, 1099–1107. [CrossRef]

131. Xiao, K.; Xu, F.; Jiang, L.; Duan, N.; Zheng, S. Resin oxidization phenomenon and its influence factor during chromium(VI) removal from wastewater using gel-type anion exchangers. *Chem. Eng. J.* **2016**, *283*, 1349–1356. [CrossRef]

132. Kaya, A.; Onac, C.; Alpoguz, H.K.; Yilmaz, A.; Atar, N. Removal of Cr(VI) through calixarene based polymer inclusion membrane from chrome plating bath water. *Chem. Eng. J.* **2016**, *283*, 141–149. [CrossRef]

133. Valentín-Reyes, J.; García-Reyes, R.B.; García-González, A.; Soto-Regalado, E.; Cerino-Córdova, F. Adsorption mechanisms of hexavalent chromium from aqueous solutions on modified activated carbons. *J. Environ. Manag.* **2019**, *236*, 815–822. [CrossRef] [PubMed]

134. Tran, H.N.; Nguyen, D.T.; Le, T.G.; Tomul, F.; Lima, E.C.; Woo, S.H.; Sarmah, A.K.; Nguyen, H.Q.; Nguyen, P.T.; Nguyen, D.D.; et al. Adsorption mechanism of hexavalent chromium onto layered double hydroxides-based adsorbents: A systematic in-depth review. *J. Hazard. Mater.* **2019**, *373*, 258–270. [CrossRef] [PubMed]

Article

Beyond Operational Energy Efficiency: A Balanced Sustainability Index from a Life Cycle Consideration

Ming Hu [1,2]

1   School of Architecture, Planning & Preservation, University of Maryland, College Park, MD 20742, USA;
    mhu2008@umd.edu
2   Faculty of Architecture, Civil Engineering and Applied Arts, University of Technology, Rolna 43,
    40-555 Katowice, Poland

Abstract: Most deep energy renovation projects focus only on an operating energy reduction and disregard the added embodied energy derived from adding insulation, window/door replacement, and mechanical system replacement or upgrades. It is important to study and address the balance and trade-offs between reduced operating energy and added embodied energy from a whole life cycle perspective to reduce the overall building carbon footprint. However, the added embodied energy and related environmental impact have not been studied extensively. In response to this need, this paper proposes a holistic sustainability index that balances the trade-off between reduced operating energy and added embodied energy. Eight case projects are used to validate the proposed method and calculation. The findings demonstrate that using a balanced sustainability index can reveal results different from a conventional operating energy-centric approach: (a) operating energy savings can be offset by the embodied energy gain, (b) the operating energy savings do not always result in a life cycle emissions reduction, and (c) the sustainability index can vary depending on the priorities the decision makers give to operating carbon, embodied carbon, and operating cost. Overall, the proposed sustainability score can provide us with a more comprehensive understanding of how sustainable the renovation works are from a life cycle carbon emissions perspective, providing a more robust estimation of global warming potential related to building renovation.

Keywords: sustainability index; life cycle consideration; operational energy efficiency

Citation: Hu, M. Beyond Operational Energy Efficiency: A Balanced Sustainability Index from a Life Cycle Consideration. *Sustainability* **2021**, *13*, 11263. https://doi.org/10.3390/su132011263

Academic Editors: Carlos Morón Fernández and Daniel Ferrández Vega

Received: 31 August 2021
Accepted: 5 October 2021
Published: 13 October 2021

Publisher's Note: MDPI stays neutral with regard to jurisdictional claims in published maps and institutional affiliations.

1. Introduction

In Europe, the existing building stock is more than 50 years old, and about 40% of the existing residential buildings were constructed before the 1960s, when building regulations for energy consumption were limited [1]. Consequently, around 75% of the existing building stock in the European Union (EU) is energy inefficient [2]. In the United States, most existing houses were built before the establishment of the Building Energy Codes Program in 1992 by the U.S. Department of Energy, Washington, DC, United States [3]. These older buildings represent about 68% of the national residential building stock and are typically energy inefficient due to air leakage and inadequate insulation [2]. The National Renewable Energy Laboratory has identified approximately 34.5 million homes with wood studs that have no wall insulation [4]. Overall, in both the United States and Europe, a large portion of residential buildings will need some type of renovation, retrofit, or upgrade in the next five to 10 years.

There have been large investments in energy efficiency-related renovation in the global market. In the period 2012–2016, in the EU, more than EUR 200 billion were invested in energy renovations for residential buildings. In the next decades, energy renovation will become the key determiner for achieving the carbon-neutral goal. The European Climate Foundation has outlined three key areas for the building industry to maintain its trajectory toward zero emissions. One of the areas is reducing energy demand—specifically the operating energy demand—through renovation of the building stock [5]. In the United

Sates, more than USD 279 billion could be invested in building retrofits, resulting in more than USD 1 trillion in energy savings over 10 years, equal to a savings of about 30% of the annual electricity used in the United States (Rockefeller Foundation 2012, New York City, NK, The United States) [6]. The above figures highlight the significant impact building retrofits already have within the EU and American economies. There is space for tremendous potential and growth in both the European Union and the United States.

Currently, the building sector responds to the need for energy retrofits by focusing on an operating energy use reduction. Increasing numbers of companies have announced their commitment to the net zero carbon goal, based on the assumption that operating energy savings leads to an overall carbon emissions reduction and a healthy environment. However, the question remains: Is the current energy-centric renovation approach sustainable? Several studies have looked at the relation between operating energy and embodied energy. Dodoo et al. [7] found that an increase in the thermal mass in the building envelope reduced the cooling load, and hence the operating energy demand; however, such a renovation increases embodied energy considerably. Ellura et al. [8] adopted a life cycle approach for studying net zero energy building, and results showed that when the addition of embodied energy was included in the whole life cycle emission count, the net zero building performance largely shifted away from the nearly zero energy goal. Hu studied energy-efficient renovations in comparison to existing buildings, and showed that the new construction had greater environmental impact potential due to the new building materials added. Such results raised concerns of focusing only on an operating energy reduction while overlooking the added environmental impact [9].

In recent years, there have been studies focusing on the trade-off between embodied and operating carbon. Crawford et al. [10] studied the impact of different building materials of eight residential construction assemblies; a theoretical generic building was used as a base building. Rossello-Batel et al. [11] studied the relation between reduced heating demand and the embodied energy of different building typologies and building envelope options. They found that adding additional insulation in the façade can reduce energy demand to one third of the existing condition while having the highest increase in embodied energy. Stephan et al. [12] also found an increase in insulation in passive houses could reduce the heating demand in the winter, but such a decrease was offset by the higher embodied energy embedded in the insulation materials. With the increase in research on the relation between embodied and operating carbon, most studies were conducted on theoretical conditions using simulated data. Studies using data from actual renovated buildings are limited due to inaccessibility of the data.

The importance of understanding the trade-off between operating and embodied energies and their related carbon emissions has been gradually recognized by practitioners and researchers. Consequently, creating a comprehensive and holistic measurement of sustainability for building energy retrofits has become an emerging research topic. However, there have been very few studies and efforts on this topic, and proposed measurements vary greatly. For example, Bakar et al. [13] proposed using an energy efficiency index as an indicator for measuring building energy performance. Such an index is calculated as the ratio of the energy input to the factor related to the energy-using component. The embodied energy was included as one related factor and measured by the weight of the raw material used. Varusha et al. [14] suggested using the EE factor to quantify the trade-off between the embodied and operating energies of a building. The EE factor is calculated as the ratio of operating energy to embodied energy of a proposed building design against the ratio of a base building based on the ASHRAE 2016 benchmark. Triana et al. [15] proposed a sustainability index in the building life cycle energy use that includes life cycle energy consumption, life cycle carbon emissions, thermal comfort hours, and cost of the building energy life cycle. Those four values are added together and then divided by four to get the sustainability index. However, there is no sustainability index proposed especially for a building retrofit yet.

To respond to such a research gap, the aim of this study is to reveal the importance of considering embodied energy in current energy retrofit practices since the most energy-efficient building is not necessarily the most sustainable building. Consequently, a comprehensive measure of the sustainability of a renovation project is proposed to measure the effectiveness of a building energy retrofit by integrating the life cycle assessment. This paper uses eight actual energy retrofit projects to demonstrate the trade-off between operating energy and embodied energy. The renovation techniques applied to the eight buildings include a building envelope retrofit, a building heating and ventilation system renovation and upgrades, a lighting system upgrade, and other renovation techniques. This study contributes mainly to the body of knowledge of sustainability by (1) highlighting the importance of embodied energy consideration in energy retrofit projects, (2) presenting a new measure for a sustainability index for renovation projects, and (3) testing the proposed sustainability index and evaluating the sensitivity of the results by applying them to real projects.

2. Method

2.1. Studied Buildings

2.1.1. Selection of Buildings

The selection of buildings was largely based on the data availability. Often the actual building energy use data after renovation were not easy to obtain. Eight projects were used in this study to control the variables of building size, age, building system used, and local climate condition. All eight buildings are located in the same city and were built around similar periods, with the energy retrofits mainly focused on the buildings' heating system and exterior façade. Demonstrated in Figure 1 the eight buildings were part of the "European cities serving as Green Urban Gate towards leadership in sustainable energy" (EU_GUGLE) project [16]. The project aimed to demonstrate the feasibility of a nearly zero energy building renovation target; it started in 2013 and lasted for six years. Six pilot cities from Italy, Austria, Finland, Denmark, Estonia, and Slovakia participated in the project. Around 200,000 m^2 of gross floor area was renovated and targeted primary (source) energy savings of up to 82%. Eight buildings in Tampere, Finland, participated in EU_GUGLE. All eight buildings are in the Tammela district, a traditional residential district close to the city center and railway station. There is a total of around 299,000 m^2 of existing building stock in Tammela district. The current average energy use intensity is 213 kWh/m^2, and the target intensity is 160 kWh/m^2 [17], around a 25% operating energy reduction.

Figure 1. Studied project location map.

2.1.2. Case Building Physical Characteristics

The studied buildings were built between 1961 and 1980; the renovations were completed between 2014 and 2019. Seven buildings have six floors and one has four floors—only the four-story building was built in the first half of the 1960s. All eight buildings selected reflected a typical building constructed in the 1960s and 1970s before building energy regulations were enforced in Finland [18]. According to Niemelä et al., 2017, Finnish multi-family buildings constructed in the first half of the 1960s were normally built on-site with a tile building façade and "bookshelf-type" framework [19]. Beginning in the late 1960s, prefabricated large concrete panels became the main construction type [19]. In fact, all the studied buildings had prefabricated concrete panels. Table 1 shows the main characteristics of the studied buildings—the buildings' initial conditions without any energy performance improvement interventions.

Table 1. Physical characteristics of the studied case buildings.

	Case 1	Case 2	Case 3	Case 4	Case 5	Case 6	Case 7	Case 8
Construction year	1961	1968	1970	1971	1974	1978	1973	1980
Renovation year	2017	2017	2017	2017	2015	2014	2017	2017
Gross area (m^2)	1960	3693	5395	5554	2488	3024	4117	6060
# of units	72	94	148	61	70	108	60	72
# of floors	4	6	6	6	6	6	6	6
U-value of envelope (W/m^2 K) [18]								
External wall	0.81	0.81	0.81	0.81	0.81	0.34	0.81	0.34
Floor	0.47	0.47	0.47	0.47	0.47	0.38	0.47	0.38
Roof	0.47	0.47	0.47	0.47	0.47	0.26	0.47	0.26
Door	2.2	2.2	2.2	2.2	2.2	1.4	2.2	1.4
Window	1.7	1.7	1.7	1.7	1.7	1.7	1.7	1.7
Ventilation								
Type	Mech exh	Mech exh	Mech exh	Mech exh	Mech exh	Mech exh	Mech exh	Mech exh
Heat recovery efficiency	0	0	0	0	0	0	0	0
Air exchange rate	0.5	0.5	0.5	0.5	0.5	0.5	0.5	0.5
Room air temperature setpoint	22	22	22	22	22	22	22	22
Envelope area (m^2)	3938	10,738	6239	9582	15,750	6510	11,550	6345
Window area (m^2)	1181	3758	2496	2875	3937	2278	3465	1903

2.1.3. Building Envelope and Materials Used

The primary existing structural systems and materials used in the studied buildings are listed in Table 2. As illustrated in Figure 2a, during the renovation, additional insulation was attached directly to the existing external wall structure; then, new plaster was painted over the insulation for protection. For calculation, the same renovated wall type was used for all the case buildings, but the thickness of the added insulation varied according to the information found on the project website and provided by the project team. Not all case buildings had renovated roofs. For those buildings that did, three additional layers were added: a new polyurethane insulation layer, a light gravel layer, and lightweight concrete (functions as insulation). In addition, a new double layer bitumen membrane was added for waterproofing (refer to Figure 2b and Table 2). The existing exterior windows and doors were mainly made of wooden frames, and all case buildings had them replaced with more energy-efficient exterior windows and doors.

Table 2. Existing structural and envelope systems and materials used.

	Existing Building [20]	
	1960s	1970–1980
External wall	Concrete structure wall + mineral wool insulation (+/−100 mm) + prefabricated panel (+/−50 mm) or brick.	Concrete structure wall + mineral wool insulation (+/−100 mm) + prefabricated panel (+/−50 mm).
Roof system	Sloped roof, open-air strike, wooden frame as structure, seamed tin roof as roofing material.	Flat roof, hollow concrete slab (+/−250 mm) + foam-based insulation (+/−125 mm) + roof plaster.
Floor system	Load-bearing concrete hollow core slab.	Load-bearing concrete hollow core slab.
Internal wall	Brick wall, plastered and painted.	Made of a bent sheet metal element with strips about 50 cm wide. Sheet metal element, plastic coated, partially tiled. Plated, wood, or metal frame.
Window	Wood window frame, some glass panels are operable.	Wood window frame, some glass panels are operable.
Exterior door	Wooden door with glass opening.	Replace with new door.

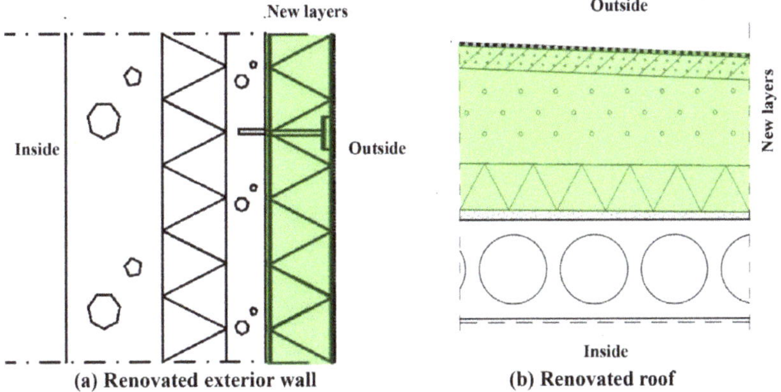

Figure 2. Building envelope renovation. (**a**) is the exterior wall details, (**b**) is roof details.

2.1.4. Building Service Systems

The information on existing building service systems—including heating, ventilation, and plumbing systems—was provided by the project team and extracted from the project website [16]. The renovations applied to each case building were also obtained from the project report, which is publicly accessible information [16]. The breakdown description for each individual building can be found in Table 3 and is explained in the following section.

2.2. Renovation Strategies and Measures

The technical improvements applied to the projects to reduce the operating energy demand are listed below as R1 through R15. Tables 3 and 4 list the applied renovation techniques for each case building.

Table 3. Renovation techniques.

R1	Exterior Wall: Additional Mineral Wool Insulation (+/−200 mm) + Plaster Render
R2	Roof: additional light gravel layer (+/−200 mm) + additional polyurethane insulation (+/−200 mm) + lightweight concrete (+/−40 mm) + bitumen membrane waterproofing
R3	Replace existing windows with energy-efficient ones
R4	Replace existing doors with energy-efficient ones
R5	Renew the thermostat radiator valves and adjust the heating network
R6	Add heat recovery from exhaust air in the ventilation system
R7	Air-to-air heat pump
R8	Air-to-water heat pump
R9	Ground-source heat pump (for heating and cooling)
R10	Connect to municipal district heating network
R11	Replace existing lighting with LED lights
R12	Add meter to monitor the water consumption
R13	Remote energy use monitoring
R14	Replace water faucets with more energy-efficient ones
R15	Replace existing elevator with more energy-efficient ones
NC	No change

Table 4. Applied renovation techniques for each individual case building.

	Case 1	Case 2	Case 3	Case 4	Case 5	Case 6	Case 7	Case 8
External wall	R1	R1	NC	NC	NC	R1	NC	R1
Roof system	R2	R2	NC	NC	NC	R2	NC	R2
Exterior window	R3	R3	R3	R3		R3	R3	R3
Exterior door	R3	R3	R3	R3	NC	R3	R3	R3
Internal wall	NC	NC	NC	NC	NC	NC	NC	NC
Floor system	NC	NC	NC	NC	NC	NC	NC	NC
Lighting	R11	R11				R11	R11	R11
Heating system	R9	R7, R10	R5, R8, R10	R5, R9	R8, R9, R10	R5, R9, R10	R5, R7, R9, R10	R5, R7, R10
Ventilation	R6	R6	R6	R6	R6	R6	R6	R6
Building management	R13	R13	R13	R13	R13	R13	R13	R13
Water supply	NC	NC	R14	NC	NC	NC	R12	R12, R14
Vertical transportation	NC	NC	R15	NC	NC	NC	NC	NC

2.3. Embodied Energy and Carbon Emissions Calculation

The software One Click LCA™, developed by the Finnish private company Bionova Ltd., was chosen for this study [21]. The software complies with EN 15987 and EN 15804 standards [17], and EN 15804 is a guideline for Environmental Product Declarations based on the ISO 14044 standard. One Click LCA includes the building material database, which is European original and Finland specific [22]. In this project, life cycle carbon emissions were calculated for individual case projects. To normalize the added embodied energy, only renovated components were included: the building envelope, heating/ventilation system, and lighting system. Structural systems and other building service systems were excluded, as they were not changed. Furniture and interior finishes were excluded as well. The life stages included in this study were A through C. As illustrated in Figure 3, stage A is the product and construction stage and includes A1 through A5. A1 through A3 is usually called "cradle to gate," and A1 through A5, "cradle to site." Stage B is the use stage, and stage C is the end-of-life stage. A1 through C3 are typically named "cradle to grave." The data used to create the LCA model were extracted from original construction documents provided by the project team and the author's visual inspection on site. Information extracted from the EU-GUGLE website, publications, presentations, and other available information can be found online. A life span of 50 years was used for the calculation, and the product service life was set as the default; for example, wood panels, the roof, and windows were set to be replaced once during the building's lifetime (50 years), and doors were set to be changed twice during the building's lifetime. In addition, in the One Click LCA database, transportation carbon was included during the production stage [23].

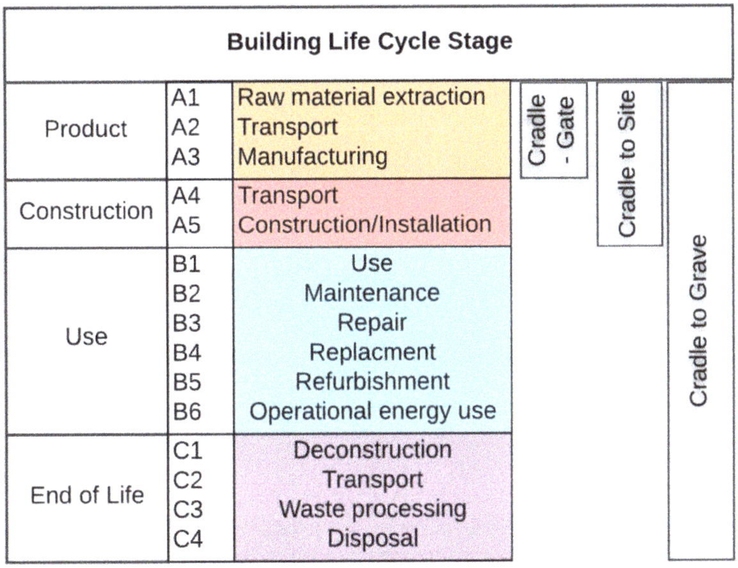

Building Life Cycle Stage						
Product	A1	Raw material extraction	Cradle - Gate	Cradle to Site	Cradle to Grave	
	A2	Transport				
	A3	Manufacturing				
Construction	A4	Transport				
	A5	Construction/Installation				
Use	B1	Use				
	B2	Maintenance				
	B3	Repair				
	B4	Replacment				
	B5	Refurbishment				
	B6	Operational energy use				
End of Life	C1	Deconstruction				
	C2	Transport				
	C3	Waste processing				
	C4	Disposal				

Figure 3. Building life cycle stages.

2.4. Proposed Sustainability Index

In this study, sustainability of the renovation project was measured by the balance (trade-off) between reduced carbon emissions (through operating energy savings) and added carbon emissions (through added building materials and systems). The framework proposed by Moran et al., 2017 was adopted and modified to calculate the sustainability of a retrofit solution, using Equation (1) through Equation (4) [23]:

$$SC_n = \frac{a\text{OES}_n + b\text{EMB}_n + c\text{ECO}_n}{k} \tag{1}$$

where a, b, and c are weighting factors for each of the respective categories; k is $\sum(a,b,c)$; OES_n is the life cycle carbon reduction due to the operating energy savings of case project n, measured by the operational cost savings (CO_2-eq/m^2); EMB_n is the life cycle carbon increase due to the embodied energy added, measured by the carbon emissions equivalent (CO_2-eq/m^2); and ECO_n is the economic impact of case project n, measured by the operational cost savings (USD/m^2). The calculation of OES_n, EMB_n, and ECO_n can be expressed mathematically as Equation (2) through Equation (4):

$$\text{OES}_n = \sum_{m=1}^{q} \left[\left(\frac{oes_{m,n}}{\left(\frac{\sum_{n=1}^{p} oes_{m,n}}{p} \right)} \right) w_m \right] \tag{2}$$

$$\text{EMB}_n = \sum_{m=1}^{q} \left[\left(\frac{emb_{m,n}}{\left(\frac{\sum_{n=1}^{p} emb_{m,n}}{p} \right)} \right) w_m \right] \tag{3}$$

$$\text{EMB}_n = \sum_{m=1}^{q} \left[\left(\frac{eco_{m,n}}{\left(\frac{\sum_{n=1}^{p} eco_{m,n}}{p} \right)} \right) w_m \right] \tag{4}$$

where $oes_{m,n}$ is the life cycle carbon reduction indicator m for case project n; m stands for the different operating energy, electricity, and district heating; $emb_{m,n}$ is the life cycle carbon increase indicator m for case project n; m represents the building elements, such as the exterior wall and windows; $eco_{m,n}$ is the economic indicator m for case project n; w_m is the weighting applied for each indicator depending on the category's importance; q is the number of the indicators evaluated in each carbon emissions reduction, carbon emissions increase, and economic category; and p is the total floor area measured. The sum of the weightings ($\sum(w_m)$) applied to indicators in each category must add up to 1.

As can be seen from Equation (1), each of the three categories for the sustainability score can be given a different weighting (a, b, c) depending on the importance of the category. The importance of the categories for different stakeholders involved in the energy retrofit projects can vary. For example, the energy supplier and building operators' priority is most likely the operating energy savings. However, for environmental protection agencies that are involved in permit review, the added embodied carbon is equally critical since it can lead to unintended environmental impacts. For building owners, the operational cost may be the primary reason for choosing retrofit solutions. In Section 3 the impact of different weightings on the sustainability index are demonstrated.

3. Findings

Table 5 shows the basic data for the studied case buildings. We will first discuss the results of the overall life cycle carbon emissions and contributors to added embodied carbon. Then, we will explain the results of applying the proposed sustainability index to the studied cases. Figure 4 illustrates the normalized whole life cycle carbon emissions of the studied case buildings (Figure 4a is the normalized carbon emission by floor area and Figure 4b is the normalized carbon emission by unit counts), and Figure 5 shows the trade-off between the reduced operating carbon and added embodied carbon.

Table 5. Case building data.

Case #	Size (m^2)	# of Units	# of Floors	Energy Intensity (After)	Energy Intensity (Before)	Operational Cost
				(kWh/m^2)	(kWh/m^2)	($/m^2/Year)
1	1960	20	4	72	291	6
2	3693	94	6	94	174	8
3	5395	148	6	125	190	10
4	5554	61	6	61	172	5
5	2488	70	6	70	187	6
6	3024	108	6	73	130	6
7	4117	60	6	24	140	2
8	6060	72	6	72	137	6

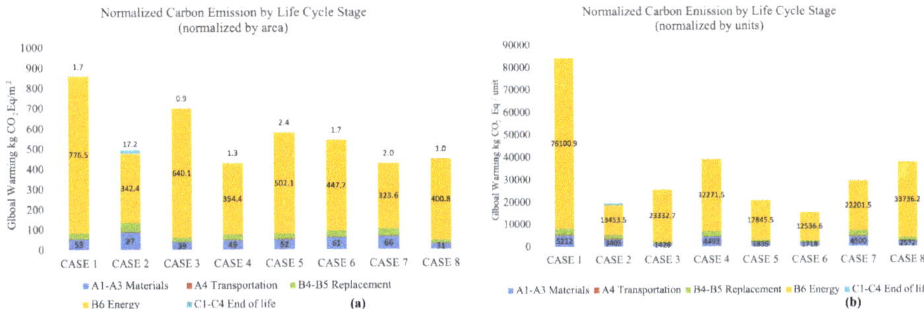

Figure 4. Normalized carbon emissions by life cycle stage.

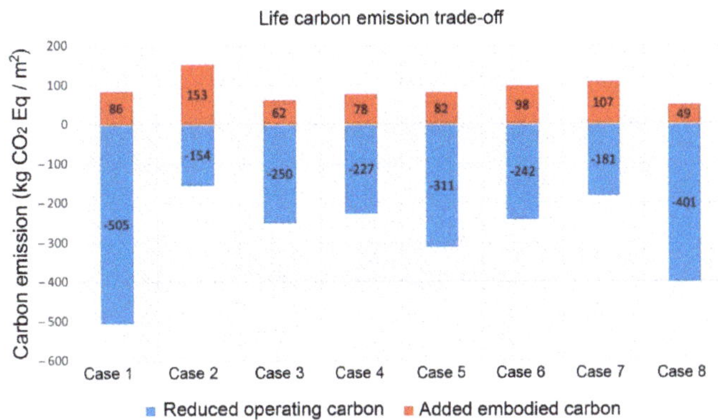

Figure 5. Life cycle carbon emissions trade-off.

3.1. Life Cycle Carbon Emissions

As showed in Figure 4, case 1 ranked first, with the highest life cycle emissions due to high energy use during the B6 stage, followed by case 5 and case 3 as the top life cycle carbon emitters, also due to the high energy demand in their use life stage. For all case buildings, the B6 stage was the dominant life cycle stage for carbon emissions, contributing around 57–83% of the total life cycle carbon emissions. These findings validate the importance of further reducing the use stage energy demand through a deep energy retrofit. Figure 4 also shows that the second highest life cycle stage contributing to life cycle carbon emissions was A1–A3, the product stage, or "cradle to gate" [24]. The fractional contributions from the remaining life cycle stages were negligible. When we normalized the life cycle emissions by the number of units included in the studied buildings (refer to Figure 4b), the results were different: Case 1 ranked first with the highest life cycle carbon emissions per unit, more than 50% higher than the emissions from case 4 (ranked second) and 60% higher than case 8 (ranked third). Since we can use the number of units as a proxy for the number of occupants in the building, we can interpret the normalized results by unit numbers as an indicator of inequality regarding the carbon emissions burden that each occupant imposes on the larger environment. For example, case building 4 has fewer occupants (refer to Table 4); however, each occupant is responsible for higher life cycle carbon emissions compared to other case buildings' occupants (except case 1). In the future, we suggest using the number of units to normalize the life cycle carbon emissions, which can better reveal the inequality among different buildings and building occupants. Despite the difference, normalized life cycle carbon emissions by unit also demonstrate that the B6 use phase is the dominant life cycle stage and needs further reductions.

As illustrated in Figure 5, for embodied carbon, case 1 had the highest increase per floor area, followed by case 2 and case 6. For the offset embodied carbon through an operating energy reduction, case 2 had the highest life cycle carbon reduction, followed by case 1 and case 7. Regarding the balance between a carbon increase and offset, all case buildings had negative life cycle carbon, which is an indicator that an energy retrofit is effective in reducing the life cycle carbon emissions of existing buildings. However, if we only look at the offset carbon emissions through an operating energy use reduction, case 1 ranks first, followed by cases 2 and 6. Despite the highest life cycle carbon emissions (refer to Figure 4), as showed in Figure 5, case 1 had the highest life cycle carbon reduction compared to the base condition before the energy retrofit. These different ranking results illustrate how evaluating sustainability using different portions of the life cycle of carbon can have different results; hence, the decisions based on the analysis results may vary.

3.2. Embodied Carbon: Building Materials and Systems

A1–A3 had the second highest life cycle stage contributing to life cycle emissions; this contribution was mostly related to the selection and production of building materials, components, and systems (refer to Figure 4). Then, we looked at the life cycle carbon derived from building materials and systems during the production and construction life stages. Figure 6a shows normalized carbon emissions (per floor area) during A1–A5 stages, which we defined as embodied carbon. The external wall was the dominant category contributing to embodied carbon, and the elevator core ranked second. The only outlier was case 2, where the elevator and roofing systems were the main contributors to the added embodied carbon. Figure 6b shows that the results of normalized carbon by unit counts were different: Case 1 ranked first with the highest per unit carbon emissions from building systems and materials used for the energy retrofit, followed by case 7 and case 4. Again, these different results demonstrate the need to potentially use occupants or unit counts as a normalization unit.

3.3. Sustainability Index Calculation and Visualization

The sustainability score was first calculated using equally weighted OES, EMB, and ECO. As illustrated in Figure 7, the X-axis represents the sustainability score (unitless), and the Y-axis is the energy use intensity after renovation, measured in kWh/m^2. The size of the bubble represents the total floor area of the case building, measured in m^2. For example, case 8 had a gross floor area of 6060 m^2, represented by the biggest circle. In general, the higher the sustainability score, the less the overall life cycle carbon emissions emitted from the renovated building, and the more sustainable the building.

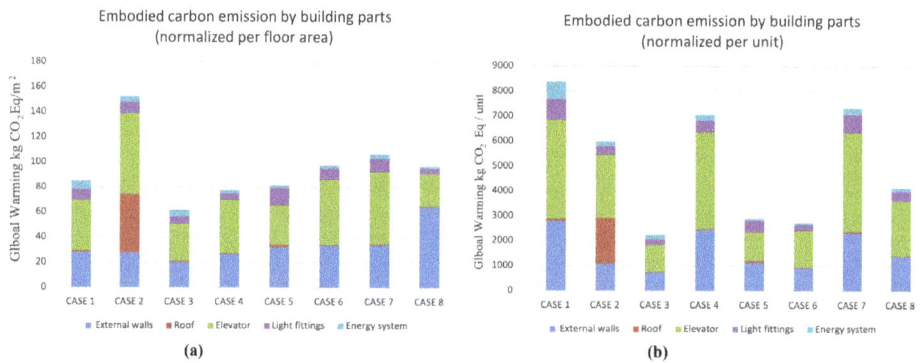

Figure 6. Embodied carbon emissions by element.

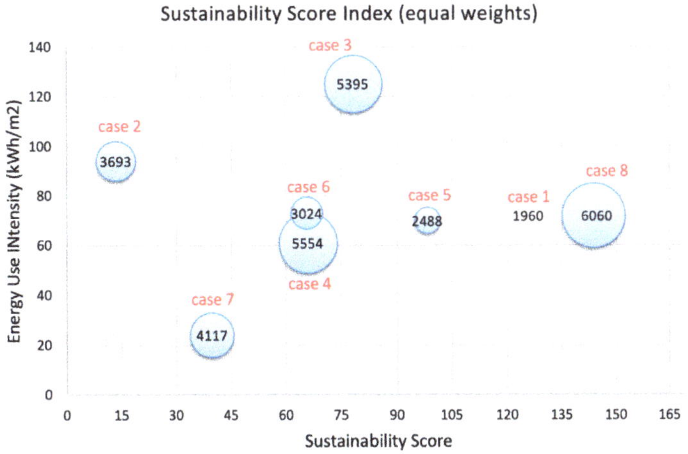

Sustainability Score Index (equal weights)

Figure 7. Sustainability score.

Figure 7 demonstrates two important findings. First, size is not correlated with the sustainability score, as the largest building (case 8) and the smallest building (case 1) had similar sustainability scores that were extremely different from the other cases. Second, energy use intensity (EUI) might not be a good measure of sustainability. In current practice, EUI is often used to measure the energy efficiency of a building and consequently the sustainability of the building. If EUI is a good indicator of sustainability, then the EUI (Y-axis) and sustainability score (X-axis) would be negatively correlated. However, they do not appear to have a clear correlation. For instance, case 7 had the lowest EUI but also the lowest sustainability score. This can be explained by the trade-off between reduced carbon emissions and increased embodied carbon. As shown in Table 5, case 2 had the highest embodied emissions increase with the second lowest operating carbon reduction. Therefore, low EUI does not necessarily mean more sustainability. In addition, case 3 had the highest EUI after renovation, which may indicate that case 3 still has much space to improve the current energy performance. However, case 3 also had the third highest sustainability score due to its poor previous energy performance followed by a large operating energy reduction achieved through the renovation (refer to Table 5). Therefore, the added embodied carbon (from the renovation) was well adjusted by the large offset of an operating carbon reduction.

As shown in Table 6, the proposed sustainability score can provide us with a more comprehensive understanding of how sustainable the renovation works are from a life cycle carbon emissions perspective, providing a more robust estimation of global warming potential related to building renovation. Only focusing on the operating energy may provide an incomplete, sometimes even opposite, interpretation to measure the effectiveness of a building energy retrofit, which is clearly demonstrated in case 7. Case 7 had the lowest energy use intensity after renovation. Based on the current commonly used measurements and criteria, it is considered very energy efficient and even has the potential to achieve net zero energy if there are renewal energy sources onsite, such as solar or wind energy. However, case 7 ranked the second lowest for the proposed sustainability score, mainly due to the trade-off between added embodied energy and reduced operating energy. From a long-term perspective, case 7 can produce more life cycle carbon than the other cases, and a large portion of such emissions are embedded in the building materials, components, and systems used in the renovation.

Table 6. Sustainability score.

	OES	EMB	ECO	EUI (After)	EUI (Before)	SC	SC Ranking
	(CO_2-eq/m^2)	(CO_2-eq/m^2)	(\$/m^2)	(kWh/m^2)	(kWh/m^2)		
Case 1	−505	86	6	72	291	141	2
Case 2	−154	153	8	94	174	13	8
Case 3	−250	62	10	125	190	78	4
Case 4	−227	78	5	61	172	66	5
Case 5	−311	82	6	70	187	98	3
Case 6	−242	98	6	73	130	66	5
Case 7	−181	107	2	24	140	40	7
Case 8	−401	49	6	72	137	144	1

4. Discussion and Sensitivity Analysis

4.1. Sensitivity Analysis

As demonstrated from this study, EUI is not always the best measure of sustainability because of its emphasis on operating energy. The overall goal of sustainable development is to reduce carbon emissions, mitigating the impact on climate change. Reducing the operating energy demand is only one important part—integrating the counting of added embodied carbon emissions due to an energy retrofit can generate a more holistic view of sustainability. Further, the evaluation of sustainability is not only impacted by the trade-off between operating carbon emissions and embodied carbon emissions but is also influenced by the stakeholders' priorities.

To test the validity of the sustainability score calculation, we used three sets of different weights to represent different stakeholders' values. For building owners or operators, operating energy savings is prioritized and related to operation cost savings; therefore, we gave the higher weights to OES and ECO to represent the building operators' view. The results are illustrated in Figure 8a. The second set of weights was for climate change policymakers. We assumed they will prioritize a life cycle carbon emissions reduction; thus, the balance between OES and EMB should be weighted equally but higher than ECO. The results are illustrated in Figure 8b. The third set of weights represented current common practices of sustainable building renovations, which are primarily related to an operating energy reduction (measured by EUI), so we gave much higher weights to OES and lower and equal weights to EMB and ECO. The results are illustrated in Figure 8c.

The results demonstrated that the sustainability index can vary depending on the priorities the decision makers give to operating carbon, embodied carbon, and operating cost. Furthermore, how sustainability is measured can have a determining impact on whether a life cycle carbon emissions reduction can be achieved. A policy maker focused on climate change and overall carbon neutrality will have a much different evaluation of sustainability compared to building owners, operators, and designers (as demonstrated in Figure 8a–c), as Figure 8a,c are more similar to each other than they are to Figure 8b. This can be problematic if the existing sustainability measurement continues to be used in the building and construction industry, as it focuses on operating energy use. There are some countries integrating embodied energy and carbon calculations and considerations in building codes, such as Norway Standard 3720 [25], but an overwhelming majority of countries have not made the embodied carbon calculation a mandatory requirement for new construction and renovation.

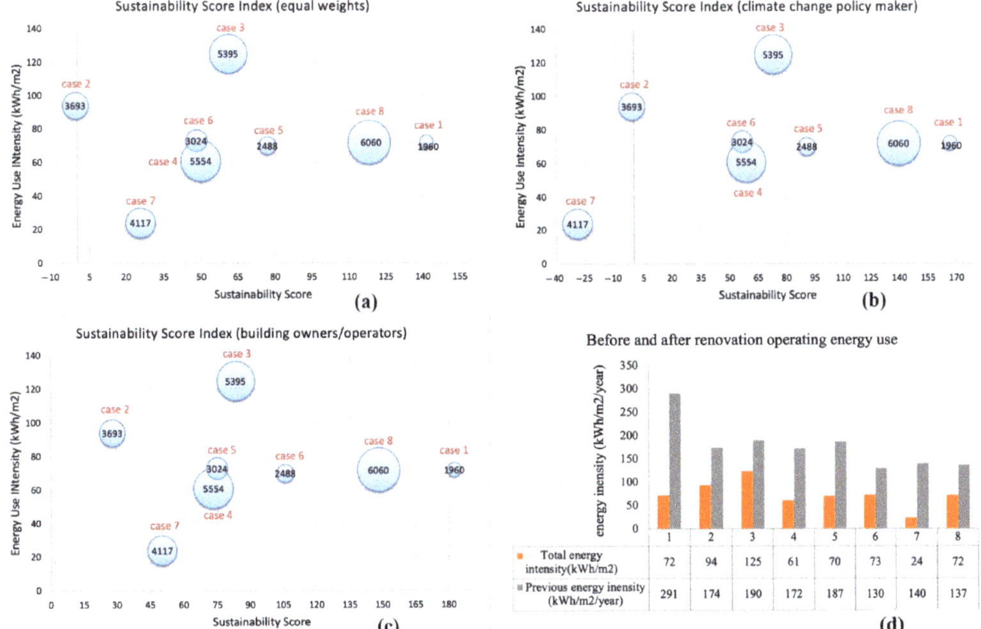

Figure 8. Sensitivity analysis. (**a**) is the sustainability score with equal weights; (**b**) is the sustainability score by climate change policy maker; (**c**) is the sustainability score by building owners; (**d**) is the before and after renovation operating energy use.

4.2. Significance of the Study

The significant features of this study, compared to previous studies on the sustainability of renovation projects, are as follows:

- Uses actual energy performance data (before and after renovation) of eight finished deep energy retrofit projects;
- Uses country-specific life cycle material inventory, and the life cycle impact model was built based on actual building construction documents, historical records, and on-site measurements and investigations;
- Offers a comprehensive view of the trade-off between an operational carbon reduction and an embodied carbon increase;
- Proposes a holistic sustainability index—including the measure of operating carbon, embodied carbon, and operating cost—tested on eight built projects.

4.3. Limitations of the Study

There are three limitations of this study: First, since the LCA database we used is location-specific, we were unable to verify the results using other LCA tools because each tool uses a different database for evaluating the life cycle environmental impact and carbon emissions. Second, we were unable to get further information on why certain renovation techniques were applied to certain buildings but not others; therefore, the motivation for the renovation strategies was not clear to us. Third, structural and other building systems were excluded from this study; the inclusion of the whole building system might reveal additional insights on the source of embodied carbon emissions derived from all building materials and systems.

5. Conclusions

This study investigated the trade-off between a reduced operational carbon reduction and added embodied carbon emissions of eight deep energy retrofit projects. The analysis of the case buildings demonstrated that the current practice, which focuses solely on an operational energy use reduction, may not result in the most sustainable solution if the embodied energy and related carbon emissions are not carefully counted. Moreover, using the proposed sustainability index, integrating both operational carbon and embodied carbon, we found that the large operating energy reduction can be offset by the added embodied energy, and the renovated building with the lowest EUI can be less sustainable than buildings with higher EUIs. In addition, this study revealed that the sustainability score varies based on stakeholders' perspectives. If embodied carbon emissions are not included in the consideration of energy retrofit planning, it is impossible for building owners to reduce the additional emissions once the renovation is finished. Consequently, addressing embodied carbon emissions should go hand in hand with deep operating energy retrofit initiatives to achieve a comprehensive sustainable result. In the future, we recommend that embodied carbon be included in building codes and regulations and that it be used together with EUI to measure the effectiveness of a building energy retrofit.

Funding: This research was funded by Finland Fulbright Foundation and the APC was funded by Faculty of Architecture, Civil Engineering and Applied Arts, University of Technology.

Institutional Review Board Statement: No Institutional Review Board required for this study.

Informed Consent Statement: Not applicable.

Data Availability Statement: Data available on request due to restrictions eg privacy or ethical. The data presented in this study are available on request from the corresponding author. The data are not publicly available due to privacy of the project owners.

Acknowledgments: I thank School of Architecture, Planning and Preservation at University of Maryland, Faculty of Built Environment, Tampere University and VTT Technical Research Centre of Finland for their support.

Conflicts of Interest: The authors declare no conflict of interest.

References

1. Zuhaib, S.; Hajdukiewicz, M.; Keane, M.; Goggins, J. Facade modernisation for retrofitting existing buildings to achieve nearly zero energy buildings. 2016.
2. Felius, L.C.; Dessen, F.; Hrynyszyn, B.D. Retrofitting towards energy-efficient homes in European cold climates: A review. *Energy Effic.* **2020**, *13*, 101–125. [CrossRef]
3. Antonopoulos, C.A.; Metzger, C.E.; Zhang, J.; Ganguli, S.; Baechler, M.C.; Nagda, H.; Desjarlais, A. *Wall Upgrades for Residential Deep Energy Retrofits: A Literature Review (No. PNNL-28690)*; Pacific Northwest National Lab. (PNNL): Richland, WA, USA, 2019.
4. Wilson, E.J.; Harris, C.B.; Robertson, J.J.; Agan, J. Evaluating energy efficiency potential in low-income households: A flexible and granular approach. *Energy Policy* **2019**, *129*, 710–737. [CrossRef]
5. European Climate Foundation. Zero Carbon Buildings 2050. Available online: https://cedelft.eu/publications/zero-carbon-buildings-2050/ (accessed on 24 May 2021).
6. Rockefeller Foundation. United States Building Energy Efficiency Retrofits. Available online: https://www.rockefellerfoundation.org/report/united-states-building-energy-efficiency-retrofits/ (accessed on 2 February 2021).
7. Dodoo, A.; Gustavsson, L.; Sathre, R. Effect of thermal mass on life cycle primary energy balances of a concrete-and a wood-frame building. *Appl. Energy* **2012**, *92*, 462–472. [CrossRef]
8. Cellura, M.; Guarino, F.; Longo, S.; Mistretta, M. Energy life-cycle approach in Net zero energy buildings balance: Operation and embodied energy of an Italian case study. *Energy Build.* **2014**, *72*, 371–381. [CrossRef]
9. Hu, M. Balance between energy conservation and environmental impact: Life-cycle energy analysis and life-cycle environmental impact analysis. *Energy Build.* **2017**, *140*, 131–139. [CrossRef]
10. Crawford, R.H.; Czerniakowski, I.; Fuller, R.J. A comprehensive framework for assessing the life-cycle energy of building construction assemblies. *Archit. Sci. Rev.* **2010**, *53*, 288–296. [CrossRef]
11. Rosselló-Batle, B.; Ribas, C.; Moià-Pol, A.; Martínez-Moll, V. An assessment of the relationship between embodied and thermal energy demands in dwellings in a Mediterranean climate. *Energy Build.* **2015**, *109*, 230–244. [CrossRef]
12. Stephan, A.; Crawford, R.H.; De Myttenaere, K. A comprehensive assessment of the life cycle energy demand of passive houses. *Appl. Energy* **2013**, *112*, 23–34. [CrossRef]

13. Bakar, N.N.A.; Hassan, M.Y.; Abdullah, H.; Rahman, H.A.; Abdullah, M.P.; Hussin, F.; Bandi, M. Energy efficiency index as an indicator for measuring building energy performance: A review. *Renew. Sustain. Energy Rev.* **2015**, *44*, 1–11. [CrossRef]
14. Venkatraj, V.; Dixit, M.K.; Yan, W.; Lavy, S. Evaluating the impact of operating energy reduction measures on embodied energy. *Energy Build.* **2020**, *226*, 110340. [CrossRef]
15. Triana, M.A.; Lamberts, R.; Sassi, P. Sustainable energy performance in Brazilian social housing: A proposal for a Sustainability Index in the energy life cycle considering climate change. *Energy Build.* **2021**, *242*, 110845. [CrossRef]
16. EU-GUGLE. Project Description. Available online: http://eu-gugle.eu/project/ (accessed on 24 May 2021).
17. EU-GUGLE. A Sustainable Renovation Pilot Project for Smarter Cities. Available online: https://www.eurac.edu/en/research/technologies/renewableenergy/conferences/Documents/4.%20Morishita%20Naomi.pdf (accessed on 24 May 2021).
18. Hirvonen, J.; Jokisalo, J.; Heljo, J.; Kosonen, R. Towards the EU emissions targets of 2050: Optimal energy renovation measures of Finnish apartment buildings. *Int. J. Sustain. Energy* **2019**, *38*, 649–672. [CrossRef]
19. Niemelä, T.; Kosonen, R.; Jokisalo, J. Cost-effectiveness of energy performance renovation measures in Finnish brick apartment buildings. *Energy Build.* **2017**, *137*, 60–75. [CrossRef]
20. Available online: https://hometalkoot.fi/kerrostalo (accessed on 5 February 2021).
21. OneClick LCA. We are One Click LCA. Available online: https://www.oneclicklca.com/about-one-click-lca/ (accessed on 2 February 2021).
22. Petrovic, B.; Myhren, J.A.; Zhang, X.; Wallhagen, M.; Eriksson, O. Life cycle assessment of building materials for a single-family house in Sweden. *Energy Procedia* **2019**, *158*, 3547–3552. [CrossRef]
23. Moran, P.; O'Connell, J.; Goggins, J. Sustainable energy efficiency retrofits as residential buildings move towards nearly zero energy building (NZEB) standards. *Energy Build.* **2020**, *211*, 109816. [CrossRef]
24. Jiménez-González, C.; Curzons, A.D.; Constable, D.J.; Cunningham, V.L. Cradle-to-gate life cycle inventory and assessment of pharmaceutical compounds. *Int. J. Life Cycle Assess.* **2004**, *9*, 114–121. [CrossRef]
25. Norway Standard 3720:2018. Available online: https://www.standard.no/en/webshop/productcatalog/productpresentation/?ProductID=992162 (accessed on 27 May 2021).

sensors

Article

Development of Condition Assessment Index of Ballast Track Using Ground-Penetrating Radar (GPR)

Fiseha Nega Birhane [1,2], Yeong Tae Choi [2,*] and Sung Jin Lee [2]

[1] Department of Transportation System Engineering, KRRI School, University of Science & Technology, 217, Gajeong-ro, Yuseong-gu, Daejeon 34113, Korea; fish@krri.re.kr
[2] Korea Railroad Research Institute, 176, Cheoldo-bangmulgwan-ro, Uiwang 16105, Korea; geolsj@krri.re.kr
* Correspondence: yeongtaechoi@krri.re.kr

Abstract: The condition of the ballast is a critical factor affecting the riding quality and the performance of a track. Fouled ballast can accelerate track irregularities, which results in frequent ballast maintenance requirements. Severe fouling of the ballast can lead to track instability, an uncomfortable ride and, in the worst case, a derailment. In this regard, maintenance engineers perform routine track inspections to assess current and future ballast conditions. GPR has been used to assess the thickness and fouling levels of ballast. However, there are no potent procedures or specifications with which to determine the level of fouling. This research aims to develop a GPR analysis method capable of evaluating ballast fouling levels. Four ballast boxes were constructed with various levels of fouling. GPR testing was conducted using a GSSI (Geophysical Survey Systems, Inc.) device (400, 900, 1600 MHz), and a KRRI (Korea Railroad Research Institute) GPR device (500 MHz), which was developed for ballast tracks. The dielectric permittivity, scattering of the depth (thickness) values, signal strength at the ballast boundary, and area of the frequency spectrum were compared against the fouling level. The results show that as the fouling level increases, the former two variables increase while the latter two decrease. On the basis of these observations, a new integrated parameter, called a ballast condition scoring index (BCSI), is suggested. The BCSI was verified using field data. The results show that the BCSI has a strong correlation with the fouling level of the ballast and can be used as a fouling-level-indicating parameter.

Keywords: track inspection; ballast fouling; ballast thickness; dielectric permittivity; signal strength; time and frequency domains; ballast condition scoring index (BCSI); KRRI; GSSI

Citation: Birhane, F.N.; Choi, Y.T.; Lee, S.J. Development of Condition Assessment Index of Ballast Track Using Ground-Penetrating Radar (GPR). *Sensors* **2021**, *21*, 6875. https://doi.org/10.3390/s21206875

Academic Editors: Carlos Morón Fernández and Daniel Ferrández Vega

Received: 11 September 2021
Accepted: 14 October 2021
Published: 16 October 2021

Publisher's Note: MDPI stays neutral with regard to jurisdictional claims in published maps and institutional affiliations.

1. Introduction

The ballast layer plays an important role in keeping the sleepers in the correct position, in supporting repeated train loads, and in transferring the train load to the substructures. It also provides free drainage of water and prevents the growth of vegetation. Sound ballast materials should be strong, stable, and drainable. Railway ballast is subjected to both traffic loads and environmental changes and, as the ballast bed deforms and degrades, it adversely affects the performance of the railway track [1–5].

Ballasts become fouled because of the fragmentation of the ballast materials, the surface infiltration of weathered particles and coal droplets, and the upward migration of fines from the subgrade and sleeper wear [2–4,6]. When fouling becomes severe, excess pore water pressure is generated under cyclic loading that, in turn, degrades the track resiliency and stability in undrained conditions [2,4].

As a result, routine track inspections to access the ballast condition are an important aspect of track operations. Cost-effective tracks are needed as the demand for faster and heavier trains increases. To meet these needs, it is essential to improve the monitoring methods and the assessment of the track performance [1].

Railway track condition assessments mainly consist of monitoring the geometric parameters of the track using inspection vehicles. The geometric parameters are related to the positioning and wearing of the rail [7]. Track geometry deformation (i.e., track settlement) is the final phenomenon, arising because of various causes, such as track stiffness differences and subgrade defects, among others [1,8]. Nondestructive tests are widely used to assess the condition of the ballast and subgrade. Common nondestructive methods are visual inspections, the light-falling weight deflection method (LFWD), and electromagnetic methods (e.g., GPR) [1,2]. A lightweight deflectometer (LWD) involves dropping a known weight from a specified height onto a circular plate over soil and measuring the vertical surface deflection and the impact load. The LWD can also be used to determine the stiffness of the track [7,9,10].

Ground penetrating radar (GPR) is currently used more often to assess track ballast conditions. GPR uses electromagnetic signals to detect subsurface features, into which it transmits the electromagnetic pulses into the ground and measures the reflected signals. The changes in the dielectric properties of different materials and, hence, the layers, alter the electromagnetic wave propagation. Such a feature, along with its ability to acquire high-speed data without disturbing train operations, makes GPR a desirable inspection tool [2,5,6,8,11,12]. Nevertheless, GPR is not yet used in a systematic manner for railway track characterization or rehabilitation planning [8], one of the reasons being the lack of a simple and reliable method. With regard to this concern, numerous studies have been carried out. This study has been conducted with the aim of resolving this issue as well.

Conventional GPR analysis primarily involves the interpretation of the signal in the time domain, and is generally performed in specific locations or for a particular purpose, such as subgrade characterization or fouling detection [2,8,13]. Interpreting radar profiles is still a challenging and time-consuming task that, therefore, requires significant experience by GPR experts. Furthermore, there are no potent procedures to find the level and type of fouling quantitatively. Recently, many studies have focused on identifying the thicknesses and fouling levels of railway ballast by means of a frequency-domain analysis, in addition to a time-domain analysis [2,8,11,13–17].

Fontul et al. [8] attempted to identify infrastructure changes by adopting short- and long-sliding windows. For these sliding windows, they calculated the area under the signal in the time and frequency domains. The peaks of these areas, and the peaks of the differences of the areas between the short and long sliding windows, were utilized to detect anomalies in ballast tracks.

Anbazhagan et al. [16], and Alani et al. [18], showed that as the fouling level increases, the electromagnetic wave velocity decreases and the dielectric constant increases. De Chiara [19] points out that the presence of moisture can significantly increase the dielectric constant of a medium. In addition, they showed that the dielectric permittivity can vary, not only according to the frequency of the GPR, but also according to the antenna configuration of the device.

Roberts et al. [20] adopt a void-scattering amplitude envelope approach to assess the condition of the ballast. After comparing the energy between the clean and fouled sections, they showed that the loss of energy is higher in fouled ballast, as it has fewer void spaces.

Most GPR applications achieve the detection and localization of buried features and discontinuities and/or the estimation of the electromagnetic properties of materials by performing signal processing tasks in the time domain. However, time-domain methods do not account for the frequency-dispersive properties of media, and do not consider the signal phase [8]. In the frequency domain, it is also easy to filter out electromagnetic interference.

Various attempts have been made to detect the fouling level of ballast using a frequency spectrum analysis, in which the time-domain GPR data are converted to the frequency domain using the Fourier transform [8,13,21]. Silvast et al. [21] showed that the area of a signal in the frequency domain is higher in clean ballast compared to fouled ballast. Shao et al. [13] used a 800 MHz antennae to evaluate the track at the network level. They proposed an automatic ballast condition assessment method using the maximum peak in the frequency domain.

Birhane et al. [11] developed a 500 MHz GPR device (referred to as a KRRI GPR device). The device has eight channels and can eliminate sleeper noise at the GPR surveying stage. In addition, an algorithm that automatically detects the thickness of the ballast layer was developed. It was found that the caliber of the thickness detection algorithm differed in the clean and fouled ballast sections of the track.

In summary, although various studies have found correlations between GPR parameters, such as the dielectric permittivity and the area of the frequency spectrum and the fouling level, there are no potent procedures by which to ascertain the level of fouling. The determination of the level and type of fouling of the ballast layer from GPR analysis remains challenging.

This paper aims to assess the fouling level of the ballast using both time- and frequency-domain parameters. The correlation of the fouling level of the ballast with the time- and frequency-domain parameters, such as the dielectric permittivity, signal strength at the bottom ballast boundary where materials change into reinforced trackbed or sublayer soil, scattering of the ballast thickness values, and the area of the frequency spectrum, is investigated thoroughly.

2. Methods and Tools for GPR Survey

2.1. Experimental Setup

Birhane et al. [11] pointed out that fouled ballast shows more signal scattering, especially at the boundary of the ballast and reinforced subgrade. In this study, to assess the fouling level of the ballast meticulously, further GPR surveys were conducted on lab-built and actual ballast tracks.

In the lab, the ballast tracks were simulated using four ballast boxes with distinct fouling levels. The four ballast boxes represent the stages of fouling on a ballast track: (1) clean (Box 1); (2) partially fouled (Box 2); (3) fully fouled (Box 3); and (4) Box 4, which has increasing fouling material content according to the depth, as illustrated in Table 1. The fouling material is added to the ballast aggregate in all cases at predetermined ratios and mixed before the boxes are filled.

Table 1. Test Track Ballast Boxes with Various Fouling Levels.

	Box 1	Box 2	Box 3	Box 4
Ratio of fouling material to clean ballast	0	0.5	1.0	0–1.0
Fouling index	0.4	34.23	51.15	30.08
Description	Clean	Partially fouled	Fully fouled	Partially fouled (layer fouled) * 5th (0.00), 4th (0.25), 3rd (0.50), 2nd (0.75), 1st layer (1.00) *

* Bottom to upper layer (1st layer = bottom, 5th layer = surface layer), () = ratio of fouling material to clean ballast.

The ballast tracks consisted of 2 m × 2 m × 0.6 m wooden frames. The front side wall was composed of a transparent material to facilitate visual inspections. Figure 1 illustrates the construction procedure of the ballast boxes. Using an excavator, the gravel and fouling materials were mixed. This mixed material was then poured into the boxes, as shown in Figure 1.

(**a**) Build wooden box   (**b**) Mix with fouled materials   (**c**) Construct ballast layer

(**d**) Box 1   (**e**) Box 2   (**f**) Box 3   (**g**) Box 4

Figure 1. Construction of ballast boxes.

Figure 2 shows the gradation of the ballast aggregates and the fouled materials. Only 0.2% of the ballast aggregates can pass through 4.75 mm and 0.075 mm sieves. In contrast, in the fouling silt material, 100% and 1.9% of the material can pass through the 4.75 mm and 0.075 mm sieves, respectively. The fouling index can be calculated by Equation (1), suggested by Selig and Waters [22]:

$$FI = P_\%^4 + P_\%^{200}$$ (1)

where $P_\%^4$ is the percentage by mass of the sampled ballast material finer than the 4.75 mm (No. 4) sieve, and $P_\%^{200}$ is the percentage by mass finer than the 0.075 mm (No. 200) sieve.

The corresponding fouling indices of each of the ballast boxes were 0.4, 34.23, 51.15, and 30.08 for Boxes 1 to 4, respectively, as shown in Table 1.

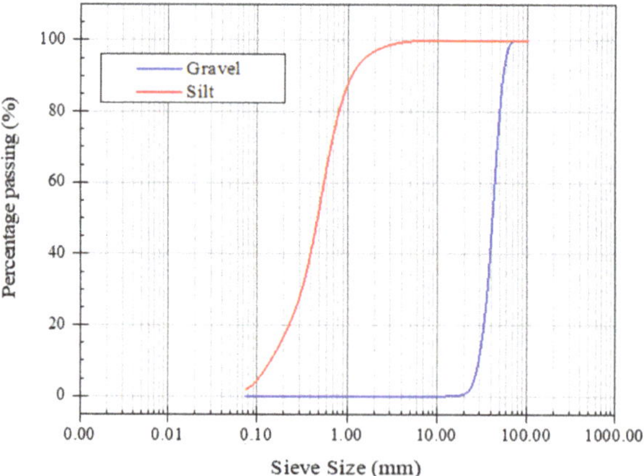

Figure 2. Grain size distribution of the ballast gravel and fouling silt.

2.2. Surveying Equipment

A KRRI (Korea Railroad Research Institute, Uiwang-si, South Korea) GPR device with an operating frequency of 500 MHz, developed for ballast track surveys, and commercial GSSI (Geophysical Survey Systems, Inc., Nashua, NH, USA) devices, with operating frequencies of 400, 900, and 1600 MHz, were used for laboratory tests. The KRRI GPR device was developed for field surveys, especially of ballast tracks. Accordingly, the KRRI GPR device was used for verification tests on the Gyeongbu high-speed railway [11].

In order to avoid the boundary effects of the ballast boxes, the centerlines of the boxes were surveyed. The test setup and operation using GSSI and KRRI devices are illustrated in Figures 3 and 4, respectively. The KRRI device has eight channels, four on the left four on the right. Thus, in order to scan the centerline only one side antenna (channel-0) was used during the lab test (pleases see Figure 4b).

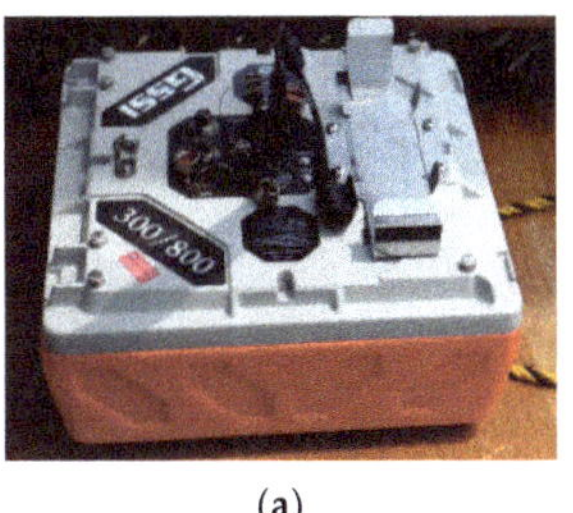
(a)

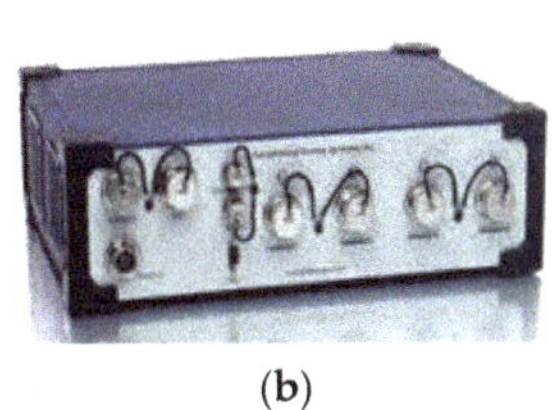
(b)

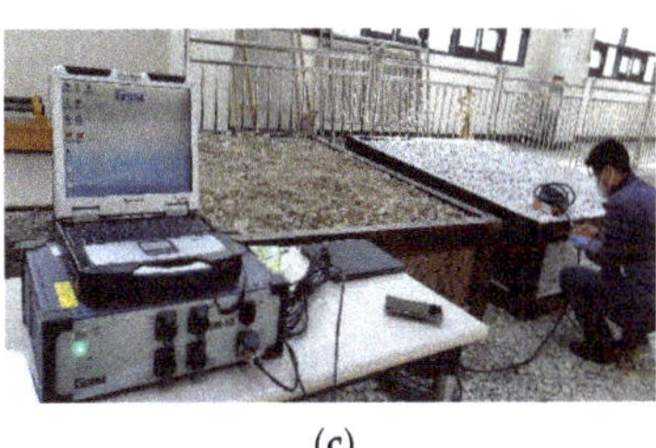
(c)

Figure 3. GSSI GPR device: (**a**) GSSI antenna; (**b**) GSSI SIR-30 controller; and (**c**) GSSI ballast survey.

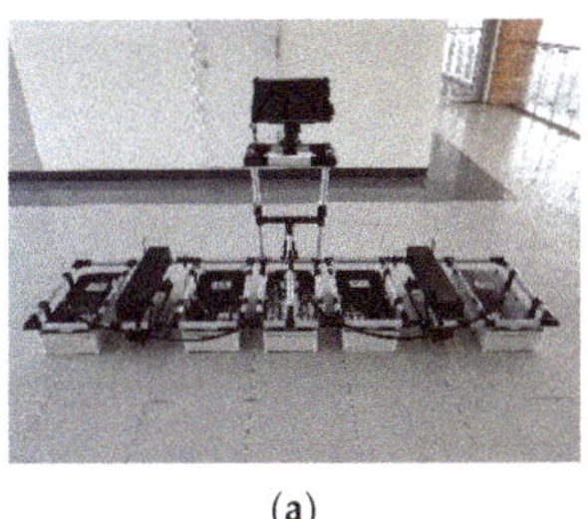
(a)

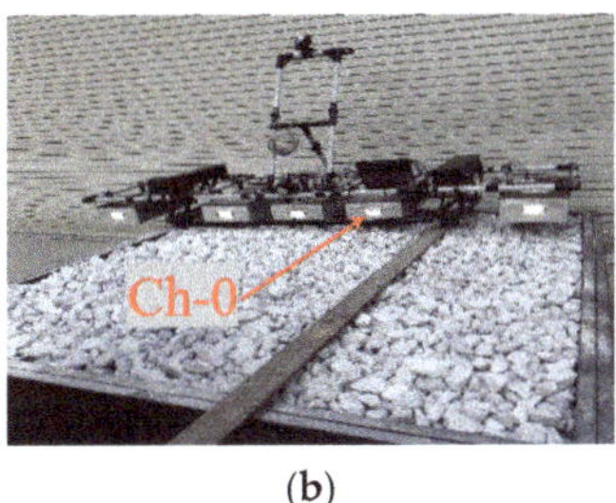

(b)

(c)

Figure 4. (**a**) KRRI GPR device; (**b**) KRRI ballast survey in the lab; and (**c**) KRRI ballast survey on the Gyeongbu HSR line.

3. Development of the Ballast Condition Scoring Index

3.1. Preprocessing of Collected GPR Signals

The GPR dataset should undergo preprocessing procedures before further analysis is carried out because it is always possible for the GPR signals to have noise from the surrounding environment [11,12,23–25]. For the KRRI device, an in-house program was used [11], while Prism2 was used for the data processing with the GSSI devices. The general preprocessing procedures applied in this paper are explained below [11,23]:

(1) Frequency filtering (high-pass and low-pass filter) to remove higher and lower frequency noises.

- High-pass filtering: the cutoff frequency is one-quarter (1/4) of the operating frequency
- Low-pass filtering: the cutoff frequency in seven-fourths (7/4) of the operating frequency

(2) Time zero correction to remove the air ground interface (i.e., the gap between the GPR antenna and the surface of the medium)

(3) Gain: a linear gain function was applied to compensate for the natural attenuation of the signal

- A linear gain function, with a slope ranging from 0.5 to 5%, can be used depending on the attenuation of the signal.

Figure 5 shows preprocessed B-scan images obtained from the GSSI and KRRI devices. As shown in the figure, it is difficult to discern the fouling level of the ballast from the B-scan image alone. Both the GSSI and the KRRI devices can detect the boundary between the ballast layer and the sublayer. The boundary in Figure 5 would be assumed to exist at the location where the GPR signal jumps. B-scan images can hardly deliver information other than the boundary.

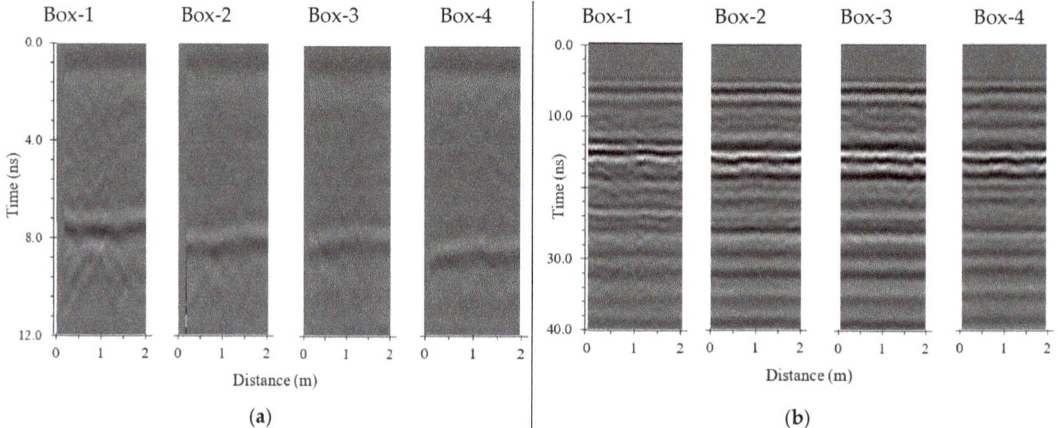

Figure 5. (**a**) B-scan 900 MHz GSSI device, and (**b**) B-scan 500 MHz KRRI device using Channel 0.

In order to evaluate the ballast condition with GPR signals, the relevant parameters should be investigated. On the basis of the literature and experience with GPR, four parameters were selected: the dielectric permittivity; signal strength at the boundary; standard deviation (i.e., signal scattering); and area of the frequency spectrum. The standard deviation comes from the calculation of the scattering of thickness values generated by the ballast thickness detection algorithms suggested by Birhane et al. [11]. The area of the frequency spectrum can be calculated by converting the GPR signals into frequency domain images. On the basis of the investigation of each parameter, a new integrated parameter, called the ballast condition scoring index (BCSI), is suggested. The details are illustrated below.

3.2. Dielectric Permittivity Variation

The dielectric permittivity is a measure of the speed at which an electromagnetic signal passes through media. The dielectric permittivity can indicate the fouling level of the ballast by comparing the measured value with that of clean ballast. The dielectric permittivity constant (ε) can be calculated at a known depth (or thickness) of the medium using Equation (2):

$$\zeta = \left(\frac{C \times \Delta t}{2D} \right)^2 \tag{2}$$

where C is the speed of light 3×10^8 m/s; Δt is the time it takes for the GPR signal to reach the boundary of the ballast (s); and D is the depth (thickness) of the ballast (m).

Several studies have found that the dielectric permittivity increases as the fouling level increases [16,18,19,26]. Figure 6 shows that the dielectric permittivity calculated at the ballast boxes varies according to the fouling index. The dielectric permittivity of Box 4 is indicated with solid markers and is different from the others (hollow markers). As illustrated in Table 1, the fouling level of Box 4 increases as the depth increases, whereas the others remain constant within each layer. For this reason, Box 4 appears not to follow the trend according to the fouling index. Therefore, the results of Box 4 are included in the graphs but are not included in any further analysis in order to develop the ballast condition scoring index.

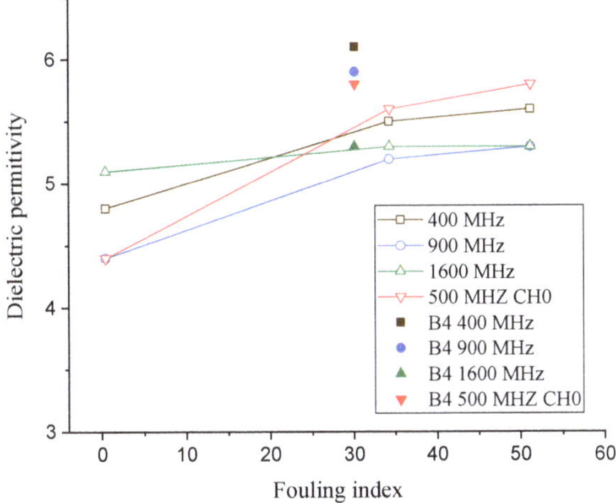

Figure 6. Dielectric permittivity vs. the fouling level at various frequencies.

As anticipated, the more fouled ballast boxes have higher dielectric permittivity levels. The dielectric permittivity values ranged from 4.4 for Box 1 (clean), to 5.8 for Box 3 (fully fouled). The range of the dielectric permittivity of the ballast layers matches those from the literature [16,18,19,26,27]. Note that the dielectric permittivity of Box 4 with layered fouling is the highest, in all GPR frequencies. This observation indicates that multiple layers of materials could reduce the speed of the electromagnetic waves and raise the dielectric permittivity of the media.

In order to use the dielectric permittivity as a ballast-fouling-level-indicating parameter, the thickness of the ballast must be known beforehand. This limits the applications of the dielectric permittivity to assess the fouling level.

3.3. Boundary Estimation and Variation

The automatic thickness detection algorithm, by Birhane et al. [11], showed contrasting results between the good condition and the fouled condition of ballast track. The good condition, i.e., ballast that was either clean or had little fouling, shows fairly continuous thicknesses, as shown in Figure 7a. However, the poor ballast condition, i.e., fouled ballast, resulted in scattered thickness detection according to the algorithm, as shown in Figure 7b.

This feature implies that the scattering of the thickness values might indicate the fouling level. Scattering, in this case, can be expressed quantitatively as the standard deviation of the ballast thickness. In this regard, the standard deviation of the depth (thickness) value generated by the depth detection algorithm (DVS) is selected as a GPR parameter.

As mentioned in Section 3.2, this stems from the fact that clean ballast is uniformly graded ballast, with more air voids and different physical and electromagnetic behaviors between the ballast and the subgrade, meaning that there are large differences in the dielectric constant. This makes the boundary more apparent for the thickness detection algorithm, whereas fouled ballast has finer particles filling the air voids and smaller aggregates segregating to the bottom section. This causes the electromagnetic characteristics of the ballast layer to resemble those of the subgrade. It is also important to reiterate that the presence of finer particles results in greater energy attenuation, that makes the boundary more obscure [20].

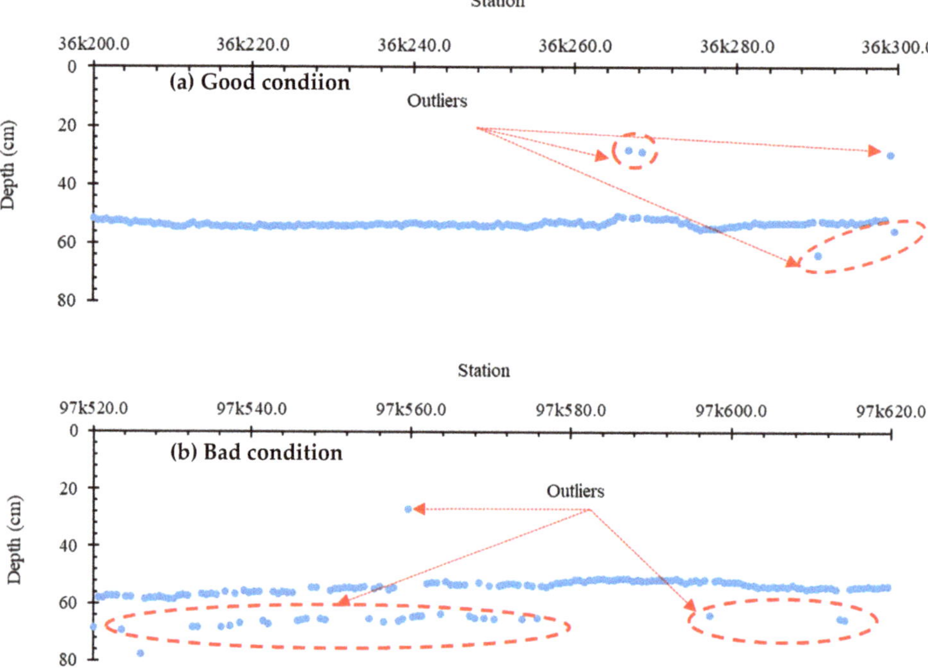

Figure 7. Ballast thickness detection (**a**) at a good site, and (**b**) at a poor site.

The standard deviation (signal scattering) of the thickness values generated by the depth (thickness) detection algorithm (DVS) was compared with the fouling level at various frequencies. As shown in Figure 8, except for 400 MHz, when the fouling level changes from a clean to a fouled condition, the inverse of the standard deviation of the depth (thickness) values from the algorithm (1/DVS) decreases. However, when the fouling index rises from 34.23 to 51.15 (Box 2 to Box 3), 1/DVS increases instead of decreasing.

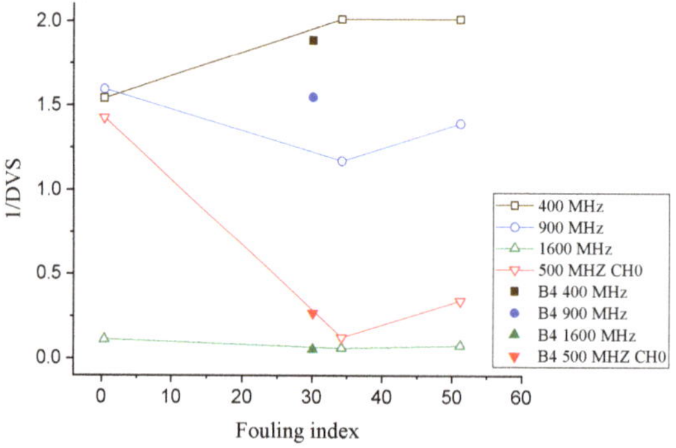

Figure 8. Inverse of the standard deviation of the thickness values generated by the according algorithm vs. the fouling level at various frequencies.

Although the inverse of DVS seems to contradict the anticipated trend, when the fouling level rises above a certain level, it is significant for distinguishing a good condition ballast from a fouled and/or semi-fouled ballast. This was verified by the analysis of the results of the GPR scan on the Gyeongbu HSR line; details are presented in Section 3.6.

3.4. Signal Strength

The signal strength at the boundary of the ballast is another GPR parameter that can indicate the fouling level of the ballast layer. Fouled ballast materials (i.e., fine particles) tend to absorb electromagnetic waves and, thus, the reflected signals become weak. This phenomenon results in lower energy at the boundary [20]. Clean ballast is uniformly graded and contains more air voids and, therefore, shows a clearer difference between the clean ballast layer and the subgrade (here, the bottom of the box). The contrasting difference at the boundary gives rise to a large reflection. The large reflection results in a strong echo back to the GPR antenna [28].

The amplitudes of the signals at the ballast boundary are compared against the fouling level (see Figure 9). Because the amplitudes in both the time domain and frequency domain vary significantly due to different frequencies and devices, an adjustment should be applied for a reasonable comparison. The adjustment factors (refer to Table 2) are determined by the amplitude of the strongest signals of each frequency and are utilized to normalize the signal strength. The strength used in this section is the amplitude of the reflected signal at the boundary, where the KRRI algorithm calculates the boundary [11].

Table 2. Adjustment Factors to Normalize the Signal Strength.

Frequency	Adjustment Factor	Remark
400 MHz	15×10^3	GSSI
900 MHz	3×10^3	GSSI
1600 MHz	6×10^3	GSSI
500 MHz	9×10^3	KRRI

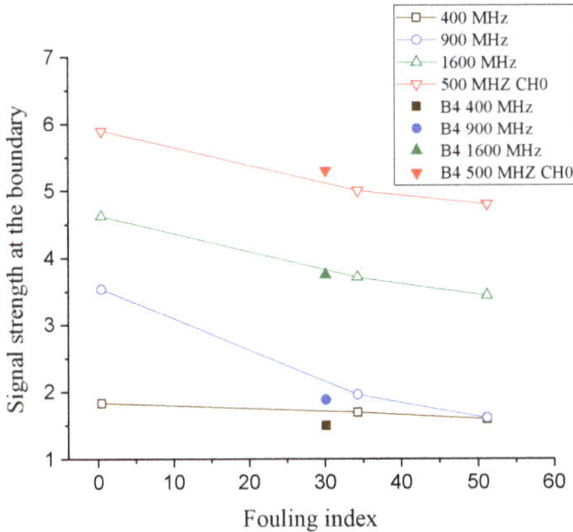

Figure 9. Signal strength at the boundary vs. fouling level with various frequencies.

As shown in Figure 9, the more the ballast is fouled, the greater the energy loss and signal attenuation and, hence, the signal strength at the boundary is lower. As stated earlier, this is mainly caused by the loss of energy, which is higher in fouled ballast, as there are fewer void spaces [20].

3.5. Area of the Frequency Spectrum of the Signal

Silvast et al. [21] compared the area of a signal in the frequency domain along fouled and clean ballast sections and showed that the area is significantly larger in clean ballast.

Therefore, the area bounded by the frequency spectrum is used as the frequency-domain analytic parameter to assess the fouling level.

In this study, the average of the area bounded by the frequency spectrum of the signal is calculated for each box and compared with regard to the fouling level. Similar to the signal strength, the adjustments have been made for the calculated areas, as shown in Table 3.

Table 3. Adjustment factors to normalize the area of the frequency domain.

Frequency	Adjustment Factor	Remarks
400 MHz	4×10^{10}	GSSI
900 MHz	2×10^{9}	GSSI
1600 MHz	2×10^{10}	GSSI
500 MHz	4×10^{9}	KRRI

As shown in Figure 10, the more the ballast is fouled, the smaller the area bounded by the frequency spectrum of a signal becomes. This is also in agreement with earlier reports in the literature [8,21].

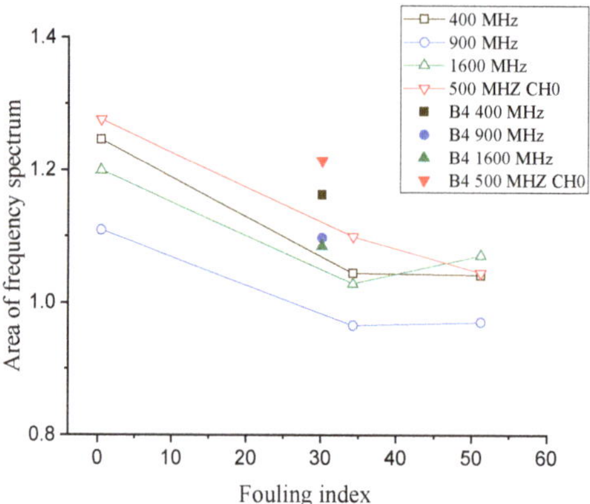

Figure 10. Area of the signal under the frequency domain vs. the fouling level at various frequencies.

3.6. Ballast Condition Scoring Index (BCSI) vs. the Fouling Level

After thoroughly analyzing the correlations among the four parameters above with the fouling level, a new integrated parameter called the ballast condition scoring index (BCSI) was suggested to indicate the fouling level. The BCSI is a function of the signal strength at the ballast boundary (SSb), the depth (thickness) value scattering (DVS), and the area of the frequency spectrum of the signal (A_{fft}). It is defined as Equation (3).

$$\text{BCSI} = \left(\text{SSb} + \frac{1}{\text{DVS}} \right)^{A_{fft}} \tag{3}$$

To come up with Equation (2), various combinations of the three parameters, starting from direct superimposition and simple multiplication, were investigated and their performances were compared. Accordingly, the function in the form of Equation (2) gives the best result. The dielectric permittivity is excluded because the thickness of the ballast should be known beforehand in order to use the dielectric permittivity. The BCSI is a coherent parameter to indicate the fouling level of the ballast and is, theoretically, more accurate and stable for the fouling analysis, as it includes both the time-domain and the frequency-domain analyses. The GPR parameters, especially the ones related to the time-domain analysis, can be affected by a number of factors in addition to the fouling level, such as the thickness of the ballast layer, the moisture content, and the type of fouling material. For example, the signal strength can be affected by the thickness of the ballast, i.e., even if the condition of the ballast is identical, as the thickness of the ballast increases, the energy level at the bottom boundary will be lower because of the higher level of attenuation [29].

Furthermore, it is reported in the literature that the moisture level influence in a frequency-domain analysis is minor compared to that in a time-domain analysis [8]. Therefore, combining the time domain and the frequency domain into a single integrated parameter enables the stabilization of the parameter and uses their combined advantages to assess the condition of the ballast, which is tampered with by multiple variables.

The correlation between the fouling level and the ballast condition scoring index (BCSI) is presented in Figure 11. A higher scoring index means a better ballast condition. Once more, all of the frequencies show a trend similar to those of the dielectric permittivity, the signal strength at the boundary, and the area of the frequency-domain graph.

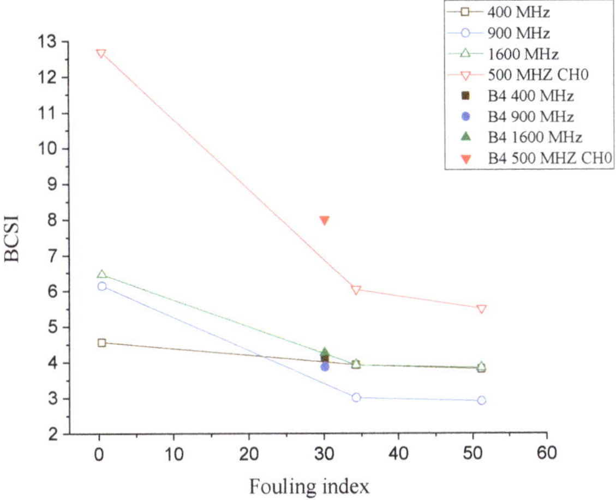

Figure 11. Ballast condition scoring index (BCSI) vs. the fouling level at various frequencies.

The results show that the parameters can clearly distinguish between clean and fouled ballast. Clean ballast has low dielectric permittivity, high signal strength at the bottom boundary, low scattering of the thickness value, and a higher area of the frequency spectrum and, thus, results in a high scoring index.

The KRRI 500 MHz GPR device appears to discern fouling levels with the BCSI. For the GSSI device, 900 and 1600 MHz antennas seem to have the capability to figure out the fouling index; however, a 400 MHz frequency would not be appropriate to evaluate the fouling level of the ballast layer in operating lines. Birhane et al. [11] proved that, in actual tracks, the GPR data collected using commercial GPR devices shows significant noise around and under sleepers. The KRRI GPR device, on the other hand, which, unlike the commercial devices, has a transmitter and receiver antennae at separate boxes, can successfully eliminate the noise, especially at the lower section of the ballast layer. The arrangement of the antennae, and the magnetic sensors that detect fasteners on the sleepers, enable the KRRI device to scan the ballast in between the sleeper, thereby significantly minimizing the interference of the sleepers. In this sense, the KRRI GPR device is used for verification in operating lines.

4. Verification of the Ballast Condition Scoring Index on a High-Speed Railway

In order to apply the suggested index, BCSI, to the maintenance of operating lines, the BCSI needed to be verified through field testing. For this purpose, the Gyeongbu high-speed railway was selected, and more than 10 km of track was scanned using a KRRI GPR device at various sites.

The test pits were excavated at randomly selected sections of bridges, tunnels, and embankments along the track. Because of the lack of an appropriate sampling technique, a quantitative fouling index could not be obtained on the Gyeongbu HSR line. Instead, the ballast condition was carefully assessed visually at selected excavation pits. After visually inspecting the excavated pits, the condition of the ballast was qualitatively graded according to the fouling level category suggested by Selig and Waters [22] (refer to Table 4 and Figure 12). On the basis of the description of the ballast condition, and the graphical description shown in Figure 12, the fouling index (FI) was estimated in a quantitative manner. Table 5 illustrates the major sites for the GPR survey, which were, in this case, the Banwall, Baebang, Pyeongtaek, and Gemeo sites. The details of the ballast thickness, a description of the fouling levels, and other factors pertaining to the test pits are summarized in Table 5.

Table 4. Fouling Index and Corresponding Ballast Condition Categories.

Fouling Index [1]	Description of Ballast Condition
<1	Clean
1 to <10	Moderately clean
10 to <20	Moderately fouled (Partially fouled)
20 to <40	Fouled
>40	Highly fouled (Fully fouled)

[1] Fouling index category [22].

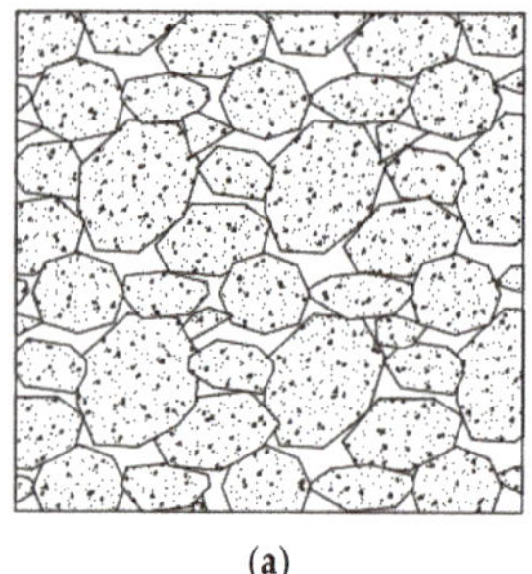

(a)

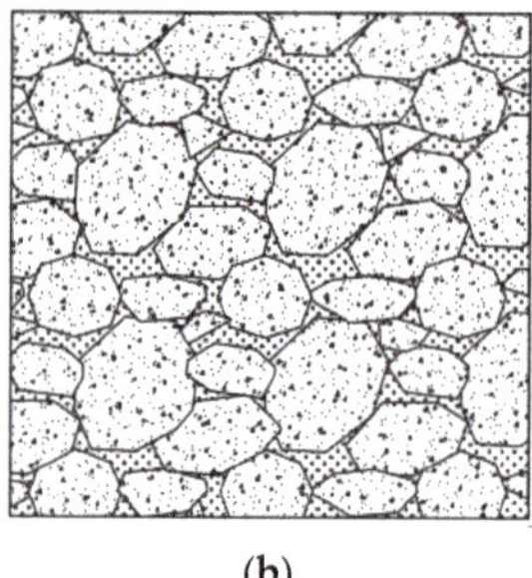

(b)

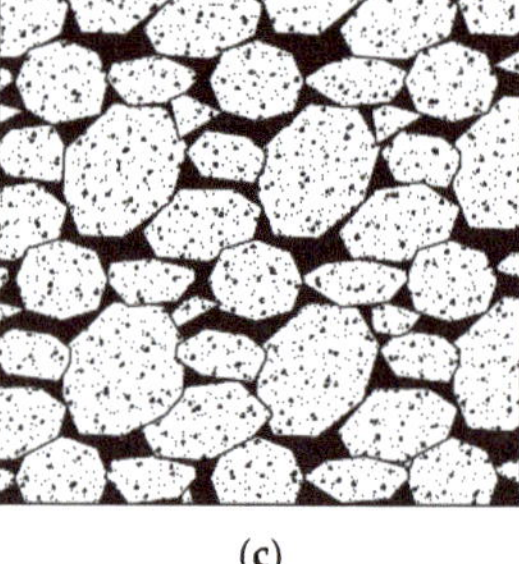

(c)

Figure 12. Fouling levels of ballast: (**a**) clean ballast; (**b**) partially fouled ballast; and (**c**) fully fouled ballast.

The relationships among the fouling levels of the ballast along the Gyeongbu HSR line and the dielectric permittivity, signal strength at the boundary, area of the frequency domain, and the standard deviation of the ballast thickness value are summarized in Figure 13. For all parameters, a 20m average length was taken so as to obtain the representative values of a certain location. As the fouling level of the ballast increases, the dielectric permittivity and the standard deviation of the thickness values from the algorithm increase, whereas the signal strength at the ballast boundary, and that of the area of the frequency spectrum, decrease.

Table 5. Condition of the Ballast Layer at Selected Trial Pits of the Gyeongbu HSR Line.

Site	Track	Location	Thickness (cm)	Description of the Fouling Level	Category According to Selig and Waters		Fouling Index (Fi)
Banwall	T1	36k260.0	53	Very good condition with slight fouling	Moderately clean	1 to <10	5
		36k294.0	48	Heavy fouling boundary at the bottom	Fouled	20 to <40	30
		36k397.0	53	Good to moderate fouling	Moderately clean	1 to <10	10
Baebang	T2	97k210.0	>36	Highly fouled—cannot be hand excavated beyond 36 cm.	Highly fouled	$\geq$40	40
		97k530.0	>53	Moisture detected on lower section.	Highly fouled	$\geq$40	40
Geomo	T1	241k295.0	56	Highly fouled—cannot be excavated beyond 36 cm.	Moderately fouled	10 to <20	20
		241k350.0	>54	Lower section has moisture.	Highly fouled	$\geq$40	40
		241k370.0	56	Moderate fouling.	Fouled	20 to <40	30
		241k460.0	69	Serious fouling.	Moderately fouled	10 to <20	20
		241k530.0	>62	Cannot be hand excavated beyond 54 cm.	Highly fouled	$\geq$40	40
	T2	241k347.0	60	Fouling moderate-serious. Moisture presence at the bottom section.	Moderately fouled	10 to <20	20
		241k370.0	53	Moderate fouling	Fouled	20 to <40	30
		241k405.0	>35	Serious fouling/not to bottom	Moderately fouled	10 to <20	20
		241k446.0	70	Moderate fouling	Moderately fouled	10 to <20	20

The GPR parameters are closely correlated with the fouling level along the HSR line; however, the correlation is not as strong as those from ballast box tests. This occurs because the ballast boxes exist in a well-controlled environment, whereas on the actual track, the thickness of the ballast layer, the moisture content, the type of fouling materials, and the fouling level vary along operating lines.

The ballast condition scoring index (BCSI) has a strong correlation with the fouling level, where a good ballast condition has a higher scoring index. A comparison of Figure 13 with Figure 14 shows the integrated parameters. The BCSI performs much better in indicating the fouling level of the ballast layer than the individual parameters. As shown in the figures, an exponential regression was performed on all the variables in Equation (2), including the BCSI. The coefficient of determination (R^2) was 0.15, 0.11, 0.84, and 0.86 for the signal strength at the boundary, area of the frequency spectrum, the inverse of the standard deviation of the depth (thickness) values from the algorithm (1/DVS), and the ballast condition scoring index (BCSI), respectively.

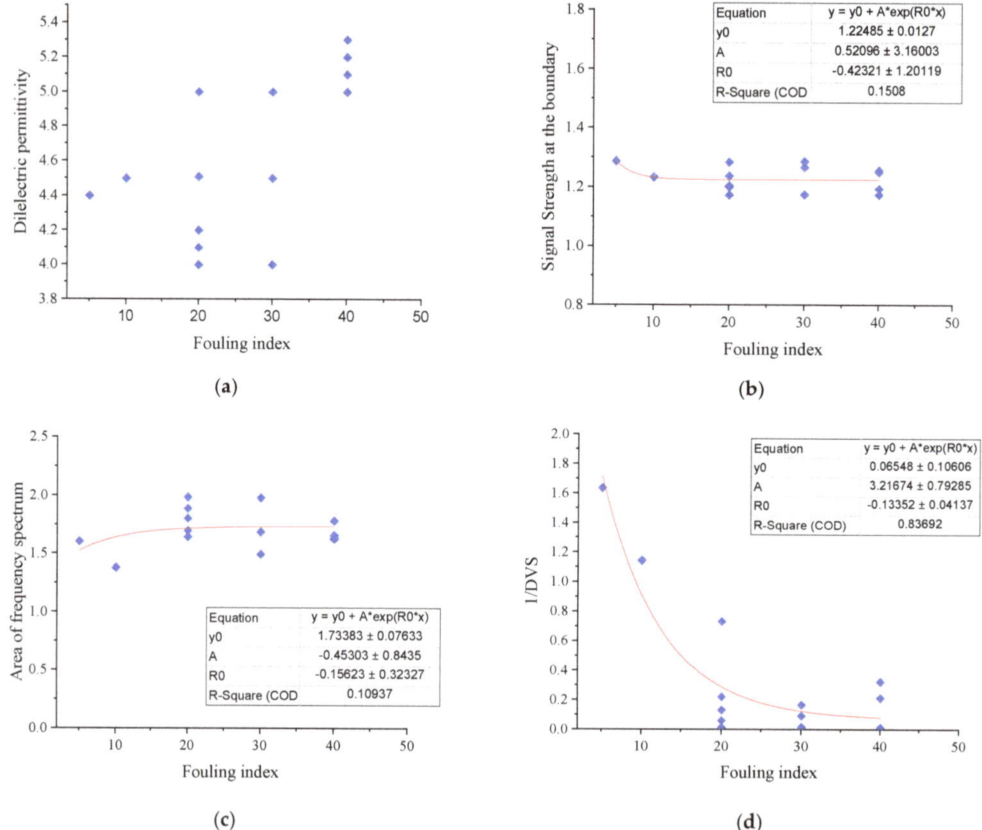

Figure 13. Relationships between the qualitative fouling level of the ballast and (**a**) the dielectric permittivity; (**b**) signal strength at the boundary; (**c**) area of the frequency spectrum; and (**d**) inverse of the standard deviation of the depth (thickness) values.

The results, both in the lab-built boxes and the actual track, show that the correlation of the fouling level with the BCSI is not only stronger, but also more stable, than with the individual parameters. This demonstrates that the BCSI can be used as a tool to indicate the possible fouled section of the track, where detailed investigation is needed. It can also be expanded and/or improved by including more fouling level informative time-domain and frequency-domain parameters, which might be investigated in future works, i.e., the BCSI can be continuously updated to include additional parameters that can possibly indicate the fouling level of ballast. With further verification, and the possible inclusion of more fouling-level-indicating parameters, the BCSI can be refined into a more robust and reliable ballast-fouling assessment index. Given the ability of GPR devices to be easily mounted on track-inspection or commercial trains, the BCSI can also serve as a swift real-time ballast-condition-assessment and monitoring tool.

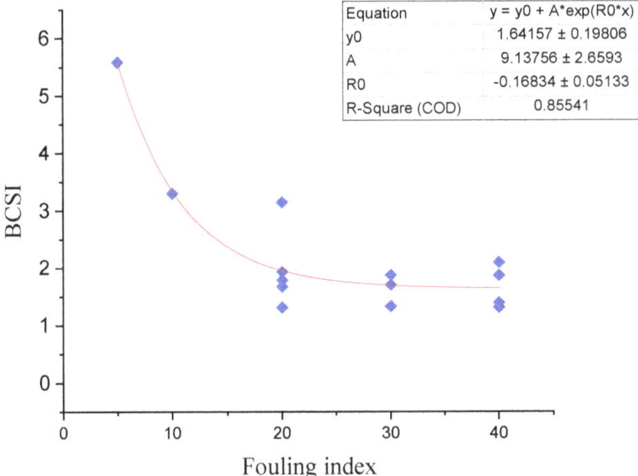

Figure 14. Relationship between the fouling level of the Gyeongbu HSR and the BCSI.

5. Conclusions

Track irregularities are the major parameters when determining the need for track maintenance. Fouling of the ballast is considered one of the most important factors causing track irregularities or the settlement of the ballast track. Many researchers have attempted to develop technologies to assess ballast conditions, and GPR has served as a suitable nondestructive tool.

This paper aims to develop an index and/or a method to evaluate the ballast condition in a practical and efficient manner. To do this, four ballast boxes were built with different fouling levels, and GPR signals were then collected. A GSSI GPR device with operating frequencies of 400, 900, and 1600 MHz, and a KRRI GPR device with an operating frequency of 500 MHz, were utilized to develop an index able to assess the ballast-fouling level.

The fouling levels of the ballast were compared with four parameters (dielectric permittivity, signal strength at the boundary, standard deviation (scattering) of the thickness, and the area of the frequency domain signal). The comparison shows that clean ballast has low dielectric permittivity, high signal strength at the bottom boundary, low scattering of the thickness value, and a larger area of the frequency-domain signal. On the basis of this observation, a new integrated parameter called the ballast condition scoring index (BCSI) is suggested.

The ballast condition scoring index (BCSI) can discern different fouling levels because it has a good correlation with the fouling level of an actual railway track line operating in Korea (the Gyeongbu high-speed railway). Accordingly, it can be used as a tool to indicate the possible fouled section of the track, where detailed investigation is needed. It can also be improved by including more fouling-level indicative, time-domain, and frequency-domain parameters.

Since GPR devices can easily be mounted on track-inspection or commercial trains, the BCSI can also serve as a swift real-time ballast condition assessment and monitoring tool.

Author Contributions: Conceptualization, F.N.B., Y.T.C. and S.J.L.; methodology, F.N.B., Y.T.C. and S.J.L.; software, F.N.B.; validation, F.N.B., Y.T.C. and S.J.L.; formal analysis, F.N.B.; investigation, F.N.B. and Y.T.C.; resources, F.N.B., Y.T.C. and S.J.L.; data curation, F.N.B., Y.T.C. and S.J.L.; writing—original draft preparation, F.N.B.; writing—review and editing, F.N.B. and Y.T.C.; visualization, F.N.B., Y.T.C. and S.J.L.; supervision, Y.T.C. and S.J.L.; project administration, F.N.B., Y.T.C. and S.J.L.; funding acquisition, Y.T.C. and S.J.L. All authors have read and agreed to the published version of the manuscript.

Funding: This work was supported by the National Research Foundation of Korea (NRF) grant, funded by the Korean government (MSIT) (Development of GPR test method for investigating the condition of ballast and trackbed, 2019R1A2C200744212).

Acknowledgments: The authors express gratitude for the funding from the National Research Foundation of Korea (NRF).

Conflicts of Interest: The authors declare no conflict of interest.

References

1. Esveld, C. *Modern Railway Track*; Digital Edition; MRT-Productions: Zaltbommel, The Netherlands, 2014.
2. Indraratna, B.; Salim, W.; Rujikiatkamjorn, C. *Advanced Rail Geotechnology—Ballasted Track*; CRC Press: London, UK, 2011.
3. Bassey, D.; Ngene, B.; Akinwumi, I.; Akpan, V.; Bamigboye, G. Ballast Contamination Mechanisms: A Criterial Review of Characterisation and Performance Indicators. *Infrastructures* **2020**, *5*, 2412–3811. [CrossRef]
4. Sussmann, T. Source of Ballast Fouling and Influence Considerations for Condition Assessment Criteria. *Transp. Res. Rec. J. Transp. Res. Board* **2012**, *2289*, 87–94. [CrossRef]
5. Smith, M.R.; Collis, L. Railway track ballast. In *Aggregates: Sand, Gravel and Crushed Rock Aggregates for Construction Purposes*; Geological Society of London: London, UK, 2001; pp. 285–289.
6. Bold, R.D. *Non-Destructive Evaluation of Railway Trackbed Ballast*; The University of Edinburgh: Edinburgh, UK, 2011.
7. Park, B.S.; Choi, Y.T.; Hwang, S.H. Ballasted Track Status Evaluation Based on Apparent Track Stiffness Index. *Appl. Sci.* **2020**, *14*, 4729. [CrossRef]
8. Fontul, S.; Paixão, A.; Solla, M.; Pajewski, L. Railway Track Condition Assessment at Network Level by Frequency Domain Analysis of GPR Data. *Remote Sens.* **2018**, *4*, 559. [CrossRef]
9. Kuttah, D. Determining the resilient modulus of sandy subgrade using cyclic light weight deflectometer test. *Transp. Geotech.* **2021**, *27*, 100482. [CrossRef]
10. Sulewska, M.J.; Bartnik, G. Application of the Light Falling Weight Deflectometer (LFWD) to Test Aggregate Layers on Geosynthetic Base. *Procedia Eng.* **2017**, *189*, 221–226. [CrossRef]
11. Birhane, F.N.; Choi, Y.T.; Lee, S.J. Development of GPR Device and Analysis Method to Detect Thickness of Ballast Layer. *J. Korean Soc. Railw.* **2020**, *23*, 269–278. [CrossRef]
12. Daniels, D.J. *Ground Penetrating Radar*, 2nd ed.; Institution of Engineering and Technology: London, UK, 2004.
13. Shao, W.; Bouzerdoum, A.; Phung, S.L.; Su, L.; Indraratna, B.; Rujikiatkamjorn, C. Automatic Classification of Ground-Penetrating-Radar Signals for Railway-Ballast Assessment. *IEEE Trans. Geosci. Remote Sens.* **2011**, *49*, 3961–3972. [CrossRef]
14. Al-Qadi, I.; Xie, W.; Roberts, R. Optimization of antenna configuration in multiple-frequency ground penetrating radar system for railroad substructure assessment. *NDT E Int.* **2010**, *43*, 20–28. [CrossRef]
15. Leng, Z.; Al-Qadi, I. Dielectric Constant Measurement of Railroad Ballast and Application of STFT for GPR Data Analysis. In *Non-Destructive Testing in Civil Engineering*; Laboratoire central des ponts et chaussées: Nantes, France, 2009.
16. Anbazhagan, P.; Naresh, P.; Bharatha, T. Identification of type and degree of railway ballast fouling using ground coupled GPR antennas. *J. Appl. Geophys.* **2016**, *126*, 183–190. [CrossRef]
17. De Bold, R.; O'Connor, G.; Morrissey, J.; Forde, M. Benchmarking large scale GPR experiments on railway ballast. *Constr. Build. Mater.* **2015**, *92*, 31–42. [CrossRef]
18. Alani, A.B.; Fabio, T.; Luca, B.C.; Alessandro, C.; Brancadoro, M.G.; Amir, M. Railway ballast condition assessment using ground-penetrating radar—An experimental, numerical simulation and modelling development. *Constr. Build. Mater.* **2017**, *140*, 508–520.
19. De Chiara, F.; Fontul, S.; Fortunato, E. GPR Laboratory Tests For Railways Materials Dielectric Properties Assessment. *Remote Sens.* **2014**, *6*, 9712–9728. [CrossRef]
20. Roberts, R.; Rudy, J.; Al-Qadi, I.; Tutumluer, E.; Boyle, J. Railroad Ballast Fouling Detection Using Ground Penetrating Radar—A New Approach Based on Scattering from Voids. In Proceedings of the 9th European Conference on NDT (ECNDT 2006), Berlin, Germany, 25–29 September 2006.
21. Silvast, M.; Levomaki, M.; Nurmikolu, A.; Noukka, J. NDT Techniques in Railway Structure Analysis. In Proceedings of the 7th World Congress on Railway Research, Montreal, QC, Canada, 4–8 June 2006.
22. Selig, T.E.; Waters, M.J. *Track Geotechnology and Substructure Management*; Thomas Telford Publishing: London, UK, 1994.

23. Jol, H. Data Processing And Interpretation. In *Ground Penetrating Radar: Theory and Application*; Elsevier: Amsterdam, The Netherlands, 2009; pp. 405–413.
24. Kim, J.-H.; Cho, S.-J.; Yi, M.-J. Removal of ringing noise in GPR data by signal processi. *Geosci. J.* **2007**, *11*, 75–81. [CrossRef]
25. Szymczyk, M.; Szymczyk, P. Preprocessing of GPR data. *Image Process. Commun.* **2013**, *18*, 83–90. [CrossRef]
26. Sadeghi, J.; Motieyan, M.; Mollazadeh, M.; Yousefi, B.; Zakeri, J. Improvement of railway ballast maintenance approach, incorporating ballast geometry and fouling conditions. *J. Appl. Geophys.* **2018**, *151*, 263–273. [CrossRef]
27. Hugenschmidt, J. Railway track inspection using GPR. *J. Appl. Geophys.* **2000**, *43*, 147–155. [CrossRef]
28. Lai, W.L.; Poon, C.S. GPR data analysis in time-frequency domain. In *The Railway Track and Its Long Term Behaviour*; Tzanakakis, K., Ed.; Springer: Dordrecht, The Netherlands, 2013; pp. 362–366.
29. Leucci, G. Ground Penetrating Radar: The Electromagnetic Signal Attenuation and Maximum Penetration Depth. *Sch. Res. Exch.* **2008**, *2008*, 1–7. [CrossRef]

Article

Strength and Acid Resistance of Ceramic-Based Self-Compacting Alkali-Activated Concrete: Optimizing and Predicting Assessment

Hassan Amer Algaifi [1,*], Mohammad Iqbal Khan [2,*], Shahiron Shahidan [1], Galal Fares [2], Yassir M. Abbas [2], Ghasan Fahim Huseien [3,*], Babatunde Abiodun Salami [4] and Hisham Alabduljabbar [5]

1   Faculty of Civil Engineering and Built Environmental, Universiti Tun Hussein Onn Malaysia, Parit Raja 86400, Malaysia; shahiron@uthm.edu.my

2   Department of Civil Engineering, College of Engineering, King Saud University, Riyadh 11421, Saudi Arabia; galfares@ksu.edu.sa (G.F.); yabbas@ksu.edu.sa (Y.M.A.)

3   Department of Building, School of Design and Environment, National University of Singapore, Singapore 117566, Singapore

4   Interdisciplinary Research Center for Construction and Building Materials, King Fahd University of Petroleum and Minerals, Dhahran 31261, Saudi Arabia; salami@kfupm.edu.sa

5   Department of Civil Engineering, College of Engineering, Prince Sattam bin Abdulaziz University, Alkharj 11942, Saudi Arabia; h.alabduljabbar@psau.edu.sa

*   Correspondence: enghas78@gmail.com (H.A.A.); miqbal@ksu.edu.sa (M.I.K.); bdggfh@nus.edu.sg (G.F.H.)

Citation: Algaifi, H.A.; Khan, M.I.; Shahidan, S.; Fares, G.; Abbas, Y.M.; Huseien, G.F.; Salami, B.A.; Alabduljabbar, H. Strength and Acid Resistance of Ceramic-Based Self-Compacting Alkali-Activated Concrete: Optimizing and Predicting Assessment. *Materials* **2021**, *14*, 6208. https://doi.org/10.3390/ma14206208

Academic Editor: Luigi Coppola

Received: 14 September 2021
Accepted: 15 October 2021
Published: 19 October 2021

Publisher's Note: MDPI stays neutral with regard to jurisdictional claims in published maps and institutional affiliations.

Abstract: The development of self-compacting alkali-activated concrete (SCAAC) has become a hot topic in the scientific community; however, most of the existing literature focuses on the utilization of fly ash (FA), ground blast furnace slag (GBFS), silica fume (SF), and rice husk ash (RHA) as the binder. In this study, both the experimental and theoretical assessments using response surface methodology (RSM) were taken into account to optimize and predict the optimal content of ceramic waste powder (CWP) in GBFS-based self-compacting alkali-activated concrete, thus promoting the utilization of ceramic waste in construction engineering. Based on the suggested design array from the RSM model, experimental tests were first carried out to determine the optimum CWP content to achieve reasonable compressive, tensile, and flexural strengths in the SCAAC when exposed to ambient conditions, as well as to minimize its strength loss, weight loss, and UPVL upon exposure to acid attack. Based on the results, the optimum content of CWP that satisfied both the strength and durability aspects was 31%. In particular, a reasonable reduction in the compressive strength of 16% was recorded compared to that of the control specimen (without ceramic). Meanwhile, the compressive strength loss of SCAAC when exposed to acid attack minimized to 59.17%, which was lower than that of the control specimen (74.2%). Furthermore, the developed RSM models were found to be reliable and accurate, with minimum errors (RMSE < 1.337). In addition, a strong correlation ($R > 0.99$, $R^2 < 0.99$, adj. $R^2 < 0.98$) was observed between the predicted and actual data. Moreover, the significance of the models was also proven via ANOVA, in which p-values of less than 0.001 and high F-values were recorded for all equations.

Keywords: mathematical assessment; optimization; self-compacting alkali-activated concrete; granulated blast furnace slag; ceramic tile waste; durability; strength; microstructure

1. Introduction

Owing to the rapid development in the construction industry, concerns about air pollution and industrial waste material have increased. As such, researchers have shifted their attention to developing green and sustainable concrete. For example, ceramic waste was utilized either as a partial replacement of aggregates [1] or cement [2] in a concrete matrix. It was found that CWP was able to produce green concrete with improved mechanical properties [3].

On the other hand, alkali-activated concrete (AAC) has emerged as one of the top trending concrete alternatives, particularly within the scientific community, owing to environmental considerations [4–7]. Free-cement-based alkali-activated concrete was developed using waste binders such as silica fume (SF), granulated blast furnace slag (GBFS), fly ash (FA), and palm oil fuel ash (POFA) [8–11]. The efficiency of the developed AAC was experimentally investigated and assessed based on the outcome of the mechanical properties and durability [12–14]. However, such classical and traditional methods require a large number of experiments to obtain the optimum content of the involved chemical reaction parameters. In addition to the variation in the experimental tests, the variable interaction investigation is also challenging via experimentation alone. Consequently, in recent years, many researchers have shifted their attention to adopting an optimization and predictive technique prior to conducting experimental studies [15,16]. For example, He et al. [17] used a simplex-centroid design method to optimize and assess the effects of ternary binders, including calcium aluminate cement (CAC), GBFS, and soda lime glass powder (GP), on the properties of alkali-activated mortar. The results of their study revealed that the optimum CAC content was higher than 10%, while the GBFS content should be lower than 5%. In addition, the GP content ranged from 77% to 90%. Recently, Dave et al. [18] successively optimised the mix of the proportions of the alkali-activated composite in an appropriate and systematic way using the Taguchi method. In another study by Koči et al. [19], an alkali-activated aluminosilicate composition involving ceramic powder, water, sodium silicate, and siliceous sand was also optimised to effectively and quickly maximize the compressive strength using the downhill simplex method. Next, the partial replacement of FA by GBFS in AAC was theoretically evaluated using the Taguchi method in a study by Mehta et al. [20]. Based on their findings, the compressive strength of FA-based ACC increased with an increase in the GBFS content, owing to further polymerization products and the available calcium-based hydration product. In addition to the Taguchi method, the optimum content of ladle furnace slag in FA-based alkali-activated cement was obtained and assessed using another optimization method, namely response surface methodology (RSM) in a study by Pinheiro et al. [21]. The effect of waste glass powder on the flexural and compressive strengths of the alkali-activated material was also investigated using the RSM model [22].

In addition, Cong et al. [23] employed both experimental works and statistical approaches to investigate several properties of GBFS/FA-based alkali-activated concrete, including the modulus of elasticity, dynamic response under impact loads, direct tensile stress-strain relationship, and compressive stress-strain relationship. The predicted results revealed that the fly ash-based AAC-containing slag exhibited superior elasticity and ductility stress. In a previous study, a random forest (RF) approach was utilized by Gomaa et al. [24] to predict the fresh and hardened properties of fly ash-based alkali-activated concrete cured in an oven under ambient and moist conditions. The obtained results showed that the RF model was able to generate predictions with a higher accuracy than that of the experimental works. In the same context, the compressive strength of slag-based geopolymer concrete incorporating silica fume (SF) and natural zeolite (NZ) was estimated using a gene expression programming (GEP) model. The accuracy of the GEP model in predicting the compressive strength of GGBS-based geopolymer concrete under different amounts of SF, NZ, time, GBFS, and sodium hydroxide solution [25]. In particular, the respective values of R^2 for both the training and validation phases were 0.918 and 0.94, respectively, indicating that both the experimental and predicted results were considered close and satisfactory. Other predictive and optimization methods are also available in the literature [26,27].

In the present study, both an experimental and a mathematical assessment using the RSM were taken into account to optimize and predict the optimal content of ceramic waste in GBFS-based self-compacting alkali-activated concrete. This is because the utilization of ceramic waste in AAC is still in its infancy [28,29]. The aim of this study was to provide in-depth information to promote the use of ceramic waste in the construction industry.

To implement the aims of this study, experimental tests were first conducted based on the suggested design array of the central composite design (CCD). The obtained results were analyzed and assessed using the RSM models. Furthermore, ANOVA and error-statistics validation methods, such as *p*-value, F-value, and RMSE, were taken into account to evaluate the accuracy and sensitivity of the models. Moreover, the correlation statistics indicators, such as R^2, adjusted R^2, predicted R^2, and R, were also computed to assess the fitness between the actual and predicted data. Finally, several microstructure tests, that is, FE-SEM, XDR, and XRF, were conducted to support the present findings.

2. Experimental and Mathematical Works

2.1. Design of Experiment Using Response Surface Methodology

Response surface methodology, specifically a face-centered central composite design (FC-CCD), was considered in the design experiments through five sequential steps. In the first step, the number of experiments was suggested and determined using the FC-CCD. In the second step, the experimental results were collected and analyzed using Design-Expert software (Version 12, Minneapolis MN 55413, U.S.). A numerical model was constructed using a second-order polynomial equation. Subsequently, the accuracy and performance of the models were verified using analysis of variance (ANOVA). Finally, the optimum CWP/GBFS ratio was obtained using the desirability function.

Two independent variables and two levels were used in this study. The GBFS replaced by the CWP was the first independent variable (d_0), whereas time (d_1) was the second. As shown in Figure 1, FC-CCD involves three types of points, namely factorial points, axial points, and central points, according to their locations. For example, the square vertices point with a coded value of -1 and $+1$ are related to the factorial point. Moreover, the points situated at the centre of each face and far from the centre of the square with a distance of $\pm\alpha$ referred to the axial or star points, while the central point situated in the centre of the square had a value of zero.

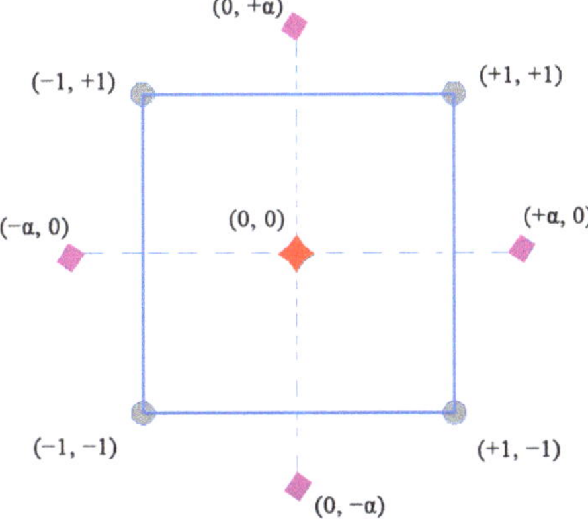

Figure 1. Schematic representation of FC-CCD.

Consequently, the suggested number of experimental tests was 13 according to Equation (1), where Z_0 and Z are the real values of the independent variables at the centre point and the real value of the independent value, respectively, and α and R are the step change value and the coded value of the independent variable, respectively. It should be noted that the main goal of using five central points was to evaluate the pure error in the proposed models. Similarly, the relationship between the coded values and the actual values of the experimental tests is expressed using Equation (2), where m is the number of centre points. Moreover, Equation (3) represents the general form of the second-order polynomial equation which was used to develop the model, where β_i denotes the linear coefficients, β_0 corresponds to the intercept of the model, and β_{ii} denotes the quadratic coefficients.

$$Q = 2^n + 2n + m \tag{1}$$

$$R = \frac{Z - Z_0}{\alpha} \tag{2}$$

$$Y = \beta_0 + \sum_i^k \beta_i d_i + \sum_i^k \beta_{ii} d_i^2 + \sum_{ij}^k \beta_{ij} d_{ij} \tag{3}$$

Herein, two RSM models, namely Model 1 and Model 2, were developed, as shown in Table 1. The compressive strength (*CS*), tensile strength (*TS*), and flexural strength (*FS*) of the GBFS-based SCACC containing CWP exposed to ambient conditions was evaluated in the first model at time intervals of 3 to 56 days. The second model was developed to determine the optimal strength loss (*SL*), weight loss (*WL*), and ultrasonic pulse velocity (*UPVL*) of the acid-treated SCACC at time intervals of 90 to 360 days.

Table 1. The suggested design array of experimental work based on FC-CCD.

Run No.	Coded Value		Real Value			FC-CCD Division
			Model 1 and 2 CWP/GBFS (%)	Model 1 Time (Days)	Model 2 Time (Days)	
1	−1	−1	10	3	90	
2	1	−1	80	3	90	Factorial
3	−1	1	10	56	360	points (2n)
4	1	1	80	56	360	
5	1	0	10	29.5	225	
6	−1	0	80	29.5	225	Axial points
7	0	−1	45	3	90	(2n)
8	0	1	45	56	360	
9	0	0	45	29.5	225	
10	0	0	45	29.5	225	
11	0	0	45	29.5	225	Centre points
12	0	0	45	29.5	225	
13	0	0	45	29.5	225	

In the same context, the accuracy of the developed equations was verified using an ANOVA. ANOVA is an integrated mathematical and statistical tool used to verify the significance and strength of a proposed model using several statistical validation indicators, such as p-value, F-value, adequate precision, R^2, predicted R^2, and adjusted R^2. In particular, the developed equation could be classified as significant if the p-value was less than 0.005 and the F-value was high (>1). Similarly, the value of the adequate precision should be greater than four to demonstrate that the model has less error prediction. Moreover, a strong correlation between the actual and estimated results could be obtained if R^2 was greater than 0.7. In addition, the model could be used for further prediction when the differences between the predicted R^2 and adjusted R^2 were less than 0.2.

To determine and assess the statistical validation indicators, several parameters should be first calculated beforehand, such as the total sum of squares (SS_T), degree of freedom (DF), and residual sum of square (SS_R). Both the SS_R and SS_T were obtained using Equations (4) and (5), respectively, where Y_A and Y_P denote the actual and predicted values, respectively, and $\overline{Y_A}$ is the average actual value. Consequently, R^2 was determined using Equation (6).

$$SS_R = \sum_{i=1}^{n} \left(Y_A - Y_P\right)^2 \tag{4}$$

$$SS_T = \sum_{i=1}^{n} \left(\overline{Y}_A - Y_A\right)^2 \tag{5}$$

$$R^2 = 1 - \frac{SS_R}{SS_T}$$

$$= 1 - \frac{\sum_{i=1}^{n} (Y_A - Y_P)^2}{\sum_{i=1}^{n} \left(\overline{Y}_A - Y_A\right)^2} \tag{6}$$

Similarly, the degree of freedom for the variance of dependent variables (DF_T) and the degree of freedom for the residual error (DF_R) were calculated to obtain the adjusted R^2 using Equation (7). Meanwhile, the predicted R^2 was determined using Equation (8), where W refers to the estimated residual sum of squares without the i_{th}. In addition, Equation (9) represents the adequate precision (SN), which is also essential for assessing the signal-to-noise ratio, where σ^2 refers to the residual mean square and p is the number of terms in the model.

$$R^2_{adj} = 1 - \frac{SS_R/DF_R}{SS_T/DF_T} \tag{7}$$

$$R^2_{pred} = 1 - \frac{W}{SS_T} \tag{8}$$

$$SN = \frac{\max(Y_p) - \min(Y_p)}{\sqrt{\frac{p\sigma^2}{n}}} \tag{9}$$

2.2. Material and Mixing Process

In the present experimental work, two types of waste materials were adopted in addition to the alkali-activated binder, GBFS and CWP. The GBFS was collected from a local iron industry (Ipoh, Malaysia) and used as the main source of calcium and silica oxides in the preparation of the SCAAC specimens without applying any laboratory treatment. It is well known that GBFS exhibits cementing and pozzolanic properties and reacts hydraulically with water. As for its physical properties, the colour of GBFS is off-white. The medium particle size was estimated to be 12.8 μm from a particle-size analysis. From the Brunauer, Emmett and Teller (BET) test, the specific surface area and specific gravity of the GBFS specimen were found to be 13.6 m^2/g and 2.89, respectively. From the viewpoint of chemistry, the chemical composition of both the CWP and the GBFS was determined using X-ray fluorescence (XRF) spectroscopy. The main composition of GBFS was calcium oxide (51.8% weight), followed by silica and aluminium (41.7 %). This result is in line with several other research results which reported that the synthesis of an alkali-activated paste is greatly dependent on the concentrations of CaO, SiO$_2$, and Al$_2$O$_3$ oxides. These compounds are involved in the development of the N,C-(A)-S-H gels during geopolymerization.

Ceramic waste was collected from a tile ceramic waste industry (Johor, Malaysia). The ceramic waste was primarily processed using laboratory treatment prior to SCAAC specimen preparation. At an earlier stage, the ceramic waste was crushed using a crushing machine and sieved through a sieve with a mesh size of 600 μm. Subsequently, the obtained ceramic was ground for 6 h to produce a fine CWP of light grey colour with a particle size of 35 μm, a specific surface area of 12.2 m^2/g, and a specific gravity of 2.61. Based on the chemical composition analysis, the silica and aluminium contents in the CWP were approximately 84.8% of the total element weight. Unlike GBFS, CWP presented a very low calcium oxide content. Moreover, neither CWP nor GBFS possessed a large amount of potassium oxide (K$_2$O). On the other hand, CWP contained 13.5% sodium oxide (Na$_2$O). Although this goes beyond the focus of the present study, there is evidence that the activation of the alkaline and geopolymerization depends significantly on the Na$_2$O and K$_2$O contents. It is also interesting to note that the CWP and GBFS conformed to ASTM C618, in which both materials exhibited extremely low loss of ignition values.

For the alkaline-activator solution preparation, two types of solutions were prepared and mixed: sodium silicate (NS) and sodium hydroxide (NH). Both NH and NS were supplied by QREC (Selangor, Malaysia). From the viewpoint of the analytical chemical grade, the NH exhibited a purity of 98%, while the NS solution contained Na_2O (14.70 wt.%), SiO_2 (29.5 wt.%), and H_2O (55.80 wt.%). The molarity of the NaOH solution was maintained at 2M during the preparation of the NH solution. Following this, the solution was left to cool for 24 h prior to the addition of sodium silicate solution to prepare the final alkaline solution with a modulus solution ratio of 1.2 (SiO_2/Na_2O). The ratio of sodium silicate to sodium hydroxide (NS:NH) was maintained at 0.75 for all mixtures. In addition, both the local river sand, with a specific gravity of 2.62, and crushed granite were utilized as fine and coarse aggregates, respectively, to prepare the SCAAC specimens. The measured specific gravity and water absorption of the crushed granite were found to be 2.67 and 0.51%, respectively. To sustain the ratio of the alkaline solution to the binder (S:B), the supplied river sand was used under a saturated surface dry condition for all mixes.

For the design of the experiments, the SCAAC mixtures were coded and represented as shown in Table 1. It should be noted that the ratios of solution to binder (S:B) and sodium silicate to sodium hydroxide (NS:NH), the sodium hydroxide molarity, the solution modulus ($SiO_2:Na_2O$), and the content of the binder, river, and crushed granite were fixed at 0.50, 0.75, 2 M, 1.2, 484 kg/m^3, 844 kg/m^3, and 756 kg/m^3, respectively, for all of the self-compacting alkali-activated mixtures. Several steps were taken into account when preparing the SCAAC specimen. First, both the CWP and GBFS were mixed and blended in a concrete mixer for 3 min. In this step, a dry condition was considered to ensure a uniform mixture between the coarse and fine aggregates. Subsequently, the alkaline solution was added to the mixture for 6 min at moderate speed. Several techniques were used to evaluate the required initial properties of the SCAAC mixtures, such as the filling and penetration capabilities, as well as the segregation resistance.

2.3. Test Methods

The European guidelines for self-compacting concrete, EFNARC (European Federation for Specialist Construction Chemicals and Concrete Systems) were adopted in this study to evaluate the fresh performance of the SCAAC mixtures. It was found that the inclusion of CWP as a GBFS replacement from 0% to 10%, 45%, and 80% led to an increase in the slum flow value from 560 mm to 605, 640, and 750 mm, respectively, while the T50 test showed an enhanced flow of fresh concrete and a time reduction from 6 s to 5.2, 5, and 3 s, respectively. An L-box test improved the workability in which the H2 to H1 ratio increased from 0.78 to 0.95.

According to ASTM C109, ASTM C496, and ASTM C78, the compressive strength (*CS*), splitting tensile (*TS*), and flexural strength (*FS*) tests were conducted at time intervals of 3, 7, 28, and 56 days. Samples of cubic (100 × 100 × 100) mm, cylindrical (200 × 100) mm, and prism (100 × 100 × 500) mm dimensions were fabricated for all the *CS*, *TS*, and *FS* tests, respectively. The average values from all the tests on all three specimen types were collected and compared to those of the control specimen prepared without any CWP content.

The durability and life service of the proposed concrete were measured in an aggressive environment by exposing the SCAAC specimen to a sulphuric acid solution. Deionized water was used to prepare the sulphuric acid (initial concentration is 96%) solution (H_2SO_4) at a concentration of 10%. The cured specimen at day 28 was evaluated with regard to the *CS*, weight, and ultrasonic pulse velocity (*UPVL*) before further immersion in the solution. To maintain the pH of the solution, a fresh H_2SO_4 solution was replaced every three months throughout the experiment. After 90, 180, and 360 days from the immersion date, the concrete specimen was examined, and several indicators were investigated for the performance assessment, such as compressive strength loss (*SL*), weight loss (*WL*), and *UPVL* according to the ASTM C267 specifications [30], as well as the surface and edge deterioration.

2.4. Microstructure

In this study, three microstructure tests were considered to evaluate the effect of the CWP content on the proposed concrete performance, including X-ray diffraction (XRD), Fourier transform infrared (FTIR) spectroscopy, and scanning electron microscopy (SEM). To analyse the obtained results, the XRD data were connected to Jade software (CA, USA), where inspection of the disordered stage of the alkali-activated pastes was enabled and carried out in the 2θ range of 5–90° at a step size of 0.02° and a 0.5 s/step scanning speed. The procedure involved positioning the sample on a brass stub holder fixed with carbon tape, prior to exposure to infrared radiation for five minutes for drying purposes. The sample was subsequently covered with a layer of gold using a blazer-sputtering coater. Patterns were documented at 20 kV and 1000× magnification. The vibration modes of the underlying chemical structures of the alkali-activated pastes were determined using an FTIR spectrophotometer. To assess the surface morphology of the tested alkali-activated pastes, the samples were coated in advance using a gold sputtering coater machine and tested using the SEM instrument by applying a similar magnification to the images.

3. Results and Discussion

3.1. Predicted Equations for Mechanical Properties and their Validation

Based on the present findings, six predictive quadratic equations were developed. In particular, model 1 reflects the mechanical properties of SCAAC exposed to ambient conditions, including compressive, flexural, and tensile strengths. The second model represents the behaviors of the SCAAC upon exposure to an acid attack, including compressive strength loss, weight loss, and UPVL. Indeed, these equations are important because the relationship between the involved parameters and responses can be quickly predicted and evaluated. The accuracy of these equations was evaluated using several statistical validation parameters, as listed in Table 2. For instance, the correlation between the predicted and actual data was assessed using the R^2. It can be seen that the R^2 values of the compressive, tensile, and flexural strengths were higher than 0.99. Similarly, the R^2 values of compressive strength loss, weight loss, and UPVL were greater than 0.99. It was also found that the differences between the adj. R^2 and predicted R^2 values were lower than 0.056 for all data sets. Such results are considered an indicator of a strong correlation between the actual and estimated data. This is in line with Mohammed et al. [31], who stated that a model could be considered accurate and reliable if the differences between the adjusted and predicted R^2 were lower than 0.2. In the same regard, a good agreement and fitness between the actual and estimated results was also proven, in which the ratio between R^2 and adj. R^2 was found to be high. This fact is consistent with Mohammed et al. [32], who demonstrated that the best fitness could be achieved if the R^2 to adj. R^2 ratio was high. In addition, the R values of the mechanical properties of SCAAC were higher than 0.995. Similarly, the R values of the strength loss, weight loss, and UPVL were 0.998, 0.9988, and 0.9999, respectively. As such, the developed models could be utilized for the purpose of prediction, which is in good agreement with previous studies. For example, Khan et al. [15] highlighted that the correlation between the predicted and the actual results could be considered as strong if the R value ranges were between 1 and 0.8, whereas moderate correlation could be achieved when the R value was greater than 0.5, but less than 0.8. This was also confirmed by Carrillo et al. [33]. In the same context, it can be seen that the value of adeq. precision was greater than four, indicating a perfect fit of the proposed models. Moreover, the adequacy and efficiency of the proposed models were verified using an error statistical indicator, namely RMSE. The implementation of statistical error indicators is necessary to assess the prediction error. It was observed that the RMSE values were less than 1.337 for all of the datasets, indicating high accuracy.

Table 2. Predictive equations of SCAAC properties.

Model	Item (MPa)	Predicted Equations and Related Statistics Indicators					
Model 1	Compressive strength	R = 0.995	$R^2 = 0.992$	$R^2_{adj} = 0.987$	$R^2_{predicted} = 0.931$	Adeq. Precision 47.938	RMSE 1.337
		$CS = 47.11 - 18.57d_0 + 8.84d_1 - 3.05d_0d_1 - 1.31d_0^2 - 3.29d_1^2$					
	Tensile strength	R = 0.998	$R^2 = 0.997$	$R^2_{adj} = 0.995$	$R^2_{predicted} = 0.971$	Adeq. Precision 74.47	RMSE 0.0631
		$TS = 4.62 - 1.36d_0 + 0.55d_1 - 0.125d_0d_1 - 0.26d_0^2 - 0.43d_1^2$					
	Flexural strength	R = 0.996	$R^2 = 0.992$	$R^2_{adj} = 0.987$	$R^2_{predicted} = 0.931$	Adeq. Precision 49.11	RMSE 0.0349
		$FS = 1.49 - 0.39d_0 + 0.29d_1 - 0.05d_0d_1 + 0.09d_0^2 - 0.21d_1^2$					
Model 2	Strength loss	R = 0.998	$R^2 = 0.997$	$R^2_{adj} = 0.995$	$R^2_{predicted} = 0.974$	Adeq. Precision 83.192	RMSE 1.022
		$SL = 29.26 - 16.95d_0 + 19.38d_1 - 12.30d_0d_1 - 75.1d_0^2 - 0.0008d_1^2$					
	weight loss	R = 0.9988	$R^2 = 0.9978$	$R^2_{adj} = 0.996$	$R^2_{predicted} = 0.977$	Adeq. Precision 88.7	RMSE 0.0269
		$WL = 0.625 - 0.5d_0 + 0.51d_1 - 0.35d_0d_1 + 0.07d_0^2 - 0.0002d_1^2$					
	UPVL	R = 0.9999	$R^2 = 0.9999$	$R^2_{adj} = 0.9998$	$R^2_{predicted} = 0.9987$	Adeq. Precision 362.72	RMSE 0.0625
		$UPVL = 7.39 - 5.09d_0 + 4.33d_1 - 3.26d_0d_1 - 0.325d_0^2$					

For the same considerations, ANOVA was used to investigate the applicability of the models using the *p*-value and F-value, as shown in Table 3. The *p*-value is considered to be one of the best mathematical methods for verifying the significance of a regression coefficient. When the *p*-value was smaller than 0.005, the model was significant. In contrast, an insignificant model was obtained when the *p*-value was greater than 0.005. The *p*-values of all models were found to be less than 0.0001, confirming that the models were significant. Similarly, the F-value was also used to assess the significance of the mean value variance. As is well known, a model can be considered significant if the F-value is higher, whereas a lower value indicates an insignificant model. In the developed models, it can be inferred that all the models were reliable and applicable for prediction owing to the high F-values. This inference is in good agreement with Ray et al. [34], who evaluated the predictive models of the compressive and splitting tensile strengths of concrete incorporating fine glass aggregate and condensed milk can fibers using F-values. Based on their findings, their models were considered accurate and significant owing to their high F-values.

After verifying the developed equations, the optimum replacement percentage of GBFS by CWP was investigated using multi-objective optimization based on a desirability function. Indeed, the desirability function (*DR*) has been widely used owing to its proven ability to determine the optimal content that either maximizes or minimizes the target output. Equation (10) describes the desirability function, where *n* refers to the number of independent and dependent variables. Two independent variables, namely time and the CWP/GBFS ratio, were considered; however, the focus of the current study is more focused on the latter (the CWP/GBFS ratio). Therefore, two scenarios were considered to determine the optimum percentage of CWP. First, the optimum percentage of CWP was obtained on the basis of the maximum compressive, tensile, and flexural strengths achieved, whereas the second scenario aimed at obtaining the optimal content with a minimum strength loss.

In general, it was found that the mechanical strength decreased with increasing ceramic content. In contrast, the durability properties, such as the weight and strength loss, as well as the UPVL, of the SCAAC exposed to acid attack improved. Based on Figure 2a,b, the optimum value of the CWP/GBFS ratio that satisfied both the strength and durability was 31%. This is because, in terms of mechanical properties, the decrease in compressive strength was 16% compared to that of the control specimen (without ceramic), which is considered an acceptable result. From the viewpoint of durability, the strength loss minimized to 59.17%, which is lower than that of the control specimen (74.2%). In other words, both the mechanical properties and the durability of the SCAAC could be considered acceptable when the replacement of the GBFS by the CWP was lower than 31%. Beyond this value, despite the fact that the durability properties of the SCAAC exposed to acid attack increased with an increase in the CWP content, a significant reduction in the compressive, tensile, and flexural strength was recorded.

$$DR = (d_1 \times d_2 \times d_3 \times \ldots\ldots d_n)^{(1/n)} \tag{10}$$

Table 3. Validation of the proposed models using ANOVA.

Model	Type	ANOVA	Term					
			Model	d_0	d_1	$d_0\,d_1$	$d_0{}^2$	$d_1{}^2$
Model 1	CS	*p*-value	<0.0001	<0.0001	<0.0001	0.0085	0.2357	0.0141
		F-value	185.37	730.08	165.57	13.13	1.68	10.54
		Sig.	Y	-	-	-	-	-
	TS	*p*-value	<0.0001	<0.0001	<0.0001	0.0131	0.0007	<0.0001
		F-value	491.75	1930.24	325.15	10.90	32.73	90.37
		Sig.	Y	-	-	-	-	-
	FS	*p*-value	<0.0001	<0.0001	<0.0001	0.0335	0.0089	<0.0001
		F-value	186.05	542.72	308.71	6.96	12.86	71.70
		Sig.	Y	-	-	-	-	-
Model 2	SL	*p*-value	<0.0001	<0.0001	<0.0001	<0.0001	0.0003	0.9992
		F-value	564.82	1042.85	1363.63	366.61	43.58	1.1E-06
		Sig.	Y	-	-	-	-	-
	WL	*p*-value	<0.0001	<0.0001	<0.0001	<0.0001	0.0067	0.9941
		F-value	635.05	1324.13	1399.85	434.40	14.44	0.0001
		Sig.	Y	-	-	-	-	-
	UPVL	*p*-value	<0.0001	<0.0001	<0.0001	<0.0001	0.0002	1.0
		F-value	10641.2	26589.5	19250.2	7308.08	50.0	0.0
		Sig.	Y	-	-	-	-	-

On the other hand, the influence of the CWP:GBFS ratio content on the mechanical properties and durability of SCAAC was clearly observed. As illustrated in Figure 3a, a sharp slope is observed, indicating a significant impact. This fact is consistent with the ANOVA results in which the *p*-values of compressive, tensile, and flexural strength were less than 0.0001 and the F-value was high, confirming that the addition of ceramic adversely affected the mechanical properties of the SCAAC. In other words, the mechanical properties of SCAAC decreased with increasing CWP content.

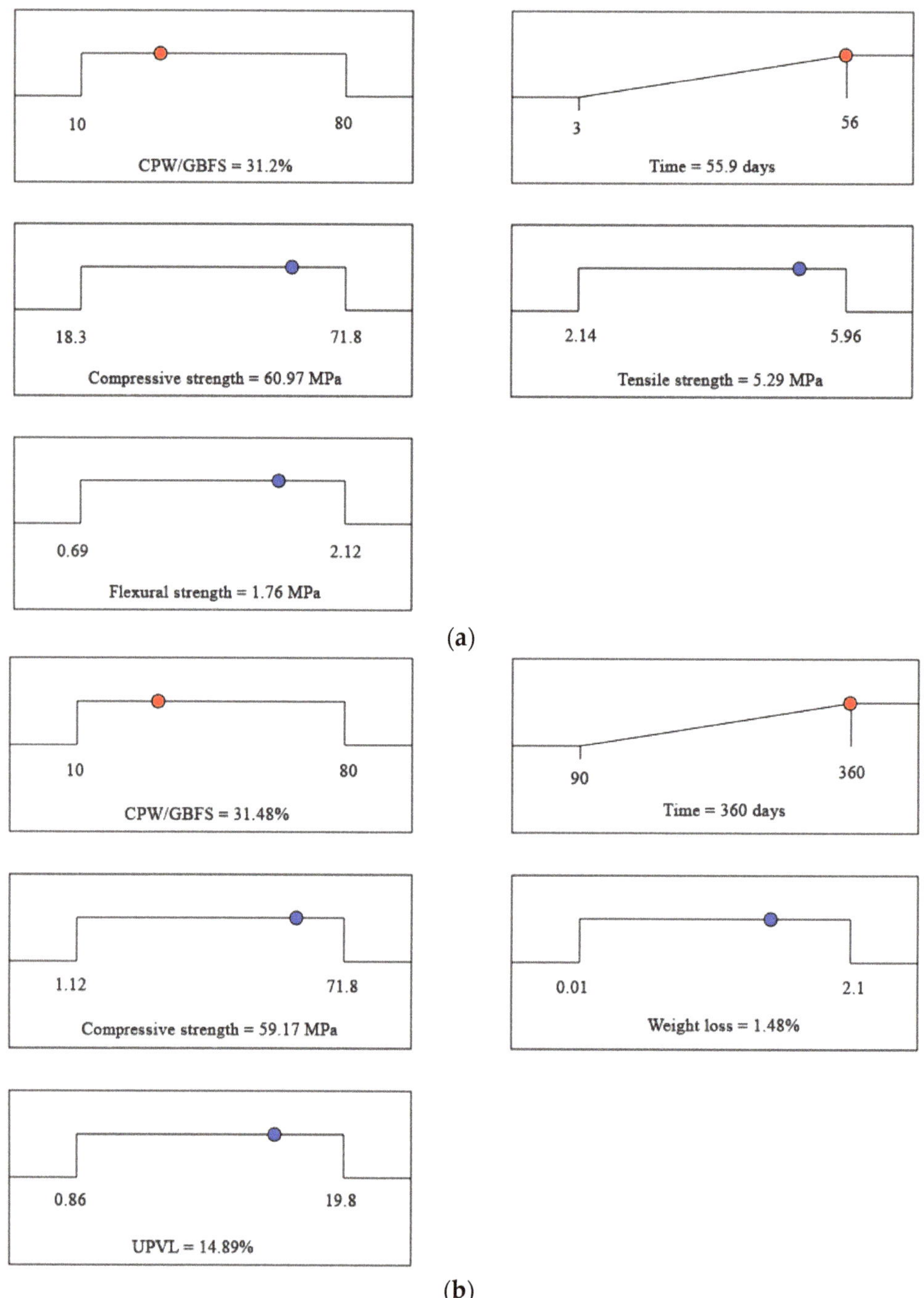

Figure 2. Optimum replacement percentage of GBFS with CWP in the aspects of (**a**) mechanical properties and (**b**) durability.

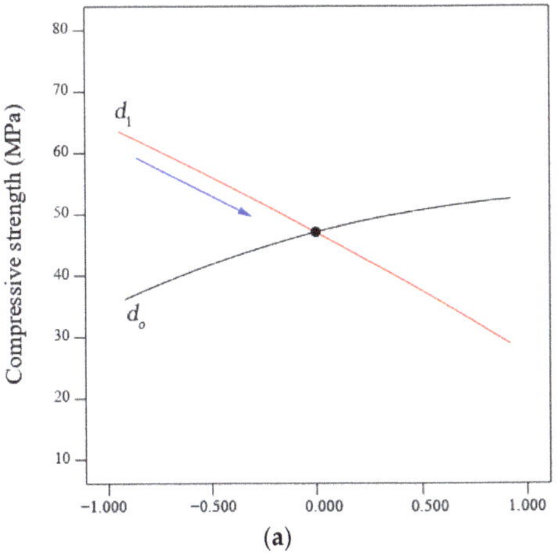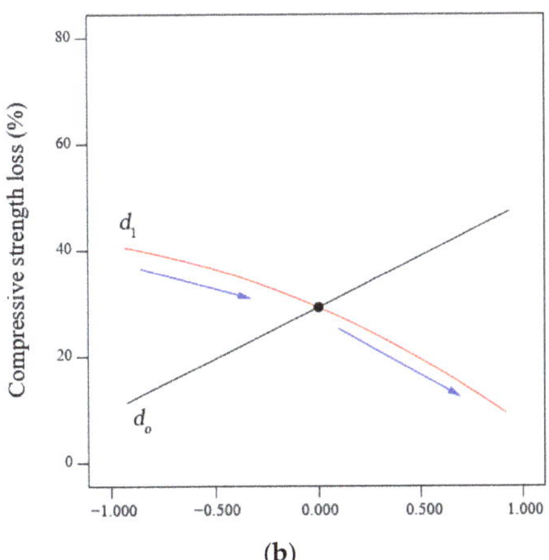

(a)

(b)

Figure 3. Perturbation curves for the (**a**) compressive strength of SCAAC (Model 1) and (**b**) strength loss of SCAAC when exposed to acid attack (Model 2).

In contrast, the strength loss, weight loss, and UPVL improved with an increase in CWP content. This fact can be observed in Figure 3b, in which the slope gradient for the ceramic content (d_1) was high, confirming that the addition of CWP was also significant in terms of durability. These results were also confirmed by ANOVA. According to Tian et al. [35], the significance and sensitivity of a linear variable were proven with a p-value of less than 0.5. In the present study, the p-values for strength loss, weight loss, and UPVL were also less than 0.0001a and the F-value was high, indicating a significant interaction of CWP with respect to durability.

3.2. Effect of CWP Incorporation on the Mechanical Properties of AAC

Figure 4 illustrates the effects of the CWP incorporation as the GBFS replacement and curing age on the evolution of the compressive strength, tensile strength, and flexural strength of the SCAAC. As shown in Figure 4a, the inclusion of CWP in the alkali-activated matrix as a part of the binder negatively affected the compressive strength of the SCAAC exposed to ambient conditions. In contrast, the results showed that the development of CS monotonically improved with the increase in age. In general, at both the early and later ages, the compressive strength was highly influenced by the CWP content. At 28 days of curing age, the results showed that the increasing of the CWP content from 0% to 80% led to the compressive strength reduction from 70.1 MPa to 26.2 MPa, respectively.

The results of the splitting tensile strength of the SCAAC specimen prepared with various levels of GBFS replacement with CWP are also presented in Figure 4b. The results revealed a decreasing trend in tensile strength with increasing CWP content. However, all the proposed concrete specimens demonstrated an increase in strength development with increasing curing age. The highest splitting tensile strength value of 5.96 MPa was achieved by the control sample, and this value dropped to 2.93 MPa with an increase in the level of CWP to 80%. For flexural strength, a similar trend was observed, and the value of the strength decreased with an increase in the CWP content from 0% to 80% as the GBFS replacement. For example, with the increasing of the content of the CWP from 0 to 80%, the value of the flexural strength dramatically dropped from 2.2 to 1.19 MPa, respectively.

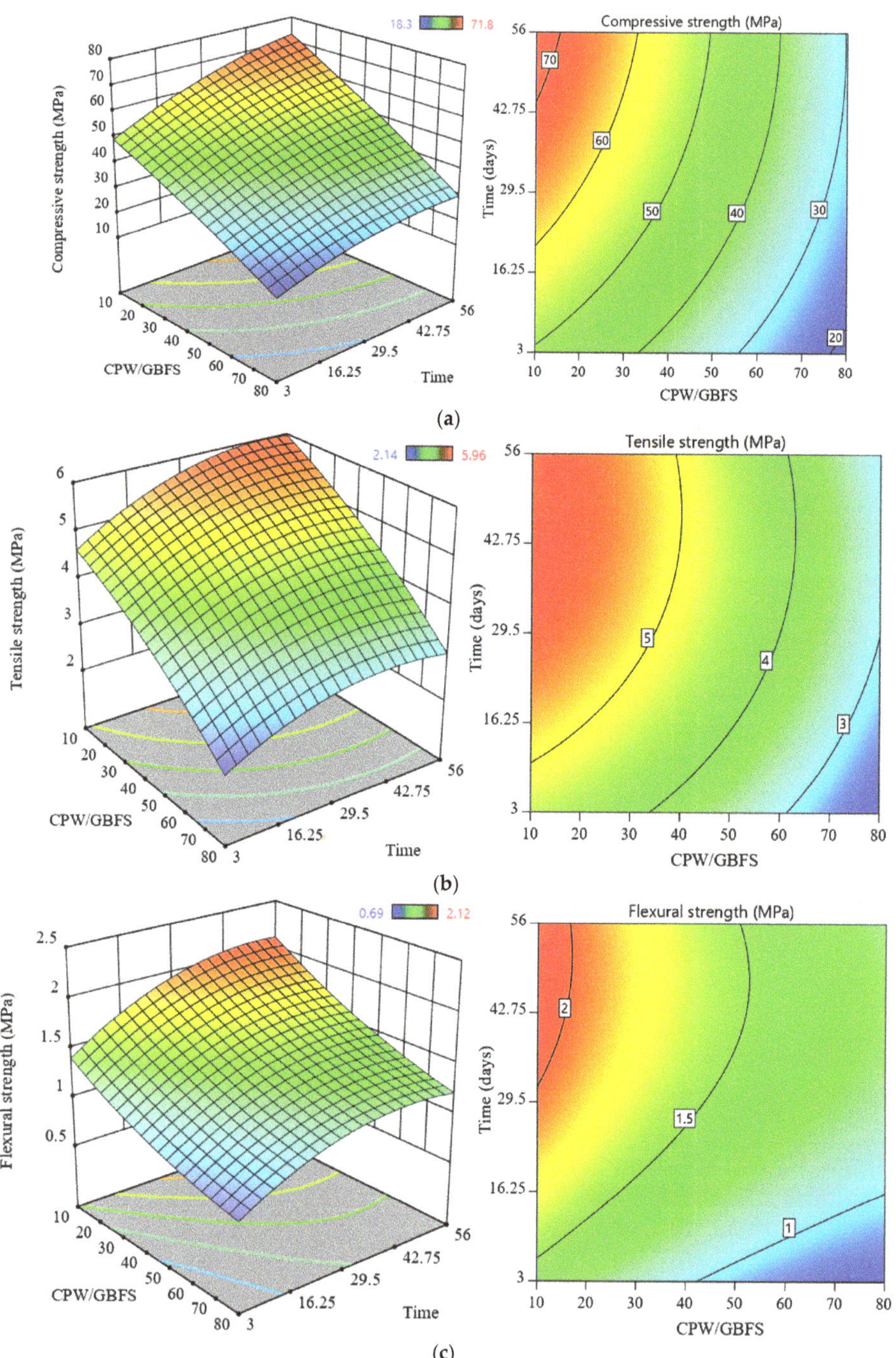

Figure 4. Strength evolution in the proposed concrete: (**a**) CS, (**b**) TS, and (**c**) FS.

In general, the decrease in the strength of the proposed alkali-activated concrete incorporating ceramic is due to the limited amount of CaO present in the mix, in which the ratio of SiO_2 to CaO is not balanced but increases steadily. A high content of CWP but a low content of CaO could negatively affect the chemical reaction rate and thus the productivity of C-(A)-S-H. In other words, the amount of C-(A)-S-H and the concrete strength are linearly dependent on the CaO content. This is consistent with a study by Rashad [36]. According to their study, the alkali-activated concrete strength decreased with increasing aluminosilicate content in materials such as fly ash. In the present study, the decrease in the CWP/GBFS-based alkali-activated concrete strength may be attributed to several reasons. For example, the chemical compositions of CWP and GBFS play a significant role in adversely affecting the alkali activation process of the binder. In addition, the decrease in the reaction rate of the CWP is considered as the second reason for the strength reduction [37]. Third, a high amount of CWP content could also lead to lower compactness and density of concrete, and thus, reduced strength. In the same context, the low sodium hydroxide concentration (2 M) is another reason for countering the lack of CaO content in the present study. Thus, it can be inferred that the compressive strength development in the SCAAC could only increase with the formation of the C-S-H and C-A-S-H gels as well as the N-A-S-H [29,38].

In the same regard, X-ray diffraction (XRD) analysis was also taken into account to analyse and support the results of the proposed SCAAC containing a high amount of CWP (45% and 80%). As shown in Figure 5, the hydrotalcite ($Mg_6Al_2CO_3OH_{16}.4H_2O$) phase emerged at a peak around 10° [39]. In addition, a reduction in the C-S-H peak intensity was observed at 30° with increasing CWP concentration. The C-S-H is most probably substituted by hydrotalcite at the peak 43° with the replacement percentage of GBFS by CWP of 80%. Similarly, calcite was found to be substituted by quartz at peaks at 17°, 21°, 28°, 38.5°, and 51° when the amount of CWP was higher than 45%. It is also remarkable to note that increasing the CWP level leads to a decreased C-S-H product and an increased non-reacted silicate quantity, which confirmed the previous results where the concrete strength dropped from 70.1 to 26.2 MPa after 28 days. This is also in line with the previous literature [40], in which the productivity of the C-S-H gel and the calcite formulation could be diminished owing to the low CaO content. Moreover, the formation of C-S-H gel, N-A-S-H gel, and calcite was also detected [41].

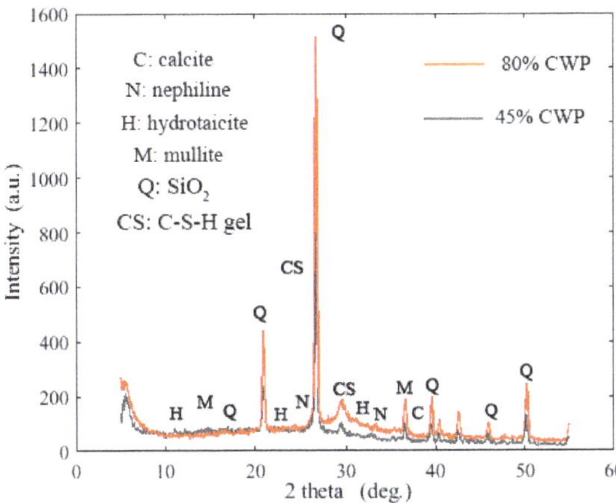

Figure 5. XRD results for the effect of GBFS replacement with CWP in SCAACs.

On the other hand, the surface morphology of the SCAACs incorporating different replacement percentages of GBFS with CWP was also examined and analysed after 28 days using SEM analysis to support the present finding. In particular, concrete samples with replacement percentages of 45% and 80% were considered. Based on Figure 6a, the concrete specimen containing 45% CWP was found to be dense, and the quantity of both the partial reaction and the unreacted particles was small. The surface of the second concrete specimen

(80% CWP) exhibited a large amount of partly reacted and unreacted particles, as shown in Figure 6b. As such, it can be inferred that poor morphology, a high degree of porosity, and a high quantity of unreacted quartz (SiO_2) were clearly observed when the value of the CWP increased from 45% to 80%. In other words, a reduction in the GBFS and a high amount of CWP had a negative effect on the C-(A)-S-H gel formulation. In addition, it leads to the generation of non-reacted particles, including quartz, and a more partially reacted gel such as mullite [42].

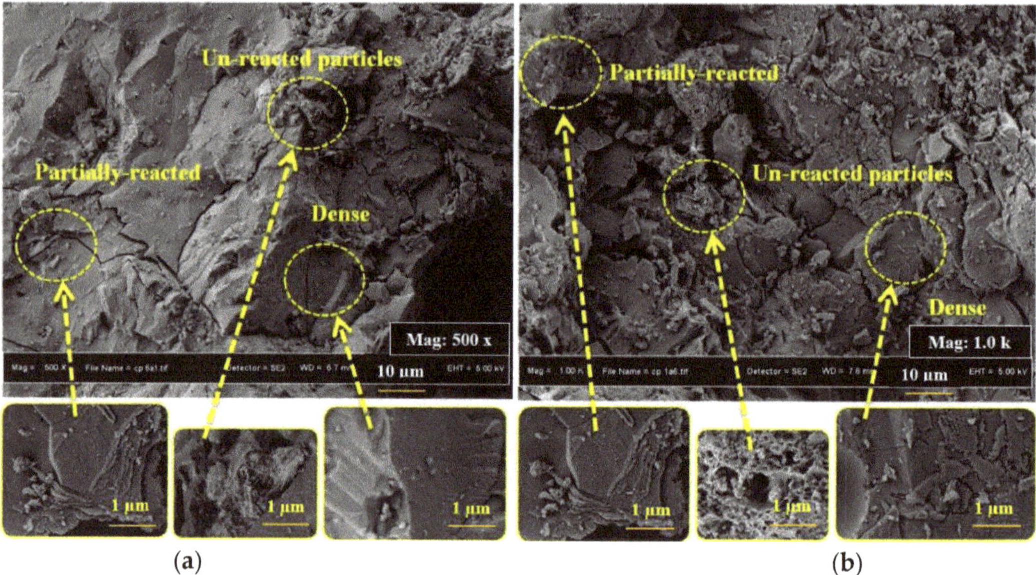

Figure 6. Surface morphology of SCAACs containing CWP at (**a**) 45% and (**b**) 80%.

In addition, Figure 7 shows the FTIR analysis results of the SCAACs prepared with various levels of CWP. The strength results indicated that the inclusion of CWP in the alkali-activated matrix reduced the specimen strength. This is because the amounts of dissolved CaO and Al_2O_3 decrease as the concentration of GBFS is minimized. Similarly, it was found that the lack of dissolved Al_2O_3 leads to a diminished extent of silicate polymerization [43]. It is well known that the FTIR test aims to analyse the bonding vibrations within the concrete specimen, which reflect the compressive strength, specifically through the occurrence of an FTIR spectral band of Si-O-Al. It can be seen that the band frequency increased when the concentration of CWP increased from 45% to 80%. In addition, the increase in the FTIR vibration frequency was due to the decrease in the Al_2O_3 level in the SCAAC mixes.

3.3. Effect of CWP Incorporation on the Durability of Acid-Treated AAC

The durability performance of the proposed SCAAC in an aggressive environment was evaluated, and a sulphuric acid environment was considered for this purpose. The losses in strength, weight, and UPV were measured after 3, 6, and 12 months from the immersion date in the 10% H_2SO_4 solution. The effects of the GBFS replacement with the CWP on the durability of the SCAAC in terms of strength, weight, and UPV losses were recorded, and the results are presented in Figure 8. For the strength loss, it was observed that the inclusion of CWP resulted in an enhanced durability in the prepared specimen and a reduced strength loss percentage from 74% to 13.3% with an increase in the CWP content from 0% to 80% for the specimen exposed to acid solution for 12 months (Figure 8a). Likewise, the weight loss decreased gradually with increasing replacement level of the

GBFS by the CWP in the binder matrix, as shown in Figure 8b. For example, the weight loss percentage dropped from 2.2% to 0.39% when the CWP content increased from 0% to 80%. This is because the formation of excessive undesirable gypsum in the control mix led to softening of the binder, and therefore, the surface of the concrete significantly deteriorated. In particular, the paste was separated from the aggregate and thus reduced the cross-section of the concrete samples. In contrast, the AAC incorporating a high amount of CWP resists binder deterioration and protects the concrete surface owing to the limited amount of calcium ions. This is also in line with previous studies [44,45]. On the other hand, the UPV test was conducted on the specimens exposed to acid solution for 12 months to measure the internal deterioration of the mixes. Figure 8c presents the loss in UPV percentage results, which indicated a loss percentage drop from 20.2% to 3.1% with an increase in the replacement level of the GBFS by the WCP from 0% to 80%.

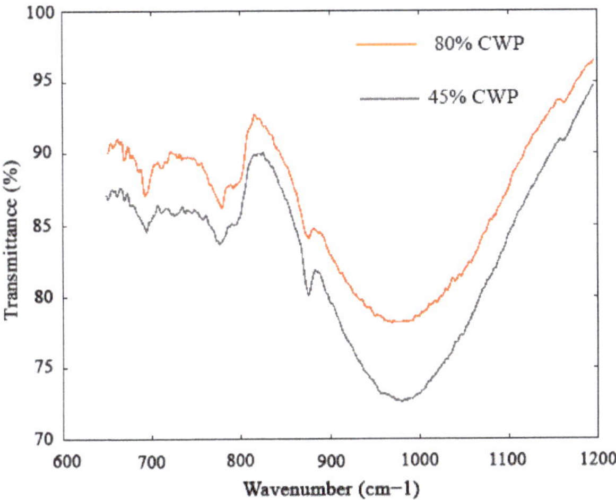

Figure 7. FTIR results of the alkali-activated binder prepared with high amount of CWP.

Moreover, the formation of cracks and surface deterioration decreased as the CWP content increased, as shown in Figure 9. Based on the visual examination, the CWP/GBFS-based AAC appeared to be in good condition compared to that of the control mix. The worsening of the GBFS-based ACC prepared with 45% and 80% CWP was less than that of the OPC. The cross section of the GFBS-based ACC containing 80% after 3, 6, and 12 months was almost the same. On the other hand, with the increase in exposure time, a significant reduction in the cross-section of the control mix was observed. In particular, crack formation occurred after three months and extended to six months. Then, a part of the concrete was removed, resulting in a reduction in the cross-section of the OPC owing to the softening of the paste around the aggregates. In addition, this was due to the high amount of calcium in the control specimen without ceramic incorporation, which resulted in a high amount of gypsum product. A concrete specimen with a high CWP content exhibited a high resistance to sulphuric acid attack. In other words, the enhancement in the durability of the proposed concrete to an acidic environment is attributed to the reduction in CaO and $Ca(OH)_2$ levels accompanied by an increase in CWP content. It is well known that the high availability of SO_4^{-2} in sulphuric acid can react with $Ca(OH)_2$ in concrete to form gypsum ($CaSO_4.2H_2O$). This product exhibited several undesirable results, in which an expansion could be generated in the concrete matrix to form further cracks. Consequently, such negative results could prompt and increase the deterioration of alkali-activated concrete [26,46]. This is also in good agreement with Allahverdi and Skvara [47], who utilized 50% GBFS in sodium-silicate-activated specimens for exposure to an acid solution. The results showed that the presence of Na^+ and Ca^{++} negatively affected the acid resistance of the concrete. In particular, the degradation of the gel product was accelerated owing to the exchange reaction between the OH ions with Ca^{++} and Na^+. Moreover, the composition and structure of the Al_2O_3-SiO_2 network underwent changes due to de-alumination [48].

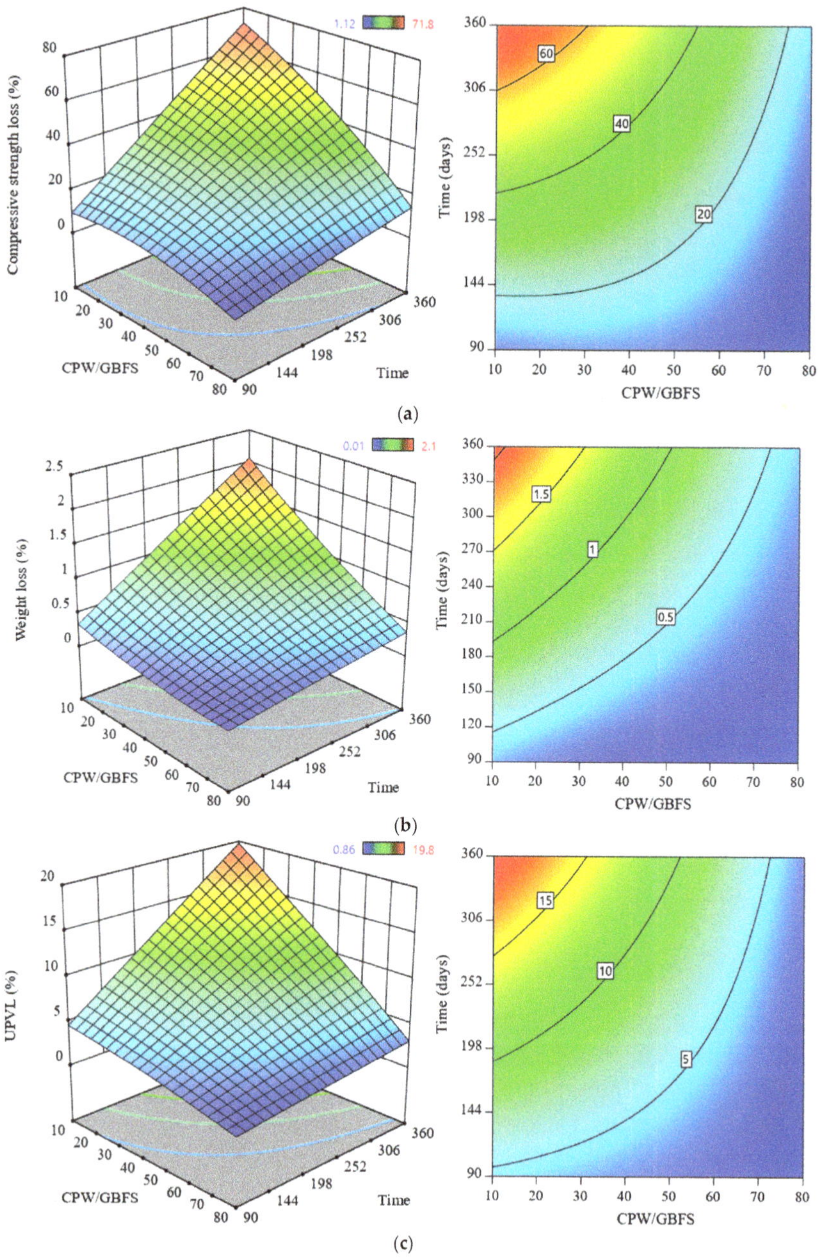

Figure 8. Durability of SCAAC exposed to acid: (**a**) strength loss, (**b**) weight loss, and (**c**) UPVL.

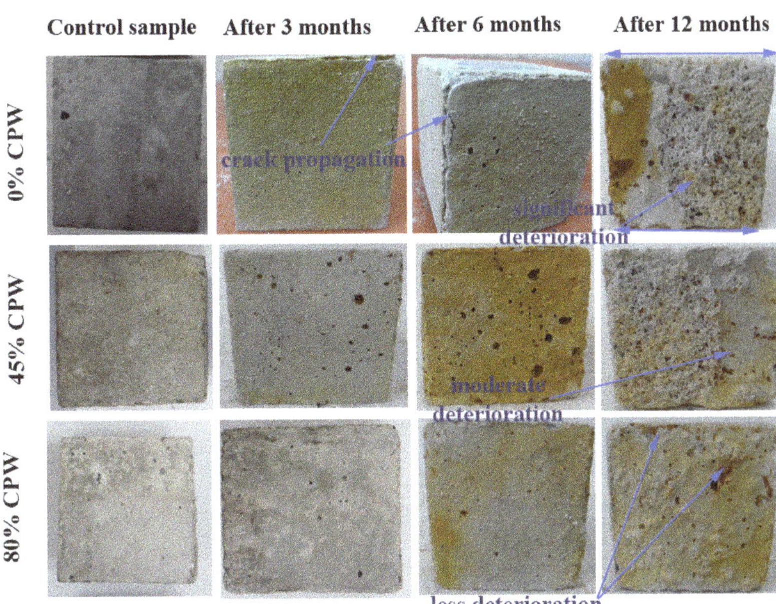

Figure 9. Surface texture of the proposed SCAAC containing 0, 50, and 70% CWP exposed to 10% H₂SO4 after different immersion periods.

4. Conclusions

Self-compacting alkali-activated concrete has been widely applied and experimentally evaluated using different binders, such as POFA, FA, and GBFS. However, the utilization of CWP as a binder in GBFS-based SCAACs is still limited. In the present study, an optimization method was considered to assess and obtain the optimal CWP content to satisfy both the strength and durability aspects. Based on the visual examination, predicted, and actual results, it can be concluded that:

- As a benefit of the RSM model, the number of experiments was found to be small enough (13) to develop a predictive equation.
- RSM proved its ability to assess the behavior of SCAAC incorporating CWP, in which the appropriate correlation between the actual data and the predicted data was achieved. In particular, the R, R^2, and adjusted R^2 values were higher than 0.95.
- The significance of the developed models was also proven by the high F-value and p-value of less than 0.0001. In addition, the proposed models can accurately predict the behavior of the SCAAC with minimum errors (RMSE < 1.337).
- The high replacement of GBFS by CWP prevented the deterioration of the concrete surface owing to the limited amount of calcium ions.
- The optimum replacement percentage of GBFS by CWP was 31%, in which a reasonable decrease in the compressive strength (16%) was obtained in addition to the minimized strength loss of the SCACC when exposed to an acid attack of only 59.17% (compared to 74.2% for the control specimen).
- The strength and weight loss of the SCAAC significantly decreased with the increase in the amount of CWP, specifically, 45% and 80%, respectively.

Therefore, it could be concluded that CWP could serve as an excellent strategic material to address serious environmental issues and enhance the durability of SCAAC-exposed acid attacks.

Author Contributions: Methodology, M.I.K., G.F.H. and H.A.A.; software, H.A.A.; investigation, G.F.H. and H.A.A.; writing—original draft preparation, H.A.A.; writing—review and editing, G.F., B.A.S., H.A., Y.M.A. and M.I.K.; supervision, M.I.K. and S.S. All authors have read and agreed to the published version of the manuscript.

Funding: This research received no external funding.

Institutional Review Board Statement: Not applicable.

Informed Consent Statement: Not applicable.

Data Availability Statement: The data presented in this study are available on request from the corresponding author. The data are not publicly available due to data privacy.

Acknowledgments: The authors extend their appreciation to the Deanship of Scientific Research at King Saud University for funding this work through the research group project No. RGP-VPP-105.

Conflicts of Interest: The authors declare no conflict of interest.

References

1. Horňáková, M.; Lehner, P. Relationship of Surface and Bulk Resistivity in the Case of Mechanically Damaged Fibre Reinforced Red Ceramic Waste Aggregate Concrete. *Materials* **2020**, *13*, 5501. [CrossRef]
2. Barreto, E.; Stafanato, K.; Marvila, M.; de Azevedo, A.; Ali, M.; Pereira, R.; Monteiro, S. Clay Ceramic Waste as Pozzolan Constituent in Cement for Structural Concrete. *Materials* **2021**, *14*, 2917. [CrossRef]
3. Hilal, N.; Saleh, R.D.; Yakoob, N.B.; Banyhussan, Q.S. Utilization of ceramic waste powder in cement mortar exposed to elevated temperature. *Innov. Infrastruct. Solut.* **2021**, *6*, 35. [CrossRef]
4. Khan, N.N.; Saha, A.K.; Sarker, P.K. Evaluation of the ASR of waste glass fine aggregate in alkali activated concrete by concrete prism tests. *Constr. Build. Mater.* **2021**, *266*, 121121. [CrossRef]
5. Hafez, H.; Kassim, D.; Kurda, R.; Silva, R.V.; de Brito, J. Assessing the sustainability potential of alkali-activated concrete from electric arc furnace slag using the ECO2 framework. *Constr. Build. Mater.* **2021**, *281*, 122559. [CrossRef]
6. Amer, I.; Kohail, M.; El-Feky, M.; Rashad, A.; Khalaf, M.A. A review on alkali-activated slag concrete. *Ain Shams Eng. J.* **2021**, *12*, 1475–1499. [CrossRef]
7. Cai, R.; Ye, H. Clinkerless ultra-high strength concrete based on alkali-activated slag at high temperatures. *Cem. Concr. Res.* **2021**, *145*, 106465. [CrossRef]
8. Ganesh, A.C.; Muthukannan, M. Development of high performance sustainable optimized fiber reinforced geopolymer concrete and prediction of compressive strength. *J. Clean. Prod.* **2021**, *282*, 124543. [CrossRef]
9. Wasim, M.; Ngo, T.D.; Law, D. A state-of-the-art review on the durability of geopolymer concrete for sustainable structures and infrastructure. *Constr. Build. Mater.* **2021**, *291*, 123381. [CrossRef]
10. Shahmansouri, A.A.; Nematzadeh, M.; Behnood, A. Mechanical properties of GGBFS-based geopolymer concrete incorporating natural zeolite and silica fume with an optimum design using response surface method. *J. Build. Eng.* **2021**, *36*, 102138. [CrossRef]
11. Rahman, S.K.; Al-Ameri, R. A newly developed self-compacting geopolymer concrete under ambient condition. *Constr. Build. Mater.* **2021**, *267*, 121822. [CrossRef]
12. Pascual, A.B.; Tognonvi, T.M.; Tagnit-Hamou, A. Optimization study of waste glass powder-based alkali activated materials incorporating metakaolin: Activation and curing conditions. *J. Clean. Prod.* **2021**, 127435. [CrossRef]
13. Zhang, B.; He, P.; Poon, C.S. Optimizing the use of recycled glass materials in alkali activated cement (AAC) based mortars. *J. Clean. Prod.* **2020**, *255*, 120228. [CrossRef]
14. Sood, D.; Hossain, K. Optimizing Precursors and Reagents for the Development of Alkali-Activated Binders in Ambient Curing Conditions. *J. Compos. Sci.* **2021**, *5*, 59. [CrossRef]
15. Khan, M.A.; Memon, S.A.; Farooq, F.; Javed, M.F.; Aslam, F.; Alyousef, R. Compressive Strength of Fly-Ash-Based Geopolymer Concrete by Gene Expression Programming and Random Forest. *Adv. Civ. Eng.* **2021**, *2021*, 6618407. [CrossRef]
16. Bin Ahmed, F.; Biswas, R.K.; Ahsan, K.A.; Islam, S.; Rahman, R. Estimation of strength properties of geopolymer concrete. *Mater. Today: Proc.* **2021**, *44*, 871–877. [CrossRef]
17. He, P.; Zhang, B.; Lu, J.; Poon, C.S. A ternary optimization of alkali-activated cement mortars incorporating glass powder, slag and calcium aluminate cement. *Constr. Build. Mater.* **2020**, *240*, 117983. [CrossRef]
18. Dave, S.V.; Bhogayata, A.; Arora, N.K. Mix design optimization for fresh, strength and durability properties of ambient cured alkali activated composite by Taguchi method. *Constr. Build. Mater.* **2021**, *284*, 122822. [CrossRef]
19. Kočí, V.; Koňáková, D.; Pommer, V.; Keppert, M.; Vejmelková, E.; Černý, R. Exploiting advantages of empirical and optimization approaches to design alkali activated materials in a more efficient way. *Constr. Build. Mater.* **2021**, *292*, 123460. [CrossRef]
20. Mehta, A.; Siddique, R.; Ozbakkaloglu, T.; Shaikh, F.; Belarbi, R. Fly ash and ground granulated blast furnace slag-based alkali-activated concrete: Mechanical, transport and microstructural properties. *Constr. Build. Mater.* **2020**, *257*, 119548. [CrossRef]
21. Pinheiro, C.; Rios, S.; da Fonseca, A.V.; Jimenez, A.M.F.; Cristelo, N. Application of the response surface method to optimize alkali activated cements based on low-reactivity ladle furnace slag. *Constr. Build. Mater.* **2020**, *264*, 120271. [CrossRef]

22. Zhang, L.; Yue, Y. Influence of waste glass powder usage on the properties of alkali-activated slag mortars based on response surface methodology. *Constr. Build. Mater.* **2018**, *181*, 527–534. [CrossRef]
23. Cong, X.; Zhou, W.; Elchalakani, M. Experimental study on the engineering properties of alkali-activated GGBFS/FA concrete and constitutive models for performance prediction. *Constr. Build. Mater.* **2020**, *240*, 117977. [CrossRef]
24. Gomaa, E.; Han, T.; El Gawady, M.; Huang, J.; Kumar, A. Machine learning to predict properties of fresh and hardened alkali-activated concrete. *Cem. Concr. Compos.* **2021**, *115*, 103863. [CrossRef]
25. Shahmansouri, A.A.; Bengar, H.A.; Ghanbari, S. Compressive strength prediction of eco-efficient GGBS-based geopolymer concrete using GEP method. *J. Build. Eng.* **2020**, *31*, 101326. [CrossRef]
26. Mohammed, B.S.; Haruna, S.; Wahab, M.M.B.A.; Liew, M. Optimization and characterization of cast in-situ alkali-activated pastes by response surface methodology. *Constr. Build. Mater.* **2019**, *225*, 776–787. [CrossRef]
27. Mohammadi, F.; Mohammadi, T. Optimal conditions of porous ceramic membrane synthesis based on alkali activated blast furnace slag using Taguchi method. *Ceram. Int.* **2017**, *43*, 14369–14379. [CrossRef]
28. Huseien, G.F.; Sam, A.R.M.; Shah, K.W.; Mirza, J. Effects of ceramic tile powder waste on properties of self-compacted alkali-activated concrete. *Constr. Build. Mater.* **2020**, *236*, 117574. [CrossRef]
29. Huseien, G.F.; Sam, A.R.M.; Shah, K.W.; Mirza, J.; Tahir, M.M. Evaluation of alkali-activated mortars containing high volume waste ceramic powder and fly ash replacing GBFS. *Constr. Build. Mater.* **2019**, *210*, 78–92. [CrossRef]
30. ASTM C267. *Standard Test Methods for Chemical Resistance of Mortars, Grouts, and Monolithic Surfacing and Polymer Concretes*; ASTM International: West Conshohocken, PA, USA, 2012.
31. Mohammed, B.S.; Khed, V.C.; Nuruddin, M.F. Rubbercrete mixture optimization using response surface methodology. *J. Clean. Prod.* **2018**, *171*, 1605–1621. [CrossRef]
32. Mohammed, M.K.; Al-Hadithi, A.I.; Mohammed, M.H. Production and optimization of eco-efficient self compacting concrete SCC with limestone and PET. *Constr. Build. Mater.* **2019**, *197*, 734–746. [CrossRef]
33. Carrillo, J.; Ramirez, J.; Lizarazo-Marriaga, J. Modulus of elasticity and Poisson's ratio of fiber-reinforced concrete in Colombia from ultrasonic pulse velocities. *J. Build. Eng.* **2019**, *23*, 18–26. [CrossRef]
34. Ray, S.; Haque, M.; Ahmed, T.; Nahin, T.T. Comparison of artificial neural network (ANN) and response surface methodology (RSM) in predicting the compressive and splitting tensile strength of concrete prepared with glass waste and tin (Sn) can fiber. *J. King Saud Univ. Eng. Sci.* **2021**. [CrossRef]
35. Tian, Z.; Zhang, Z.; Zhang, K.; Tang, X.; Huang, S. Statistical modeling and multi-objective optimization of road geopolymer grouting material via RSM and MOPSO. *Constr. Build. Mater.* **2021**, *271*, 121534. [CrossRef]
36. Rashad, A.M. Properties of alkali-activated fly ash concrete blended with slag. *Iran. J. Mater. Sci. Eng.* **2013**, *10*, 57–64.
37. Puertas, F.; Martinez-Ramirez, S.; Alonso, S.; Vázquez, T. Alkali-activated fly ash/slag cements: Strength behaviour and hydration products. *Cem. Concr. Res.* **2000**, *30*, 1625–1632. [CrossRef]
38. Huseien, G.F.; Sam, A.R.M.; Shah, K.W.; Asaad, M.A.; Tahir, M.M.; Mirza, J. Properties of ceramic tile waste based alkali-activated mortars incorporating GBFS and fly ash. *Constr. Build. Mater.* **2019**, *214*, 355–368. [CrossRef]
39. Mozgawa, W.; Deja, J. Spectroscopic studies of alkaline activated slag geopolymers. *J. Mol. Struct.* **2009**, *924*, 434–441. [CrossRef]
40. Ravikumar, D.; Peethamparan, S.; Neithalath, N. Structure and strength of NaOH activated concretes containing fly ash or GGBFS as the sole binder. *Cem. Concr. Compos.* **2010**, *32*, 399–410. [CrossRef]
41. Ismail, I.; Bernal, S.A.; Provis, J.L.; San Nicolas, R.; Hamdan, S.; van Deventer, J.S. Modification of phase evolution in alkali-activated blast furnace slag by the incorporation of fly ash. *Cem. Concr. Compos.* **2014**, *45*, 125–135. [CrossRef]
42. Huseiena, G.F.; Ismaila, M.; Tahirb, M.; Mirzac, J.; Husseina, A.; Khalida, N.H.; Sarbinia, N.N. Effect of Binder to Fine Agregate Content on Performance of Sustainable Alkali Activated Mortars Incorporating Solid Waste Materials. *Chem. Eng.* **2018**, *63*, 667–672.
43. Huseien, G.F.; Ismail, M.; Tahir, M.M.; Mirza, J.; Khalid, N.H.A.; Asaad, M.A.; Husein, A.A.; Sarbini, N.N. Synergism between palm oil fuel ash and slag: Production of environmental-friendly alkali activated mortars with enhanced properties. *Constr. Build. Mater.* **2018**, *170*, 235–244. [CrossRef]
44. Mehta, A.; Siddique, R. Sulfuric acid resistance of fly ash based geopolymer concrete. *Constr. Build. Mater.* **2017**, *146*, 136–143. [CrossRef]
45. Ariffin, M.; Bhutta, M.; Hussin, M.; Tahir, M.M.; Aziah, N. Sulfuric acid resistance of blended ash geopolymer concrete. *Constr. Build. Mater.* **2013**, *43*, 80–86. [CrossRef]
46. Zhang, W.; Yao, X.; Yang, T.; Zhang, Z. The degradation mechanisms of alkali-activated fly ash/slag blend cements exposed to sulphuric acid. *Constr. Build. Mater.* **2018**, *186*, 1177–1187. [CrossRef]
47. Allahverdi, A.; Skvara, F. Sulfuric acid attack on hardened paste of geopolymer cements—Part 1. Mechanism of corrosion at relatively high concentrations. *Ceram. Silik.* **2005**, *49*, 225.
48. Gu, L.; Bennett, T.; Visintin, P. Sulphuric acid exposure of conventional concrete and alkali-activated concrete: Assessment of test methodologies. *Constr. Build. Mater.* **2019**, *197*, 681–692. [CrossRef]

materials

Article

Effects of HPMC on Workability and Mechanical Properties of Concrete Using Iron Tailings as Aggregates

Xiaowei Gu [1], Xiaohui Li [1,*], Weifeng Zhang [1], Yuxin Gao [2,3], Yaning Kong [3], Jianping Liu [4] and Xinlong Zhang [5]

[1] Science and Technology Innovation Center of Smart Water and Resource Environment, Northeastern University, Shenyang 110819, China; guxiaowei@mail.neu.edu.cn (X.G.); wfzhang@stumail.neu.edu.cn (W.Z.)

[2] College of Materials Science and Engineering, Chongqing University, Chongqing 400045, China; gaoyuxin@cscec.com

[3] China West Construction Academy of Building Materials, Chengdu 610015, China; kongyaning@cscec.com

[4] School of Architecture and Civil Engineering, Shenyang University of Technology, Shenyang 110870, China; pliu@sut.edu.cn

[5] Key Lab of Structures Dynamic Behaviour and Control of the Ministry of Education, Harbin Institute of Technology, Harbin 150090, China; 21b933082@stu.hit.edu.cn

[*] Correspondence: lixiaohui@stumail.neu.edu.cn; Tel.: +86-18602440326

Abstract: Iron ore tailings (IOTs) are gradually used as building materials to solve the severe ecological and environmental problems caused by their massive accumulation. However, the bulk density of IOT as aggregate is too large, which seriously affects the concrete properties. Therefore, in this paper, the effect of hydroxypropyl methylcellulose (HPMC) on the workability, mechanical properties, and durability of concrete prepared from IOT recycled aggregate was studied. The action mechanism of HPMC on the workability and the mechanical properties of the IOT concrete was analyzed by mercury intrusion porosimetry (MIP) and scanning electron microscope (SEM). The results show that HPMC can effectively improve the segregation problem caused by the sinking and air entrainment of IOT aggregate and improve the crack resistance of concrete with little effect on its compressive strength and electric flux. These results are due to the air-entraining thickening effect of HPMC, which improves the slurry viscosity, hinders the sinking of aggregate, and improves the workability. At the same time, HPMC film, after concrete hardening, will bridge the slurry and aggregate through physical and chemical effects, hinder the propagation of microcracks, and improve the crack resistance.

Keywords: HPMC; iron tailings; workability; mechanical properties

Citation: Gu, X.; Li, X.; Zhang, W.; Gao, Y.; Kong, Y.; Liu, J.; Zhang, X. Effects of HPMC on Workability and Mechanical Properties of Concrete Using Iron Tailings as Aggregates. *Materials* **2021**, *14*, 6451. https://doi.org/10.3390/ma14216451

Academic Editor: Jorge de Brito

Received: 10 October 2021
Accepted: 24 October 2021
Published: 27 October 2021

Publisher's Note: MDPI stays neutral with regard to jurisdictional claims in published maps and institutional affiliations.

1. Introduction

With the continuous development of the global social economy, the consumption of concrete as a building material is increasing rapidly at the rate of 4.4 billion tons per year [1]. More than 75% of concrete raw materials are composed of aggregates [2]. The acquisition of natural aggregates has a serious impact on the ecological environment. As a developing country, the vigorous development of infrastructure has promoted China's rapid development. The overuse of land, minerals, water, and other non-renewable precious natural resources in the short term will damage the survival of our future generations. Currently, mineral admixtures such as fly ash (FA) [3], silica fume [4], coal bottom ash [5], and ground granulated blast furnace slag (GGBFS) [6] have been used as partial replacement of cement in concrete, which not only cuts carbon emission footprints but also reduces costs of production. Therefore, an eco-friendly production mode must be adopted to realize the sustainable development of the concrete industry.

Iron has always played an essential role in the development of modern human civilization [7]. After the iron is extracted from the iron ore by various beneficiation processes in the concentrator, a large number of iron tailings are discharged to the IOT pond for stacking [8]. China's cumulative stockpile of tailings reached 60 billion tons, including 475 million tons of IOTs based on the 2018–2019 Chinese development report. IOTs have become one of the most critical solid wastes in the world [9]. Their massive accumulation will not only cause land pollution, poor water quality, and aggravation of dust but also threaten human health [10]. The managers of mining enterprises need to spend more resources to maintain IOT ponds every year [11]. More seriously, the existence of tailing dams is a potential safety hazard, and the human life safety events caused by the tailings dam breaks around the world are increasing every year [9]. With extensive research on the recycling of IOT, the recycled IOT is classified into four categories at home and abroad: (i) extracting valuable metals in IOT [12]; (ii) used for a backfill material [13]; (iii) turned into a building material [14,15]; (iv) reclaiming farmland on the tailings dumping ground. The large-scale application of iron tailings in concrete building materials has become the focus of all major sectors of society.

Compared with the natural aggregate, the IOT aggregate in concrete is made from the waste stone stripped from the open-pit mine through crushing and screening [16]. The IOT coarse aggregate has many edges and corners [17]. At the same time, the IOT fine aggregate has many powders during the crushing, which further increases the IOT specific surface area, increases the water demand, and leads to poor workability of IOT concrete [18].

Cellulose ethers (CEs) have been widely used in modern building products due to the function of preventing uncontrolled water loss into porous matrix [19]. HPMC, a type of CH, has been used because of its better effect on improving the workability of concrete [20]. The water retention of HPMC mainly hinders the migration of water molecules between pores through the action of internal groups. Pourchez et al. reported that CE slows down the hydration process of cement, which helps to reduce the loss of water [21]. Brumaud et al. revealed how CE slows down the cement hydration process [22]. Weyer et al. showed that it is necessary to control the CE content [23]. Only adding an appropriate amount of CE can slow down the hydration of cement.

Until now, researchers have studied the mechanism of different kinds of CEs in paste and mortar, but there are few reports of HPMC on concrete, especially the concrete used by IOTs as an aggregate. This paper focuses on the effect of HPMC on the workability of full IOT aggregate concrete. Moreover, the influence mechanism of HPMC on the mechanical properties of IOT concrete was also clarified.

2. Materials and Methods

2.1. Raw Materials

Cement (P.O. 42.5) is the main binder produced by Benxi Yuntian Cement Co., Ltd. (Benxi, China) according to GB/T 175-2007 [24]. The physical and chemical properties of cement are presented in Table 1.

Table 1. Physical and chemical properties of cement.

Chemical Composition	(wt.%)	Physical Properties	Value
SiO_2	28.16	Specific surface area (m^2/kg)	400
Al_2O_3	7.44	Water requirement of consistency/%	26.0
Fe_xO_y [25]	2.76	Initial setting/min	185
CaO	54.86	Final setting/min	260
Na_2O	0.03	3 d flexural strength/MPa	4.1
MgO	4.36	28 d flexural strength/MPa	8.2
SO_3	2.24	3 d compressive strength/MPa	21.0
Loss	2.67	28 d compressive strength/MPa	49.4

The coarse and fine aggregates are formed by the IOT waste rock mechanism. The particle size of the coarse aggregate is 5–20 mm, while the fineness modulus of the fine aggregate is 2.2. The gradation compositions of coarse and fine aggregates are presented in Figure 1, respectively. Table 2 displays the aggregates' basic properties. The coarse and fine aggregates were evaluated according to GB/T 14685-2011 and GB/T 14684-2011, respectively. Polycarboxylic acid high-performance water reducer (water reduction rate: 34%) was used to ameliorate the workability of concrete supplied by the Changyuan admixture technology Co., Ltd. (Jilin, China). The NDJ viscosity of HPMC was 196,000 mPa·s. and was purchased from Huzhou, Zhejiang.

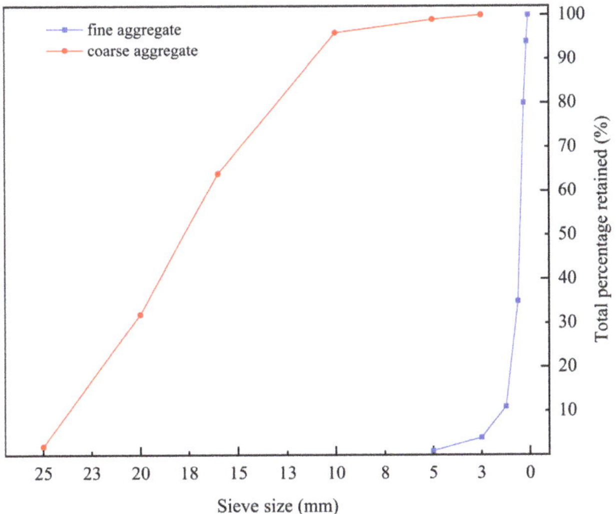

Figure 1. The gradation compositions of coarse and fine aggregates.

Table 2. The basic properties of aggregates.

	Coarse Aggregate	Fine Aggregate
Apparent density (kg/m^3)	2630	2590
Bulk density (kg/m^3)	1490	1540
Sediment percentage/%	0.3	2.4
Clay lump/%	0	0.8
Void ratio/%	43	41
Strength crushing index/%	7.4	/
Needle flake particle content/%	3	/

2.2. Mixture Proportions

The mixture design of IOT concrete in this paper is shown in Table 3. In HPMC0, the HPMC is not added in concrete. In HPMC1, the HPMC is added into concrete. HPMC and an appropriate amount of water were added to the agitator cup and stirred at the speed of 3000 rpm for 180 s. HPMC is 0.1% by mass of cement [26].The characteristic of fly ash admixture is shown in Table 4.

Table 3. Mixture proportions.

Symbol	C (kg/m^3)	FA (kg/m^3)	FAs (kg/m^3)	CAs (kg/m^3)	Water (kg/m^3)	SPs (kg/m^3)	HPMC (kg/m^3)	w/b
HPMC0	380	40	875	945	165	4	/	0.4
HPMC1	380	40	875	945	165	4	0.38	0.4

Note: Cement (C); Fly ash (FA); Fine Aggregates (FAs); Coarse Aggregates (CSa); Superplasticisers (SPs); Hydroxypropyl methyl cellulose (HPMC); water to binder (w/b).

Table 4. The characteristic of fly ash admixture.

Test Items	Detection Result
Water demand ratio/%	87
Loss on ignition/%	1.21
Water content/%	0.5
Sulfur trioxide/%	0.23
Free calcium oxide content/%	0.52
SiO_2/wt%	60.1
Al_2O_3/wt%	25.1
Fe_2O_3/wt%	6.7

2.3. Specimen Preparation

The SJD60 concrete horizontal-axle mixer was selected for concrete preparation. The cement and fly ash were dry mixed for one minute, the aggregate was added for another minute, and water and SPs were added for another two minutes. Specimens of 100 mm cubes, 150 mm cubes, and 100 × 400 mm prisms were prepared for each concrete mixture. The 100 mm and 150 mm cubes were used to determine the compressive strength and splitting tensile strength tests. The prisms were used to evaluate the flexural strength test. The permeability test took a 50 mm thick round pie sample from a 100 × 200 mm cylinder. The samples were cured under ambient conditions for 1 d, demolded, and further cured in a standard curing tank at a temperature of 20 ± 2 °C and humidity of more than 95% until the aging.

2.4. Test Methods

2.4.1. Fresh Properties

The slump, expansion, and entrained air content of the IOT concrete were measured. The slump and expansion tests were conducted by ASTM C143-15a [27]. The entrained air content of concrete mixtures was measured using the water column method according to the British Standard EN 12350-7 [28]. For the slump and entrained air content experiments, each group of experiments obtained the average value and the result by measuring three values. The average of the two outermost expansion diameters was taken as the final result for the expansion experiment.

2.4.2. Mechanical Properties

The mechanical properties of IOT concrete were determined using a universal testing machine with a loading capacity of 2000 kN following the Chinese Standard GB/T50081-2002 [29]. The loading rates applied in the compressive, splitting tensile, and flexural strengths were 0.6 MPa/s, 0.06 MPa/s, and 0.06 MPa/s, respectively. The flexural strength was measured with a four-point bending test. The results of the above three strength tests were determined by the average value of three measured results.

2.4.3. Permeability Test

To determine the chloride ions charge in the concrete specimen, the Chloride penetration tests were conducted according to ASTM C1202 [30]. The cylindrical sample, 100 × 200 mm, was cut into 100 two-part pieces of 100 × 100 mm. We then took 50 mm thick samples from the two parts respectively and used the surface of the first cut as the test surface. After cutting, the sample shall be vacuum water retained. The cathode solution is 3% NaCl solution, and the anode solution is 0.3 mol/l NaOH solution. The DC power supply for six hours can stably output 60 V voltage, the accuracy can meet the requirements of 0.1 V, and the current is in the range of 0–10 A. Use a data logger to record the total charge passed in 6 h every 30 min. Table 5 shows the evaluation of chloride ion permeability grade.

Table 5. Evaluation of chloride ion permeability grade.

Chloride Permeability	Charge (Coulomb)	Typical Concrete
High	>4000	High w/c ratio (>0.6)
Moderate	2000–4000	Moderate w/c ratio (0.4–0.5)
Low	1000–2000	Low w/c ratio (<0.4)
Very low	100–1000	Latex-modified concrete, internally sealed concrete
Negligible	<100	Polymer impregnated concrete, polymer concrete

2.4.4. Mercury Intrusion Porosimetry

The mercury intrusion porosimetry (MIP) testing instrument is AutoPore IV 9500, produced by the Micrometrics Instrument Corporation. The relationship between the applied pressure and the cylindrical pore diameter is described by the Washburn equation [31]. The total porosity and the pore size distribution could be obtained from MIP. The threshold pore diameter above which minimal intrusion occurs is still appropriate as a comparative index for the connectivity of different samples [32]. The threshold pore diameter is closely related to the permeability and diffusivity of cementitious materials and could be considered an indicator of material durability [33]. After curing to the corresponding age, the cube was sliced and soaked in absolute ethanol for seven days to terminate hydration. The samples were then dried in a 50 °C oven for two days.

2.4.5. Scanning Electron Microscopy-Backscattered Electron-Image Analysis (SEM-BSE-IA)

The slices were cut, immersed in isopropanol, and dried in an oven. Then, the slices were immersed in epoxy resin under vacuum for about 30 min and placed for hardening for one day. The hardened sample was first polished on sandpaper (320#) so that the sample surface was completely exposed, and the sample edge could be seen under the irradiation of a fluorescent lamp. Then, the samples were polished to 1 μm until the sample surface was smooth and flat on the polishing machine, using diamond suspension as the polishing solution and gasoline as a lubricant. The sandpaper grades used ranged from 800#, 1000#, 1200#, 1500#, 2000#, 2500#, to 4000#. According to the sample surface conditions and each grinding condition, the grinding time ranged from 20 to 60 s. In order to avoid the influence of residual diamond impurities and improve the efficiency of sample preparation, the samples need to be cleaned with ultrasonic in isopropyl alcohol solution during polishing. The polishing disc needed to be cleaned with a special detergent and brush. After polishing, the samples were stored in a vacuum dryer for 2 days to remove the residual organic solvent and avoid polluting the probe inside the scanning electron microscope (SEM). The samples were sprayed with carbon to form a carbon layer with a thickness of about 10 nm on the surface before the SEM test. The backscattered electron (BSE) images were obtained by FEI FEG quanta 650 field emission environmental scanning electron microscope with an accelerating voltage of 20 kV and a beam spot diameter of 3.0 nm.

BSE was used to evaluate the properties of ITZ, and the image analysis was carried out to quantify the microstructure. The ImageJ software was used to remove the aggregate. For image binarization, a 5 μm strip was used. The proportion of the object image was obtained by dividing the number of pixels in the gray range of the object image by the total number of pixels in the strip. The multiples and resolutions of the images used were consistent.

3. Results and Discussion

3.1. Workability

The iron ore tailings are generated by the waste rock mechanism stripped from the open-pit mine. Compared with ordinary natural aggregate, the IOT has a larger bulk density due to the iron content inside, leading to the natural sinking and difficulty of mortar binding with the aggregate [16]. When IOT is used as a fine aggregate, the specific surface area is large, and the water absorption is high due to the internal stone powder, which worsens the workability of the IOT concrete. The fineness modulus of raw materials used to prepare concrete is positively correlated with the workability of concrete [34].

The addition of HPMC improves the fluidity of concrete, which is mainly due to the action of HPMC's chemical groups to improve the water retention effect of concrete so that more mortar can well wrap the aggregate (Figure 2, Table 6). As the entrained air content experiment results show, in addition to ameliorating the water retention of concrete, the HPMC also introduces a large number of bubbles, resulting in a significant increase in the entrained air content of concrete from 1.6% to 4.5%. Chen [26] also found that the addition of HPMC can effectively improve the water retention performance of concrete, resulting in significant improvement in workability.

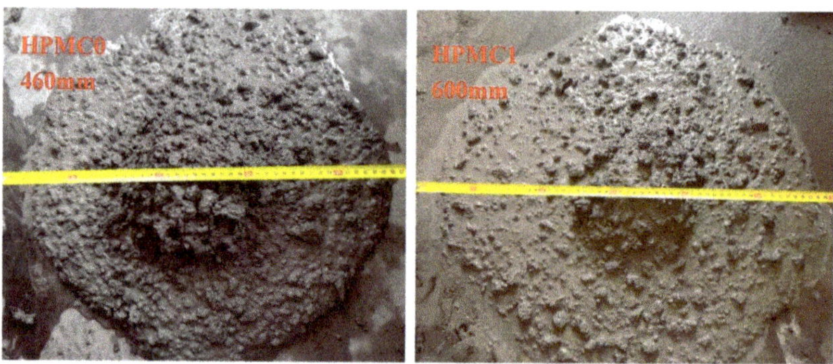

Figure 2. Changes of concrete workability before and after adding HPMC.

Table 6. Multiple indexes of concrete workability.

Symbol	Slump (mm)	Expansion (mm)	Entrained Air Content (%)
HPMC0	220	460	1.6
HPMC1	240	600	4.5

3.2. Mechanical Properties

3.2.1. Compressive Strength

The results of compressive strength tests of IOT concrete with and without HPMC are presented in Figure 3. The compressive strength of HPMC1 was lower than that of HPMC0, regardless of curing ages. The reduction rate at 3 and 28 days was 11.5% and 3.1%, respectively. The macro strength of the concrete mainly depends on the micropore structure distribution and porosity [35]. Although the addition of HPMC makes the water retention effect of concrete better, it increases the internal porosity while the compressive strength is decreased. Furthermore, when the concrete is mixed with water, the HPMC latex particles are adsorbed on the cement surface to generate a latex film, reducing the cement hydration and compressive strength [20]. The particles' pore plugging induces a decrease in the water transport properties of the cement matrix [36]. With the continuous hydration, the decreased range of compressive strength is decreasing, indicating that the increasing hydration products are making up for the loss of compressive strength caused by HPMC. This corresponds to the above experimental results of entrained air content, which proves that introducing many bubbles weakens the compressive strength of the IOT.

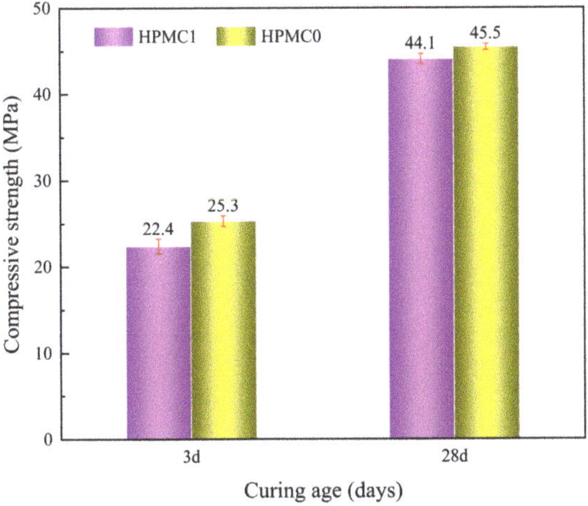

Figure 3. Development of compressive strength at different curing ages.

3.2.2. Splitting Tensile Strength

The splitting tensile strength test was presented in Figure 4. The results indicate that the splitting tensile strength increased when adding HPMC at all curing ages. The growth rate of splitting tensile strength at 3 and 28 days was 7.3% and 10.6%, respectively. When HPMC is added, the methoxy and hydroxypropyl groups on the molecular chains react with Ca^{2+} and Al^{3+} to form a viscous gel and fill in the concrete internal void, playing a role of flexible reinforcement, thereby improving the compactness of concrete and increasing the splitting tensile strength. Moreover, smaller IOT powder enhances the bond between the aggregate and the matrix, improving the splitting tensile strength [37].

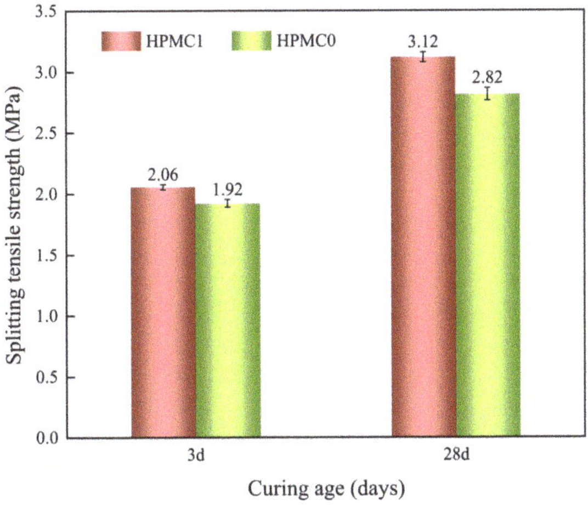

Figure 4. Splitting tensile strength of concrete with increasing age.

3.2.3. Flexural Strength

The flexural strength test results of IOT concrete with and without HPMC are displayed in Figure 5. The flexural strength of HPMC1 was lower than that of HPMC0 at 3 and 28 days curing age. The reduction range of flexural strength was 6.06% and 6.12%, respectively. It can be observed that the flexural strength is not sensitive to the addition of HPMC. The addition of HPMC improves the water retention effect of IOT concrete, but the internal porosity also increases, resulting in reduced flexural strength. HPMC latex particles form latex film by adsorbing on the cement surface, reducing the cement hydration and flexural strength.

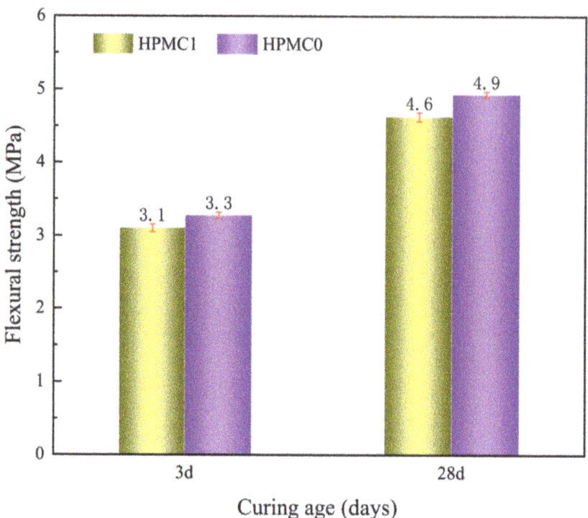

Figure 5. Development of flexural strength of IOT concrete at different curing ages.

3.3. Chloride Penetration Test

The chloride ion flux results for HPMC0 and HPMC1 are displayed in Figure 6. The HPMC1 shows higher charges passed compared to HPMC0 at 3 and 28 days. The charges passed were greater than 4000 C at three days, which was classified as high (seen in Table 5) according to ASTM C1202 standard. At the initial stage of hydration, a large number of bubbles were introduced by HPMC, increasing the internal porosity and resulting in poor resistance to chloride ion permeability. The coulomb charge in HPMC0 and HPMC1 specimens reduced significantly after 28 days. The charges passed were within the 1000–2000 C range, which was classified as low. With the continuous hydration reaction, more hydration products fill the internal pores, which reduces the porosity, optimizes the pore size distribution, and reduces the chloride ion electric flux significantly. Pierre [38] reported that HPMC would plug part of the cement paste porosity, which increases the apparent viscosity and leads to a substantial decrease in the material's apparent permeability. It has also been shown that the retardation mechanism results mainly from the adsorption of HPMC onto the surface of cement grains [39]. The addition of HPMC increases the entrained air content of concrete, and the introduction of many bubbles increases the internal pores, which do not block the passage of chloride ions.

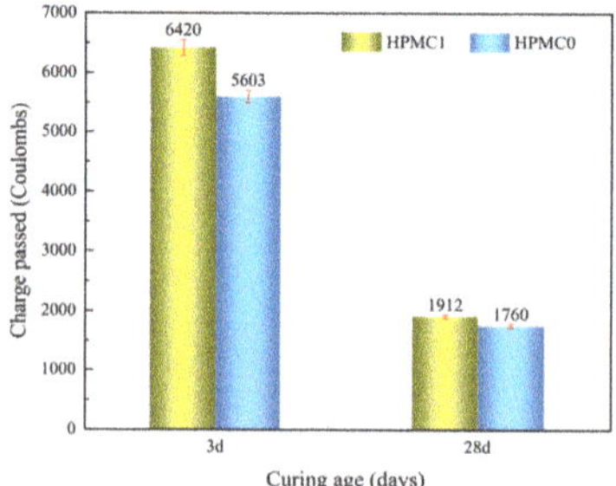
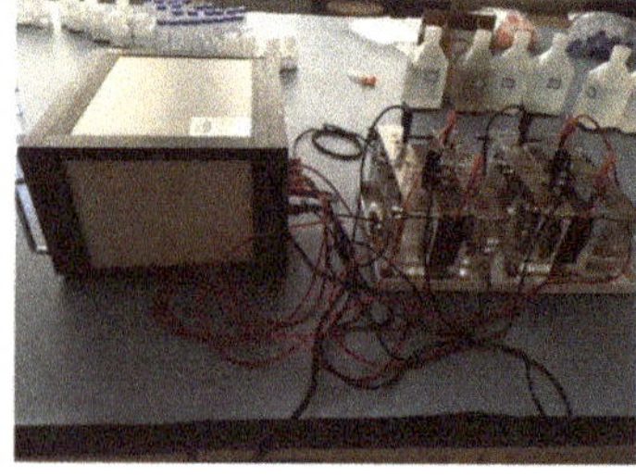

Figure 6. The charge passed of concrete with and without HPMC at different curing ages.

3.4. MIP

The differential pore size distribution and cumulative porosity of HPMC0 and HPMC1 are shown in Figures 7 and 8. The harmful pores and total porosity of HPMC0 and HPMC1 at 3 and 28 days are as shown in Table 7. The addition of HPMC can significantly increase the pore volume and the total porosity after curing for three days. The addition of HPMC introduces many bubbles, which increases the entrained air content and the porosity of IOT concrete. Figure 7 demonstrated that the cumulative pore volume increased from 0.025 mL/g to 0.06 mL/g. As generally accepted, when the pores inside the concrete increase more than 20 nm or even close to 100 nm, they can reduce the properties of concrete, e.g., mechanical properties, permeability, and durability. Hence, when adding HPMC, the number of harmful pores (pore size in the range of 50 nm to 200 nm) showed a slight increase, which means that its compressive strength, flexural strength, and chloride resistance were reduced [40].

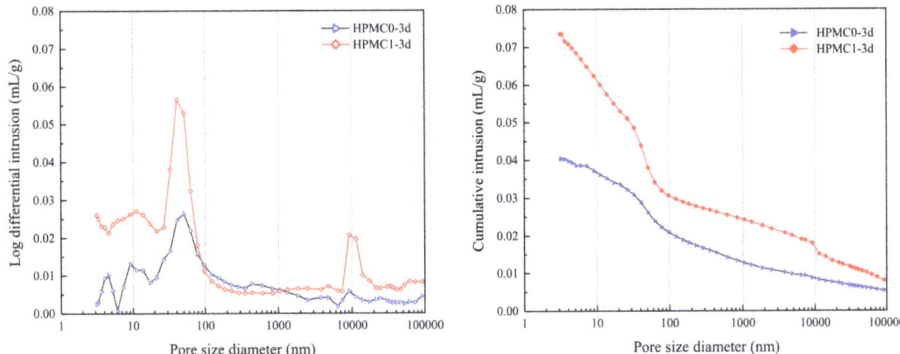

Figure 7. The pore structure of HPMC0 and HPMC1 at 3 days.

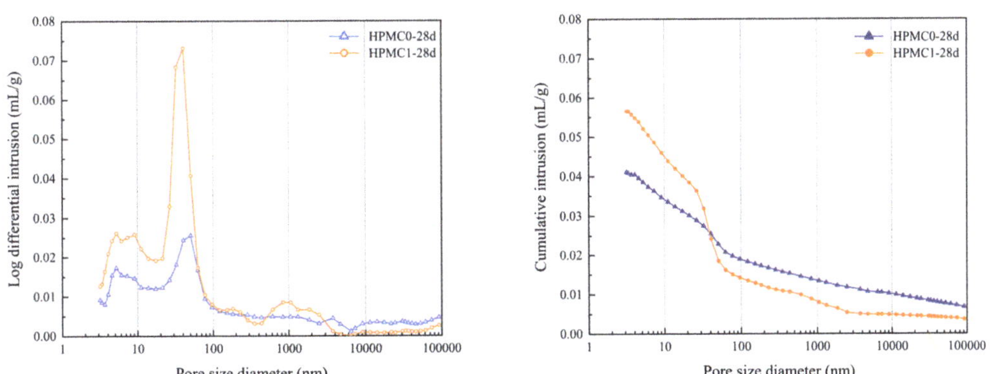

Figure 8. The pore structure of HPMC0 and HPMC1 at 28 days.

Table 7. Harmful pores and total porosity of HPMC0 and HPMC1.

Symbol	Harmful Pores (50–200 nm) (mL/g)	Total Porosity (%)
HPMC0-3d	0.105	9.51
HPMC1-3d	0.137	15.25
HPMC0-28d	0.077	9.33
HPMC1-28d	0.097	12.44

After curing for 28 days, the harmful pores and the total porosity for HPMC0 and HPMC1 have decreased compared to those cured for only three days (as shown in Table 7). The total porosity from 3 days to 28 days was reduced by 0.18% and 2.81%, respectively, indicating that the hydration products filled the voids and refined the pore structure with the increase of curing age.

The MIP test results confirm the overall performance; as the entrained air content of concrete increases, the compressive and flexural strength decreases, and the chloride ion electric flux increases after adding HPMC.

3.5. SEM-BSE-IA

Under the condition of constant stress, the different elastic and shear moduli of aggregate and matrix lead to the different strains. Currently, ITZ plays a bridging role between the aggregate and the matrix. Even if each phase in concrete has a high hardness, if ITZ can not effectively transfer stress, the concrete strength will be reduced [41]. Therefore, ITZ is considered the weakest link of concrete strength as it leads to the strength of concrete being lower than that of aggregate and cement [42].

Microcracks and high porosity in ITZ are the key factors affecting the mechanical properties of concrete. Large size CH will weaken the van der Waals force, decreasing the CH interlayer bonding ability, which is not conducive to the mechanical properties of concrete [43]. After adding HPMC, as Figure 9 shows, the ITZ contained many pores, the increase of porosity made the combination of aggregate and matrix relatively loose, slender ettringite was generated, and large CH crystals were enriched to form the preferred orientation. The ITZ without HPMC was fairly dense, and the enrichment volume of CH crystals formed near the aggregate was small. The SEM-BSE-IA test results confirmed that the addition of HPMC increased the pores in ITZ, loosened the structure, and worsened the mechanical properties.

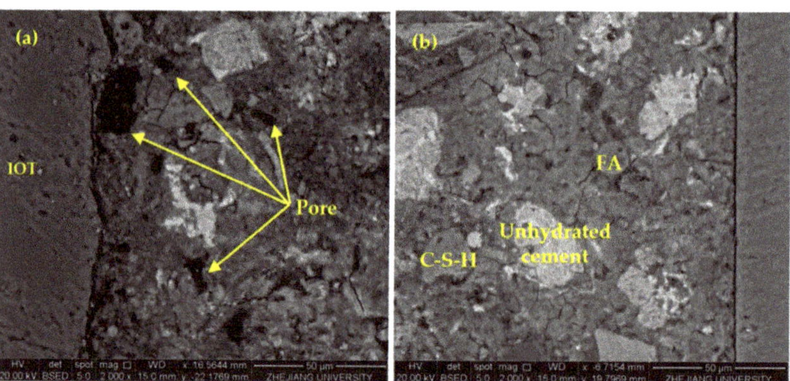

Figure 9. The interface transition zone (ITZ) of HPMC1 (**a**) and HPMC0 (**b**) at 28 days.

Figure 10 analyzed the porosity and unhydrated rate in the ITZ of HPMC0 and HPMC1, and the results are shown in Tables 8 and 9. In freshly mixed concrete, the water film is formed on the surface of the large aggregate particles so that the water–cement ratio in the ITZ is much larger than that of the cement–mortar matrix. The water–cement ratio is significant, and there are plate-like calcium hydroxide crystals and the acicular column ettringite enrichment near the aggregate surface, forming a framework with more pores than the cement–mortar matrix. Therefore, the closer to the aggregate, the higher the porosity of the ITZ, the higher the degree of hydration, and the lower the non-hydration rate.

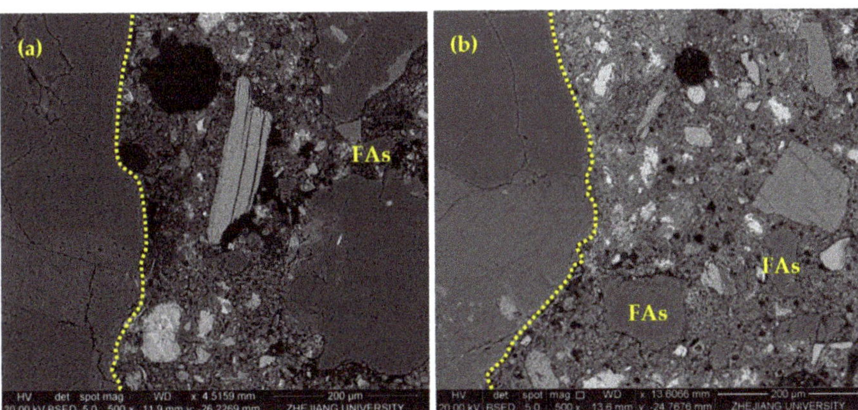

Figure 10. Quantitative calculation of the interfacial transition zone: HPMC1 (**a**) and HPMC0 (**b**) at 28 days.

Table 8. BSE quantitative calculation of HPMC0.

Distance from Aggregate/μm	Porosity/%	Unhydrated Rate/%
5	43.27	15.96
10	30.24	34.95
15	20.33	44.17
20	13.95	52.68

Table 9. BSE quantitative calculation of HPMC1.

Distance from Aggregate/μm	Porosity/%	Unhydrated Rate/%
5	59.44	15.50
10	53.18	19.55
15	52.85	20.07
20	47.85	24.06

4. Conclusions

In this study, the effect of HPMC on the workability of IOT concrete was studied. Moreover, the influence mechanism of HPMC on the mechanical properties of IOT concrete was further clarified. Based on the experimental results, several conclusions can be drawn:

(1) Due to the iron content in tailings aggregate, the density is relatively large, which leads to the aggregate sink and separation. The surface activation of HPMC can induce air, retain water and thicken to improve the viscosity of mortar and aggregate, prevent tailings aggregate from sinking, and improve the workability of concrete.

(2) Although HPMC can improve workability, it is not conducive to the development of compressive and flexural strength. However, it can improve the splitting tensile strength, which shows that the addition of HPMC can improve crack resistance.

(3) When the concrete is hardened, the air entrainment of HPMC will form pores and reduce the compressive strength and flexural strength. At the same time, HPMC reacts with Ca^{2+} and Al^{3+} to form membrane structure, which plays a bridging role, hinders the propagation of microcracks, improves the splitting tensile strength of concrete, and then ameliorates the crack resistance of concrete.

(4) The ITZ of ordinary C30 concrete is the key to affecting the concrete's strength and its weakest link. The addition of HPMC increases more bubbles in the ITZ and reduces the compactness, which is not conducive to the development of strength. However, the final strength grade is consistent with that without HPMC. The chloride ion flux is

about 1800, which does not affect the durability. Therefore, it is an effective measure to improve the workability of IOT concrete by adding HPMC.

Author Contributions: Conceptualization, X.G., X.L.; data curation, W.Z., Y.G., and Y.K.; writing of the original draft, X.L., W.Z., and Y.K.; writing of review and editing, W.Z., J.L., and X.Z. All authors have read and agreed to the published version of the manuscript.

Funding: This research was funded by the National Key Research and Development Plan of China (2019YFC1907202), the Research and Development Project, Liaoning (2020020307-JH1/103-02), and the Fundamental Research Funds for the Central Universities (N2101019).

Institutional Review Board Statement: Not applicable.

Informed Consent Statement: Not applicable.

Data Availability Statement: Not applicable.

Acknowledgments: Not applicable.

Conflicts of Interest: The authors declare no conflict of interest.

References

1. Zhao, J.; Ni, K.; Su, Y.; Shi, Y. An evaluation of iron ore tailings characteristics and iron ore tailings concrete properties. *Constr. Build. Mater.* **2021**, *286*, 122968. [CrossRef]
2. Shi, C.; Meyer, C.; Behnood, A. Utilization of copper slag in cement and concrete. *Resour. Conserv. Recy.* **2008**, *52*, 1115–1120. [CrossRef]
3. Liu, X.; Ni, C.; Meng, K.; Zhang, L.; Liu, D.; Sun, L. Strengthening mechanism of lightweight cellular concrete filled with fly ash. *Constr. Build. Mater.* **2020**, *251*, 118954. [CrossRef]
4. Adil, G.; Kevern, J.T.; Mann, D. Influence of silica fume on mechanical and durability of pervious concrete. *Constr. Build. Mater.* **2020**, *247*, 118453. [CrossRef]
5. Muthusamy, K.; Rasid, M.H.; Jokhio, G.; Budiea, A.M.A.; Hussin, M.W.; Mirza, J. Coal bottom ash as sand replacement in concrete: A review. *Constr. Build. Mater.* **2020**, *236*, 117507. [CrossRef]
6. Shen, D.; Jiao, Y.; Kang, J.; Feng, Z.; Shen, Y. Influence of ground granulated blast furnace slag on early-age cracking potential of internally cured high performance concrete. *Constr. Build. Mater.* **2020**, *233*, 117083. [CrossRef]
7. Argane, R.; Benzaazoua, M.; Hakkou, R.; Bouamrane, A. Reuse of base-metal tailings as aggregates for rendering mortars: Assessment of immobilization performances and environmental behavior. *Constr. Build. Mater.* **2015**, *96*, 296–306. [CrossRef]
8. Zuccheratte, A.C.V.; Freire, C.B.; Lameiras, F.S. Synthetic gravel for concrete obtained from sandy iron ore tailing and recycled polyethyltherephtalate. *Constr. Build. Mater.* **2017**, *151*, 859–865. [CrossRef]
9. Che, T.K.; Pan, B.F.; Sha, D.; Lu, J.L. Utilization of iron tailings as fine aggregates in low-grade cement concrete pavement. *IOP Conf. Ser. Mat. Sci. Eng.* **2019**, *479*, 012053. [CrossRef]
10. Zhao, Y.; Zhang, Y.; Chen, T.; Chen, Y.; Bao, S. Preparation of high strength autoclaved bricks from hematite tailings. *Constr. Build. Mater.* **2012**, *28*, 450–455. [CrossRef]
11. Li, T.; Wang, S.; Xu, F.; Meng, X.; Li, B.; Zhan, M. Study of the basic mechanical properties and degradation mechanism of recycled concrete with tailings before and after carbonation. *J. Clean. Prod.* **2020**, *259*, 120923. [CrossRef]
12. Tang, C.; Li, K.; Ni, W.; Fan, D. Recovering Iron from Iron Ore Tailings and Preparing Concrete Composite Admixtures. *Miner.* **2019**, *9*, 232. [CrossRef]
13. Yang, C.M.; Cui, C.; Qin, J. Recycling of low-silicon iron tailings in the production of lightweight aggregates. *Ceram. Int.* **2015**, *41*, 1213–1221. [CrossRef]
14. Mendes, B.C.; Pedroti, L.G.; Fontes, M.; Ribeiro, J.C.L.; Vieira, C.M.F.; Pacheco, A.A.; de Azevedo, A.R. Technical and environmental assessment of the incorporation of iron ore tailings in construction clay bricks. *Constr. Build. Mater.* **2019**, *227*, 116669. [CrossRef]
15. Wang, Z.; Xu, C.; Wang, S.; Gao, J.; Ai, T. Utilization of magnetite tailings as aggregates in asphalt mixtures. *Constr. Build. Mater.* **2016**, *114*, 392–399. [CrossRef]
16. Tan, Y.; Zhu, Y.T.; Xiao, H.L. Evaluation of the Hydraulic, Physical, and Mechanical Properties of Pervious Concrete Using Iron Tailings as Coarse Aggregates. *Appl. Sci.* **2020**, *10*, 2691. [CrossRef]
17. Lv, X.; Lin, Y.; Chen, X.; Shi, Y.; Liang, R.; Wang, R.; Peng, Z. Environmental impact, durability performance, and interfacial transition zone of iron ore tailings utilized as dam concrete aggregates. *J. Clean. Prod.* **2021**, *292*, 126068. [CrossRef]
18. Shettima, A.U.; Hussin, M.W.; Ahmad, Y.; Mirza, J. Evaluation of iron ore tailings as replacement for fine aggregate in concrete. *Constr. Build. Mater.* **2016**, *120*, 72–79. [CrossRef]
19. Patural, L.; Marchal, P.; Govin, A.; Grosseau, P.; Ruot, B.; Devès, O. Cellulose ethers influence on water retention and consistency in cement-based mortars. *Cem. Concr. Res.* **2011**, *41*, 46–55. [CrossRef]

20. Bülichen, D.; Plank, J. Water retention capacity and working mechanism of methyl hydroxypropyl cellulose (MHPC) in gypsum plaster—Which impact has sulfate? *Cem. Concr. Res.* **2013**, *46*, 66–72. [CrossRef]
21. Pourchez, J.; Grosseau, P.; Ruot, B. Changes in C3S hydration in the presence of cellulose ethers. *Cem. Concr. Res.* **2010**, *40*, 179–188. [CrossRef]
22. Brumaud, C.; Baumann, R.; Schmitz, M.; Radler, M.; Roussel, N. Cellulose ethers and yield stress of cement pastes. *Cem. Concr. Res.* **2014**, *55*, 14–21. [CrossRef]
23. Weyer, H.; Müller, I.; Schmitt, B.; Bosbach, D.; Putnis, A. Time-resolved monitoring of cement hydration: Influence of cellulose ethers on hydration kinetics. *Nucl. Instrum. Methods Phys. Res. Sect. B: Beam Interact. Mater. Atoms* **2005**, *238*, 102–106. [CrossRef]
24. Chinese Standard GB/T 175. *Common Portland Cement*; Standards Press of China: Beijing, China, 2007.
25. Nergis, D.D.B.; Vizureanu, P.; Ardelean, I.; Sandu, A.V.; Corbu, O.C.; Matei, E. Revealing the Influence of Microparticles on Geopolymers' Synthesis and Porosity. *Materials* **2020**, *13*, 3211. [CrossRef] [PubMed]
26. Chen, N.; Wang, P.; Zhao, L.; Zhang, G. Water Retention Mechanism of HPMC in Cement Mortar. *Materials* **2020**, *13*, 2918. [CrossRef] [PubMed]
27. ASTM C143-15a. *Standard Test Method for Slump of Hydraulic-Cement Concrete*; ASTM International: West Conshohocken, PA, USA, 2015.
28. EN12390-part7. *Testing Fresh Concrete-Air Content-Pressure Methods*; BSI Standards Publication: Brussels, Belgium, 2009.
29. Chinese Standard GB/T 50081. *Standard for Test Method of Mechanical Properties on Ordinary Concrete*; Standards Press of China: Beijing, China, 2002.
30. ASTM C1202. *Standard Test Method for Electrical Indication of Concrete's Ability to Resist Chloride Ion Penetration*; ASTM International: West Conshohocken, PA, USA, 2012.
31. Washburn, E.W. Note on a Method of Determining the Distribution of Pore Sizes in a Porous Material. *Proc. Natl. Acad. Sci. USA* **1921**, *7*, 115–116. [CrossRef]
32. Diamond, S. Reply to the discussion by S. Chatterji of the paper "Mercury porosimetry—An inappropriate method for the measurement of pore size distributions in cement-based materials". *Cem. Concr. Res.* **2001**, *31*, 1659. [CrossRef]
33. Garboczi, E.J.; Bentz, D.P. Percolation aspects of cement paste and concrete-properties and durability. *Aci Special Publ.* **1999**, *189*, 147–164.
34. Arum, C.; Owolabi, A.O. Suitability of iron ore tailings and quarry dust as fine aggregates for concrete production. *J. Appl. Sci. Technol.* **2012**, *17*, 46–52.
35. Kumar, R.; Bhattacharjee, B. Porosity, pore size distribution and in situ strength of concrete. *Cem. Concr. Res.* **2003**, *33*, 155–164. [CrossRef]
36. Marliere, C.; Mabrouk, E.; Lamblet, M.; Coussot, P. How water retention in porous media with cellulose ethers works. *Cem. Concr. Res.* **2012**, *42*, 1501–1512. [CrossRef]
37. Li, G.; Chen, Y.J.; Di, H. The Study on used Properties of Mine Tailings Sand. *Adv. Mater. Res.* **2014**, *859*, 87–90. [CrossRef]
38. Pierre, A.; Perrot, A.; Picandet, V.; Guevel, Y. Cellulose ethers and cement paste permeability. *Cem. Concr. Res.* **2015**, *72*, 117–127. [CrossRef]
39. Betioli, A.; Gleize, P.; Silva, D.; John, V.M.; Pileggi, R. Effect of HMEC on the consolidation of cement pastes: Isothermal calorimetry versus oscillatory rheometry. *Cem. Concr. Res.* **2009**, *39*, 440–445. [CrossRef]
40. Fan, Y.-F.; Luan, H.-Y. Pore structure in concrete exposed to acid deposit. *Constr. Build. Mater.* **2013**, *49*, 407–416. [CrossRef]
41. Malachanne, E.; Jebli, M.; Jamin, F.; Garcia-Diaz, E.; El Youssoufi, M.-S. A cohesive zone model for the characterization of adhesion between cement paste and aggregates. *Constr. Build. Mater.* **2018**, *193*, 64–71. [CrossRef]
42. Ke, Y.; Ortola, S.; Beaucour, A.-L.; Dumontet, H. Identification of microstructural characteristics in lightweight aggregate concretes by micromechanical modelling including the interfacial transition zone (ITZ). *Cem. Concr. Res.* **2010**, *40*, 1590–1600. [CrossRef]
43. Aitcin, P.-C. Concrete structure, properties and materials. *Can. J. Civ. Eng.* **1986**, *13*, 499. [CrossRef]

Article

Electrochemical Study of Clean and Pre-Corroded Reinforcements Embedded in Mortar Samples with Variable Amounts of Chloride Ions

María de las Nieves González [1], María Isabel Prieto [1,*], Alfonso Cobo [1] and Fernando Israel Olmedo [2]

[1] Escuela Técnica Superior de Edificación, Universidad Politécnica de Madrid, Avda. Ramiro de Maeztu, 7, 28040 Madrid, Spain; mariadelasnieves.gonzalez@upm.es (M.d.l.N.G.); alfonso.cobo@upm.es (A.C.)
[2] Valladares Ingeniería S.L, C/Julián Camarillo, 42, 28037 Madrid, Spain; fiolmedoz@gmail.com
* Correspondence: mariaisabel.prieto@upm.es; Tel.: +34-910-675-4341

Abstract: The present study investigates the possibility of re-surfacing previously corroded reinforcements and the suitability of the two electrochemical techniques that are widely used to determine the state of corrosion of steel (the corrosion potential E_{corr} and the corrosion rate i_{corr}). In order to test this, 32 pre-corroded B500SD reinforcing steel bars have been used for one year, where half of the bars have been cleaned to eliminate corrosion products. The other half have been maintained with the generated corrosion products. Subsequently, the bars have been embedded in cement mortar samples with variable amounts of chloride ion, and E_{corr} and i_{corr} have been measured for 250 days. The results showed that it is not possible to rework the reinforcement without removing corrosion products and that it is not possible to predict the passive or active state of steel by measuring E_{corr} only.

Keywords: repassivation; corrosion; reinforcement; electrochemical techniques

Citation: González, M.d.l.N.; Prieto, M.I.; Cobo, A.; Olmedo, F.I. Electrochemical Study of Clean and Pre-Corroded Reinforcements Embedded in Mortar Samples with Variable Amounts of Chloride Ions. *Materials* **2021**, *14*, 6883. https://doi.org/10.3390/ma14226883

Academic Editor: Emilia Morallon

Received: 2 October 2021
Accepted: 12 November 2021
Published: 15 November 2021

Publisher's Note: MDPI stays neutral with regard to jurisdictional claims in published maps and institutional affiliations.

1. Introduction

Reinforcement corrosion is accepted as the main cause of the reduction in the service life of reinforced concrete structures (RCSs) [1–3]. The enormous economic impact of this problem, due to the direct and indirect costs involved, has led to a vast development of new technologies and materials with the purpose of increasing the durability of RCSs [4–7]. In the USA, direct costs due to corrosion of RCSs infrastructure are estimated at 0.25% of GNP, which corresponds to USD 16.6 billion per year [8].

The steel reinforcements embedded in the RCS are in a passive state that are protected against corrosion. This protection is due to the existence of a passive layer formed at the steel/concrete interface, which is self-healing and very thin around 10 nm [9]. Its formation and stability are guaranteed by the high alkalinity of the concrete, usually between the range of pH 13–14, and by the existence of an appropriate electrochemical potential [10].

The loss of the passivity of the RCS reinforcement is due, in most cases, to the following factors: the presence of despassivating ions, essentially chlorides, in sufficient quantity to locally break up the passivating layers [11]; or the decrease in the pH of the concrete, due to the effect of the CO_2 present in the atmosphere [12–14]. In addition to the triggering factor that induces corrosion, the environment in which the structure is located determines the variables that most significantly influence its behavior. In the case of marine environments, the chloride diffusion coefficient, the concentration of chlorides on the surface and the thickness of the rebar coating are the most determining parameters to be able to evaluate its behavior, while in the case of underground tunnels, although corrosion is induced by chlorides, the formation of NaCl crystals is a parameter to take into account [15,16].

For decades, numerous investigations have been carried out to explain the role of the factors that trigger corrosion of rebars, especially chloride ions [17,18], whose mechanism can be seen in the Figure 1.

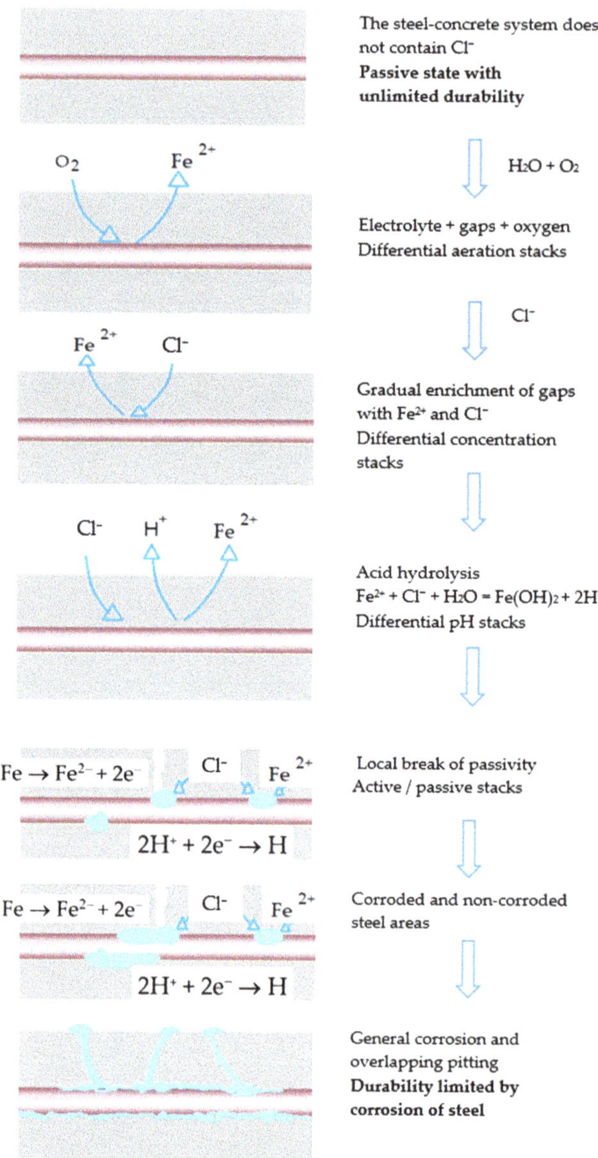

Figure 1. Chloride corrosion mechanism in reinforced concrete structures.

The critical threshold of chloride ions (C_{crit}) needed to despassivate the steel is one of the critical parameters in the prediction and evaluation of corrosion [17]. The knowledge of C_{crit} is fundamental when establishing the requirements to achieve structures with sufficient durability and to evaluate the service life or residual life of existing structures [18]. The onset of corrosion due to the effect of chlorides in the concrete reinforcements has been related to an increase in the ratio $[Cl^-]/[OH^-]$ above a certain value, which was initially set by Haussman at 0.6 [19–24]. However, it was later found that it can be affected by other parameters such as the amount of air retained in the holes of the steel-concrete interface or the presence of defects [23,25–30]. Numerous studies depict that there is a significant dispersion in C_{crit} values. In most cases, the limits are set in relation to the weight of the cement: For example, for concrete with a water/cement ratio of 0.5 mainly located in marine environments in the intertidal zone in Europe with a limit value of 0.5% suggested [31]. In Spain, the mandatory regulations establish a durability strategy based on the service life of the structure, but always with a limit of Cl^- of 0.4% [32]. In the United States, the thresholds are set according to the exposure class of the structure, ranging from 0.15% to 0.30%, and may reach 1% for structures located in dry environments [33]. At present, there are numerous investigations that model the corrosion behavior of the reinforcements embedded in the reinforced concrete and the influence of the concrete quality, the concentration of chlorides and the coating and the cracking, on the rate of corrosion, under different environmental conditions [34–37]. In addition, these investigations, validated on the corresponding experimental works, have allowed to simulate the influence that the size of the pits have on the geometry of the fissures and how the accumulation of oxide in the pits influence the adhesion between concrete and steel [38].

The realization of potential maps according to ASTM C876-09 [39] is the electrochemical technique most commonly used to diagnose the risk of corrosion in RCSs [2,40]. However, the results of the evaluation of RCSs with corroded reinforcements may indicate different degrees of corrosion or probability of corrosion depending on the technique used for corrosion assessment [41,42]. It is generally accepted that the measurements of corrosion potential (E_{corr}) should be completed with other procedures [40,43]. The measurement of corrosion rate is a quantitative technique that is known for decades [44]. However there have been found large variations in the values of corrosion rate for narrow ranges of values of corrosion potential [45].

Taking into account the previous premises, the aims of the work are to verify the possibility of re-surfacing previously corroded reinforcements and to verify the suitability of two electrochemical techniques widely used to determine the corrosion status of reinforcement by correlating the E_{corr} and the i_{corr} measurements.

2. Experimental Program

In this study, two issues related to corrosion of RCS reinforcements have been investigated: (i) The possibility of re-surfacing previously corroded reinforcements and (ii) the verification of the suitability of two electrochemical techniques widely used to determine the corrosion status of reinforcements. In order to test this, 32 B500SD reinforcing steel bars of 6 mm diameter and 120 mm length have been used. The chemical composition of the steel is shown in Table 1. The analyses have been carried out with an Optical Emission Spectrometer by Arc/Spark model SPECTROMAX.

Table 1. Chemical composition of the steel used.

Element	C	Si	Mn	P	S	Cr	Ni	Cu	Mo
Composition (%)	0.21	0.22	0.72	<0.01	0.022	0.05	0.09	0.08	<0.05

The rebars have been embedded in cement mortar samples of $80 \times 55 \times 20$ mm^3 with a cement/sand/water dosage of 1/3/0.5. CEM I 42.5R Portland cement according to RC-16 standard [46], siliceous sand with a maximum size of 0.4 mm and drinking water supplied by Canal de Isabel II in Madrid were used. The physical characteristics and chemical composition of the cement and the sand used can be seen in Table 2.

Table 2. Physical characteristics and chemical composition of cement and sand.

		Physical Characteristics		Chemical Composition	
Cement	Blaine specific surface area	414 m^2/kg		SO$_3$	3.40%
	Density	3.15 g/cm^3		Cl$^-$	0.01%
	Initial setting time	108 min		Calcination loss	1.72%
	Final setting time	160 min		Insoluble residue	0.40%
Sand	Sand equivalent	78		S, SO$_3$, Cl$^-$ and low specific weight particles	0.00%
	Real density	2.619 g/cm^3		Fine	0.78%
	Normal absorption coefficient	15%			
	Saturated surface dry density	2.630 g/cm^3			
	Clay clumps	0.01%			
	Coefficient of type of course aggregate	0.26%			
	Soft particles	0.93%			

The steel rebars were embedded in the specimens, leaving 5 cm outside, in which the risk of differential aeration at the triple atmosphere/mortar/steel interface was eliminated by means of adhesive tape. The specimens were demolded after 24 h and cured in a wet chamber for 28 days at a temperature of 20 ± 2 °C and a humidity above 90%. Subsequently, they were subjected to a constant anodic polarization of 20 µA/cm^2 during one year. In the photograph (Figure 2a), one of the specimens can be seen after the passage of the electric current. The samples were mixed with an addition of 2% of Cl$^-$ in relation to the weight of the cement to ensure that the anodic polarization would lead to corrosion of the reinforcements and would not cause the electrolysis of the water in the pore network.

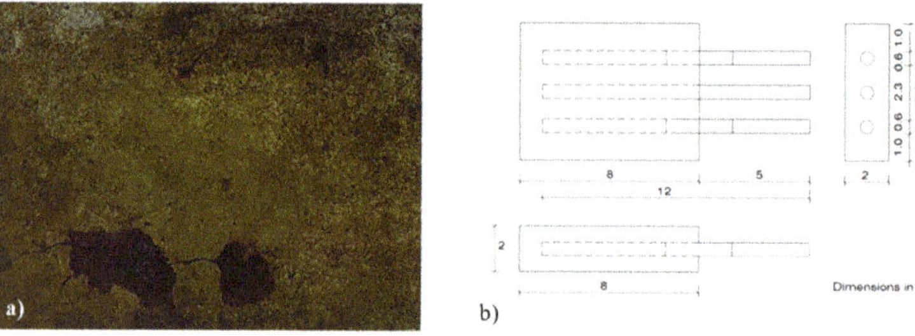

Figure 2. Specimens used in the study of corrosion. (**a**) Symptoms in pre-corroded specimens (first phase); (**b**) Scheme of the specimens with variable amounts of chloride ion (second phase).

Once the pre-corrosion of the rebars was complete, the mortar samples were then broken and the rebars removed. Half of the rebars (16 pieces) were chosen at random, and the corrosion products were completely removed by dissolving the iron oxides and hydroxides in 50% HCl inhibited with 4g/l urotropin (hexamethylenetetramine). With the clean rebars (CLN), 8 specimens similar to those of the first phase were performed, but with variable amounts of chloride ion of 0.0%; 0.2%; 0.4%; 0.6%; 0.8%; 1.0%; 1.5%; and 2.0% in relation to the weight of cement and with the incorporation of a third rebar located in the center to facilitate the electrochemical measurements (Figure 2b). The same process has been carried out with the rebars that hold the corrosion products. Then the 16 specimens were kept in a wet chamber for 250 days at a temperature of 20 ± 2 °C and a humidity above 90%. During this time the corrosion potential (E$_{corr}$), and corrosion rate were periodically recorded using an AUTOLAB/PGSTAT302N potentiostat in which the outer bars were used as working electrodes and the central bar as an auxiliary electrode

(Figure 3). The reference electrode used was $Cu/CuSO_4$. The corrosion rate was measured by the corrosion current density (i_{corr}) obtained by measuring the polarization resistance (R_p) by the Stern and Geary equation [47]:

$$i_{corr} = \frac{B}{R_p} \qquad (1)$$

where the value of 26 mV for the constant B was chosen [48]. The value of R_p was obtained by applying polarization of 10 mV and measuring the current response after 15 s. The reference values for predicting the corrosion state as a function of E_{corr} and i_{corr} are shown in Table 3 [39,49].

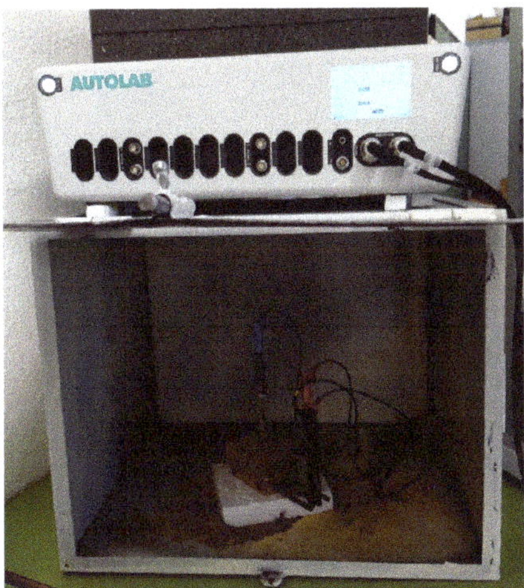

Figure 3. Performing electrochemical measurements.

Table 3. Reference values for E_{corr} and i_{corr}.

Measurement	Risk	Values
E_{corr} (mV)	High >90% Uncertainty Low <10%	$E_{corr} < -350$ $-350 < E_{corr} < -200$ $E_{corr} > -200$
i_{corr} ($\mu A/cm^2$)	Active state High corrosion Low corrosion Passive state	$i_{corr} > 1\,\mu A/cm^2$ $0.5\,\mu A/cm^2 < i_{corr} < 1\,\mu A/cm^2$ $0.1\,\mu A/cm^2 < i_{corr} < 0.5\,\mu A/cm^2$ $i_{corr} < 0.1\,\mu A/cm^2$

With respect to the thresholds of the E_{corr} value there is a very broad consensus throughout the scientific and technical community, most likely due to the enormous diffusion at the international level of the ASTM C 876-09 standard. Regarding i_{corr} thresholds, it is widely accepted that $i_{corr} < 0.1\,\mu A/cm^2$ corresponds to steel in a passive state, and above this value steel corrodes and $i_{corr} > 1\,\mu A/cm^2$ is very dangerous [42,44,50–53].

3. Results and Discussion

Figure 4 shows the evolution in time of the i_{corr} of all the analyzed rebars. Each of the data have been obtained as the arithmetic mean of the two rebars in the same state and embedded in the same specimen. The abbreviation CLN indicates that the rebar has been cleaned while the abbreviation COR indicates that the rebar keeps the corrosion products. The final numbers indicate the amount of Cl^-, expressed as a percentage by weight of cement, present in the specimens. It can be seen that, in most cases, the starting point is an

i_{corr} that decreases in a very similar way during the first 35 days of exposure. After this period of exposure, the i_{corr} remains at very stable values most probably because of the humidity of the environment remaining constant. Figure 5 shows the values corresponding to the i_{corr} of all the rebars after 250 days.

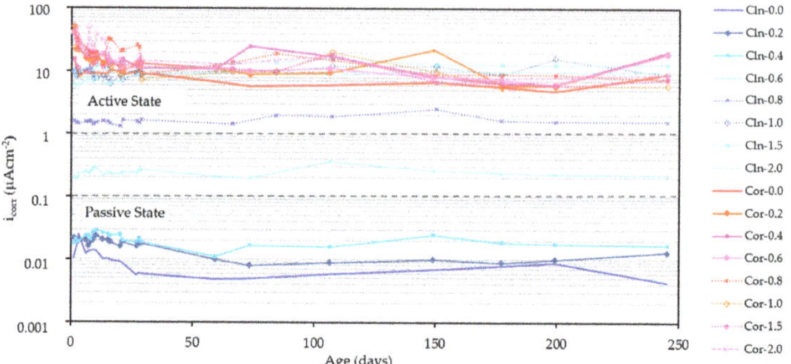

Figure 4. Evolution of i_{corr} over time for all the rebars studied.

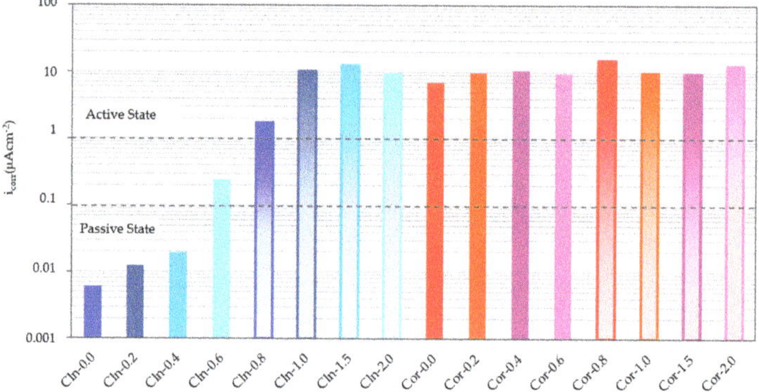

Figure 5. i_{corr} measurement of all rebars after 250 days.

If the clean rebars (CLN) are analyzed, the analysis depicted in Figures 4 and 5 indicate that only the rebars embedded in mortar samples of up to 0.4% of Cl^- in weight of cement are kept in a passive state. These results are in agreement with the indications of the EHE and ACI [32,33,54]. The embedded rebars in specimens with 0.6% Cl^- are maintained with low corrosion levels. Rebars in specimens with 0.8% Cl^- are maintained with i_{corr} between 1 and 2 $\mu A/cm^2$. With higher amounts of Cl^-, the rebars maintain i_{corr} at approximately 10 $\mu A/cm^2$. Similar results were obtained in previous investigations, observing that a greater amount of chloride ion present in the mortar, generates a higher rate of corrosion in the rebars [49,55,56].

All the rebars that have kept the corrosion products (COR) exhibit very high i_{corr} values (10 $\mu A/cm^2$) regardless of the amount of Cl^- present in the mortar. Moreover, the i_{corr} of the bars does not depend on the amount of chloride ion, which proves the impossibility of reworking reinforcements with thick corrosion products when these are not eliminated. This is because the i_{corr} in the reinforcements that corrode in an active state are sufficient to maintain an acid pH in the steel/corrosion products interface within such an alkaline medium as concrete, so that, once corrosion has been triggered, Cl^- are not necessary to maintain it [49,55,57]. Similar results were obtained by Miranda et al., when they observed that the higher the degree of pre-corrosion of the steel, the higher its corrosion rate, even in chloride-free environments [56,57].

The evolution of the E_{corr} in the same period of time can be seen in Figure 6. Figure 7 depicts in more detail the E_{corr} values reached by all the bars after 250 days.

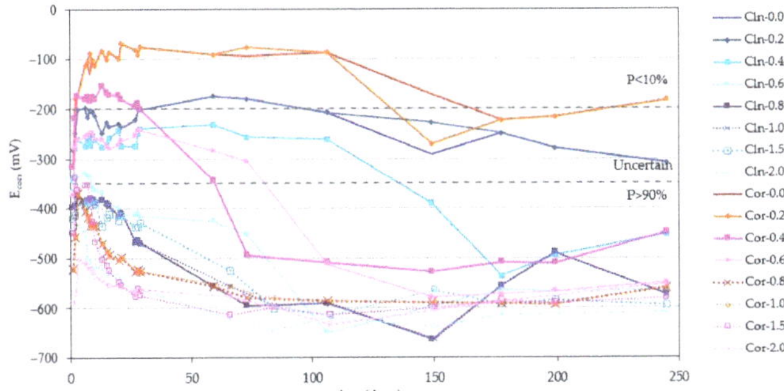

Figure 6. Evolution of the E_{corr} over time for all the rebars studied.

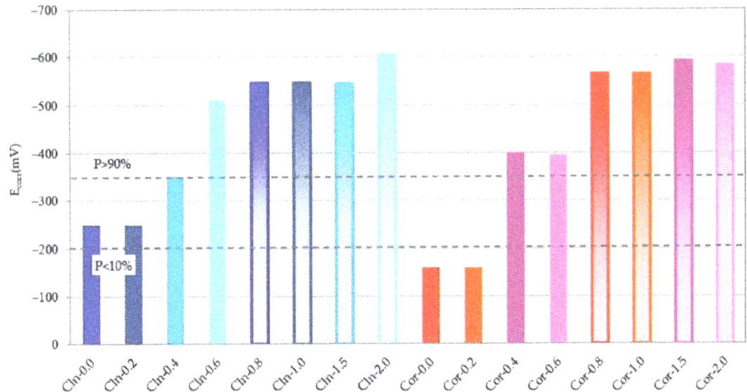

Figure 7. Measurement of the E_{corr} of all rebars after 250 days.

If we compare the evolution in time of the graphs in Figures 4 and 6, we can see that the E_{corr} suffers a greater variation than the i_{corr}, reaching, in some cases (corroded rebar introduced in a specimen with 0.4% of Cl^-), values, during the whole studied period, typical of the passive, uncertain and active state, since the measurement of the potential for corrosion varies enormously depending on various factors, such as temperature and humidity [58]. It can also be seen that at 250 days, the E_{corr} value depends much more on the amount of Cl^- in the specimen than on the passive or active state of the rebar, so that the E_{corr} of rebars in specimens containing the same amount of Cl^- are very similar, regardless of whether the rebar has been embedded in the clean specimen or with the corrosion products (Figures 6 and 7).

If the values of E_{corr} and i_{corr} are represented in a system of axes together with the thresholds that delimit the passive and active states, it is possible to check the validity of the E_{corr} measurements as a predictor of the corrosion state of a rebar (Table 3). Choosing as ordinate axis the value of E_{corr} and as x-axis the value of i_{corr} and marking by vertical and horizontal lines the thresholds of these values, the space is divided into 6 quadrants (Figure 8).

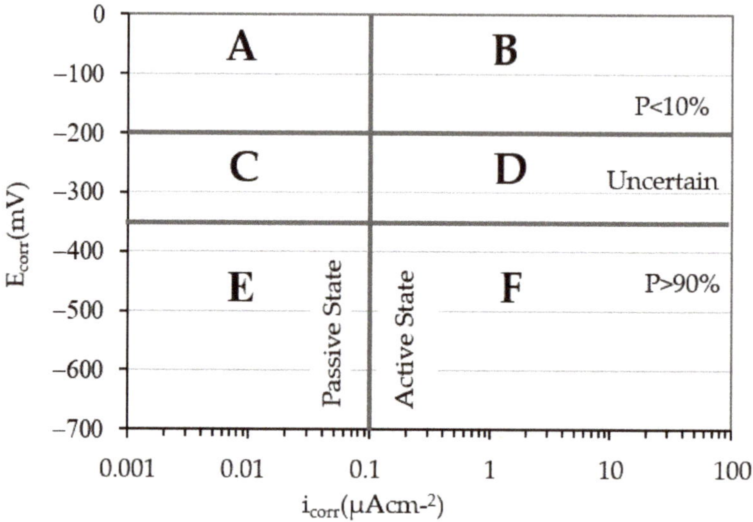

Figure 8. Joint representation of E_{corr} and i_{corr} in the plane.

In quadrants A and F, corrosion states coincide that predict the measurements of the E_{corr} and i_{corr}: Quadrant A corresponds to measurements that indicate the passive state while the measurements of quadrant F indicate the active state. In contrast, quadrants B and E define corrosion states with total discrepancy between the prediction of the E_{corr} and i_{corr}. Quadrants C and D define corrosion states where the E_{corr} predicts an uncertain state.

Figure 9 depicts the E_{corr} and i_{corr} values for each of the specimens obtained in all measurements. The graphs on the left column correspond to the clean rebars (CLN) while the right column corresponds to the values of the rebars with corrosion products (COR). In the two graphs located in the same row, the amount of Cl^- present in the test pieces coincides. Figure 10 shows the values of all the rebars together.

The analysis of all the graphs in Figure 9, shows that only the predictions of the E_{corr} and i_{corr} for rebars with an ongoing corrosion process and containing at least 0.8% Cl^- coincide. In these cases, all the data points of the graphs are located in quadrant F indicated in Figure 8, regardless of whether the rebars are clean or with corrosion products. In addition, the E_{corr} value becomes more electronegative as the amount of Cl^- present in the specimen increases and regardless of the state of the rebar. This trend is more pronounced for very low amounts of Cl^- and does not depend on the corrosion state of the reinforcement.

The analysis in Figure 10 allows verification of (i) the rebars in passive state ($i_{corr} < 0.1$ $\mu A/cm^2$) show a large number of E_{corr} values corresponding to high probabilities of corrosion (quadrant A) or uncertain states (quadrant C), and (ii) the rebars in active state ($i_{corr} > 0.1$ $\mu A/cm^2$) that show a large number of E_{corr} values corresponding to low probabilities of corrosion (quadrant F) or uncertain states (quadrant D). These results show the impossibility of predicting the passive or active state of the steel only by means of E_{corr} measurements [58].

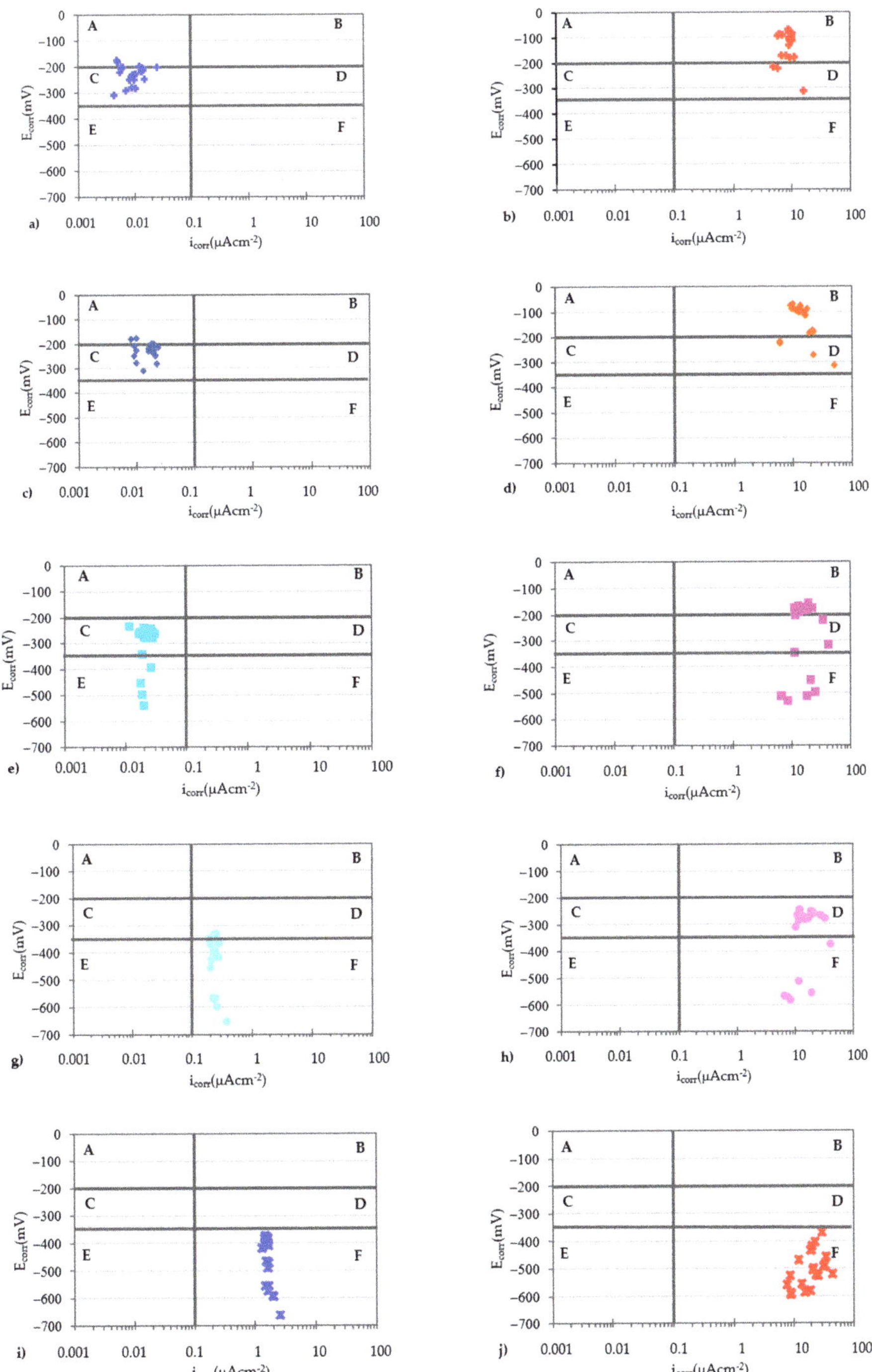

Figure 9. *Cont.*

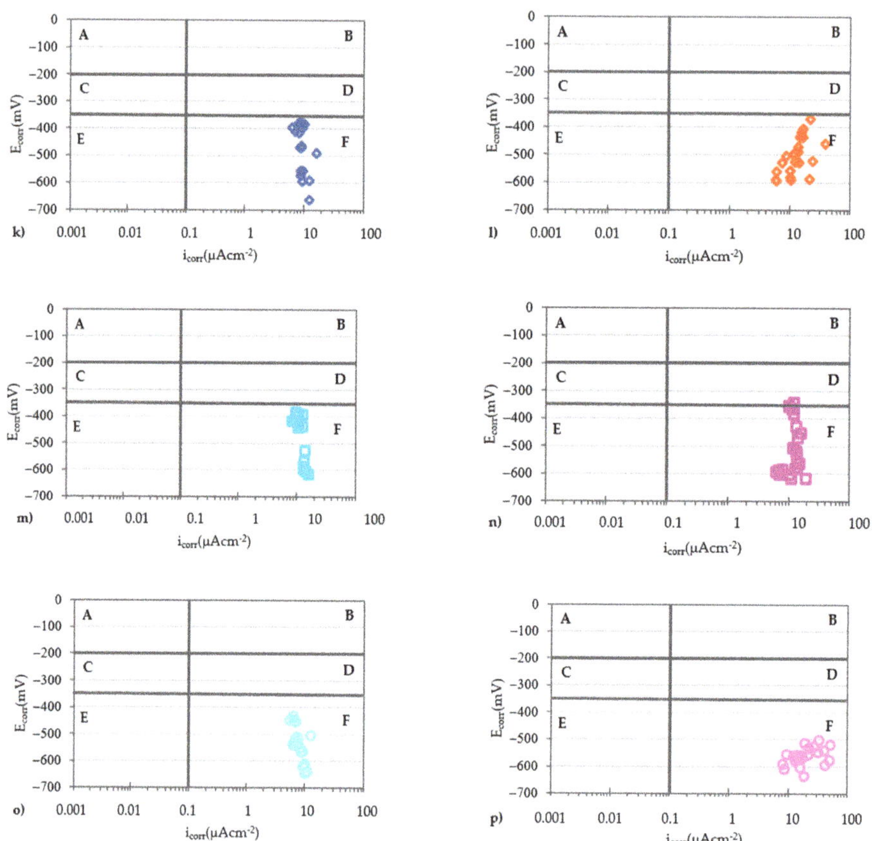

Figure 9. E_{corr} and i_{corr} values for each of the rebars. (**a**) Cln-0.0; (**b**) Cor-0.0; (**c**) Cln-0.2; (**d**) Cor-0.2; (**e**) Cln-0.4; (**f**) Cor-0.4; (**g**) Cln-0.6; (**h**) Cor-0.6; (**i**) Cln-0.8; (**j**) Cor-0.8; (**k**) Cln-1.0; (**l**) Cor-1.0; (**m**) Cln-1.5; (**n**) Cor-1.5; (**o**) Cln-2.0; (**p**) Cor-2.0.

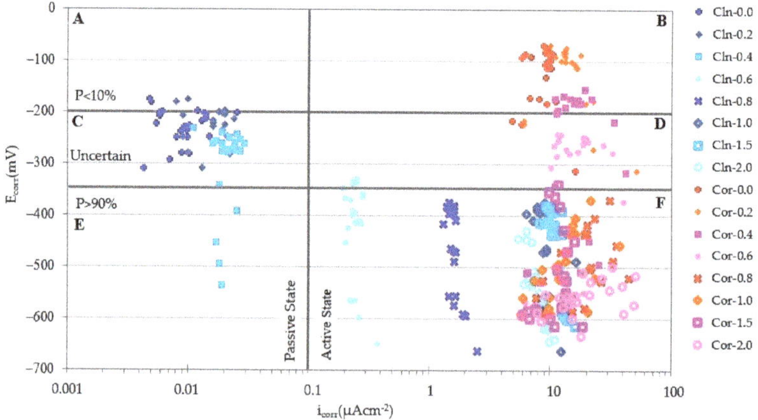

Figure 10. E_{corr} and i_{corr} values for all rebars.

4. Conclusions

Previously corroded reinforcing steel bars have been soaked in mortar samples with varying contents of Cl- in two different states: previously pickled and with the corrosion products. They have been kept for 250 days in a humid chamber, and E_{corr} and i_{corr} have been measured and obtained with the following conclusions being drawn:

- All the rebars that have kept the corrosion products showed a very high i_{corr} and a similar value, approximately 10 $\mu A/cm^2$, regardless of the number of chlorides present in the specimen.
- Clean rebars (CLN) embedded in specimens with 0.6 Cl^- remain with uncertain i_{corr} (values between 0.1 and 1.0 $\mu A/cm^2$).
- The values of the E_{corr} measurement depended more on the amount of Cl^- present in the specimen than on the passive or active state of the rebars.
- Only the predictions of the E_{corr} and i_{corr} coincided in bars embedded in specimens with at least a 0.8% Cl^- by weight cement ratio, regardless of whether the rebar is clean or maintains the corrosion products.
- The low correlation of the results obtained in the E_{corr} and i_{corr} in different situations, makes it impossible to predict the passive or active state of the steel solely based on E_{corr} measurements.
- To repair a concrete structure corroded by the effect of chloride ions, the concrete that surrounds the rebars must be removed so that all the corrosion products generated on the surface of the rebars can be eliminated. If the complete removal of the corrosion products is not achieved, even if a repair mortar is placed on it, the rebars will remain in active state.

Author Contributions: Conceptualization, A.C. and M.d.l.N.G.; methodology, A.C. and M.I.P.; software, F.I.O.; validation, M.d.l.N.G. and M.I.P.; formal analysis, M.I.P. and F.I.O.; investigation, A.C.; resources, M.d.l.N.G.; data curation, F.I.O.; writing—original draft preparation, A.C.; writing—review and editing, M.I.P.; visualization, M.d.l.N.G.; supervision, A.C.; project administration, M.d.l.N.G. All authors have read and agreed to the published version of the manuscript.

Funding: This research received no external funding.

Institutional Review Board Statement: Not applicable.

Informed Consent Statement: Not applicable.

Data Availability Statement: Data available on request.

Conflicts of Interest: The authors declare no conflict of interest.

References

1. Slater, J.E. *Corrosion of Metals in Association with Concrete*; ASTM-STP: Philadelphia, PA, USA, 1983; Volume 818.
2. Flis, J.; Pickering, H.W.; Osseo-Asare, K. Interpretation of Impedance Data for Reinforcing Steel in Alkaline Solutions Containing Chlorides and Acetates. *Electrochim. Acta* **1998**, *43*, 1921–1929. [CrossRef]
3. Page, C.L. Mechanism of corrosion protection in reinforced concrete marine structures. *Nature* **1975**, *258*, 514–515. [CrossRef]
4. Hope, B.B.; Page, J.A.; Ip, A.K.C. Corrosion rates of steel in concrete. *Cem. Concr. Res.* **1986**, *16*, 771–781. [CrossRef]
5. Federal Highway Administration. *6 FHWA Research Engineer Wins Arthur S. Flemming Award Promoted Advanced Bridge Inspection Technologies*; Federal Highway Administration: Washington, DC, USA, 2001.
6. BRE Group. *Report No 4*; Building Research Establishment Ltd.: Watford, UK, 2001.
7. Angst, U.M. Challenges and opportunities in corrosion of steel in concrete. *Mater. Struct.* **2018**, *51*, 4. [CrossRef]
8. Pillai, R.G.; Trejo, D. Surface condition effects on critical chloride threshold of steel reinforcement. *ACI Mater. J.* **2005**, *102*, 103–109.
9. Gancedo, J.R.; Alonso, C.; Andrade, C.; Gracia, M. AES study of the passive layer formed on iron in saturated $Ca(OH)_2$ solutions. *Corrosion* **1989**, *45*, 976–977. [CrossRef]
10. Torbati-Sarraf, H.; Poursaee, A. Corrosion of coupled steel with different microstructures in concrete environment. *Constr. Build. Mater.* **2018**, *167*, 680–687. [CrossRef]
11. González, J.A.; Otero, E.; Feliú, S.; Bautista, A.; Ramírez, E.; Rodríguez, P.; López, W. Some considerations on the effect of chloride ion son the corrosion of Steel reinforcements embedded in concrete structures. *Mag. Concr. Res.* **1998**, *50*, 189–199. [CrossRef]

12. Galán, I.; Andrade, C.; Castellote, M. Natural and accelerated CO_2 binding Kinetics in cement paste at different relative humidity. *Cem. Concr. Res.* **2013**, *49*, 21–28. [CrossRef]

13. Revert, A.B.; De Weerdt, K.; Hornbstel, K.; Geiker, M.R. Carbonation-induced corrosion: Investigation of the corrosion onset. *Constr. Build. Mater.* **2018**, *162*, 847–856. [CrossRef]

14. Piqueras, M.A.; Company, R.; Jódar, L. Numerical analysis and computing of free boundary problems for concrete carbonation chemical corrosion. *J. Comput. Appl.* **2018**, *336*, 297–316. [CrossRef]

15. Tristan Senga Kiesse, T.S.; Bonnet, S.; Amiri, O.; Ventura, A. Analysis of corrosion risk due to chloride diffusion for concrete structures in marine environment. *Mar. Struct.* **2020**, *73*, 102804. [CrossRef]

16. Li, C.; Chen, Q.; Wang, R.; Wu, M.; Jiang, Z. Corrosion assessment of reinforced concrete structures exposed to chloride environments in underground tunnels: Theoretical insights and practical data interpretations. *Cem. Concr. Compos.* **2020**, *112*, 103652. [CrossRef]

17. Angst, U.; Elsener, B.; Larsen, C.K.; Vennesland, O. Critical chloride content in reinforced concrete—A review. *Cem. Concr. Res.* **2009**, *39*, 1122–1138. [CrossRef]

18. Wang, Y.; Liu, C.; Wang, Y.; Li, Q.; Liu, Z. Investigation on chloride threshold for reinforced concrete by a test method combining ANDT and ACMT. *Constr. Build. Mater.* **2019**, *214*, 158–168. [CrossRef]

19. Page, C.L.; Lambert, P.; Vassie, P.R.W. Investigation of reinforcement corrosion: 1. The pore electrolyte phase in chloride-contaminated concrete. *Mater. Struct.* **1991**, *24*, 243–252. [CrossRef]

20. Lambert, P.; Page, C.L.; Vassie, P.R.W. Investigation of reinforcement corrosion: 2. Electrochemical monitoring of steel in chloride-contaminated concrete. *Mater. Struct.* **1991**, *24*, 351–358. [CrossRef]

21. Oh, B.H.; Jang, S.Y.; Shin, Y.S. Experimental investigation of the threshold chloride concentration for corrosion initiation in reinforced concrete structures. *Mag. Concr. Res.* **2003**, *55*, 117–124. [CrossRef]

22. Kayyali, O.A.; Haque, M.N. The ratio of Cl^-/OH^- in chloride contaminated concrete. A most important criterion. *Mag. Concr. Res.* **1995**, *47*, 235–242. [CrossRef]

23. Ann, K.T.; Song, H.W. Chloride threshold level for corrosion of steel in concrete. *Corros. Sci.* **2007**, *49*, 4113–4133. [CrossRef]

24. Haussman, D.A. Steel corrosion in concrete; How does it occur? *Mater. Prot.* **1967**, *6*, 19–23.

25. Rossi, E.; Polder, R.; Copuroglu, O.; Nigland, T.; Savija, B. The influence of defects at the steel/concrete interface for chloride-induced pitting corrosion of naturally deteriorated 20 years old specimens studied through X-ray Computed Tomography. *Constr. Build. Mater.* **2020**, *235*, 117474. [CrossRef]

26. Shi, J.; Ming, J. Influence of defects at the steel-mortar interface on the corrosion behavior of steel. *Constr. Build. Mater.* **2017**, *136*, 118–125. [CrossRef]

27. Soylev, T.A.; Francois, R. Quality of steel-concrete interface and corrosion of reinforcing steel. *Cem. Concr. Res.* **2003**, *33*, 1407–1415. [CrossRef]

28. Mohammed, T.U.; Otsuki, N.; Hamada, H.; Yamaji, T. Chloride-induced corrosion of steel bars in concrete with presence of gap at steel-concrete interface. *ACI Mater. J.* **2002**, *99*, 149–156.

29. Kenny, A.; Katz, A. Steel-concrete interface influence on chloride threshold for corrosion–Empirical reinforcement to theory. *Constr. Build. Mater.* **2020**, *244*, 118376. [CrossRef]

30. Alhozaimy, A.; Hussain, R.R.; Al-Negheimish, A. Electro-chemical investigation for the effect of rebar source and surface condition on the corrosion rate of reinforced concrete structures under varying corrosive environments. *Constr. Build. Mater.* **2020**, *244*, 118317. [CrossRef]

31. Duracrete. *General Guidelines for Durability Design and Redesign*; CUR: Gouda, Belgium, 2000.

32. de Fomento, M. *Instrucción de Hormigón Estructural EHE-08*; Madrid, Spain, 2008. Available online: https://www.mitma.gob.es/recursos_mfom/1820100.pdf (accessed on 20 October 2021).

33. American Concrete Institute. *ACI 318-19 Building Code Requirements for Structural Concrete and Commentary*; American Concrete Institute: Farmington Hills, MI, USA, 2019.

34. Ožbolt, J.; Balabanic, G.; Kušter, M. 3D Numerical modelling of steel corrosion in concrete structures. *Corros. Sci.* **2011**, *53*, 4166–4177. [CrossRef]

35. Chauhan, A.; Sharma, U.K. Crack propagation in reinforced concrete exposed to non-uniform corrosion under real climate. *Eng. Fract. Mech.* **2021**, *248*, 107719. [CrossRef]

36. Chen, J.; Zhang, W.; Tang, Z.; Huang, Q. Experimental and numerical investigation of chloride-induced reinforcement corrosion and mortar cover cracking. *Cem. Concr. Compos.* **2020**, *111*, 103620. [CrossRef]

37. Xu, F.; Xiao, Y.; Wang, S.; Li, W.; Liu, W.; Du, D. Numerical model for corrosion rate of steel reinforcement in cracked reinforced concrete structure. *Constr. Build. Mater.* **2018**, *180*, 55–67. [CrossRef]

38. Fekak, F.E.; Garibaldi, L.; Elkhalfi, A.; Alami, E.E. A numerical study of pitting corrosion in reinforced concrete structures. *J. Build. Eng.* **2021**, *43*, 102789.

39. ASTM C876-09 Standard. *Standard Test Method for Half-Cell Potentials of Uncoated Reinforcing Steel in Concrete*; ASTM International: West Conshohocken, PA, USA, 2009. [CrossRef]

40. Videm, K. *Corrosion of Reinforcement in Concrete. Monitoring, Prevention and Rehabilitation. (EFC 25)*; RC Press: London, UK, 1998; Chapter 10, pp. 104–121.

41. Ismail, M.; Ohtsu, M. Corrosion rate of ordinary and high-performance concrete subjected to chloride attack by AC impedance spectroscopy. *Constr. Build. Mater.* **2006**, *20*, 458–469. [CrossRef]
42. Reou, J.S.; Ann, K.Y. Electrochemical assessment on the corrosion risk of steel embedment in OPC concrete depending on the corrosion detection techniques. *Mater. Chem. Phys.* **2009**, *113*, 78–84. [CrossRef]
43. Andrade, C.; Keddam, M.; Nóvoa, X.R.; Pérez, M.C.; Rangel, C.M.; Takenouti, H. Electrochemical behavior of steel rebars in concrete: Influence of environmental factors and cement chemistry. *Electrochim. Acta* **2001**, *46*, 3905–3912. [CrossRef]
44. González, J.A.; Miranda, J.M.; Feliú, S. Considerations on reproducibility of potential and corrosion rate measurements in reinforced concrete. *Corros. Sci.* **2004**, *46*, 2467–2485. [CrossRef]
45. Sagüés, A. Corrosion Measurement Techniques for Steel in Concrete. In *Corrosion-National Association of Corrosion Engineers Annual Conference*; NACE: Houston, TX, USA, 1993.
46. Ministerio de la Presidencia. *RC-16, Instrucción para la Recepción de Cementos*; Boletín Oficial del Estado: Madrid, Spain, 2016.
47. Stern, M.; Geary, A.L. Electrochemical polarization I. A theoretical analysis of the shape of the polarization curves. *J. Electrochem. Soc.* **1957**, *104*, 56–63. [CrossRef]
48. Bastidas, D.M.; González, J.A.; Feliú, S.; Cobo, A.; Miranda, J.M. A quantitative study of concrete-embedded steel corrosion using potentiostatic pulses. *Corrosion* **2007**, *63*, 1094–1100. [CrossRef]
49. Prieto, M.I.; Cobo, A.; Rodríguez, A.; González, M.N. The efficiency of surface-applied corrosion inhibitors as a method for the repassivation of corroded reinforcement bars embedded in ladle furnace slag mortars. *Constr. Build. Mater.* **2014**, *54*, 70–77. [CrossRef]
50. Browne, R.D.; Geoghegan, M.P.; Baker, A.F. Corrosion Monitoring of Steel in Concrete. In *Corrosion of Reinforcement in Concrete Construction*; Crane, A.P., Ed.; Ellis Horwood Ltd.: Chichester, UK, 1983; Chapter 13; pp. 193–222.
51. Elsener, B.; Andrade, C.; Gulikers, J.; Polder, R.; Raupach, M. Half-cell potential measurements—Potential mapping on reinforced concrete structures. *Mater. Struct.* **2003**, *36*, 1–11. [CrossRef]
52. Andrade, C.; Alonso, C. Test methods for on-site corrosion rate measurements of steel reinforcement in concrete by means of the polarization resistance method. *Mater. Struct.* **2004**, *37*, 623–643. [CrossRef]
53. Network, D. *Manual de Inspección, Evaluación y Diagnóstico de Corrosión en Estructuras de Hormigón Armado*; CYTED Programe: Rio de Janeiro, Brazil, 1998; ISBN 980-296-541-3.
54. *Eurocode 2: Design of Concrete Structures—Part 1-1: General Rules and Rules for Buildings*; CEN: Brussels, Belgium, 2004.
55. Prieto, M.I.; Cobo, A.; Rodríguez, A.; Calderón, V. Corrosion behavior of reinforcements bars embedded in mortar specimens containing ladle furnace slag in partial substitution of aggregate and cement. *Constr. Build. Mater.* **2013**, *38*, 188–194. [CrossRef]
56. Martinez, M.J.; Lopez, M.; Cantero, D.; Rodríguez, J. Influence of the previous state of corrosion of rebars in predicting the service life of reinforced concrete structures. *Constr. Build. Mater.* **2018**, *188*, 915–923. [CrossRef]
57. Miranda, J.M.; González, J.A.; Cobo, A.; Otero, E. Several question about electrochemical rehabilitation methods for reinforced concrete structures. *Corros. Sci.* **2006**, *48*, 2172–2188. [CrossRef]
58. Macdonald, D.D.; Qiu, J.; Zhu, Y.; Yang, J.; Engelhardt, G.R.; Sagüés, A. Corrosion of rebar in concrete. Part I: Calculation of the corrosion potential in the passive state. *Corros. Sci.* **2020**, *177*, 109018. [CrossRef]

Article

Experimental Study on Solidification of Pb²⁺ in Fly Ash-Based Geopolymers

Fang Liu [1], Ran Tang [1], Baomin Wang [2],* and Xiaosa Yuan [1]

[1]   Shaanxi Key Laboratory of Safety and Durability of Concrete Structures, Xijing University,
      Xi'an 710123, China; liufang_winter@163.com (F.L.); tangran627@163.com (R.T.);
      yuanxiaosa2009@163.com (X.Y.)

[2]   School of Civil Engineering, Dalian University of Technology, Dalian 116023, China

*   Correspondence: wangbm@dlut.edu.cn; Tel.: +86-0411-8470-7101

Abstract: Fly ash from the incineration of domestic waste contains heavy metals, which is harmful to the environment. To reduce and prevent their contamination, heavy metal ions need to be sequestered. In this study, the geopolymer prepared by fly ash, a kind of power plant waste, is used to cure the heavy metal Pb²⁺, and to investigate the effect of different concentrations of Pb²⁺ on the compressive strength of the solidified body at different ages; the curing effect is judged by the toxic leaching concentration of heavy metals; the resistance of the solidified body to immersion is evaluated by comparing the change in strength before and after leaching; the fly ash-based geopolymer solidified body is compared with the cement solidified body in terms of curing effectiveness; the properties of the geopolymer and its mechanism of curing heavy metals are explored by microscopic tests. The results show that the fly ash-based geopolymer solidified body has good resistance to immersion; the optimum curing concentration of Pb²⁺ in fly ash-based geopolymers is 2.0%; compared to pure geopolymers, the strength of the solidified body at 28 d decreases by only 13.0%, and the leaching concentration of Pb²⁺ is 4.73 mg·L⁻¹, which meets the specification requirements; the curing effect of the fly ash-based geopolymer is better than the cement solidified body; the microscopic test results indicate that the curing of Pb²⁺ by the fly ash-based geopolymer is a combination of both chemical bonding and physical fixation.

Keywords: fly ash; geopolymer; solidification; heavy metal; leaching concentration

Citation: Liu, F.; Tang, R.; Wang, B.; Yuan, X. Experimental Study on Solidification of Pb²⁺ in Fly Ash-Based Geopolymers. *Sustainability* **2021**, *13*, 12621. https://doi.org/10.3390/su132212621

Academic Editors: Carlos Morón Fernández and Daniel Ferrández Vega

Received: 20 October 2021
Accepted: 12 November 2021
Published: 15 November 2021

Publisher's Note: MDPI stays neutral with regard to jurisdictional claims in published maps and institutional affiliations.

1. Introduction

With the development of society, the discharge of domestic waste increases rapidly. Proper disposal of these wastes has become an important task. At present, there are three common methods: landfilling, composting, and incineration [1], of which incineration is becoming the main method because of its high efficiency, speed, significant volume reduction, and the possibility of converting waste into other energy sources; however, after waste incineration, most of the heavy metals such as Pb, Cr, and Zn will go into the fly ash that can cause irreversible damage to the environment and humans if handled improperly.

An important means of dealing with heavy metal waste is to cure them to reduce or even prevent their contamination. The solidified cement is commonly used for this purpose, but it has obvious drawbacks, including high concentrations of toxic leaching of heavy metals, poor durability, and poor resistance to acidic soils and rainfall [2,3].

The geopolymer [4] is a new type of inorganic polymeric material with high strength and durability. It is prepared by the polymerization of silica-aluminous raw materials, such as metakaolin [5,6], slag [7], kaolinite [8], and fly ash [9,10], under the solubilization of alkali activator to form an amorphous silica-aluminum compound. Andini, S. [11] investigated the effect of different curing temperatures on the performance of geopolymers synthesized from fly ash, and determined the optimum curing temperature. Duxson, P. [12] studied the mechanical properties of geopolymer with different types of alkali activators and different silica-aluminum ratios in relation to age, and found that the sodium alkali activator is more effective than the potassium alkali activator in stimulating geopolymers.

The internal structure of geopolymers exhibits a three-dimensional mesh-like structure similar to that of zeolites that makes it uniquely suited to the solidification of almost all heavy metal ions [13–16].

As reported in [17,18], geopolymers can efficiently bind heavy metals in their matrix structure; the reaction mechanism could be physical or chemical bonding. Studies by Van Jaarsveld et al. [19] suggested that Pb might be chemically bonded within the aluminosilicate matrix, but the exact form was unclear. In contrast, Palacios and Palomo [20] suggested that Pb was solidified to insoluble Pb_3SiO_5 in NaOH-activated fly ash cementitious materials.

Therefore, a geopolymer was prepared using waste fly ash as a matrix in this study to investigate its curing of heavy metals and the mechanism. Additionally, the heavy metal Pb^{2+} was solidified through the geopolymer to study the effect of its concentration on the compressive strength of the solidified body at different ages, and on the toxic leaching concentration of heavy metals, and to compare the curing effect of the fly ash-based geopolymer with that of the cement solidified body. Furthermore, the properties of the geopolymer and its mechanism of curing heavy metals were explored through microscopic tests (XRD, FT-IR, and SEM) to investigate the reaction process, structural morphological changes, and curing mechanisms at a microscopic level.

2. Raw Materials and Experimental Procedures

2.1. Raw Materials

2.1.1. Fly Ash

The ultra-fine fly ash used in this study came from Guodian Zhuanghe power plant in Liaoning Province. The chemical composition of this fly ash is given in Table 1, and its main technical indicators are listed in Table 2.

Table 1. Chemical composition of fly ash.

SiO_2	Al_2O_3	Fe_2O_3	CaO	TiO_2	K_2O	MgO
48.67%	29.16%	8.98%	4.73%	2.75%	2.66%	0.74%

Table 2. Summary of main technical indicators of fly ash.

Item	Test Result	Standard of Class I Fly Ash
Water content	0.72%	$\leq 1\%$
Water requirement ratio	93.8%	$\leq 95\%$
Fineness	2.3%	$\leq 12\%$
Strength activity index	78.1%	$\geq 70\%$

2.1.2. Water Glass

The sodium water glass ($Na_2O \cdot nSiO_2$) was obtained from Usolf Chemical Technology Co., Ltd. in Shandong Province. The specific parameters are listed in Table 3.

Table 3. Relevant parameters of water glass.

Name	Content
Na_2O	8.3%
SiO_2	26.5%
H_2O	65.2%
Modulus	3.3
Appearance	Transparent and viscous liquid

2.1.3. Other Raw Materials and Chemical Reagents

(1) NaOH

It was analytical pure level that was used to adjust the modulus of the water glass to configure a compound alkali activator required for the experiment.

(2) H_2SO_4, HNO_3

They were analytical pure levels that were used to prepare leaching solutions in simulated acidic environments, and to leach fly ash-based geopolymer samples in preparation for subsequent heavy metal leaching concentration tests.

(3) Heavy metal salt

The aim of this study is to research the solidification of Pb^{2+}; therefore, nitrate ($Pb(NO_3)_2$, analytical pure level) was chosen to avoid the interference of anions in heavy metal salts.

(4) Water

The deionized water used in this test was supplied by Bonuo Chemical Reagent Factory.

2.2. Preparation of the Fly Ash-Based Geopolymer

In this test, Pb^{2+} was introduced in the form of $Pb(NO_3)_2$, while its content (in terms of mass of Pb^{2+} as a percentage of solid mass) was set at six levels of 0.5%, 1.0%, 1.5%, 2.0%, 2.5%, and 3.0%, respectively. The experimental ratios of materials used for the fly ash-based geopolymer are listed in Table 4.

Table 4. Material proportioning for fly ash-based geopolymer solidified body.

No.	Fly Ash/g	Water Glass/g	NaOH/g	Water/g	$Pb(NO_3)_2$/g	Pb^{2+} Content/%
1	1000.00	299.17	47.80	104.94	8.00	0.50
2	1000.00	299.17	47.80	104.94	15.99	1.00
3	1000.00	299.17	47.80	104.94	23.99	1.50
4	1000.00	299.17	47.80	104.94	31.98	2.00
5	1000.00	299.17	47.80	104.94	39.98	2.50
6	1000.00	299.17	47.80	104.94	47.97	3.00

According to Table 4, fly ash, water glass, water, and $Pb(NO_3)_2$ were added to the mixing pot of the mechanical mixer in turn, and stirred for 3 min (30 s slow stirring and 1 min fast stirring, repeated, and then finished). Next, the mixture was poured into a $40 \times 40 \times 40$ mm mold, which had been painted in advance with a release agent, and vibrated for 2 min on a vibrating table, before sealing the surface with polyethylene film to prevent moisture evaporation. After standing for 2 h at room temperature, the mold was placed in a high temperature curing box at 65 °C for 24 h before being taken out and demolded, and then it continued to be placed in a standard curing box for closed curing until the age of 3 d, 7 d, and 28 d.

2.3. Analytical Test Methods

2.3.1. Compressive Strength Test

The compressive strength test was performed according to *Method of testing cements—Determination of strength*, GB/T17671-1999.

2.3.2. Leaching Concentration Test for Heavy Metal Ions

The test was performed according to GB5085.3-2007 and GB14569.1-2011. The preparation of the leaching solution required was carried out according to HJ/T299-2007 and HJ557-2010, simulating the acid rain environment and the normal environment, respectively; and then, the leaching concentration of heavy metals was measured by Inductive Coupled Plasma Emission Spectrometer (ICP) test according to GB/T23942-2009.

2.3.3. Microscopic Test

X-ray diffraction (XRD), Fourier-transform infrared spectroscopy (FT-IR), and scanning electron microscopy (SEM) were used to test and analyze the performance of samples of the fly ash-based geopolymer, and the heavy metal ion solidified body on a micromorphological basis.

3. Results and Discussion

3.1. Compressive Strength

The compressive strength and its trends in the curing of Pb^{2+} by the fly ash-based geopolymer are shown in Table 5 and Figure 1.

Table 5. Compressive strength of fly ash-based geopolymer solidified body.

No.	Pb^{2+} Content/%	Compressive Strength at 3 d (Strength Loss)/MPa	Compressive Strength at 7 d (Strength Loss)/MPa	Compressive Strength at 28 d (Strength Loss)/MPa
0	0.0	45.26	46.66	48.72
1	0.5	44.07 (2.6%)	45.73 (2.0%)	47.84 (1.8%)
2	1.0	42.96 (5.1%)	44.47 (4.7%)	46.67 (4.2%)
3	1.5	40.25 (11.1%)	41.99 (10.0%)	44.82 (8.0%)
4	2.0	37.54 (17.1%)	39.67 (15.0%)	42.39 (13.0%)
5	2.5	33.79 (25.3%)	36.86 (21.0%)	39.46 (19.0%)
6	3.0	27.38 (39.5%)	29.40 (37.0%)	31.67 (35.0%)

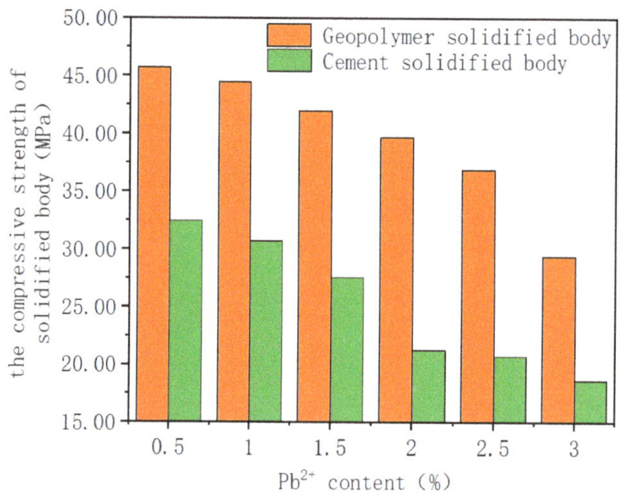

Figure 1. Comparison of the compressive strength of the geopolymer and the cement solidified body.

According to Table 5, the compressive strength of the fly ash-based geopolymer at each age decreases gradually with the increase in Pb^{2+} concentration. Additionally, at the same Pb^{2+} concentration, the internal reaction of the geopolymer is gradually complete, and the alkali activator fully acts with the increase of age, corresponding to the increasing strength of the solidified body. It is analyzed that, before Pb^{2+} concentration reaches 2.0%, the compressive strength of the solidified body at 3 d, 7 d, and 28 d can still be maintained at around 40 MPa, with good mechanical properties; after 2.0%, the strength decreases rapidly; when Pb^{2+} reaches 3.0%, the compressive strength at 3 d decreases by nearly 40%. The reason could be that the introduction of Pb^{2+} inhibits the polymerization reaction to some extent, or it has some effect on the internal structure of the solidified body.

3.2. Toxic Leaching Concentration of Heavy Metals

3.2.1. Neutral Environment

According to Table 6, in a neutral environment, the toxic leaching concentration of heavy metals in the fly ash-based geopolymer solidified body increases with the increase of Pb^{2+} content. At the same heavy metal concentration, the polymerization reaction within the solidified body tends to be thorough with the increase in age; its internal cage-like sequestration structure is more complete; the sequestration effect on heavy metal ions is more effective; the leaching concentration of the heavy metal solidified body decreases accordingly. When the heavy metal content is 2.5%, the leaching concentrations of the solidified body at 3 d and 7 d exceed the specification maximum limit (5 mg·L^{-1}), while the concentration at 28 d does not; when the heavy metal content reaches 3%, the leaching concentration of the solidified body at 28 d is only above the limit. Therefore, the optimum Pb^{2+} content ranges from 2% to 2.5% in a neutral environment.

Table 6. Results of toxic leaching concentration corresponding to different Pb^{2+} contents in a neutral environment.

No.	Pb^{2+} Content/%	Leaching Concentration of Heavy Metals in Solidified Body/mg·L^{-1}		
		3 d	7 d	28 d
1	0.5	1.05	0.81	0.53
2	1.0	1.67	1.34	1.03
3	1.5	3.53	3.09	2.82
4	2.0	4.26	3.79	3.35
5	2.5	5.38	5.03	4.87
6	3.0	6.42	5.89	5.24

3.2.2. Acid Environment

As shown in Table 7, the toxic leaching pattern of heavy metals in an acid environment is generally the same as that in a neutral environment; however, the external and internal structure of the solidified body is more significantly attacked by the acid environment than the neutral environment, so that heavy metal ions are more likely to migrate out of the internal structure of the geopolymer, resulting in higher leaching concentrations for each Pb^{2+} content and each age of the solidified body in an acid environment than in a neutral environment.

Table 7. Results of toxic leaching concentrations corresponding to different Pb^{2+} contents in an acid environment.

No.	Pb^{2+} Content/%	Leaching Concentration of Heavy Metals in Solidified Body/mg·L^{-1}		
		3 d	7 d	28 d
1	0.5	1.25	1.12	0.89
2	1.0	1.93	1.72	1.46
3	1.5	4.06	3.95	3.35
4	2.0	4.93	4.89	4.73
5	2.5	6.31	6.02	5.47
6	3.0	8.36	7.97	7.61

When Pb^{2+} reaches 2%, the leaching concentrations of the solidified body at the ages of 3 d, 7 d, and 28 d are close to the normative limit, but does not yet exceed it, while at the Pb^{2+} content of 2.5%, the limit is exceeded at all ages. To sum up, the optimum solidifying concentration range for Pb^{2+} in the fly ash-based geopolymer in an acid environment is similarly between 2% and 2.5%.

3.3. Immersion Resistance

Two groups of fly ash-based geopolymer solidified body with different Pb^{2+} doping are set. One group is used as a control group to test the strength before immersion; the other is placed in a water tank at room temperature and is flooded over the surface. The specimens are removed and dried with a rag after being immersed for 90 days, and then, they are tested for the compressive strength. Test results are listed in Table 8.

Table 8. Changes of the strength before and after immersion.

No.	Pb^{2+} Content/%	Strength before Immersion/MPa	Strength after Immersion/MPa	Strength Loss/%
1	0.5	47.84	36.31	24.10
2	1.0	46.67	37.98	18.62
3	1.5	44.82	34.59	22.82
4	2.0	42.39	35.82	15.50
5	2.5	39.46	29.65	24.85
6	3.0	31.67	21.97	30.63

According to Table 8, the strength loss of the solidified body does not exceed the limit of 25% for each Pb^{2+} content, except for the highest content (3.0%). Therefore, it can be concluded that the fly ash-based geopolymer solidified body has good resistance to soaking.

3.4. Comparison between the Fly Ash-Based Geopolymer and the Cement Solidified Body

To verify the effectiveness of the fly ash-based geopolymer in solidifying Pb^{2+}, the cement solidified body is chosen for comparison in this study. Both are set to the same water to ash ratio, and the same amount of Pb^{2+} (0.5%, 1.0%, 1.5%, 2.0%, 2.5%, and 3.0%). The compressive strength of the solidified body at 7 d, and the toxic leaching concentrations at 28 d (acid environment) are used as indicators to compare the solidifying effect of these two. The experimental results are shown in Figures 1 and 2.

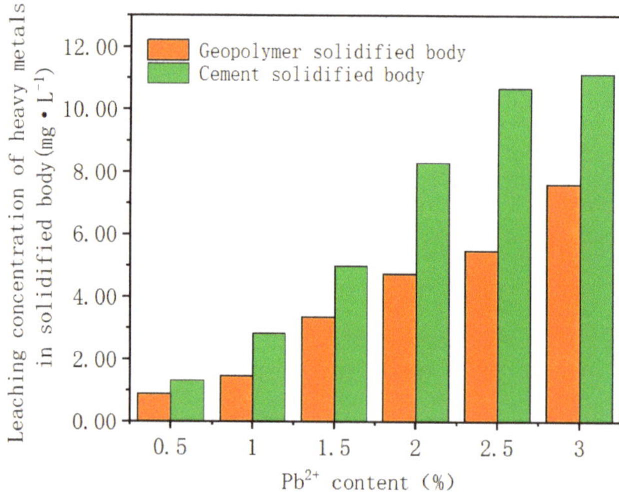

Figure 2. Comparison of heavy metal leaching concentration of geopolymer and cement solidified body.

As shown in Figure 1, the compressive strength of the fly ash-based geopolymer solidified body is significantly higher than that of the cement solidified body. It is analyzed that the compressive strengths of both tend to decrease with the increase in Pb^{2+} content. When Pb^{2+} concentration is 2%, the compressive strength of the geopolymer solidified body is still around 40 MPa; the strength decreases slowly until the Pb^{2+} concentration reaches 2%, and more rapidly after this. In contrast, the strength of the cement solidified body decreases at a relatively rapid rate all along. Therefore, the performance of the geopolymer solidified body is significantly better from the point of view of the compressive strength.

According to Figure 2, in an acid environment, the solidification of Pb^{2+} in the fly ash-based geopolymer solidified body is better than in the cement solidified body. At 2% Pb^{2+}, the leaching concentration of the geopolymer solidified body is still below 5 mg·L^{-1}, the limit value for Pb^{2+} is specified in GB5085.3-2007, whereas the cement solidified body is very close to the normative limit at 1.5% Pb^{2+}.

This is mainly due to that in an acid environment, the hydration products of cement, such as C-S-H (calcium silicate hydrate) and $Ca(OH)_2$, react with the acid to produce the corresponding calcium salts, resulting in an increase in pore space within the solidified body, which ultimately leads to the destruction of the solidified body and a reduction in its solidifying ability. By comparison, the silicon–oxygen (Si-O) and aluminum–oxygen (Al-O) bonded in the internal structure of the geopolymer are difficult to break in an acid environment, making it more resistant to acid attack.

To sum up, whether from the point of view of mechanical properties (compressive strength) or solidifying properties (leaching concentration), the performance of the fly ash-based geopolymer solidified body is better than that of the cement solidified body.

3.5. Solidifying Mechanism Analysis

3.5.1. XRD

The XRD pattern of the fly ash used for the test is shown in Figure 3. It can be seen that the main body of the fly ash is amorphous, and its crystalline substances are quartz and mullite, which are the main substances containing silicon and aluminum. The characteristic peaks of quartz appear at 22.53°, 50.71°, and 54.69°; and that of mullite at 17.15°, 27.33°, 32.67°, and 34.82°. The diffraction peaks in a "bun" shape in the 2θ angle range of 15° to 30° indicate the presence of vitreous body in the fly ash, from which the activity of the fly ash is mainly derived.

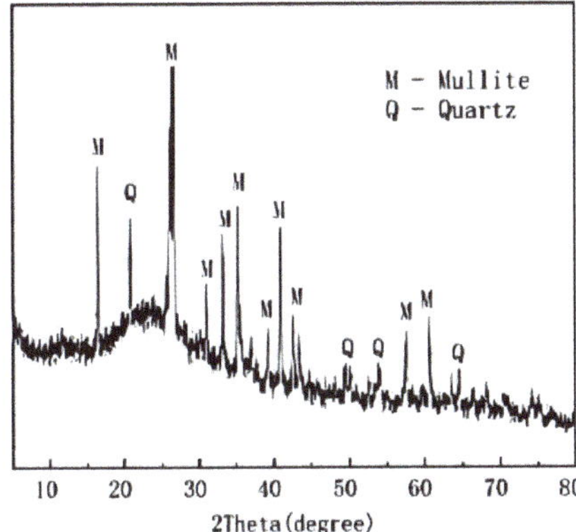

Figure 3. XRD pattern of fly ash.

The variations in the mineral phase of the fly ash-based geopolymer and its solidified body are shown in Figure 4. Comparing Figure 4 with Figure 3, the crystalline phases in the fly ash-based geopolymer are mainly mullite and quartz, with a small amount of calcium silicate present at 39.54°, and the crystalline diffraction peaks are somewhat reduced compared to that in the fly ash XRD pattern, demonstrating that the crystalline substances in the fly ash react to form amorphous substances during the alkali excitation process. Additionally, in the curve of the pure fly ash-based geopolymer, "bun"-shaped dispersion peaks appear in the interval 18° to 35 °, which are slightly shifted to the right compared to the fly ash curve, indicating that the dissolution of the vitreous body in the fly ash produces amorphous gel-like substance in the fly ash-based geopolymer; the absence of new phases and the disappearance of old phases mean that some of the crystalline particles in the fly ash are dissolved, and some are encapsulated by the gel material in the geopolymer, which corresponds to the SEM results.

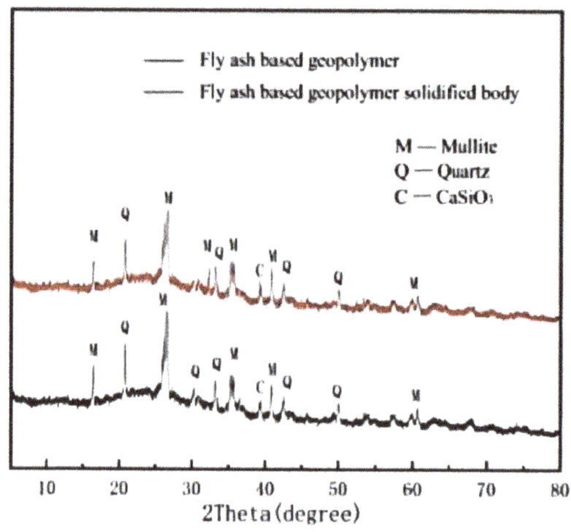

Figure 4. XRD pattern of the fly ash-based geopolymer and its solidified body.

Compared to the curve in Figure 3, the overall curve in Figure 4 is flatter, which means a relatively high degree of reaction, and a more adequate polymerization of the silica-aluminous material in the raw material with the alkali activator, resulting in a good strength of the geopolymer. Diffraction peaks of crystalline minerals associated with Pb^{2+} are not found on the curve of the solidified body doped with Pb^{2+}, and no new phases appear, which means that Pb^{2+} is not involved in the overall internal chemistry of the geopolymer and the Pb_3SiO_5 phase suggested by Palomo et al. [20] is not detected. It can be assumed that Pb^{2+} may have been bonded to the internal structural skeleton of the fly ash-based geopolymer, or the solidification of heavy metal ions by the geopolymer solidified body is a physical encapsulation.

3.5.2. SEM

Figure 5a shows a micrograph of the fly ash-based geopolymer in the early stage of the reaction. At this stage, the fly ash particles have not fully reacted; half of them are dissolved, while the other half appears partially encapsulated on the surface.

Figure 5b shows a micrograph of the fly ash-based geopolymer at a later stage of the reaction. It can be observed that the fly ash, after polymerization with the alkali activator, produces a very dense geopolymer gel (silica-aluminate gel) with a uniformly dense and amorphous outer surface; the particles in the fly ash are completely dissolved or encapsulated within the formed geopolymer gel; the beads are no longer visible compared to the pre-reaction period, but traces of their internal concavity left by the dissolution reaction can be found. Overall, the internal morphology of the fly ash-based geopolymer is dominated by a dense amorphous structure, which means that the reaction of the materials in this test is sufficient and the polymerization reaction of internal depolymerization and polycondensation is complete. As a result, the fly ash-based geopolymer obtained from the reaction possesses good physicochemical properties.

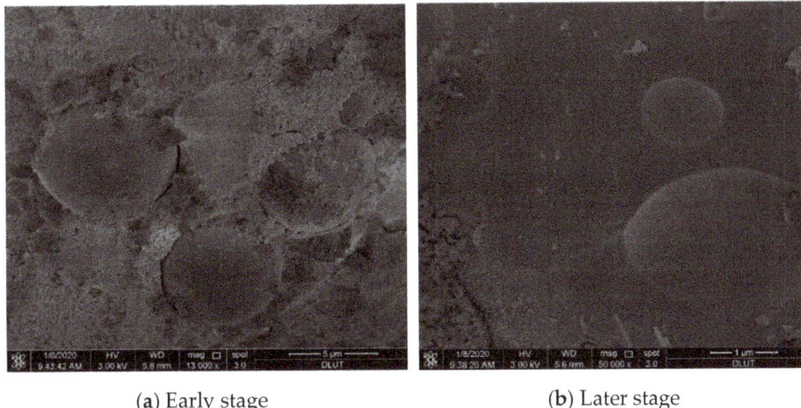

(**a**) Early stage   (**b**) Later stage

Figure 5. SEM of the fly ash-based geopolymer in the early and later stages of the reaction.

Figure 6 shows the electron micrographs of the fly ash-based geopolymer during the reaction. Compared to the early stage of the reaction in Figure 5a, the wrapping condition in Figure 6a is more complete, and the progress of the reaction can be clearly observed. Figure 6b is a partial magnification, showing that the gel in the geopolymer slowly grows and extends from the top of the particle downwards until it completely wraps around the geopolymer, forming a kind of core-shell structure. These indicate that a portion of the fly ash particles is encapsulated by the gel into a core-shell structure, except for a portion being dissolved by the alkaline solution. The outer shell of this structure extends until it completely encapsulates the geopolymer and then forms a integral whole with its surrounding dense gel-like substances, which also has a beneficial effect on the mechanical properties of the geopolymer.

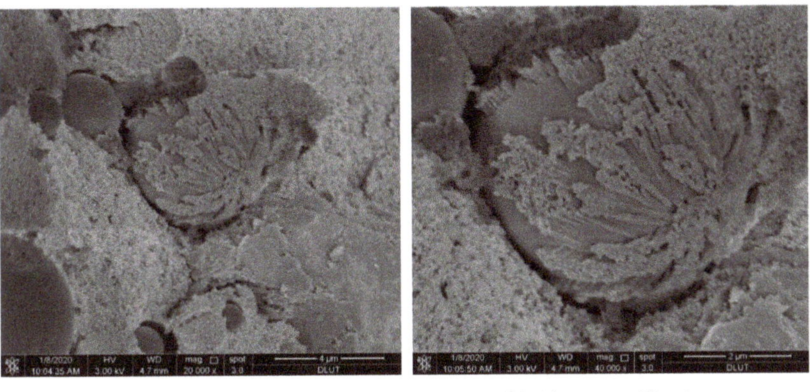

(**a**) Before magnification   (**b**) After magnification

Figure 6. SEM of the fly ash-based geopolymer during the reaction.

Figure 7a shows the state during the solidifying Pb^{2+} reaction of the fly ash-based polymer, and the fully reacted state is given in Figure 7b. Compared to Figure 7a, a more complete and dense whole is formed in the latter figure, resulting in a good performing solidified body of the fly ash-based geopolymer.

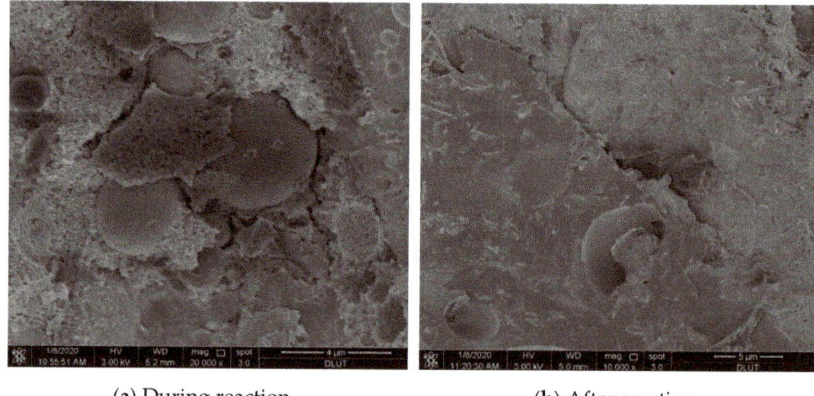

(a) During reaction   (b) After reaction

Figure 7. SEM of Pb^{2+} solidified body of the fly ash-based geopolymer.

3.5.3. FT-IR

The infrared spectrum of the fly ash-based geopolymer and its solidified body is shown in Figure 8.

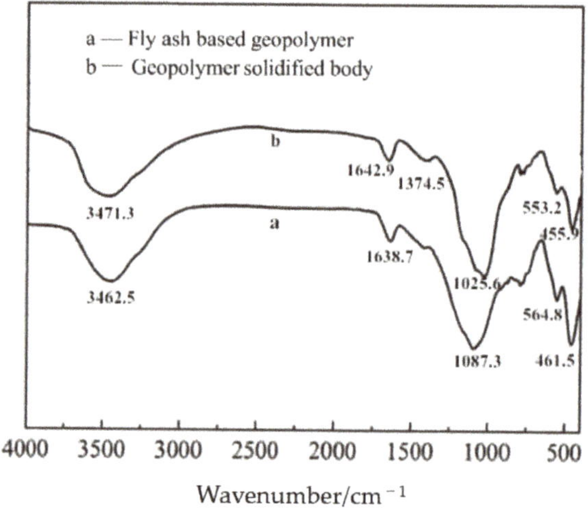

Figure 8. Infrared spectrum of the fly ash-based geopolymer and its solidified body.

According to the figure, the geopolymer exhibits O-H stretching vibration and O-H bending vibration at 3462.5 cm^{-1} and 1638.7 cm^{-1}, respectively; while the solidified body exhibits O-H stretching vibration and O-H bending vibration at 3471.3.4 cm^{-1} and 1642.9 cm^{-1}, respectively, which is surmised to be due to the presence of bound water within the geopolymer and the solidified body. The absorption peaks at 1087.3 cm^{-1} of the geopolymer and 1025.6 cm^{-1} of the solidified body in the infrared spectrogram correspond to the stretching vibration of Si-O-T (T = Si or Al). The introduction of Pb^{2+} has an effect on the stretching vibrations of Si-O-T around 1087.3 cm^{-1} in the geopolymer, indicating that it is bonded into the backbone of the geopolymer. The absorption peaks near 564.8 cm^{-1} of the geopolymer and 553.2 cm^{-1} of the solidified body correspond to the stretching vibration of Al-O-Si. The bending vibration of Si-O or Al-O of the geopolymer and its solidified body is located near 461.5 cm^{-1} and 455.9 cm^{-1}, respectively. The peak at 1374.5 cm^{-1} occurs only in the solidified body, which is analyzed as being due to the nitrate contained in the heavy metal salts when Pb^{2+} is introduced. Overall, the infrared spectrum of the two samples is generally consistent, with no significant differences in the positions of the individual absorption peaks, indicating that the main structure of the solidified body is a complex three-dimensional mesh structure formed by the polymerization of silica-oxygen and aluminum-oxygen tetrahedra, although there are some differences in its physicochemical properties after the introduction of Pb^{2+}.

4. Conclusions

In this study, the fly ash-based geopolymer was prepared to solidify Pb^{2+}; the mechanical properties and leaching concentration of fly ash-based geopolymer were studied; the effects of different concentrations of Pb^{2+} on the compressive strength and solidifying effect of the solidified body were analyzed, and were compared with the cement solidified body; the reaction process, structural morphology, and solidifying mechanism were studied by XRD, FT-IR and SEM. The main conclusions are as follows:

(1) The compressive strength of the solidified body decreases compared to that of the pure geopolymer; the higher the concentration of Pb^{2+} added, the greater the decrease in strength. The solidifying performance of the fly ash-based geopolymer solidified body in a neutral environment is better than that in an acidic environment. Taking into account the compressive strength and leaching concentration, the optimum solidifying concentration of Pb^{2+} in the fly ash-based geopolymer is 2.0%.

(2) The fly ash-based geopolymer solidified body performs better than the cement solidified body, in terms of both mechanical properties (the compressive strength) and solidifying properties (leaching concentration). At Pb^{2+} concentration of 2%, the compressive strength of the geopolymer solidified body can still reach around 40 MPa and its leaching concentration is below the limit, whereas the compressive strength of the cement solidified body is only 21.25 MPa, and the leaching concentration is already as high as $8.29 \ mg \cdot L^{-1}$.

(3) The analysis of the mineral composition, microscopic morphology, and chemical structure of the fly ash-based geopolymer and its solidified body by XRD, SEM and FTIR reveals that the fly ash-based geopolymer forms a complete whole inside with a very dense structure, and possesses good physicochemical properties. The solidifying of Pb^{2+} by the geopolymer is a combination of both chemical bonding and physical encapsulation.

Author Contributions: Data curation and investigation, R.T.; writing—original draft preparation, F.L.; writing—review and editing, X.Y.; supervision, B.W. All authors have read and agreed to the published version of the manuscript.

Funding: This work is supported by the National Natural Science Foundation of China [51878116 and 51902270]; Liaoning Province Key Project of Research and Development Plan [2020JH2/10100016]; Dalian Science and Technology Innovation Fund Project [2020JJ26SN060]; the Youth Innovation Team of Shaanxi Universities.

Institutional Review Board Statement: Not applicable.

Informed Consent Statement: Not applicable.

Data Availability Statement: The data are not to be shared.

Conflicts of Interest: The authors declare no conflict of interest.

References

1. Zhang, Y. The Status and Prediction for Solid Waste Treatment Technology in China. *Environ. Sanit. Eng.* **2000**, *2000*, 81–84. (In Chinese)
2. Bie, R.; Chen, P.; Song, X.; Ji, X. Characteristics of municipal solid waste incineration fly ash with cement solidification treatment. *J. Energy Inst.* **2016**, *89*, 704–712. [CrossRef]
3. Liu, X. Discussion on current treatment and disposal technology situation of MSW fly ash in our country. *Eng. Constr.* **2014**, *46*, 56–60. (In Chinese)
4. Zheng, J.; Qin, W. Research progress of Geopolymer Materials. *New Build. Mater.* **2002**, *2002*, 11–12. (In Chinese)
5. Lancellotti, I.; Kamseu, E.; Michelazzi, M.; Barbieri, L.; Corradi, A.; Leonelli, C. Chemical stability of geopolymers containing municipal solid waste incinerator fly ash. *Waste Manag.* **2010**, *30*, 673–679. [CrossRef] [PubMed]

6.   Lizcano, M.; Kim, H.S.; Basu, S.; Radovic, M. Mechanical properties of sodium and potassium activated metakaolin-based geopolymers. *J. Mater. Sci.* **2012**, *47*, 2607–2616. [CrossRef]

7.   Ma, S.; Chen, P.; Liu, R. Study on seawater erosion to geopolymeric materials. *New Build. Mater.* **2009**, *36*, 58–61. (In Chinese)

8.   Mo, X.; Wang, R.; Pang, J.; Su, F.; Cui, X. Study on preparation and flexural strength of geopolymer based composites materials. *J. Guangxi Univ. (Nat. Sci. Ed.)* **2009**, *34*, 183–186. (In Chinese)

9.   Guo, X.; Shi, H.; Dick, W.A. Compressive strength and microstructural characteristics of class C fly ash geopolymer. *Cem. Concr. Compos.* **2010**, *32*, 142–147. [CrossRef]

10.   Luna Galiano, Y.; Fernández Pereira, C.; Vale, J. Stabilization/solidification of a municipal solid waste incineration residue using fly ash-based geopolymers. *J. Hazard. Mater.* **2011**, *185*, 373–381. [CrossRef] [PubMed]

11.   Andini, S.; Cioffi, R.; Colangelo, F.; Grieco, T.; Montagnaro, F.; Santoro, L. Coal fly ash as raw material for the manufacture of geopolymer-based products. *Waste Manag.* **2008**, *28*, 416–423. [CrossRef] [PubMed]

12.   Duxson, P.; Mallicoat, S.; Lukey, G.; Kriven, W.; van Deventer, J. The effect of alkali and Si/Al ratio on the development of mechanical properties of metakaolin-based geopolymers. *Colloids Surf. A Physicochem. Eng. Asp.* **2007**, *292*, 8–20. [CrossRef]

13.   Bankowski, P.; Zou, L.; Hodges, R. Using inorganic polymer to reduce leach rates of metals from brown coal fly ash. *Miner. Eng.* **2004**, *17*, 159–166. [CrossRef]

14.   Zhang, J.; Provis, J.L.; Feng, D.; van Deventer, J.S. Geopolymers for immobilization of Cr^{6+}, Cd^{2+}, and Pb^{2+}. *J. Hazard. Mater.* **2008**, *157*, 587–598. [CrossRef] [PubMed]

15.   Provis, J.L.; Rose, V.; A Bernal, S.; van Deventer, J.S.J. High-Resolution Nanoprobe X-ray Fluorescence Characterization of Heterogeneous Calcium and Heavy Metal Distributions in Alkali-Activated Fly Ash. *Langmuir* **2009**, *25*, 11897–11904. [CrossRef] [PubMed]

16.   Minarikova, M.; Skvara, F. Fixation of heavy metals in geopolymeric materials based on brown coal fly ash. *Ceram. Silik.* **2006**, *50*, 200–207.

17.   Singh, R.; Budarayavalasa, S. Solidifacation and stabilization of hazardous wastes using geopolymers as sustainable binders. *J. Mater. Cycles Waste* **2021**, *23*, 1699–1725. [CrossRef]

18.   El-Eswed, B.; Yousef, R.; Alshaaer, M.; Hamadneh, I.; Al-Gharabli, S.; Khalili, F. Stabilization/solidification of heavy metals in kaolin/zeolite based geopolymers. *Int. J. Miner. Process.* **2015**, *137*, 34–42. [CrossRef]

19.   van Jaarsveld, J.; van Deventer, J. The effect of metal contaminants on the formation and properties of waste-based geopolymers. *Cem. Concr. Res.* **1999**, *29*, 1189–1200. [CrossRef]

20.   Palomo, A.; Palacios, M. Alkali-activated cementitious materials: Alternative matrices for the immobilisation of hazardous wastes: Part II. Stabilisation of chromium and lead. *Cem. Concr. Res.* **2003**, *33*, 289–295. [CrossRef]

sustainability

MDPI

Article

Exploring the Critical Barriers to the Implementation of Renewable Technologies in Existing University Buildings

Joaquín Fuentes-del-Burgo [1,*], Elena Navarro-Astor [2], Nuno M. M. Ramos [3,*] and João Poças Martins [4,*]

1   Escuela Politécnica de Cuenca, Universidad de Castilla-La Mancha, 16071 Cuenca, Spain
2   Escuela Técnica Superior de Ingeniería de Edificación, Universitat Politècnica de València, 46022 Valencia, Spain; enavarro@omp.upv.es
3   CONSTRUCT-LFC, Faculty of Engineering (FEUP), University of Porto, 4200-465 Porto, Portugal
4   CONSTRUCT-GEQUALTEC, Faculty of Engineering (FEUP), University of Porto, 4200-465 Porto, Portugal
*   Correspondence: Joaquin.fuentes@uclm.es (J.F.-d.-B.); nmmr@fe.up.pt (N.M.M.R.); jppm@fe.up.pt (J.P.M.)

Abstract: For more than a decade, the European Union has been implementing an ambitious energy policy focused on reducing CO_2 emissions, increasing the use of renewable energy and improving energy efficiency. This paper investigates the factors that hinder the application of renewable energy technologies (RETs) in existing university buildings in Spain and Portugal. Following a qualitative methodology, 33 technicians working in the infrastructure management offices of 24 universities have been interviewed. The factors identified have been classified into economic-financial, administrative and legislative barriers, architectural, urban planning, technological, networking, social acceptance, institutional and others. It is concluded that there have not been sufficient economic incentives to carry out RETs projects in this type of building. Conditioning factors can act individually or jointly, generating a greater effect. Most participants consider that there are no social acceptance barriers. Knowledge of these determinants can facilitate actions that help implement this technology on university campuses in both countries.

Keywords: barriers; energy efficiency; renewable energy technologies; university building; energy performance of buildings directive (EPDB)

Citation: Fuentes-del-Burgo, J.; Navarro-Astor, E.; Ramos, N.M.M.; Martins, J.P. Exploring the Critical Barriers to the Implementation of Renewable Technologies in Existing University Buildings. *Sustainability* **2021**, *13*, 12662. https://doi.org/10.3390/su132212662

Academic Editors: Carlos Morón Fernández and Daniel Ferrández Vega

Received: 20 October 2021
Accepted: 12 November 2021
Published: 16 November 2021

Publisher's Note: MDPI stays neutral with regard to jurisdictional claims in published maps and institutional affiliations.

1. Introduction

About 40% of the world's final energy is consumed in the building sector [1–3], while in the European Union (EU) it accounts for 40% of final energy and 36% of total CO_2 emissions from its Member States [4,5]. In view of these data, policies have been designed to promote energy saving measures that improve the energy efficiency of existing buildings [5,6].

The EU has made great efforts to meet the objectives set out in the EPDB Directive 2010/31/EU [4,7], turning energy and climate policy into cornerstones of EU policy [8].

Although it is essential to achieve the objectives set out in the Energy Performance of Buildings Directive (EPBD) Directive [4], the size of a country's real estate is much larger than that of new buildings [2,9,10]. In fact, existing buildings cover around 75% to 85% of the building stock today and in the next fifty years [11]. In addition, a many of the existing buildings consume more energy than the new ones as they do not comply with the energy requirements established in the new regulations based on the EPBD directives [12].

Publicly owned buildings represent a significant percentage of the real estate surface, around 12% in the case of the EU [4,13]. According to Bertone et al. [14], most public buildings were designed and built without taking energy consumption into account, therefore generating a significant impact on social, environmental and economic sustainability [2].

With this background, improving the energy efficiency of existing public administration is an important step towards reducing their environmental impact [2]. In Europe, there is a growing concern and awareness of using sustainable construction strategies, measures and solutions in buildings, both new and renovated [15].

Europe's public buildings include a large number of university buildings for classrooms, laboratories, offices, sports halls, university residences, etc., with high energy consumption for diverse purposes such as lighting, heating, ventilation, air conditioning and for the operation of the equipment [16–18]. Moreover, university campuses are usually large facilities, with a relevant impact on local energy consumption in the cities in which they are located. On the other hand, they are occupied and visited by large numbers of people, so any sustainability-related measures adopted in these buildings can increase the population's awareness of these issues and affect their own decisions towards implementing energy-efficient solutions elsewhere. Thus, adopting renewable technologies in university buildings can produce direct and indirect effects on the sustainability of cities, as promoted by UNSDG 11, and on the sustainability of other facilities, as promoted by UNSDG 9 [19].

Reducing energy consumption by installing RETs could be considered an energy-saving measure [20–23]. The use of these technologies in existing buildings could offset their energy balance, reducing the emission of CO_2 and other polluting gases, thus helping to achieve the European directives' objectives and reducing their environmental impact [9,24–26].

Within the framework of the Europe 2020 Strategy [27], whose objectives have been expanded with the 2030 Climate and Energy Framework [28], focusing on RETs in existing university buildings [29], the objective of this research is to identify the different types of barriers the objective of this research is to identify the different types of barriers to the implementation of RETs in existing university buildings. For this reason, semi-structured interviews have been carried out with technical professionals who work in the maintenance offices of university infrastructures in Spain and Portugal.

The structure of the article continues with the literature review, the description of the qualitative methodology used, and the analysis and discussion of the results, and it ends with the conclusions.

2. Literature Review

Despite the advantages of using RETs in new and existing buildings, a series of barriers have arisen that have slowed down their adoption, arousing the interest of scientists and politicians [22,30]. Some barriers have already been identified by the European Commission itself—pointing out the poor application of current legislation, administrative procedures, accessing existing networks or the implementation of adequate measures by the Member States to guarantee their growth, among others [31]. These factors may be specific to one type of technology, while others may be specific to a region or country [32,33].

The literature classifies the factors that influence RETs development and installation in various ways. Some classifications try to cover all types of RETs, while others focus on some type of energy and/or technology (solar, wind, photovoltaic, thermal or geothermal) or on a specific country or region [33–39].

The number of factors and their grouping varies from one author to another. Beck and Martinot [34] include economic and regulatory factors, institutional disadvantages, lack of skills and information, and technological biases. The exhaustive list by Klessmann et al. [36] encompasses economic and market factors, legal and administrative factors, related to networks, lack of experienced labor, and information and acceptance barriers. Sidiras and Koukios [35] find that the main barriers for solar thermal installations are cost, installation difficulty, lack of energy saving incentives, short product lifespan and poor quality of materials and aesthetics.

Regardless of the number of factors that condition RETs use and their classification, their multidimensional nature is outlined [37]. If these technologies are to be applied to the building sector, new conditions must appear with different weights and treatments depending on whether they are applied to new or existing building, as well as their use, due to the different energy consumption profiles [40].

A number of economic and financial factors have been pointed out. Among the economic barriers, the high cost of RETs [41–43] stands out, particularly in building-mounted systems [6,44]. Regarding financial barriers, there is the need for a high level of initial funding, fewer financial resources, the lack of adequate financing, high financing costs, the absence of adequate structures to facilitate investment, increased credit risk perception, and long investment recovery periods (long payback periods) [7,10,21,25,33,34,45].

Likewise, existing barriers relating to network connection have been documented, mainly electric power, where it is difficult to obtain access due to factors such as not being open to renewable energies, lack of transparency in procedures, and long time periods to obtain the permit or lack of network capacity [36]. Related to the above, a European Commission [46] report indicates that renewable energy producers in different Member States have problems with regulatory procedures.

Social acceptance is a factor that also influences the use of RETs. It includes multiple personal, psychological and/or contextual elements. Furthermore, perceived utility, intention to use, installation conditions, cost, trust, place and people's position in relation to the renewable energy used play an important role [38]. Moreover, not all RETs are treated equally, as there is more local opposition to the development of one type of project compared to others [39]. Social acceptance barriers have also been explained by attitudes such as "not in my back yard—NIMBY", which can be determined by feelings of fairness, justice, selfishness or ignorance [47].

Considering the set of barriers to RET implementation, their identification and classification is essential. This will allow the design of appropriate approaches and policies [24], which in turn will facilitate the installation of this type of equipment in existing university buildings.

3. Methodology

Qualitative research is a rigorous and internationally recognized method of social research that has been recently used in research related to building energy [48–52], proving its usefulness in studies on drivers and barriers to RET adoption [30].

This research is an exploratory study through semi-structured interviews, to allow the participants the freedom to give their opinions and experiences [53]. The data collected provide specific information on relevant aspects of the RETs, the adequacy of policies and the barriers that appear in this environment, instead of quantity, proportion or magnitude of factors [49,54]. The research does not intend to generalize the results among populations but rather to collect data that clarifies the study phenomenon based on the opinions and experiences of the technicians who work in the infrastructure management offices of the universities of Spain and Portugal [33,50].

These professionals have been chosen because they are the ones who intervene in the entire life cycle of any project that affects university's real estate. In addition to being responsible for real estate maintenance, they are familiar with their individual problems, the current conditions of the buildings, and factors related to their use on a daily basis. No requirements have been established regarding their academic qualification in order to participate in this research, since each university freely decides when recruiting them.

Participants' citations are identified with a number together with the letters E (Spain) or P (Portugal), to identify the country of origin [21].

The research has focused on Spain and Portugal due to the fact that they constitute a very well defined geographical and climatic unit in the Iberian Peninsula [55] and to the common characteristics that both countries share in the EU [56].

The script of questions for the interviews was developed based on a literature review [39,57]. As the number of interviews progressed, new factors appeared in participants' responses, causing questions to be added or removed from the initial questionnaire [58]. Considering that Directive 2012/47/EU was in force during the period of time in which the interviews were carried out, the questions asked were the following:

1. What is your opinion on Directive 2012/27/EU of the European Parliament where it establishes the reduction of CO_2 emissions by 20% and the increase in the use of renewable energies by 20%?

2. I am going to ask your opinion about the existence of a series of barriers that can affect the integration of renewable energy technologies in university existing buildings:
   - ○ What are the financial barriers?
   - ○ What are the technological barriers?
   - ○ What are the barriers in connecting to existing networks?
   - ○ What are the regulatory or administrative barriers?
   - ○ What are the social acceptance barriers?
   - ○ What are the architectural barriers?
   - ○ What are the urban planning barriers?

3. If you think there are other barriers, please describe them.

Two different strategies were followed for the interview format. Face-to-face interviews were carried out if the distance by road was less than 300 km, or if there was a combination of high-speed trains between Cuenca (main author's residence) and the destination city. The second strategy consisted of telephone interviews when the above conditions were not met. In the case of the Portuguese universities, there were difficulties in arranging telephone interviews with technicians, so it was decided to do all the interviews face-to-face. Table 1 shows the type of interview and the universities that took part in the research.

Table 1. Participating universities and type of interview conducted.

Country	Face-to-Face Presential	Telephone Interview
Spain	University of Castilla-La Mancha Polytechnic University of Madrid University of Valencia University of Zaragoza University Juame I University Carlos III of Madrid University of Córdoba Polytechnic University of Valencia National Distance Learning University (Madrid)	University of Oviedo University of Málaga University of Santiago de Compostela University of Vigo University of Cantabria University of Valladolid Polytechnic University of Catalonia CEU Cardenal Herrera Oria University University of Mondragón University of Las Palmas de Gran Canaria, University of La Coruña University of La Laguna (Tenerife)
Portugal	University of Porto University of Lisbon Lusófona University	

The process of finding the technicians from the different universities, establishing contact with them and interviewing them was complicated and time-consuming [21]. Hence, the data collection phase lasted for a period of 27 months [30], taking place from May 2017 to August 2019.

In total, 26 universities were contacted in Spain and six in Portugal, of which 21 in Spain and three in Portugal agreed to participate.

As mentioned, different methods were used during the data acquisition process. Sixteen individual face-to-face interviews, two group personal interviews, and twelve telephone interviews were conducted [48,52]. The reason for this mix of methods was the location of the interviewees, the availability of the participants, or the country [59].

In total, 33 technicians were interviewed: 30 from Spain and 3 from Portugal. Given the exploratory approach of the research, Guest et al. [60] (p. 79) state that "for most research projects, . . . , in order to understand common perceptions and experiences among a group of relatively homogeneous individuals, 12 interviews should suffice".

Initially, participants were asked to provide information on their age, gender, academic qualifications and years of experience in the position (see Tables 2 and 3). Twenty-eight men and five women participated. Their qualifications cover a wide range of disciplines [25]:

six are architects, seven industrial engineers, one is a mining engineer, one is a computer engineer, one is an industrial organization engineer, five are building engineers, one is a maritime engineering graduate, one is a statistics graduate, one is a political science graduate, five are industrial technical engineers, three have a baccalaureate and one has a vocational training certificate.

Table 2. Participants' age.

Age (Years)	Number of Participants
From 30 to 34	1
From 35 to 39	6
From 40 to 44	5
From 45 to 49	6
From 50 to 54	4
From 55 to 59	4
From 60 to 63	7

Table 3. Years of experience in the job.

Years of Experience in the Position	Number of Participants
Less than 5	6
From 6 to 10	7
From 11 to 15	3
From 16 to 20	6
From 21 to 25	1
From 26 to 30	7
More than 30	3

The same list of questions was used [61] in both telephone and in-person interviews. The interviews, including those conducted in Portuguese, were recorded and transcribed verbatim and fully into Spanish [61]. The average duration of the in-person interview was 32 min, the telephone interviews were approximately 26 min and the group interviews took around 39 min.

Creswell's proposal [62] was used for the analysis. It consists of six main steps: data organization and preparation for analysis; data reading repeatedly to get its general sense; coding; description of the environment; contextualization and location of links between topics; and, finally, data interpretation.

The ATLAS.ti v.8.0 program was used throughout the data coding and interpretation process.

Open coding was used, which allowed for the categorization of concepts, and the establishment of common themes and patterns through careful examination of the data [50,63]. In this phase, the use of constant comparison of the data contained in the text segments [64] was essential, which made it possible to refine the concepts, identify their properties and explore their interrelationships [65]. Common patterns appeared that formed different categories and relationships between them, and that allowed for grouping of the text segments into specific analysis constructs [65,66].

Although perceptions of the effect and importance of barriers vary among participants [24], differences between responses are also part of the results [57]. The analysis reflects what respondents consider important at a specific point in time [57].

4. Results and Discussion

4.1. Opinion on Directive 2012/27/EU of the European Parliament and of the Council of 25 October 2012

All participants are aware of the objectives of reducing total greenhouse gas emissions by 20% and achieving 20% energy consumption renewable in 2020. They are conscious of the need to reduce energy consumption and polluting emissions. Even so, they express varied opinions, considering these objectives as "very or too ambitious" (No. 1E and 8P), "necessary" (No. 18E), "very necessary" (No. 21E), "very important" (No. 28P), "praise-worthy" (No. 5E), "A utopia" (No. 3E), "difficult to achieve" (No. 5E) or "insufficient" (No. 9E). Some also emphasize the importance of increasing the implementation of RETs in university buildings. More than half believe that it will be very difficult to achieve the targets proposed in Directive 2012/27/EU [67].

Several respondents believe that Public Administrations should set an example in the implementation of these measures in its buildings, especially the University: "It is mandatory from an ethical point of view and beneficial for society. It is the responsibility of the Administrations to contribute to this change" (No. 23E). This opinion is reflected in the Energy Efficiency Plan 2011 [13], according to which the public sector should promote an exemplary role, accelerating the pace of public building renovation, leading to high levels of energy performance.

Few participants affirm that this type of policy should have begun to be applied earlier, pointing out that the targets imposed should have been higher. They recognize that these objectives seemed impossible to achieve at first, but with recent technological developments, they are now achievable. Some were unaware that the first Energy Performance of Buildings Directive (EPBD) was approved in 2002 [68], since all referred to the subsequent reworking published in 2010 [4], which expanded the objectives of the original directive [5,69].

Some interviewees believe that deadlines for reducing emissions and implementing the technologies have been short and, as there are no intermediate milestones to verify the degree of scope, policy application has not been easy. In this regard, they are unaware that the Energy Efficiency Plan 2011 provides that the European Commission would present a legal instrument requiring public authorities to renovate at least 3% of their buildings (by surface area) every year [13].

Regarding the common percentage marked throughout the EU, the differences between the countries is pointed out and how these circumstances can affect the achievement of the targets: "that is one of the problems, which is about unifying and we cannot have the same peculiarities that Norway or Germany have" (No. 4E). They identify differentiating factors such as culture, climate, existing infrastructures and/or the country's economic situation.

In their opinions, the participants acknowledge that politics and economics have affected the Directive implementation. At the political level, several Spanish technicians consider that there has been no will to achieve the objectives due to the lack of coordination between the different public administrations or the approval of laws that did not facilitate or delayed the implementation of some technologies, such as the Personal Consumption Royal Decree 900/2015 [70]: "The tax on the sun. It is that the position of these (electricity) companies and the government is detrimental to this being possible" (No. 16E).

Another factor associated with politics is the difference between the terms set by national governments, which work under the perspective of a 4-year legislature, versus the 10 or 20-year terms established in the directives. This gap means that the priorities and the degree of achievement of the results are not aligned.

It is also identified that, when establishing a common energy policy in the directives, some countries have not been aware of the economic investment necessary for its development: "My opinion, as a technical manager with 25-year experience in a public university, is that the Public Administration will find it difficult to attain those values, if not more involved: It's complicated" (No. 12E).

The participants report that in a proposal as important as this one, with such a large existing building stock, there has not been an economic policy that helps to achieve the objectives. As noted by Curtius [30], insufficient government incentives are a barrier to development in adopting integrated photovoltaic energy in buildings.

In addition to the lack of financial support, they consider that the way to achieve the targets or what to do has not been adequately specified: "One thing is to put on paper everything you want to do, and another thing is to tell us how to do it in the day to day" (No. 15E). Although the EPBD requires minimum standards in the case of major renovations of existing buildings, according to Rosenow and Kern [69], meeting the energy performance requirements of building codes in rehabilitation actions can be problematic.

Finally, all the participants point out that the recession has negatively impacted the development of both Spain and Portugal: "we came out of a major crisis with a population who had made a lot of sacrifices. Achieving everything at once is very difficult" (No. 8P). Furthermore, different member states, in particular those most affected by the 2008 international recession, have reduced or eliminated support schemes for renewable energies [71].

4.2. Barriers to the Implementation of TERs in Existing University Buildings

Participants' answers to the specific questions about factors that may act as barriers to RETs installation are detailed below. Being related to each other, some barriers do not act in isolation. The combined impact of several barriers can have a significant negative effect on the implementation of these technologies [72]. In any case, the barriers place RETs at a relative disadvantage compared to other forms of energy supply [34].

4.2.1. Economic-Financial Barriers

All those interviewed point to the economy as the main barrier that conditions RETs implementation in university buildings. This barrier has, in turn, different components shown in the responses: high investment, the cost of the projects, the country's economic situation, the economic crisis, the investment crisis, the financial assessment criteria, high repayment terms, insufficient budgets or simply the lack of financial resources. As Figure 1 shows, these could be grouped into four interrelated categories: the national economy, university budgets, the cost of facilities and financial factors.

There is a direct relationship between the different economic components mentioned. The cost of the projects is related to the investment that the universities have to make, as well as their financial capacity and/or the availability of funding to tackle them. As Nasirov et al. [33] point out, economic and financial barriers act as a key obstacle for the development of RET's projects.

Although some participants acknowledge that the cost of certain technologies, such as photovoltaic modules, has gone down, the investment required and the initial financial cost continue to be very high [25,73].

Hiring specialized personnel to maintain said facilities is also identified as a cost, which increases the repayment period. As stated in Yaqoot et al. [74], the hiring of personnel for maintenance services is considered as one more barrier.

On a financial level, these projects are usually assessed based on their repayment term or based on economic benefit criteria. As a condition of viability, it is usually imposed that the repayment be as short as possible, as participant No. 27P summarizes: "Here a great effort is made by the technicians who prepare the project to demonstrate that it can have an average payback of 7, at most 15 years for a great project". The combination of long repayment terms and low profits derived from the energy generated, in the face of high investment costs, means that university management does not consider these projects economically attractive [6,45].

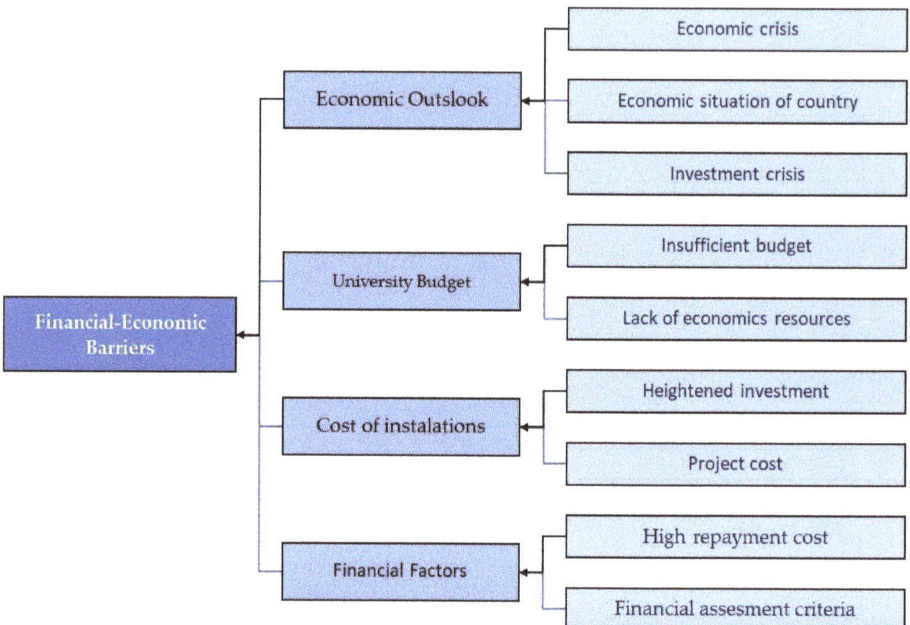

Figure 1. Factors considered as economic-financial barriers.

The payback period is directly related to the amount of energy generated by the RET, clearly a technological factor. In this sense, several variables are cited that directly condition the repayment period, such as the low performance of the facilities, the availability of the energy resource (wind, solar radiation, outdoor temperatures, etc.), the geographical location, the characteristics of the urban environment where the building is located, the amount of energy used in the building and its price in the energy marketplace.

To implement the RETs, some universities have had to request a "lease"; participate in calls for regional, state or European aid for this type of project; or seek aid through contracts with energy companies. The general opinion is that there are not enough external aids or subsidies for these actions.

More than half of the sample of Spanish public universities confirm that they do not have a budget for these projects: "the regional government considers that in the budget it gives us for the operation of the University everything is already included" (No. 2E). They relate the political decisions adopted by the autonomous governments to the economic restrictions to implement RETs.

In both countries, reduced budgets and lack of funding for universities are associated with the economic situation of the individual states and the economic crisis suffered. Furthermore, the financial crisis has also had a negative effect on the availability of funds provided by the banking sector [39].

4.2.2. Administrative and Legislative Barriers

Many interviewees who have been involved in the entire process of administrative concession, legalization and/or execution of an RET project in their universities, identify the associated bureaucracy as a barrier, although some do not consider it insurmountable. As stated in ürge-Vorsatz et al. [75], regulatory barriers make it difficult to implement RETs in buildings.

This bureaucracy depends on the country and its energy policy. The participants from Portugal do not currently acknowledge administrative or legislative barriers to RET implementation in their country. In Spain, one participant describes RET projects carried out in the first decade of the 21st century, taking advantage of the facilities and aid established in the Renewable Energy Promotion Plan and the legislation approved in those years [76].

Based on their experience, more than half of the sample acknowledge that the procedures are numerous. For some it is a serious barrier: "they are an insurmountable problem" (No. 19E), and others recognize the difficulty of the process describing it as "laborious" (No. 20E), "complex" (No. 15E), "complicated" (No. 12E) or "heavy-going" (No. 23E). Several consider that the procedures can be solved by increasing the human and technical resources of the university destined to these projects, assuming that there will be a delay in project execution.

In Spain, due to the decentralization of the administrative system, the participants recognize the problems of RET projects due to the numerous legislations that they have to consider at the national, regional and local levels [77], as well as the regulations of electric suppliers, in the case of photovoltaic installations. This was pointed out by Coenraads et al. [78] when the number of authorities involved in obtaining construction permits for these projects was high. As Byrnes et al. [43] discussed, legislative processes must be rational, both within and between the different administrations.

In addition, the lack of communication between different Public Administrations (state, regional and local) has been identified throughout the administration of tendering procedures, legalizations, etc., aggravated by the lack of interlocutors to facilitate said communication. In this sense, the impossibility of clearly identifying the professional roles involved and the lengthy procedures are regulatory deficiencies [14]. This situation restricts the development of the RETs, poses a risk to the project during the development phase [33] and may cause a significant delay in the commissioning of the facilities [32].

Another negative factor is the complexity of the legislation, which makes it difficult to interpret, as well as the rigidity associated with its compliance. In the responses of the Spanish participants, Law 9/2017 on Public Sector Contracts [79] is repeatedly cited as an added problem that involves more workload for them, as stated by technician No. 23E: "the Contract Law itself makes any type of project very heavy-going, very complicated". As new laws appear, doubts about their interpretation, the complexity of their analysis [80] and the lack of experience in their application can become an administrative barrier that limits, slows down or prevents the development of these projects. Some of these barriers appear at Figure 2.

There is unanimity that the public bidding processes for the supply or work related to the implementation of RETs in university buildings are very complex and require a trial and error process to meet all the necessary requirements.

Based on the need to reduce CO_2 emissions and improve the existing building stock, some interviewees hope that the entire bureaucratic process associated with RETs implementation will be simplified in the future.

There seems to be an important difference between the countries investigated at the energy policy level. In Portugal, the participants acknowledge that the policy favors the implementation of photovoltaic solar installations. In the case of Spain, the problems highlighted by Coenraads et al. [78] persist, and the administrative authorization procedure depends on each region and is not harmonized, with regulatory differences between the 17 autonomous communities.

The greater or lesser difficulty of drafting, processing and implementing RETs projects is linked to the applicable national legislation. In Spain, during data collection there was a change in the law that governed the production and supply of electrical energy through self-supply systems. With the first Royal Decree 900/2015 [70], the participants affirmed that the bureaucratic and technical procedures were very complicated, making it difficult to implement photovoltaic solar installations: "Legalising a photovoltaic installation is a hassle" (No. 19E). However, the new law, Royal Decree 244/2019 [81], is more flexible and does not impose as many requirements as the previous one, which changed the perception of those interviewed: "the problems have greatly diminished" (No. 20E).

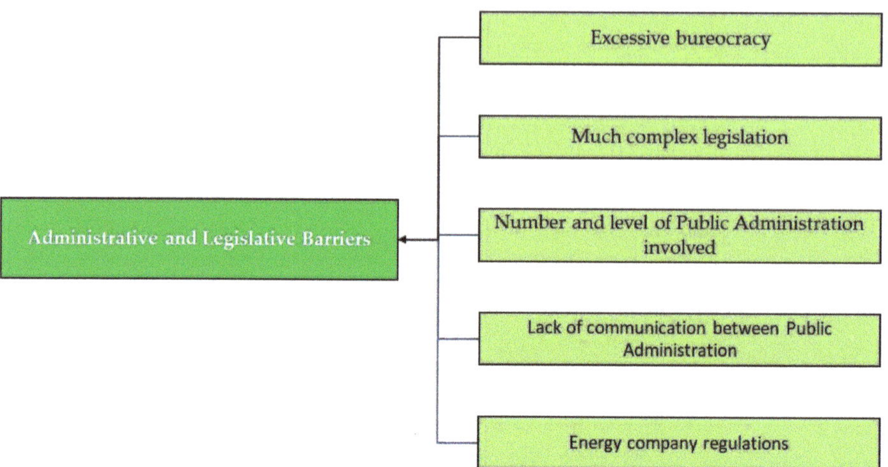

Figure 2. Factors considered administrative and legislative barriers.

The situation described is a clear example of how a law approved according to a specific energy policy can become a barrier for the development and implementation of certain RETs, such as photovoltaic installations in this case. Thus, legal instability negatively affects the development of RETs [77], as it generates a risk associated with the level of reliability and continuity of state policies [30]. Frequent policy changes aimed at promoting RETs discourage widespread adoption of technologies [82].

4.2.3. Architectural Barriers

This section includes factors that can prevent, hinder or involve an increase in the cost of execution in the implementation of RETs. Quite a few participants consider that architectural barriers are an extra cost in these projects. One even supports that "the architectural barrier hangs over the economic one, there is no architectural barrier if there is money to do the installation" (No. 18E). Thus, some architectural barriers could be eliminated with the necessary budget, transforming the architectural factor into an economic one.

More than half of the sample considers that buildings' age can be a problem for the implementation of RETs. It involves a technical complexity added to the project, which can be solved with an increase in the cost of execution, due to the works to be carried out and the auxiliary equipment to be used: "we have old buildings, not protected, but old or of poor construction quality, so we need to carry out more work than planned" (No. 31E). This is confirmed by Dadzie et al. [83], who found that investing heavily in renewable technologies when buildings are old is considered uneconomical as the cost of implementation exceeds the savings attained.

Some respondents highlight that one of the main difficulties to be solved is the integration of RETs without negative impact on the building, especially in buildings classified as heritage or those with a special design. Historic buildings impose limitations when it comes to maintaining their authenticity [84]. To solve this problem, they propose to move away from standard solutions and to carefully study the options offered by the different technologies to choose the one that best suits the building and its environment. As participant No. 25E affirms: "you have to have imagination." Each RET project in existing buildings is unique and requires a careful study of possible solutions on a case-by-case basis [22].

Factors have also been identified such as the difficulty of adapting certain technologies to buildings depending on the materials used in their construction, the future conservation of the building envelope involved, the size and adequacy of the affected technical premises, and the difficulty and/or security of access to the installation for future maintenance jobs.

They also indicate that the improper placement of solar installations on roofs has been a source of damage due to loss of waterproofing, generating humidity and leaks, which have led to conservation problems, with the corresponding economic cost. Polo and Frontini [85] point out that the origin of these problems is that installers and engineering companies try to implement the cheapest solution without taking into account the future conservation of the building.

Likewise, problems derived from the age of the existing facilities are mentioned. The implementation of RETs may involve the comprehensive remodeling of the facilities related to the new project. "Older buildings generally have embedded installations, they are not easy to register, which requires significant additional funds." (No. 20E)

Furthermore, age is directly related to the criteria used during the design and construction of buildings. Hence, adequate spaces were not considered for the laying of the facilities [25] in their conception and construction. This reality greatly conditions RETs implementation, being an added problem. As participant No. 33E highlights: "many times the buildings have an architecture or a distribution that complicates the implementation".

Another factor that limits the integration of technologies based on solar radiation appears in buildings with complex and irregular geometries, without an adequate orientation to capture the energy resource [86]. According to Dadzie et al. [83], existing buildings with complex designs make the selection and installation of these technologies difficult.

In any case, roofs are identified as the best location to install RETs (photovoltaic, thermal and/or mini-wind), due to the limitation of free space around the existing building [86] and the possibility of providing the proper orientation and inclination of photovoltaic and thermal modules to maximize energy production [84]. Despite this, they mention possible drawbacks: the roofing of the building, recommending acting only on flat roofs; the materials that make up the cover; or the existence of other facilities that occupy a lot of surface area, leaving little space available [87]. Furthermore, as the available surface area is reduced, the amount of energy generated is limited [25], affecting the repayment period of the projects.

Buildings' age could be related to administrative and legislative factors when they are classified as of heritage status. In this regard, both countries have organizations for the preservation of historical heritage, at all levels of the administration, which impose rigorous conditions to be met in projects for RETs implementation in listed buildings [88]. Some of these factors and barriers appear at Figure 3.

Directive 2010/31/EU [4] states that Member States may decide not to establish or apply the requirements set forth in its Section 1 in the case of officially protected buildings when the implementation may unacceptably alter their character or appearance. In this regulatory framework, AbdelAzim et al. [6] indicated that policies and standards related to the sustainability of buildings, in general, are more focused on new buildings, paying less attention to the existing buildings as a whole.

Given the experience with listed buildings, several researchers recognize the difficulty of taking action in compliance with the constraints, their high cost and the need to hire specialized companies. In some cases, they are discouraged because the end result has not been very good despite the effort invested. They consider that the cost becomes an economic barrier, hence any improvement in energy performance is perceived as a prohibitive investment.

Most interviewees admit that they are not considering the installation of RETs with this type of building, except those that can be installed on the roof, such as photovoltaic and solar thermal. Astiaso et al. [11] mention that not all energy efficiency interventions or thermal systems can take place in historic buildings. Technologies that can be installed out of sight of building users and visitors have to be selected [88].

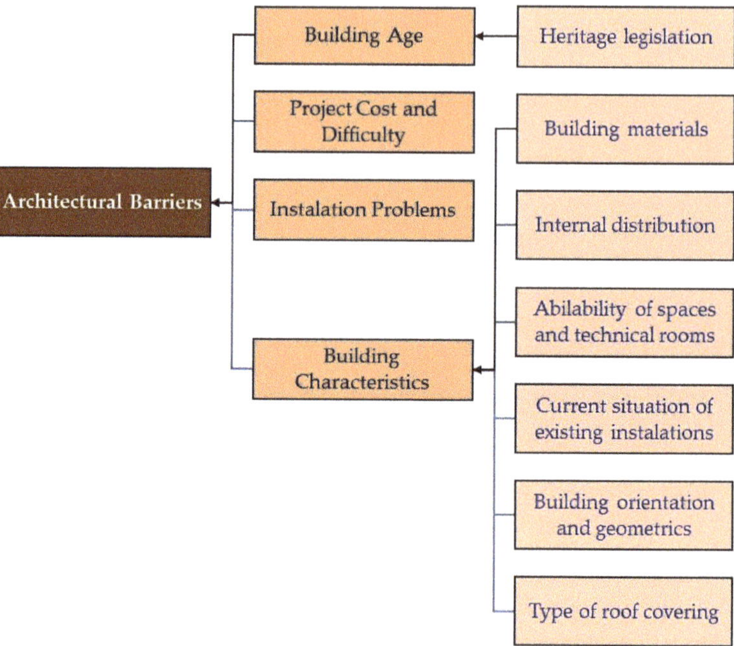

Figure 3. Factors considered as architectural barriers.

For some participants, historical heritage regulations are not a barrier as such. They prioritize the conservation of this type of building over any type of intervention that could negatively affect its historical value. Several authors recommend that historical buildings must preserve their antiquity, integrity and cultural significance as well as architectural aspects and be in harmony with the surrounding cultural heritage [86,88,89].

4.2.4. Urban Planning Barriers

Factors in this section are related to the structure and dispersion of university campuses, their situation in the city and its urban planning regulations. Due to legislative aspects, this last factor is related to administrative barriers, because as participant No. 30E points out: "the municipal ordinance is often a bit rigid, and that rigidity means we have administrative barriers." In Spain, the organizations responsible for urban planning are the autonomous communities and, basically, the local bodies [77,90]. At the urban level, factors such as aesthetics, noise or safety are factors considered by urban planning technicians of local bodies [34].

The shape and arrangement of the buildings within the university campus can either be a barrier or a facilitator for RETs implementation. Thus, in a campus with a distribution of nearby buildings of different heights, the shade generated can reduce the energy production of solar installations. This drawback does not occur on large campuses and with buildings of similar height. In addition, campuses located on the outskirts of urban centers have fewer problems regarding municipal regulations. In contrast, those located within the urban center or that have their buildings scattered across the city have their capacity to act limited to the roof.

Buildings not classified as heritage but located in a protected area such as historic neighborhoods may also be affected. Since they play an important role in cities, contributing economic and tourist value to their heritage [91], maintaining the authenticity of buildings in the processes in these neighborhoods is usually a mandatory requirement [84]. This situation increases the difficulty of the work and the extension of the administrative procedures to be done.

The general urban planning plans are identified as a barrier to carrying out some actions, as stated by technician No. 11E: "Within buildings prior to 1980, not being impossible, not being unfeasible, you can transform roofs, but if they are in urban spaces it is very difficult, you cannot transform a tile roof, because the general urban plan won't let you". Curtius [30], in his study on the implementation of photovoltaic installations, found that obtaining the necessary building permits was an important barrier because local administrations sought to preserve the original character of the urban environment.

Despite the above, several participants do not consider general urban plans an insurmountable problem for RETs implementation. In fact, they have recently noticed an evolution of these regulations to favor their use.

Some interviewees raised the need for the general urban planning plans and their subsidiary regulations to evolve at the local authority level to allow and/or facilitate both the legislative processing and the execution of RET projects. As a more ambitious proposal, some indicate that the plans should contemplate the option of integrating the district's energy generation into urban land: "sooner or later planning will conform to the fact that urban spaces can be made, or have to be left, for energy districts" (No. 25E).

Some respondents include here the negative effects that some buildings may have on others in order to obtain the energy resource, such as the generation of shade for solar radiation [84,92] or of obstacles to wind circulation: "The buildings in front of us prevent us from achieving the desired sunlight" (No. 24E). In this regard, providing existing and/or future buildings with access to solar radiation should be one more aspect to include in the consideration of ordinances and urban development plans.

4.2.5. Technological Barriers

All Portuguese participants and two-thirds of the Spaniards believe that, given the current level of RETs development, technology cannot be considered a barrier. Although renewable energies require much more advanced technology than conventional ones [93], some are convinced that the advances that are taking place will further facilitate the application of RETs in the future. Technological innovation has been a crucial variable for the removal of barriers and cost reduction of these technologies [94].

However, the technicians point out a series of factors that hinder the application of RETs in existing university buildings. One is the high energy demand they have, with their peak and valley consumption hours [95–98], and its relationship with the amount of energy produced; the availability of the energy resource; and the performance of renewable energy production systems, both electrical and thermal [25,42]. They also mention that the reliability of these technologies in supplying energy during peak periods depends on uncontrollable weather factors [32,99]. Finally, the potential risk of equipment failure to perform its task is added [73].

Thus, in the case of solar thermal installations, they report that setting up sanitary hot water production equipment in buildings with little consumption (cleaning the building and employee hygiene) generates serious problems that have forced them to be rendered inoperative. In addition, in facilities for sports centers, swimming pool water heating, etc., systems to cover solar collectors during closing months have had to be implemented or designed.

Another drawback, related to the multidimensional nature of the barriers, is the requirements imposed by the technical regulations that govern the projects of each type of installation [100]. The technical requirements contained in them may limit, condition or hinder the application in certain buildings.

The large amount of space or the high volume required for the implementation of some RETs are also pointed out. Thus, in the case of biomass heating installations, several interviewees refer to the large warehouse capacity required to have adequate autonomy [99]. They also outline its maintenance cost as a disadvantage along with the need to hire specialized operators.

Regarding low temperature geothermal installations, they report that those carried out with vertical boreholes are more efficient than those done with horizontal ones [11]. In addition, they identify the large number of necessary boreholes as an obstacle, as well as the space they occupy to avoid thermal influence between the different boreholes. They reject the use of shallow buried horizontal boreholes, due to the large surface area required for their implementation. Related to this result, Čenejac et al. [101] recommend this type of exchange only in medium-sized buildings. Regarding their application in existing buildings, several participants consider that geothermal facilities present great difficulties if there is no free space around them or, simply, "geothermal energy is not viable in existing buildings" (No. 28P).

The poor performance of some of these technologies is also pointed out as a barrier, normally focusing on photovoltaic installations. This forces an increase in the surface area to be installed when the space of the building envelope is insufficient and/or inadequate to generate all the necessary energy [87,102], although they do affirm that they can help reduce the electrical energy consumed from the network, reducing expenditure [103].

Concerning the above, they also report that the local weather and the geographical location of the buildings have a strong influence. As Sovacool [104] points out, the energy production of solar systems varies with the season; the time of day; and the presence of fog, clouds and rain. Despite the good geographical situation and the climate of the Iberian Peninsula compared to other European countries [105], all participants from the North of the Peninsula point out that factors such as the low value of solar irradiation and the reduced number of daily sunshine hours in autumn and winter prevent these facilities from being repaid over a reasonable period.

Likewise, the integration of photovoltaic modules in the building must be considered, although according to Jelle [106], these facilities are the ones that can best be incorporated into the architectural design. Thus, they can be integrated without difficulty anywhere in the envelope, with different cell options (rigid, flexible, opaque or transparent) [107], with different types of modules, over or on roofs and walls, and on opaque and semi-transparent envelope surfaces [30,102,103].

In the case of installation on the façade, interviewees affirm that energy production is influenced by their orientation and the limitations in the inclination of the modules. However, they point out that it can be a very good solution as a shading element for classroom windows, offices and laboratories [75,106]. Thus, with this application, it is possible to reduce energy consumption in the air conditioning of these spaces during the summer [108].

4.2.6. Network Connection Barriers

Direct access to the existing network infrastructure is essential for successfully implementing photovoltaic solar installation projects [72]. Concerning this barrier, participants from both countries show different perspectives. While the Portuguese technicians claim not to have had problems connecting to networks, complying with the conditions imposed in their legislation, the Spanish technicians do not show a unified response.

On one hand, around half affirm that the electricity companies themselves generate connection problems: "We are in the hands of the supply companies, which are the ones that are saying: "I'll give it to you or I won't give it to you" and, furthermore, in a relatively obscure way, because they don't give you an explanation of why not either" (No. 14E). On the other hand, a minority recognizes that either they have not had this problem or it has been resolved with the new legislation. Finally, very few Spanish participants claim to be unaware of the current legislation or not to have had previous experiences, so they do not comment on this factor.

Several affirm that there can be no technical problems when connecting photovoltaic generators to the existing electrical distribution network. Based on their experience, they conclude that companies create an administrative problem which makes connection difficult.

The numerous bureaucratic procedures imposed by electricity companies in the process of legalization and connection of photovoltaic generators are also identified. Other researchers have identified the lack of clarity in the connection procedure of different types of RETs [78,93].

Some participants have developed theories about the reasons why electricity companies make it difficult for small energy producers to implement projects. Due to this and after negative experiences with this type of project, an interviewee recommends using these facilities only for self-supply.

In addition, they describe some problems of a technological nature such as the voltage difference of the photovoltaic generator and the grid voltage; the types of connection and control devices that companies require to install, increasing the cost of the project; or the difference between the electrical diagrams projected compared to those included in the company's technical regulations.

Another determining factor is the location of the building where the photovoltaic generator is installed. On campuses found in urban areas or isolated university buildings in the city, the age of the existing network and its degree of saturation may be the limitation that forces electricity companies to impose such strict technological conditions: "distribution companies are very clear, when you introduce something or disturb the network, everything goes down the drain" (No. 10E). According to Lehmann et al. [109], the lack of capacity of the existing network can become a barrier for new projects.

On the contrary, when it comes to buildings on exempt university campuses, with their own distribution network and transformation stations, there are fewer technical problems when making the connection.

4.2.7. Social Acceptance Barriers

The majority of those in the sample do not perceive these obstacles. On the contrary, they point out that the student body, the teaching staff and the administration personnel of the universities are very aware, committed and sensitive to environmental issues. Indeed, due to this awareness, they ask for more actions of this type: "some demand that things be done and ask why they are not done" (No. 24E). Thus, what could initially have been a barrier becomes an incentive for RETs implementation, as summarized by participant No. 30E: "social non-acceptance would be doing nothing". When the RET project is considered beneficial and the community sees it in a positive light [110], it could be considered an example of PIMBY (Please in My Back Yard).

Some respondents think that this awareness is the result of training and dissemination that is being carried out in different areas of society: "the students are demanding it, they are asking for it. People are well educated, television has harped on us a lot" (No. 29E). Considering the age and education of the university community members, Bertsch et al.'s [111] words apply: education is a highly relevant socio-demographic variable in RETs' social acceptance.

However, others have encountered the rejection or mistrust of members of government teams, faculty deans or school directors regarding the usefulness of this type of project or, as participant No. 28P affirms, "there is a tendency to distrust new things". This misconception of RETs, possibly generated by a lack of understanding or knowledge of the subject, or fear of the unknown, can have a negative impact on their implementation [37,42].

Despite encountering initial reluctance in RETs implementation, they also report the change of opinion in their university management team members when they verify the return on the investments made and the benefits brought. Thus, as stakeholders become familiar with RETs and discover their advantages, their institutional development and social acceptance increases [43]. The phenomenon described by Hai [51] is reproduced (tendency to emulate when people from the same community share success stories).

Consequently, to prevent people with decision-making power from opposing these developments, some participants believe that more education on these issues should be carried out in the different governing bodies.

"The problem I see is that there is not good dissemination or awareness-raising. To me, education in saving and clean energy is essential, it's fundamental" (No. 12E).

Considering that universities are part of the Public Administration [112,113], several interviewees argue that they should be a good example to society. Furthermore, as an added benefit, some universities use RETs implementation projects to promote themselves, by improving both their institutional image and their social commitment.

Some technicians believe that this social acceptance barrier could arise from negative effects that some types of RETs may generate, such as the increase in taxes to implement them, the smell of burnt wood on campus in the case of using biomass boilers in district heating, or noise from mini wind generators installed on building roofs or on campus.

Several respondents report news read or transmitted by colleagues, of positive or negative experiences of RETs implementation in other public administration buildings. This knowledge influences their opinion on the advantages and disadvantages of certain technologies. Based on this information, they establish value judgments rather than seeking the opinion of experts, which can influence their decision-making [21,24]. The impact of failure cases is relevant, since they generate a negative opinion towards the corresponding technology, causing the technicians to discard its use.

4.2.8. Institutional Barriers

The concept of institutional barrier encompasses those factors existing in an organization that limit the acquisition of sustainable energy technologies [114].

More than half of the sample recognizes that RETs implementation in the university depends on the governing teams having the political will to present such projects to the government councils and adopt the necessary measures to carry them out. Therefore, it would be a decision-making process by multiple actors who may have different and even conflicting motivations, which can make it difficult to obtain funds or approve projects [114]. In these types of situations, although financial factors are important, the strategic nature of the action may have more of an influence on decision-making than the profitability of the projects itself [115].

Extreme opinions are found in the responses, from governing teams that have an active role, executing RETs implementation actions, to others that, despite recognizing their importance, do nothing. They claim that this inaction is due to economic factors, so this barrier is also multidimensional in nature. In any case, without the support of the university governing teams, RETs implementation programs cannot be carried out [17].

On the other hand, some universities propose to meet environmental targets by obtaining official certificates from regional governments or by trying to improve their international rankings such as the UI GreenMetric World University Ranking. In other cases, the policy consists of constructing new buildings with an almost zero energy building certificate (Net Zero Energy Building (NZEB)), such as the BREEAM, LEED or BEAM certification schemes [116,117].

Some interviewees acknowledge the difficulty of getting a rectoral team involved in a renewable energy project with a long-term return on investment since it would tie up university's financial resources for several terms. According to technician No. 13E: "They do not think long term, they are interested in the four years that they are going to be in government. They are moved more by political interests than by energy efficiency".

One strategy developed by universities in both countries consists of reinvesting the savings produced by RETs investments in projects of the same type, or in energy efficiency measures in buildings. This option constitutes a procedure for continuous improvement of the energy efficiency of buildings, avoiding dependance on energy policy changes of each government team.

However, in other universities, technicians regret not being able to implement it due to what they call an "accounting barrier" (No. 20E), preventing savings made to continue generating investments in this type of measure.

The existence of different service units related to the operation of buildings, independently of each other, is identified as a negative factor. For example, one unit is dedicated to building maintenance, another to works, another to sustainability, another to energy, etc. In these cases, the structure itself, with high segregation of functions, can act as a brake due to a lack of uniqueness of criteria focused on energy efficiency, consumption reduction and RETs integration. Furthermore, their capacity to act depends on the budget assigned to each unit, and if this is reduced, their work will be "exclusively a philosophical service" (No. 11E).

4.2.9. Other Barriers

- Lack of training and/or experience in RETs.

Some technicians admit not being experts in these technologies or lacking knowledge and experience in executing related projects. The lack of skills and information can increase uncertainty about RETs implementation and make it difficult to make decisions about their application [34]. This confirms Cooke et al. [21], who found that "ignorance and lack of understanding" act as barriers.

Although it is not something to be generalized, there seems to be a certain training deficit in the application of RETs in building construction. This situation may be caused by inadequate human resource management in those universities that have not provided adequate training in new technologies to technicians working in infrastructure management areas [41].

- Lack of information.

Several technicians indicate not having enough information to adequately evaluate all the technical, economic and financial aspects of RETs application in their buildings. The lack of adequate, specific, simple and timely information makes it difficult to make relevant decisions regarding these projects' implementation and execution [24,74,118].

Some participants say that they learn more through conversations with other colleagues and/or professors and in work meetings within the university environment. They value this source of information very much because, as participant No. 8P states: "I think that most of these people with whom I speak are from here, from the university, they do not have any commercial intention behind it. I give them more credit because I don't think there are any underlying interests behind the things they say. They are taken in good faith".

They also cite the reading of feasibility studies or proposals submitted by installation or commercial companies of these technologies. However, they manifest a certain suspicion towards its content because they lack the tools to ensure everything is correct. In addition, when they analyze the economic viability, the results do not coincide with their calculations: "when marketing companies present us with the repayment plan, they are not credible either, because there are times when we save more than we spend, which upsets me" (No. 29E).

The lack of information can condition the choice of a certain technology or negatively influence the decision-making process. As participant No. 4E acknowledges: "if we could have all the information, if we could manage all the parameters of what we want to implement, how we want to implement it, and that we are convinced that this is what we want to implement".

Added to the above is the lack of confidence when the information is provided by installation or engineering companies that do not show adequate experience or reliability: "I did not see solid, dependable technical support, which is what you want to see with initiatives of this kind, but rather, it seemed a bit like an adventure to me" (No. 5E). Thus, when in doubt, the application and viability of these technologies is dismissed.

Several participants indicate a certain degree of isolation between the different universities. They identify little or no information transfer between the university infrastructure management offices to share experiences on RETs implementation or energy efficiency measures. "We are constantly reinventing the wheel. I am here in my office, I start inventing the wheel and when I have invented the wheel I realise that someone else has already done it . . . knowledge transfer is lacking" (No. 13E).

In this regard, there is the Spanish Confederation of Rectors of State Universities (CRUE) and the commission called CRUE-Sustainability, which have different working groups. Only half of the Spanish sample claim to know their reports, so there could be a disclosure problem. This lack of information exchange between technicians from different universities can negatively affect the spread of innovations in RETs [119].

- High daily workload.

It is acknowledged that the daily workload is so high that they do not have time to think about the implementation of these facilities during the day. Some refer to the large amount of documentation to be carried out for any task, as stated by participant No. 3E: "We are overwhelmed and now everything is paperwork, paperwork, paperwork".

This excess of bureaucracy and the limitation of personnel in the infrastructure management offices make it difficult for the responsible technicians to consider the study of this type of projects. They also acknowledge that they cannot prepare all the documentation required to participate in official calls for aid for national or European renewable energy implementation projects.

- Recent actions on heating installations in buildings.

Having implemented recent renovations and/or upgrades of boiler rooms in buildings, associated with energy supply contracts with gas distribution companies for a certain period of time, appears as a limiting factor.

"When you decide on one, you have to exploit it for at least its useful life, those ten years that company X imposes on us" (No. 5E).

In this decision-making, it seems that those responsible for the university have only taken into account the economic factors and the technical advantages, prioritizing the products with the lowest initial cost [114], without considering the environmental effects arising from the continued use of conventional fossil fuel technologies [34]. This situation is favored by the direct and indirect financial support that promotes the supply of conventional energies, limiting the expansion of renewable energy sources [120].

Although RETs advantages are well recognized, some people are not inclined to use them since conventional fossil fuel technologies are better known, more available and more affordable [121].

5. Conclusions

The objectives set by the EPBD Directive, in its 2012 version, are known and generate a varied opinion among technicians who work in university infrastructure management offices, although the perception prevails that achieving them is very difficult.

Achieving the 20% reduction of CO_2 emissions and increasing the implementation of renewable energy technologies by another 20% in university's existing buildings involves the establishment of temporary milestones that have not existed previously and the provision of public funds that have not been received. The general perception is that, in the drafting of the EPDB directive, each country's economic situation and specific characteristics were not taken into account. In this case, the uniformity of criteria has set objectives that are difficult to attain for Spain and Portugal.

Although the objectives set out in the Europe 2020 Strategy are considered necessary, when the building park is of a certain age, increasing the percentage of RETs integrated into the buildings, without an economic policy that accompanies these objectives, can lead to the failure of their fulfillment. In addition, the time difference between the application of the directives (10 years) and the duration of the national governments' legislature may be a factor to consider for the application of the energy and economic policies necessary to comply with the directives. The people interviewed perceive that there have not been sufficient financial incentives to achieve the EPBD directive objectives. They highlight the persistence of the effects of the 2008 international economic downturn, which could have negatively affected the application of appropriate policies.

Based on the barriers identified in the literature review, participants' responses underline a series of factors that act as obstacles to the implementation of RETs in existing university buildings. These factors have been grouped into economic and financial, administrative and legislative, architectural, urban, technological, network-connection and institutional barriers of the university.

An important result found is that there are no social acceptance barriers. On the contrary, different groups demand greater implementation of this type of technology in university buildings.

Due to the characteristics of qualitative research, other barriers such as lack of training and/or experience, lack of information, technicians' daily workload and recent actions in engine rooms were discovered.

These factors can act either individually or jointly, in the latter case becoming real obstacles to the development of these projects. Although no attempt has been made to classify the importance of the different barriers, after analyzing the responses, the most important are economic-financial and administrative/legislative barriers.

The perception of insufficient direct support to achieve the objectives of the EPDB in public buildings, together with the difficulties generated by administrative and legal procedures, as well as the lack of aid and communication by and between the different public administrations involved in the legalization process of these projects, make it difficult to execute it. Furthermore, in the case of photovoltaic installation projects, the difficulties and barriers imposed by the electricity companies themselves must be taken into account.

A significant difference has been found in the perception of technicians from Spain and Portugal since the latter did not perceive that administrative and legislative barriers were as high in their country.

Cities are sources of energy consumption and gas emissions. Those responsible for preparing urban plans must allow for solutions that reduce or eliminate those urban criteria that prevent RETs installation in cities. Urban regulations need to evolve, including sustainability criteria that facilitate, or at least do not hinder, energy efficiency measures in actions on the outer envelope of buildings, as well as the integration of RETs that, at present, are impossible to apply in some municipalities due to their municipal regulations.

The main contribution of this paper is the identification of factors and their grouping into a series of barriers that can help to design strategies and actions that tend to eliminate them or, at least, to consider them as minimizing their impact and thus facilitating the development and implementation of RETs in existing university buildings. At the same time, this knowledge will help to achieve the UNSDG-9 and UNDSG-11 goals set out in the 2030 Agenda for Sustainable Development.

Limitations and Future Directions

During the sample preparation phase, it was impossible to achieve the same percentage of participants from both countries. The difficulties that arose during the arrangement of interviews with technicians from Portuguese universities could not be avoided. This limitation may have influenced the determination of other differences or similarities, and their effect, between the barriers and factors that affect the application of RETs in university buildings.

To take advantage of the results obtained, it would be useful to develop a quantitative investigation that would allow one to classify the different barriers in order to order them from greater to lesser influence, to establish actions or policies aimed at minimizing the negative effect of those barriers with greater importance.

Future lines of research would be to know if the barriers found affect the application of RETs in other existing buildings intended for teaching use such as schools and secondary education centers, as well as public administration buildings, at each of their levels: local, regional or state.

Public procurement processes for the supply or operation of RETs in university existing buildings are complex and need to meet many requirements set in the regulations. Hence, they generate difficulties and problems that have been identified as administrative and legislative barriers. In further research, it would be interesting to explore and determine the strategies used by universities in this respect, in order to improve them.

As the new 2030 Climate and Energy Framework raises more demanding objectives than the previous framework, this study could be replicated once the strategy has been developed, checking whether the barriers encountered continue or disappear, or whether new ones emerge.

Author Contributions: Conceptualization, J.F.-d.-B.; methodology, J.F.-d.-B. and E.N.-A.; formal analysis, J.F.-d.-B.; investigation, J.F.-d.-B., E.N.-A., N.M.M.R. and J.P.M.; writing—original draft preparation, J.F.-d.-B. and E.N.-A.; writing—review and editing, J.F.-d.-B., E.N.-A., N.M.M.R. and J.P.M. All authors have read and agreed to the published version of the manuscript.

Funding: This research received funds given by the Department of Civil Engineering and Building (University of Castilla-La Mancha), for covering part cost of manuscript's publication.

Institutional Review Board Statement: Not applicable.

Informed Consent Statement: Informed consent was obtained from all subjects involved in the study.

Data Availability Statement: Data sharing not applicable.

Acknowledgments: The authors wish to thank the support given by the University of Porto and the collaboration of the interviewees in this research.

Conflicts of Interest: The authors declare no conflict of interest.

References

1. IPCC. *Climate Change 2014: Mitigation of Climate Change. Contribution of Working Group III to the Fifth Assessment Report for the Intergobernmental Panel on Climate Change*; Cambridge University Press: New York, NY, USA, 2014.
2. Ruparathna, R.; Hewage, K.; Sadiq, R. Improving the energy efficiency of the existing building stock: A critical review of commercial and institutional buildings. *Renew. Sustain. Energy Rev.* **2016**, *53*, 1032–1045. [CrossRef]
3. Tsemekidi-Tzeiranaki, S.; Labanca, N.; Cuniberti, B.; Toleikyte, A.; Zangheri, P.; Bertoldi, P. *Analysis of the Annual Reports 2018 under the Energy Efficiency Directive—Summary Report, EUR 29667 EN.*; Publications Office of the European Union: Luxembourg, 2019.
4. European Parliament. Directive 2010/31/EU of the European Parliament and of the Council of 19 May 2010 on the energy performance of buildings (recast). *Off. J. Eur. Union* **2010**, *153*, 13–35.
5. Burman, E.; Mumovic, D.; Kimpian, J. Towards measurement and verification of energy performance under the framework of the European directive for energy performance of buildings. *Energy* **2014**, *77*, 153–163. [CrossRef]
6. AbdelAzim, A.I.; Ibrahim, A.M.; Aboul-Zahab, E.M. Development of an energy efficiency rating system for existing buildings using Analytic Hierarchy Process—The case of Egypt. *Renew. Sustain. Energy Rev.* **2017**, *71*, 414–425. [CrossRef]
7. Sandoval Fernández, P. Reto Europeo: La Eficiencia Energética en Edificios. La Nueva Directiva Comunitaria 31/2010. *Seqüência* **2011**, *32*, 55–77.
8. Liobikienė, G.; Butkus, M. The European Union possibilities to achieve targets of Europe 2020 and Paris agreement climate policy. *Renew. Energy* **2017**, *106*, 298–309. [CrossRef]
9. Ameli, N.; Brandt, N. *What Impedes Household Investment in Energy Efficiency and Renewable Energy?* Economics Department Working Papers No. 1222; ECO/WKP(2015)40; Organisation for Economic Cooperation and Development (OECD): Paris, France, 2015.
10. Ashrafian, T.; Yilmaz, A.Z.; Corgnati, S.P.; Moazzen, N. Methodology to define cost-optimal level of architectural measures for energy efficient retrofits of existing detached residential buildings in Turkey. *Energy Build.* **2016**, *120*, 58–77. [CrossRef]
11. Astiaso Garcia, D.; Cumo, F.; Tiberi, M.; Sforzini, V.; Piras, G. Cost-Benefit Analysis for Energy Management in Public Buildings: Four Italian Case Studies. *Energies* **2016**, *9*, 522. [CrossRef]
12. Gangolells, M.; Casals, M.; Forcada, N.; Macarulla, M.; Cuerva, E. Energy mapping of existing building stock in Spain. *J. Clean. Prod.* **2016**, *112*, 3895–3904. [CrossRef]
13. European Commission. *Communication from the Commission to the European Parliament, the Council, the European Economic and Social Committee and the Committee of the Regions Brussels, Energy Efficiency Plan 2011*; European Commission: Brussels, Belgium, 2011.
14. Bertone, E.; Sahin, O.; Stewart, R.A.; Zou, P.; Alam, M.; Blair, E. State-of-the-Art review revealing a roadmap for public building water and energy efficiency retrofit projects. *Int. J. Sustain. Built Environ.* **2016**, *5*, 526–548. [CrossRef]
15. Almeida, R.M.; Ramos, N.M.; Simões, M.L.; de Freitas, V.P. Energy and water consumption variability in school buildings: Review and application of clustering techniques. *J. Perform. Constr. Facil.* **2015**, *29*, 04014165. [CrossRef]

16. Erhorn-Kluttig, H. Overview. Energy Efficient University Campus Projects. The European Portal for Energy Efficiency in Buildings. 2017. Available online: http://www.buildup.eu/en/news/overview-energy-efficient-university-campus-projects-0 (accessed on 27 August 2020).

17. Leal Filho, W.; Lange Salvia, A.; do Paço, A.; Anholon, R.; Gonçalves Quelhas, O.L.; Izabela Simon Rampasso, I.; Artie Ng, A.; Balogun, A.; Kondev, B.; Londero Brandli, L. A comparative study of approaches towards energy efficiency and renewable energy use at higher education institutions. *J. Clean. Prod.* **2019**, *237*, 117728. [CrossRef]

18. Wadud, Z.; Royston, S.; Selby, J. Modelling energy demand from higher education institutions: A case study of the UK. *Appl. Energy* **2019**, *233*, 816–826. [CrossRef]

19. General Assembly of the United Nations. *Transforming Our World: The 2030 Agenda for Sustainable Development*; UN: New York, NY, USA, 2015.

20. Born, F.J.; Clark, J.A.; Johnstone, C.M.; Kelly, N.J.; Burt, G.; Dysko, A.; McDonald, J.; Hunter, I.B. On the integration of renewable energy systems within the built environment. *Build. Serv. Eng. Res. Technol.* **2001**, *22*, 3–13. [CrossRef]

21. Cooke, R.; Cripps, A.; Irwin, A.; Kolokotroni, M. Alternative energy technologies in buildings: Stakeholder perceptions. *Renew. Energy* **2007**, *32*, 2320–2333. [CrossRef]

22. Woo, J.H.; Menassa, C. Virtual Retrofit Model for aging commercial buildings in a smart grid environment. *Energy Build.* **2012**, *80*, 424–435. [CrossRef]

23. Economidou, M.; Todeschi, V.; Bertoldi, P.; Agostino, D.D.; Zangheri, P.; Castellazzi, L. Review of 50 years of EU energy efficiency policies for buildings. *Energy Build.* **2020**, *225*, 110322. [CrossRef]

24. Reddy, S.; Painuly, J.P. Diffusion of renewable energy technologies—Barriers and stakeholders' perspectives. *Renew. Energy* **2004**, *29*, 1431–1447. [CrossRef]

25. Zhang, X.; Shen, L.; Chan, S.Y. The diffusión of solar energy use in HK: What are the barriers? *Energy Policy* **2012**, *41*, 241–249. [CrossRef]

26. Aksamija, A. Regenerative design and adaptive reuse of existing commercial buildings for net-zero energy use. *Sustain. Cities Soc.* **2016**, *27*, 185–195. [CrossRef]

27. European Commission. *Europe 2020. A Strategy for Smart, Sustainable and Inclusive Growth*; Publications Office of the European Union: Brussels, Belgium, 2010.

28. Kulovesi, K.; Oberthür, S. Assessing the EU's 2030 Climate and Energy Policy Framework: Incremental change toward radical transformation? *RECIEL* **2020**, *29*, 151–166. [CrossRef]

29. Day, J.K. Survey and Interview Approaches to Studying Occupants. In *Exploring Occupant Behavior in Buildings. Methods and Challenges*; Wagner, A., O'Brien, W., Dong, B., Eds.; Springer International Publishing: Cham, Switzerland, 2017; pp. 213–238.

30. Curtius, H.C. The adoption of building-integrated photovoltaics: Barriers and facilitators. *Renew. Energy* **2018**, *126*, 783–790. [CrossRef]

31. European Commission. *Communication from the Commission to the Council and the European Parliament. The Renewable Energy Progress Report: Commission Report in Accordance with Article 3 of Directive 2001/77/EC; Article 4(2) of Directive 2003/30/EC and on the Implementation of the EU Biomass Action Plan*; European Commission: Brussels, Belgium, 2009.

32. Mirza, U.K.; Ahmad, N.; Harijan, K.; Majeed, T. Identifying and addressing barriers to renewable energy development in Pakistan. *Renew. Sust Energ Rev.* **2009**, *13*, 927–931. [CrossRef]

33. Nasirov, S.; Silva, C.; Agostini, C.A. Investors' Perspectives on Barriers to the Deployment of Renewable Energy Sources in Chile. *Energies* **2015**, *8*, 3794–3814. [CrossRef]

34. Beck, F.; Martinot, E. Renewable Energy Policies and Barriers. In *Encyclopedia of Energy*; Cleveland, C.J., Ed.; Academic Press/Elsevier Science: San Diego, CA, USA, 2004; pp. 365–383.

35. Sidiras, D.K.; Koukios, E.G. Solar systems diffusion in local markets. *Energy Policy* **2004**, *32*, 2007–2018. [CrossRef]

36. Klessmann, C.; Held, A.; Rathmann, M.; Ragwitz, M. Status and perspectives of renewable energy policy and deployment in the European Union—What is needed to reach the 2020 targets? *Energy Policy* **2011**, *39*, 7637–7657. [CrossRef]

37. Huang, S.; Lo, S.; Lin, Y. To Re-Explore the Causality between Barriers to Renewable Energy Development: A Case Study of Wind Energy. *Energies* **2013**, *6*, 4465–4488. [CrossRef]

38. Moula, M.E.; Maula, J.; Hamdy, M.; Fang, T.; Jung, N.; Lahdelma, R. Researching social acceptability of renewable energy technologies in Finland. *Int J. Sustain. Built Environ.* **2013**, *2*, 89–98. [CrossRef]

39. Eleftheriadis, I.M.; Anagnostopoulou, E.G. Relationship between Corporate Climate change disclosures and firm factors. *Bus. Strateg. Environ.* **2015**, *24*, 780–789. [CrossRef]

40. Pérez-Lombard, L.; Ortiz, J.; Pout, C. A review on building energy consumption information. *Energy Build.* **2008**, *40*, 394–398. [CrossRef]

41. Margolis, R.; Zuboy, J. *Nontechnical Barriers to Solar Energy Use: Review of Recent Literature*; National Renewable Energy Laboratory: Golden, CO, USA, 2006.

42. Ogunleye, I.O.; Awogbemi, O. Constraints to the use of solar photovoltaic as a sustainable power source in Nigeria. *AJSIR* **2010**, *2*, 11–16. [CrossRef]

43. Byrnes, L.; Brown, C.; Foster, J.; Wagner, L.D. Australian renewable energy policy: Barriers and challenges. *Renew. Energy* **2013**, *60*, 711–721. [CrossRef]

44. Sharpe, T.; Proven, G. Crossflex: Concept and early development of a true building integrated wind turbine. *Energy Build.* **2010**, *42*, 2365–2375. [CrossRef]

45. Di Giuseppe, E.; Iannaccone, M.; Telloni, M.; D'Orazio, M.; Di Perna, C. Probabilistic life cycle costing of existing buildings retrofit interventions towards nZE target: Methodology and application example. *Energy Build.* **2017**, *144*, 416–432. [CrossRef]

46. European Commission. *Report from the Commission to the European Parliament, the Council, the European Economic and Social Committee and the Committee of the Regions*; Renewable energy progress report; European Commission: Brussels, Belgium, 2015.

47. Evans, B.; Parks, J.; Theobald, K. Urban wind power and the private sector: Community benefits, social acceptance and public engagement. *J. Environ. Plan. Manag.* **2011**, *54*, 227–244. [CrossRef]

48. Baker, S.E.; Edwards, R. *How Many Qualitative Interviews is Enough? Expert Voices and Early Career Reflections on Sampling and Cases in Qualitative Research*; National Centre for Research Methods (NCRM): Southhampton, UK, 2012.

49. Galvin, R. How many interviews are enough? Do qualitative interviews in building energy consumption research produce reliable knowledge? *J. Build. Eng.* **2015**, *1*, 2–12. [CrossRef]

50. Sommerfeld, J.; Buys, L.; Vine, D. Residential consumers' experiences in the adoption and use of solar PV. *Energy Policy* **2017**, *105*, 10–16. [CrossRef]

51. Hai, M.A. Rethinking the social acceptance of solar energy: Exploring "states of willingness" in Finland. *Energy Res. Soc. Sci* **2019**, *51*, 96–106. [CrossRef]

52. Bavaresco, M.V.; D'Oca, S.; Ghisi, E.; Lamberts, R. Methods used in social sciences that suit energy research: A literature review on qualitative methods to assess the human dimension of energy use in buildings. *Energy Build.* **2020**, *209*, 1–19. [CrossRef]

53. Ugulu, A.I. Barriers and motivations for solar photovoltaic (PV) adoption in urban Nigeria. *IJSEPM* **2019**, *21*, 19–34.

54. Painuly, J.P.; Fenhann, J.V. *Implementation of Renewable Energy Technologies—Opportunities and Barriers: Summary of Country Studies*; UNEP Collaborating Centre on Energy and Environment, Risø National Laboratory: Roskilde, Denmark, 2002.

55. Font Tullot, I. *Climatología de España y Portugal. Nueva Versión*; Ediciones Universidad de Salamanca: Salamanca, Spain, 2007.

56. Sequeiros Tizón, J.S. Integración económica y comercio internacional. *REM* **2000**, *2*, 151–177.

57. Ahlborg, H.; Hammar, L. Drivers and barriers to rural electrification in Tanzania and Mozambique–Grid-extension, off-grid, and renewable energy technologies. *Renew. Energy* **2014**, *61*, 117–124. [CrossRef]

58. Caven, V. Agony aunt, hostage, intruder or friend? The multiple personas of the interviewer during fieldwork. *Intang. Cap.* **2012**, *8*, 548–563.

59. Denzin, N.K.; Lincoln, Y.S. *The SAGE Handbook of Qualitative Research*, 4th ed.; Sage: Thousand Oaks, CA, USA, 2011.

60. Guest, G.; Bunce, A.; Johnson, L. How many interviews are enough? An experiment with data saturation and variability. *Field Methods* **2006**, *18*, 59–82. [CrossRef]

61. Von Wirth, T.; Gislason, L.; Seidl, R. Distributed energy systems on a neighborhood scale: Reviewing drivers of and barriers to social acceptance. *Renew. Sustain. Energy Rev.* **2018**, *82*, 2618–2628. [CrossRef]

62. Creswell, J. *Research Design: Qualitative, Quantitative, and Mixed Methods Approaches*; Sage: Thousand Oaks, CA, USA, 2009.

63. Blismas, N.G.; Dainty, A.R.J. Computer-Aided qualitative data analysis: Panacea or paradox? *BRI* **2003**, *31*, 455–463. [CrossRef]

64. Bowen, G.A. Preparing a qualitative research-based dissertation: Lessons learned. *Qual. Rep.* **2005**, *10*, 208–222. [CrossRef]

65. Tesch, R. *Qualitative Research. Analysis Types & Software Tools*; Routledge: New York, NY, USA, 2013.

66. Charmaz, K. *Constructing Grounded Theory. A Practical Guide through Qualitative Analysis*; Sage: London, UK, 2006.

67. European Union. Directive 2012/27/EU of the European Parliament and of the Council of 25 October 2012 on energy efficiency, amending Directives 2009/125/EC and 2010/30/EU and repealing Directives 2004/8/EC and 2006/32/EC. *Off. J. Eur. Union.* **2012**, *315*, 1–56.

68. European Union. Directive 2002/91/EC of the European Parliament and of the Council of 16 December 2002 on the energy performance of buildings. *Off. J. Eur. Union* **2003**, *L1*, 65–71.

69. Rosenow, J.; Kern, F. EU energy innovation in policy: The curious case of energy efficiency. In *Research Handbook on EU Energy Law and Policy*; Leal-Arcas, R., Wouters, J., Eds.; Edward Elgar Publishing Limited: Cheltenham, UK; Northampton, MA, USA, 2017; pp. 501–517.

70. Ministerio de Industria, Energía y Turismo. *Real Decreto 900/2015, de 9 de Octubre, Por el Que se Regulan las Condiciones Administrativas', Técnicas y Económicas de las Modalidades de Suministro de Energía Eléctrica con Autoconsumo y de Producción con Autoconsumo*; Ministerio de Industria, Energía y Turismo: Madrid, Spain, 2015; pp. 94874–94917.

71. Fernandez, R.M. Conflicting energy policy priorities in EU energy governance. *J. Environ. Stud. Sci.* **2018**, *8*, 239–248. [CrossRef]

72. Abdmouleh, Z.; Alammari, R.A.M.; Gkastli, A. Review of policies encouraging renewable energy integration & best practices. *Renew. Sustain. Energy Rev.* **2015**, *45*, 249–262.

73. Owen, A.D. Renewable energy: Externality costs as market barriers. *Energy Policy* **2006**, *34*, 632–642. [CrossRef]

74. Yaqoot, M.; Diwan, P.; Kandpal, T.C. Review of barriers to the dissemination of decentralized renewable energy systems. *Renew. Sustain. Energy Rev.* **2016**, *58*, 477–490. [CrossRef]

75. Ürge-Vorsatz, D.; Danny Harvey, L.D.; Mirasgedis, S.; Levine, M.D. Mitigating CO_2 emissions from energy use in the world's buildings. *BRI* **2007**, *35*, 379–398.

76. Espejo Marín, C. La energía solar fotovoltaica en España. *Nimbus* **2004**, *13–14*, 5–31.

77. Martínez Montes, G.; Serrano López, M.M.; Rubio Gámez, M.C.; Menéndez Ondina, A. An overview of renewable energy in Spain: The small hydro-power case. *Renew. Sustain. Energy Rev.* **2005**, *9*, 521–534. [CrossRef]

78. Coenraads, R.; Reece, G.; Voogt, M.; Ragwitz, M.; Held, A.; Resch, G.; Faber, T.; Haas, R.; Konstantinaviciute, I.; Juraj Krivošík, J.; et al. *Progress, Promotion and Growth of Renewable Energy Sources and Systems*; Final report. TREN/D1/42-2005/S07.56988; Ecofys Netherlands: Utrecht, The Netherlands, 2008.
79. Jefatura del Estado. *Ley 9/2017, de 8 de Noviembre, de Contratos del Sector Público, Por la Que se Transponen al Ordenamiento Jurídico Español las Directivas del Parlamento Europeo y del Consejo 2014/23/UE y 2014/24/UE, de 26 de Febrero de 2014*; Boletín Oficial del Estado, núm. 272, de 09 de Noviembre de 2017, Referencia: BOE-A-2017-12902; Jefatura del Estado: Madrid, Spain, 2017.
80. Hirst, E. Improving energy efficiency in the USA: The federal role. *Energy Policy* **1991**, *19*, 567–577. [CrossRef]
81. Ministerio de Transición Ecológica. *Real Decreto 244/2019, de 5 de Abril, Por el Que se Regulan las Condiciones Administrativas, Técnicas y Económicas del Autoconsumo de Energía Eléctrica*; Boletín Oficial del Estado, núm. 83, de 6 de abril de 2019; Ministerio de Transición Ecológica: Madrid, Spain, 2019; pp. 35674–35719.
82. Kammen, D.M.; Sunter, D.A. City-Integrated renewable energy for urban sustainability. *Science* **2016**, *352*, 922–928. [CrossRef]
83. Dadzie, J.; Runeson, G.; Ding, G.; Bondinuba, F.K. Barriers to Adoption of Sustainable Technologies for Energy-Efficient Building Upgrade—Semi-Structured Interviews. *Buildings* **2018**, *8*(4), 57. [CrossRef]
84. Aronova, E.; Radovic, G.; Murgul, V.; Vatin, N. Solar Power Opportunities in Northern Cities (Case Study of Saint-Petersburg). *Appl. Mech. Mater.* **2014**, *587*, 348–354. [CrossRef]
85. Polo López, C.S.; Frontini, F. Energy efficiency and renewable solar energy integration in heritage historic buildings. *Energy Procedia* **2014**, *48*, 1493–1502. [CrossRef]
86. Webb, A.L. Energy retrofits in historic and traditional buildings: A review of problems and methods. *Renew. Sustain. Energy Rev.* **2017**, *77*, 748–759. [CrossRef]
87. Papamanolis, N. An overview of solar energy applications in buildings in Greece. *Int J. Sustain. Energy* **2016**, *35*, 814–823. [CrossRef]
88. Hayter, S.J.; Kandt, A. *Renewable Energy Applications for Existing Buildings (No. NREL/CP-7A40-52172)*; National Renewable Energy Laboratory (NREL): Golden, CO, USA, 2011.
89. De Santoli, L.; Mancini, F.; Clemente, C.; Lucci, S. Energy and tecnological refurbishment of the School of Architecture Valle Giulia, Rome. *Energy Procedia* **2017**, *133*, 382–391. [CrossRef]
90. De Guerrero Manso, C. La clasificación del suelo urbano en el contexto urbanístico actual de regeneración de la ciudad. *Rev. Aragonesa De Adm. Pública* **2010**, *37*, 139–185.
91. Vieites, E.; Vassileva, I.; Arias, J.E. European initiatives towards improving the energy efficiency in existing and historic buildings. *Energy Procedia* **2015**, *75*, 1679–1685. [CrossRef]
92. Imenes, A.G. Performance of BIPV and BAPV installations in Norway. In Proceedings of the 43rd IEEE Photovoltaic Specialists Conference (PVSC), Portland, OR, USA, 5–10 June 2016; pp. 3147–3152.
93. Aranda Usón, A.; Ortego Bielsa, A. *Integración de Energías Renovables en Edificios*; Prensas Universitarias: Zaragoza, Spain, 2011.
94. Verbruggen, A.; Fischedick, M.; Moomaw, W.; Weir, T.; Nadaï, A.; Nilsson, L.J.; Nyboer, J.; Sathaye, J. Renewable energy costs, potentials, barriers: Conceptual issues. *Energy Policy* **2010**, *38*, 850–861. [CrossRef]
95. Lund, P.D.; Lindgren, J.; Mikkola, J.; Salpakari, J. Review of energy system flexibility measures to enable high levels of variable renewable electricity. *Renew. Sustain. Energy Rev.* **2015**, *45*, 785–807. [CrossRef]
96. Boydak, Ö. Commercial Buildings Energy Consumption Survey (CBECS) and Its Comparison with Turkey Applications. *J. Clean. Energy Technol.* **2017**, *5*, 69–72. [CrossRef]
97. Touzani, S.; Granderson, J.; Fernandes, S. Gradient boosting machine for modeling the energy consumption of commercial buildings. *Energy Build.* **2018**, *158*, 1533–1543. [CrossRef]
98. Mbungu, N.T.; Bansal, R.C.; Naidoo, R.; Miranda, V.; Bipath, M. An optimal energy management system for a commercial building with renewable energy generation under real-time electricity prices. *Sustain. Cities Soc.* **2018**, *41*, 392–404. [CrossRef]
99. Ellabban, O.; Abu-Rub, H.; Blaabjerg, F. Renewable energy resources: Current status, future prospects and their enabling technology. *Renew. Sustain. Energy Rev.* **2014**, *39*, 748–764. [CrossRef]
100. Liu, F.; Meyer, A.S.; Hogan, J.F. *Mainstreaming Building Energy Efficiency Codes in Developing Countries*; The World Bank: Washington, DC, USA, 2010.
101. Čenejac, A.R.; Bjelaković, R.M.; Anđelković, A.S.; Djaković, D.D. Covering of heating load of object by using ground heat as a renewable energy source. *Therm. Sci.* **2012**, *16*, 225–235. [CrossRef]
102. Scognamiglio, A.; Røstvik, H.N. Photovoltaics and zero energy buildings: A new opportunity and challenge for design. *Prog. Photovolt.* **2013**, *21*, 1319–1336. [CrossRef]
103. Samir, H.; Ali, N.A. Applying building-integrated photovoltaics (BIPV) in existing buildings, opportunities and constraints in Egypt. *Procedia Environ. Sci.* **2017**, *37*, 614–625. [CrossRef]
104. Sovacool, B.K. The intermittency of wind, solar, and renewable electricity generators: Technical barrier or rhetorical excuse? *Util. Policy* **2009**, *17*, 288–296. [CrossRef]
105. Ordoñez García, J.; Jadraque Gago, E.; Alegre Bayo, J.; Martínez Montes, G. The use of solar energy in the buildings construction sector in Spain. *Renew. Sustain. Energy Rev.* **2007**, *11*, 2166–2178. [CrossRef]
106. Jelle, B.P. Building Integrated Photovoltaics: A Concise Description of the Current State of the Art and Possible Research Pathways. *Energies* **2016**, *9*, 21. [CrossRef]
107. Pandey, A.K.; Tyagi, V.V.; Selvaraj, J.A.; Rahim, N.A.; Tyagi, S.K. Recent advances in solar photovoltaic systems for emerging trends and advanced applications. *Renew. Sustain. Energy Rev.* **2016**, *53*, 859–884. [CrossRef]

108. Peng, C.; Huang, Y.; Wu, Z. Building-integrated photovoltaics (BIPV) in architectural design in China. *Energy Build.* **2011**, *43*, 3592–3598. [CrossRef]
109. Lehmann, P.; Creutzig, F.; Ehlers, M.H.; Friedrichsen, N.; Heuson, C.; Hirth, L.; Pietzcker, R. Carbon lock-out: Advancing renewable energy policy in Europe. *Energies* **2012**, *5*, 323–354. [CrossRef]
110. Jobert, A.; Laborgne, P.; Mimler, S. Local acceptance of wind energy: Factors of success identified in French and German case studies. *Energy Policy* **2007**, *35*, 751–760. [CrossRef]
111. Bertsch, V.; Hall, M.; Weinhardt, C.; Fichtner, W. Public acceptance and preferences related to renewable energy and grid expansion policy: Empirical insights for Germany. *Energy* **2016**, *114*, 465–477. [CrossRef]
112. Ramírez Brouchoud, M.F. Las reformas del Estado y la administración pública en América Latina y los intentos de aplicación del New Public Management. *Estud. Políticos* **2009**, *34*, 115–141.
113. Tardío Pato, J.A. ¿Tiene sentido que las universidades públicas dejen de ser administraciones públicas en las nuevas leyes del sector público y de procedimiento administrativo común? *Doc. Adm.* **2015**, *2*, 12. [CrossRef]
114. Wang, L.; Morabito, M.; Payne, C.T.; Robinson, G. Identifying institutional barriers and policy implications for sustainable energy technology adoption among large organizations in California. *Energy Policy* **2020**, *146*, 111768. [CrossRef]
115. Cooremans, C. The role of formal capital budgeting analysis in corporate investment decision-making: A literature review. In Proceedings of the ECEEE 2009 Summer Study, La Colle sur Loup, France, 1–6 June 2009; pp. 237–245.
116. Wells, L.; Rismanchi, B.; Aye, L. A review of Net Zero Energy Buildings with reflections on the Australian context. *Energy Build.* **2018**, *158*, 616–628. [CrossRef]
117. Goh, C.S.; Jack, L.; Bajracharya, A. Qualitative study of sustainability policies and guidelines in the built environment. *J. Leg. Aff. Disput. Resolut. Eng. Constr.* **2020**, *12*, 04520016. [CrossRef]
118. Alam, S.S.; Nor, N.F.M.; Ahmad, M.; Hashim, N.H.N. A Survey on Renewable Energy Development in Malaysia: Current Status, Problems and Prospects. *Environ. Clim. Technol.* **2016**, *17*, 5–17. [CrossRef]
119. Foxon, T.J.; Gross, R.; Chase, A.; Howes, J.; Arnall, A.; Anderson, D. UK innovation systems for new and renewable energy technologies: Drivers, barriers and systems failures. *Energy Policy* **2005**, *33*, 2123–2137. [CrossRef]
120. Kofoed-Wiuff, A.; Sandholt, K.; Marcus-Møller, C. *Renewable Energy Technology Deployment (RETD): Barriers, Challenges and Opportunities: A Synthesis of Various Studies on Barriers, Challenges and Opportunities for Renewable Energy Deployment*; EA Energy Analyses: Copenhagen, Denmark, 2006.
121. Mann, S.; Harris, I.; Harris, J. The development of urban renewable energy at the existential technology research center (ETRC) in Toronto, Canada. *Renew. Sustain. Energy Rev.* **2006**, *10*, 576–589. [CrossRef]

Article

Effect of Manufactured Sand with Different Quality on Chloride Penetration Resistance of High–Strength Recycled Concrete

Jincai Feng [1,*], Chaoqun Dong [1], Chunhong Chen [1], Xinjie Wang [1] and Zhongqiu Qian [2]

[1] Department of Civil Engineering, Changzhou University, Changzhou 213164, China; cczudcq@163.com (C.D.); chench@cczu.edu.cn (C.C.); wangxinjie@cczu.edu.cn (X.W.)
[2] Jiangsu Nigao Science & Technology Co., Ltd., Changzhou 213141, China; qianzhongqiu@126.com
* Correspondence: hjxfjc@cczu.edu.cn

Abstract: High–strength manufactured sand recycled aggregate concrete (MSRAC) prepared with manufactured sand (MS) and recycled coarse aggregate (RCA) is an effective way to reduce the consumption of natural aggregate resources and environmental impact of concrete industry. In this study, high–, medium– and low–quality MS, which were commercial MS local to Changzhou and 100% by volume of recycled coarse aggregate, were used to prepare MSRAC. The quality of MS was determined based on stone powder content, methylene blue value (MBV), crushing value and soundness as quality characteristic parameters. The variation laws of compressive strength and chloride penetration resistance of high–strength MSRAC with different rates of replacement and different qualities of MS were explored. The results showed that for medium– and low–quality MS, the compressive strength of the MSRAC increased first and then decreased with increasing rate of replacement. Conversely, for high–quality MS, the compressive strength gradually increased with increasing rate of replacement. The chloride diffusion coefficient of MSRAC increased with decreasing MS quality and increasing rate of replacement. The chloride diffusion coefficient of MSRAC basically met the specifications for 50–year and 100–year design working life when the chloride environmental action was D and E. To prepare high–strength MSRAC, high–quality MS can 100% replace RS (river sand), while rates of replacement of 50–75% for medium–quality MS or 25–50% for low–quality MS are proposed. Scanning Electron Microscope (SEM) images indicated that an appropriate amount of stone powder is able to improve the compressive strength of RAC, but excessive stone powder content and MBV are unfavorable to the compressive strength and chloride penetration resistance of RAC.

Keywords: manufactured sand; chloride penetration resistance; recycled aggregate concrete; high–strength concrete

Citation: Feng, J.; Dong, C.; Chen, C.; Wang, X.; Qian, Z. Effect of Manufactured Sand with Different Quality on Chloride Penetration Resistance of High–Strength Recycled Concrete. *Materials* **2021**, *14*, 7101. https://doi.org/10.3390/ma14227101

Academic Editor: Jean-Marc Tulliani

Received: 7 October 2021
Accepted: 13 November 2021
Published: 22 November 2021

1. Introduction

The traditional cement concrete industry is unprepossessing due to its high consumption of natural resources and energy, and major environmental damage [1]. In 2020, China produced 2.5 billion m^3 of concrete, which consumed nearly 1.65 billion tons of sand and gravel, including a large number of natural river sand (RS). As a non–renewable resource, RS is overconsumed. Long–term mining of RS has caused serious threats to the safety and stability of bridges and river banks, the ecological system of origin, and the health of local residents [2–4]. In addition, China has promulgated a policy of restriction and prohibition of mining RS. This further leads to less and less high–quality RS being available for the concrete industry, and the rising cost of RS. Therefore, it is imperative to find alternative materials for RS in order to meet the goal of sustainable development in the concrete industry.

Manufactured sand (MS) is a kind of fine aggregate with particle size less than 4.75 mm that is obtained by crushing and screening rocks. It is regarded as a potential alternative material for RS [5]. However, due to its preparation process, MS has defects such as irregular particle shape, unreasonable gradation, high stone powder content, large methylene blue value (MBV), etc. These defects also lead to the performance differences (workability, mechanical properties, and durability) between MS concrete and RS concrete [6–9]. Many scholars have studied the influence of MS characteristics on the performance of concrete. Shen et al. [10,11] studied the effects of MS of different crushing values, stone powder content and the MBV on performance of C50 MS concrete. It was found that the stone powder content and MBV had the greatest impact on the workability of the concrete. When the stone powder content and MBV remained unchanged, the compressive strength was negatively correlated with the crushing value. Liu et al. [12] studied the influence of MMBV on the durability of medium– and low–strength MS concrete. MMBV is the multiplication of stone powder content and MBV. MMBV has shown an effect on the workability of medium– and low–strength concrete. With increasing MMBV, concrete workability decreases. The compressive strength and chloride penetration resistance of medium–strength concrete were more affected than those of low–strength concrete. Chen et al. [13] studied the coupling effect of different stone powder content and MBV in the performance of C40 concrete, and they calculated the coupling coefficient of stone powder content and MBV on concrete performance. Li et al. [14] studied the effect of stone powder content on the durability of MS fully replaced medium–strength marine concrete. MS with low crushing value and MBV was selected to prepare marine concrete. The optimum content of stone powder was about 7.5%. Li et al. [15] used MS of different crushing value and stone powder content to prepare pavement concrete. The results of research showed that with increasing stone powder content, the compressive strength of pavement concrete improved. The lower the crushing value, the higher the compressive strength. To sum up, the MS characteristics have significant effects on the mechanical properties and durability of low– and medium–strength concrete. However, there are few studies on the effects of MS characteristics on performance of high–strength concrete.

In recent years, the market share of MS has been increasing [16]. Hence, it is generally believed that MS is a low–quality substitute for RS and is only suitable for the preparation of medium– and low–strength concrete [17]. This can also be found from the above studies. However, Shen et al. [18] proved that MS with low crushing value, stone powder content, and MBV can be used to completely replace RS in the preparation of ultra–high–strength concrete, and the compressive strength at 28 d exceeded 130 MPa. This indicates that MS has the potential to be used to prepare high–strength concrete. In fact, on the basis of the main quality characteristic parameters of MS, including stone powder content, MBV, crushing value and soundness, the quality of MS can be divided into categories I, II and III, corresponding to high, medium and low quality, respectively, according to the Chinese standard [19]. Table 1 presents the quality of the MS used in some studies according to the Chinese standard. It can be found that in the past, most studies changed one or two characteristic quality parameters to study their influence on concrete performance. However, they did not reckon with the influence of overall MS quality on concrete performance. The category of a certain quality characteristic parameter cannot be used to determine the overall quality of MS. Hence, the overall quality of MS needs to be comprehensively determined on the basis of multiple quality characteristic parameters. As shown in Table 1, the stone powder content, crushing value, and MBV of the MS used in reference [18] are determined to be class I. The MS can be regarded as being of high quality, and is able to completely replace RS in the preparation of high–strength concrete. On the other hand, for medium– and low–quality MS, the optimum rate of replacement for the preparation of high–strength concrete remains unclear. In addition, few studies about manufactured sand recycled aggregate concrete (MSRAC) have been reported, especially with respect to the effect of different levels of MS quality on the durability of high–strength RAC.

Table 1. Quality of MS used in some references.

Reference	Stone Powder Content/%	MBV	Crushing Value/%	Soundness/%
Shen et al. [10]	III	I	III	–
	III	I	I	–
	III	III	I	–
	I	I	II	–
Shen et al. [11]	III	I	III	–
	III	I	I	–
	III	III	I	–
	I	I	I	–
	III	III	II	–
	II	III	III	–
	I	I	I	–
	I	I	II	–
Li et al. [14]	I	I	I	I
	II	I	I	I
	III	I	I	I
Shen et al. [18]	I	I	I	–
	I	I	I	–
	II	I	II	–
Tian et al. [20]	I	–	–	–
	III	–	–	–

Note: I, II and III represent high, medium and low quality, respectively.

The research and use of RAC in China started later than that in developed countries. At present, the use rate of waste concrete in China is low, at only about 10%. A large amount of waste concrete has piled up, resulting in serious land pollution [20–22]. The application of RAC contributes to solving this problem. Many studies have investigated the durability of RAC [23–25]. Chloride penetration resistance is one of the most important durability properties of RAC. Mata et al. [26] found that RAC had higher chloride permeability than NAC. The lower resistance of RAC to chloride ion penetration compared with NAC can be attributed to the old interfacial transition zone and cement paste attached to the recycled aggregate, which make RAC more penetrable than NAC. Due to the presence of old mortar, RAC has multiple interfaces, such as new mortar–old mortar, new mortar–aggregate, and old mortar–aggregate, leading to RAC having more complex and heterogeneous material properties. The old mortar layer is the fundamental reason for the poor durability of RAC [27]. Loose and porous old mortar and the weak interfacial transition zone provide a large number of channels for chloride ions to enter the concrete [28,29]. Chloride ions are present and migrate in the concrete in the free and chemically or physically bound states, deteriorating the interface structure of the concrete and causing the corrosion of steel bars. This leads to the cracking and destruction of reinforced concrete structures, and continuously reduces the bearing capacity of concrete structures. Especially in concrete structures exposed to marine environments, deicing salt environments and coastal underground environments, the chloride penetration resistance of the concrete protective layer is particularly important.

Therefore, in order to improve the use rate of MS and promote the application of medium– and low–quality MS in chloride ion environments, it is necessary to clarify the influence of different levels of MS quality on the chloride penetration resistance of high–strength RAC at different rates of replacement. In this experiment, high–, medium– and low–quality MS are used to prepare MSRAC. The effects of MS of different quality on the compressive strength and chloride penetration resistance of high–strength RAC with different rates of replacement are explored in order to define the optimum rate of replacement of MS of different quality. The microstructure of MSRAC was observed using

Scanning Electron Microscope (SEM) technology to clarify the influence of different levels of MS quality on MSRAC.

2. Experimental Section

2.1. Materials

Type P.O. 52.5 Portland cement was used. Commercially available fly ash and silica fume were used as mineral additives. The composition of fly ash and silica is shown in Table 2. The properties of the cement are shown in Table 3.

Table 2. The composition of fly ash and silica.

Composition (%)	CaO	SiO_2	Al_2O_3	FeO_3	MgO	MnO	K_2O	TiO_2	SO_3
Fly ash	3.72	51.50	29.33	3.77	1.16	0.22	1.70	0.98	1.69
Silica	0.27	87.03	1.12	0.97	0.88	0.14	–	–	0.86

Table 3. The properties of cement.

Cement Sort		P.O. 52.5
Blaine fineness (m^2/kg)		382
Normal consistency (%)		27.8
Soundness		Qualified
Initial setting time (min)		187
Final setting time (min)		258
Flexural strength (MPa)	3 d	6.2
	28 d	9.3
Compressive strength (MPa)	3 d	30.8
	28 d	55.7

River sand and three different levels of MS quality were used, which were denoted as RS, MS1, MS2 and MS3, respectively (Figure 1). The MS were commercially manufactured locally in Changzhou. The characteristics of the MS were tested according to GB/T 14684–2011 [19]. The specific results are shown in Table 3. The gradation curve of MS is shown in Figure 2. Stone powder was used to adjust the stone powder content of the MS. The particle size distribution and XRD results of the stone powder are shown in Figures 3 and 4, respectively. Stone powder content, MBV, crushing value and soundness were selected as parameters for comprehensively determining the quality of the MS. It can be seen from Table 4 that, according to GB/T 14684–2011 [19], MS1, MS2 and MS3 can be categorized as high–quality, medium–quality and low–quality MS. Because its MBV > 1.4 and it has a high stone powder content, MS3 can even be classified as a substandard product.

Figure 1. Three levels of MS quality: (**a**) MS1; (**b**) MS2; (**c**) MS3.

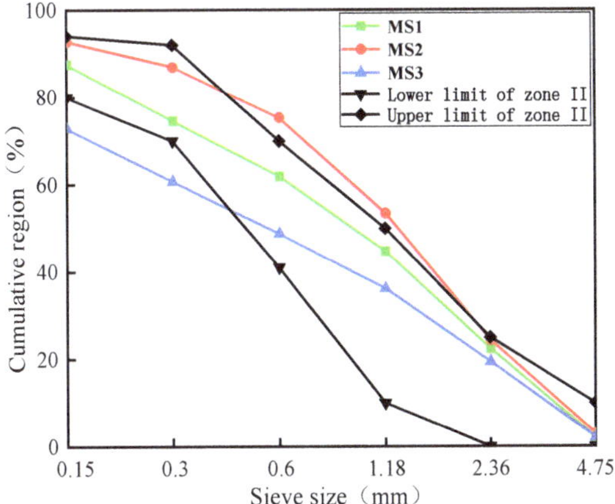

Figure 2. Gradation curves of three MS.

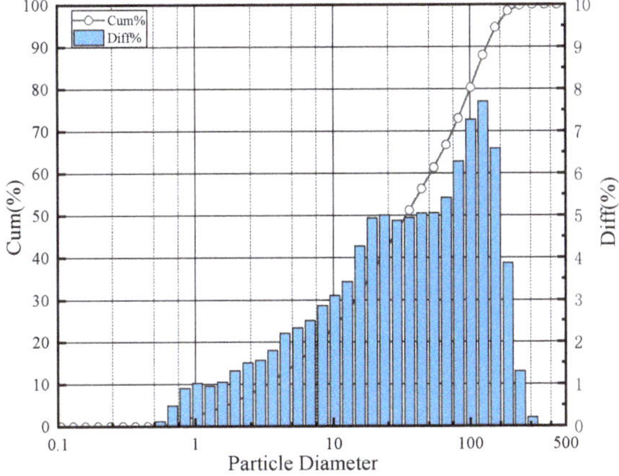

Figure 3. Particle size distribution of stone powder.

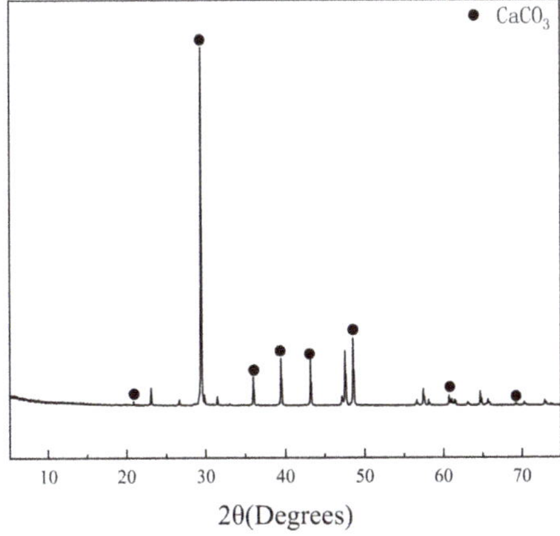

Figure 4. XRD pattern of stone powder.

Table 4. Physical properties of MS.

MS	Standard Values—I	MS1	Standard Values—II	MS2	Standard Values—III	MS3
Fineness modules	–	2.8	–	3.2	–	2.3
Apparent density (kg/m^3)	–	2737	–	2746	–	2654
Stone powder content (%)	0–10	7.0	0–10	10.2	10–15	15.6
MBV(g/kg)	0–0.5	0.5	0.5–1.0	0.9	1.0–1.4	1.7
Crushing value (%)	0–20	15.2	20–25	20.8	25–30	29.0
Soundness (%)	0–8	6.4	0–8	7.8	8–10	10.3
Water absorption(%)	–	1.2	–	1.8	–	2.0

Notes: Standard values—I, II and III represent high, medium and low quality, respectively.

The coarse aggregate consisted of a 5–20 mm continuous gradation of RCA, which was tested according to GB/T 25177–2010 [30]. The specific results are shown in Table 5.

Table 5. Physical properties of RCA.

Coarse Aggregate	Close Packing Density (kg/m^3)	Apparent Density (kg/m^3)	Crushing Value (%)	Water Absorption (%)
RCA	1477	2620	16%	6.2%

The admixture was a polycarboxylate superplasticizer, and the water reduction rate was 40%. The mixed water was tap water.

2.2. Concrete Mix Design

The gradation of the high–quality MS was reasonable. The values of the quality characteristic parameters—stone powder content, crushing value and soundness—were close to those of natural sand, and the MBV was low. Two rates of replacement of 50% and 100% were selected. The performance of medium–quality MS was slightly worse than that of natural sand, and rates of replacement of 50%, 75% and 100% were selected. The optimum rate of replacement for low–quality MS was the research focus, so four rates of replacement of 25%, 50%, 75% and 100% selected. The designed strength of the concrete was 80 MPa, and the water–binder ratio was 0.25. The amount of additional water was calculated on the basis of the water absorption of RCA and MS. The control group was RAC, which was prepared entirely using RS. To make the performance of RAC comparable, the designed strength, water–binder ratio, and the amount of RCA and cementitious material of the MSRAC were consistent with those in the RAC. The amount of superplasticizer in the RAC was adjusted according to the actual mixing situation, and the amount in MSRAC was the same as in RAC. Table 6 shows details of the proportions in the concrete mixture.

Table 6. Concrete mix design (kg/m^3).

Sample	RS	MS	RCA	Cement	Fly Ash	Silica Fume	Water	Additional Water	Superplasticizer	Fresh Concrete Fluidity
RAC	609	0	1095	438	88	58	146	68	5.8	Good
MS1–50	304	318	1095	438	88	58	146	72	5.8	Good
MS1–100	0	636	1095	438	88	58	146	76	5.8	Slightly viscous
MS2–50	304	319	1095	438	88	58	146	73	5.8	Slightly viscous
MS2–75	152	478	1095	438	88	58	146	75	5.8	Slightly viscous
MS2–100	0	638	1095	438	88	58	146	78	5.8	Extremely viscous
MS3–25	456	154	1095	438	88	58	146	72	5.8	Slightly good
MS3–50	304	308	1095	438	88	58	146	74	5.8	Slightly viscous
MS3–75	152	462	1095	438	88	58	146	77	5.8	Extremely viscous
MS3–100	0	616	1095	438	88	58	146	80	5.8	Extremely viscous

2.3. Methods

According to the proportions shown in Table 5, the mixture was mixed with the secondary mixing process proposed by Tam et al. [31]. The specific process used is shown in Figure 5. The fresh concrete was loaded into the mold, vibrated for 10–20 s, and demolded after 24 h. Each sample was placed in a standard curing room at a temperature of $20 \pm 2\,°C$ and a relative humidity of 95% for 28 d according to GB/T 50081–2002 [32].

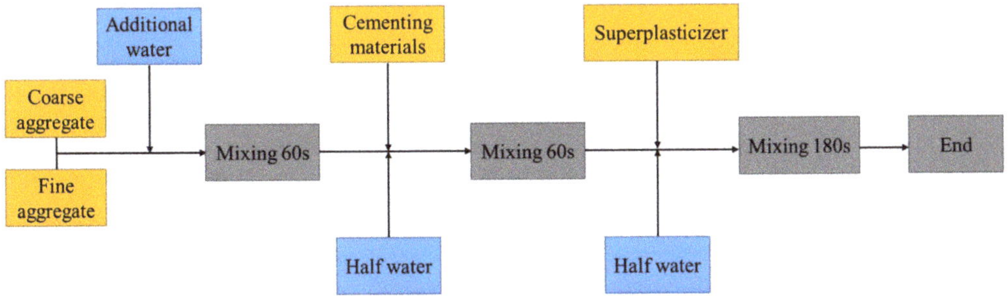

Figure 5. Flow diagram of the concrete secondary mix process.

The particle size distribution of the stone powder was determined using a laser granulometer SALD–2201 and without using a solvent. The XRD pattern of the stone powder was determined using a D8 ADVANCE from Bruker, which was used for conventional phase analysis and semi–quantitative analysis of polycrystalline samples, determination and correction of cell parameters, determination of grain size and crystallinity, and indexing of unknown polycrystalline samples. The test angle was 6–75 degrees. The anticathode was a copper target.

The slump of fresh concrete was tested according to GB/T 50080–2016 [33]. Concrete fluidity was observed when performing the concrete slump test. Compressive strength was determined by testing 150 mm $\times$ 150 mm $\times$ 150 mm cubes according to GB/T50081–2002 [32]. Three test blocks were tested for each proportion.

The chloride penetration resistance was determined by means of the method for testing the rapid chloride ion migration coefficient (RCM) and was represented by the chloride diffusion coefficient according to the GB/ T 50082–2009 [34]. The test block was placed at the bottom of the rubber sleeve, and it was tightened up and down. The rubber sleeve containing the test block was installed in the test tank, and the anode plate was installed. Then, 300 mL 0.3 mol/L NaOH solution was injected into the rubber sleeve, and 12 L 10% NaCl solution was added into the cathode. The continuous voltage of the test was 60 V. The specimen size was ø100 mm $\times$ 50 mm. Three test blocks were tested for each proportion.

The contents of free and bound chloride ions were determined according to SL352–2006 [35]. After the RCM test, the specimens were divided into two halves along the radius, and the mortar specimens at depths of 2.5 mm, 7.5 mm, 12.5 mm and 17.5 mm were taken from the bottom of specimens. These mortars were used to test the free and total chloride ions at different concrete depths. The free chloride ion content is represented by c_f. The total chloride ion content is represented by c_t. The bound chloride ion content is represented by c_b and is equal to c_t minus c_f. The chloride binding capacity is represented by R is equal to c_b divided by c_f.

The SEM observations were performed by means of a Zeiss SU–PRA55 scanning electron microscope. SEM images were obtained under an accelerating voltage of 20 kV. Before the observation, gold was sputtered on the sample.

3. Results and Discussion

3.1. Workability of High–Strength MSRAC

Figure 6 illustrates the results of the slump tests for the fresh concrete. Table 5 shows the fluidity of the fresh concrete. It can be seen that with increasing MS rates of replacement, slump gradually decreased, and the fluidity of the fresh concrete deteriorated. The slump of MSRAC was lower than that of RAC. When the rate of replacement of MS2 exceeded 75%, the slump decreased significantly and the fluidity of the concrete deteriorated. When the rate of replacement of MS3 exceeded 50%, the slump decreased sharply, and the fresh concrete had great viscosity and almost no fluidity. This phenomenon could be due to the quality of the MS. The stone powder content and MBV of the MS were higher than the RS, and increased with decreasing MS quality. Since the specific surface area of stone powder was much higher than that of MS and RS, the water consumption required for wrapping inevitably increased with increasing stone powder content, thereby reducing the workability of the fresh concrete. Studies have shown that MBV is positively correlated with the mud powder content in MS. The higher the mud powder content, the higher the MBV [36]. Mud powder absorbed a large amount of free water, reducing the mixing water. As a result, the slump of the fresh concrete decreased and the workability deteriorated.

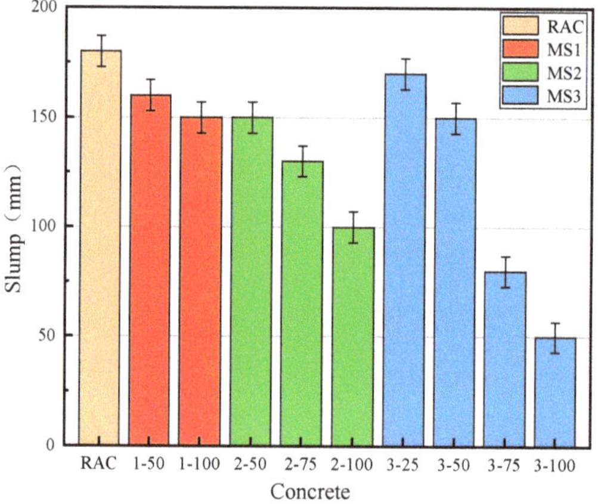

Figure 6. Slump of concrete.

3.2. 28 d Compressive Strength of High—-Strength MSRAC

Figure 7 shows the 28 d compressive strength of concrete prepared with three different levels of MS quality at different rates of replacement. It can be seen that the compressive strength of MSRAC increased with increasing rate of replacement with MS1. Specifically, the 28 d compressive strengths of concrete containing 50% and 100% MS1 were 2.9% and 6.4% higher than that of RAC, respectively. Due to the low crushing value of MS1, the "interlocking effect" between MS particles took place fully, based on the grade coordination. In addition, an appropriate amount of stone powder in MS1 improved the particle packing density of the concrete, which increased the density of the interface transition zone between the slurry and the RAC and improved the compressive strength [37]. Additionally, the limestone particles in the stone powder played a nucleation role in the formation of $Ca(OH)_2$ and C–S–H during the early stage of cement hydration, accelerating the hydration of C3S minerals [38] and improving the compressive strength of the concrete [39].

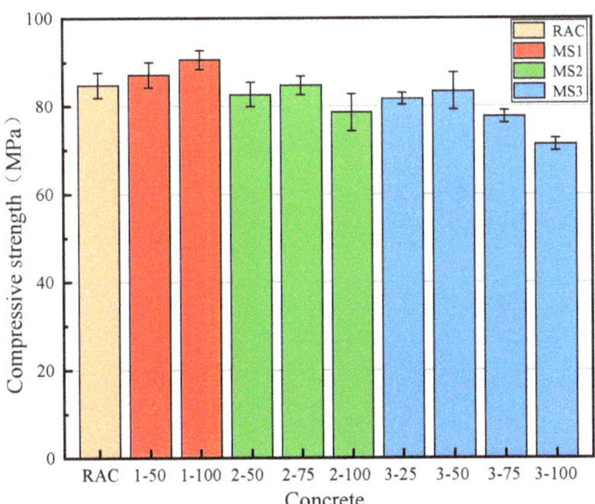

Figure 7. 28 d compressive strength of concrete.

However, the 28 d compressive strength of the MSRAC first increased, and then subsequently decreased with increasing rates of replacement with MS2 and MS3. The variations in the compressive strength of concrete containing 50% and 75% MS2 were not obvious, but compressive strength decreased sharply when MS2 replacement reached 100%. The compressive strength of the MSRAC with 100% MS2 was 6.2 MPa lower than that of RAC. For the MSRAC with MS3, the variations in compressive strength of the concrete containing 25% and 50% MS3 were not obvious, but compressive strength decreased sharply when MS3 replacement reached 75%. The compressive strengths of concrete containing 75% and 100% MS3 were 11.1% and 16.3% lower than that of RAC, respectively. The quality of MS2 and MS3 was inferior to that of MS1 and RS, leading to a lower compressive strength in the concrete. In addition, the excessive stone powder content in MS2 and MS3 was responsible for higher water absorption and the slump decrease. This led to a lower amount of water being available for hydration, as well as a more porous cementitious matrix. The densest accumulations of aggregates in the concrete were destroyed, and the interfacial transition zone of the RAC was deteriorated, resulting in decreased concrete strength. In summary, in terms of the compressive strength of the MSRAC, the optimal rate of replacement of high–quality MS was 100%, the optimal rate of replacement of medium quality MS was 50–75%, and the optimal rate of replacement of low–quality MS was 25–50%.

3.3. Chloride Diffusion Coefficient of High—-Strength MSRAC

Figure 8 shows the chloride diffusion coefficient of MSRAC. It can be seen that the worse the MS quality and the higher the rate of replacement, the higher the chloride diffusion coefficient. When the replacement of MS was 100%, the maximum chloride diffusion coefficient occurred in the concrete containing MS3, and the minimum chloride diffusion coefficient occurred in the concrete containing MS1. For the MSRAC with MS1, the chloride diffusion coefficient of MS1–50 was slightly lower than that of RAC, while it increased by 14.3% when the rate of replacement with MS1 was 100%. For the MSRAC with MS2, the chloride diffusion coefficient of the concrete was greater than that of RAC. The chloride diffusion coefficients of MSRAC containing 50%, 75% and 100% MS2 were 17.9%, 7.1% and 21.4% higher than that of RAC, respectively. For the MSRAC with MS3, the chloride diffusion coefficient increased with increasing rate of MS3 replacement. The chloride diffusion coefficient of MSRAC containing 100% MS3 was 50% higher than that of RAC.

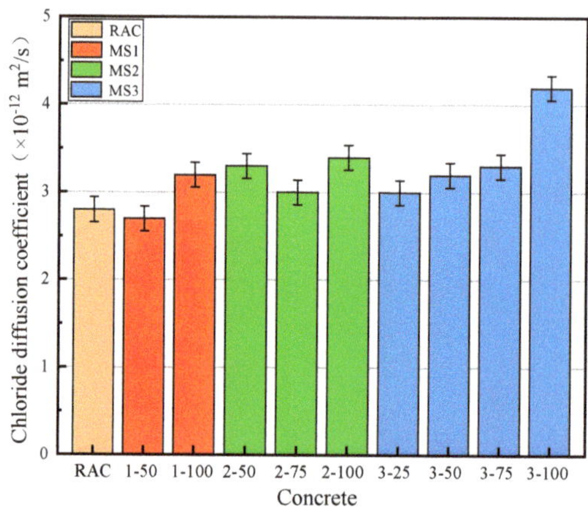

Figure 8. Chloride diffusion coefficient of concrete.

In RAC, the mortar and interfacial transition zone were weak areas that favored chloride penetration. On the one hand, due to the small particle size of stone powder, the stone powder was able to fill in the pores of the mortar and the interfacial transition zone, resulting in a closer accumulation of concrete particles. The number of open pores decreased and the chloride diffusion coefficient of the concrete decreased [40]. On the other hand, the mud powder content, represented by MBV, also affected the chloride penetration resistance of the concrete. When the mud powder content increased, the mud powder absorbed water, which was subsequently no longer available for cement paste hydration and workability, resulting in greater porosity, promoting chloride penetration [36]. When the positive effect is greater than the negative effect, the chloride diffusion coefficient increases. High–strength concrete contains a large amount of cementitious materials. Therefore, the density of concrete is high, and the accumulation of particles is close. The positive effect of stone powder as microfiller was not strong. The excess stone powder exerted a negative effect similar to that of mud powder. With decreasing MS quality, the stone powder content and MBV increased, resulting in an increase in the chloride diffusion coefficient.

The chloride diffusion coefficient limitation of concrete according to GB/T 50476–2019 [41] is shown in Table 7. The chloride diffusion coefficients of RAC prepared with medium– and high–quality MS at different rates of replacement all met the design requirements of the specification. Therefore, the maximum rate of replacement of medium– and high–quality MS based on their chloride penetration resistance was 100%.

Table 7. The chloride diffusion coefficient limitation of concrete [41].

Environment Action Classification	Design Working Life	
	50 Years	100 Years
D [a]	$< 10 \times 10^{-12}$ m^2/s	$< 7 \times 10^{-12}$ m^2/s
E [b]	$< 6 \times 10^{-12}$ m^2/s	$< 4 \times 10^{-12}$ m^2/s

[a] Atmospheric region (mild salt fog): Atmosphere at sea more than 15 m above average water level; Outdoor environment on land within 100–300 m of the tidal shoreline; Mild sputtering environment of deicing salt; Contact with high concentration of chloride ion water and a dry–wet alternate environment. [b] Atmospheric region (heavy salt fog): Atmosphere at sea less than 15 m above average water level; Outdoor environment on land within 100 m of the tidal line and 15 m below the sea level; Tide and splash zones; Direct contact with de–icing salt solution environment; Heavy sputtering by deicing salt solution or heavy salt fog environment; Contact with high concentration of chloride solution and dry–wet alternate environment.

The chloride diffusion coefficient of MS3–100 was 4.2×10^{-12} m^2/s. Therefore, through data fitting of MS3 series' chloride diffusion coefficient, Equation (1) was obtained. It can be seen that the rate of replacement of low–quality MS had a quadratic parabolic relationship with the chloride diffusion coefficient, as shown in Figure 9. Substituting $D_{RCM} = 4 \times 10^{-12} m^2/s$ into Equation (1), the maximum rate of replacement of low–quality MS based on chloride penetration resistance was found to be 96.0%.

$$y = 1.48571 \times 10^{-4}x^2 - 0.00246x + 2.86571, \ x \in [0, 100], \ R^2 = 0.933 \tag{1}$$

where y is the chloride diffusion coefficient, and x is the rate of replacement.

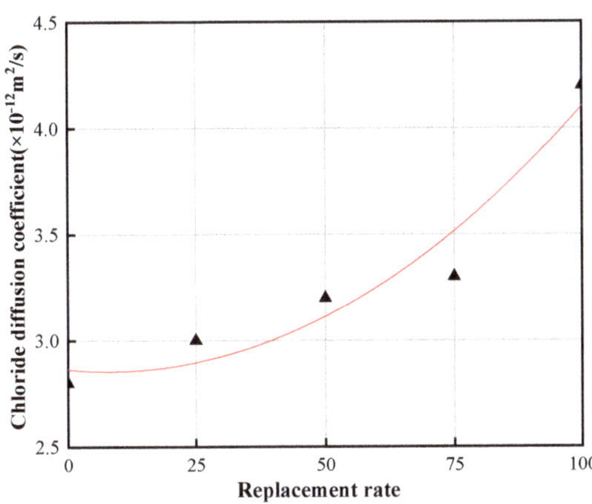

Figure 9. Fitted graph of chloride diffusion coefficient of MS3 series concrete.

3.4. Free and Bound Chloride Ion Content of High–Strength MSRAC

Figure 10a shows the relationship between the free chloride ion content of concrete and sample depth. It can be seen that the free chloride ion contents of RAC and MS1–50 at different depths were similar and were the lowest, which is due to the dense structures of RAC and MS1–50, restricting the invasion of chloride ions. The free chloride ion content of MS3–100 at each depth was the highest. This is because the concentrations of mud powder and stone powder in MS3 were significantly greater than those in RS, MS1 and MS2. These fine particles absorbed free water, reduced workability and cement hydration, increased porosity, and provided channels for chloride ion penetration. Therefore, the free chloride ion content of MS3–100 was the highest. Free chloride ion is the main cause of steel corrosion and bearing capacity decline in concrete. The passive film on the surface of steel bars breaks, and pitting corrosion occurs as a result of chloride ion corrosion, resulting in the generation of ferric hydroxide, the volume of which expands by 3–4 times, resulting in expansion and cracking of concrete protective layer [42].

Chloride ion binding in concrete is the phenomenon whereby chloride ion binds with cement hydration products, including through physical adsorption and chemical bonding. The physical adsorption of chloride ions is due to electrostatic or van der Waals forces attached to the pore wall or hydration products. This combination is unstable, and chloride ions can easily become free [43]. Chemical bonding is the interaction between chloride ions and hydration products in the form of chemical bonds. On the one hand, chloride ions react with calcium aluminate hydrates (C–A–H) to form Friedel's salt ($3CaO \bullet Al_2O_3 \bullet CaCl_2 \bullet 10H_2O$). On the other hand, they react with calcium hydroxide in the pores to form expansive compound salt ($CaCl_2 \bullet Ca(OH)_2 \bullet 2H_2O$) [44]. Figure 10b,c show the trends in the variations of bound chloride ion content and chloride ion binding rate with sample depth. It can be seen that the bound chloride ion content and chloride ion binding rate at the same depth increased with decreasing MS quality and increasing rate of replacement. This is mainly due to the high stone powder and mud powder content. These powders absorbed free water, which was then no longer available for cement paste hydration and workability, subsequently resulting in greater porosity, destroying the densest accumulation of aggregates in the concrete and resulting in an increase in the porosity and specific surface area of mortar. The binding ability of the concrete to chloride ions was improved, leading to high bound chloride ion content and chloride ion binding rate. Additionally, it can be seen from Figure 10b that there was a decrease in the bound chloride ion content with increasing sample depth. In Figure 10c, the chloride ion binding rate increased with increasing sample depth. The reason for this is that the penetration depth of chloride ion into concrete is limited. With increasing sample depth, the free chloride ion content decreased, resulting in less bound chloride ions. Finally, the free chloride ion content gradually became equal to the bound chloride ion content. Therefore, the bound chloride ion content decreased, and the chloride binding rate increased. Previous studies have shown that the bound chloride ion content in concrete is unstable. Due to the action of environmental chloride ion concentration, acidification, chemical corrosion, external electric field, and temperature, the stability of bound chloride ion content can change, releasing free chloride ions [45,46]. Therefore, the influence of bound chloride ion content should be considered in the evaluation of the chloride penetration resistance of MSRAC.

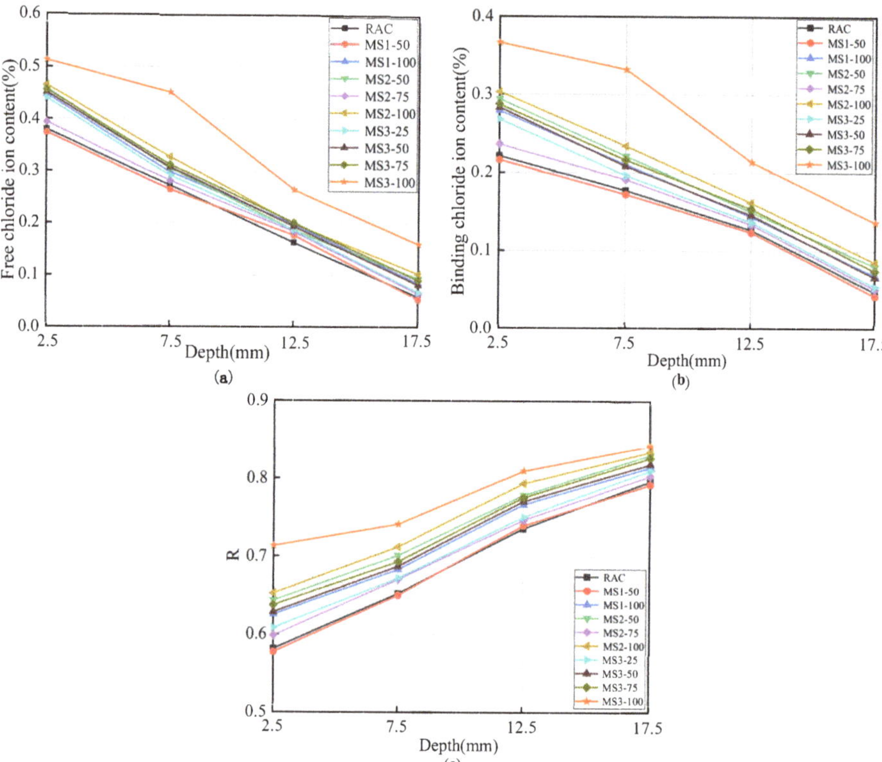

Figure 10. Relationship between chloride content and diffusion depth of concrete: (**a**) free chloride ion content; (**b**) bound chloride ion content; (**c**) chloride binding rate.

Overall, the chloride ion content and chloride ion binding rate of the MSRAC were higher than those of the RAC. The lower the quality of MS and the higher the rate of replacement, the higher the chloride ion content and binding rate were. In addition, the chloride ion content of concrete was positively correlated with chloride diffusion coefficient.

Izquierdo et al. [47] showed that the critical free chloride ion content (in the mass fraction of cementitious material) of reinforcement ranged from (0.497 ± 0.26)% to (0.569 ± 0.177)%, and the critical total chloride ion content (in the mass fraction of cementitious material) ranged from (0.632 ± 0.112)% to (0.771 ± 0.346)%. It can be concluded from Figure 10a,b that at different sample depths, only the free chloride ion content of MS3–100 at 2.5 mm reached the critical value. The total chloride ion content of MS1–100, MS2–50, MS2–100 and MS3 series at 2.5 mm reached the critical value.

3.5. SEM Analysis

Figure 11 presents the SEM images of RAC and MSRAC with medium– and low–quality MS before the RCM test. Compared to RAC (Figure 11a), it can be seen that the microstructure of the MSRAC prepared with MS was looser, with MS3–100 being the looses. From Figure 11b–e, it can be seen that the microstructure of MSRAC became increasingly loose, accompanied by the formation of micro–cracks and pores, with increasing rate of replacement of medium– and low–quality MS. The bonding between mortar and aggregate was fragile, and the interfacial transition zone was fuzzy. Additionally, there were some micro–cracks around the interfacial transition zone. In this experiment, the quality and rate of replacement of MS were the determinants of MSRAC performance. The gradual deterioration of the microstructure indicated that the compressive strength and chloride resistance penetration of the MSRAC deteriorated with decreasing MS quality and increasing rate of replacement. This verified the macro law at the micro level.

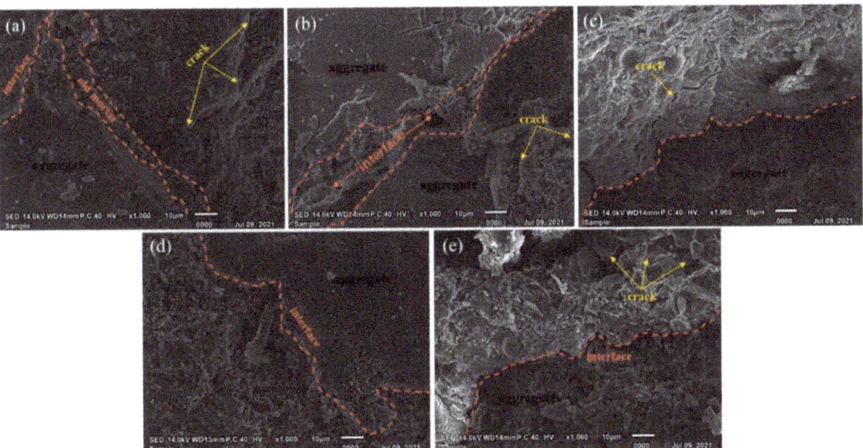

Figure 11. SEM images of RAC and MSRAC with medium– and low–quality MS before the RCM test: (**a**) RAC; (**b**) MS2–50; (**c**) MS2–100; (**d**) MS3–25; (**e**) MS3–100.

Figure 12 shows the SEM images of the RAC and MSRAC with medium– and low–quality MS after the RCM test. Compared to Figure 11, it can be seen that after the RCM test, the microstructure of the concrete had deteriorated further, resulting in more microcracks and pores, and the surface became rougher. Large amounts of stone and mud powder can be observed on the surface of MS2–100, MS3–25 and MS3–100. These fine particles absorbed free water for hydration, leading to blurring and refining of the interfacial transition zone, resulting in more microcracks in the interfacial transition zone. These microcracks decreased the combination of aggregate and mortar and provided more channels for chloride ion penetration. In addition, there were many white salt crystals on the concrete surface after the RCM test, especially on the surface of MS3–100. This indicated that after entering the concrete, the chloride ions underwent a chemical combination reaction with the hydration products, forming expansive compound salt. These two reactions consumed the hydroxyl ion inside the concrete. The pH value of the pore solution decreased, resulting in the decomposition of C–S–H. This deteriorated the pore structure of the concrete and decreased the compactness and stability of the concrete [48,49]. This also explained the reason for which MS3–100 possessed the highest chloride diffusion coefficient and chloride ion content.

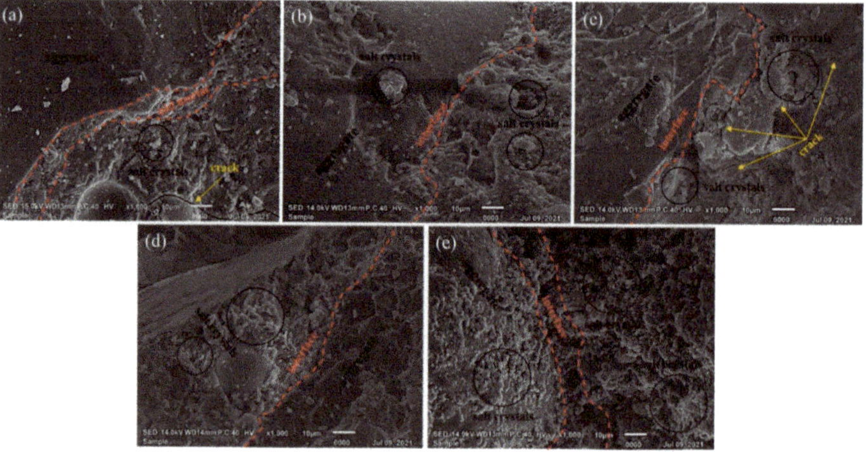

Figure 12. SEM images of RAC and MSRAC with medium– and low–quality MS after the RCM test: (**a**) RAC; (**b**) MS2–50; (**c**) MS2–100; (**d**) MS3–25; (**e**) MS3–100.

4. Conclusions

Three quality levels of MS were used to prepare MSRAC with different rates of replacement. The effects of the three kinds of MS with different rates of replacement on the compressive strength and chloride ion permeability of high–strength RAC were investigated. The following conclusions can be drawn from the results of this study:

(1) The worse the quality of the MS, the lower the compressive strength of the MSRAC. The compressive strength of MSRAC with high–quality MS increased with increasing rate of replacement. However, for medium– and low–quality MS, the compressive strength of MSRAC first increased and then decreased with increasing rate of replacement.

(2) The chloride diffusion coefficient, and the free and bound chloride ion content of MSRAC increased with decreasing MS quality and increasing MS rate of replacement. Except for MS3–100, the chloride diffusion coefficient was lower than the standard limits of a working life of 50 years and 100 years when the chloride environmental action grade is D and E. The chloride diffusion coefficient of MSRAC with MS3 showed a quadratic parabola relationship with the rate of replacement. The maximum rate of replacement of MS3 in 100 years under environmental action level E was 96.0%.

(3) In light of the compressive strength and chloride penetration resistance, high–quality MS was able to completely replace RS, while the rates of replacement for medium– and low–quality MS were 50–75% and 25–50% in high–strength MSRAC, respectively.

(4) SEM analysis showed that an appropriate amount of stone powder can fill the micro cracks and interfacial transition zone of RAC. This can improve the compressive strength of the RAC. Conversely, excessive stone powder and mud powder absorbed free water when the concrete mixing is insufficient, resulting in the blurring of the interfacial transition zone and an increase in the number of microcracks and pores, having an unfavorable effect on compressive strength and chloride penetration resistance.

Author Contributions: Conceptualization, J.F.; funding acquisition, J.F.; formal analysis, C.D.; methodology, C.D.; writing—original draft preparation, C.D.; investigation, C.C.; supervision, C.C.; writing—review and editing, X.W.; project administration, X.W.; data curation, Z.Q.; resources, Z.Q. All authors have read and agreed to the published version of the manuscript.

Funding: This research was funded by the National Natural Science Foundation of China (52108190), Prospective Joint Research Project of Jiangsu Province (BY2020471).

Institutional Review Board Statement: Not applicable.

Informed Consent Statement: Not applicable.

Data Availability Statement: The data presented in this study are available on request from the corresponding author.

Acknowledgments: The authors hereby sincerely acknowledge the funder for their support.

Conflicts of Interest: The authors declare that they have no known competing financial interests or personal relationships that could have appeared to influence the work reported in this paper.

Abbreviations

MS	manufactured sand
NAC	natural aggregate concrete
RCA	recycled coarse aggregate
SEM	Scanning Electron Microscope
RCM	rapid chloride ions migration coefficient
C_3S	Ca_3SiO_5
RAC	recycled aggregate concrete
MSRAC	manufactured sand recycled aggregate concrete
MMBV	multiplication of stone powder content and MBV
MBV	methylene blue value
RS	river sand
C–S–H	calcium silicate hydrates

References

1. Shen, W.; Cao, L.; Li, Q.; Zhang, W.; Wang, G.; Li, C. Quantifying CO_2 emissions from China's cement industry. *Renew. Sustain. Energ. Rev.* **2015**, *50*, 1004–1012. [CrossRef]
2. Raman, S.N.; Ngo, T.; Mendis, P.; Mahmud, H.B. High–strength rice husk ash concrete incorporating quarry dust as a partial substitute for sand. *Constr. Build. Mater.* **2011**, *25*, 3123–3130. [CrossRef]
3. Manning, D.; Vetterlein, J. *Exploitation and Use of Quarry Fines. Report No: 087. MIST2/DACM/01*; Mineral Solutions Limited: Manchester, UK, 2004.
4. Thomas, T.; Dumitru, I.; van Koeverden, M. Manufactured Sand. National Test Methods and Specification Values. Available online: https://www.ccaa.com.au (accessed on 7 April 2016).
5. Gonçalves, J.P.; Tavares, L.M.; Toledo Filho, R.D.; Fairbairn, E.M.; Cunha, E.R. Cunha, Comparison of natural and manufactured fine aggregates in cement mortars. *Cem. Concr. Res.* **2007**, *37*, 924–932. [CrossRef]
6. Mo, K.H.; Alengaram, U.J.; Jumaat, M.Z.; Liu, M.Y.; Lim, J. Assessing some durability properties of sustainable lightweight oil palm shell concrete incorporating slag and manufactured sand. *J. Clean. Prod.* **2016**, *112*, 763–770. [CrossRef]
7. Xu, W.; Wen, X.; Wei, J.; Xu, P.; Zhang, B.; Yu, Q.; Ma, H. Feasibility of kaolin tailing sand to be as an environmentally friendly alternative to river sand in construction applications. *J. Clean. Prod.* **2018**, *205*, 1114–1126. [CrossRef]
8. Nanthagopalan, P.; Santhanam, M. Fresh and hardened properties of self–compacting concrete produced with manufactured sand. *Cem. Concr. Compos.* **2011**, *33*, 353–358. [CrossRef]
9. Cortes, D.D.; Kim, H.K.; Palomino, A.M.; Santamarina, J.C. Rheological and mechanical properties of mortars prepared with natural and manufactured sands. *Cem. Concr. Res.* **2008**, *38*, 1142–1147. [CrossRef]
10. Shen, W.; Liu, Y.; Wang, Z.; Cao, L.; Wu, D.; Wang, Y.; Ji, X. Influence of manufactured sand's characteristics on its concrete performance. *Constr. Build. Mater.* **2018**, *172*, 574–583. [CrossRef]
11. Shen, W.; Yang, Z.; Cao, L.; Cao, L.; Liu, Y.; Yang, H.; Lu, Z.; Bai, J. Characterization of manufactured sand: Particle shape, surface texture and behavior in concrete. *Constr. Build. Mater.* **2016**, *114*, 595–601. [CrossRef]
12. Liu, Z.A.; Zhou, M.K.; Li, B.X. Relationships between modified methylene blue value of microfines in manufactured sand and concrete properties. *J. Wuhan Univ. Technol.–Mater. Sci. Ed.* **2016**, *31*, 574–581. [CrossRef]

13. Chen, X.; Yuguang, G.; Li, B.; Zhou, M.; Li, B.; Liu, Z.; Zhou, J. Coupled effects of the content and methylene blue value (MBV) of microfines on the performance of manufactured sand concrete. *Constr. Build. Mater.* **2020**, *240*, 117953. [CrossRef]
14. Li, B.X.; Yin, L.Y.; Feng, Z.H.; Fan, L.L.; Ye, X.S. Study on the effects of rock dust content on the durability of C60 manufactured sand marine concrete. *Concrete* **2017**, *10*, 169–173. (In Chinese)
15. Li, B.X.; Ke, G.J.; Zhou, M.K. Influence of manufactured sand characteristics on strength and abrasion resistance of pavement cement concrete. *Constr. Build. Mater.* **2011**, *25*, 3849–3853. [CrossRef]
16. Ji, T.; Chen, C.Y.; Zhuang, Y.Z.; Chen, J.F. A mix proportion design method of manufactured sand concrete based on minimum paste theory. *Constr. Build. Mater.* **2013**, *44*, 422–426. [CrossRef]
17. Westerholm, M.; Lagerblad, B.; Silfwerbrand, J.; Forssberg, E. Influence of fine aggregate characteristics on the rheological properties of mortars. *Cem. Concr. Compos.* **2008**, *30*, 274–282. [CrossRef]
18. Shen, W.G.; Liu, Y.; Cao, L.H. Mixing design and microstructure of ultra high strength concrete with manufactured sand. *Constr. Build. Mater.* **2017**, *143*, 312–321. [CrossRef]
19. GB/T 14684–2011. *Sand for Construction*; Standardization Administration of the People's Republic of China: Beijing, China, 2011.
20. Tian, Y.; Jiang, J.; Wang, S.; Yang, T.; Qi, L.; Geng, J. The mechanical and interfacial properties of concrete prepared by recycled aggregates with chloride corrosion media. *Constr. Build. Mater.* **2021**, *282*, 122653. [CrossRef]
21. Hao, L.; Liu, Y.; Wang, W.; Zhang, J.; Zhang, Y. Effect of salty freeze–thaw cycles on durability of thermal insulation concrete with recycled aggregates. *Constr. Build. Mater.* **2018**, *189*, 478–486. [CrossRef]
22. Zhao, Y.X.; Zeng, W.L.; Zhang, H.R. Properties of recycled aggregate concrete with different water control methods. *Constr. Build. Mater.* **2017**, *152*, 539–546. [CrossRef]
23. Bravo, M.; De Brito, J.; Pontes, J.; Evangelista, L. Durability performance of concrete with recycled aggregates from construction and demolition waste plants. *Constr. Build. Mater.* **2015**, *77*, 357–369. [CrossRef]
24. Kapoor, K.; Singh, S.P.; Singh, B. Durability of self–compacting concrete made with Recycled Concrete Aggregates and mineral admixtures. *Constr. Build. Mater.* **2016**, *128*, 67–76. [CrossRef]
25. Mardani–Aghabaglou, A.; Yüksel, C.; Beglarigale, A.; Ramyar, K. Improving the mechanical and durability performance of recycled concrete aggregate–bearing mortar mixtures by using binary and ternary cementitious systems. *Constr. Build. Mater.* **2019**, *196*, 295–306. [CrossRef]
26. Matar, P.; Barhoun, J. Effects of waterproofing admixture on the compressive strength and permeability of recycled aggregate concrete. *J. Build. Eng.* **2020**, *32*, 101521.
27. Duan, Z.H.; Poon, C.S. Properties of recycled aggregate concrete made with recycled aggregates with different amounts of old adhered mortars. *Mater. Des.* **2014**, *58*, 19–29. [CrossRef]
28. Dodds, W.; Christodoulou, C.; Goodier, C.; Austin, S.; Dunne, D. Durability performance of sustainable structural concrete: Effect of coarse crushed concrete aggregate on rapid chloride migration and accelerated corrosion. *Constr. Build. Mater.* **2017**, *155*, 511–521. [CrossRef]
29. Guo, H.; Shi, C.; Guan, X.; Zhu, J.; Ding, Y.; Ling, T.C.; Zhang, H.; Wang, Y. Durability of recycled aggregate concrete—A review. *Cem. Concr. Compos.* **2018**, *89*, 251–259. [CrossRef]
30. GB/T 25177–2010. *Recycled Coarse Aggregate for Concrete*; Standardization Administration of the People's Republic of China: Beijing, China, 2010.
31. Tam, V.; Tam, C. Assessment of durability of recycled aggregate concrete produced by two–stage mixing approach. *J. Mater. Sci.* **2007**, *42*, 3592–3602. [CrossRef]
32. GB/T 50081–2002. *Standard for Test Method of Mechanical Properties on Ordinary Concrete*; Standardization Administration of the People's Republic of China: Beijing, China, 2002.
33. GB/T 50080–2016. *Standard for Test Method of Performance on Ordinary Fresh Concrete*; Standardization Administration of the People's Republic of China: Beijing, China, 2016.
34. GB/T 50082–2009. *Standard Test Method for Long–Term Performance and Durability of Ordinary Concrete*; Standardization Administration of the People's Republic of China: Beijing, China, 2009.
35. SL 352–2006. *Test Code for Hydraulic Concrete*; China Water Conservancy Press: Beijing, China, 2006.
36. Wang, J.; Niu, K.; Tian, B.; Sun, L. Effect of methylene blue (MB)–value of manufactured sand on the durability of concretes. *J. Wuhan Univ. Technol.–Mater. Sci. Ed.* **2012**, *27*, 1160–1164. [CrossRef]
37. Tragardh, J. Microstructural feature sand related properties of self–compacting concrete. In Proceedings of the First International RILEM Symposium on Self–compacting Concrete, Cachan Cedex: RILEM, Stockholm, Sweden, 13–15 September, 1999; pp. 175–186.
38. Bonavetti, V.; Donza, H.; Rahhal, V.; Irassar, E. Influence of initial curing on the properties of concrete containing limestone blended cement. *Cem. Concr. Res.* **2000**, *30*, 703–708. [CrossRef]
39. Bonavetti, V.L.; Rahhal, V.F.; Irassar, E.F. Studies on the carboaluminate formation in limestone filler–blended cements. *Cem. Concr. Res.* **2001**, *31*, 853–859. [CrossRef]
40. Li, B.X.; Wang, J.L.; Zhou, M.K. Effect of limestone fines content in manufactured sand on durability of low– and high–strength concretes. *Constr. Build. Mater.* **2009**, *23*, 2846–2850. [CrossRef]
41. GB/T 50476–2019. *Standard for Design of Concrete Structure Durability*; Standardization Administration of the People's Republic of China: Beijing, China, 2019.

42. Tu, X.; Zhu, B.J.; Zhu, J. Two–dimensional numerical simulation approach for corrosion and cracking of reinforced concrete induced bychloride. *Concrete* **2018**, *9*, 23. (In Chinese)
43. Su, L.; Niu, D.T.; Luo, Y.; Huang, D.G.; Luo, D.M. Inner Chloride ion diffusion and capillary water absorption properties of fly ash coral aggregate concrete. *J. Build. Mater.* **2021**, *1*, 77–86. (In Chinese)
44. Li, H.; Guo, Q.J.; Wang, J.B.; Zhang, K.F. Meso–/Micro–Structure of Interfacial Transition Zone and Durability of Recycled Aggregate Concrete: A Review. *Mater. Rep.* **2020**, *34*, 13050–13057. (In Chinese)
45. Byung, H.O.; Seung, Y.J. Effects of material and environmental parameters on chloride penetration profiles in concrete structures. *Cem. Concr. Res.* **2007**, *37*, 47–53.
46. Han, S.H. Influence of diffusion coefficient on chloride ion penetration of concrete structure. *Constr. Build. Mater.* **2007**, *21*, 370–378. [CrossRef]
47. Izquierdo, D.; Alonso, C.; Andrade, C.; Castellote, M. Potentiostatic determination of chloride threshold values for rebar depassivation Experimental and statistical study. *Electrochim. Acta* **2004**, *49*, 2731–2739. [CrossRef]
48. Andrés, V.Z.Y.; Javier, Z.C.; Antonio, D.M.Á. Chloride Penetration and Binding in Recycled Concrete. *J. Mater. Civil. Eng.* **2008**, *20*, 449–455.
49. Zuquan, J.; Xia, Z.; Tiejun, Z.; Jianqing, L. Chloride ions transportation behavior and binding capacity of concrete exposed to different marine corrosion zones. *Constr. Build. Mater.* **2018**, *177*, 170–183. [CrossRef]

 *applied sciences*

Article

Evaluation of Cement Performance Using Industrial Byproducts Such as Nano MgO and Fly Ash from Greece

Panagiota P. Giannakopoulou [1], Aikaterini Rogkala [1], Paraskevi Lampropoulou [1], Maria Kalpogiannaki [1] and Petros Petrounias [1,2,*]

[1] Section of Earth Materials, Department of Geology, University of Patras, 26504 Patras, Greece; peny_giannakopoulou@windowslive.com (P.P.G.); k.rogkala@upatras.gr (A.R.); p.lampropoulou@upatras.gr (P.L.); maria.kalpogiannaki@gmail.com (M.K.)
[2] Centre for Research & Technology Hellas (CERTH), Chemical Process & Energy Resources Institute, Maroussi, 15125 Athens, Greece
* Correspondence: Geo.plan@outlook.com

Abstract: The need for environmentally friendly construction materials is growing more and more these days. This paper investigates byproducts from Greece, such as magnesite tailings from Evoia and fly ash from Kardia (Ptolemais), in order to evaluate their suitability as cement additives. For this purpose, the raw materials were tested and studied regarding their mineralogical and chemical components for their morphological characteristics. Different cement specimens of various mixtures of raw materials were produced and tested. These raw materials are considered suitable for cement additives. The effect of nano MgO content seems to have played a more critical role in the physicomechanical performance of produced cement compared to that of the fly ash content. Furthermore, more satisfactory results in the physicomechanical properties of the produced cement gave samples of group II containing 3–4% of nano MgO. Nano MgO content up to 4% seems to have negative influence on the compressive strength of the produced cement, simultaneously reducing its durability. The increase of nano MgO content leads to the increase of the expansion of the produced cement specimens. In the early stage, the expansion rate was intensively larger. With the consumption of nano MgO, the expansion in the later stage gradually slowed down and tended to stabilize.

Keywords: magnesite tailings; nano MgO; fly ash; cement

Citation: Giannakopoulou, P.P.; Rogkala, A.; Lampropoulou, P.; Kalpogiannaki, M.; Petrounias, P. Evaluation of Cement Performance Using Industrial Byproducts Such as Nano MgO and Fly Ash from Greece. *Appl. Sci.* **2021**, *11*, 11601. https://doi.org/10.3390/app112411601

Academic Editors: Carlos Morón Fernández and Daniel Ferrández Vega

Received: 22 October 2021
Accepted: 4 December 2021
Published: 7 December 2021

Publisher's Note: MDPI stays neutral with regard to jurisdictional claims in published maps and institutional affiliations.

1. Introduction

Cement is widely used as a basic component of concrete. Due to the rapid construction development, the demand for cement and natural aggregates has, exceptionally, been increased. More specifically, in 2014, about 40 billion tons of aggregates and 4 billion tons of cement were required for constructions all over the world [1,2]. As a result, a great amount of carbon dioxide (CO_2) is released in the air. This is the reason why several researchers, i.e., [3,4], have turned their attention to producing environmentally friendly applications, and at the same time, they have reduced CO_2 emissions during their production. The reaction of cement and water produces volume shrinkage, which may lead to structural cracking of cement-based materials [5]. This is detrimental not only for the mechanical properties and durability of cement-based materials, but it may also lead to shorter service life of constructions, resulting in a great loss of natural resources. The problem of shrinkage of cement is becoming more and more serious, as by increasing the concrete strength, the amount of cement per unit volume of concrete increases, too [6,7]. Cementitious concrete is the most-used manmade material among all, and comprises a mixture of mortar, aggregates, and water [2,4,8]. The main component of concrete is the material which binds the aggregate particles together, commonly comprising a mixture of cement and water [4,8]. Concrete structures can be described as a three-phase system composed of hardened cement paste, aggregate, and the interface between aggregate particles and cement paste [8,9].

One of the effective ways to reduce cracking is by adding expansion agent in cement concrete or by using the expansion generated during the hydration process of expansion agent to compensate the shrinkage of cement concrete [5,6]. Nowadays, the expansion agents mainly used in several applications primarily include sulfoaluminate-type, CaO-type, and MgO-type expansion agents. In comparison with the first two kinds of expansion agents, MgO expansion agent (nano MgO) displays a lot of advantages, including less water requirement for hydration, stable hydration product $Mg(OH)_2$, adjustable design of expansion process, and so on, in order to widely to be used in modern concrete [10]. Since the 1970s, numerous researchers have begun to study MgO expansion agent (MEA), which has been applied to a wide range of buildings, such as dams. The expansion produced by MEA hydration is used to compensate the temperature drop shrinkage and dry shrinkage of mass concrete, effectively improving the crack resistance and durability of the structure [11,12].

Cements with high MgO content have gained more and more popularity in the last decade, perhaps due to the augmented concern about climate change, with the intention and need of reducing the CO_2 emissions regarding the production of conventional Portland cements. Some authors believe that it is possible to produce such type of cements with a high MgO content and reduced CO_2 emissions [13]. In recent decades, the major motivation for the development and uptake of MgO-based cements has been driven from an environmental standpoint. The lower temperatures required for the production of MgO compared to the conversion of $CaCO_3$ to PC and the energy savings associated with this reduced temperature have led researchers to envision MgO-based cements as being central to the future of ecofriendly cement production. Equally, the ability of MgO to absorb CO_2 from the atmosphere to form a range of carbonates and hydroxycarbonates lends itself well to the discussion of "carbon-neutral" cements, which could potentially absorb close to as much CO_2 during their service life as was emitted during their manufacture. These two interconnected aspects have led to a recent explosion in interest, both academic and commercial, in the area of MgO-based cements.

Magnesium is the eighth most abundant element in the Earth's crust, at ~2.3% by weight, present in a range of rock formations such as dolomite, magnesite, and silicate. Magnesium is also the third-most abundant element in solution in seawater, with concentrations of ~1300 ppm [5]. The mineral magnesite comprises a widely used source of magnesium oxide (MgO), because magnesium is considered a critical element by the EU. However, alternative sources, such as several Mg-rich silicates, may become more relevant in the near future. The transition of magnesite to magnesia has been achieved by heat treatment at temperatures above 600 °C. This transition from carbonate to oxide by release of carbon dioxide (CO_2) initially produces a porous microstructure with low bulk density. Moreover, this material is characterized as a strong reactive due to its low magnesia particle sizes with high surface areas which can be retained if not heated further. Such high surface area variants of magnesia are known as "active magnesia" in cement-related research. Exposing this porous form to higher temperatures (>1200 °C) solidifies the structure by a sintering mechanism and drastically reduces the reactive surface area as well as the affected sites that promote wetting and dissociation of magnesia in acid phosphate solutions. Magnesia-based cements, by definition, use MgO as a building block rather than the CaO which represents more than 60% of the elemental composition of PC. Due to the substantially different chemistry of MgO compared to that of CaO, one cannot simply change the feedstock for conventional Ca-based cements to produce a directly corresponding material using the same infrastructure. Comparing the respective (MgO, CaO)–Al_2O_3–SiO_2 ternary phase diagrams, vast differences in chemistry and phase formation are shown. More specifically, no magnesium silicate phases are formed in high temperatures that have hydraulic properties akin to those formed in the calcium-rich region of the CaO–SiO_2–Al_2O_3 system: Ca_3SiO_5, Ca_2SiO_4, and $Ca_3Al_2O_6$ are critical hydraulic phases in PC, but have no magnesian analogues.

Currently, the MEA used in the market is mainly produced by the calcining of magnesite. However, with the continuous mining of mineral resources, the problems of resource exhaustion and environmental pollution are becoming more and more threatening. Following governments' instructions regarding the mining of nonrenewable mineral resources, magnesite resources are decreasing [14] while, at the same time, a large amount of magnesite is discarded annually. These gradually accumulated large amounts of magnesite not only occupy cultivated land but also cause waste of resources. Magnesite is a raw material used in different industrial applications. It is used mainly after heat treatment for the production of caustic magnesia and dead burned magnesia, whilst untreated magnesite has a few industrial applications too (e.g., in the production of fertilizers, electrodes, and environmental protection). The main magnesite deposits in Greece appear in Chalkidiki and Evoia Island. Regarding the case of Evoia's magnesite deposits, they have been started to be exploited since 1893 mainly for the production of magnesia-rich raw materials and refractory final products. In this area, the proven, probable, and possible reserves have been approximately estimated as 35, 45, and 60 million tons, respectively [15]. The concession areas cover ~404 km^2 while all mines are within a 14 km radius around the refractory industry plants. The last is 2 km away from the loading port. According to the current activities, the annual capacity is 120,000 MT of caustic magnesia and dead burned magnesia, while above 250,000 tons of stockpile waste have been disposed of in the area during last year's activities. They consist of the fine ore (-40 mm) which is rejected before the beneficiation processing and specifically before the hand-sorting stage of magnesite. This tailing contains significant amounts of magnesite, and in this work, the feasibility of its utilization as MEA is studied. In the Greek market, materials of similar quality are found, but they are produced from natural mineral raw materials and not from byproducts.

The real contribution of such initiatives does not result in a notable decrease in negative externalities. This fact can be assigned to a limited acceptance of "greener" building materials [16]. The relatively poor adoption of modified binders represents a barrier towards a more sustainable construction industry [17]. However, the change of this paradigm needs to be completed not only by the technical parameters but also must be combined with other scientific disciplines involved in the sustainability principles [18]. In a nutshell, new design and development materials must go hand in hand with corresponding leadership, convincing communication, and complex assessment of ecofriendly materials to overcome major barriers in the conservative building industry. Numerous researchers who study green construction materials and applications often use micropetrographic analytical methods [19].

The aim of this study is to evaluate the impact of MEA combined with fly ash on final physicomechanical performance of the produced mortar, providing a new way to solve the environmental pollution of magnesite tailings and fly ashes. For this scope, magnesite tailings and fly ash from Greece were used where the influence of different percentages of MEA combined with fly ash on the physicomechanical performance of the produced cements were studied.

2. Geological Setting of Magnesite Tailings

Magnesite tailings are located in central and northern Evoia and are occupied by Jurassic ophiolitic rocks such as harzburgites and basalts (Figure 1). These rocks overlay a mélange (the Pagondas complex), comprising Upper Triassic to Jurassic sedimentary and volcanic rocks, which was developed as an accretionary prism [20]. The entire sequence of the ophiolitic rocks and the mélange is thrust on the Pelagonian carbonate platform. The ultramafic rocks are covered by Pliocene terrigenous and minor marine and lacustrine sediments, and Quaternary deposits [21]. The overthrust of the Evoia ophiolite is structurally overlain by a dismembered metamorphic sole which constitutes evidence for a late Upper Jurassic emplacement age [22,23]. Magnesite is presented as cryptocrystalline, mostly nodular, and appears as open space fillings along joints of ultramafic rocks, forming stockworks, as veins controlled by brittle faults, indicating that magnesite was deposited during opening of the veins in an extensional geodynamic regime, and as nodular forms that resulted from replacement of ultramafic rock. Those nodules are mainly centimeter-sized and contain relics of chromite. Magnesite also predates deposition of dolomite and silica polymorphs where a vertical mineral zonation is evident: magnesite modal percentages decrease towards the surface, whereas dolomite and quartz gradually become more abundant. Moreover, vuggy silica textures are presented in the uppermost parts. Brecciation of early-formed magnesite and cementation by magnesite deposited at a later stage is also observed. Both the deposits of stockwork type of mineralization and the main veins have not been deformed. Significant displacements by younger faults affecting the deposits are not observed.

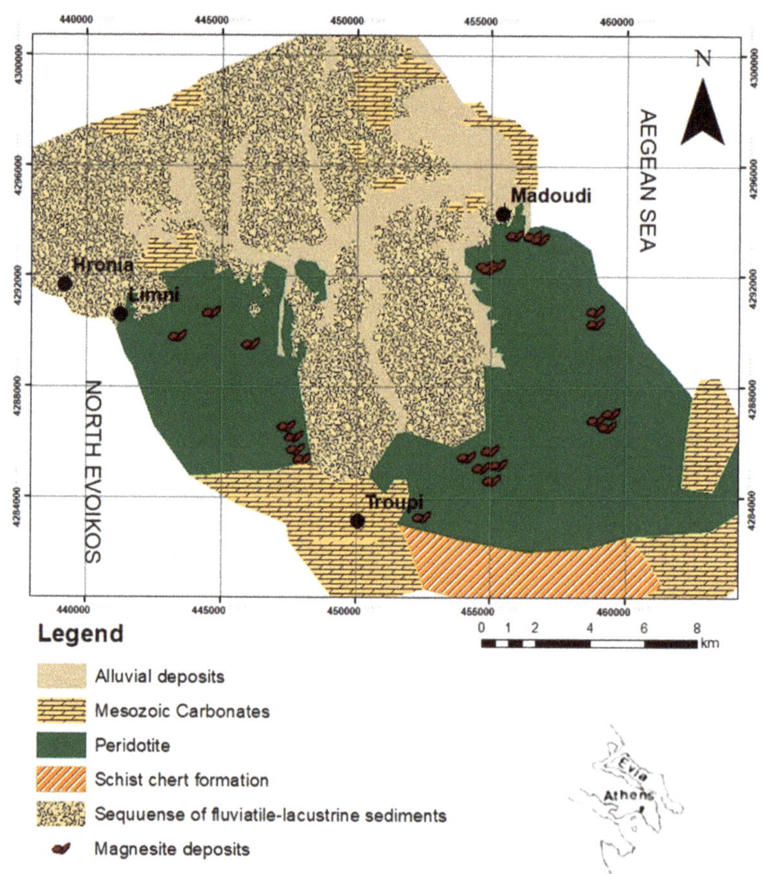

Figure 1. Modified geological map of central and northern Evoia [24] using ArcMap GIS mapping.

3. Materials and Methods

3.1. Materials

In order to investigate the influence of byproducts on the final mortar behavior, providing a new way to solve the environmental pollution of magnesite tailings, magnesite tailings from Evoia (Greece) were collected and used, in order to identify their influence on various produced cements performance. The samples were subsequently crushed, in order to be suitable for all the engineering tests and processing, which were performed according to European and International standards. Additionally, fly ash obtained from a

Kardia lignite-fired power plant in Ptolemais was used in this study in order to investigate its behavior and properties in the produced cements with MEA. Fly ash, along with bottom ash, constitutes the inorganic solid residue after coal combustion. Greek lignite yields, on average, around 30% inorganic content (fly and bottom ash), which is the basic problem because of the enormous amounts produced from lignite combustion. The cement was P·II42.5 Portland cement produced by Titan Co., Ltd. (GREECE). It is a composite cement that completes the range of Portland cements. Its enhanced properties offer high levels of early strength which make it ideal for voluminous concreting and prefabrication work. The used cement was TITAN CEM II 42,5 N which is a composite Portland-type cement with pozzolana/limestone. The pH value of water was 7.0.

3.2. Methods

3.2.1. Methods for Raw Materials

The petrographic features of raw materials were examined using a combination of methods. The mineralogical and textural characteristics of the magnesite tailings were examined in thin sections using a polarizing microscope (Leica Microsystems Leitz Wetzlar, Germany) (University of Patras). The mineralogical composition of the studied samples was determined with X-ray diffraction using a Bruker D8 advance diffractometer with Ni-filtered CuKα radiation. Random powder mounts were prepared by gently pressing the powder into the cavity holder. The scanning area for bulk mineralogy of specimens covered the 2θ interval 2–70°, with a scanning angle step size of 0.015° and a time step of 0.1 s. The mineral phases were determined using the DIFFRACplus EVA 12® software (Bruker-AXS, Fitchburg, MA, USA) based on the ICDD Powder Diffraction File of PDF-2 2006 (University of Patras) while the semi-quantitative calculations were performed by TOPAS 3.0 ® software (TOPAS MC Inc., Oakland, CA, USA), based on the Rietveld method (wt% phase error <1). Morphological texture of raw materials, as well as the chemical composition of minerals, were examined using a scanning electron microscope (SEM) (JEOL JSM-6300 SEM) equipped with an energy dispersive spectrometer (EDS) (Tokyo, Japan). The chemical composition of minerals was determined using natural and synthetic standards and 20 kV accelerating voltage with 10 nA beam current. The samples were carbon-coated to acquire conductance (Forth-ICE-HT, University of Patras). The bulk chemical composition was determined using a Bruker S4-Pioneer XRF wavelength dispersive spectrometer. The spectrometer was fitted with an Rh tube, five analyzing crystals, namely: LIF200, LIF220, LIF420, XS-55, and PET, and the detectors were a gas-flow proportional counter, a scintillation detector, or a combination of the two. Samples were analyzed at 60 kV and 45 mA tube-operating conditions (University of Patras).

The magnesite tailings were crushed by large-, medium-, and small-sized jaw crushers in turn, then milled in a ball mill for 1.0 h. Material retained in each sieve was weighed and expressed as a percentage of the whole sample by using sieves which reached up to 10 μm. The test adopted the method of calcination after briquetting. After stirring 50 g of sieved tailings and 2.0% water and holding the pressure at 6 MPa for 5 s, a cube of 5 cm $\times$ 5 cm $\times$ 1 cm was made. Then, it was calcined in a box furnace. The calcination was performed on average 1000 °C for 1.0 h, in order for carbonates and/or other clay minerals of tailings to decompose. The calcined tailing was cooled rapidly in the air and passed through a square whole sieve of 0.08 (or 0.1) mm after being milled for 3 min by vibration mill (Western Greece Region, University of Patras).

3.2.2. Methods for the Produced Cement

In order to investigate the consistency and the fluidity of the produced cement specimens, the below process was followed. Based on the mix ratio provided in Equation (1), the test method of cement paste consistency was carried out in accordance with GB/T1346-2011, "Test Methods for Water Requirement of Normal Consistency, Setting Time and Soundness of the Portland Cement". Therefore, the GB/T2419-2005, "Test Method for Fluidity of Cement Mortar", was used. The control sample was composed of 70% ordinary

Portland cement and 30% fly ash. Under dry conditions, nano MgO and fly ash were mixed into the cement to prepare the samples for experiments. The water–binder ratio of all the samples was 0.28. The mixing ratios of the samples are shown in Table 1. Three samples were used for each mix ratio group. The sample size was 25 mm × 25 mm × 280 mm (Western Greece Region).

Table 1. Mixing ratio of cement mortar.

Sample Code	OP Cement	Fly Ash	Nano MgO	Water
Fk1	345	151	0	140
Fk2	340	149	5	140
Fk3	336	147	10	140
Fk4	330	145	15	140
Fk5	325	143	20	140
Fk6	323	141	25	140
Fk7	320	139	30	140
Fk8	310	137	35	140

In order to investigate the expansion of the produced cement specimens, a cuboid of 40 mm × 40 mm × 160 mm was used, with a sand ratio of 1:3 and water cement ratio of 0.5. In order to measure the change of its length, two nail heads were embedded at the end. The substitution rate of fly ash for cement was 0% and 30%, and that of MEA for cement was 0%, 4%, 8%, 12%, and 16%, respectively. The calcination temperature of MEA was 1100 °C, and the holding time was 1 h. The specimen was demolded after standard curing 24 h after forming, and its initial length (L0) was measured after precuring in 20 °C water for 2 h (Western Greece Region). Then, the test pieces were placed into water at 20 °C for long-term curing, and the length (L1) was measured after the corresponding age. The calculation formula of the expansion rate (φ) of the test piece is

$$\varphi = (L1 - L0)/L \tag{1}$$

where L is the effective length of the mortar specimen, taking 150 mm.

The porosity and pore size distribution of mortar were measured by GT-60 mercury intrusion tester produced by Canta Instruments Co., Ltd. (USA) (Western Greece Region).

Regarding the investigation of the compressive strength of the produced cements, three samples were used for each set of mix ratios. The size of the sample was 40 mm × 40 mm × 160 mm. The specimens were cured for 28 and 365 days, autoclave cured, and then their compressive strength and flexural strength were measured. The autoclave curing was performed by curing the sample in a 216 °C and 2.0 MPa saturated steam for 6.0 h after curing in water at 20 °C for 365 days. The model of the autoclave was YZF-2A, which has an inner diameter of 160 mm, volume of 0.0085 m^3, and maximum allowable steam pressure of 25 MPa. The flexural and compressive strengths of the specimens were tested using a compression bending machine and a universal testing machine, respectively. The loading speed of the compressive strength test was controlled at 2.4 ± 0.2 kN/s, and the compressed area of the specimens was 1600 mm^2. The loading speed of the flexural strength test was automatically controlled at 50 N/s. The specimens used for the compressive strength test were adopted after the flexural strength test. The strength of each group of specimens was taken as the arithmetic mean of the measured values of the three specimens (University of Patras).

Atomic force microscopy (AFM) test was taken with digital instruments (University of Patras). AFM height images, including root mean square and roughness, were conducted in order to investigate the morphology of the produced cement.

4. Results and Discussion

4.1. Raw Materials

The MgO content of the magnesite tailings is approximately 40 wt%, while low amounts of silica (5 wt%) and calcium oxides (2.85 wt%) complete the bulk composition (Table 2). The chemistry of this material corresponds to the mineralogical composition of Figures 2 and 3, where magnesite, dolomite quartz, and serpentine were detected. Moreover, the microscopic study using a polarizing and electron microscope assumes the coexistence of magnesite with significant amounts of other crystalline phases. The last reduces the purity in magnesium of the studied materials which is usually demanded for their uses of high economic value (e.g., refractory production). Nevertheless, these wastes are immediately presented clear and free of possible harmful elements.

Table 2. Chemical composition of magnesite tailing (%).

Material	Chemical Compositions						
	CaO	MgO	Al_2O_3	SiO_2	Fe_2O_3	Loss	Total
Magnesite tailing	2.85	40.35	0.03	5.00	0.05	50.75	99.03

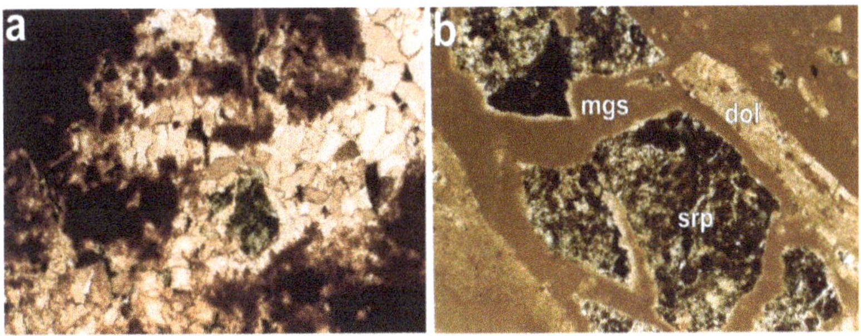

Figure 2. (**a**). Microphotographs where fragments of serpentine (srp), the replacement of magnesite (mgs) by coarse grain dolomite (dol), are observed (XPL, ×68); (**b**) fragments of serpentine in the magnesite vein. In the contact of magnesite with the serpentine fragment, dolomite is formed (XPL, ×68).

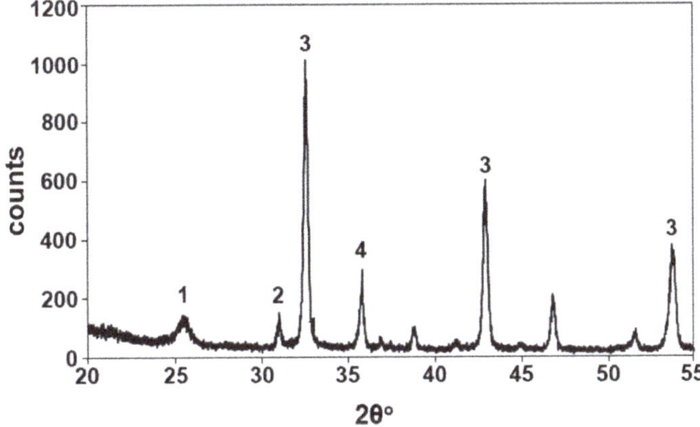

Figure 3. XRD pattern of magnesite tailing from Evoia. 1: quartz, 2: dolomite, 3: magnesite, 4: serpentine.

As it is shown in Figure 2a,b, fragments of serpentine are present in the microstructure of magnesite. Moreover, in Figure 2a, the replacement of magnesite by the coarse grain dolomite can also be observed. In general, the studied magnesite tailings presented are suitable to be used in a variety of applications in which an augmented content of nano MgO is needed.

From the conducted microscopic observation and analyses of the studied tailings, it was shown that although these tailings constitute a rich in magnesium appearance containing over 45% of magnesite, it seems very difficult to obtain a high-quality product (bipolar magnesium) with over 90% MgO. The reason why this happens is due to the nature of this appearance, and more specifically calcium silicate impurities, as are presented in Table 2 and Figure 2. Thus, the mineralogy and the adhesion of the present minerals to the magnesite are responsible for the characterization of the magnesite tailings as a poor appearance for immediate exploitation, despite its MgO content. Magnesite tailings contain in their matrix 79% magnesite, 3% quartz, 3% dolomite, and 15% serpentine (Figure 3). It has been proven that magnesium refractory materials consist of 90% MgO and calcium/SiO_2 ratio of >2.85. In order to use magnesite tailings as refractories, they should contain similar characteristics to those of physical magnesites, such as low content of impurities as CaO, SiO_2, FeO (tot), and B_2O_3, and high amounts of MgO.

This conclusion is obvious as the tailings contain thin veins of serpentine and dolomite and very fine distributed quartz, not only in the dolomite veins, but also in the mass of magnesite. The coexistence of these minerals makes their release very difficult and, therefore, the appearance should be processed and enriched with high-cost processes by spending significant amounts of energy. This is exactly the main reason why these tailings are not easy to use in other applications, something that the present study fully investigates when filling the existing research gaps regarding the use of tailings deriving from magnesites from Greece (as these tailings which are used for the production of environmentally friendly cements have not yet been tested). Additionally, similar raw materials have been used in mortars and have been tested regarding the influence of their MgO content on their final performance [6], where it was found that certain MgO content seems to positively affect the final behavior of the produced mortars. The chemical composition of the used cement is shown in Table 3 and its mineral composition is shown in Figure 4. The main mineral composition of the used cement is C_3S, C_2S, C_3A, and C_4AF, with a small amount of gypsum and bassanite (Figure 4).

Table 3. Chemical compositions of cement (%).

Material	Chemical Compositions									
	CaO	MgO	Al_2O_3	SiO_2	Fe_2O_3	SO_3	Na_2O	K_2O	Loss	Total
Cement	63.16	1.43	4.75	19.47	3.43	2.68	0.28	0.62	3.26	99.08

Nano magnesium oxide is the product produced when the magnesite is heated to 1100 °C; whereupon most of the CO_2 is removed, while 2–7% CO_2 remains, depending on the amount burned [25]. The burned magnesium oxide produced in this study by the thermal decomposition of magnesite at a relatively low temperature, ~1000 °C, is porous, of low bulk density, and chemically active. It is very easily hydrated to $Mg(OH)_2$ or carbonized to $MgCO_3$ even with the humidity or CO_2 of the atmosphere. XRD results indicate the new major phase of periclase to coexist with low amounts of brucite and portlandite (or lime), and the primary remaining magnesite and quartz (Figure 5). The same diffractogram, by its low peak intensities, reveals the fine-grained character of this material. Scarcely, in the microstructure of this calcined product, enriched microregions with the $CaO/Ca(OH)_2$ phases were observed (Figure 6a,b).

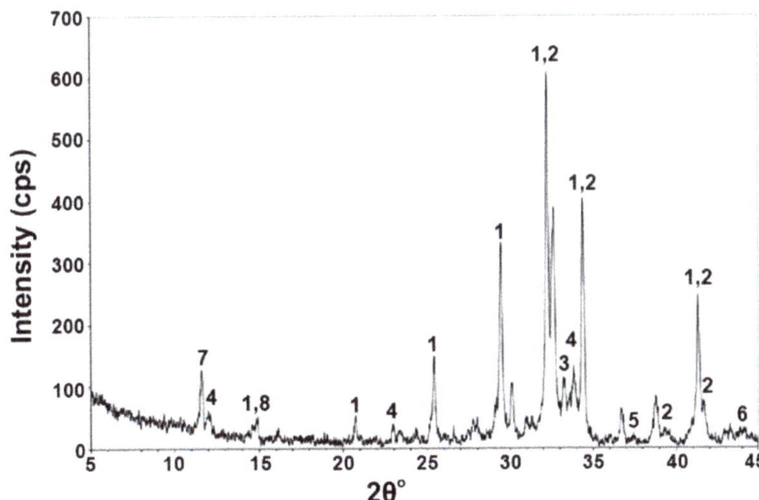

Figure 4. XRD pattern of P 52.5 Portland cement produced by Titan Co., Ltd. (Athens, Greece) where 1: C_3S, 2: C_2S, 3: C_3A, 4: C_4AF, 5: CaO_f, 6:MgO, 7: $CaSO_4$-$2H_2O$-Gypsum, 8: $CaSO_4$ $0.5H_2O$-Bassanite.

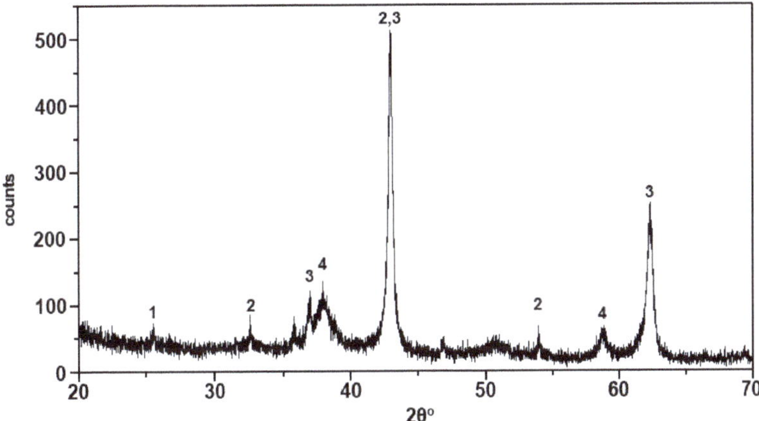

Figure 5. X-ray pattern of bulk mineral composition of final product. 1: Quartz, 2: magnesite, 3: periclase, 4: brucite.

In this study, fly ash was also used, as numerous researchers have used fly ash as an additive for the production of cements of normal and of increased mechanical strength. For this reason, fly ash derived from a Kardia lignite-fired power plant in Ptolemais Basin was collected and studied in order to investigate the behavior and physicomechanical properties of the produced cement in combination with MEA. The characterization of the tested fly ash, which includes mineralogical composition, morphological determination, and chemical composition, revealed that fly ash is enriched in CaO (Table 4). The increased specific surface of its particles makes fly ash capable of restraining elements and participating in chemical reactions due to the microporous minerals included, such as clays and micas [26,27]. Furthermore, the SiO_2 presence constitutes the main carrier of pozzolanic properties, making it useful in cement as an additive. The mineralogical analysis results are shown in Table 3, while powder X-ray diffraction patterns are shown in Figure 7. The basic components of Kardia fly ash are lime, anhydrite, calcite, gehlenite, and less quartz, plagioclase, and hematite. Micas appear in minor amounts. According to the XRPD results, in the fly ash, amorphous phase ranges to percentages of 25 wt%. Lime anhydrite and quartz appear as the major crystalline phases. Anorthite, portlandite, and gehlenite were also detected in lower amounts.

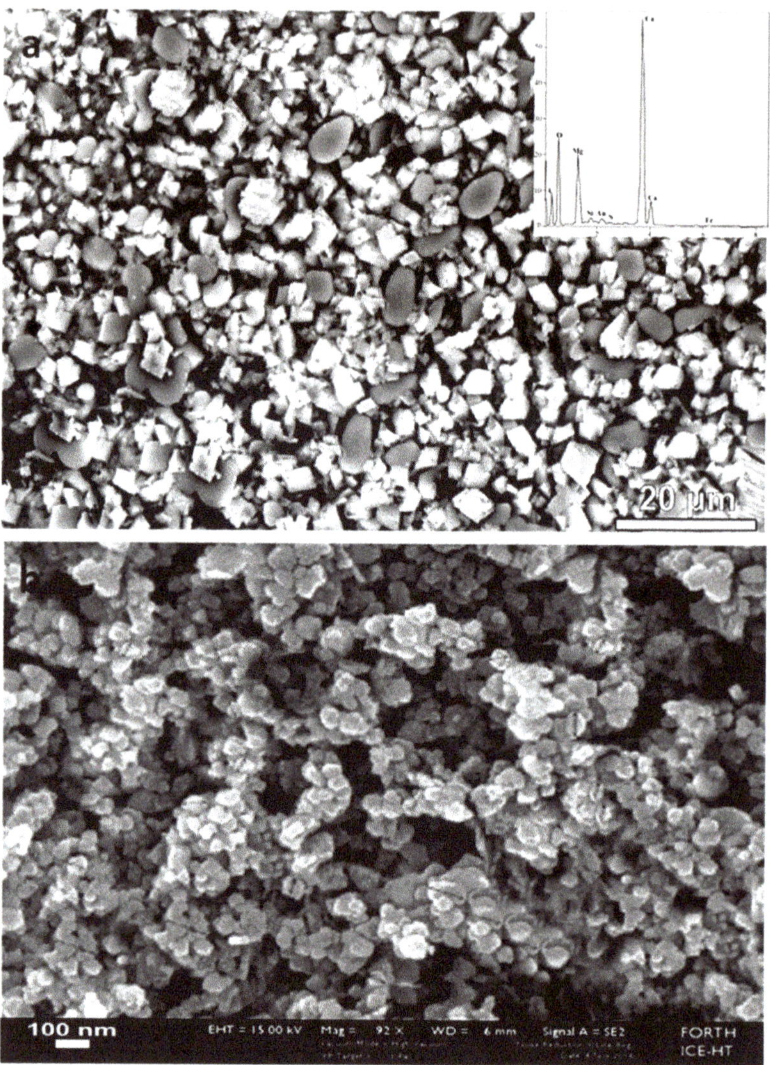

Figure 6. (a,b) Backscattered image indicating microregion in the matrix of the final product enriched in lime/portlandite (white), and periclase (grey).

Table 5 presents the chemical composition of the investigated fly ash, and the XRD pattern of Figure 7 shows the mineralogical composition of fly ash. The presence of silica dioxide in calcium-rich fly ashes is important even in small amounts [28]. It has been demonstrated that when incorporating high-calcium fly ashes in cementitious systems, the soluble silica present in the ash significantly affects their hydration. Although fly ash rich in active silica is generally more preferable, this does not exclude the possibility of achieving higher reaction rates and superior performance when utilizing different ashes [25]. The investigated fly ash presents $SiO_2 + Al_2O_3 + Fe_2O_3$ less than 70% as it is required for Class C classification according to ASTM standards (ASTM 618C) (Table 5). $SiO_2 + Al_2O_3 + Fe_2O_3$ content also has to be >50%, which is not the case in Greek fly ash due to its high CaO content. However, several authors [25,26] classify Greek fly ash in Class C, taking into consideration the significantly high CaO concentration. It should be mentioned that one major problem regarding fly ash use is the heterogeneity of its chemical composition derived from the chemical composition of feed lignite, the combustion conditions, and the amount of coexcavated inorganic strata that led to combustion. The variability of these factors excludes a constant chemical composition of fly ash for further utilization.

Table 4. Mineral composition (%) of fly ash from a Kardia power plant.

Kardia	
Minerals	**(%)**
Quartz	8
Anhydrite	15
Lime	30
Plagioclase	3
Calcite	14
Gehlenite	10
Hematite	2
Mica	1
Clay minerals	-
Amorphous	17

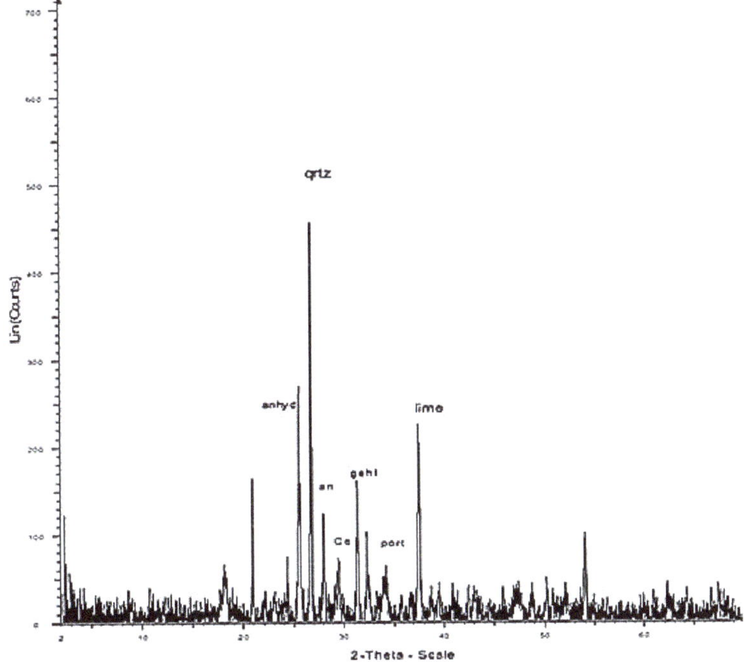

Figure 7. XRD pattern of fly ash from a Kardia power plant: Qrtz: quartz, Cc: calcite, an: anorthite, anhyd: anhydrite, gehl: gehlenite, port: portlandite, lime: lime.

Table 5. Chemical composition (in %) of the fly ash under study (bdl: below detection limit).

Samples	Kardia
Oxides	
CaO	53.00
SiO_2	17.00
Fe_2O_3	12.05
Al_2O_3	5.40
SO_3	8.00
K_2O	2.02
MgO	0.60
TiO_2	1.00
Cr_2O_3	0.19
P_2O_5	0.26
MnO	0.14
SrO	0.09
NiO	0.09
Na_2O	Bdl
ZrO_2	0.06
CuO	0.04
ZnO	0.03
Rb_2O	0.02

As for the morphological features of the tested fly ash, properties of plerospheres, cenospheres (hollow spheres), and agglomerates, which result from lignite combustion, were observed. A plerosphere constitutes a cenosphere that may encapsulate a mass of microspheres and is generally silica-coated (Figure 8a–c).

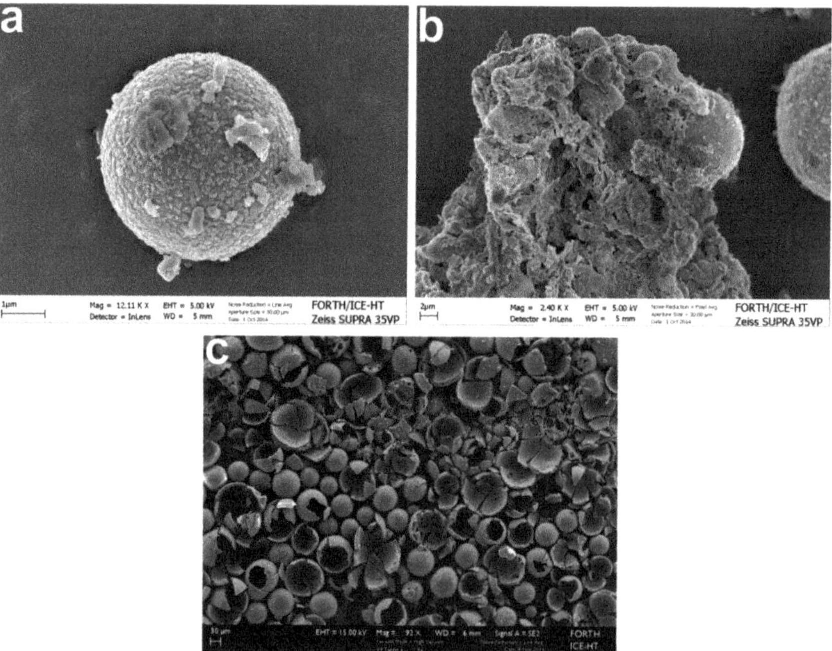

Figure 8. Fly ash particles from a Kardia power plant: (**a–c**): Cenospheres after releasing their gases, silica-coated perfect sphere, and with smaller particles on its surface.

The authors, after having taken into account the above results of the used raw materials, concluded that the raw materials are individually suitable to be used as raw materials for cement production of high mechanical, as well as of high physical, characteristics. Each of the basic components of cement exhibits a capable behavior, as it is shown that by varying their relative percentages, the properties of cement can be modified. Similar raw materials have been investigated and tested by numerous researchers to be used as cement additives and they are considered as suitable for this type of application. However, the exact proportions of raw materials used in order to produce such green applications has not been identified yet and is something that this study investigates.

4.2. Factors Influencing the Behavior of the Products

MEA is typically produced by calcining magnesite at a temperature ranging from 900 °C to 1200 °C, which is much lower than the cement clinker temperature, and as a result provides higher hydration reactivity in comparison to the dead burnt periclase. In this study, MEA was used after its calcining in 1100 °C by staying at this temperature for one hour and was mixed with fly ash to study their effect on the final behavior of the produced cements (Figure 6). The mineral composition analysis results of MEA produced by calcining magnesite tailings at 1100 °C for 1.0 h, presented in Figure 5, and the comparison of minerals before and after calcination, in Figures 3 and 5, when the calcination temperature and holding time were up to 1100 °C for 1 h, show that carbonate and clay minerals were decomposed, or only a trace amount could not be displayed on the XRD pattern (Figure 5). In addition, the XRD pattern shows that the main minerals obtained after 1 h of heat preservation were periclase (MgO), as well as small amounts of portlandite, brucite, and initial quartz. This mineralogical modification during calcining appears to have a significant effect on the activity and expansion performance. This is exactly shown by the diagram of Figure 9, which shows that a small percentage of calcining magnesite tailings at 1100 °C for 1.0 h performs satisfactorily.

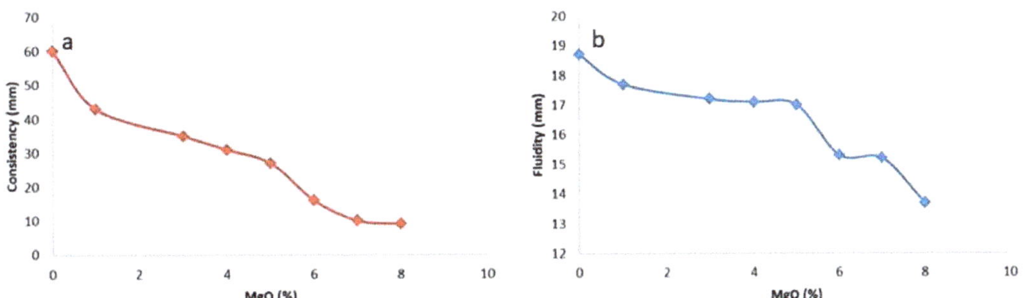

Figure 9. Influence of MgO on the (**a**) consistency and (**b**) fluidity of cement.

4.2.1. The Effect of Nano MgO on the Consistency of the Produced Cement

The influence of nano MgO, as well as of fly ash in certain contents, as they are displayed in Table 1, seems to be intensively strong for the total of the tests of the produced cement. However, the most significant influence presented is the nano MgO content. The effect of nano MgO on the consistency of the produced cement paste is presented in Figure 9a. The observation of this figure shows that the increased amount of nano MgO results in the gradual decrease of its consistency. Several researchers have reached similar conclusions [14,28]. In the case that the content of nano MgO was 1%, the sinking depth of the test cone was 42 mm, while when the content of nano MgO was 3%, the consistency of cement was 35 mm. When the content of nano MgO was more than 4%, the consistency value of the specimen decreased more significantly.

The general behavioral trend of the studied mixtures could be defined into three (3) different groups regarding the total of their properties, as is presented in Figure 9a,b. It can be observed that group I consists of FK1 and FK2 samples, group II of FK3–FK5, and group III of FK6–FK8. This discrimination of clear limits reveals that there are certain critical values of MgO and fly ash content which determine the final cement properties. When studying the diagrams of Figure 1 combined with Table 1, and knowing the raw materials used as waste materials of other processes (tailings) and fly ash, it was observed that high content of fly ash (Kardia, Greece) combined with low content of MgO provides high values of consistency and fluidity, and, on the other hand, low content of fly ash and high content of nano MgO provides lower values in consistency and fluidity; something that seems to be defined mainly by the nano MgO content and, more specifically, by their hydration activity.

4.2.2. The Effect of Nano MgO on the Fluidity of the Produced Cement

The way in which the fluidity of the produced cement was influenced by the content of nano MgO is depicted in the diagram of Figure 9b, and as is shown in this figure, the nano MgO content plays a severe role. It can be observed that by the increase of nano MgO content, the fluidity is decreased almost linearly, which is consistent with the law of cement paste consistency. Xiaoyan Wang [29] investigated the effect of nano MgO on the fluidity of oil-well cement paste and found a similar phenomenon. The effect of nano MgO content on the fluidity of cement paste was relatively obvious, and the fluidity of cement paste decreased with the increase in nano MgO content. When the nano MgO content was 1.0%, the fluidity of the oil-well cement slurry worsened because the water demand of nano MgO increased and generated $Mg(OH)_2$, which also caused the consistency of the cement paste to become worse. The nano MgO has a small bulk density, which may affect the bulk density of the freshly mixed specimens. Therefore, it may disrupt the uniform particle assembly. Furthermore, it has a large specific surface area and high hydration activity, and as a result, it reacts with water rapidly. In fresh cement paste, nano MgO consumed more free water for hydration and covering its surface layer [30–32], thus reducing the water–cement ratio and leading to reduced fluidity and consistency. Xue Zhang found that when a high-MgO admixture of cement was used or excess MgO was used to mix high-performance concrete, a higher water–solid ratio was required [33,34]. Ye et al. [35], when studying the performance of cement-based materials made by cement paste, including nano MgO oxide and fly ash deriving from their surrounding areas, found similar trends to the results of our study which contains cements made by nano MgO oxides and fly ash deriving from Greece.

4.2.3. The Effect of Nano MgO on the Expansion of the Produced Cement

In the present study, the ability of cement paste to expand was examined, as it is a particularly important parameter of its basic characteristics. The present experimental study used numerous samples for a significant period of time up to 400 days, making a systematic study of the raw materials which showed that a specific content of nano MgO and fly ash in powder as specific additives can be decisively active in the controlled shrinkage and expansion of the cement. This achieves the desired degree of volume dimension, while compensating for the shrinkage effects, avoiding the volume shrinkage of the cement. The present raw materials, as they do not present metallic elements, are not feasible and will affect any future Fe reinforcements. Its dilating action seems to begin in the first 8 days and ends depending on the ratio of nano MgO/fly ash from 8 to 400 days. Samples containing low percentage of MgO immediately complete their expansion, while mixtures with an increased percentage in nano MgO continue to evolve 400 days later. The expansion of the cement paste has been remarkably improved due to the nano MgO addition; nevertheless, the accurate limits of the MgO content added have not yet been defined. The expansion performance increases as the nano MgO content increases, following the extension of the curing age. Firstly, the expansion of the cement paste containing nano MgO oxide was significantly higher than that of the cement paste without MgO oxide. This may have happened because the nano MgO particles were small, and $Mg(OH)_2$ crystals were quickly formed due to hydration, and as a result, this gives higher expansion to the cement paste. Furthermore, Mg(OH)$_2$ crystals grow around the pore rims and near the MgO particles, and the increased volume of Mg(OH)$_2$ crystals squeezes the surrounding cement paste and causes expansion [6,30,35,36]. The expansion of the tested cement paste samples are presented in Figure 10. As can be observed from the diagram (Figure 10), as the nano MgO rises, the expansion increases. The expansion of the sample without nano MgO was 364 μm/m after autoclave curing, while the expansion amounts were 1060, 1700, 2500, 3300, and 4000 μm/m when nano MgO was 4, 5, 6, 7, and 8%, respectively, after autoclave treatment. Several scientists have obtained similar results [6,35]. It is worth mentioning that in the expansion test, a clear discrimination among the three groups is observed, characterized by clear limits in the used mixtures of nano MgO/fly ash. In the present systematic study of the effect of additives on the behavior of the produced cement, representative samples from the two groups were studied, giving the most satisfactory results in order to identify the effect of raw materials on the final behavior of cement.

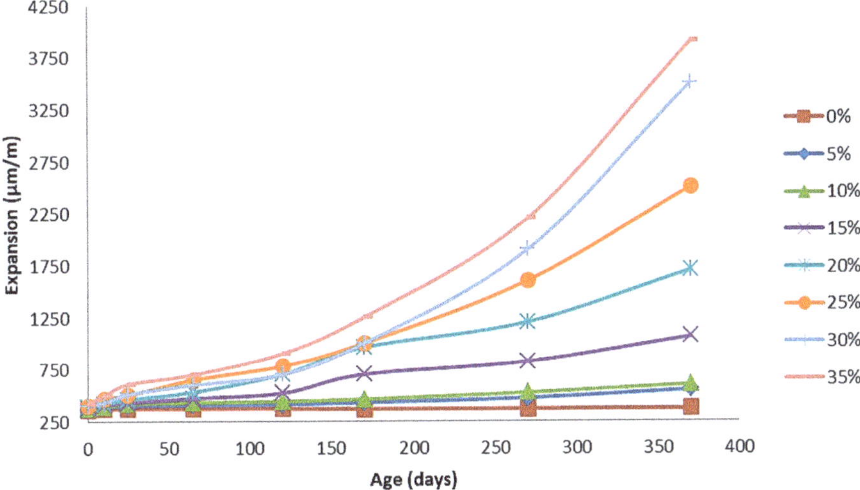

Figure 10. Effect of nano MgO content on expansion performance of cement paste.

4.2.4. The Effect of Nano MgO/Fly Ash on the Cement Porosity

As shown in Figure 11, we observed different pore distribution in the cement paste directly depending on the content of nano MgO/fly ash. Pore distribution and total porosity are factors that affect the final strength, permeability, and volume changes in the hardened cement paste. The pore size distribution and porosity of cement paste is affected by the nano MgO, as can be seen in Figure 11. The total porosity of the cement decreased as the curing age increased. The porosity of the produced cement containing 8% of nano MgO oxide was higher than that of cement containing 4% of nano MgO oxide, which may be due to the excessive content of nano MgO oxide, which led to harmful expansion and increased the harmful pore size of mortar. This observation may be due to the presence of MgO as well as the process of MgO hydration; there are three main stages that may occur at the same time but can be dominant at different times. In the first two stages, reactive nano MgO particles dissolve and leave pores in the whole system, and then minor Mg(OH)$_2$ crystals grow freely to fill these pores without generating crystal growth pressure. The third stage is dominated by confined crystallization, and expansion is mainly caused by microcracks induced by the crystal growth pressure in the whole system. Hydration of MgO makes the pore structure partially dense or increases the tortuosity of pores, which will improve durability; furthermore, if high expansion occurs, high porosity will reduce durability. The strength of hardened cement is considered to be its most important property. Each of the basic components of cement presents a specific behavior where, by varying their relative percentages, the mechanical properties of the cement can be modified. It should also be noted that as the total porosity of the cement decreased as the curing age increased, the cement became denser, and the strength became higher. This is exactly what is shown in Figure 12, where the division into three groups with clear boundaries of specific ratios of nano MgO/fly ash is clearly shown. The clear separation of the three groups, from the beginning of the study, with the factors affecting the cement is obviously shown in Figure 12, which is directly related to the total of the results given so far. For example, the differentiation of the two groups in terms of porosity and its distribution (Figure 11) is in complete agreement with the groups of Figure 12, and the distribution of pores is affected by the ratio of water to cement and the degree of hydration of the cement. Large pores (>50 nm) affect the compressive strength, while smaller ones have a greater effect on drying shrinkage and creep.

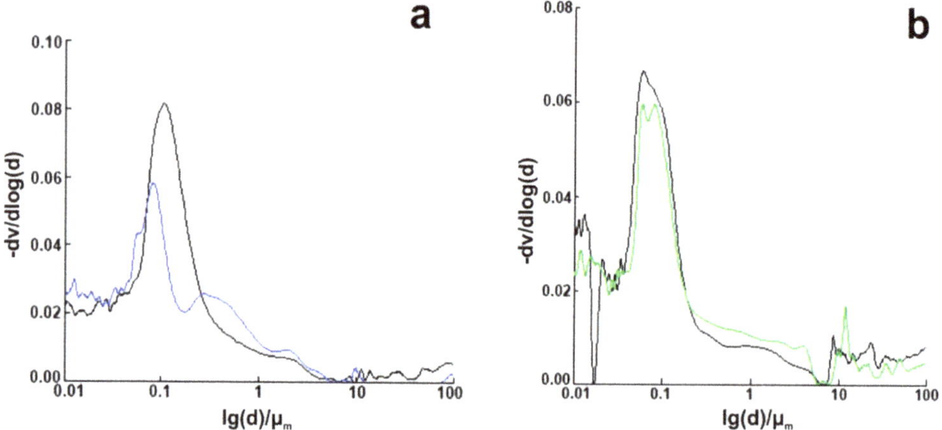

Figure 11. Pore size distribution of mortars containing MEA: (**a**) 4% MEA; (**b**) 8% MEA.

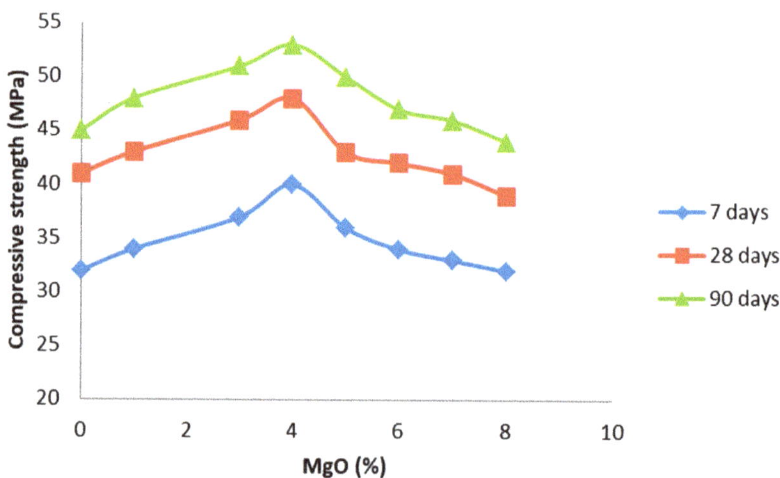

Figure 12. The compressive strength of cement paste samples with MgO.

4.2.5. The Effect of Nano MgO on the Mechanical Performance of the Produced Cement

Moreover, the physical interpretation of the diagrams of Figures 9 and 10 is harmonious with the final behavior and strength of the hardened cement directly dependent on the mineral raw materials, where they participate in the composition of the cement (Table 1). In general, the cement strength seems to show increasing trends in the tests of 7, 28, and 90 days for contents of nano MgO up to 4% of the volume of cement, probably because the microstructure improved after nano MgO incorporation. With the extension of the curing age, the compressive strength of all samples gradually increased, most likely because fly ash underwent a pozzolanic reaction, which improved the interface transition zone of the cement [37]. However, we can observe that when the content of MgO was up to 4%, the compressive strength values, regardless of the curation days, show reducing trends, which is probably due to the development of a strong network of microcracks in the structure of the mortar, displaying its properties. We found similar conclusions from the test of expansion. In addition, hydration of C–S–H in these critical samples seems to have been accelerated, leading to the loss of cohesive forces within the cement structure.

4.2.6. The Effect of Microroughness on the Cement Performance

The AFM technique constitutes a very useful tool for examining the surface morphology of various materials such as cement. The initial grouping into three subgroups is also verified by the study of the microroughness of the cements through the AFM method. As shown in Figure 13, the samples of group I, representing the lowest percentage in MEA, show low microroughness and therefore may constitute a determining factor in the low cohesion of the cement paste, as the microtopography is likely to create limited development of cohesive forces within the structure of the cement paste. In contrast to the above, in group II, there is increased microroughness due to the increase in the size of MEA, where it creates increased cohesive forces within the cement paste. This increase in the microroughness, in addition to the mechanical cohesion, is likely to create additional van der Waals forces between the nano MgO and the cement minerals due to this increase in the specific surface area. Finally, group III is presented with a particularly high microroughness compared to the previous groups. However, this increase seems to overcome the mechanical forces and the van der Waals forces, so it breaks into bonds, creating a number of microcracks. This is supported by both the results of the uniaxial compressive strength and the petrographic study of the produced cements.

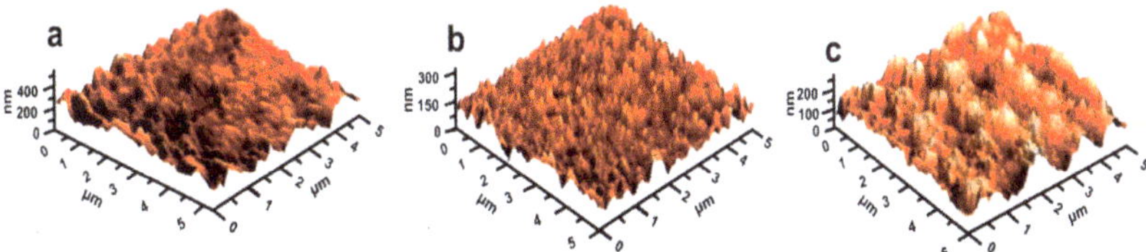

Figure 13. The AFM results of cement made by nano MgO and fly ash additives: (**a**), group I, (**b**) group II, (**c**) group III.

5. Conclusions

This paper examines byproducts (magnesite tailings and fly ash) to evaluate their suitability as cement additives, with the ultimate goal of the production of environmentally friendly cements from now on. The main conclusions of this study are given in the below remarks:

- Byproducts from different regions of Greece, such as magnesite tailings from Evoia and fly ash from Kardia (Ptolemais), which have been combined together for first time, are suitable reinforcement additives.
- The effect of nano MgO content is the most critical parameter for the physicomechanical performance of produced cement, compared to that of the fly ash content.

- More satisfactory results in the physicomechanical properties of the produced cement were from samples of group II, containing 3–4% of nano MgO. Nano MgO content above 4% seems to have a negative influence on the compressive strength of the produced cement, simultaneously reducing its durability. Therefore, the amount of nano MgO determines the physicomechanical performance of the produced cement.
- The increased microroughness of the cements is in accordance with the mechanical behavior of the produced cements.

Future scope: Our research team, following the basic principles of the circular economy, aims to produce a green concrete in the future using exclusively industrial byproducts for both cement and aggregates in mixtures capable of withstanding uniaxial loads and temperature changes.

Author Contributions: Conceptualization, P.P.G. and P.P.; methodology, P.P.G., P.P., A.R., P.L. and M.K.; software, M.K. and P.P.; investigation, P.P.G., P.P., A.R. and P.L.; resources, P.P.G., P.P. and P.L.; data curation, P.P.G., P.P., A.R. and P.L.; writing—original draft preparation, P.P.G. and P.P.; writing—review and editing, P.P.; visualization, P.P.; supervision, P.P. All authors have read and agreed to the published version of the manuscript.

Funding: This research received no external funding.

Institutional Review Board Statement: Not applicable.

Informed Consent Statement: Not applicable.

Conflicts of Interest: The authors declare no conflict of interest.

Abbreviations

AFM	Atomic force microscopy
ASTM	International standards worldwide
dol	Dolomite
EDS	Energy dispersive spectrometer
MEA	MgO expansion agent
mgs	Magnesite
OP	Ordinary Portland
PC	Portland cement
srp	Serpentine
XRD	X-ray diffraction
XRF	X-ray fluorescence
XRPD	X-ray powdered diffraction

References

1. Freedonia. World Construction Aggregates-Demand and Sales Forecasts, Market, Share, Market, Size, Market, Leaders. In *Industry Study No. 3389*; The Freedonia Group: Cleveland, OH, USA, 2016; p. 390.
2. USGS. Commodity Statistics and Information Mineral. In *Yearbooks*; USA Geological Survey: Washington, DC, USA, 2015.
3. Petrounias, P.; Rogkala, A.; Giannakopoulou, P.P.; Lampropoulou, P.; Xanthopoulou, V.; Koutsovitis, P.; Koukouzas, N.; Lagogiannis, I.; Lykokanellos, G.; Golfinopoulos, A. An Innovative Experimental Petrographic Study of Concrete Produced by Animal Bones and Human Hair Fibers. *Sustainability* **2021**, *13*, 8107. [CrossRef]
4. Petrounias, P.; Giannakopoulou, P.P.; Rogkala, A.; Lampropoulou, P.; Tsikouras, B.; Rigopoulos, I.; Hatzipanagiotou, K. Petrographic and Mechanical Characteristics of Concrete Produced by Different Type of Recycled Materials. *Geosciences* **2019**, *9*, 264. [CrossRef]
5. José, N.; Ahmed, H.; Miguel, B.; Luís, E.; Jorge, D.B. Magnesia (MgO) Production and Characterization, and Its Influence on the Performance of Cementitious Materials: A Review. *Materials* **2020**, *13*, 4752. [CrossRef]
6. Ye, Y.; Liu, Y.; Shi, T.; Hu, Z.; Zhong, L.; Wang, H.; Chen, Y. Effect of Nano-Magnesium Oxide on the Expansion Performance and Hydration Process of Cement-Based Materials. *Materials* **2021**, *14*, 3766. [CrossRef]
7. Shayanfar, M.A.; Farnia, S.M.H.; Ghanooni-Bagha, M.; Massoudi, M.S. The Effect of Crack Width On Chloride Threshold Reaching Time in Reinforced Concrete Members. *Asian J. Civ. Eng.* **2020**, *21*, 625–637. [CrossRef]
8. Jackson, N. *Civil Engineering Materials*; Macmillan Press Ltd.: London, UK, 1981.
9. Neville, A.M. *Properties of Concrete, ELSB*, 5th ed.; Pearson Education Publishing Ltd.: London, UK, 2005.

10. Kabir, H.; Hooton, R.D. Evaluating Soundness of Concrete Containing Shrinkage-Compensating MgO Admixtures. *Constr. Build. Mater.* **2020**, *253*, 119141. [CrossRef]
11. Ye, Q.; Chen, H.; Wang, Y.; Wang, S.; Lou, Z. Effect of MgO and gypsum content on long-term expansion of low heat Portland slag cement with slight expansion. *Cem. Concr. Compos.* **2004**, *26*, 331–337. [CrossRef]
12. Lou, Z.; Ye, Q.; Chen, H.; Wang, Y.; Shen, J. Hydration of MgO in clinker and its expansion property. *J. Chin. Ceram. Soc.* **1998**, *26*, 430–436. [CrossRef]
13. Walling, S.; Provis, J. Magnesia-based cements: A journey of 150 years, and cements for the future? *Chem. Rev.* **2016**, *116*, 4170–4204. [CrossRef] [PubMed]
14. Liu, Z.; Wang, S.; Huang, J.; Wei, Z.; Guan, B.; Fang, J. Experimental Investigation on the Properties and Microstructure of Magnesium Oxychloride Cement Prepared with Caustic Magnesite and Dolomite. *Constr. Build. Mater.* **2015**, *85*, 247–255. [CrossRef]
15. Dimopoulos, I.; Anastassakis, G. Recovery of magnesite from fine waste material rejected before hand-sorting. In Proceedings of the XV Balkan Mineral Processing Congress, Sozopol, Bulgaria, 12–16 June 2013; Volume 1, pp. 213–216.
16. Martek, I.; Hosseini, M.R.; Shrestha, A.; Edwards, D.J.; Durdyev, S. Barriers inhibiting the transition to sustainability within the Australian construction industry: An investigation of technical and social interactions. *J. Clean. Prod.* **2019**, *211*, 2812–2892. [CrossRef]
17. Kuppig, V.D.; Cook, Y.C.; Carter, D.A.; Larson, N.J.; Williams, R.E.; Dvorak, B.I. Implementation of sustainability improvements at the facility level: Motivations and barriers. *J. Clean. Prod.* **2016**, *139*, 15291–15538. [CrossRef]
18. Murtagh, N.; Scott, L.; Fan, J.L. Sustainable and resilient construction: Current status and future challenges. *J. Clean. Prod.* **2020**, *268*, 10. [CrossRef]
19. Vishwakarma, V.; Uthaman, S. Environmental impact of sustainable green concrete. In *Smart Nanoconcretes and Cement-Based Materials: Properties, Modelling and Applications*; Elsevier: Amsterdam, The Netherlands, 2019; pp. 241–255.
20. Robertson, A.H.E. Origin and emplacement of an inferred late Jurassic subduction-accretion complex, Euboea, eastern Greece. *Geol. Mag.* **1991**, *128*, 27–41. [CrossRef]
21. Katsikatsos, G. *Geological Map, Sheet LIMNI*; Scale 1:50.000,1.G.M.E.; Institute of Geology and Mineral Exploration: Athens, Greece, 1980.
22. Spray, J.G.; Roddick, J.C. Petrology and 40Ar/39Ar Geochronology of some Hellenic Sub-Ophiolite Metamorphic Rocks. *Contrib. Mineral. Petrol.* **1980**, *72*, 43–55. [CrossRef]
23. Thuizat, R.; Whitechurch, H.; Montigny, R.; Juteau, T. K-Ar dating of some infra-ophiolitic metamorphic soles from the Eastern Mediterranean: New evidence for oceanic thrustings before obduction. *Earth Planet. Sci. Lett.* **1981**, *52*, 302–310. [CrossRef]
24. Gartzos, E. Carbon and oxygen irotope constraints on the origin of magnesite deposits, North Evia (Greece). *Schweiz. Mineral. Petrogr. Mitt.* **1990**, *70*, 67–72.
25. Mehta, P.K. Pozzolanic and cementitious byproducts as mineral admixtures for concrete—-A critical review. In Proceedings of the 1st International Conference on the Use of Fly Ash, Silica Fume, Slag, and Natural Pozzolans in Concrete, ACI SP-79, Detroit, MI, USA, July 1983; p. 1.
26. Rai, P.; Qiu, W.; Pei, H.; Chen, J.; Ai, X.; Liu, Y.; Ahmad, M. Effect of Fly Ash and Cement on the Engineering Characteristic of Stabilized Subgrade Soil: An Experimental Study. *Geofluids* **2021**, *2021*, 1–11. [CrossRef]
27. Koukouzas, N.K.; Zeng, R.; Perdikatsis, V.; Xu, W.; Kakaras, E.K. Mineralogy and geochemistry of Greek and Chinese coal fly ash. *Fuel* **2006**, *85*, 2301–2309. [CrossRef]
28. Gao, P.; Lu, X.; Geng, F.; Li, X.; Hou, J.; Lin, H.; Shi, N. Production of MgO-type Expansive Agent in Dam Concrete by Use of Industrial By-Products. *Build. Environ.* **2008**, *43*, 453–457. [CrossRef]
29. Wang, X. Investigations of Nano-Materials Modified Cement Slurry(Stone) Structure and Properties. Master's Thesis, China University of Petroleum, Beijing, China, 2016; pp. 37–38.
30. Ding, W. Study on the Properties of Cement Paste and Mortar with MgO Expansive Admixture. Master's Thesis, Southwest University of Science and Technology, Mianyang, China, 2016; pp. 36–37.
31. Mo, L.; Deng, M. Thermal behavior of cement matrix with high-volume mineral admixtures at early hydration age. *Cem. Concr. Res.* **2006**, *36*, 1992–1998.
32. Mo, L.; Min, D.; Tang, M. Effects of calcination condition on expansion property of MgO-type expansive agent used in cement-based materials. *Cem. Concr. Res.* **2010**, *40*, 437–446. [CrossRef]
33. Zhang, X. Microscopic Characterization and Early age Performance of MgO and Nano MgO High Performance Concrete. Master's Thesis, Shandong University, Jinan, China, 2019; pp. 33–34.
34. Ding, W.; Tan, K.; Liu, L.; Tang, K.; Zhao, C. Influence research of light-burned magnesia on autogenous shrinkage and pore structure of steam-cured cement paste. *China Concr. Cem. Prod.* **2016**, *5*, 21–26.
35. Ye, Q.; Yu, S.; Zhang, Z.; Zhang, Q.; Shi, T. Effect of Nano-MgO on the expansion and strength of hardened cement paste. *J. Build. Mater.* **2017**, *20*, 765–769.
36. Zhou, L.; Wang, Y.; Zhang, G.; Zhang, F. Research progress in the effect of Nano-MgO on the properties of cement-based materials. *China Concr. Cem. Prod.* **2019**, *5*, 13–18.
37. Chen, H. Expansive property and pore structure characterisitics of magnesium oxide slight expansive cement- fly ash binding material. *J. Chin. Ceram. Soc.* **2005**, *33*, 516–519.

sustainability

Article

COVID-19 Experience Transforming the Protective Environment of Office Buildings and Spaces

Panupant Phapant [1], Abhishek Dutta [2] and Orathai Chavalparit [2,*]

1   Interdisciplinary Program of Environment, Development and Sustainability, Graduate School, Chulalongkorn University, Bangkok 10330, Thailand; panupant.p@gmail.com
2   Department of Environmental Engineering, Faculty of Engineering, Chulalongkorn University, Bangkok 10330, Thailand; duttabob@gmail.com
*   Correspondence: orathai.c@chula.ac.th; Tel.: +66-815-536-884

Abstract: The COVID-19 pandemic has affected human life in every possible way and, alongside this, the need has been felt that office buildings and workplaces must have protective and preventive layers against COVID-19 transmission so that a smooth transition from 'work from home' to 'work from office' is possible. However, a comprehensive understanding of how the protective environment can be built around office buildings and workspaces, based on the year-long experience of living with COVID-19, is largely absent. The present study reviews international agency regulation, country regulation, updated journal articles, etc., to critically understand lessons learned from the COVID-19 pandemic and evaluate the expected changes in sustainability requirements of office buildings and workplaces. The built environment, control environment, and regulatory environment around office buildings and workplaces have been put under test on safety grounds during the pandemic. Workers switched over to safely work from home. Our findings bring out the changes required to be affected in the three broad environmental dimensions to limit their vulnerability status experienced during the pandemic. Office building designs should be fundamentally oriented to provide certain safety protective measures to the workers, such as touch-free technologies, open working layouts, and workplace flexibilities to diminish the probability of getting infected. Engineering and administrative control mechanisms should work in a complementary way to eliminate the risk of disease spread. Country regulation, agency regulations, and operational guidelines need to bring behavioral changes required to protect workers from the COVID-19 pandemic.

Keywords: COVID-19; office buildings and workplaces; office-built environment; control environment; regulatory environment

Citation: Phapant, P.; Dutta, A.; Chavalparit, O. COVID-19 Experience Transforming the Protective Environment of Office Buildings and Spaces. *Sustainability* **2021**, *13*, 13636. https://doi.org/10.3390/su132413636

Academic Editor: John Rennie Short

Received: 28 October 2021
Accepted: 3 December 2021
Published: 9 December 2021

Publisher's Note: MDPI stays neutral with regard to jurisdictional claims in published maps and institutional affiliations.

1. Introduction

The COVID-19 outbreak is one of the most significant global pandemics in human history. As the World Health Organization (WHO) recently announced, the COVID-19 pandemic is slowly becoming an inescapable part of human life and work [1,2]. WHO data revealed that, as of September 2021, more than two hundred and thirty million people had been infected by and almost five million have died from COVID-19 [3]. Even as humans are looking to minimize COVID-19's impact on daily life through specific vaccine availabilities, different evolving virus variants have held the human world in the doldrums. This pandemic has created significant challenges in public health and other various areas, including politics, economics, education, and social behavior. The outbreak has resulted in a global economic decline, and economic growth in every global region was forecasted to decrease significantly. For instance, the World Bank predicted the economic growth of Asia and the Pacific to be a mere 0.5%, and South Asia's economy to contract by 2.7%, Europe and Central Asia's by 4.7%, and Latin America's by 7.2% compared to the onset of the pandemic [4]. Even after a year of struggling to cope with COVID-19 and vaccine introductions, the International Monetary Fund's outlook for the world's economy indicates very high uncertainty and uneven economic recovery during 2021 [5].

The modes of transmission for the SARS-CoV-2 (COVID-19 causing) virus are similar to those of the flu, SARS-CoV-1, MERS, and influenza viruses, which include contact, droplet, airborne, fomite, fecal–oral, blood borne, mother-to-child, and animal-to-human transmission [6]. Pamidimukkala et al. [7] concluded that the virus can be transmitted from person-to-person and causes symptoms that include fever, dry cough, fatigue, and shortness of breath. Due to its strong infectious nature, extended incubation period, difficulty in detection, and vagueness in transmission modes, COVID-19 is becoming a very difficult disease to control [8]. The pandemic has resulted in the enactment of social distancing policies. It has dramatically affected many workforces and workplaces as governments worldwide have implemented various social restrictions to curb infection rates, such as closing schools and asking people to work from home [9]. Many companies have been forced to shut down, affecting many countries' economies, while employees whose jobs have been retained have experienced heightened concerns for their mental health and physical wellbeing. People worldwide are panicking about the virus's fast transmission rates and are isolating themselves from office buildings, and a significantly large percentage of every country's population is spending their time indoors only. Nevertheless, after months of strict quarantine, a reopening of society is inevitable. Many countries are planning exit strategies to progressively lift lockdown measures without leading to an increase in the number of COVID-19 cases [10]. Due to the virus's ongoing persistence, engineers and scientists are trying their best to innovate and discover new cures to fight against this virus [11–14]. Whether official or residential buildings are considered, they should all be equipped with multiple protective layers to contain the spread of viruses, to the point where current pandemics will slowly disappear [15]. By implementing multiple protective layers against viruses, humans will create an environment in which the transmission and mutation of fast-adapting viruses will be minimized.

COVID-19 has affected human life in every possible way; in turn, human patterns that are changing because of COVID-19 could lead to insights that are useful in the fight against COVID-19 [16]. Ever since COVID-19 surfaced, people have started deliberately altering building infrastructures, house interiors, and lifestyles in ways to provide some protective layers against the virus [17]. Fear of COVID-19 has redirected customers' preferences from hypermarkets to decentralized commercial places for quicker shopping. Decentralization, redistribution, and restructuring have become key design concerns since COVID-19. Even the vision of cities' futures has changed in the minds of city dwellers, who are now favoring six-lane cycling tracts over six-lane highways as their favorite option of commuting to and from offices and schools [18].

In their review paper, Honey-Roses et al. (2020) [19] concluded that the long battle with COVID-19 will make the world rethink the design of urban public spaces to serve people's best interests. Sepe (2021) [20] went one step further and pointed out that the COVID-19 pandemic has raised questions about the validity of the principles and practices of the 'Charter of Public Space' being followed by several countries when designing public spaces. Dogan et al. (2020) [21] indicated a positive correlation between meteorological variables, and ambient air quality ($PM_{2.5}$, PM_{10}) with COVID-19 cases, while Shahzad et al. (2021) [22] showed that air quality and temperature positively correlated with COVID-19 death cases. Emmanuel et al. (2020) [23] reemphasized the importance of introducing fresh air into crowded and poorly ventilated buildings [7].

2. Workplace Regulations to Mitigate COVID-19

Since workers can come into close contact with colleagues and visitors coming from different localities with various COVID-19 pandemic infectivity levels, virus exposure can occur at any time within the workplace [24]. Therefore, regulatory guidelines and directives are essential in fighting the COVID-19 pandemic in workplace settings. Policies aimed at protecting workers also prevent community transmission of the virus and protect national economies by maintaining open and safe workplaces [25]. Apart from international health regulations, countries also need to implement pandemic control regulations and directives that reflect their specific needs and legislative backgrounds. Sector-specific guidelines can give further clarity to business operators and benefit different sectors of the economy. Both governments and industry associations must actively support devising industry-specific risk and safety guidelines that workers need to follow. Tokazhanov et al. (2020) [26] noted that existing building sustainability ratings based on "environmental impact" and "energy performance" required total overhauling as the current pandemic has exposed their inadequacies. Jiang et al. (2021) [27] pointed out that drastic changes in the spatial distribution of living spaces will be required to control the spread of the epidemic in the community. The Architectural Society of China (2020) [28] emphasized the need for improving buildings' health performances as buildings are inevitably linked to their communities in epidemic prevention and control measures. Awada et al. (2021) [29] confirmed that the current layout of buildings cannot effectively withstand the spread of COVID-19 and therefore require changes. Valizadeh et al. (2021) [30] further suggested that alteration of the indoor environment is required as most infections occur indoors.

The COVID-19 pandemic has forced businesses to allow officials to stay home and remain active for official engagements through the online presence. Experts are apprehensive that working from home will not be productive as workers are away from their coworkers and managers [31]. Additionally, studies have shown that workers cannot disengage from official work in time when working from home, which has led to increased worker stress and decreased productivity [32]. In fact, both managers and workers are unhappy with the continuance of working from home. Currently, many countries are in the middle of their second or third waves of the pandemic, and yet both government and private establishments are already planning for the eventual reopening of offices to employees. Consequently, the persistence of the coronavirus pandemic has led to reconsiderations of existing office buildings and official workplaces; both office buildings and workplaces need to become more resilient against COVID-19 disease spread for the eventual return of workers. Companies are preparing for the return of employees to their offices with extensive safety arrangements to prevent COVID-19 transmission. At this juncture, it is essential to review how office buildings and workplaces can build up protective and preventive layers against COVID-19 transmission among employees and workers.

In that context, by the literature review, this paper intends to address the following questions: (1) what are the deficiencies in office buildings and their layouts in regards to the infection spreading propensity during the COVID-19 pandemic; (2) to what extent do COVID-19 regulations and guidelines bring about safety to officials; and (3) what are some emerging solutions that will make workplaces safer against COVID-19 when considering office building structural designs and workplace layouts, the virus spread elimination, capacity control, and effective environmental regulations? The answer to these questions will be essential for policymakers for managing a smooth transition from 'work from home' to 'work from the office' as usual.

3. Methodology

COVID-19 related scientific literature grew much faster than any other pandemic the world has witnessed so far. This paper gathered data and information through reviewing the literature. The literature review took place during June–July 2020 and focused mainly on the Web of Science (WOS), Google Scholar, and PubMed databases for journal paper search. The key words used for searching were "office environment and COVID-19", "office regulations and COVID-19", "office layouts and COVID-19", and "office safety and COVID-19". A total of 112 journal papers were retrieved using this method. On reviewing the above journal papers, 19 articles were excluded because of content similarity or they dealt with other aspects of the COVID-19 outbreak not covered by the objectives of the present study. The current article is a narrative review of the 112 papers related to the protective and preventive office environment relevant to the COVID-19 pandemic. This study also used important information provided by credible blog sites.

4. Measure/Regulation to Contain the Spread of the COVID-19 Disease in Offices

SARS-CoV-2, the virus that causes COVID-19, attacks the respiratory tract of infected individuals, causing breathing troubles, fevers, and coughs. Infection with this virus primarily causes respiratory illness that ranges from no symptoms, mild to severe symptoms, or even death. According to current knowledge about SARS-CoV-2, transmission primarily occurs with close contact with an infected person or coughing and sneezing. Viral transmission rates depend on the amount of viable virus being shed and expelled by the infected person, the types of contact an infected person has with others, the settings where exposure occurs, and what preventative measures are in place [25]. Similar to regular flu viruses, SARS-CoV-2 droplets are released by infected persons through coughing or exhalation and spread by touch or inhalation. Touch transmission occurs when individuals touch contaminated surfaces and then their face, nose, eyes, or mouth, while airborne transmission occurs when individuals merely breathe in air that contains viral droplets. Indoor climate-controlled environments exacerbate airborne viral transmissions, especially when ventilation systems are malfunctioning or simply inadequate [33]. As such, strategies to mitigate airborne transmission, such as de-densification, will not be adequate to contain virus spread. At the end of the day, managers and employees will need foolproof, tailor-made COVID-19 protocols that only take the regulations and guidelines produced from the international and national levels as basic considerations in order to protect their wellbeing during this pandemic emergency.

4.1. Agency Regulations

International organizations/agencies, such as the WHO and International Labour Organization (ILO), have released various health regulations to counter the spread of COVID-19 disease in workplaces. These regulations have emphasized cleaning working surfaces, maintaining good hygiene, regular hand washing, promoting good respiratory hygiene, and developing well-defined meeting protocols in workplaces [34]. The WHO [25,34–38] and ILO [39–41] have also provided national and local governments, employers, workers and their representatives, and occupational health services with practical guidance on preventing COVID-19 outbreaks at work by minimizing exposure to and transmission of SARS-CoV-2 among workers [25]. The USA Occupational Safety and Health Administration (OSHA) [42,43] also published regulations related to infection prevention and industrial hygiene practices for the workplace. OSHA regulations emphasize employers' responsibility to provide employees with a workplace free from recognized hazards that can cause death or serious physical harm [43]. In response to the COVID-19 pandemic, the Centers for Disease Control and Prevention (CDC) [44–47] issued updated employer regulations to ensure safe and healthy workplaces. The CDC made it compulsory for employers to regularly assess workplace hazards and implement an effective hazards control system [44]. Similarly, the European Centre for Disease Prevention and Control (ECDC) [48–51] recommended the installation of heating, ventilation, and air conditioning systems (HVAC) for minimizing airborne transmission of COVID-19 in workplaces and measures which reduce the risk of airborne SARS-CoV-2 transmission [52]. Additionally, the American Society of Heating, Refrigerating and Air-Conditioning Engineers (ASHRAE) recommends various strategies to fight against COVID-19 in workplaces, especially ventilation improvement [53–58].

Supplementary Table S1 shows a summary of the preventive and protective recommendations against COVID-19 established by various international organizations. The hierarchy of controls—that is, the levels of risk management—comprises of five groups which include, in order from most to least effective, elimination, substitution, engineering control, administrative and organization control, and personal protective equipment (PPE). Elimination is the most effective strategy; however, it is not always possible to completely eliminate hazards in any operations, and therefore a combination of the other measures is required. Engineering controls involve considering the details of building facilities and equipment that can be improved to reduce exposure to hazards without reliance on worker behavior.

4.2. Country Regulations

Besides workplace guidelines from various international organizations, different countries have also presented their own regulations and measures to contain the spread of the COVID-19 disease in offices and workplaces. Although a country's success in containing the pandemic is not the outcome of solely effective workplace regulations, successful pandemic containment is almost impossible without the latter. The Bloomberg COVID-19 Resilience Ranking provides a serviceable snapshot of COVID-19 prevention successes of 53 countries around the world. In the ranking, countries listed with effective regulations vis-à-vis achieved success in containing COVID-19 were New Zealand, Singapore, Australia, Israel, South Korea, Finland, Norway, Denmark, Mainland China, and Hong Kong [59]. The examples of various country regulations are as follows:

Singapore [60,61]: Work-from-home became the default mode of work in this country during the COVID-19 pandemic; penalties were implemented for anyone violating this policy. Employers were instructed to take care of their workers and workplaces, especially individuals who became unwell in the workplace. To fulfill this, employers were instructed to implement safe management measures such as monitoring plans and safe management officers. Personal hygiene, safe distancing, and mental health measures were also essential strategies in this country's success in containing the pandemic.

South Korea [62]: The government of the Republic of Korea launched the "All about Korea's Response to COVID-19" initiative to help employers and employees maintain good hygiene and social distancing in the workplace. Strategies included avoiding crowded work environments and implementing work-from-home and flexible work schedules.

Finland, Norway, and Denmark: These Nordic countries all had high rankings for best COVID-19 management by implementing advisories for employers and employees [63–66], recommending remote work [67], and re-opening offices [68].

China: The State Council for the People's Republic of China published guidelines on staying safe from COVID-19 in workplaces for both employers and employees. Strategies included [69] guidelines on the commute to work, using stairs (avoiding elevators) upon arrival at work, guidelines in the office, and guidelines for after work. Holiday practices were also outlined [70].

Hong Kong: The Centre for Health Protection, Department of Health, advised for the prevention of COVID-19 in the workplace and businesses (Interim) [71] by maintaining good personal and environmental hygiene in workplaces, maintaining toilet hygiene, and maintaining the hygiene of overnight rooms

4.3. Operational Guidelines

The risk of infection spread across employees in a workplace also depends on the characteristics of the working environment and workplace operational guidelines. For example, Dyal et al. (2020) [72] reported the ordeal of many workers in the meat processing sector when group infections occurred in this sector's closed factory environments which often lack adequate ventilation. Agostinho (2020) [73] pointed out that COVID-19 spread quickly in the meat processing sector because of its intrinsic processing requirement which prevented coworkers from maintaining a six-foot safe working distance between one another. Furthermore, Van Den Berg et al. (2021) [74] emphasized the importance of maintaining physical distancing between 3 versus 6 feet in public schools with mandatory face mark wearing. The data reveals that lower distancing can be applied in a school setting. Therefore, Ijaz et al. (2021) [75] emphasized the necessity of framing policies and working protocols exclusively for meat production and supply chain to curtail disease spread.

Elsewhere, health sector workers were the most vulnerable to COVID-19 during the pandemic's peak periods in some respective countries due to the sector's intrinsic work patterns [76]. Iavicoli et al. (2021) [77] presented a general workplace risk assessment framework based on three intrinsic requirements: exposure, proximity, and aggregation. Unsurprisingly, health and social workers had higher risk infectivity based on this assessment than workers in education, arts, entertainment, and recreation. This fact supports the notion that workers in different sectors cannot be effectively protected from the COVID-19 pandemic by a single safety guideline or protocol [78]. Instead, the understanding of sector-specific risk factors and workplace characteristics must be a prerequisite for developing effective, tailor-made safety guidelines and protocols for a particular workplace. Finally, workers must receive timely communications about updated guidelines and protocols, such that strategies are correctly followed and problematic misperceptions between managers and workers can be avoided [79].

5. Effects of COVID-19 Pandemic on Office and Workplace Sustainability

In the long duration of the COVID-19 pandemic, since its outbreak, the pandemic has provided varied office working experiences that have given a greater understanding of workplace vulnerabilities against disease transmission. COVID-19 has affected the working style of professionals, irrespective of whether they worked from home or in commercial office spaces. In some industrial operations, where working from home was not possible, business owners and facility managers had to make arrangements and implement engineering and administrative measures to ensure the safety of on-site work. Workplace layouts were adjusted to provide some distance between neighboring workstations and ventilation systems were improved. Administrative instructions were issued to employees to maintain distancing requirements from other colleagues while at the office. All these measures acted as collaborative protective layers against the spread of COVID-19 to make office buildings and workplaces coherent working places once again.

5.1. Engineering Control Environment in Workplaces

Many workers converted their homes to makeshift offices during the COVID-19 pandemic. Working from home has given workers flexibility in performing official work and, more importantly, protection from virus infection. As managers and workers have become more experienced in handling their own safety during the pandemic, their desire to return to regular office-working arrangements also increased. However, with the dangers of the virus still present, office workplaces need to be modified to provide workers with a sense of security such that their normal productivity could resume [80–82]. The dangers of COVID-19 should compel office managers to develop the built environment around offices with protective layers to prevent infection spread among workers. In this context, experiences of living with COVID-19 have provided many practical insights.

Workplaces are vulnerable areas where rapid multiplication of COVID-19 cases or infection clustering is possible. Although allowing employees to work from home can negate the hazards of the workplace, most workforces include a large portion of employees that cannot work from home due to the criticality of their sectors. On top of that, whether workers work in a factory or plush office chamber, their exposure to infection hazards also put their family members and coworkers at risk [83]. Considering how infected workers can quickly spread pandemic viruses to their urban residential centers, the workplace must have protective layers to protect workers from virus transmission [84].

Günther (2020) [85] reported on the evidence of how a lack of infection control efforts from management in a German food processing factory led to the continuous exposure of workers to recirculated infected air, resulting in a rapid increase in new COVID-19 cases. Elsewhere, Herstein et al. (2021) [86] reported on a similar super-spreading event of COVID-19 in a meat processing factory in Nebraska, USA, once again due to a lack of infection controls. Investigations into these rapid virus transmission cases in factories revealed the absence of three important infection control measures: lack of fresh air circulation in a closed space, non-existent physical distancing between coworkers during working situations, and lack of administrative controls. It is essential for management to develop well-thought-out infection control plans to protect workers from the COVID-19 disease, especially where risk propensity is inherently high, such as open office space environments [87]. Management's advanced preparedness to prevent COVID-19 and other pandemic diseases will be very relevant for their workers' safety. Infection prevention should take top priority in any manager's plans against COVID-19. Successful infection prevention hinges on correctly assessing the varying degrees of risk relevant to different industrial and office setups and managing these risks with effective control strategies [88]. Managers will be challenged to understand the infection-spreading gateways in their respective controlled areas and then implement the right tools and techniques to close these gateways. In the case of the earlier-mentioned meat factories, workers commented that managers allowed for closed spaces with reduced fresh air circulation, leading to the rapid spread of infection [89]. Workers also noted that the virus's longer surviving ability in a low-temperature closed room was not considered before allowing workers to be present [85]. Employees are demanding to return to offices that are more adaptive against COVID-19. Adaptive office facilities may include flexible desk arrangements, improved touch-free technologies, less crowding made possible by employee roster presence plans, frequent sanitization, and flexibility of switching to remote working arrangements as needed [90].

Building owners and policymakers must collaborate with the office manager to implement the infection protective and preventive strategies by following agency regulations and country measures. Furthermore, they have to take control actions against COVID-19, which may include (a) implementing engineering controls (such as ventilation) to prevent airborne virus transmission and (b) implementing administrative controls and COVID-19 protocols that apply to everyone present in the workplace.

5.1.1. Ventilation

Literature on COVID-19 supports airborne transmission of the SARS-CoV-2 virus, mainly by respiratory droplets smaller than 5 microns. However, although larger respiratory droplets (i.e., more than 10 microns) cannot be suspended in the air for much time, they can still cause disease transmission when they accumulate on and contaminate surfaces. There is some contradicting opinion regarding the duration of aerosol viruses staying active while afloat, how fresh air dilution affects viral loads, and how far viruses can travel while airborne. However, some existing case studies undeniably confirm the risk of airborne COVID-19 transmission. Aerosol SARS-CoV-2 viruses are a genuine concern for workers in closed office spaces. Evidence has pointed out that aerosol viruses can remain active in the air for some time and make a direct airborne journey into workers' respiratory tracts [91].

Research from different countries during the first COVID-19 waves highlighted the possible respiratory spread of the novel coronavirus under low temperature and low relative humidity (RH) conditions [92,93]. Low RH and temperature environments supported the rapid transmission of the novel coronavirus across the population [94]. Management's responsibility will be to implement engineering controls that prevent the spread of COVID-19 in workplaces, be it by installing or maintaining adequate air replenishing systems or climate control systems. In cases where engineered measures are inadequate or impractical to install, managers must look into other ways to prevent disease transmission in their workplaces.

5.1.2. Distancing and Building Layout

The built environment around offices needs to be reshaped and adapted under the backdrop of the COVID-19 pandemic [95]. The pandemic has shown that conventional office design concepts have been too narrowly focused on productivity and economic efficiency. Although high-rise commercial buildings can accommodate huge worker populations, they also expose more people to a higher risk of contamination through human contact. Interior office spaces that opted for less square footage per worker for the sake of cost-effectiveness have inversely become the least desirable option under COVID-19 [96,97]. State-of-the-art high-speed elevators in sophisticated multistory office buildings have suddenly become places of high infection spread propensity; employees now perceive offices with sprawling staircases as safer and more beneficial to their health [98].

The attitude of office managers towards open office plans changed fast when they had to reduce virus propagation among office workers during COVID-19 [80]. Open office layouts can crowd and confine many employees into a shared space, thereby facilitating aerial viral transmission when infected coworkers cough and expel coronavirus droplets. The SARS-CoV-2 virus can be present in aerosols for up to three hours, during which time people can be infected [99,100]. As society develops an in-depth understanding of the plausible modes of transmission for the SARS-CoV-2 viruses, more and more employees find existing open space offices unacceptable [101]. Literature on COVID-19 also supports single-occupancy office rooms and even suggests keeping rooms empty for some time, depending on airflow rate and room size, before new employees enter [102].

5.1.3. Facilities and Equipment

Of course, devices with touch-free technology can altogether eliminate physical contact by workers. Demand to make day-to-day office operations touch-free will be increasing [103]. Before the COVID-19 pandemic, employees had to touch many facilities to operate them. Day-to-day office work involved touching devices such as biometric fingerprint scanners, computers, printers, tea and coffee vending machines, and many more. With the coronavirus situation, employees have become more reluctant to operate office facilities via touch. After COVID-19, there will be a demand for offices where employees rarely need to touch items with their hands. At the same time, lifts and coffee vending machines can be operated using personal smartphones [104]. Offices will need to be equipped with touch-free technologies as employees return to the workplace [105]. Management must plan to use affordable, available, and workable technologies to drastically reduce office touchpoints to allay workers' fears of infection. Otherwise, workers may become hesitant in regularly attending offices that overly rely on touch-based operations.

Individuals can be infected with COVID-19 by touching different virus-contaminated surfaces in the office. Van Doremalen et al. (2020) [106] indicated that the active period of SARS-CoV-2 viruses vary based on the type of surface they are present on. SARS-CoV-2 remains active comparatively longer (2 to 3 days in laboratory condition) on plastic and stainless-steel surfaces but much shorter (1 day) on cardboard surfaces. Copper surfaces, despite being metal, provide the minimum life of viruses. Laboratory experiments have also shown that a single contaminated door handle can cause rapid multiplication of infected persons in an office within hours [107]. Surfaces that can retain viruses longer need to become less frequent touch points. Workers need to be aware of the risk profile of their immediate office environment and become more strategic in keeping infection possibilities at a low level [108].

5.1.4. Water and Sanitary System Safety

During the temporary building closures due to COVID-19, the water tank of the buildings are likely to have microbiological contaminants [109], which may take a long period to recover to the normal levels. Furthermore, increased legionella [110] can cause health hazards for the building's occupants.

5.2. Administrative and Organizational Control Environment in Workplaces

It is of particular importance for office employees to be well aware of COVID-19 safety precautions such as mask-wearing. It has been found that the doffing of COVID-19 protective equipment is common in workplaces, which invites infection risks to all who are present [111]. Additionally, workplace visitors must also be delicately managed to prevent external virus sources from penetrating the workplace's disease-prevention layers.

To effectively combat COVID-19, every employee in a typical workplace must be unitedly involved in following protocols designed to prevent disease transmission. Lapses or suppression of disease symptoms by a single worker can put entire workplace populations at risk of infection that may lead to the closure of the entire office, depending on local statutory guidelines [112]. As such, office administrators must frequently and effectively communicate basic COVID-19 prevention guidelines to all employees. More importantly, they must take on the challenge of maintaining employees' motivation in adhering to enforced guidelines and systems [113].

5.3. Vulnerabilities of Office Buildings and Workplaces

For workers to return to their respective workplaces, different pandemic-resilient requirements must be ingrained into the collective system of building architectures, layouts, safety measures, and many more. If these much-required protective layers are successfully executed in office and factory environments, they can ensure the sustainability of workplaces during the pandemic. The key vulnerabilities in the implementation of preventive COVID-19 regulations, engineering control, and administrative control have been identified and summarized in Table 1. Each vulnerability needs to be addressed to minimize disease transmission risks among employees and built antivirus protective layers.

Table 1. Vulnerabilities of office buildings and workplaces based on experiences with COVID-19.

Category	Subcategory	Vulnerabilities	References
Regulatory Environment	International Agency Regulations	• Virus exposure can occur anytime at workplaces where no regulations can help. • Various health regulations from the various international agencies to be followed in workplaces are unknown. • A simple translation of the requirements documented at the international, national, and sectoral levels to office-specific guidelines and protocols may not be adequate.	[24,114,115]
	Country Regulations	• National level COVID-19 guidelines may have some mismatches with international guidelines. • Country regulations may reflect national COVID-19 resiliency but are not office specific. • Country regulations only serve as general reflections of COVID-19 threats.	[59,115]
	Operational Guidelines	• Guidelines are rapidly changing and evolving, which may lead to confusion. • Workers may not receive updated COVID-19 operating procedures and guidelines in a timely fashion.	[77–79,116]
Engineering Controls Environment	Ventilations	• Filtered air or fresh air supplies are not available in some offices and workplaces. • Lack of infection controls through engineering means in office spaces. • Management may not know proper indoor temperature and humidity control protocols that make the COVID-19 disease less infectious.	[86,91,92,94]

Table 1. *Cont.*

Category	Subcategory	Vulnerabilities	References
Engineering Controls Environment	Distancing and Building Layout	• Indoor offices are rigid and unadaptable structures that prohibit workers from maintaining a safe distance from one another. • Existing open plan office interior designs deemed no longer appropriate due to high COVID-19 transmission propensity. • High-rise commercial buildings have more risk of contamination.	[15,80,96–98,100]
	Facilities and Equipment	• Day-to-day work requires touching of different gadgets and surfaces, increasing the chance of virus transmission. • Hands-free cultures are generally absent and difficult to introduce into traditional offices. • Automated office environments with less physically operated systems are absent. • Prevailing work cultures are not resilient in preventing the spread of infectious diseases such as COVID-19. • Historically, office equipment and facilities were not designed for preventing infectious diseases. • SARS-CoV-2 has been discovered on surfaces in the isolation area where suspected patients were being held.	[108,117–119]
Administrative and Organizational Controls Environment	Water and Sanitary System Safety	• Stagnant water can lead to poor water quality, notably microbiology and legionella, which can harm building occupants.	[109,110]
	Workplace Safety Measures	• Employees doffing COVID-19 protective gear in the workplace enhances the risk for colleagues.	[111]
	Cleaning and Disinfection	• Typical office space cleaning policies and procedures are inadequate at preventing infection risks.	[88]
	Emergency Plan Development	• Office surveillance systems based on staff attendance are almost useless during the COVID-19 pandemic and require upgrades. • Employees suppressing disease symptoms create risk for entire offices.	[86,112]
	Business Continuity Guideline	• Management is not adequately assessing the carrying degrees of risk relevant to different industrial and office setups.	[88]
	Education and Communication	• Less motivated employees may avoid being trained on COVID-19 protocols or following the protocols.	[113]

6. Protective and Preventive Layers for Office Buildings and Workplaces

The current pandemic has shown the need for the built environment to be unfriendly towards severe acute respiratory syndrome coronavirus 2 (SARS-CoV-2) types of viruses [15]. A building's resiliency against the coronavirus will become a design consideration alongside productivity and aesthetics. To prevent COVID-19, especially when workers are returning to their respective workplaces, policymakers and building owners have to analyze, prepare, and implement various strategies for the safety of the workers. Based on the lessons learned and the vulnerabilities exposed during the pandemic stint of more than a year, a number of anti-COVID-19 considerations will be required for offices and workplaces for the safety of the workers. These requirements can be grouped under three broad headings: (1) regulatory environment, (2) engineering control environment, and (3) administrative and organization control environment, as shown in Table 2.

Table 2. Summary of office buildings and workplaces requirements to safeguard against COVID-19.

Category	Subcategory	Requirements
Regulatory Environment	Agency Regulations	• Various international agency regulations that provide practical updated workplace guidelines to protect workplaces against COVID-19 [79].
	Country Regulations	• COVID-19 guidelines differ across different countries, reflecting the differences in their health care facilities and strengths. • In some instances, country-specific COVID-19 regulations may deviate from IHF regulations [114].
	Operational Guidelines	• Constant updates of regulatory frameworks in workplaces are needed. Workplaces must be able to adapt their operational guidelines to encompass the latest regulatory updates at the national and international levels [78]. • Regulations should bring about behavioral changes required to protect workers from the COVID-19 pandemic [83]. • Specific behavior standards should be developed and implemented for each work sector based on general guidelines established by the health authority and with the participation of business chambers, trade unions, and government authorities [120]. • Workplace policies and protocols should be ready at the organization level to ensure employee safety and security during COVID-19 [79,121].
Engineering Control Environment	Ventilation	• Increase fresh air supply and spent air exhaust rates in and out of workplaces through demand-controlled ventilation systems [56]. • The disabling of air recirculation systems should be strictly controlled [122]. • High-Efficiency Particulate Air (HEPA) filters must be installed to allow absolute clean air inflow [123]. • Control inflow rates of filtered air into workplaces to prevent virus-friendly environments (control indoor humidity and temperature) [33]. • Revamp industrial hygiene (IH) and occupational and environmental health and safety (OEHS) design to include considerations for virulent pathogens [88].
	Distancing and Building Layout	• Designs must fundamentally consider health and sustainability [15]. • Working places should be more functional, healthier, and morally fulfilling [124]. • People in tall buildings may be more at risk [98]. • Install physical barriers in workplaces made of plexiglasses, polycarbonate, and tempered glass [72]. • Attention must be paid to avoid overcrowded interior spaces [125,126]. • Office spaces should become more varied, with fewer desks and more choices of spaces to meet, eat, exercise, and unwind [127]. • Opt for internal layouts with adequately open working spaces and allocate spaces for individuals [128].
	Facilities and Equipment	• Provide personal digital terminals (tablets, laptops, desktop PCs, etc.) [33]. • Automate bathroom facilities (toilets, etc.) [129]. • Opt for devices with touch-free technology such as automated infrared temperature checks, facial recognition admittance devices, keycard swipers, and voice-activated elevators. [130]. • Consider selecting materials for surfaces (such as office desks, etc.) that sustain a shorter half-life of the SARS-CoV-2 virus [131].
	Water and Sanitary Ware System	• During COVID-19 temporary building closure, routine flushing and shock disinfection were indicated as possible microbiological and legionella risk control methods [109].

Table 2. *Cont.*

Category	Subcategory	Requirements
Administrative Control Environment	Workplace Safety Measures	• Together, source control and pathway control play an essential role in minimizing high-risk situations [132] • Implement an efficient surveillance system that can track infection and transmission trends amongst employees and communication thereof to employees [133].
	Cleaning and Disinfection	• Cleaning detergent and disinfectants are recommended to be made available at strategic places, especially at the frequent touch surfaces [134].
	Emergency Plan Development	• Implement an efficient surveillance system that can track infection and transmission trends amongst employees and communicate to employees [133].
	Business Continuity Guideline	• Establish teams of workplace experts that can enforce effective control measures [94]. • Allow for flexible working models, such as working from home [135].
	Education and Communication	• Train workers to develop acquaintance with COVID-19 guidelines [39].

6.1. Workplace Regulations

The WHO's International Health Regulations (2005) (IHR) is an international agreement, agreed upon by 196 countries, with a scope to prevent, protect against, control, and provide a public health response to the international spread of disease. This regulation came into effect on 15 June 2007. The IHR addresses how different countries can take action against the spread of diseases such as COVID-19 or other public health issues. Therefore, in a sense, individual countries' COVID-19 health-protective guidelines must follow IHR regulations. Unfortunately, some countries were reported to have deviated from the IHR during COVID-19 [114].

Many international and national agencies, led by the WHO, have proactively issued COVID-19 safety guidelines for workplaces and are continually updating these guidelines as their understanding of the SARS-CoV-2 virus matures with time. The objectives of any of these agencies' guidelines are to provide national frameworks and organizational guidelines for the safe return of workers into workplaces. The authors of [136], reported that the COVID-19 guidelines of six countries during their initial outbreaks differed and reflected these nations' different healthcare strengths explicitly. The authors of [137] also pointed out the ad hocism of national and local COVID-19 guidelines across their 20-country study.

All regulations and guidelines, be they international or national, are fundamentally meant to prevent pandemic spread [138]. However, they may not address specific requirements for successful implementation in particular workplaces. For example, national or international guidelines may not specifically address how ventilation systems should work for the consideration of assembly line workers.

As different sectors of the economy reopen, countries need to develop sector-specific guidelines with collaboration between business chambers, workers' bodies, and government authorities so that benchmark standards can be implemented and followed by individual organizations [120]. Individual organizations should be ready to ensure employees' safety and security during COVID-19 [79,122]. Office managers should actively execute in-depth workplace analyses with expert input to develop the essential guidelines suitable for their workers. Constant updating of workplace guidelines according to the international and national guideline updates is required [78]. It must be noted that regulations cannot immediately bring about behavioral changes in workers [83]. Similarly, written policies or regulations alone cannot bring about the required behavioral changes [83]. Dennerlein et al., 2020 [139], therefore, indicated the necessity of multidimensional workplace regulatory frameworks that cover human factors and ergonomic principles, because, ultimately, employees must be cognizant of their responsibility towards their health.

6.2. Engineering Control Measure Environment

Efficient source control can play an important role in controlling infectious spreads in high-risk places [132], eliminating workers' risks of contracting pandemic diseases. Administrative actions that serve this purpose may include controlling office overcrowding, eliminating PPE doffing (such as masks, shields, etc.), and banning entry for visitors unwilling to maintain social distancing [140]. Source control may be especially difficult in sprawling office spaces, but surveillance systems may help by tracking and prohibiting possible protocol violations at their nascent stages. Bashir et al. (2020) [141] pointed out that even simple, low-cost end-to-end IoT architectures can aid in the surveillance of office environments to uphold standard operating procedures (SOP). With surveillance systems implemented, employers can ensure the strength of source control measures and quickly communicate about infection trends to inform employees of their actions [133].

Effective air cleaning, filtration, and ventilation systems are forms of engineering controls that can block virus entry pathways and maintain a healthy workplace environment [113]. The task of HVAC systems will be to prevent or eliminate airborne aerosols laden with micro respiratory droplets less than 5 microns in size from office environments. The epidemic task force of ASHRAE (American Society of Heating, Refrigerating and Air-Conditioning Engineers) suggested running office HVAC systems with at least minimum outdoor airflow rates to flush airborne viruses out of office spaces. HVACs should remain in operation until fresh air is replenished in every single employee's space within an office. Air cleaners and/or filters need to be utilized in conjunction with HVACs to prevent the recycling of unwanted aerosols back into the room [123]. ASHRAE also suggests using MERV (Minimum Efficiency Reporting Value) 13 or better rated air filters that can stop >95% of 0.30–1.0 micron-sized pathogens that are present in the air [56]. HVACs should also help maintain healthy workplace temperature and relative humidity (RH) that are unfriendly towards disease spread; in regard to this, research is still ongoing to establish the exact temperature and RH ranges that can best subdue the COVID-19 virus [142].

Health and sustainability should be at the core of any office building's architecture [15]. The COVID-19 pandemic made it starkly evident that designers and planners did not put enough importance on the issue of infectious disease control in building design. Before the pandemic, office buildings were developed with primary emphasis on productivity and aesthetics without sufficient consideration for infectious disease control features [143]. Therefore, office environments must become more versatile, infection resistant, and morally fulfilling for workers to achieve higher productivity [124]. Tall and crowded office buildings fitted with multiple fast-moving elevators are no longer looked upon proudly by employees due to their fear of virus infection. Instead, employees prefer lower-levelled office buildings with sprawling staircases [98]. Designers must also use building materials with a shorter half-life for the SARS-CoV-2 virus [131,144].

Open office layouts used to be much-touted for their productivity-enhancement capacity. However, their vulnerability to disease transmission was exposed during COVID-19, particularly by one case in Zurich where an entire team working in an open office became infected [87]. In India, workers wary of virus infection were reluctant to attend open-concept information technology call center offices. As such, designers must alter open plan layouts to make them more disease resilient [145]. Architectural innovations will be needed to create open office layouts that respect personal spaces and promote employee morale even during pandemics [146]. In addition, physical barriers made of plexiglass, polycarbonate, or tempered glass placed in refreshment rooms and meeting rooms can further reduce the possibility of close contact between employees [72].

Touch-free technologies can also mitigate the spread of virulent diseases in the office. Virus-laden surfaces, such as elevator keys, doorknobs, attendance registering keyboards, or even electrical switches, are prominent modes of disease transfer. Network-connected devices (Internet of Things, IoT networks) can help employees minimize physical contact with high-touch surfaces when passing through entry security systems, operating elevators, accessing personal cabins, or using office equipment (computers, lights, fans, etc.) [130,147,148].

6.3. Administrative and Organization Control Measure Environment

Regardless of what sorts of management-enforced risk elimination and engineering controls are implemented in the workspace, they will be ineffective in keeping out COVID-19 unless efficient administrative follow-up is performed. Perhaps an organization's greatest asset against COVID-19 is motivated employees. Efficient administration can motivate employees to cooperate and maintain management-enacted guidelines and protocols for COVID-19 disease control. Without employee motivation, individual workers who defy protocols can undermine all management efforts in containing COVID-19 [149]. To this effect, the office administration's ability to involve workers in the safety planning and development of COVID-19 guidelines and protocols, such as workplace safety measures, cleaning and disinfection, emergency plan development, business continuity guidelines, and education and communication, will be immense. Administrators must not tire employees with rigorous COVID-19 protocol training but instead develop engaging and practical teaching sessions to prevent workers' inhibition against following the laid down protocols.

7. Conclusions

The current COVID-19 pandemic has brought different changes in our society. The continuance of the pandemic has forced humanity to think of erecting protective and preventive barriers that will allow people to return to everyday life. The rapid spread of COVID-19 forced the office and factory workers to leave their working premises and opt for work from home. However, offices and factories must function normally so as to maintain the economic health of any country. More resilient office and factory buildings with safety assurances are needed for workers to walk back to their offices and factories. The long COVID-19 stretch has exposed different workplace-related vulnerabilities that are required to be plugged in for workers to return to their workplaces. In this direction, this paper considered the vulnerabilities experienced in offices and working places under three major categories, the built environment, control environment, and regulatory environment, and then assessed methods to make workplaces conducive towards eliminating and controlling COVID-19 disease spread.

The COVID-19 pandemic made it clear that the built environment of offices was grossly incapable of controlling infectious disease spread as health and sustainability were not the fundamental part of the design aspects. Open space office layouts may provide cost efficiency as they enhance workers' productivity but can be an option during the COVID-19 pandemic. Office facilities and equipment must be equipped with more and more touch-free technologies. The control environment in offices needs to be reoriented, with the main focus on eliminating sources of infection and blocking the pathways of COVID-19 disease spread. Both engineering and administrative measures must work in a complementary mode to make the COVID-19 disease control environment fully effective in offices and factories. Regulations and guidelines enacted at the international or national level are essential in the context of the office environment to prevent pandemic spread. Agency regulations and country regulations are the important sources of updated research-based understanding that can be used for developing operational guidelines to protect workplaces against COVID-19. Specific behavior standards should be developed and implemented for each work sector based on general guidelines established by the health authority. The concerted efforts of business chambers, trade unions, and government regulations should bring about behavioral changes required to protect workers from the COVID-19 pandemic. Finally, the study suffers from a few limitations. The study entirely depended on the literature published for a limited period, starting from the origin of COVID-19. We found that number of research studies focusing on vulnerabilities of office buildings due to COVID-19 diseases were comparatively less in number than residential building vulnerabilities.

Supplementary Materials: The following are available online at https://www.mdpi.com/article/10.3390/su132413636/s1, Table S1: Preventive and Protective Measures Against COVID-19 From International Agencies.

Author Contributions: Conceptualization, P.P., A.D. and O.C.; methodology, A.D. and O.C.; validation, O.C.; formal analysis, P.P. and A.D.; resources, P.P. and A.D.; data curation, O.C.; writing-original draft preparation, P.P. and A.D.; writing-review and editing, A.D. and O.C.; visualization, P.P.; project administration, P.P.; funding acquisition, P.P. and O.C. All authors have read and agreed to the published version of the manuscript.

Funding: This research was funded by "The 100th Anniversary Chulalongkorn University Fund for Doctoral Scholarship".

Institutional Review Board Statement: Not Applicable.

Informed Consent Statement: Not Applicable.

Data Availability Statement: Not Applicable.

Acknowledgments: This research was (partially) supported by the Ratchadapisek Sompoch Endowment Fund (2021), Chulalongkorn University (764002-ENV). We acknowledge the funding support provided by the Second Century Fund (C2F) from Chulalongkorn University awarded to Abhishek Dutta's postdoctoral fellowship.

Conflicts of Interest: The authors declare no conflict of interest.

References

1. United Nations Conference on Trade and Development (UNCTAD). *International Production beyond the Pandemic: UNCTAD World Investment Report*; United Nations Publications: New York, NY, USA, 2020.
2. Wang, Z.; Tang, K. Combating COVID-19: Health equity matters. *Nat. Med.* **2020**, *26*, 458. [CrossRef] [PubMed]
3. World Health Organization. WHO Coronavirus Disease (COVID-19) Dashboard. WHO Coronavirus Disease (COVID-19) Dashboard. 12 June 2021. Available online: https://covid19.who.int/# (accessed on 1 October 2021).
4. Bank, T.W. The Global Economic Outlook during the COVID-19 Pandemic: A Changed World. 8 June 2020. Available online: https://www.worldbank.org/en/news/feature/2020/06/08/the-global-economic-outlook-during-the-covid-19-pandemic-a-changed-world (accessed on 31 July 2021).
5. International Monitory Fund (IMF). Managing Divergent Recoveries, World Economic Outlook. 2021. Available online: https://www.imf.org/en/Publications/WEO/Issues/2021/03/23/world-economic-outlook-april-2021 (accessed on 31 July 2021).
6. Bazaid, A.S.; Aldarhami, A.; Binsaleh, N.K.; Sherwani, S.; Althomali, O.W. Knowledge and practice of personal protective measures during the COVID-19 pandemic: A cross-sectional study in Saudi Arabia. *PLoS ONE* **2020**, *15*, e0243695. [CrossRef] [PubMed]
7. Pamidimukkala, A.; Kermanshachi, S. Impact of COVID-19 on field and office workforce in construction industry. *Proj. Leadersh. Soc.* **2021**, *2*, 100018. [CrossRef]
8. Alenezi, M.N.; Al-Anzi, F.S.; Alabdulrazzaq, H. Building a sensible SIR estimation model for COVID-19 outspread in Kuwait. *Alex. Eng. J.* **2021**, *60*, 3161–3175. [CrossRef]
9. Yifang, X.U.; Jiannan, C.A.I.; Shuai, L.I.; Qiang, H.E.; Siyao, Z.H.U. Airborne infection risks of SARS-CoV-2 in US Schools and impacts of different intervention strategies. *Sustain. Cities Soc.* **2021**, *74*, 103188.
10. D'Angelo, D.; Sinopoli, A.; Napoletano, A.; Gianola, S.; Castellini, G.; del Monaco, A.; Fauci, A.J.; Latina, R.; Iacorossi, L.; Salomone, K.; et al. Strategies to exiting the COVID-19 lockdown for workplace and school: A scoping review. *Saf. Sci.* **2021**, *134*, 105067. [CrossRef]
11. Al-Humairi, S.N.S.; Kamal, A.A.A. Opportunities and challenges for the building monitoring systems in the age-pandemic of COVID-19: Review and prospects. *Innov. Infrastruct. Solut.* **2021**, *6*, 79. [CrossRef]
12. Draghi, M. We face a war against coronavirus and must mobilise accordingly. *Financ. Times.* **2020**, *25*.
13. Kindervag, J. Cybersecurity Lessons from the COVID-19 Pandemic. 2020. Available online: https://www.securityroundtable.org/cybersecurity-lessons-from-the-coronavirus (accessed on 1 July 2021).
14. Ahlefeldt, F. Antivirus Architecture as Urban Design. 2020. Available online: https://fritsahlefeldt.com/2020/04/28/antivirus-architecture-as-urban-design/ (accessed on 1 July 2021).
15. Megahed, N.A.; Ghoneim, E.M. Antivirus-built environment: Lessons learned from COVID-19 pandemic. *Sustain. Cities Soc.* **2020**, *61*, 102350. [CrossRef] [PubMed]

16. Bereitschaft, B.; Scheller, D. How Might the COVID-19 Pandemic Affect 21st Century Urban Design, Planning, and Development? *Urban Sci.* **2020**, *4*, 56. [CrossRef]
17. Dietz, L.; Horve, P.F.; Coil, D.A.; Fretz, M.; Eisen, J.A.; van den Wymelenberg, K. 2019 novel coronavirus (COVID-19) pandemic: Built environment considerations to reduce transmission. *Msystems* **2020**, *5*, e00245. [CrossRef]
18. Anderson, J.; Rainie, L.; Vogels, E.A. Experts Say the 'New Normal' in 2025 Will Be Far More Tech-Driven, Presenting More Big Challenges. Available online: https://www.pewresearch.org/internet/2021/02/18/experts-say-the-new-normal-in-2025-will-be-far-more-tech-driven-presenting-more-big-challenges (accessed on 1 July 2021).
19. Honey-Rosés, J.; Anguelovski, I.; Chireh, V.K.; Daher, C.; van den Bosch, C.K.; Litt, J.S.; Mawani, V.; McCall, M.K.; Orellana, A.; Oscilowicz, E.; et al. The impact of COVID-19 on public space: An early review of the emerging questions—design, perceptions and inequities. *Cities Health* **2020**. [CrossRef]
20. Sepe, M. COVID-19 pandemic and public spaces: Improving quality and flexibility for healthier places. *Urban Des. Int.* **2021**, *26*, 159–173. [CrossRef]
21. Doğan, B.; Ben Jebli, M.; Shahzad, K.; Farooq, T.H.; Shahzad, U. Investigating the Effects of Meteorological Parameters on COVID-19: Case Study of New Jersey, United States. *Environ. Res.* **2020**, *191*, 110148. [CrossRef]
22. Shahzad, K.; Farooq, T.H.; Doğan, B.; Hu, L.Z.; Shahzad, U. Does environmental quality and weather induce COVID-19: Case study of Istanbul, Turkey. *Environ. Forensics* **2021**. [CrossRef]
23. Emmanuel, U.; Osondu, E.D.; Kalu, K.C. Architectural design strategies for infection prevention and control (IPC) in health-care facilities: Towards curbing the spread of COVID-19. *J. Environ. Health Sci. Eng.* **2020**, *18*, 1699–1707. [CrossRef] [PubMed]
24. Asian Development Bank (ADB). Protecting the Safety and Well-Being of Workers and Communities from COVID-19. 2020. Available online: https://www.adb.org/sites/default/files/publication/614811/safety-well-being-workers-communities-covid-19.pdf (accessed on 31 July 2021).
25. World Health Organization. Preventing and Mitigating COVID-19 at Work: Policy Brief. 19 May 2021. Available online: https://www.who.int/publications/i/item/WHO-2019-nCoV-workplace-actions-policy-brief-2021-1 (accessed on 25 June 2021).
26. Tokazhanov, G.; Tleuken, A.; Guney, M.; Turkyilmaz, A.; Karaca, F. How is COVID-19 Experience Transforming Sustainability Requirements of Residential Buildings? A Review. *Sustainability* **2020**, *12*, 8732. [CrossRef]
27. Jiang, J.; Chen, M.; Zhang, J. How does residential segregation affect the spatiotemporal behavior of residents? Evidence from Shanghai. *Sustain. Cities Soc.* **2021**, *69*, 102834. [CrossRef]
28. Architectural Society of China. *Draft of Assessment Standard for Healthy Building, in T/ASC 02—2016*; China Construction Industry Press: Beijing, China, 2020; pp. 163–166.
29. Awada, M.; Becerik-Gerber, B.; Hoque, S.; O'Neill, Z.; Pedrielli, G.; Wen, J.; Wu, T. Ten questions concerning occupant health in buildings during normal operations and extreme events in-cluding the COVID-19 pandemic. *Build. Environ.* **2021**, *188*, 107480. [CrossRef]
30. Valizadeh, J.; Hafezalkotob, A.; Alizadeh, S.M.S.; Mozafari, P. Hazardous infectious waste collection and government aid distribution during COVID-19: A robust mathematical leader-follower model approach. *Sustain. Cities Soc.* **2021**, *69*, 102814. [CrossRef] [PubMed]
31. Collins, J.H. The benefits and limitations of telecommuting. *Def. AR J.* **2009**, *16*, 55.
32. Eddleston, K.A.; Mulki, J. Toward Understanding Remote Workers' Management of Work–Family Boundaries: The Complexity of Workplace Embeddedness. *Group Organ. Manag.* **2015**, *42*, 346–387. [CrossRef]
33. Cortiços, N.D.; Duarte, C.C. COVID-19: The impact in US high-rise office buildings energy efficiency. *Energy Build.* **2021**, *249*, 111180. [CrossRef]
34. World Health Organization. Considerations for Public Health and Social Measures in the Workplace in the Context of COVID-19. 10 May 2020. Available online: https://www.who.int/publications/i/item/considerations-for-public-health-and-social-measures-in-the-workplace-in-the-context-of-covid-19 (accessed on 31 July 2021).
35. World Health Organization. Getting Your Workplace Ready for COVID-19: How COVID-19 Spreads. 19 March 2020. Available online: https://apps.who.int/iris/bitstream/handle/10665/331584/WHO-2019-nCov-workplace-2020.2-eng.pdf?sequence=1&isAllowed=y (accessed on 5 May 2020).
36. World Health Organization. Water, Sanitation, Hygiene, and Waste Management for the COVID-19 Virus (Interim Guidance). 23 April 2020. Available online: https://apps.who.int/iris/handle/10665/331846 (accessed on 5 July 2020).
37. World Health Organization. Roadmap to Improve and Ensure Good Indoor Ventilation in the Context of COVID-19. 1 March 2021. Available online: https://www.who.int/publications/i/item/9789240021280 (accessed on 5 July 2021).
38. World Health Organization. Considerations for Implementing and Adjusting Public Health and Social Measures in the Context of COVID-19. Interim guidance. 14 June 2021. Available online: https://www.who.int/publications/i/item/considerations-in-adjusting-public-health-and-social-measures-in-the-context-of-covid-19-interim-guidance (accessed on 25 June 2021).
39. International Labour Organization. A Safe and Healthy Return to Work during the COVID-19 Pandemic. May 2020. Available online: https://www.ilo.org/wcmsp5/groups/public/---ed_protect/---protrav/---safework/documents/instructionalmaterial/wcms_745541.pdf (accessed on 25 June 2021).
40. International Labour Organization. Safe Return to Work: Ten Action Points. May 2020. Available online: https://www.ilo.org/wcmsp5/groups/public/---ed_protect/---protrav/---safework/documents/instructionalmaterial/wcms_745541.pdf (accessed on 25 June 2021).

41. International Labour Organization. Prevention and Mitigation of COVID-19 at Work, Action Checklist. 9 April 2020. Available online: https://www.ilo.org/wcmsp5/groups/public/---ed_protect/---protrav/---safework/documents/instructionalmaterial/wcms_741813.pdf (accessed on 25 June 2021).
42. Occupational Safety and Health Administration. Protecting Workers: Guidance on Mitigating and Preventing the Spread of COVID-19 in the Workplace. 2021. Available online: https://www.osha.gov/coronavirus/safework (accessed on 9 August 2021).
43. Occupational Safety and Health Administration, U.S., Department of Labor. Protecting Workers: Guidance on Mitigating and Preventing the Spread of COVID-19 in the Workplace. 10 June 2021. Available online: https://www.osha.gov/coronavirus/safework (accessed on 3 August 2021).
44. Centers for Disease Control and Prevention. COVID-19 Employer Information for Office Buildings 2021. Available online: https://www.cdc.gov/coronavirus/2019-ncov/community/office-buildings.html (accessed on 31 July 2021).
45. Centers for Disease Control and Prevention. Guidance for Reopening Buildings After Prolonged Shutdown or Reduced Operation. 22 September 2020. Available online: https://www.cdc.gov/nceh/ehs/water/legionella/building-water-system.html (accessed on 9 April 2021).
46. Centers for Disease Control and Prevention. Guidance for Businesses and Employers Responding to Corona. 8 March 2021. Available online: https://www.cdc.gov/coronavirus/2019-ncov/community/guidance-business-response.html (accessed on 25 June 2021).
47. Centers for Disease Control and Prevention. Cleaning and Disinfecting Your Facility. 15 June 2021. Available online: https://www.cdc.gov/coronavirus/2019-ncov/community/disinfecting-building-facility.html (accessed on 25 June 2021).
48. European Centre for Disease Prevention and Control. Guidelines for the Implementation of Non-Pharmaceutical Interventions Against COVID-19. 24 September 2020. Available online: https://www.ecdc.europa.eu/sites/default/files/documents/covid-19-guidelines-non-pharmaceutical-interventions-september-2020.pdf (accessed on 1 July 2021).
49. European Centre for Disease Prevention and Control. Disinfection of Environments in Healthcare and Non-Healthcare Settings Potentially Contaminated with SARS-CoV-2. 26 March 2020. Available online: https://www.ecdc.europa.eu/sites/default/files/documents/Environmental-persistence-of-SARS_CoV_2-virus-Options-for-cleaning2020-03-26_0.pdf (accessed on 1 July 2021).
50. European Centre for Disease Prevention and Control. Heating, Ventilation, and Air-Conditioning Systems in the Context of COVID-19: First Update. 10 November 2020. Available online: https://www.ecdc.europa.eu/sites/default/files/documents/Heating-ventilation-air-conditioning-systems-in-the-context-of-COVID-19-first-update.pdf (accessed on 1 July 2021).
51. European Centre for Disease Prevention and Control. Guideline for the Use of Non-Pharmaceutical Measures to Delay and Mitigate the Impact of 2019-nCoV. 10 February 2020. Available online: https://www.ecdc.europa.eu/sites/default/files/documents/novel-coronavirus-guidelines-non-pharmaceutical-measures_0.pdf (accessed on 1 July 2021).
52. European Centre for Disease Prevention and Control. All Resources on COVID-19. Available online: https://www.ecdc.europa.eu/en/covid-19/all-reports-covid-19 (accessed on 25 June 2021).
53. American Society of Heating. Refrigerating and Air-Conditioning Engineers. ASHRAE Position Document on Infectious Aerosols. 14 April 2020. Available online: https://www.ashrae.org/file%20library/about/position%20documents/pd_infectiousaerosols_2020.pdf (accessed on 25 June 2021).
54. American Society of Heating. Refrigerating and Air-Conditioning Engineers. Pandemic COVID-19 and Airborne Transmission, Environmental Health Committee (EHC) Emerging Issue Brief. 17 April 2020. Available online: https://www.ashrae.org/file%20library/technical%20resources/covid-19/eiband-airbornetransmission.pdf (accessed on 25 June 2021).
55. American Society of Heating. Refrigerating and Air-Conditioning Engineers. COVID-19: One Page Guidance Documents. 2021. Available online: https://www.ashrae.org/technical-resources/covid-19-one-page-guidance-documents (accessed on 25 June 2021).
56. American Society of Heating. Refrigerating and Air-Conditioning Engineers. ASHRAE EPIMEDIC TASK FORCE. 6 January 2021. Available online: https://www.ashrae.org/file%20library/technical%20resources/covid-19/core-recommendations-for-reducing-airborne-infectious-aerosol-exposure.pdf (accessed on 25 June 2021).
57. American Society of Heating. Refrigerating and Air-Conditioning Engineers. Guidance for Re-Opening Buildings. 14 September 2021. Available online: https://www.ashrae.org/file%20library/technical%20resources/covid-19/guidance-for-re-opening-buildings.pdf (accessed on 25 July 2021).
58. American Society of Heating. Refrigerating and Air-Conditioning Engineers. Guidance for Building Operation during the COVID-19 Pandemic. 2020. Available online: https://www.ashrae.org/file%20library/technical%20resources/ashrae%20journal/2020journaldocuments/72-74_ieq_schoen.pdf (accessed on 25 June 2021).
59. Chang, R.V.K.; Munoz, M.; Tam, F.; Makol, M.K.; The COVID Resilience Ranking. The Best and Worst Places to Be in COVID: US. Sinks in Ranking. Bloomberg. 28 July 2020. Available online: https://www.bloomberg.com/graphics/covid-resilience-ranking/ (accessed on 9 August 2021).
60. Ministry of Manpower, S. Requirements for Safe Management Measures at the Workplace. 10 June 2021. Available online: https://www.mom.gov.sg/covid-19/requirements-for-safe-management-measures (accessed on 25 June 2021).
61. Singapore Business Federation. Guide on Business Continuity Planning for COVID-19. 2020. Available online: https://www.enterprisesg.gov.sg/-/media/esg/files/covid-19/guide-on-business-continuity-planning-for-covid.pdf?la=en (accessed on 31 July 2021).

62. Ministry of Foreign Affairs, Republic of Korea. All About Korea's Response to COVID-19. 7 October 2020. Available online: https://www.mofa.go.kr/eng/brd/m_22591/view.do?seq=35&srchFr=&srchTo=&srchWord=&srchTp=&multi_itm_seq=0&itm_seq_1=0&itm_seq_2=0&company_cd=&company_nm=&page=1&titleNm= (accessed on 31 July 2021).
63. Finnish Institute of Occupational Health. Guidelines for Workplaces to Prevent Coronavirus Infection. 22 April 2021. Available online: https://hyvatyo.ttl.fi/en/koronavirus/en/fioh-guidelines-for-workplaces-to-prepare-for-the-coronavirus-epidemic/ (accessed on 25 June 2021).
64. Denmark, W. Prevent the Spread of Coronavirus. 2021. Available online: https://www.sst.dk/en/English/Corona-eng (accessed on 25 June 2021).
65. Norwegian Institute of Public Health (NIPH). Hand Hygiene, Cough Etiquette, Cleaning and Laundry—Advice and Information to the Genral Public. 4 March 2021. Available online: https://www.fhi.no/en/op/novel-coronavirus-facts-advice/facts-and-general-advice/hand-hygiene-cough-etiquette-face-masks-cleaning-and-laundry (accessed on 25 June 2021).
66. Norwegian Institute of Public Health (NIPH). Cleaning for COVID-19—Advice for Sectors Outside the Healthcare Service. 22 March 2021. Available online: https://www.fhi.no/en/op/novel-coronavirus-facts-advice/advice-and-information-to-other-sectors-and-occupational-groups/cleaning-and-disinfection/ (accessed on 25 June 2021).
67. Finnish Institute of Occupational Health. Guidelines for Remote Work. 5 March 2021. Available online: https://www.ttl.fi/en/guidelines-for-remote-work/ (accessed on 25 June 2021).
68. Mia Boesen, S.P. Guidance for Employers in Denmark. 2021. Available online: https://www.twobirds.com/en/news/articles/2020/denmark/covid-19-guidance-for-employers-in-denmark (accessed on 25 June 2021).
69. The State Council of the People's Republic of China. China How to Stay Safe from COVID-19 at Your Work. 2020. Available online: http://covid-19.chinadaily.com.cn/a/202002/25/WS5e67679fa31012821727ded6.html (accessed on 26 June 2021).
70. The State Council of the People's Republic of China. China WHO Recommends 10 Basic Personal Prevention Measures Against COVID-19. 25 February 2020. Available online: https://english.www.gov.cn/news/topnews/202002/29/content_WS5e59a8cfc6d0c201c2cbd36f.html (accessed on 26 June 2021).
71. Centre for Health Protection. Health Advice on Prevention of Coronavirus Disease (COVID-19) in Workplace (Interim). 14 April 2021. Available online: https://www.chp.gov.hk/files/pdf/nid_guideline_workplace_eng.pdf (accessed on 26 June 2021).
72. Dyal, J.W. COVID-19 among workers in meat and poultry processing facilities—19 states, April 2020. *Morb. Mortal. Wkly. Rep.* **2020**, *69*, 557–561. [CrossRef] [PubMed]
73. Agostinho, S. Coronavirus Outbreak Continues to Infect Workers at Both Companies Despite All Measures on the Ground. 2020. Available online: https://www.jornalvalorlocal.com/nuacutemero-de-casos-de-covid-jaacute-chega-aos-26-na-sonae-e-aos-129-na-avipronto.html (accessed on 31 July 2021).
74. Berg, P.V.D.; Schechter-Perkins, E.M.; Jack, R.S.; Epshtein, I.; Nelson, R.; Oster, E.; Branch-Elliman, W. Effectiveness of 3 Versus 6 ft of Physical Distancing for Controlling Spread of Coronavirus Disease 2019 among Primary and Secondary Students and Staff: A Retrospective, Statewide Cohort Study. *Clin. Infect. Dis.* **2021**, *73*, 1871–1878. [CrossRef] [PubMed]
75. Ijaz, M.; Yar, M.K.; Badar, I.H.; Ali, S.; Islam, S.; Jaspal, M.H.; Hayat, Z.; Sardar, A.; Ullah, S.; Guevara-Ruiz, D. Meat Production and Supply Chain Under COVID-19 Scenario: Current Trends and Future Prospects. *Front. Veter-Sci.* **2021**, *8*, 660736. [CrossRef]
76. Adams, J.G.; Walls, R.M. Supporting the Health Care Workforce during the COVID-19 Global Epidemic. *JAMA* **2020**, *323*, 1439. [CrossRef] [PubMed]
77. Iavicoli, S.; Boccuni, F.; Buresti, G.; Gagliardi, D.; Persechino, B.; Valenti, A.; Rondinone, B.M. Risk assessment at work and prevention strategies on COVID-19 in Italy. *PLoS ONE* **2021**, *16*, e0248874. [CrossRef] [PubMed]
78. Spinazzè, A.; Cattaneo, A.; Cavallo, D.M. COVID-19 Outbreak in Italy: Protecting Worker Health and the Response of the Italian Industrial Hygienists Association. *Ann. Work. Expo. Health* **2020**, *64*, 559–564. [CrossRef]
79. Wong, E.; Ho, K.F.; Wong, S.Y.; Cheung, A.W.; Yeoh, E. Workplace safety and coronavirus disease (COVID-19) pandemic: Survey of employees. *Bull. World Health Organ.* **2020**. preprint. [CrossRef]
80. Molla, R. This Is the End of the Office as We Know It. 2020. Available online: https://www.vox.com/recode/2020/4/14/21211789/coronavirus-office-space-work-from-homedesign-architecture-real-estate (accessed on 29 July 2021).
81. Aday, S.; Aday, M.S. Impact of COVID-19 on the food supply chain. *Food Qual. Saf.* **2020**, *4*, 167–180. [CrossRef]
82. Braier, A.; Garrett, M.; Smith, S.; Datar, A.; Eggers, W.D. Designing Adaptive Workplaces, How the Public Sector Can Capitalize on Lessons Learned from COVID-19. Deloitte Insights. 10 February 2021. Available online: https://www2.deloitte.com/us/en/insights/industry/public-sector/designing-for-adaptive-work-in-the-public-sector.htm (accessed on 5 July 2021).
83. Semple, S.; Cherrie, J.W. COVID-19: Protecting Worker Health. *Ann. Work. Expo. Health* **2020**, *64*, 461–464. [CrossRef]
84. Kurgat, E.K.; Sexton, J.D.; Garavito, F.; Reynolds, A.; Contreras, R.D.; Gerba, C.P.; Leslie, R.A.; Edmonds-Wilson, S.L.; Reynolds, K.A. Impact of a hygiene intervention on virus spread in an office building. *Int. J. Hyg. Environ. Health* **2019**, *222*, 479–485. [CrossRef] [PubMed]
85. Günther, T.; Czech-Sioli, M.; Indenbirken, D.; Robitaille, A.; Tenhaken, P.; Exner, M.; Ottinger, M.; Fischer, N.; Grundhoff, A.; Brinkmann, M.M. SARS–CoV–2 outbreak investigation in a German meat processing plant. *EMBO Mol. Med.* **2020**, *12*, e13296. [CrossRef] [PubMed]
86. Herstein, J.J.; Degarege, A.; Stover, D.; Austin, C.; Schwedhelm, M.M.; Lawler, J.V.; Lowe, J.J.; Ramos, A.K.; Donahue, M. Characteristics of SARS-CoV-2 transmission among meat processing workers in Nebraska, USA, and ef-fectiveness of risk mitigation measures. *Emerg. Infect. Dis.* **2021**, *27*, 1032. [CrossRef]

87. Weissberg, D.; Böni, J.; Rampini, S.K.; Kufner, V.; Zaheri, M.; Schreiber, P.W.; Abela, I.A.; Huber, M.; Sax, H.; Wolfensberger, A. Does respiratory co-infection facilitate dispersal of SARS-CoV-2? investigation of a super-spreading event in an open-space office. *Antimicrob. Resist. Infect. Control.* **2020**, *9*, 191. [CrossRef]
88. Zisook, R.E.; Monnot, A.; Parker, J.; Gaffney, S.; Dotson, S.; Unice, K. Assessing and managing the risks of COVID-19 in the workplace: Applying industrial hygiene (IH)/occupational and environmental health and safety (OEHS) frameworks. *Toxicol. Ind. Health* **2020**, *36*, 607–618. [CrossRef] [PubMed]
89. Bromage, E. The Risks—Know Them—Avoid Them. 2020. Available online: https://www.erinbromage.com/post/the-risks-know-them-avoid-them (accessed on 9 August 2021).
90. Consulting, D. Office Layouts for the Post COVID-19 Workplace. 2020. Available online: www.condecosoftware.com/blog/office-design-post-covid-19-workplace/ (accessed on 31 July 2021).
91. Judson, S.D.; Munster, V.J. Nosocomial Transmission of Emerging Viruses via Aerosol-Generating Medical Procedures. *Viruses* **2019**, *11*, 940. [CrossRef]
92. Falagas, M.E.; Theocharis, G.; Spanos, A.; Vlara, L.A.; Issaris, E.A.; Panos, G.; Peppas, G. Effect of meteorological variables on the incidence of respiratory tract infections. *Respir. Med.* **2008**, *102*, 733–737. [CrossRef] [PubMed]
93. Mäkinena, T.M.; Juvonen, R.; Jokelainen, J.; Harju, T.H.; Peitso, A.; Bloigu, A.; Silvennoinen-Kassinen, S.; Leinonen, M.; Hassi, J. Cold temperature and low humidity are associated with increased occurrence of respiratory tract infec-tions. *Respir. Med.* **2009**, *103*, 456–462. [CrossRef] [PubMed]
94. Moriyama, M.; Hugentobler, W.J.; Iwasaki, A. Seasonality of Respiratory Viral Infections. *Annu. Rev. Virol.* **2020**, *7*, 83–101. [CrossRef]
95. Priday, C. Architecture after Coronavirus. 2020. Available online: https://exepose.com/2020/05/05/architecture-after-coronavirus (accessed on 31 July 2021).
96. Volini, E. Returning to Work in the Future of Work. 2020. 15 May 2020. Available online: https://www2.deloitte.com/us/en/insights/focus/human-capital-trends/2020/covid-19-and-the-future-of-work.html (accessed on 29 July 2021).
97. Muggah, R.; Ermacora, T. Opinion: Re-Designing the COVID-19 City. 2020. Available online: https://www.npr.org/2020/04/20/839418905/opinion-redesigning-the-covid-19-city (accessed on 29 July 2021).
98. Gormley, M.; Coronavirus: People in Tall Buildings May Be More at Risk. 20 September 2020. Available online: https://www.fastcompany.com/90488321/live-in-a-tall-building-you-may-be-more-at-risk-of-contracting-coronavirus (accessed on 29 July 2021).
99. Klompas, M.; Baker, M.A.; Rhee, C. Airborne transmission of SARS-CoV-2: Theoretical considerations and available evi-dence. *JAMA* **2020**, *324*, 441–442. [CrossRef] [PubMed]
100. Franklin, N. Offices in the Post Lockdown Era Will Focus on What They Are Good at 16 June 2020. Available online: https://workplaceinsight.net/offices-in-the-post-lockdown-era-will-focus-on-what-they-are-good-at/ (accessed on 29 July 2021).
101. Zhang, J. Integrating IAQ control strategies to reduce the risk of asymptomatic SARS CoV-2 infections in classrooms and open plan offices. *Sci. Technol. Built Environ.* **2020**, *26*, 1013–1018. [CrossRef]
102. Augenbraun, B.L.; Lasner, Z.D.; Mitra, D.; Prabhu, S.; Raval, S.; Sawaoka, H.; Doyle, J.M. Assessment and mitigation of aerosol airborne SARS-CoV-2 transmission in laboratory and office environments. *J. Occup. Environ. Hyg.* **2020**, *17*, 447–456. [CrossRef] [PubMed]
103. Diebner, R.; Silliman, E.; Ungerman, K.; Vancauwenberghe, M. Adapting Customer Experience in the Time of Coronavirus. 2020. Available online: https://www.mckinsey.com/business-functions/marketing-and-sales/our-insights/adapting-customer-experience-in-the-time-of-coronavirus (accessed on 31 July 2021).
104. Kretchmer, H. COVID-19: Is This What the Office of the Future Will Look Like 2020. Available online: https://www.weforum.org/agenda/2020/04/covid19-coronavirus-change-office-work-homeworking-remote-design/ (accessed on 31 July 2021).
105. Billington, L. Understanding the Touchless Workplace. 15 April 2020. Available online: https://www.gensler.com/blog/understanding-the-touchless-workplace (accessed on 31 July 2021).
106. Van Doremalen, N.; Morris, D.H.; Holbrook, M.G.; Gamble, A.; Williamson, B.N.; Tamin, A.; Harcourt, J.L.; Thornburg, N.J.; Gerber, S.I.; Lloyd-Smith, J.O.; et al. Aerosol and surface stability of SARS-CoV-2 as compared with SARS-CoV-1. *N. Engl. J. Med.* **2020**, *382*, 1564–1567. [CrossRef]
107. Reynolds, K.A.; Beamer, P.I.; Plotkin, K.R.; Sifuentes, L.Y.; Koenig, D.W.; Gerba, C.P. The healthy workplace project: Reduced viral exposure in an office setting. *Arch. Environ. Occup. Health* **2016**, *71*, 157–162. [CrossRef]
108. Capolongo, S.; Rebecchi, A.; Buffoli, M.; Appolloni, L.; Signorelli, C.; Fara, G.M.; D'Alessandro, D. COVID-19 and Cities: From Urban Health strategies to the pandemic challenge. A Decalogue of Public Health opportunities. *Acta Biomed.* **2020**, *91*, 13–22. [PubMed]
109. Ye, C.; Xian, X.; Bao, R.; Zhang, Y.; Feng, M.; Lin, W.; Yu, X. Recovery of microbiological quality of long-term stagnant tap water in university buildings during the COVID-19 pandemic. *Sci. Total. Environ.* **2021**, *806*, 150616. [CrossRef] [PubMed]
110. Liang, J.; Swanson, C.S.; Wang, L.; He, Q. Impact of building closures during the COVID-19 pandemic on Legionella infection risks. *Am. J. Infect. Control* **2021**, *49*, 1564–1566. [CrossRef]
111. Baloh, J.; Reisinger, H.; Dukes, K.; Da Silva, J.P.; Salehi, H.P.; Ward, M.; E Chasco, E.; Pennathur, P.R.; Herwaldt, L. Healthcare Workers' Strategies for Doffing Personal Protective Equipment. *Clin. Infect. Dis.* **2019**, *69* (Suppl. S3), S192–S198. [CrossRef] [PubMed]

112. McMillan, L.L.P. Terminations of Employees Breaching COVID-19 Protocols. Canada. 2020. Available online: https://iclg.com/briefing/15267-canada-terminations-of-employees-breaching-covid-19-protocols (accessed on 9 August 2021).

113. Dehghani, F.; Omidi, F.; Yousefinejad, S.; Taheri, E. The hierarchy of preventive measures to protect workers against the COVID-19 pandemic: A review. *Work* **2020**, *67*, 771–777. [CrossRef] [PubMed]

114. Habibi, R.; Burci, G.L.; de Campos, T.C.; Chirwa, D.; Cinà, M.; Dagron, S.; Eccleston-Turner, M.; Forman, L.; O Gostin, L.; Meier, B.M.; et al. Do not violate the International Health Regulations during the COVID-19 outbreak. *Lancet* **2020**, *395*, 664–666. [CrossRef]

115. Sinha, R.; Bhati, D. Ivermectin for COVID-19: Mismatch between Global and INDIAN Policies. 19 May 2021. Available online: https://www.downtoearth.org.in/news/health/ivermectin-for-covid-19-mismatch-between-global-and-indian-policies-76984 (accessed on 30 July 2021).

116. Chikermane, G.; Agrawal, R. *Regulatory Changes in India in the Time of COVID-19: Lessons and Recommendations. ORF Special Report No. 110*; Observer Research Foundation: New Delhi, India, 2020; Available online: https://www.orfonline.org/research/regulatory-changes-in-india-in-the-time-of-covid19-67604/ (accessed on 30 July 2021).

117. Boland, B.; Smet, A.D.; Palter, R.; Sanghvi, A. Reimagining the Office and Work Life after COVID-19. 8 June 2020. Available online: www.mckinsey.com/business-functions/organization/our-insights/reimagining-the-office-and-work-life-after-covid-19 (accessed on 29 July 2021).

118. Alter, L. Architecture after the Coronavirus. 2020. Available online: https://www.treehugger.com/green-architecture/architecture-after-coronavirus.html (accessed on 30 July 2021).

119. Guo, Z.-D.; Wang, Z.-Y.; Zhang, S.-F.; Li, X.; Li, L.; Li, C.; Cui, Y.; Fu, R.-B.; Dong, Y.-Z.; Chi, X.-Y.; et al. Aerosol and Surface Distribution of Severe Acute Respiratory Syndrome Coronavirus 2 in Hospital Wards, Wuhan, China, 2020. *Emerg. Infect. Dis.* **2020**, *26*, 1583–1591. [CrossRef]

120. Economic Commission for Latin America and the Caribbean (ECLAC). COVID-19 Special Report No.4: Sectors and Businesses Facing COVID-19: Emergency and Reactivation. 2021. Available online: https://repositorio.cepal.org/bitstream/handle/11362/45736/5/S2000437_en.pdf (accessed on 31 July 2021).

121. Tran, B.X.; Hoang, M.T.; Pham, H.Q.; Hoang, C.L.; Le, H.T.; A Latkin, C.; Ho, C.S.; Ho, R.C. The operational readiness capacities of the grassroots health system in responses to epidemics: Implications for COVID-19 control in Vietnam. *J. Glob. Health* **2020**, *10*, 011006. [CrossRef] [PubMed]

122. Persily, A.K.; E Chapman, R.; Dols, S.J.E.W.S.; Lavappa, H.D.P.; Rishing, A.; Emmerich, S.J.; Davis, H.; Rushing, A.S. *Building Retrofits for Increased Protection Against Airborne Chemical and Biological Releases*; National Institute of Standards and Technology: Gaithersburg, MD, USA, 2007.

123. Christopherson, D.A.; Yao, W.C.; Lu, M.; Vijayakumar, R.; Sedaghat, A.R. High-Efficiency Particulate Air Filters in the Era of COVID-19: Function and Efficacy. *Otolaryngol. Neck Surg.* **2020**, *163*, 1153–1155. [CrossRef] [PubMed]

124. Alraouf, A.A. The new normal or the forgotten normal: Contesting COVID-19 impact on contemporary architecture and urbanism. *Archnet-IJAR* **2021**, *15*, 167–188. [CrossRef]

125. Mudditt, J. Pandemic-Proofing Offices Could Involve Short-Term Fixes, New Working Patterns and Long-Term Design UPGRADES that Put Hygiene at the Heart of Workplace Planning. 2020. Available online: https://www.bbc.com/worklife/article/20200514-how-the-post-pandemic-office-will-change (accessed on 9 August 2021).

126. Budds, D. Design in the Age of Pandemics, throughout History, How We Design and Inhabit Physical Space Has Been a Primary Defense Against Epidemics, Curbed. 2020. Available online: https://www.curbed.com/2020/3/17/21178962/design-pandemics-coronavirus-quarantine (accessed on 31 July 2021).

127. Buxton, P. Rethink: Working Miracles, the Office Design Revolution. 22 May 2020. Available online: https://www.ribaj.com/intelligence/post-pandemic-design-offices-bco-alexi-marmot-cushman-and-wakefield-hassell (accessed on 9 August 2021).

128. Wilson, M. Our Offices Will Never Be the Same after COVID-19. Here's What They Could Look Like. 2020. Available online: https://www.fastcompany.com/90488060/our-offices-will-never-be-the-same-after-covid-19-heres-what-they-could-look-like (accessed on 9 August 2021).

129. Pocock, L.; Shams, R. The post COVID world: Architecture, internal spaces, public spaces and streetscapes. *Middle East J. Bus.* **2020**, *15*, 20–25.

130. Carino, M.M. Offices Prepare for Post-Virus Return to Work. 2020. Available online: https://www.marketplace.org/2020/04/22/how-will-office-spaces-change-after-covid19/ (accessed on 9 August 2021).

131. Cirrincione, L.; Plescia, F.; Ledda, C.; Rapisarda, V.; Martorana, D.; Moldovan, R.E.; Theodoridou, K.; Cannizzaro, E. COVID-19 Pandemic: Prevention and Protection Measures to Be Adopted at the Workplace. *Sustainability* **2020**, *12*, 3603. [CrossRef]

132. Brosseau, L.M.; Rosen, J.; Harrison, R. Selecting controls for minimizing SARS-CoV-2 aerosol transmission in workplaces and conserving respiratory protective equipment supplies. *Ann. Work. Expo. Health* **2021**, *65*, 53–62. [CrossRef]

133. Plantes, P.J.; Fragala, M.S.; Clarke, C.; Goldberg, Z.N.; Radcliff, J.; Goldberg, S.E. Model for mitigation of workplace transmission of COVID-19 through population-based testing and sur-veillance. *Popul. Health Manag.* **2021**, *24* (Suppl. S1), S-16–S-25. [CrossRef]

134. Pan, L.; Wang, J.; Wang, X.; Ji, J.S.; Ye, D.; Shen, J.; Li, L.; Liu, H.; Zhang, L.; Shi, X.; et al. Prevention and control of coronavirus disease 2019 (COVID-19) in public places. *Environ. Pollut.* **2021**, *292*, 118273. [CrossRef] [PubMed]

135. Helmold, M. *New Work, Transformational and Virtual Leadership*; Springer: Singapore, 2021.

136. Yoo, J.Y.; Dutra, S.V.O.; Fanfan, D.; Sniffen, S.; Wang, H.; Siddiqui, J.; Song, H.-S.; Bang, S.H.; Kim, D.E.; Kim, S.; et al. Comparative analysis of COVID-19 guidelines from six countries: A qualitative study on the US, China, South Korea, the UK, Brazil, and Haiti. *BMC Public Health* **2020**, *20*, 1853. [CrossRef] [PubMed]
137. Lavizzari, A.; Klingenberg, C.; Profit, J.; Zupancic, J.A.F.; Davis, A.S.; Mosca, F.; Molloy, E.J.; Roehr, C.C.; The International Neonatal COVID-19 Consortium. International comparison of guidelines for managing neonates at the early phase of the SARS-CoV-2 pandemic. *Pediatric Res.* **2021**, *89*, 940–951. [CrossRef] [PubMed]
138. Johansen, T.B.; Astrup, E.; Jore, S.; Nilssen, H.; Dahlberg, B.B.; Klingenberg, C.; Berg, A.S.; Greve-Isdahl, M. Infection prevention guidelines and considerations for paediatric risk groups when reopening primary schools during COVID-19 pandemic, Norway, April 2020. *Eurosurveillance* **2020**, *25*, 2000921. [CrossRef] [PubMed]
139. Dennerlein, J.T.; Burke, L.; Sabbath, E.L.; Williams, J.A.R.; Peters, S.; Wallace, L.; Karapanos, M.; Sorensen, G. An Integrative Total Worker Health Framework for Keeping Workers Safe and Healthy during the COVID-19 Pandemic. *Hum. Factors J. Hum. Factors Ergon. Soc.* **2020**, *62*, 689–696. [CrossRef] [PubMed]
140. Brooks, J.T.; Butler, J.C. Effectiveness of mask wearing to control community spread of SARS-CoV-2. *JAMA* **2021**, *325*, 998–999. [CrossRef]
141. Bashir, A.; Izhar, U.; Jones, C. IoT-based COVID-19 SOP compliance and monitoring system for businesses and public offices. *Eng. Proc.* **2020**, *2*, 8267.
142. Dutta, A.; Jinsart, W. Air Quality, Atmospheric Variables and Spread of COVID-19 in Delhi (India): An Analysis. *Aerosol Air Qual. Res.* **2021**, *21*, 200417. [CrossRef]
143. Forsyth, A. What Role Do Planning and Design Play in a Pandemic? 19 March 2020. Available online: https://www.gsd.harvard.edu/2020/03/what-role-do-planning-and-design-play-in-a-pandemic-ann-forsyth-reflects-on-covid-19s-impact-on-the-future-of-urban-life/ (accessed on 31 July 2021).
144. Spolidoro, B. Healthy Buildings: How Architecture Can Defend Us from COVID-19. 2020. Available online: https://www.workdesign.com/2020/05/healthy-buildings-how-architecture-can-defend-us-fromcovid-19/ (accessed on 31 July 2021).
145. Bahadursingh, N. Ways COVID-19 Will Change Architecture. 2020. Available online: https://architizer.com/blog/inspiration/industry/covid19-city-design/ (accessed on 31 July 2021).
146. Alhusban, A.A.; Alhusban, S.A.; Alhusban, M.A. How the COVID 19 pandemic would change the future of architectural design. *J. Eng. Des. Technol.* epub ahead of printing. **2021**. [CrossRef]
147. Wainwright, O. Smart Lifts, Lonely Workers, No Towers or Tourists: Architecture after Coronavirus. 2020. Available online: https://www.theguardian.com/artanddesign/2020/apr/13/smart-lifts-lonely-workers-no-towers-architecture-after-covid-19-coronavirus (accessed on 9 August 2021).
148. Nichols, M. An IoT-Enabled Office Is a Must, Post-COVID. 2020. Available online: https://connectedremag.com/smart-buildings/iot/an-iot-enabled-office-is-a-must-post-covid/ (accessed on 31 July 2021).
149. Michaels, D.; Wagner, G.R. Occupational Safety and Health Administration (OSHA) and Worker Safety during the COVID-19 Pandemic. *JAMA* **2020**, *324*, 1389. [CrossRef]

MDPI
St. Alban-Anlage 66
4052 Basel
Switzerland
www.mdpi.com

MDPI Books Editorial Office
E-mail: books@mdpi.com
www.mdpi.com/books